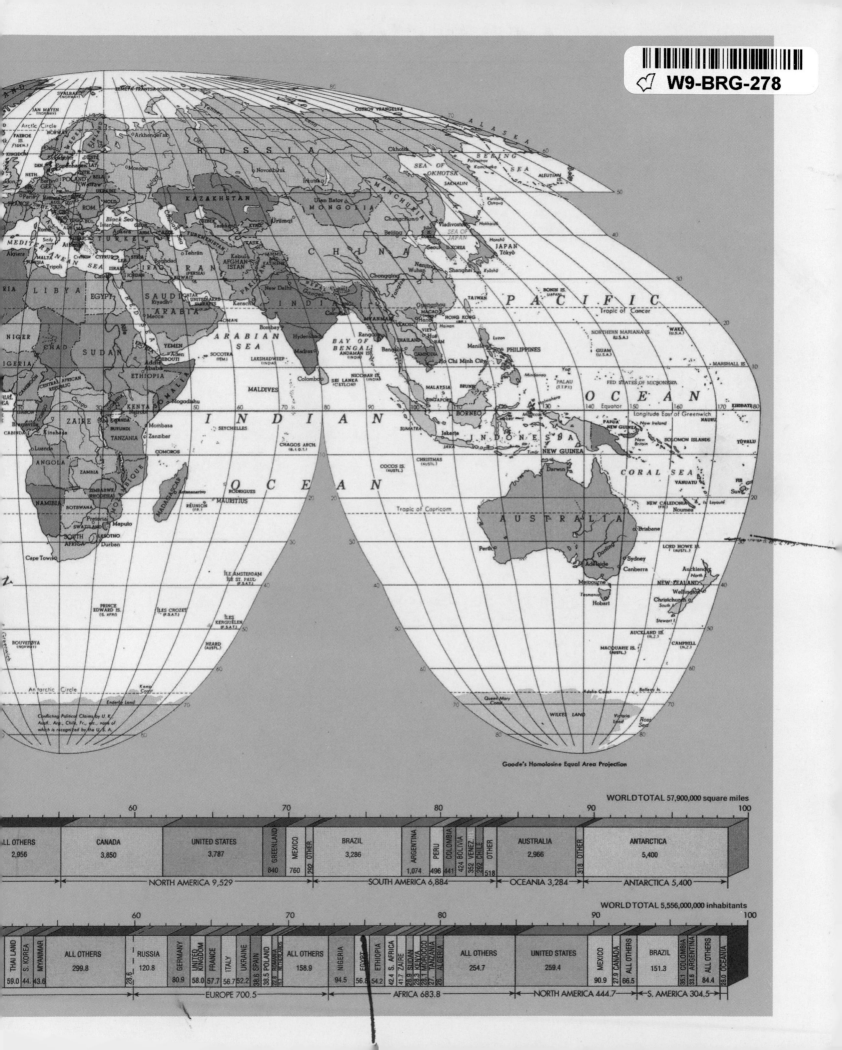

Goode's Homolosine Equal Area Projection

WORLD TOTAL 57,900,000 square miles

| | 60 | | 70 | | 80 | | 90 | | 100 |

| ALL OTHERS 2,956 | CANADA 3,850 | UNITED STATES 3,787 | GREENLAND 840 | MEXICO 760 | OTHER 282 | BRAZIL 3,286 | ARGENTINA 1,074 | PERU 496 | COLOMBIA 441 | BOLIVIA 424 | VENEZ. 352 | CHILE 292 | OTHER 518 | AUSTRALIA 2,966 | OTHER 318 | ANTARCTICA 5,400 |

NORTH AMERICA 9,529 — SOUTH AMERICA 6,884 — OCEANIA 3,284 — ANTARCTICA 5,400

WORLD TOTAL 5,556,000,000 inhabitants

| | 60 | | 70 | | 80 | | 90 | | 100 |

| THAILAND 59.0 | S. KOREA 44.1 | MYANMAR 43.6 | ALL OTHERS 299.8 | RUSSIA 120.8 | 28.6 | GERMANY 80.9 | UNITED KINGDOM 58.0 | FRANCE 57.7 | ITALY 56.7 | UKRAINE 52.2 | SPAIN 38.6 | POLAND 38.5 | ROMANIA 22.8 | NETHERLANDS 15.4 | ALL OTHERS 158.9 | NIGERIA 94.5 | EGYPT 56.8 | ETHIOPIA 54.2 | S. AFRICA 42.4 | ZAIRE 41.7 | SUDAN 28.9 | KENYA 28.3 | MOROCCO 28.1 | TANZANIA 27. | ALGERIA 26. | ALL OTHERS 254.7 | UNITED STATES 259.4 | MEXICO 90.9 | CANADA 27.9 | ALL OTHERS 66.5 | BRAZIL 151.3 | COLOMBIA 35.1 | ARGENTINA 33.6 | ALL OTHERS 84.4 | OCEANIA 28.0 |

EUROPE 700.5 — AFRICA 683.8 — NORTH AMERICA 444.7 — S. AMERICA 304.5

IMPORTANT:

HERE IS YOUR REGISTRATION CODE TO ACCESS
YOUR PREMIUM McGRAW-HILL ONLINE RESOURCES.

For key premium online resources you need THIS CODE to gain access. Once the code is entered, you will be able to use the Web resources for the length of your course.

If your course is using **WebCT** or **Blackboard**, you'll be able to use this code to access the McGraw-Hill content within your instructor's online course.

Access is provided if you have purchased a new book. If the registration code is missing from this book, the registration screen on our Website, and within your WebCT or Blackboard course, will tell you how to obtain your new code.

Registering for McGraw-Hill Online Resources

TO gain access to your McGraw-Hill web resources simply follow the steps below:

1. USE YOUR WEB BROWSER TO GO TO: **http://www.mhhe.com/environmentalscience**

2. CLICK ON **FIRST TIME USER**.

3. ENTER THE REGISTRATION CODE* PRINTED ON THE TEAR-OFF BOOKMARK ON THE RIGHT.

4. AFTER YOU HAVE ENTERED YOUR REGISTRATION CODE, CLICK **REGISTER**.

5. FOLLOW THE INSTRUCTIONS TO SET-UP YOUR PERSONAL UserID AND PASSWORD.

6. WRITE YOUR UserID AND PASSWORD DOWN FOR FUTURE REFERENCE.
   KEEP IT IN A SAFE PLACE.

TO GAIN ACCESS to the McGraw-Hill content in your instructor's **WebCT** or **Blackboard** course simply log in to the course with the UserID and Password provided by your instructor. Enter the registration code exactly as it appears in the box to the right when prompted by the system. You will only need to use the code the first time you click on McGraw-Hill content.

Thank you, and welcome to your McGraw-Hill Online Resources!

* YOUR REGISTRATION CODE CAN BE USED ONLY ONCE TO ESTABLISH ACCESS. IT IS NOT TRANSFERABLE.

0-07-244000-7 ENGER: ENVIRONMENTAL SCIENCE, 9E

REGISTRATION CODE

dedication-58905639

Eldon D. Enger
Delta College

Bradley F. Smith
Western Washington University

ENVIRONMENTAL
Science

A Study of Interrelationships
NINTH EDITION

Mc Graw Hill **Higher Education**

Boston   Burr Ridge, IL   Dubuque, IA   Madison, WI   New York   San Francisco   St. Louis
Bangkok   Bogotá   Caracas   Kuala Lumpur   Lisbon   London   Madrid   Mexico City
Milan   Montreal   New Delhi   Santiago   Seoul   Singapore   Sydney   Taipei   Toronto

ENVIRONMENTAL SCIENCE: A STUDY OF INTERRELATIONSHIPS
NINTH EDITION

Published by McGraw-Hill, a business unit of The McGraw-Hill Companies, Inc., 1221 Avenue of the Americas, New York, NY 10020. Copyright © 2004, 2002, 2000, 1998, 1995, 1992, 1989, 1986, 1983 by The McGraw-Hill Companies, Inc. All rights reserved. No part of this publication may be reproduced or distributed in any form or by any means, or stored in a database or retrieval system, without the prior written consent of The McGraw-Hill Companies, Inc., including, but not limited to, in any network or other electronic storage or transmission, or broadcast for distance learning.

Some ancillaries, including electronic and print components, may not be available to customers outside the United States.

 This book is printed on recycled, acid-free paper containing 10% postconsumer waste.

International   1 2 3 4 5 6 7 8 9 0 DOW/DOW 0 9 8 7 6 5 4 3
Domestic   1 2 3 4 5 6 7 8 9 0 DOW/DOW 0 9 8 7 6 5 4 3

ISBN 0–07–244000–7
ISBN 0–07–121454–2 (ISE)

Publisher: *Margaret J. Kemp*
Senior developmental editor: *Kathleen R. Loewenberg*
Executive marketing manager: *Lisa L. Gottschalk*
Lead project manager: *Peggy J. Selle*
Lead production supervisor: *Sandy Ludovissy*
Senior media project manager: *Tammy Juran*
Media technology producer: *Renee Russian*
Designer: *David W. Hash*
Cover/interior designer: *Jamie E. O'Neal*
Cover photo: *Clyde Butcher,* Title: *Gaskin Bay, Ten Thousand Islands, Everglades National Park, Florida*
Back cover images: ©*SFWMD: Our Mission is to manage and protect water resources of the region by balancing and improving water quality, flood control, natural systems, and water supply.*
Senior photo research coordinator: *Lori Hancock*
Photo research: *LouAnn K. Wilson*
Supplement producer: *Brenda A. Ernzen*
Compositor: *GAC–Indianapolis*
Typeface: *10/12 Times*
Printer: *R. R. Donnelley Willard, OH*

The credits section for this book begins on page C-1 and is considered an extension of the copyright page.

Library of Congress Cataloging-in-Publication Data

Enger, Eldon D.
    Environmental science : a study of interrelationships. — 9th ed. / Eldon D. Enger, Bradley F. Smith.
    p.   cm.
    Includes index.
    ISBN 0–07–244000–7 (acid-free paper)
    1. Environmental sciences. I. Smith, Bradley Fraser. II. Title.

GE105 .E54   2004
363.7—dc21   2002075341
                                    CIP

INTERNATIONAL EDITION ISBN 0–07–121454–2
Copyright © 2004. Exclusive rights by The McGraw-Hill Companies, Inc., for manufacture and export. This book cannot be re-exported from the country to which it is sold by McGraw-Hill. The International Edition is not available in North America.

www.mhhe.com

To Judy, my wife and friend, for
sharing life's adventures.
Eldon Enger

———

For Morgan and the summers on Beavertail,
where the roots of environmental
concern took hold.
Brad Smith

brief contents

contents

New for Chapter 1
• *expanded coverage on sustainable development and Agenda 21.*

New for Chapter 2
• *new section — "Do we consume too much?"*
• *more information on the Global Reporting Initiative*
• *new Global Perspective reading on the gray whales of Neah Bay*

New for Chapter 3
• *expanded section on public perceptions of environmental risks*
• *more information on extended product responsibility*

PART *ii*

Ecological Principles and
Their Application   64

"Fixing" Nature?: Restoration
and the Florida Everglades   65

CHAPTER 4
Interrelated Scientific Principles: Matter, Energy, and Environment   66

New for Chapter 4
• *rewritten section on the scientific method*
• *additional material and an illustration on isotopes*
• *new section on the molecular nature of matter*
• *clarification usage of the words molecule, compound, ions*
• *new material on endothermic and exothermic reactions*
• *new content on chemical reactions for photosynthesis and respiration*

CHAPTER 5
Interactions: Environment and Organisms   82

New for Chapter 5
• *new section — Genes, Populations, and Species*
• *new herbicide resistance illustration*
• *new section on polyploidy*
• *biogeochemical cycles now in nutrient cycles section*
• *new material on the operation of the carbon cycle in aquatic systems*

CHAPTER 6
Kinds of Ecosystems and Communities   108

New for Chapter 12
• *rewritten section on plantation forestry*
• *clarified definitions of marine, brackish, and freshwater*
• *updated figure on forms of pollution*
• *new figure on change in forest areas*
• *new figure on trends in world fish production, including capture and aquaculture*
• *rewritten section on aquaculture*
• *new material on the African Eurasian waterbird agreement*
• *new coverage on the role of keystone species and how their elimination alters ecosystems*
• *updated references to President Bush's administrative policy*

CHAPTER 13
Land-Use Planning

New for Chapter 13
• *new section on smart growth*
• *new table on state comprehensive growth legislation*

CHAPTER 14
Soil and Its Uses

New for Chapter 14
• *table 14.1, Percentage of Land Suitable for Agriculture, has been updated*
• *information on land degradation has been expanded*

CHAPTER 15
Agricultural Methods and Pest Management

New for Chapter 15
• *new Environmental Close-Up reading on industrial livestock production*
• *reorganization of section on integrated pest management*
• *updated information on genetically modified organisms*

CHAPTER 16
Water Management

New for Chapter 16
• *new Global Perspective reading on the dead zone of the Gulf of Mexico*
• *expanded coverage on global water issues*
• *more information on water treatment in the Salina Valley, California*
• *expanded material on chemicals entering drinking water sources*
• *more discussion of the New York City water supply*
• *additional text on water diversions and extractions*
• *expanded material on wetlands*
• *new table on the population of the world's ten largest watersheds*
• *new table on international water disputes*
• *new coverage of the National Research Council's report on Envisioning
the Agenda for Water Resource Research in the 21st Century*

PART v
Pollution and Policy   384
Environmental Policy: Pragmatic
or Polluted?   385

CHAPTER 17
Air Quality Issues

CONTENTS

preface

Why "A Study of Interrelationships?"

Environmental science is an interdisciplinary field. Because environmental disharmonies occur as a result of the interaction between humans and the natural world, we must include both when seeking solutions to environmental problems. It is important to have a historical perspective, appreciate economic and political realities, recognize the role of different social experiences and ethical backgrounds, and integrate these with the science that describes the natural world and how we affect it. *Environmental Science: A Study of Interrelationships* incorporates all of these sources of information when discussing any environmental issue. Furthermore, the authors have endeavored to present a balanced view of issues, diligently avoiding personal biases and fashionable philosophies.

Environmental Science: A Study of Interrelationships is intended as a text for a one-semester, introductory course for students with a wide variety of career goals. They will find it interesting and informative. The central theme is interrelatedness. No text of this nature can cover all issues in depth. What we have done is to identify major issues and give appropriate examples that illustrate the complex interactions that are characteristic of all environmental problems. Many facts are presented in charts, graphs, and figures that help to illustrate the scope of environmental issues. This is not the core of the text, however, since the facts will change.

Organization and Content

This book is divided into five parts and 20 chapters. It is organized to provide an even, logical flow of concepts along with clear illustrations of the major environmental issues of today. In this ninth edition, each part opens with a guest author's essay highlighting an environmental issue close to their home.

Part 1 establishes the theme of the book by looking at the kinds of **environmental issues** typical of different regions of North America. In each region, the specific issues selected involve scientific, social, political, and economic components typical of environmental problems. Chapter 2 focuses on the philosophical base needed to examine environmental issues by discussing various ethical and moral stands that shape how people approach environmental issues. Chapter 3 introduces economic issues and the concept of risk analysis. Both of these topics will be also brought up at several points later in the text.

Part 2 provides an understanding of the **ecological principles** that are basic to organism interactions and the flow of matter and energy in ecosystems. The nature of food chains and how they affect the flow of matter and energy are discussed. Other topics include the efficiency of energy flow through ecosystems, the intricacies of organism-to-organism interaction, and the creative role of natural selection in shaping ecological relationships. Principles of population structure and organization are also developed in this section, with particular attention to the implications of these principles to the growth and impact of human populations.

Part 3 focuses on **energy.** A major emphasis is on the historically important, nonrenewable fossil fuels that have stimulated economic success of the developed economies of the world. Renewable sources of energy are discussed, but with the recognition that they currently are a small part of the world energy picture. Weapons production and nuclear power plants use enormous amounts of energy that can be released from the nucleus of the atom. Both of these uses have caused fear among the public related to the dangers of radiation and the adequacy of waste disposal. These issues are discussed in this section.

Part 4 emphasizes the impact of **human activity on natural ecosystems.** As human populations grow and technology changes, the magnitude of human actions becomes more apparent. The natural ecosystems on land and water are modified to meet human needs. The heavy use of pesticides in agriculture is discussed in this section.

Part 5 deals with the major types of **pollution.** Pollution affects the health and welfare of humans and other organisms. Air pollution, solid waste, and hazardous and toxic substances are discussed in this section. The cost of pollution cannot always be measured in financial terms but may be reflected in the mental and physical health of the populace. Ultimately, governments must address environmental concerns and develop policy to address the concerns. Increasingly, the concerns are international in scope and require negotiations between governments with very different economic conditions and concerns.

What Makes This Text Unique?

This text is written with the student in mind. Both authors have many years of teaching experience and use their knowledge of what helps students learn to shape the text. All aspects of the text: writing style, illustrations, review materials, and boxed readings are designed to be informative without being overly complex. Many of the factual details are included in illustrations and tables rather than in the narrative of the text. For the person who wants facts, they are present but do not obscure the general concepts and principles being described.

Often the clearest way to present information is with an illustration. Each drawing, chart, graph, or photograph is designed to help students visualize an idea or concept, or create a mental picture that enhances the written text. The review materials at the end of each chapter are designed as learning tools. Review questions, vocabulary lists, and concept maps are all useful aids to help students assess whether they have a firm grasp of the content of the chapter.

The authors work very hard to present a balanced, unbiased presentation of the material. It is not the purpose of a textbook to tell you what to think. The purpose of a text is to provide access to information and the conceptual framework needed to understand complex issues so that you can understand the nature of environmental problems and formulate your own views.

Special Features and Learning Aids

1. A **world map** with political boundaries can be found on the inside front cover. We believe that this will help the reader to more fully understand and appreciate global environmental issues. Each of the five parts of the text begins with a **guest essay** that places the upcoming chapters in context for the reader by describing a current environmental issue.

2. Each chapter begins with a set of **learning objectives,** an **outline,** and a **conceptual diagram,** all of which give the student a broad overview of the interrelated forces that are involved in the material to be discussed. The student is encouraged to refer to these resources while reading and reviewing the chapter.

3. Chapters conclude with an Issues—Analysis **case study,** a **summary,** a list of **key terms, review questions, critical thinking questions,** a list of topics that correspond to specific **Internet links** on the accompanying website, and a **new feature—concept mapping.** This new exercise helps to reinforce understanding of basic concepts and principles through creating concept maps from a list of key words. Combined with the introductory conceptual diagram at the beginning of the chapter, these mapping exercises help to illustrate the connections between environmental principles, issues, and possible solutions.

4. To dramatize and clarify text material, each chapter includes a number of **tables, charts, graphs, maps, drawings,** or **photographs.** Each illustration has been carefully chosen to provide a pictorial image or an organized format for showing detailed information, which helps the reader comprehend the chapter material by reinforcing the written word in the text.

5. Each chapter also includes **boxed readings.** Each is an in-depth consideration of a specific situation, an alternative viewpoint, or a wider worldview of the issues discussed in the chapter.

6. The text concludes with two **appendices** that deal with the following topics: critical thinking, and the periodic table of the elements. In addition, there is a complete **glossary** and an **index.** A table of **metric conversions** is located on the inside back cover.

New to This Edition!

Concept Mapping A new learning activity that involves the student in constructing a concept map has been added at the end of each chapter. This activity is designed to help students see how concepts and ideas are related to one another.

Guest Authors Each Part Opener in this edition features a guest author's essay that describes an environmental issue in the author's own backyard. These essays help point out the different types of concerns developing in various parts of the United States, and critical thinking questions help illustrate that these problems often have no easy solution.

Over 100 Reviewers As with previous editions, reviewers' suggestions are incorporated into the text, either through small changes in text or figures to improve clarity and accuracy, or by providing the most up-to-date information available.

Chapter Updates Several chapters have had *major changes* in content or emphasis:

- Chapter 1 has expanded coverage on **sustainable development** and **Agenda 21** to provide current information.
- Chapter 2 has a new section entitled **"Do We Consume Too Much?"** to highlight the economic and ecological impact of consumption.
- Chapter 4 has a completely rewritten section on the **scientific method** to better describe the process of science.
- Chapter 5 has a new section on **genes, populations, and species** to help the reader see the connections between these topics.
- Chapter 6 has new material on both **Mediterranean and dry tropical forests ecosystems,** as suggested by reviewers.
- Chapters 7 and 8 have been updated with recent information on **human population.**
- Chapter 12 has completely revised sections on **plantation forestry and aquaculture.**
- Chapter 13 has a new section on **smart growth** that highlights new trends in **urban land-use planning.**
- Chapter 15 has a rewritten section on **genetically modified organisms,** requested by reviewers.
- Chapter 16 has new information that emphasizes the need for careful **management of water resources.**

- Chapter 17 has been restructured to reflect **current ways of classifying** the different kinds of **air pollutants.**
- Chapter 18 has a new section that better describes the various **ways wastes are categorized.**
- Chapter 19 has new content on **hazardous wastes** and **brownfields.**
- Chapter 20 includes new material on **ecoterrorism.**

New Readings   There are seven all new boxed readings, chosen to complement new text content.

New Art   There are **over 20 new figures and tables,** and many other pieces have been revised.

Useful Supplements

- **Instructor's Manual.** Available to instructors via the accompanying website, this valuable resource contains chapter overviews, key concepts and terms, answers to review questions, suggested classroom activities, and a unique "Resource Locator" that pulls together appropriate material from numerous sources to use with each chapter.
- **Instructor's Testing and Resource CD-ROM.** This cross-platform CD-ROM provides a wealth of resources for the instructor. Supplements featured on this CD-ROM include the lab manual and a computerized test bank using Brownstone Diploma@ testing software to quickly create customized exams. This user-friendly program allows instructors to search for questions by topic, format, or difficulty level; edit existing questions or add new ones; and scramble questions and answer keys for multiple versions of the same test. Other assets on the Instructor's Testing and Resource CD-ROM are grouped within easy-to-use folders.
- **Transparencies.** A set of 100 transparencies is available to users of the text. These acetates include key figures from the text, including new art from this edition.
- **Essential Study Partner CD-ROM.** A complete, interactive student study tool, this CD features animations, videos, and learning activities. From quizzes to interactive diagrams, you'll find that there has never been a better study partner to ensure the mastery of core concepts. Best of all, it's FREE with a new textbook purchase.
- **Digital Content Manager CD-ROM.** This multimedia collection of visual resources allows instructors to use art from the text in multiple formats to create customized classroom presentations, visually based tests or quizzes, dynamic course website content, or attractive printed support materials. The digital assets on this cross-platform CD-ROM are grouped within the following easy-to-use folders:
  - *Illustrations and Photos*   All of the line drawings from the text and hundreds of photos are in ready-to-use digital files.
  - *PowerPoint Lecture Outline*   Ready-made presentations combine art from the text with customized lecture notes covering all 20 chapters.
  - *Tables*   Every table that appears in the text is provided in electronic form.
  - *Active Art*   These special art pieces consist of key images from the text that are converted to a format that allows instructors to break the art down into core elements and then group the various pieces and create customized images. This is especially helpful with difficult concepts, which can be presented step by step.
  - *Animations*   Numerous full-color animations illustrating many different concepts covered in the study of environmental science are provided. The visual impact of motion will enhance classroom presentations and increase comprehension.
- **Online Learning Center** (http://www.mhhe.com/environmentalscience). This comprehensive website offers numerous resources for both students and instructors:

  Student Resources—Everything you need in one place:
  - Study questions
  - Labeling exercises
  - Practice quizzing
  - Hyperlinks on chapter topics
  - Guide to electronic research
  - Regional perspectives (case studies)
  - Environmental issues world map
  - Key-term flashcards
  - How to write a paper
  - Metric equivalents and conversion tables
  - Career information
  - PowerWeb's hundreds of current articles and daily news items integrated into each chapter on the OLC
  - How to contact your government officials

  Instructor Resources—In addition to all of the above, you'll receive:
  - Instructor's manual with a supplements resource chart for each chapter
  - Interactive lecture outlines on PowerPoint
  - Answers to critical thinking questions
  - PageOut (create your own course website)

Related Titles of Interest

1. *Field and Laboratory Activities Manual, 7ed.* (0-07-290913-7) by Enger and Smith
2. Interactive World Issues: Of Place and Planet CD-ROM (0-07-255648-X), Cambridge Studios
3. *Annual Editions: Environment 02/03* (0-07-250682-5), John L. Allen, editor
4. *Taking Sides: Clashing Views on Controversial Environmental Issues, 9ed.* (0-07-303184-4), Theodore D. Goldfarb, editor

5. *Sources: Notable Selections in Environmental Studies, 2ed.* (0-07-303186-0), Theodore D. Goldfarb, editor

6. The Dushkin *Student Atlas of Environmental Issues* (0-697-36520-4), John Allen, University of Connecticut, editor

7. *Life Science Living Lexicon* (CD = 0-697-37993-0; Print = 0-697-12133-X) by William Marchuk

8. *You Can Make a Difference: Be Environmentally Responsible* (0-07-292416-0) by Judy Getis

9. *Environmental Ethics: Divergence and Convergence* (0-07-006180-7) by Botzler and Armstrong

10. *Environmental Problem-Solving: A Case Study Approach* (0-07-027686-2) by Isobel W. Heathcote

11. *Eyewitness World Atlas CD-ROM* (0-07-233220-4), published by Dorling-Kindersley

Acknowledgements

The creation of a textbook requires a dedicated team of professionals who provide guidance, criticism, and encouragement. It is also important to have open communication and dialog to deal with the many issues that arise during the development and production of a text. Therefore, we would like to thank Marge Kemp, Kathy Loewenberg, Heather Wagner, Peggy Selle, Marilynn Taylor, Jamie O'Neal, and Lou Ann Wilson for their suggestions and kindnesses. We would also like to acknowledge the attendees of the McGraw-Hill Environmental Science Symposium in Bellingham, Washington. Matthew Laposata, Allan Matthias, Kenneth Rhinehart, Royal Berglee, Nicole Welch, Paul Rowland, Lynnette Danzl-Tauer, Catherine Christie, Linda Fitzhugh, David Charlet, David Hassenzahl, and Stacy Smith were all generous in their sharing of ideas and gave us a renewed sense of dedication to providing the very best textbook we can for their use in the classroom. Finally, we'd like to thank our many colleagues who have reviewed all, or part, of *Environmental Science: A Study of Interrelationships*. Their valuable input has continued to shape this text and help it meet the needs of instructors around the world.

Eldon D. Enger
Bradley F. Smith

Francis O. Adeola
University of New Orleans
John V. Aliff
Georgia Perimeter College
David F. Arieti
Waubonsee Community College, Oakton Community College
Joe Arruda
Pittsburg State University
Ellen Baker
Santa Monica College
Aaron Barkatt
The Catholic University of America
George C. Boone
Susquehanna University
Kathleen T. Brown
GMC—Augusta Community College
Britt A. Bunyard
Ursuline College
Ray D. Burkett
Southwest Tennessee Community College
Pat Calie
Eastern Kentucky University
Diane V. Carlton
State University of New York, Oneonta
John A. Conners
Alderson-Broaddus College
Terence H. Cooper
University of Minnesota
Kimberly Dawn Cutshaw
Tusculum College
Steven Dahlberg
Concordia College
Elizabeth A. Desy
Southwest State University
Craig Diamond
Florida State University
Jerald J. Dosch
College of Visual Arts
Diane Dudzinski
Washington State Community College
Leslie Fay
Rock Valley College
David K. Ferris
University of South Carolina, Spartanburg
Michael Fidanza
Pennsylvania State University, Reading
Kenneth Finkelstein
Suffolk University
David G. Fisher
Maharishi University of Management
Linda Mueller Fitzhugh
Gulf Coast Community College
Dennis M. Forsythe
The Citadel, Charleston Southern University
Chris Fox
Community College of Baltimore County, Catonsville
Stephanie Smith Freese
Northeast Mississippi Community College
Allan A. Gahr
Gordon College
Sandi B. Gardner
Triton College
Paul J. Gier
Huntingdon College
Marcia L. Gillette
Indiana University, Kokomo
David Goldblum
University of Wisconsin, Whitewater

Gian Gupta
University of Maryland Eastern Shore
Clark G. Haberman
Central Community College, Hastings
Judy C. Hardin
Tusculum College
Graham C. Hickman
Texas A & M University, Corpus Christi
W. Wyatt Hobach
University of Nebraska, Kearney
Robert E. Holtz
Concordia University
Alan R. Holyoak
Manchester College
Robert M. Hordon
Rutgers University
John C. Inman
Presbyterian College
M. Latif A. Khan
Georgia Military College
Ned J. Knight
Linfield College
John Koscelny
Coffeyville Community College
David Kowalewski
Alfred University
Thomas L. Kramer
Muscatine Community College
Steve LaDochy
California State University, Los Angeles
William R. Lamnula
Nazareth College
Matthew Laposata
Kennesaw State University
Peter V. Lindeman
Edinboro University of Pennsylvania
Matthew J. Lindstrom
Siena College
David A. Lovejoy
Westfield State College
Paul E. Lutz
Lenoir-Rhyne College
John S. Mackiewicz
State University of New York, Albany
Michael J. Manetas
Humboldt State University
Heidi Marcum
Baylor University
Robert J. Mason
Temple University
John Mbonifor
Saint Paul's College
Marlene McCall
Community College of Allegheny County
Sheila G. Miracle
Southeast Community College
Kiran Misra
Edinboro University of Pennsylvania

Anita T. Morzillo
Southern Illinois University, Carbondale
Catherine Mossman
University of Wisconsin, Parkside
Victor I. Okereke
State University of New York, Morrisville
Dale Patterson
Truett-McConnell College
James L. Pazun
Pfeiffer University
Mary Lou Peltier
Saint Martin's College
Murray P. Pendarvis
Our Lady of the Lake College
Barbara Yohai Pleasants
Iowa State University
John M. Pleasants
Iowa State University
Usha Rao
St. Joseph's University
Paul A. Rollinson
Southwest Missouri State University
Elaine J. Ross
Kendall College
Michael Ross
College of Saint Benedict/Saint John's University
Wesley Rouse
Hope International University
Robert M. Sanford
University of Southern Maine
Todd L. Scarlett
University of South Carolina
Fred Schindler
Indian Hills Community College
Ronald R. Schultz
Florida Atlantic University
Benjamin B. Steele
Colby-Sawyer College
David A. Steffy
Jacksonville State University
Maura O. Stevenson
Community College of Allegheny College
Jonathan S. Taylor
California State University, Fullerton
Olli H. Tuovinen
Ohio State University
Marge Welch
Southwestern Community College
Sally Welch
Marygrove College
Donald L. Williams
Sterling College
Jeffrey S. Wooters
Pensacola Junior College
Leonard C. Yannielli
Naugatuck Valley Community College
Xiaoqi Zhang
University of Missouri, Rolla

guided tour

The features of this book are *unique*:

The organization and principle features of this book were planned with the students' wholistic learning and comprehension in mind.

World Map

A **world map** with political boundaries can be found on the inside front cover. This will help the reader to more fully understand and appreciate global environmental issues.

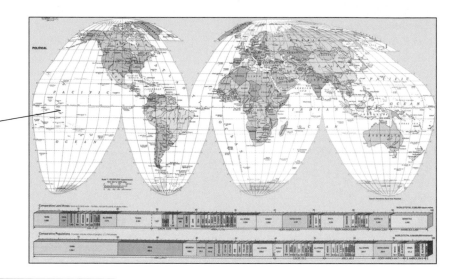

Deer Hunting Within the City of Brotherly Love?

Anne Todd Bockarie

Philadelphia University/Pennsylvania

The industrial Northeast of the United States has had a long history of human intervention in nature. Our footprint on the land has resulted in radical changes in how plants and animals interact. As cities, suburbs, and industry grew, large expanses of forests were cut into small patches. These activities have affected our relationship to white-tailed deer, a keystone species of these forests. Deer were a critical source of food and clothing as Europeans settled the eastern seaboard. Logging, agriculture, and extensive hunting almost wiped out the population by 1930. People were concerned about the low numbers of deer, so the state of Pennsylvania embarked on a restocking program to bring the deer back for recreational hunting and to maintain their critical role in the forest ecosystem.

The program was successful, and today, we have the opposite problem—too many deer. Along the entire East Coast of the United States, deer numbers have surged due to the nutritious food resources found in millions of well-tended lawns, gardens, and farms; the loss of natural predators; and wildlife professionals managing the herd for hunting. The impact on humans of so many deer has been more than 100,000 cases of Lyme disease in the Northeast being reported to the Centers for Disease Control and Protection since 1982. Ticks on deer and mice transmit a bacterium, or spirochete, that causes Lyme disease into the bloodstream when they bite humans. Lyme disease can be treated with antibiotics, but it is difficult to diagnose, and symptoms may last months or years after the initial bite if the infection is not caught in its early stages. The auto accident rate also increased as more drivers struck deer as they crossed roads. Researchers estimate that in 1995, motor vehicle accidents involving deer resulted in 211 deaths, 29,000 injuries, and approximately $1 billion in property damage.

Philadelphia's Fairmount Park Commission recently had to decide what to do about the overpopulation of deer in the 3642 hectares (9000 acres) of parks in the city. It was believed that the deer were eating all the native ground and understory plants, thereby inhibiting the forest from regenerating. Every time a tree would sprout, the deer would browse the seedling. As the plants disappeared, so did the habitat needed for songbirds that use the parks for food and cover on their migration north in the spring. However, part of what makes the park experience unique in Philadelphia is being able to walk through dense woods and see a herd of deer or other wildlife. Most park users want to see and watch deer. The commission hired a scientist to estimate the number of deer in the largest park. He found 60 deer per square mile. It is estimated that the park could feed eight deer per square mile without a negative impact on the forest.

What are the options? The park could be fenced, but it is over 607 hectares (1500 acres) in size and has many entrances, roads, railroad lines, bridges, and sewer and water pipes running through it. The deer in the park could be hunted or culled, but over 1 million people use the area throughout the day and evening, so safety would be a concern. Park officials could use birth control (immunocontraception) on the deer, but this method is very expensive; the park's budget has stayed the same over the past 15 years, and the number of employees has been reduced from 1000 to less than 250. The commission held a public hearing to discuss the options and the concerns of citizens around the park. Here are some of the comments:

"Deer have brothers, fathers, sisters, and mothers. I object on moral and ethical grounds."

"Why are only lethal methods approved by the Pennsylvania Game Commission? Seven percent of Pennsylvanians are hunters, yet they control this state agency and its policies."

"At Fire Island in New York, there has been a 17 percent reduction in herd size using birth control in the last seven years."

"The American Psychological Association has shown that exposure to violence leads to violent actions in children. We are the City of Brotherly Love. The message [of culling the deer] is it's okay to kill."

"Every relevant federal, state, and local agency has been consulted on this issue and supports the cull. One must look at the ecological big picture. The experts have given us the results."

"Our Park Friends Group has seen deer ravage the vegetation. Every tree we plant is eaten. Who is standing up for the vegetation in the park?"

The Fairmount Park Commission decided that the best option would be to hire a professional sharpshooter to cull the herd and impose a curfew at night to keep park users out of harm's way. Protesters demonstrated against this option, and several were arrested when they went into the park and tried to stop the cull.

What Do You Think?

1. Should we leave nature to manage itself, even if it means starvation or the severe decline of a species?
2. In 1842, the U.S. Supreme Court declared that wildlife should be held in trust for all citizens, so how do we resolve the conflict between wildlife protectors and hunters?
3. Do plants and animals have the same inherent rights as humans?
4. Should we support hunting for recreational or management purposes?
5. How much would you be willing to pay to control white-tailed deer in a city with contraceptives?

New!

Guest authors contribute **essays** on environmental issues close to home. Appearing with each Part Opener, these essays are accompanied by critical thinking questions and present real problems with no easy solution.

Human Population Issues

CHAPTER 8

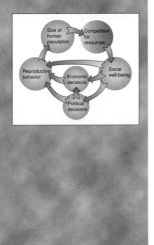

Objectives

After reading this chapter, you should be able to:

- Apply some of the principles discussed in chapter 7 to the human population.
- Differentiate between birthrate and population growth rate.
- Describe the current population situation in the United States.
- Explain why the age distribution and the status and role of women affect population growth projections.
- Recognize that countries in the developed world are experiencing an increase in the average age of their populations.
- Recognize that most countries of the world have a rapidly growing population.
- Describe the implications of the demographic transition concept.
- Understand how an increasing world population will alter the worldwide ecosystem.
- Recognize that rapid population growth and poverty are linked.
- Explain why less-developed nations have high birthrates and why they will continue to have a low standard of living.
- Recognize that the developed nations of the world will be under greater pressure to share their abundance.

Conceptual Study Aids

Each chapter begins with a set of **learning objectives**, an **outline**, and a **conceptual diagram**, all of which give the student a broad overview of the interrelated forces that are involved in the material to be discussed. Students are encouraged to refer to these resources while reading and reviewing the chapter.

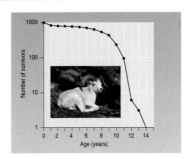

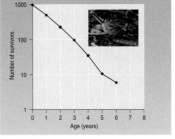

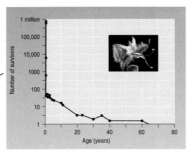

figure 7.2 **Types of Survivorship Curves** (a) The Dall sheep is a large mammal that produces relatively few young. Most of the young survive, and survival is high until individuals reach old age, when they are more susceptible to predation and disease. (b) The curve shown for the white-crowned sparrow is typical of that for many kinds of birds. After a period of high mortality among the young, the mortality rate is about equal for all ages of adult birds. (c) Many small animals and plants produce enormous numbers of offspring. Mortality is very high in the younger individuals, and few individuals reach old age.

Quality Art

To dramatize and clarify text material, each chapter includes a number of **tables, charts, graphs, maps, drawings,** or **photographs.** Every illustration has been carefully chosen to provide a pictorial image or an organized format for showing detailed information, helping the reader better comprehend the chapter material.

Population Density and Spatial Distribution

Because of such factors as soil type, quality of habitat, and availability of water, organisms normally are distributed unevenly. Some populations have many individuals clustered into a small space, while other populations of the same species may be widely dispersed. **Population density** is the number of organisms per unit area. For example, fruit-fly populations are very dense around a source of rotting fruit, while they are rare in other places. Similarly, humans are often clustered into dense concentrations we call cities, with lower densities in rural areas. When the population density is too great, all individuals within the population are injured because they compete severely with each other for necessary resources. Plants may compete for water, soil nutrients, or sunlight. Animals may compete for food, shelter, or nesting sites. In animal populations, overcrowding might cause some individuals to explore and migrate into new areas. This movement from densely populated locations to new areas is called **dispersal.** It relieves the overcrowded conditions in the home area and, at the same time, increases the population in the places to which they migrate. Often, it is juvenile individuals that relieve overcrowding by leaving. The pressure to migrate from a population (**emigration**) may be a result of seasonal reproduction leading to a rapid increase in population size or environmental changes that intensify

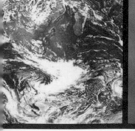

New! Concept Mapping

Chapters conclude with an **"Issues—Analysis"** case study, a **summary,** a list of **key terms, review questions, critical thinking questions,** and a new feature—**concept mapping.** In Chapter 1 the value and nature of concept maps are explained. Examples and guidelines for constructing concept maps are also given. Each subsequent chapter provides a list of terms for students to complete their own mapping. An integrated study tool, concept mapping helps readers make the important "connections" that foster real learning.

Concept Map

The construction of a concept map is a technique that helps students recognize how separate concepts are related to one another. Some concept maps may be simple, orderly lists. Others may form networks of connections that help to show how ideas are linked. It is important to understand that there is not just one way in which things can be put together. The examples show two different ways the same concepts can be organized. (Take another look at figure 1.2. It is a variety of concept map.)

Construct a concept map to show relationships among the following concepts:

biology	observation	scientific method
experiment	science	theory
hypothesis		

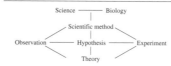

Example 1

Example 2

Interactive Exploration

Check out the website at **http://www.mhhe.com/environmentalscience** and click on the cover of this textbook for quizzing, career information, case studies, and hot links for the following topics:

General Environmental Sites

General Ecology Sites

Environmental and Ecological Organization Sites

History of Environmental Studies

Miscellaneous Environmental Resources

Introductory Materials and Governmental Sites

Science as a Process

Introductory Sites

Writing Papers and Study Tips

Glossaries and Dictionaries

Careers in Science

Utility and Organizational Sites

Global Ecology

Shaping U.S. Environmental Policy as the New Century Begins

During the presidential campaign of 2000, the economy, health care, and education were primary issues, and environmental issues were secondary. However, as the new administration began to formulate policy, national and global environmental issues became increasingly visible. Public debates about drilling in the Arctic National Wildlife Refuge, global warming, the Kyoto Protocol, and the environmental concerns surrounding world trade made the environment front-page news.

Rather than attempting to pass judgment on the Bush administration's policies, it is perhaps more beneficial to look at the polarization that exists over the policies. The statements that follow present widely different perspectives on the significance of policy decisions of the Bush administration. One presents the views of the administration, the other the views of environmental organizations.

The Bush Administration's Environmental Record
(Published by the Bush Administration)

Preamble

President Bush has articulated a vision for environmental protection that focuses on results: Cleaner air, water and land, and healthier people and ecosystems. To achieve these goals, we need a strong and growing economy. Our environmental policies thus recognize the importance of a robust economy, which provides the public and private resources needed to make new investments in environmental conservation. The President sees these goals as complementary, rather than competing—strong economic growth and strong environmental protection can and must go hand-in-hand. In his first year, President Bush made significant progress toward achieving each of these goals.

Brownfields Cleanup—Bringing New Life to Abandoned Sites in Our Cities and Towns: Fulfilling an important campaign commitment, President Bush signed historic legislation that will result in more cleanup and redevelopment of contaminated industrial sites, improving the environment, protecting public health, creating jobs, and revitalizing communities.

Clear Skies—A Clean Air Act for the 21st Century: President Bush's initiative would dramatically improve air quality by cutting power plant's emissions of three critical pollutants by 70 percent—more than any other presidential clean air initiative. This historic legislative proposal would bring clean air to American communities faster, more reliably, and more effectively than the current Clean Air Act.

Energy Bill—Promoting Clean, Affordable, Reliable Energy for America: President Bush has prepared the first national energy policy in years, and is working with Congress to pass legislation that will promote affordable, reliable, and clean energy that is essential to America's security, environmental quality, and economic growth.

Land Conservation—Working in Partnership with States: President Bush has pushed to fully fund the Land and Water Conservation Fund, and worked successfully with Congress to significantly increase its funding. President Bush has also used the Land and Water Conservation Fund to increase support for partnerships for cooperative conservation, and has requested $100 million in FY '03 funding for a new Cooperative Conservation Initiative.

Global Environment—A Realistic, Growth-Oriented Approach to Climate Change: The President has committed America to a new strategy to meet the challenge of long-term global climate change by reducing the greenhouse gas intensity of our economy by 18 percent over the next 10 years. This goal is supported by a broad range of domestic and international climate change initiatives, including $4.5 billion in FY '03 funding for climate change, as well as $178 million for the Global Environment Facility and $50 million to help conserve tropical forests through programs like debt-for-nature swaps.

Budget: President Bush's $44.4 billion FY '03 environment and natural resources budget request is the highest ever—$1.4 billion, or 3 percent, higher than FY '02 enacted. The President's budget proposal provides $4.1 billion, the highest level ever for EPA's operating program, and provides the highest level ever for EPA state program grants, $1.2 billion.

The Bush Budget: Bad News for Our Environment and Our Health (Published by the Sierra Club, with comments from the Natural Resources Defense Council)

Preamble

As this is written on the eve of Earth Day 2002, our nation's environmental landscape is changing for the worse. Agencies throughout the Bush administration are taking explicit directions from big corporate polluters, allowing these corporations to rewrite the agency rules that give life to America's environmental laws.

It is not news that the Bush administration has an anti-environmental tilt. In fact, the early months of this presidency were defined in part by overwhelming public disapproval of the administration's positions on arsenic in drinking water, drilling in the Arctic National Wildlife Refuge, and carbon dioxide pollution from power plants. Since September 11, however, the environmental assault has a quietly intensified, bolstered by a growing critical mass of presidential appointees at key federal agencies actively pursuing an anti-environment agenda, emboldened by the president's surge in popularity, and unchecked by news media distracted by the war on terrorism.

Energy Research Cuts: The Bush Administration has proposed cutting energy efficiency research and development by 27% overall, with over 50% cut in some specific programs in FY '03. These cuts would hamstring efforts to improve efficiency in homes, vehicles, businesses, and industry.

President Bush has proposed cutting renewable energy research and development programs by 36% in FY '03. This cut could slow the development of key renewable energy technologies.

Environmental Protection Agency Cuts: Overall President Bush is cutting $500 million from the EPA's budget including a cut of $158 million from the EPA's efforts to enforce laws that keep polluters from fouling the air we breathe and the water we drink. In addition to these cuts, his budgetary sleight of hand shifts money to states, crippling the federal government's ability to enforce fair and consistent environmental standards.

Interior Department Cuts: The President's budget includes numerous examples where he shifts money away from conserving landscapes and wildlife and instead uses the money for mining and oil drilling on our public lands. In addition, it tilts the balance from experienced federal oversight and lets individual states decide whether to protect wildlife and open space.

The President cuts the U.S. Fish and Wildlife Service budget by $168 million, slashing money dedicated to protecting wildlife habitat, wetlands restoration, and endangered species.

Web Integration

In a section at the end of each chapter, **Interactive Exploration** lists important topics for which there are hyperlinks available on the accompanying website. Already researched and validated, these links are a helpful study tool for students.

Relevant Box Readings

Each chapter also includes **boxed readings.** These provide an in-depth consideration of a specific situation that is relevant to the content, an alternative viewpoint, or a wider worldview of the issues discussed in the chapter.

Appendices

The text concludes with two **appendices** that offer some valuable information on critical thinking, and the periodic table of the elements. A chart on metric conversion can be found on the inside back cover. In addition, there is a complete **glossary** and **index.**

Traditionally, elements are represented in a shorthand form by letters. For example, the formula for water, H_2O, shows that a molecule of water consists of two atoms of hydrogen and one atom of oxygen. These chemical symbols for each of the atoms can be found on any periodic table of the elements. Using the periodic table, we can determine the number and position of the various parts of atoms.

Notice that atoms number 3, 11, 19, and so on are in column one. The atoms in this column act in a similar way since they all have one electron in their outermost layer. In the next column, Be, Mg, Ca, and so on act alike because these metals all have two electrons in their outermost electron layer. Similarly, atoms number 9, 17, 35, and so on all have seven electrons in their outer layer.

Knowing how fluorine, chlorine, and bromine act, you can probably predict how iodine will act under similar conditions. At the far right in the last column, argon, neon, and so on all act alike. They all have eight electrons in their outer electron layer. Atoms with eight electrons in their outer electron layer seldom form bonds with other atoms.

Periodic Table of the Elements

Representative Elements (s Series)

Key

1
Hydrogen
H
1.0079

Atomic Number
Name
Symbol
Atomic Weight

Representative Elements (p Series)

Transition Metals (d Series of Transition Elements)

Inner Transition Elements (f Series)

*Lanthanides

| 58 Cerium Ce 140.12 | 59 Praseodymium Pr 140.907 | 60 Neodymium Nd 144.24 | 61 Promethium Pm 144.913 | 62 Samarium Sm 150.35 | 63 Europium Eu 151.96 | 64 Gadolinium Gd 157.25 | 65 Terbium Tb 158.925 | 66 Dysprosium Dy 162.50 | 67 Holmium Ho 164.930 | 68 Erbium Er 167.26 | 69 Thulium Tm 168.934 | 70 Ytterbium Yb 173.04 | 71 Lutetium Lu 174.97 |

** Actinides

| 90 Thorium Th 232.038 | 91 Protactinium Pa (231) | 92 Uranium U 238.03 | 93 Neptunium Np (237) | 94 Plutonium Pu 244.064 | 95 Americium Am (243) | 96 Curium Cm (247) | 97 Berkelium Bk (247) | 98 Californium Cf 242.058 | 99 Einsteinium Es (254) | 100 Fermium Fm 257.095 | 101 Mendelevium Md 258.10 | 102 Nobelium No 259.10 | 103 Lawrencium Lr 260.105 |

A-2

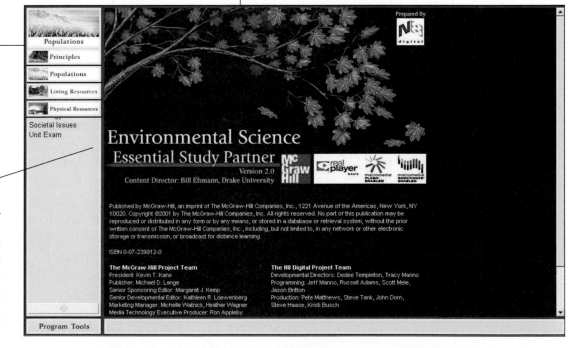

The Essential Study Partner (ESP) for Environmental Science is a student tutorial CD containing high-quality 3-D animations, interactive study activities, illustrated overviews of key topics in environmental science, and self-quizzes and exams for each important unit.

Digital Content Manager

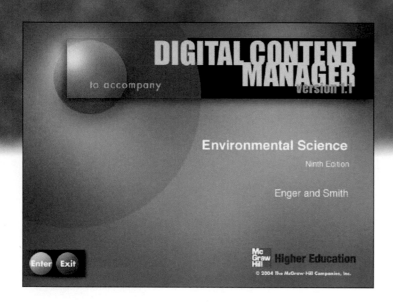

Call the McGraw-Hill Customer Service Department at 800-338-3987, or contact your local sales representative to obtain this valuable teaching aid.

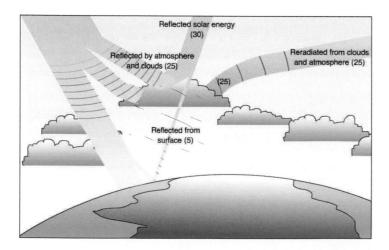

This multimedia collection of visual resources allows instructors to utilize artwork from the text in multiple formats to create customized classroom presentations, visually based tests and/or quizzes, dynamic course website content, or attractive printed support materials. The digital assets on this cross-platform CD-ROM are grouped within the following easy-to-use folders:

- **Art Library.** Full-color digital files of all illustrations in the book.

- **Photo Library.** Hundreds of discipline appropriate photos in digital files.

- **PowerPoint Lecture Outline.** Ready-made presentations combine art from the text with customized lecture notes, covering all 20 chapters.

- **Table Library.** Every table that appears in the text is provided in electronic form.

- **Active Art.** These special art pieces consist of key images from the text that are converted to a format that allows instructors to break the art down into core elements and then group the various pieces and create customized images. This is especially helpful with difficult concepts; they can be presented step by step.

- **Animations Library.** Numerous full-color animations illustrating many different concepts covered in the study of environmental science are provided. The visual impact of motion will enhance classroom presentations and increase comprehension.

The Online Learning Center

www.mhhe.com/environmentalscience

Imagine the advantages of having so many learning and teaching tools in one place—all at your fingertips.

Students, you'll appreciate extensive self-quizzing opportunities, interactive activities, case studies, career information and related web links, in addition to PowerWeb, which gives you access to hundreds of current articles and daily news items, and Access Science—the equivalent of an encyclopedia on-line.

Instructors, you'll want to take advantage of classroom activities, answers to critical thinking questions, an Instructor's Manual with a supplements chart for each chapter, case studies, and access to the PageOut: Course Website Development Center.

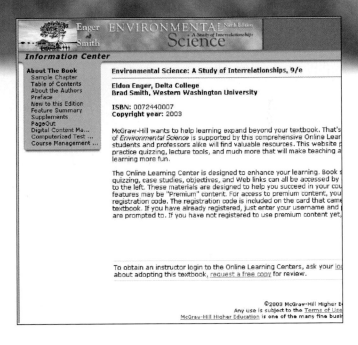

Contact your
McGraw-Hill
sales representative
for more information or
visit *www.mhhe.com*

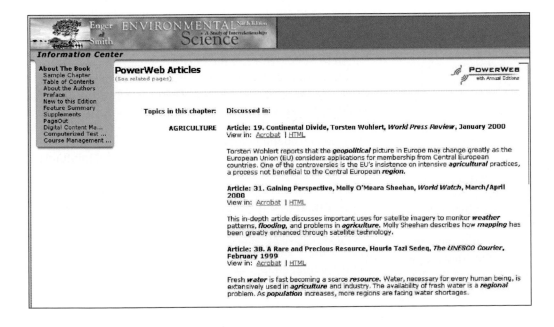

about the authors

Eldon D. Enger

Eldon D. Enger is a professor emeritus of biology at Delta College, a community college near Saginaw, Michigan. He received his B.A. and M.S. degrees from the University of Michigan. Professor Enger has over 30 years of teaching experience, during which he has taught biology, zoology, environmental science, and several other courses. He has been very active in curriculum and course development. Recent activities include the development of a learning community course in stream ecology, and a plant identification course. He was also involved in the development of an environmental regulations course, and an environmental technician curriculum.

Professor Enger is an advocate for variety in teaching methodology. He feels that if students are provided with varied experiences, they are more likely to learn. In addition to the standard textbook assignments, lectures, and laboratory activities, his classes are likely to include writing assignments, student presentation of lecture material, debates by students on controversial issues, field experiences, individual student projects, and discussions of local examples and relevant current events. Textbooks are valuable for presenting content, especially if they contain accurate, informative drawings and visual examples. Lectures are best used to help students see themes and make connections, and laboratory activities provide important hands-on activities.

Professor Enger has been a Fulbright Exchange Teacher to Australia and Scotland, received the Bergstein Award for Teaching Excellence and the Scholarly Achievement Award from Delta College, and participated as a volunteer in an Earthwatch Research Program in Costa Rica, the Virgin Islands, and Western Australia. In 2002, he was a member of a People to People delegation to South Africa, which involved learning about issues and challenges concerning resource management in South Africa. He has also visited New Zealand, New Guinea, Fiji, Puerto Rico, Mexico, Canada, Morocco, many areas in Europe, and much of the United States. During these travels, he has spent considerable time visiting coral reefs, ocean coasts, mangrove swamps, alpine tundra, prairies, tropical rainforests, cloud forests, deserts, temperate rainforests, coniferous forests, deciduous forests, and many other special ecosystems. This extensive experience provides the background to look at environmental issues from a broad perspective.

Professor Enger is married, has two adult sons, and enjoys a variety of outdoor pursuits such as kayaking, cross-country skiing, hiking, hunting, fishing, beekeeping camping, and gardening.

Bradley F. Smith

Bradley F. Smith is the dean of Huxley College of Environmental Studies at Western Washington University in Bellingham, Washington. Prior to assuming the position as dean in 1994, he served from 1991 to 1994 as the first director of the Office of Environmental Education for the U.S. Environmental Protection Agency in Washington, D.C. Dean Smith also served as the acting president of the National Environmental Education and Training Foundation in Washington, D.C., and as a special assistant to the EPA administrator.

Before moving to Washington, D.C., Dean Smith was a professor of political science and environmental studies for 15 years, and the executive director of an environmental education center and nature refuge for five years.

Dean Smith has considerable international experience. He was a Fulbright Exchange Teacher to England and worked as a research associate for Environment Canada in New Brunswick. He is a frequent speaker on environmental issues worldwide and serves on the International Scholars Program for the U.S. Information Agency. He also served as a U.S. representative on the Tri-Lateral Commission on Environmental Education with Canada and Mexico. In 1995, he was awarded a NATO fellowship to study the environmental problems associated with the closure of former Soviet military bases in Eastern Europe. Dean Smith is an adjunct professor at Far Eastern State University in Vladivostok, Russia, and is a member of the Russian Academy of Transport. He also serves as a commissioner for the International Union for the Conservation of Nature. He is a frequent speaker at universities in China.

Nationally, Dean Smith serves as a member/advisor for many environmental organizations' board of directors, advisory councils, and executive committees, including the President's Council for Sustainable Development (Education Task Force), the Science Advisory Boards for MOTE Marine Laboratory in Sarasota, Florida, and the Center for Sustainable Futures in Vermont. In 2002, he was appointed by the govenor of the state of Washington to chair the States Sustainable Advisory Committee.

Dean Smith holds B.A. and M.A. degrees in political science and public administration and a Ph.D. from the School of Natural Resources and Environment at the University of Michigan.

Dean Smith lives with his wife, Daria, daughter, Morgan, son, Ian, and English setter, Skye, along Puget Sound south of Bellingham. He is an avid outdoor enthusiast.

ENVIRONMENTAL
Science

A Study of Interrelationships

PART

i

Interrelatedness

Deer Hunting Within the City of Brotherly Love?

Anne Todd Bockarie

Philadelphia University/Pennsylvania

The industrial Northeast of the United States has had a long history of human intervention in nature. Our footprint on the land has resulted in radical changes in how plants and animals interact. As cities, suburbs, and industry grew, large expanses of forests were cut into small patches. These activities have affected our relationship to white-tailed deer, a keystone species of these forests. Deer were a critical source of food and clothing as Europeans settled the eastern seaboard. Logging, agriculture, and extensive hunting almost wiped out the population by 1930. People were concerned about the low numbers of deer, so the state of Pennsylvania embarked on a restocking program to bring the deer back for recreational hunting and to maintain their critical role in the forest ecosystem.

The program was successful, and today, we have the opposite problem—too many deer. Along the entire East Coast of the United States, deer numbers have surged due to the nutritious food resources found in millions of well-tended lawns, gardens, and farms; the loss of natural predators; and wildlife professionals managing the herd for hunting. The impact on humans of so many deer has been more than 100,000 cases of Lyme disease in the Northeast being reported to the Centers for Disease Control and Protection since 1982. Ticks on deer and mice transmit a bacterium, or spirochete, that causes Lyme disease into the bloodstream when they bite humans. Lyme disease can be treated with antibiotics, but it is difficult to diagnose, and symptoms may last months or years after the initial bite if the infection is not caught in its early stages. The auto accident rate also increased as more drivers struck deer as they crossed roads. Researchers estimate that in 1995, motor vehicle accidents involving deer resulted in 211 deaths, 29,000 injuries, and approximately $1 billion in property damage.

Philadelphia's Fairmount Park Commission recently had to decide what to do about the overpopulation of deer in the 3642 hectares (9000 acres) of parks in the city. It was believed that the deer were eating all the native ground and understory plants, thereby inhibiting the forest from regenerating. Every time a tree would sprout, the deer would browse the seedling. As the plants disappeared, so did the habitat needed for songbirds that use the parks for food and cover on their migration north in the spring. However, part of what makes the park experience unique in Philadelphia is being able to walk through dense woods and see a herd of deer or other wildlife. Most park users want to see and watch deer. The commission hired a scientist to estimate the number of deer in the largest park. He found 60 deer per square mile. It is estimated that the park could feed eight deer per square mile without a negative impact on the forest.

What are the options? The park could be fenced, but it is over 607 hectares (1500 acres) in size and has many entrances, roads, railroad lines, bridges, and sewer and water pipes running through it. The deer in the park could be hunted or culled, but over 1 million people use the area throughout the day and evening, so safety would be a concern. Park officials could use birth control (immunocontraception) on the deer, but this method is very expensive; the park's budget has stayed the same over the past 15 years, and the number of employees has been reduced from 1000 to less than 250. The commission held a public hearing to discuss the options and the concerns of citizens around the park. Here are some of the comments:

"Deer have brothers, fathers, sisters, and mothers. I object on moral and ethical grounds."

"Why are only lethal methods approved by the Pennsylvania Game Commission? Seven percent of Pennsylvanians are hunters, yet they control this state agency and its policies."

"At Fire Island in New York, there has been a 17 percent reduction in herd size using birth control in the last seven years."

"The American Psychological Association has shown that exposure to violence leads to violent actions in children. We are the City of Brotherly Love. The message [of culling the deer] is it's okay to kill."

"Every relevant federal, state, and local agency has been consulted on this issue and supports the cull. One must look at the ecological big picture. The experts have given us the results."

"Our Park Friends Group has seen deer ravage the vegetation. Every tree we plant is eaten. Who is standing up for the vegetation in the park?"

The Fairmount Park Commission decided that the best option would be to hire a professional sharpshooter to cull the herd and impose a curfew at night to keep park users out of harm's way. Protesters demonstrated against this option, and several were arrested when they went into the park and tried to stop the cull.

What Do You Think?

1. Should we leave nature to manage itself, even if it means starvation or the severe decline of a species?

2. In 1842, the U.S. Supreme Court declared that wildlife should be held in trust for all citizens, so how do we resolve the conflict between wildlife protectors and hunters?

3. Do plants and animals have the same inherent rights as humans?

4. Should we support hunting for recreational or management purposes?

5. How much would you be willing to pay to control white-tailed deer in a city with contraceptives?

Environmental Interrelationships

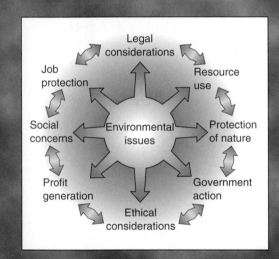

CHAPTER 1

Objectives

After reading this chapter, you should be able to:

- Understand why environmental problems are complex and interrelated.
- Realize that environmental problems involve social, ethical, political, and economic issues, not just scientific issues.
- Understand that acceptable solutions to environmental problems are not often easy to achieve.
- Understand that all organisms have an impact on their surroundings.
- Understand what is meant by an ecosystem approach to environmental problem solving.
- Recognize that different geographic regions have somewhat different environmental problems, but the process for resolving them is the same and involves compromise.

Chapter Outline

The Field of Environmental Science

The Interrelated Nature of Environmental Problems

Environmental Close-Up: *Science Versus Policy*

Global Perspective: *Fish, Seals, and Jobs*

An Ecosystem Approach

Regional Environmental Concerns

Environmental Close-Up: *Headwaters Forest*

    The Wilderness North

    The Agricultural Middle

Environmental Close-Up: *The Greater Yellowstone Ecosystem*

    The Dry West

    The Forested West

    The Great Lakes and Industrial Northeast

    The Diverse South

The Field of Environmental Science

Environmental science is an interdisciplinary area of study that includes both applied and theoretical aspects of human impact on the world. Since humans are generally organized into groups, environmental science must deal with politics, social organization, economics, ethics, and philosophy. Thus, environmental science is a mixture of traditional science, individual and societal values, and political awareness. (See figure 1.1.)

Although environmental science as a field of study is evolving, it is rooted in the early history of civilization. Many ancient cultures expressed a reverence for the plants, animals, and geographic features that provided them with food, water, and transportation. These features are still appreciated by many modern people. Although the following quote from Henry David Thoreau (1817–1862) is over a century old, it is consistent with current environmental philosophy:

> I wish to speak a word for Nature, for absolute freedom and wildness, as contrasted with a freedom and culture merely civil . . . to regard man as an inhabitant, or a part and parcel of Nature, rather than a member of society.

The current interest in the state of the environment began with philosophers like Thoreau and scientists like Rachel Carson and received emphasis from the organization of the first Earth Day on April 22, 1970. Subsequent Earth Days reaffirmed this commitment. As a result of this continuing interest in the state of the world and how people both affect it and are affected by it, environmental science is now a standard course or program at many colleges. It is also included in the curriculum of high schools. Most of the concepts covered by environmental science courses had previously been taught in ecology, conservation, biology, or geography courses. Environmental science incorporates the scientific aspects of

these courses with input from the social sciences, such as economics, sociology, and political science, creating a new interdisciplinary field.

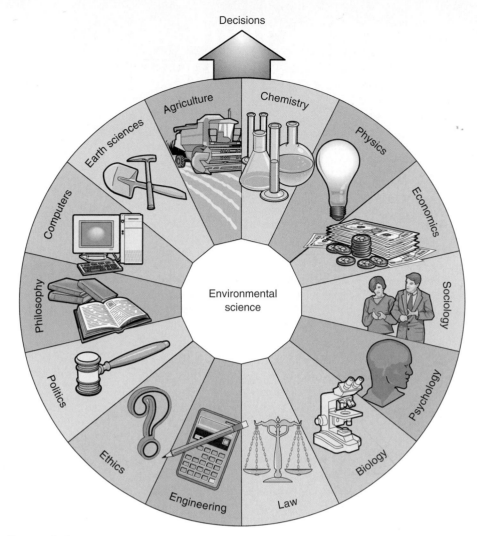

figure 1.1 **Environmental Science** The field of environmental science involves an understanding of scientific principles, economic influences, and political action. Environmental decisions often involve compromise. A decision that may be supportable from a scientific or economic point of view may not be supportable from a political point of view without modification. Often political decisions relating to the environment may not be supported by economic analysis.

The Interrelated Nature of Environmental Problems

Environmental science is by nature an interdisciplinary field. The word *envi-ronmental* is usually understood to mean the surrounding conditions that affect people and other organisms. In a broader definition, **environment** is everything that affects an organism during its lifetime. In turn, all organisms including people affect many components in their environment. (See figure 1.1.) From a human perspective, environmental issues involve concerns about science, nature, health, employment, profits, politics, ethics, and economics.

Most social and political decisions are made with respect to political jurisdictions, but environmental problems do not necessarily coincide with these artificial political boundaries. For example,

environmental CLOSE-UP

Science Versus Policy

Scientific knowledge and government policy do not always agree. The scientific community can advise governments but cannot insist that certain policies be adopted. Governments may halt some scientific research because they control funding sources, or they may introduce regulations that make continuing the research difficult. For example, much federal money was spent on alternative energy research during the Carter presidency, but many of these projects were not in favor during the Reagan and George H. W. Bush presidencies. Funding was reduced and much of the research into alternative fuels stopped. Conversely, the passage of the 1990 Clean Air Act during the Bush presidency mandated that alternative-fuel automobiles be used in some cities with severe air pollution problems.

Government policy may be contrary to prevailing scientific opinion for economic or political reasons. For many years during the Reagan administration, most scientists in the United States and Canada agreed that the burning of high-sulfur coal and other acid-producing fuels was responsible for acid rain, which was leading to the deaths of lakes in parts of Canada and the northeastern United States. The administration continued to insist that the information was not conclusive and that the problem should be studied in greater detail.

The Clinton and, to a larger degree, the George W. Bush administrations faced a similar dilemma regarding the debate over global warming. While the scientific community strongly supported the position that global warming is real and that it is due in large part to human causes, Clinton, and especially Bush were cautious in moving too fast to reduce emissions in the United States. This is in part due to the potential economic consequences involved. It is difficult to separate scientific knowledge, governmental policy, and economic policy. Can you identify similar examples of this debate in your community? Can you explain how industry, government, consumers, and environmental issues are interrelated? Why is this the case?

---

air pollution may involve several local units of government, several states or provinces, and even different nations. The forest fires that raged in Mexico in 1998 had a severe impact on air quality in Texas. Air pollution generated in China affects air quality in western coastal states in the United States and in British Columbia, Canada. On a more local level, the air pollution problems in Juarez, Mexico, are also problems in El Paso, Texas. But the issue is more than air quality and human health. Lower wage rates and less strict environmental laws have influenced some U.S. industries to move to Mexico for economic advantages. Mexico and many other developing nations are struggling to improve their environmental image and need the money generated by foreign investment to improve the conditions and the environment in which their people live.

Air pollutants produced in the major industrial regions of the United States drift across the border into Canada, where acid rain damages lakes and forests. A long-standing dispute exists between the United States and Canada over this issue. Canada claims that the United States should be doing more to reduce emissions that cause acid rain, and the United States claims it is doing as much as it can. In another example, farmers who use water from the Colorado River for irrigation reduce the quality and quantity of water entering Mexico. This causes political friction between Mexico and the United States.

The issue of declining salmon stocks in the Pacific Northwest of the United States and British Columbia, Canada, is another example of political friction over a shared natural resource. It has been calculated that on the U.S. side of the salmon issue alone, there are five federal cabinet-level departments, two federal agencies, five federal laws in question as well as numerous tribal treaties, commissions, and court decisions. All of this is in addition to many state-level departments, commissions, and rulings. If all of this were not sufficient, international bodies such as the United Nations and international treaties impact the fate of the salmon. Considering all this complexity, it is not surprising that the salmon is in such a dangerous status. (See figure 1.2.)

Because of all these political, economic, ethical, and scientific links, solving environmental problems is complicated. Environmental problems seldom have simple solutions. However, international organizations, such as the International Joint Commission, have had major bearing on the quality of the environment over broad regions of the world.

The International Joint Commission was established in 1909, when the Boundary Waters Treaty was signed between the United States and Canada. The treaty was established in part to provide that the "boundary waters and waters flowing across the boundary shall not be polluted on either side to the injury of health or property of the other." The commission has been instrumental in identifying areas of concern and encouraging the cleanup of polluted sites that affect the quality of the Great Lakes and other boundary waters. In general, the two governments have

Harvest

Overfishing has contributed to the decline of many fish populations. Often this exploitation is caused by fishery managers trying to access harvestable hatchery salmon or other abundant fish in areas that contain depleted wild salmon populations.

Hatcheries

Hatchery fish that augment harvest levels can interbreed with wild fish, resulting in the loss of genetic diversity. Hatchery fish can also spread disease and compete with wild fish for food and habitat.

Habitat

Rural and Urban: Salmon face multiple, complex threats in the developed lower regions of watersheds. Problems include low water flows, pollution, degraded physical habitats, and migration barriers such as culverts.

Forests: Improper forest practices and road construction and maintenance are the biggest threat to salmon in the upper watershed. Department of Natural Resources in Washington receives 12,000 applications for forest practices annually.

Hydropower

Dams can block fish migration to and from the ocean, kill fish passing through turbines, delay migration, and increase predation. Dams can also cause inadequate flow downstream.
There are 1,018 dams on Washington rivers. The Columbia River hosts 150 hydroelectric projects and 250 reservoirs — more than half the length of the river is blocked to salmon and steelhead.

figure 1.2 **The Four H's: Human Activities That Affect Wild Salmon Survival** The interrelated nature of environmental problems is evident in the diminishing numbers of wild salmon in the Pacific Northwest of the United States and in British Columbia, Canada. This diagram portrays the plight of wild salmon in the state of Washington, but the issue is regionwide.

Source: Washington Department of Natural Resources Newsletter, Winter 1998, Department of Natural Resources, Olympia, Washington.

listened to the commission's advice and have responded by initiating cleanup activities.

The first worldwide meeting of heads of state directed to concern for the environment took place at the Earth Summit, formally known as the United Nations Conference on Environment and Development (UNCED) in Rio de Janeiro in 1992. Most countries have also signed agreements on **sustainable development** and biodiversity. The policy statements on sustainable development at the UNCED were identified as Agenda 21. Agenda 21 is a comprehensive plan of action to be taken globally, nationally, and locally by organizations of the UN system, national governments, and major groups in every area in which humans impact the environment.

More than 178 governments at the 1992 conference adopted Agenda 21, the Rio Declaration on Environment and Development, and the Statement of Principles for the Sustainable Management of Forests to ensure effective follow-up of UNCED. The Commission on Sustainable Development (CSD) was created in 1993 to monitor and report on implementation of the agreements at the local, national, regional, and international levels. It was agreed that a five-year review of Earth Summit progress would be made in 1997 by the United Nations General Assembly meeting in special session. The Fifty-fifth General Assembly session decided in 2000 that the CSD would serve as the central organizing body for the 2002 World Summit on Sustainable Development, which was held in Johannesburg, South Africa. The Fifty-fifth General Assembly session also noted that, because of globalization, external factors have become critical in determining the success or failure of developing countries in their national sustainable development efforts.

In 1997, representatives from 125 nations met in Kyoto, Japan, for the Third Conference of the United Nations Framework Convention on Climate Change. This conference, commonly referred to as the Kyoto Conference on Climate Change, resulted in commitments from the participating nations to reduce their overall emissions of six greenhouse gases (linked to global warming) by at least 5 percent below 1990 levels and to do so between the years 2008 and 2012. The Kyoto Protocol, as the agreement was called, was viewed by many as one of the most important steps to date in environmental protection and international diplomacy. It may be years before we will know if all countries that signed these agreements will meet their commitments to environmental improvement, but they have at least stated their intention to do so.

The United Nations, through the United Nations Educational, Scientific, and Cultural Organization (UNESCO) and the United Nations Environment Programme (UNEP), has supported many environmental programs. A recent undertaking is the International Environmental Education Programme (IEEP). This program recognizes the need for both formal environmental education in schools and the informal education that occurs through the media and groups of interested citizens. Conferences on environmental education were first held during the 1970s and continue to the present.

Global Perspective

Fish, Seals, and Jobs

In 1995, the Canadian government announced a moratorium on cod fishing along the east coast of Canada. The cod industry contributes $700 million a year and 31,000 jobs to the Canadian economy, primarily in Newfoundland. At the same time, the government announced that it would begin a program to encourage the harvesting of harp seals by helping develop markets for seal products. Environmental groups opposed the harvesting of the seals.

How do all these pieces fit together? It is thought that the low numbers of cod in the North Atlantic are partially the result of an increasing population of harp seals. While overfishing, larger nets, and other factors have also contributed to the decline of the cod, it is true that harp seals feed on fish that could have been harvested. The current harp seal population along the Atlantic Coast more than doubled since the 1970s to 4.8 million in 1997 and was projected to reach 6 million by 2002 if there were no hunts. The increase in harp seals is at least partly the result of actions during the 1970s by environmental groups that sought to stop the killing of seals because they considered the harvesting method inhumane. The traditional method involves clubbing the young seals to death. In 1996, the Canadian government increased the seal hunt quota from 186,000 to 275,000.

Concern is also growing about the health of the harp seal population. The Northwest Atlantic harp seal population migrates annually between Greenland and Canada. It is hunted during the summer months in Greenland and, in the spring, along Canada's east coast. In setting its Total Allowable Catch (TAC) (275,000 in 1997, 1998, 1999, 2000, and 2001.) it has been argued that the Canadian government did not completely account for the number of animals harvested in the increasing and largely unregulated Greenland summer hunt, which now takes up to 80,000 harp seals per year. In actuality in 2000, the number of animals harvested was far below the TAC primarily because of low prices for the pelts.

The following two websites can provide additional information to this complicated and interrelated issue.

Fisheries and Oceans, Canada—www.ncr.dfo.ca/home_e.htm
International Marine Mammal Association—www.imma.org

In 1997, scientists from seven countries met in St. John's, Newfoundland, to consider the interactions between harp seals and fisheries in the Northwest Atlantic. The conclusions of the meeting supported the earlier findings of the dramatic population growth among the seals and also noted that the animals were growing more slowly and the pregnancy rate was lower than in the 1980s. These are the effects you would likely see when food becomes more difficult to find.

The St. John's meeting reinforced the fact that the diet of harp seals in the nearshore waters of the Labrador-Newfoundland shelf is dominated by Arctic cod. It was not, however, determined whether or not harp seals were affecting commercial fishing stocks—and Atlantic cod, in particular—on the Labrador-Newfoundland shelf. This was because there is a need for an estimate of the amount of juvenile cod in both inshore and offshore areas and for an assessment of the amounts of cod that are being taken by the other important predators such as Greenland halibut, whales, and seabirds.

It appears that what might be seen initially as a number of isolated and unrelated factors are really issues interrelated in a way that affects the economy of an entire region.

An Ecosystem Approach

The natural world is organized into interrelated units called ecosystems. An **ecosystem** is a region in which the organisms and the physical environment form an interacting unit. Weather affects plants, plants use minerals in the soil and affect animals, animals spread plant seeds, plants secure the soil, and plants evaporate water, which affects weather.

Ecosystems sometimes have fairly discrete boundaries, as is the case with a lake, island, or biosphere. Sometimes the boundaries are indistinct, as in the transition from grassland to desert. Grassland gradually becomes desert, depending on the historical pattern of rainfall in an area.

An ecosystem approach requires a look at the way the natural world is organized. Where do the rivers flow? What are the prevailing wind patterns? What are the typical plants and animals in the area? How does human activity affect nature? The task of an environmental scientist is to recognize and understand the natural interactions that take place and to integrate these with the uses humans must make of the natural world.

To illustrate the interrelated nature of environmental issues, we will look at several regions of North America and highlight some of the key features and issues of each.

Regional Environmental Concerns

No region is free of environmental concerns. Most regions tend to focus on specific, local environmental issues that apply directly to them. For example, protecting endangered species is a concern in many parts of the world. In the Pacific

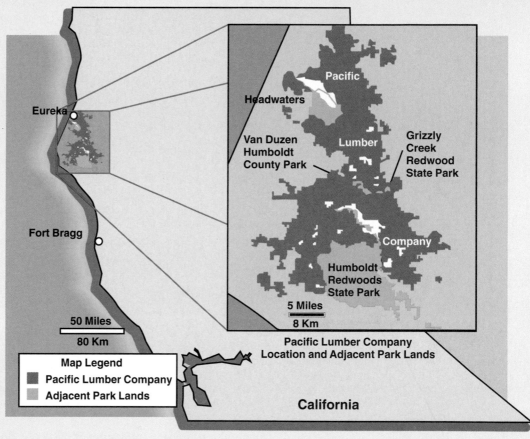

environmental CLOSE-UP

Headwaters Forest

The Headwaters Forest in Humbolt County, California, is the last ancient redwood forest remaining in private ownership. The old-growth redwoods are centuries-old trees, distinguished from second-growth trees, which have regenerated after logging. The term *old-growth* can apply to groves of trees or individual trees. The owner of this forest until 1999 was the Pacific Lumber Company (PL). Given the unique nature of the old growth and the local ecosystem, a great deal of public attention was focused on the fate of lands that were in private hands. Pacific Lumber Company purchased the forest to lumber it, not to preserve it. The fate of the forest was never certain during a decade of almost constant and bitter controversy between those who wanted to log the forest and those who wanted to preserve it.

In 1996, a federal and state agreement was entered into with Pacific Lumber. This agreement committed the federal and state governments to provide $380 million for the purchase of the largest grove of old-growth redwood still in private hands anywhere in the world—the Headwaters Grove on PL lands. In addition, the agreement specified that PL would develop a Habitat Conservation Plan and a Sustained Yield Plan for the remainder of its lands—approximately 81,000 hectares (200,000 acres). In 1999, with a final cost of $450 million, the Pacific Lumber Company and federal and state government agencies signed the Headwaters Forest Agreement. The controversy, however, did not end with the signing.

On the one hand, the agreement was hailed as a landmark. It was proclaimed as a compromise that every "reasonable" person should be able to accept. It protects every extensive tract of old-growth redwood remaining in the possession of Pacific Lumber Company, which means every tract of biological importance. It places heavier protections on salmon-spawning streams running through PL property than are applied to any private timberland in California. It also mandates more extensive precautions against stream siltation and landsliding than are currently in effect anywhere in the state. The agreement includes a covenant that guarantees these protections will run with the land—PL cannot void its habitat conservation plan by selling its land to another company. At the same time, the agreement provides PL with a predictable annual level of timber harvest on which to base its economic calculations.

Source: *California Resources Agency, 1998.*

Not all, however, were totally pleased with the agreement. Opponents to the agreement argued that the Habitat Conservation Plan gives PL too much latitude in logging sensitive habitat. All habitat conservation plans are little more than an end run around the Endangered Species Act and threaten the health of the forest, critics of the agreement stated. It was also argued that the agreement left little buffer between PL's ongoing logging operations and the public land. Apparently not all the "reasonable" people were in total agreement with the plan.

What do you think?

Should public tax dollars be spent to acquire properties such as the Headwaters Forest?

Is compromise possible on such a divisive issue?

Do you or your family have redwood furniture or a redwood deck?

What is the connection? Is there one?

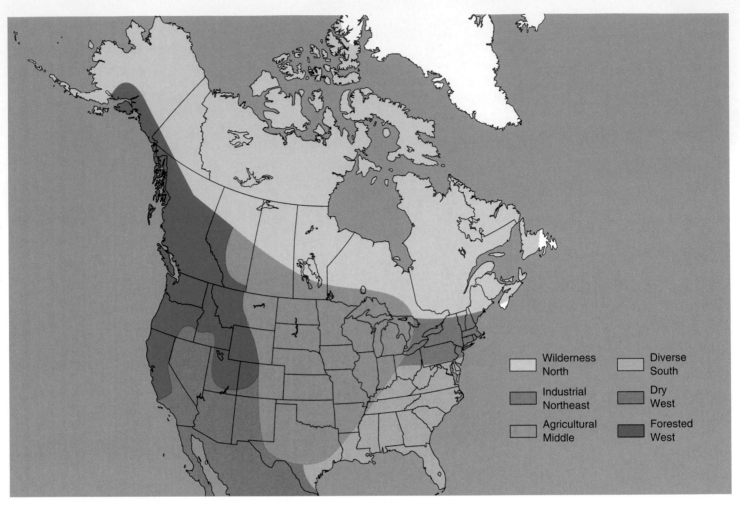

figure 1.3 **Regions of North America** Because of natural features of the land and the uses people make of the land, different regions of North America face different kinds of environmental issues. Certainly within each region people face a large number of specific issues, but certain kinds of issues are more important in some regions than others.

Northwest, for example, an endangered species known as the northern spotted owl depends on undisturbed mature forests for its survival. Development and logging may conflict with the survival of the owl. In most metropolitan areas, the problem of endangered species is purely historical, since the construction of cities has destroyed the previously existing ecosystem. Here we present a number of regional vignettes to illustrate the complexity and interrelatedness of environmental issues. (See figure 1.3.)

The Wilderness North

Much of Alaska and Northern Canada can be characterized as **wilderness**—areas with minimal human influence. Much of this land is owned by governments, not by individuals, so government policies have a large effect on

what happens in these regions. These areas have important economic values in their trees, animals, scenery, and other natural resources. Exploitation of the region's natural resources involves significant trade-offs. Usually, a portion of the natural world is altered permanently, but the area altered is so small that many people consider it insignificant. Because of the severe climate, northern wilderness areas tend to be very sensitive to insults and take a long time to repair damage done by unwise exploitation. Mining, oil exploration, development of hydroelectric projects, and harvesting of timber all require roads and other human artifacts, involve the insertion of new technologies into native culture, and generate economic benefits.

In the past, many short-term political and economic decisions failed to look at long-term environmental impli-

cations. Today, however, people are concerned about these remaining wilderness areas. Politicians are more willing to look at the scientific and recreational values of wilderness as well as the economic value of exploitation.

Native people, who consider much of this region to be their land, have become increasingly sophisticated in negotiating with state, provincial, and federal governments to protect rights they feel they were granted in treaties. They are sensitive to changes in land use or government policy that would force changes in their traditional way of life.

Concerned citizens, business interests, and environmental activists have become increasingly sophisticated in influencing decisions made by government. The process of compromise is often difficult and does not always assure

Walrus harvesting

A clear-cut forest

Grizzly bear fishing for salmon

figure 1.4   **The Wilderness North** Protection of wilderness is a major issue of this region. The major points of conflict involve the government role in managing these lands and wildlife, the protection of the rights and beliefs of native people, and the desire of many to exploit the mineral and other resources of the region.

A well-kept farm

Agricultural chemicals

Barges loaded with grain

figure 1.5   **The Agricultural Middle** The rich soil resource of this region has been converted to managed agricultural activity. The use of pesticides and fertilizer and exposure of the land to erosion cause concern about pollution of surface and groundwater. Most farmers still maintain that these practices are essential in modern agriculture and that they can be used safely and with minimal pollution.

wise decisions, but most governments now realize they must listen to the concerns of their citizens and balance economic benefits with social and cultural benefits. (See figure 1.4.)

The Agricultural Middle

The middle of the North American continent is dominated by intensive agriculture. This means that the original, natural ecosystems have been replaced by managed agricultural enterprise. It is important to understand that this area was at one time wilderness. Today, you would need to search very hard to find regions of true wilderness in Iowa, Indiana, or southern Manitoba. Some special areas have been set aside to preserve fragments of the original natural plant and animal associations, but most of the land has been converted to agriculture wherever practical.

The economic value generated by this use of a rich soil resource is tremen-

dous, and most of the land is privately owned. Governments cannot easily control what happens on these privately held lands. But governments indirectly encourage certain activities through departments of agriculture that encourage agricultural research, grant special subsidies to farmers in the form of guaranteed prices for their products and other special payments, and develop markets for products. Yet because the economic risks involved in farming are great, the number of farmers constantly declines. There are a number of reasons for farm failures, including drought, disease, lack of markets, increasing labor shortages, and fuel and equipment costs.

One of the major, nonpoint pollution sources (pollution that does not have an easily identified point of origin) is agriculture. Air pollution in the form of dust is an inevitable result of tilling the land. Soil erosion occurs when soil is exposed to wind and moving water and leads to siltation of rivers, impoundments, and lakes. Fertilizers and other agricultural

chemicals blow or are washed from the areas where they are applied. Nutrients washed from the land enter rivers and lakes where they encourage the growth of algae, lowering water quality. The use of pesticides causes concern about human exposure, effects on wild animals that are accidentally exposed, and residues in foods produced.

Since many communities in this region rely on groundwater for drinking water, the use of fertilizers and pesticides, and their potential for entering the groundwater as a result of unwise or irresponsible use, is a consumer issue. In addition, many farmers use groundwater for irrigation, which lowers the water table and leaves less groundwater for other purposes.

In an effort to stay in business and preserve their way of life, farmers must use modern technology. Careful use of these tools can reduce their impact; irresponsible use causes increased erosion, water pollution, and risk to humans. (See figure 1.5.)

environmental CLOSE-UP

The Greater Yellowstone Ecosystem

In 1872, the U.S. government established Yellowstone National Park as the world's first national park. It was an expansive area that protected unique natural features such as geysers, hot springs, rivers, lakes, and mountains. It was also a preserve for many kinds of wildlife such as grizzly bears, elk, moose, and bison. At the time it was established, the park was thought to be of adequate size to protect the scenic resources and the wildlife. Since that time, the lands surrounding the park have been converted to a variety of uses, including cattle grazing, timber production, hunting, and mining.

Fortunately, most of the lands surrounding Yellowstone National Park and the adjacent Grant Teton National Park are still under government control as national forests, national wildlife refuges, and other state, local, or federal entities. Some of the park wildlife, particularly the grizzly bear and bison, often wander across the park

boundaries. The grizzly in particular needs large regions of wilderness to survive as a species.

Many people assert that it is essential that these lands be integrated into a Greater Yellowstone Ecosystem management plan encompassing about 7.3 million hectares (18 million acres). The plan is based on more natural boundaries than the original boundaries established in 1872. This would require changes in the way much of the land surrounding Yellowstone is being used. The trade-offs are significant. Logging, mining, hunting, and grazing would be stopped or significantly reduced. This would result in a loss of jobs in those industries. Proponents argue that additional jobs would be created in the tourist and related service industries. The advantages, they argue, would be equal to or greater than the economic losses caused by stopping current uses. Individual and group decisions result in organizational policies and consumer behavior that support or weaken the ecosystem.

The Dry West

Where rainfall is inadequate to support agriculture, ranching and raising livestock are possible. This is true in much of the drier portions of western North America. Because much of the land is of low economic value, most is still the property of government, which encourages its use by providing water for livestock and irrigation at minimal cost, offering low rates for grazing rights, and encouraging mining and other development.

Many people believe that government agencies have seriously mismanaged these lands. They assert that the agencies are controlled by special interest groups and powerful politicians sensitive to the demands of ranchers, that they subsidize ranchers by charging too little for grazing rights, and that they allow destructive overgrazing because of the economic needs of ranchers. Ranchers argue that they require access to government-owned land, cannot afford significantly increased grazing fees, and that changing

government policies would destroy a way of life that is important to the regional economy.

Water is an extremely valuable resource in this region. It is needed for municipal use and for agriculture. Many areas, particularly the river valleys, have fertile soils that can be used for intensive agriculture. Cash crops such as cotton, fruits, and vegetables can be grown if water is available for irrigation. Because water tends to evaporate from the soil rapidly, long-term use of irrigated lands often results in the buildup of salts in the soil, thus reducing fertility. Irrigation water flowing from fields is polluted by agricultural chemicals that make it unsuitable for other uses such as drinking. As cities in the region grow, an increasing conflict arises between urban dwellers who need water for drinking and other purposes, and ranchers and farmers who need the water for livestock and agriculture. Increased demand for water will result in shortages, and decisions will have to be made about who

will ultimately get the water and at what price. If the urban areas get the water they want, some farmers and ranchers will go out of business. If the agricultural interests get the water, urban growth and development will have to be limited and expensive changes will have to be made to conserve domestic water use.

Because population density is low in most of this region, much of the land has a wilderness character. Increasingly, a conflict has developed between the economic management of the land for livestock production and the desire on the part of many to preserve the "wilderness." Designating an area as wilderness means that certain uses are no longer permitted. This offends individuals and groups who have traditionally used the area for grazing, hunting, and other pursuits. A long history of use and abuse of this land by overgrazing, modification to encourage plants valuable for livestock, and the introduction of grasses for livestock has significantly altered the region so that it cannot truly be called

Irrigation water and electrical generation from Glen Canyon Dam

Wilderness area

Overgrazed land

Bryce Canyon

figure 1.6 **The Dry West** Water is a key issue in this region. Both city dwellers and rural ranchers and farmers need water, and conflict results when there is not enough water to satisfy the desires of all. In addition, much of the land in this region is owned by the government. This raises concerns about how the government manages the land and how government policy affects the people of the region.

wilderness. The low population density does, however, provide a remoteness and natural character that many seek to preserve. (See figure 1.6.)

The Forested West

The coastal areas and mountain ranges of the western United States and Canada receive sufficient rainfall for coniferous forests to dominate as vegetation. Since most of these areas are not suitable for farmland, they have been maintained as forests with some grazing activity in the more open forests. Governments and large commercial timber companies own large sections of these lands. Government forest managers (U.S. Forest Service, Bureau of Land Management, Environment Canada, and various state and provincial departments) historically have sold timber-cutting rights at a loss and are thought by many to be too interested in the production of forest prod-

ucts at the expense of other, less tangible values. In 1993, the U.S. Forest Service was directed to stop below-cost timber sales.

This policy change has become a major issue in the old-growth forests of the Pacific Northwest where timber interests maintain that they must have access to government-owned forests in order to remain in business. Many of these areas have significant wilderness, scenic, and recreational value. Environmental interests point out that it makes no sense to complain about the destruction of tropical rainforests in South America while North America makes plans to cut large areas of previously uncut, temperate rainforest. Are the intangible values of preserving an ancient forest ecosystem as important as the economic values provided by timber and jobs?

Environmental interests are concerned about the consequences logging

would have on organisms that require mature, old-growth forests for their survival. Grizzly bear habitat in Alaska and British Columbia could be altered significantly by logging; the northern spotted owl has become a symbol of the conflict between logging and preservation in Oregon and Washington; and preservation of coastal redwood forests has become an issue in northern California. (See figure 1.7.) The issue of government ownership of large areas of land and the policy of multiple uses of land is not new in the West.

The Great Lakes and Industrial Northeast

While much of the West and Central regions of North America are characterized by low population densities and small towns, major portions of the Great Lakes and Northeast are dominated by large metropolitan complexes

Cut logs being hauled

Native elk

U.S. Forest Service ownership

figure 1.7 **The Forested West** The cutting of forested areas for timber production destroys the previous ecosystem. Some see the trees as a valuable resource that provides jobs and building materials. Others see the forest ecosystem as a natural resource that should be preserved. In addition, government ownership of much of this land has generated considerable political debate about what the appropriate use of the land should be.

that generate social and resource needs that are difficult to satisfy. Many of these older cities were formed around industrial centers that have declined, leaving behind poverty, environmental problems in abandoned industrial sites, and difficulties with solid waste disposal, air quality, and land-use priorities. Interspersed among the major metropolitan areas are small towns, farmland, and forests.

One of the major resources of the region is water transport. The Great Lakes and eastern seacoast are extremely important to commerce; ships can travel throughout the area by way of the St. Lawrence Seaway and the Great Lakes through a series of locks and canals that bypass natural barriers. Because of the importance of shipping in this region, harbors have been constructed and waterways have been deepened by dredging. The waterways are maintained at considerable government expense.

One of the greatest problems associated with the industrial uses of the Great Lakes and East Coast is contamination of the water with toxic materials. In some cases, unthinking or unethical individuals have dumped toxins directly into the water. In other cases, small, accidental spills or leaks over long periods of time have contaminated the sediments in harbors and bays.

A major concern about these pollutants is that they bioaccumulate (see chapter 15) in the food chain. The concentrations of some chemicals in the fat tissue of top predators, such as lake trout and fish-eating birds, can be a million times higher than the concentration in the water. Because of this, government agencies have issued consumption advisories for some fish and shellfish in contaminated areas. Since many kinds of fish can swim great distances, advisories for the Great Lakes warn against eating

Inner-city decay in Chicago Harbor in Duluth, Minnesota Central Park in New York City

figure 1.8 **The Great Lakes and the Industrial Northeast** Industry, waterways, and population centers are defining elements of this region. The historically extensive use of the Great Lakes and coastal areas of the Northeast for industry, because of the ease of providing water transportation, has resulted in many older cities with poor land-use practices. Rebuilding cities, providing recreational opportunities for urban dwellers, and repairing previous environmental damage are important issues. The water resources of the region provide transportation, recreation, and industrial opportunities.

certain fish taken anywhere within the lakes, not just from the site of contamination. Similarly, Chesapeake Bay has been subjected to years of thoughtless pollution, resulting in reduced fish and shellfish populations and advisories against consuming some organisms taken from the bay.

Water always generates considerable recreational value. Consequently, conflicts arise between those who want to use the water for industrial and shipping purposes and those who wish to use it for recreation. Due to the fact that so much of the North American population is concentrated in this region, the economic value of recreational use is extremely high. Consumer pressure is great to clean up contaminated sites and prevent the pollution of new ones. Contaminated areas do not enhance tourism or quality of life.

Most of these older, large cities had no plan to shape their growth. As a result, open space for people is limited and urban dwellers have few opportunities to interact with the natural world. Children who grow up in these cities often do not know that milk comes from a cow—they

have never seen, smelled, or touched a cow. Consequently, urban people have difficulty understanding the feeling rural people have for the land. These urban dwellers may never have an opportunity to experience wilderness. Their major environmental priorities are cleaning up contaminated sites, providing more parks and recreation facilities, reducing air and water pollution, and improving transportation. (See figure 1.8.)

The Diverse South

In many ways, the South is a microcosm of all the regions previously discussed. The petrochemical industry dominates the economies of Texas and Louisiana, and forestry and agriculture are significant elements of the economy in other parts of the region. Major metropolitan areas thrive, and much of the area is linked to the coast either directly or by the Mississippi River and its tributaries. The environmental issues faced in the South are as diverse as those in the other regions.

Some areas of the South (particularly Florida) have had extremely rapid

population growth, which has led to groundwater problems, transportation problems, and concerns about regulating the rate of growth. Growth means money to developers and investors, but it requires municipal services, which are the responsibility of local governments. Too many people and too much development also threaten remaining natural ecosystems.

Poverty has been a problem in many areas of the South. This creates a climate that encourages state and local governments to accept industrial development at the expense of other values. Often, jobs are more important than the environmental consequences of the jobs; low-paying jobs are better than no jobs.

The use of the coastline is of major concern in many parts of the South. The coast is a desirable place to live, which may encourage unwise development on barrier islands and in areas that are subject to flooding during severe weather. In addition, industrial activity along the coast has resulted in the loss of wetlands. (See figure 1.9.)

Miami metropolitan area

Everglades

Chemical plant on lower Mississippi

figure 1.9 **The Diverse South** Poverty has been a historically important problem in the region. Often the creation of jobs was considered more important than the environmental consequences of those jobs. The use of coastal areas for industry has resulted in pollution of coastal waters. The heavy use of the Mississippi River for transportation and industry has caused pollution problems. In addition, the desirable climate in the South has resulted in intense pressure to develop new housing for those who want to move to the region. Unwise development of housing on fragile coastal sites has resulted in damage to buildings by storms and the actions of the oceans. This causes intense debate on land use.

Summary

Artificial political boundaries create difficulties in managing environmental problems because most environmental units, or ecosystems, do not coincide with political boundaries. Therefore, a regional approach to solving environmental problems, one that incorporates natural geographic units, is ideal. Each region of the world has certain environmental issues that are of primary concern because of the mix of population, resource use patterns, and culture.

Environmental problems become issues when there is disagreement. This inevitably leads to a confrontation between groups that have different views on the consequences of an environmental problem. Many social, economic, ethical, and scientific issues shape a person's opinions. The process of environmental decision making must account for all of these issues when viewing an acceptable compromise.

Environmental problems are people problems. They occur because the uses of natural resources, which some people feel are justified, result in a diminished environment for others in the region. Environmental problems are defined by the person who perceives the problem. When perceptions differ, conflict occurs. Environmental decisions inevitably involve economic consequences because someone is receiving value from the resources being used or someone perceives an economic loss because a use has been withdrawn.

• Some argue that economic consequences should not be important when making environmental decisions; others argue that economic considerations can resolve all environmental issues.

- Some argue that regulation is necessary to protect resources; others argue that regulation hinders valuable use of resources.
- Some consider nonhuman organisms as important as humans; others feel that humans have a primary place in nature.
- Some are against change; others recognize that change must occur if negative consequences are to be prevented.
- Some believe that environmental responsibility rests on each decision maker, whether at home, in the workplace, or in the community. Each hour and dollar the consumer spends involves environmental consequences. How do you feel about this statement?

With all these differing opinions, compromise is the only way to resolve the conflicts. The social institution of government must play a role. Economic evaluation is important. Recognition of the validity of opposing points of view is essential. The field of environmental science seeks to find that middle ground.

Key Terms

ecosystem *8*
environment *5*

environmental science *5*
sustainable development *7*

wilderness *10*

Review Questions

1. Describe why finding solutions to environmental problems is so difficult. Do you think it has always been as complicated?
2. Describe what is meant by an ecosystem approach to environmental problem solving. Is this the right approach?
3. List two key environmental issues for each of the following regions: the wilderness North, the agricultural middle, the forested West, the dry West, the Great Lakes and industrial Northeast, and the South. How are the issues changing?
4. Define environment and ecosystem and provide examples of these terms from your region.
5. Describe how environmental conflicts are resolved.
6. Select a local environmental issue and write a short essay presenting all sides of the question. Is there a solution to this problem?

Critical Thinking Questions

1. Imagine you are a U.S. congressional representative from a western state and a new wilderness area is being proposed for your district. Who might contact you to influence your decision? What course of action would you take? Why?
2. How do you weigh in on the issue of jobs or the environment? What limits do you set on economic growth? Environmental protection?
3. Imagine you are an environmentalist in your area who is interested in local environmental issues. What kinds of issues might these be?
4. Imagine that you lived in the urban East and that you were an advocate of wilderness preservation. What disagreements might you have with residents of the wilderness North or the arid West. How would you justify your interest in wilderness preservation to these residents?
5. You are the superintendent of Yellowstone National Park and want to move to an ecosystem approach to managing the park. How might an ecosystem approach change the current park? How would you present your ideas to surrounding landowners?
6. Look at the issue of global warming from several different disciplinary perspectives—economics, climatology, sociology, political science, agronomy. What might be some questions that each discipline could contribute to our understanding of global warming?

Concept Map

The construction of a concept map is a technique that helps students recognize how separate concepts are related to one another. Some concept maps may be simple, orderly lists. Others may form networks of connections that help to show how ideas are linked. It is important to understand that there is not just one way in which things can be put together. The examples show two different ways the same concepts can be organized. (Take another look at figure 1.2. It is a variety of concept map.)

Construct a concept map to show relationships among the following concepts:

biology	observation	scientific method
experiment	science	theory
hypothesis		

Example 1

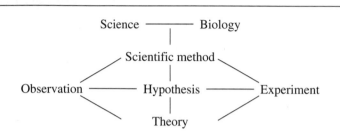

Example 2

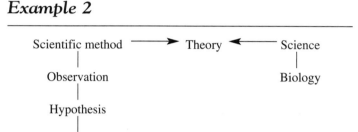

Interactive Exploration

Check out the website at **http://www.mhhe.com/environmentalscience** and click on the cover of this textbook for quizzing, career information, case studies, and hot links for the following topics:

General Environmental Sites

General Ecology Sites

Environmental and Ecological Organization Sites

History of Environmental Studies

Miscellaneous Environmental Resources

Introductory Materials and Governmental Sites

Science as a Process

Introductory Sites

Writing Papers and Study Tips

Glossaries and Dictionaries

Careers in Science

Utility and Organizational Sites

Global Ecology

Environmental Ethics

Objectives

After reading this chapter, you should be able to:

- Differentiate between ethics and morals.
- Define personal ethics.
- Explain the connection between material wealth and resource exploitation.
- Describe how industry exploits resources and consumes energy to produce goods.
- Explain how corporate behavior is determined.
- Describe the influential power that corporations wield because of their size.
- Explain why governmental action was necessary to force all companies to meet environmental standards.
- Describe the factors associated with environmental justice.
- Describe what has been the general attitude of consumers and business toward the environment.
- Explain the relationship between economic growth and environmental degradation.
- List three conflicting attitudes toward nature.

Chapter Outline

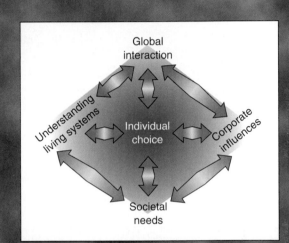

Views of Nature

The most beautiful object I have ever seen in a photograph in all my life, is the planet Earth seen from the distance of the moon, hanging in space, obviously alive. Although it seems at first glance to be made up of innumerable separate species of living things, on closer examination every one of its things, working parts, including us, is interdependently connected to all the other working parts. It is, to put it one way, the only truly closed ecosystem any of us know about.

—Lewis Thomas

There are no passengers on Spaceship Earth. We are all crew members.

—Buckminster Fuller

One of the marvels of recent technology is that we can see the Earth from the perspective of space, a blue sphere unique among all the planets in our solar system. (See figure 2.1.) Looking at ourselves from space, it becomes obvious, says ecologist William Clark of Harvard University, that only as a global species, "pooling our knowledge, coordinating our actions, and sharing what the planet has to offer—do we have any prospect for managing the planet's transformation along pathways of sustainable development."

Many people see little value in an undeveloped river and feel it is unreasonable to leave it flowing in a natural state. It could be argued that rivers throughout the world have been "controlled" to provide power, irrigation, and navigation at the expense of the natural world. It could also be argued that to not use these resources would be wasteful.

In the U.S. Pacific Northwest, there is a conflict over the value of old-growth forests. Economic interests want to use the forests for timber production and feel that to not do so would cause economic hardship. They argue that the trees are going to die anyway and they might as well be used for the betterment of the human community. Others feel that all the

figure 2.1 **The Earth as Seen from Space** Political, geographical, and nationalistic differences among humans do not seem so important from this perspective. In reality, we all share the same "home."

living things that make up the forest have a value we do not yet appreciate. Removing the trees would destroy something that took hundreds of years to develop and may never be replaced.

Interactions between people and their environment are as old as human civilization. The problem of managing those interactions, however, has been transformed today by unprecedented increases in the rate, scale, and complexity of the interactions. At one time, pollution was viewed as a local, temporary event. Today, pollution may involve several countries—as with the concern over acid deposition in Europe and in North America—and will affect multiple generations. The debates over chemical and radioactive waste disposal are examples of the increasingly international nature of pollution. For example,

many European countries are concerned about the transportation of radioactive and toxic wastes across their borders. What were once straightforward confrontations between ecological preservation and economic growth now involve multiple linkages that blur the distinction between right and wrong. For example, the enhanced greenhouse effect is thought to result from energy consumption, agricultural practices, and climatic change.

Many people believe that we have entered an era characterized by global change that stems from the interdependence between human development and the environment. They argue that self-conscious, intelligent management of the earth is one of the greatest challenges facing humanity as we begin the twenty-first century. To meet this challenge, they believe, a new environmental ethic must evolve.

Environmental Ethics

Ethics is one branch of philosophy. Ethics seeks to define fundamentally what is right and what is wrong, regardless of cultural differences. For example, most cultures have a reverence for life and hold that all humans have a right to live. It is considered unethical to deprive an individual of life.

Morals differ somewhat from ethics because morals reflect the predominant feelings of a culture about ethical issues. For example, in almost all cultures, it is certainly unethical to kill someone; however, when a country declares war, most of its people accept the necessity of killing the enemy. Therefore, it is a moral thing to do even though ethics says that killing is wrong. No nation has ever declared an immoral war.

Environmental issues require a consideration of ethics and morals. For example, because there is enough food in the world to feed everyone adequately, it is unethical to allow some people to starve while others have more than enough. However, the predominant mood of those in the developed world is one of indifference. They don't feel morally bound to share what they have with others. In reality, this indifference says that it is permissible to allow people to starve. This moral stand is not consistent with a purely ethical one.

As we can see, ethics and morals are not always the same; thus, it is often difficult to clearly define what is right and what is wrong. Some individuals view the world's energy situation as serious and have reduced their consumption. Others do not believe there is a problem and so have not modified their energy use. Still others do not care what the situation is. They will use energy as long as it is available.

Other issues are population and pollution. Is it ethical to have more than two children when the world faces overpopulation? Should an industry persuade legislators to vote no on a particular bill because it might reduce profits, even though its passage would improve the environment? The stand we take on such issues often depends on our position. An industrial leader, for example, would probably not look upon pollution as negatively as someone who participates in outdoor activities. In fact, many business leaders view the behavior of active preservationists as immoral because it restricts growth and, in some cases, causes unemployment.

Most ethical questions are very complex. Ethical issues dealing with the environment are no different. It is important to explore environmental issues from several points of view before taking a stand.

When we take an ethical stand, we become open to attack from those who disagree with our stand. Often, individuals are portrayed as villains for pursuing a course of action they consider righteous.

Environmental ethics is a topic of applied ethics that examines the moral basis of environmental responsibility. In these environmentally conscious times, most people agree that we need to be environmentally responsible. Toxic waste contaminates groundwater, oil spills destroy shore lines, and fossil fuels produce carbon dioxide, thus adding to global warming. The goal of environmental ethics, then, is not to convince us that we should be concerned about the environment—many already are. Instead, environmental ethics focuses on the moral foundation of environmental responsibility and how far this responsibility extends. There are three primary theories of moral responsibility regarding the environment. Although each supports environmental responsibility, their approaches are different.

The first of these theories is **anthropocentric,** or human-centered. Environmental anthropocentrism is the view that all environmental responsibility is derived from human interests alone. The assumption here is that only human beings are morally significant organisms and have a direct moral standing. Since the environment is crucial to human well-being and human survival, we have a duty toward the environment, that is, a duty that is derived from human interests. This involves the duty to ensure that the Earth remains environmentally hospitable for supporting human life and that its beauty and resources are preserved so that human life on earth continues to be pleasant. Some have argued that our environmental duties are derived both from the immediate benefit that people receive from the environment and from the benefit that future generations of people will receive. But critics have maintained that since future generations of people do not yet exist, then, strictly speaking, they cannot have rights any more than a dead person can have rights. Nevertheless, both parties to this dispute acknowledge that environmental concern derives solely from human interests.

A second theory of moral responsibility to the environment is **biocentric.** According to the broadest form of the life-centered theory, all forms of life have an inherent right to exist. Some biocentric thinkers give species a hierarchy of values. Some, for example, believe we have greater responsibility to protect animal species than plant species. Others determine the rights of various species depending on the harm they do to humans. For example, they see nothing wrong in killing pest species such as rats or mosquitoes. Some go further and believe that each individual organism, not just each species, has a basic right to survive. Individuals who support the animal rights movement tend to place more value on the individuals of animal species than on plant species. Trying to decide what types of species or individuals should be protected from early extinction or death resulting from human activities is an ethical dilemma. It is hard to know where to draw the line and be ethically consistent.

The third approach to environmental responsibility, called **ecocentrism,** maintains that the environment deserves direct moral consideration and not one that is merely derived from human (and animal) interests. In ecocentrism, it is suggested that the environment has direct rights, that it qualifies for moral personhood, that it is deserving of a direct duty, and that it has inherent worth. The environment, by itself, is considered to be on a moral par with humans.

The position of ecocentrism is the view advocated by the ecologist and writer Aldo Leopold in his book *A Sand County Almanac* (1949). In response to the relentless destruction of the landscape, *A Sand County Almanac* redefined the relationship between human- kind and the Earth. Leapold devoted an entire chapter of his book to "The Land Ethic."

> All ethics so far evolved rest upon a single premise: that the individual is a member of a community of interdependent parts. . . . The land ethic simply enlarges the boundaries of the community to include soils, waters, plants, and animals, or collectively the land . . . a land ethic changes the role of *Homo sapiens* from conqueror of the land- community to plain member and citizen of it. It implies respect for his fellow-members, and also respect for the community as such.

> It is inconceivable to me that an ethical relation to land can exist without love, respect, and admiration for land, and a high regard for its value. By value, I of course mean something far broader than mere economic value; I mean value in the philosophical sense.

What Leopold put forth in "The Land Ethic" was viewed by many as a radical shift in how humans perceive themselves in relation to the environment. Originally we saw ourselves as conquerors of the land. Now, according to Leopold, we need to see ourselves as members of a community that also includes the land and the water.

Leopold also wrote that "a thing is right when it tends to preserve the integrity, stability, and beauty of the biotic community. It is wrong when it tends otherwise. . . . We abuse land because we regard it as a commodity belonging to us. When we see land as a community to which we belong, we may begin to use it with love and respect."

Preservation

Development

Conservation

Recreation

figure 2.2 **The Views of Nature** Individuals envision the same resources used differently.

As traditional political and nationalistic boundaries begin to fade or shift globally, new variations of environmental thought and ethics are also evolving. Some of the new thoughts on environmental ethics are founded on an awareness that humanity is part of nature and that nature's many parts are interdependent. In any natural community, the well-being of the individual and of each species is tied to the well-being of the whole. In a world increasingly without environmental borders, nations, like individuals, should have a fundamental ethical responsibility to respect nature and to care for the Earth, protecting its life-support systems, biodiversity, and beauty and caring for the needs of other countries and future generations.

Environmental ethicists argue that to consider environmental protection as a "right" of the planet is a natural extension of the concept of human rights. Many also argue that an environmental ethic considers one's actions toward the environment as a matter of right and wrong, rather than one of self-interest.

Environmental Attitudes

There are many different attitudes about the environment, most of which fall under one of three headings: (*a*) the development ethic, (*b*) the preservation ethic, and (*c*) the conservation ethic. Each of these ethical positions has its own code of conduct against which ecological morality may be measured. (See figure 2.2.)

The **development ethic** is based on individualism or egocentrism. It assumes that the human race is and should be the master of nature and that the Earth and its resources exist for our benefit and pleasure. This view is reinforced by the work ethic, which dictates that humans should be busy creating continual change and that things that are bigger, better, and faster represent "progress," which itself is good. This philosophy is strengthened

Naturalist Philosophers

The philosophy behind the environmental movement had its roots in the last century. Among many notable conservationist philosophers, several stand out: Ralph Waldo Emerson, Henry David Thoreau, John Muir, Aldo Leopold, and Rachel Carson.

In Emerson's first essay, *Nature,* published in 1836, he claimed that "behind nature, throughout nature, spirit is present." Emerson was an early critic of rampant economic development, and he sought to correct what he considered to be the social and spiritual errors of his time. In his *Journals,* published in 1840, Emerson stated that "a question which well deserves examination now is the Dangers of Commerce. This invasion of Nature by Trade with its Money, its Credit, its Steam, its Railroads, threatens to upset the balance of Man and Nature."

Henry David Thoreau was a naturalist who held beliefs similar to Emerson's. Thoreau's bias fell on the side of "truth in nature and wilderness over the deceits of urban civilization." The countryside around Concord, Massachusetts, fascinated and exhilarated him as much as the commercialism of the city depressed him. It was near Concord that Thoreau wrote his classic, *Walden,* which describes a year in which he lived in the country to have direct contact with nature's "essential facts of Life." In his later writings and journals, Thoreau summarized his feelings toward nature with prophetic vision:

> But most men, it seems to me, do not care for Nature and would sell their share in all her beauty, as long as they may live, for a stated sum—many for a glass of rum. Thank God, man cannot as yet fly, and lay waste the sky as well as the earth! We are safe on that side for the present. It is for the very reason that some do not care for these things that we need to continue to protect all from the vandalism of a few. (1861)

John Muir combined the intellectual ponderings of a philosopher with the hard-core, pragmatic characteristics of a leader. Muir believed that "wilderness mirrors divinity, nourishes humanity, and vivifies the spirit." Muir tried to convince people to leave the cities for a while to enjoy the wilderness. However, he felt that the wilderness was threatened. In the 1876 article entitled, "God's First Temples: How Shall We Preserve Our Forests?" published in the *Sacramento Record Union,* Muir argued that only government control could save California's finest sequoia groves from the "ravages of fools." In the early 1890s, Muir organized the Sierra Club to "explore, enjoy, and render accessible the mountain regions of the Pacific Coast" and to enlist the support of the government in preserving these areas. His actions in the West convinced the federal government to restrict development in the Yosemite Valley, which preserved its beauty for generations to come.

Aldo Leopold was another thinker as well as a doer in the early conservation field. As a philosopher, Leopold summed up his feelings in *A Sand County Almanac:*

> Wilderness is the raw material out of which man has hammered the artifact called civilization. No living man will see again the long grass prairie, where a sea of prairie flowers lapped at the stirrups of the pioneer. No living man will see again the virgin pineries of the Lake States, or the flatwoods of the coastal plain, or the giant hardwoods.

Ralph Waldo Emerson

Henry David Thoreau

John Muir

Aldo Leopold

Leopold founded the field of game management. In the 1920s, while serving in the Forest Service, he worked for the development of a wilderness policy and pioneered his concepts of game management. He wrote extensively in the *Bulletin* of the American Game Association and stated that the amount of space and the type of forage of a wildlife habitat determine the number of animals that can be supported in an area. Furthermore, he said that regulated hunting can maintain a proper balance of wildlife.

Rachel Carson

While most people talk about what's wrong with the way things are, few actually go ahead and change it. Rachel Carson ranks among those few. A distinguished naturalist and best-selling nature writer, Rachel Carson published in the *New Yorker* in 1960 a series of articles that generated widespread discussion about pesticides. In 1962, she published *Silent Spring,* which dramatized the potential dangers of pesticides to food, wildlife, and humans and eventually led to changes in pesticide use in the United States.

Although some technical details of her book have been shown to be in error by later research, her basic thesis that pesticides can contaminate and cause widespread damage to the ecosystem has been established. Unfortunately, Carson's early death from cancer came before her book was recognized as one of the most important events in the history of environmental awareness and action in the twentieth century.

by the idea that "if it can be done, it should be done" or that our actions and energies are best harnessed in creative work.

Examples of the development ethic abound. The notion that bigger is better is certainly not new to us, nor is the belief that if something can be done or built, it should be. The dream of upward mobility is embodied in this ethic. In some circles, questioning growth is considered almost unpatriotic. In the development ethic, nature has only instrumental value; that is, the environment has value only insofar as human beings economically utilize it. Only in the past fifty to one hundred years have the by-products and waste associated with development been considered.

The **preservation ethic** considers nature special in itself. Nature, it is argued, has intrinsic value or inherent worth apart from human appropriation. Preservationists have diverse reasons for wanting to preserve nature. Some hold an almost religious belief regarding nature. They have a reverence for life and respect the right of all creatures to live, no matter what the social and economic costs.

During the nineteenth century, preservationists forthrightly gave ethical and spiritual reasons for protecting the natural world. John Muir condemned the "temple destroyers, devotees of ravaging commercialism" who, "instead of lifting their eyes to the God of the mountains, lift them to the Almighty dollar." This was not a call for better cost-benefit analysis: Muir described nature not as a commodity but as a companion. Nature is sacred, Muir held, whether or not resources are scarce.

Philosophers such as Emerson and Thoreau thought of nature as full of divinity. Walt Whitman celebrated a leaf of grass as no less than the "journey-work of the stars" "After you have exhausted what there is in business, politics, conviviality, love, and so on and found that none of these finally satisfy, or permanently wear—what remains? Nature remains," Whitman wrote. These philosophers thought of nature as a refuge from economic activity, not as a resource for it.

Some preservationists' interest in nature is primarily aesthetic or recreational. They believe that nature is beautiful and refreshing and should be available for picnics, hiking, camping, fishing, or just peace and quiet.

In addition to the religious and recreational preservationists, there are also preservationists whose reasons are essentially scientific. They argue that the human species depends on and has much to learn from nature. Rare and endangered species and ecosystems, as well as the more common ones, must be preserved because of their known or assumed long-range, practical utility. In this view, natural diversity, variety, complexity, and wilderness are thought to be superior to humanized uniformity, simplicity, and domesticity. Scientific preservationists want to lock up not all the land but only what they consider important to future generations.

The third environmental ethic is referred to as the **conservation or management ethic.** It is related to the scientific preservationist view but extends the rational consideration to the entire Earth and for all time. It recognizes the desirability of decent living standards, but it works toward a balance of resource use and resource availability. The conservation ethic stresses a balance between total development and absolute preservation. It stresses that rapid and uncontrolled growth in population and economics is self-defeating in the long run. The goal of the conservation ethic is one people living together in one world, indefinitely.

Societal Environmental Ethics

Society is composed of a great variety of people with diverse viewpoints. This variety can be distilled into a set of ideas that reflect the prevailing attitudes of society. The collective attitudes can be analyzed from an ethical point of view. Western, developed societies have long acted as if the Earth has unlimited reserves of natural resources, an unlimited ability to assimilate wastes, and a limitless ability to accommodate unchecked growth.

The economic direction and rationale of developed nations have been that of continual growth. Unfortunately, this growth has not always been carefully planned or even desired. This "growth mania" has resulted in the use of our nonrenewable resources for comfortable homes, well-equipped hospitals, convenient transportation, fast-food outlets, VCRs, home computers, and battery-operated toys, among other things. In economic statistics, such "growth" measures out as "productivity." But the question arises, "What is enough?" Poor societies have too little, but rich societies never say, "Halt! We have enough." The Indian philosopher and statesman Mahatma Ghandi said, "The Earth provides enough to satisfy every person's need, but not every person's greed."

Growth, expansion, and domination remain the central sociocultural objectives of most advanced societies. **Economic growth** and **resource exploitation** are attitudes shared by developing societies. We continue to consume natural resources as if the supplies were never ending. All of this is reflected in our increasingly unstable relationship with the environment, which grows out of our tendency to take from the "common good" without regard for the future.

This attitude is deeply embedded in the fabric of our society. Since the first settlers arrived in North America, nature has been considered an enemy. Frequently, the colonists expressed their relation to the wilderness in military terms. They viewed nature as an enemy to be "conquered," "subdued," or "vanquished" by a pioneer "army." Any qualms the pioneers may have felt about invading and exploiting the wilderness were justified by religious beliefs. They were driven by what they perceived to be a "moral imperative." This attitude toward nature is still popular today. Many view wilderness solely as underdeveloped land and see value in land only if it is farmed, built upon, or in some way developed. The notion that land and wilderness should be preserved is incomprehensible to some. The thought of

Environmental Philosophy

Nature, growth, and progress are concepts that we all use but seldom define either in discussion or to ourselves. We speak about environmental ethics, environmental philosophy, ecophilosophy, and so on, but what do we put into these concepts? We say that we have a responsibility for future generations and that this is a question of morals, but how should questions about morals be decided?

We all have reasons for our opinions on these matters—whether we believe that nature shall serve humanity, that humanity shall serve nature, or something in between—but we seldom make them explicit or draw conclusions from them. What are these reasons? Are they reasonable, rational, defensible, scientifically grounded, emotional, religious? And what do we mean by saying that a reason is "rational"? What value does it have that something is scientifically grounded? What weight should emotions carry in this context? Can we reason rationally about them and come to an agreement about them, or do we have to put them aside and just stick to facts? How can we know what is right?

These are all philosophical questions, and as such, they may seem far removed from real, concrete environmental problems such as ozone holes and forest death. Can it really be meaningful to spend energy on a debate of these questions, when if not disaster then at least crisis stands at the doorway? Is it at all meaningful to philosophize about the environmental questions? Isn't it action that is needed?

Trying to answer the philosophical questions does not, of course, in itself solve any environmental problems, but on the other hand, is it questionable whether we can solve these problems without discussing them on a philosophical level? Because whether we discuss them or not, we have ideas and conceptions that guide our way of thinking, what we see as a problem, what we see as causes of problems and what we see as possible, desirable, or necessary solutions. And, quite seriously, the problems that stand before us today are hardly founded on our having thought too much.

What are your thoughts on this question? Is there already too much talk, or is it too little listening?

Source: Department of Philosophy, University of Gothenburg, Sweden. No copyright.

purposely opting to not develop a resource is considered almost a sin.

Corporate Environmental Ethics

Many tasks of industry, such as procuring raw materials, manufacturing and marketing, and disposing of wastes, are in large part responsible for pollution. This is not because any industry or company has adopted pollution as a corporate policy. Industry is naturally dirty because it consumes energy and resources. When raw materials are processed, some waste (useless material) is inevitable. It is usually not possible to completely control the dispersal of all by-products of a manufacturing process. Also, some of the waste material may simply be useless.

For example, the food-service industry uses energy to prepare meals. Much of this energy is lost as waste heat. Smoke and odors are released into the atmosphere, and spoiled food items must be discarded.

The cost of controlling waste can be very important in determining a company's profit margin. **Corporations** are legal entities designed to operate at a profit, which is not in itself harmful. The corporation has no ethics, but the people who make up the corporation are faced with ethical decisions. Ethics are involved when a corporation cuts corners in production quality or waste disposal to maximize profit. The cheaper it is to produce an item, the greater the possible profit. It is cheaper in the short run to dump wastes into a river than to install a wastewater treatment facility, and it is cheaper in the short run to release wastes into the air than it is to trap them in filters. Many people consider such pollution unethical and immoral, but some corporations think of it as just one of the factors that determines **profitability.** (See figure 2.3.) Because stockholders expect an immediate return on their investment, corporations often make decisions based on short-term profitability rather than long-term benefit to society.

The amount of profit a corporation realizes determines how much it can expand. To expand continually, a corporation increases the demand for its products through advertising. The more it expands, the more power it attains. The more power it has, the greater its influence over decision makers who can create conditions favorable to its expansion plans. The process becomes a seemingly never-ending spiral.

Nations of the world must confront the problem of corporate irresponsibility toward the environment. In business, incorporation allows for the organization and concentration of wealth and power far surpassing that of individuals or partnerships. Some of the most important decisions affecting our environment are made not by governments or the public but by executives who wield massive corporate power. Often, these executives make only minimal concessions to the public interest, while they make every effort to maximize profits.

Business decisions and technological developments have increased the exploitation of natural resources. In addition, many political and legal institutions have generally supported the development of private enterprise. They have also defended and promoted private property rights rather than social and environmental concerns. Businesses and individuals typically use loopholes, political pressure, and the time-consuming nature of legal action to circumvent or delay compliance with social or environmental regulations.

Is industry becoming more environmentally concerned? Corporations have certainly made more frequent references to worldwide environmental issues over the past several years. Is such concern only rhetoric and social marketing, or is it the beginning of a new corporate ethic? The "Valdez Principles," a tool that industry and the public can use to evaluate corporate environmental responsibility, were developed as a result of the 1989 oil spill in Alaska. These were later named the CERES Principles. (See figure 2.4.)

The Oil Protection Act of 1990, passed in the wake of the *Valdez* spill, was supposed to regulate supertankers and reduce the chances of supertanker oil spills. However, to get around the law, many oil carriers have shifted their oil transport operations to lightly regulated oil barges pulled by tugboats. This reduction in oil spill safety has led to several barge oil spills, including one in January 1996 in Rhode Island's Moonstone Bay and another in March 1997 in Texas's Galveston Bay.

In 1992, the Valdez Principles were proposed and adopted by the CERES Organization (Coalition for Environmentally Responsible Economics). This organization is an independent group of environmentalists and social investors.

The CERES Principles are a set of codes that businesses may adopt voluntarily. The codes provide environmental standards against which all companies can be assessed and compared. The 10 principles encompass a wide range of goals that include minimizing pollutants, making sustainable use of renewable resources, reducing health and safety risks

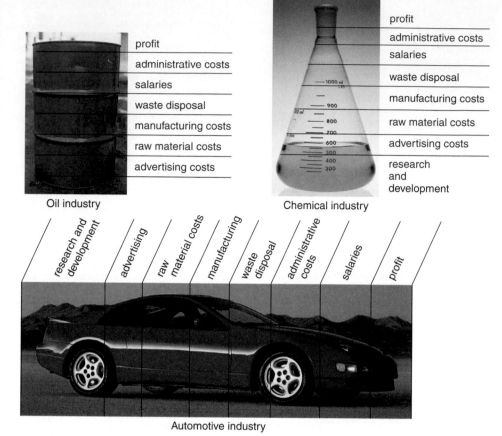

figure 2.3 **Corporate Decision Making** Corporations must make a profit. When they look at pollution control, they view its cost like any other cost: any reductions in cost increase profits.

a.   b.

figure 2.4 **CERES Principles** The 1989 oil spill in Alaska led to the development of the CERES Principles. (a) These waterfowl were victims of the spill. (b) The *Exxon Valdez*.

for employees and communities, and representing environmental interests on corporate boards. The CERES Principles are looked upon as a guide for corporate environmentalism. The goal, some argue, should be to make compliance with the CERES Principles a prerequisite for doing business.

In 1997, the Global Reporting Initiative (GRI) was established. Convened by CERES in partnership with the United Nations Environment Programme (UNEP), the GRI incorporates the active participation of corporations, nongovernmental organizations (NGOs), accountancy organizations, business associations, and other stakeholders from around the world. The mission of the GRI was to develop globally applicable guidelines for reporting on the economic,

Global Perspective

Chico Mendes and Extractive Reserves

Francisco "Chico" Alves Mendes Filho was born in 1944 in the western Brazilian Amazon. A second-generation rubber tapper, Chico was active in the rubber tappers' union for over 15 years. He and many other peasants made a living by extracting latex from rubber trees and selling it. Rubber tappers also collect and sell other natural products of the forest, such as Brazil nuts, fruits, and native medicines. Mendes was interested in preserving the portion of the Brazilian rainforest that provided their livelihood, and he supported the concept of "extractive reserves."

Extractive reserves involves setting aside land for rubber tappers, who would continue their traditional lifestyles and use the rainforest in its natural state for generations to come. This idea put Mendes in conflict with powerful people interested in clearing the rainforest to raise cattle. Most cattle-ranching operations show short-term economic gains but ultimately become uneconomical when land fertility declines from overgrazing.

In 1987, Mendes received two international environmental awards for his efforts in establishing extractive reserves. One was the Global 500 award from the United Nations Environment Programme (UNEP); the other was Ted Turner's Better World Society Environment Award.

In 1988, Chico led the Xapuri Rural Workers Union in a winning effort to stop cattle rancher Darli Alves from deforesting an area the rubber tappers wanted to make into a reserve. On December 22, 1988, as he walked from his home, Chico Mendes was shot by members of a vigilante group who supported local ranchers. In 1990, Darli Alves da Silva and his son were convicted of the murder. Alves da Silva was sentenced to 19 years in prison but escaped in 1993, leading to a three-year exhaustive manhunt. He is now in a high-security prison in Brasilia.

Before his death, Mendes had said, "I want to live to defend the Amazon." His life and death appear to have made a difference in the way the Brazilian rainforest is being used. In 1990, the Chico Mendes Extractive Reserve was established covering about 6 percent of the state of Acre in northwest Brazil.

Due to the circumstances that surrounded his death and to his role as a leader in the rubber tappers' union, his murder received international notice and caused many people to ask if the natural rainforest perhaps has as much to offer as ranches do.

While confrontations between the rubber tappers and ranchers have decreased, there have been new problems relating to mismanagement of funds in the Chico Mendes Foundation and mismanagement of the rubber tappers' cooperative that runs the Brazil-nut factory. The future of both the cooperative and the foundation remains in doubt.

---

environmental, and social performance, initially for corporations and eventually for any business, governmental, or nongovernmental organization.

The GRI's *Sustainability Reporting Guidelines* were released in draft form in 1999. The GRI guidelines represent the first global framework for comprehensive sustainability reporting, encompassing the "triple bottom line" of economic, environmental, and social issues. Twenty-one pilot test companies, numerous other companies, and a diverse array of noncorporate stakeholders commented on the draft guidelines during a pilot test period during 1999 to 2000. The final guidelines were released in June 2000.

Improved disclosure of sustainability information is an essential ingredient in the mix of approaches needed to meet the governance challenges in the globalizing economy. Today, at least 2000 companies around the world voluntarily report infor-

mation on their economic, environmental, and social policies, practices, and performance. Yet, this information is generally inconsistent, incomplete, and unverified. Measurement and reporting practices vary widely according to industry, location, and regulatory requirements. The GRI's *Sustainability Reporting Guidelines* are designed to address some of these challenges.

Practicing an environmental ethic should not interfere with corporate and other social responsibilities or obligations, though this is not always the case. It must be integrated into overall systems of belief and coordinated with economic systems. Environmental advocates, in turn, need to consider others' objectives just as they demand that others consider environmental consequences in decision making. It makes little sense to preserve the environment if that objective produces national economic collapse. Nor does it make sense to maintain stable in-

dustrial productivity at the cost of breathable air, drinkable water, wildlife species, parks, and wilderness. But to maintain profitability, influence, and freedom, businesses must be sensitive to their impact on current and future citizens, not just in terms of the price and quality of the goods they produce but also in terms of public approval of their social and political influence. A 2000 Harris Poll, for example, found that 65 percent of Americans wanted increased government spending, either national or state, to address environmental problems, even if they had to pay higher taxes. In another poll, eight of 10 Americans said they would be willing to pay extra for a product packaged with recyclable materials.

In the middle 1990s, a concept emerged called **industrial ecology** that reflects the link between the economy and the environment. This concept argues that good ecology is also good economics and that alternatives exist for corporations to provide goods and services in ways that do not destroy the environment.

One of the most important elements of industrial ecology is that, as in biological systems, it accounts for waste. Dictionaries define waste as useless or worthless material. In nature, however, nothing is eternally discarded; in various ways, all materials are reused. In our industrial world, discarding materials taken from the Earth at great cost is also generally unwise. Perhaps materials and products that are no longer in use should be termed *residues* rather than *wastes;* wastes are merely residues that our economy has not yet learned to use efficiently. A simpler way of saying this is to view a pollutant as a resource out of place. Such a statement forces us to view pollution and waste in a new way.

Environmental Justice

In 1982, a proposed PCBs landfill in Warren County, North Carolina, was protested by residents of an African-American community that was located close to the planned facility. Some observers asserted that local officials had been practicing a form of environmental racism, citing studies showing that hazardous waste treatment, storage, and disposal facilities were disproportionately located in areas occupied predominantly by minorities. Others argued that poor communities were often poorly represented in political circles, and therefore, were the easiest places in which to locate objectionable facilities. Eventually, the U.S. Environmental Protection Agency (EPA) granted the permits for this waste disposal site, which brought nationwide protests.

In response to protests stemming from this and similar incidents, President Bill Clinton issued an executive order (EO 12,898) in 1994 mandating that the EPA establish an Office of Environmental Justice. In 1998, the EPA defined **environmental justice** as fair treatment, meaning that "no group of people, including racial, ethnic, or socioeconomic groups, should bear a disproportionate share of the negative environmental consequenccs resulting from industrial, municipal, and commercial operations or the execution of federal, state, local, and tribal programs and policies." According to the EPA definition, deliberate discrimination need not be involved. Any action that affects protected groups disproportionately is in violation of EPA's rules. The difficulty arises in defining what to measure and what should be the standard of comparison.

As a first step in evaluating whether a group is unduly disadvantaged, a policy maker must consider who is affected. Most ethnic data relate to census tracts, zip codes, city boundaries, and counties. If the facility is to be located in a wealthy county but near the county border close to a poor community, how does the policy maker draw the line? Should prevailing winds be considered? Many industrial sites are located where land is cheap; people of low income may choose to live in those areas to minimize living expenses. How are these decisions weighed?

Another difficulty arises in determining whether and how particular groups will be disadvantaged. Landfills, chemical plants, and other industrial works bring benefits to some, although they may harm others. They create jobs, change land values, and generate revenues that are spent in the community. How do officials compare the benefits with the losses? How should potential health risks from a facility be compared with the overall health benefits that jobs and higher incomes bring?

The environmental justice movement is also occasionally referred to as Environmental Equity—which the EPA defines as equal protection from environmental hazards of all individuals, groups, or communities regardless of race, ethnicity, or economic status. Although environmental justice has many facets (e.g., legal, economic, and political), it may be approached appropriately in a variety of ways by the public and private sectors. In addition, the health community should naturally focus on the health aspect of environmental justice.

At its core, environmental justice means fairness. It speaks to the impartiality that should guide the application of laws designed to protect the health of human beings and the productivity of ecological systems on which all human activity, economic activity included, depends. It is emerging as an issue because studies show that certain groups of North Americans and citizens of other nations may suffer disproportionately from the effects of pollution.

Governments have established numerous laws, mandates, and directives to eliminate discrimination in housing, education, and employment, but few attempts have been made to address discriminatory environmental practices. In the United States, people of color have borne a disproportionate burden in the location of municipal landfills, incinerators, and hazardous-waste treatment, storage, and disposal facilities.

Hazardous waste sites and incinerators are not randomly located. While waste generation is correlated directly with per capita income, few toxic waste sites are located in affluent suburbs. Waste facilities are often located in communities that have high percentages of poor, elderly, young, and minority residents. Often such facilities are deliberately sited in these communities because they are seen as providing the path of least resistance and are less expensive to build because of cheaper land costs.

Questions of environmental justice extend beyond the location of toxic waste sites. Exposure to harmful pesticides and other toxic agricultural substances is a major health issue among hired farm workers, the majority of whom are people of color. There is also concern that because some Native American communities consume much greater amounts of fish from certain areas such as the Great Lakes than does the general population, they are at greater risk for dietary exposure to toxic chemicals.

Historically, the environmental movement has been a concern of middle-class whites, but there is a growing level of activism by people of color. Minority participation has broadened the debate to include many issues that were being ignored. It has also forced a dialogue about race, class, discrimination, and equity. Minorities have pushed the plight of their communities to the forefront. They have also brought a new perspective to the environmental movement and will be a part of any future environmental agenda.

Environmental Justice Highlights

1979

Houston community group filed suit over landfill: While the black community group from Houston eventually lost its court case, its work produced some of the first research on environmental justice showing that most incinerators and landfills built in the city were in black neighborhoods.

1982

More than 500 arrested in landfill fight: In mostly black Warren County, N.C., a battle over a landfill intended to hold PCB-contaminated dirt led to the arrest of hundreds including the Washington, D.C., delegate to Congress.

1983

Government report found hazardous waste bias: An analysis by the Government Accounting Office found that three of the four largest hazardous waste sites in the Southeast were in black communities.

1987

Church found environmental racism a national problem: The United Church of Christ published a report that showed that communities with hazardous waste facilities had higher percentages of minorities than those that had had no such facilities.

1989

EPA took up environmental justice: Bush administration EPA Administrator William Reilly establishes the Environmental Equity Work Group, marking the first official EPA response to the problem.

1993

EPA accepted its first civil rights complaint: The Louisiana case was filed by the Tulane (University) Environmental Law Clinic. The legal group later would file another case over the Shintech, Inc., PVC plant in rural Louisiana, which became a nationally important battle. At about the same time, the EPA formed the National Environmental Justice Advisory Council, a group of activists, local officials, and industry experts intended to serve as advisers to the EPA.

1994

President Clinton signed environmental justice executive order: The order requires all federal agencies to begin taking the issue into account. "Each Federal agency shall make achieving environmental justice part of its mission by identifying and addressing, as appropriate, disproportionately high and adverse human health or environmental effects of its programs, policies, and activities on minority populations and low-income populations."

1998

EPA released environmental justice "guidance": The new agency rules, produced with no input from states, cities, or industry, create an uproar.

Individual Environmental Ethics

The environmental movement has effectively influenced public opinion and moved the business community toward an environmental ethic. The result of this changing view of business's responsibilities will complicate business decision making through this century. More complex environmental and safety demands by the public and a broadening of horizons on the part of business will be a dominant theme of corporate life during the next decade. As human populations and economic activity continue to grow, we are facing a number of environmental problems that threaten not only human health and the productivity of ecosystems but in some cases the very habitability of the globe.

If we are to respond to these problems successfully, our environmental ethic must express itself in broader and more fundamental ways. We have to recognize that each of us is individually responsible for the quality of the environment we live in and that our personal actions affect environmental quality, for better or worse. The recognition of individual responsibility must then lead to changes in individual behavior. In other words, our environmental ethic must begin to express itself not only in national laws but also in subtle but profound changes in the ways we all live our daily lives.

Various public opinion polls conducted over the past decade have indicated that Americans think environmental problems can often be given a quick technological fix. The Roper polling organization has stated that, "They believe that cars, not drivers, pollute, so business should invent pollution-free autos. Coal utilities, not electricity consumers, pollute, so less environmentally dangerous generation methods should be found." It appears that many individuals want the environment cleaned up, but they do not want to make major lifestyle changes to make that happen.

Decisions and actions by individuals faced with ethical choices collectively determine the hopes and quality of life for everyone. As ecological knowledge and awareness begin to catch up with good intentions, people in all walks of life will need to live by an environmental ethic.

Do We Consume Too Much?

In 1994, when delegates from around the world gathered in Cairo for the International Conference on Population and Development, representatives from developing countries protested that a baby born in the United States will consume during its lifetime 20 times as much of the world's resources as an African or Indian baby. The problem for the world's environment, they argued, is overconsumption in the North,

Global Perspective

International Trade in Endangered Species

Illegal trading in rare or disappearing species as an international business turns over more than $1.6 billion a year, second only to smuggling in drugs or arms. It directly affects the populations of more than 37,000 animal and plant species and represents a severe threat to their survival.

The international groups that conduct this trading make a fortune. They buy a whole range of cheap animal and plant products, which in certain markets then sell for many times the buying price. It is estimated that each year 1.5 million caiman skins leave Brazil, Bolivia, and Paraguay, which as well as being a tragedy for the species means a loss of millions of dollars for these countries. The most important markets are to be found in the wealthy countries; the chief buyers for these wildlife products are in Japan, the United States, and the European Union. In the United States, legal trading accounts for more than $200 million, while illegal trading accounts for more than $300 million. This unregulated traffic squanders natural resources and is one of the most dangerous forms of wildlife and biosphere destruction. It does serious damage to Southern societies rich in biodiversity, as it yields no lasting profits for them and only enriches the intermediaries. It pays no taxes or customs duty. The returns for the poachers are limited, because what they receive for illegally capturing or gathering macaws, tigers, crocodiles, or orchids is very little compared to what is paid for them in the wealthy countries. Such trading threatens the sustainability of the ecosystems where the species live and perpetuates inequalities between the wealthy consumer countries and the poorer producer countries, which hardly gain anything from the trade.

The consequences of this trading are dramatic. Animals such as rhinoceroses, tigers, leopards, otters, South American caimans, macaw, some primates, butterflies, frogs, and tortoises and plants such as orchids, cacti, carnivorous plants, trees bearing precious wood like the mahogany are just a few examples affected. The list goes on.

Many animals are sacrificed for one specific product. But in the traffic of live specimens, the mortality is very high, both at the moment of capture and during shipping; many animals die so that a few can reach their destination alive. Traditional, more respectful forms of hunting and gathering are abandoned; poachers often cut down trees to reach the highest nests, taking males and females indiscriminately.

In 1973, after much debating and pressure by scientific and nongovernmental organizations, the Convention on International Trade in Endangered Animal and Plant Species (CITES) was signed in Washington, D.C., by 21 Western countries. Today, there are 125 member countries. The chief aim of CITES is to prevent illegal international trading in endangered species that are divided into three categories:

- Appendix 1 lists species in which trading is not allowed due to their imminent danger of extinction.
- Appendix 2 lists species in which trading is allowed under rigid scientific control, including those listed in Appendix 1 born in captivity.
- Appendix 3 lists species for which there are no general restrictions on trade but which have endangered populations in certain specific countries.

The convention foresees that each country should pass its own particular legislation to support and enforce the treaty's final provisions. This means that protection can vary from one member country to another. As with any law enforcement, it is not easy. Illegal trade continues to grow because the demand is there. Have you ever encountered illegally traded plants or wildlife? Have you ever asked when purchasing fish for an aquarium or perhaps a pet bird where they came from? Would you purchase a plant, fish, or perhaps jewelry with coral in it if you thought that it was traded illegally?

---

not overpopulation in the South. Do we in the North consume too much?

North Americans, only 5 percent of the world's population, consume one-fourth of the world's oil. They use more water and own more cars than anybody else. They waste more food than most people in sub-Saharan Africa eat.

As the rest of the world becomes more like America (China now consumes almost half as much meat per capita as Americans do), will something vital—water, oil, food—simply run out?

Ever since he wrote a book called *The Population Bomb* in 1968, ecologist Paul Ehrlich has argued that the American lifestyle is driving the global ecosystem to the brink of collapse. But others, including the economist Julian Simon, have argued that Ehrlich couldn't be more wrong. It is not resources that limit economic growth and lifestyles, Simon has insisted, but human ingenuity.

In 1980, the two wagered money on their competing world views. They picked something easily measurable—the value of metals—to put their theories to the test. Ehrlich predicted that world economic growth would make copper, chrome, nickel, tin, and tungsten scarcer and thus drive the prices up. Si-

mon figured human ingenuity would overcome scarcity and that the prices would go down. Ehrlich lost. By 1990, all five metals had decreased in value. Ehrlich claimed it was a result of a global recession that had reduced industrial demand for raw metals. But Simon argued that the metals decreased in price because superior materials such as plastics, fiber optics, and ceramics had been developed to replace them. The Ehrlich-Simon argument is actually an old one, and, despite the outcome of their bet, it remains unsettled. What do you think? With the question of consumption in mind, let's look at how consumption could affect several areas in the future—food, nature, oil, and water.

Food

Two centuries ago, Thomas Malthus declared that worldwide famine was inevitable as human population growth outpaced food production, In 1972, a group of scholars known as the Club of Rome predicted much the same thing for the waning years of the twentieth century. It did not happen because so far, at least—human ingenuity has outpaced population growth.

Fertilizers, pesticides, and high-yield crops have more than doubled world food production in the past 40 years. The reason 800 million people go hungry today is not that there is not enough food in the world but that they cannot afford to buy it.

It is argued that it is not a lack of resources that makes a people poor today but rather bad government. Consider, for example, Angola, a resource-rich country too wracked by civil war to exploit its wealth, and Russia, comparable to the United States in natural resources and intellectual capital but impoverished by the legacy of communism.

Norman Borlaug, who won the Nobel peace prize in 1970 for his role in developing high-yield crops, predicts that genetic engineering and other new technologies will keep food production ahead of population increases over the next half century. Perhaps not everyone will have all the meat they want, but most experts—even those who have doubts about genetic engineering—agree that enough food can be produced for the world in the twenty-first century. Whether everybody will get a fair share is much less certain.

Nature

As more people around the world achieve the American dream, they will consume more resources and generate more population. Tropical rainforests will be cut, and wilderness entombed under pavement. Mighty rivers like the Yangtze and Nile, already dammed and diverted, will become even more canal-

like. As the new century progresses, fewer and fewer of us will live on the land. Half of humanity will live in "megacities" like Tokyo and Sao Paulo, Brazil—cities of 12 million to 25 million people. Untamed nature will exist only in scattered remnants, preserved like artifacts in a museum. We will increasingly live in a world of our own making.

Oil

If everybody on Earth consumed as much oil as the average American, the world's known reserves would be gone in a decade. Even at current rates of consumption, known reserves would not last through the current century. Experts, however, are not worried. New technologies, they say, will avert a global energy crisis.

Already, oil companies have developed cheaper ways to find oil and extract it from the ground, effectively extending the supply into the twenty-second century. Still, there is a finite amount of oil on the planet, and someday it will be gone. Even before that happens, concerns about global warming may compel the world to stop burning so much fossil fuel.

The energy industry is preparing for that day by investing in technologies that will replace fossil fuels. Solar, nuclear, and wind power are all possibilities, but many experts say the most likely candidate is the fuel cell. A fuel cell is essentially a hydrogen-powered battery that produces no pollution. Its only by-product is water. And since hydrogen is the most abundant element in the universe, supply should not be a problem. This all depends, however, on the technology being developed.

Water

The world of the future may not need oil, but without water, humanity could not last more than a few days. Right now, humans use about half the planet's accessible supply of renewable, fresh water—the supply regenerated each year and available for human use. A simple doubling of agricultural produc-

tion with no efficiency improvements would push that fraction to about 85 percent. Unlike fossil fuels, which could eventually be replaced by other energy sources, there is no substitute for water.

Technologies such as desalination, which removes salt from seawater, can be used in rare circumstances. But removing salt from water takes a lot of energy and is expensive. In the Persian Gulf, one place that uses desalination, wealth makes it possible. It has been said that in the Persian Gulf, they turn "oil into water."

Some regions have already reached their water limits, with massive dams and aqueducts diverting almost every drop of water for human use. In the U.S. Southwest, the diversion is so complete that by the time the Colorado River reaches its mouth in the Sea of Cortez, it has no water in it. Much of Los Angeles' water comes from more than 300 kilometers (186 miles) away.

More than any other resource, water may limit the expansion of American consumerism during the next century. "In the next century," World Bank Vice President Ismail Serageldin predicted a few years ago, "wars will be fought over water."

The Unknown

How many people will be able to live the American way of life 50 years from now? The question is impossible to answer, many experts say, because of one major unknown, global warming. Estimates of the effect of global warming vary; however, most experts do expect some change in the Earth's average surface temperature. A warming at the lower end of the range might be barely noticeable, perhaps bringing rain instead of snow a few days each winter. But a warming at the farther end of the range could be catastrophic, shifting agricultural regions, threatening species with extinction, and pushing tropical diseases into areas where they are currently unknown. Glaciers would melt and ocean waters would expand, flooding heavily populated, low-lying places like Florida, the Netherlands, and Bangladesh.

Global Perspective

The Gray Whales of Neah Bay

In the mid-1990s, the Makah Tribe of Neah Bay in the state of Washington expressed the desire to resume their ancient tradition of hunting whales. Through the U.S. government, they petitioned the International Whaling Commission (IWC) for permission to do so. When news of this request reached the general public, it was met with controversy; some people expressed support for the Makah, while many others reacted with disapproval or anger.

The Makah were granted the legal right to hunt whales in the 1855 Treaty of Neah Bay, in which the tribe ceded the majority of their land at Cape Flattery to the U.S. government. Their fishing, sealing, and whaling rights were explicitly provided in the treaty. However, the tribe ceased their ceremonial whale hunts in the 1920s, when gray whale populations dwindled to near extinction. It was with the recovery of the gray whale populations that the Makah sought to begin whaling once again.

The controversy surrounding the decision to approve whaling is multifaceted. The tribe argues that their culture has disintegrated since the cessation of whaling; the last two generations have not known the discipline and pride that come with ceremonial hunting and have nutritionally suffered from the lack of a traditional seafood and sea mammal meat diet. Restoring the tradition of whale hunting, they claim, will result in the restoration of their culture.

Meanwhile, environmentalists concerned about the killing of gray whales argue that the tribe's intended hunting techniques are not traditional, thereby discrediting the Makah's arguments of cultural preservation. Historically, the whale would have been killed with repeated thrusts of a spear, causing internal bleeding and ultimate death. In lieu of this, the Makah, in consultation with the International Whaling Commission, intended to use a rifleshot following the initial harpooning, causing immediate death to the whale. While this method is more humane than its traditional counterpart, antiwhaling groups argue that this deviation from tradition reveals that the Makah are not, in fact, concerned with restoring culture but intend to sell the meat for economic gain. The Makah assert that they are committed to using the meat solely for ceremonial and subsistence purposes. Some activists also present an ethical side of the argument, as they contend that it is morally unacceptable to kill whales due to their demonstrated intelligence.

International policy is also entangled in this controversy. The International Whaling Commission granted a yearly quota of five whales to the Makah. This quota is accommodated under the 620 gray whales currently allotted for a five-year time frame to the indigenous people of Chukotka, Russia. Also, some environmentalists allege that the Makah have received funding from Japan and Norway to support their campaign for whaling. Both of these countries host commercial whaling industries, and allowing the tribe to resume hunting, environmentalists argue, would set an important precedent for the Japanese and Norwegians to argue for "cultural subsistence" whaling also.

Numerous national entities are involved in the policy behind this debate as well. Compiling the agreements put forth by the IWC and those reached between the National Oceanic and Atmospheric Administration (NOAA) and the Makah tribal council, the Northwest Indian Fisheries Commission put forth a management plan that explicitly defines quotas, hunting techniques, area restrictions, and monitoring and enforcement regulations. Additionally, the National Marine Fisheries Service has played an active part in defining sustainable populations of whales and other marine animals. The Makah are subject to strict regulation and would be under careful surveillance if whaling commenced.

In the fall of 1998, the Makah began their first attempt at harvesting a gray whale in the waters of Neah Bay. The tribe met with animated opposition, as citizens and activists protested on the shore of the hunting site and the Sea Shepherd Conservation Society sailed their fleet of vessels into the bay to shield whales from the hunting canoes. The presence of local and national media brought international attention to the dispute. Although numerous attempts were made, no whale was taken at the time.

The tribe made a second attempt in May 1999 and again met strong opposition, the intensity of which led to the arrests of numerous activists. However, on May 17, the hunting team succeeded in harpooning and fatally shooting a young female gray whale.

What do you think? Should the U.S. government honor its treaty with the Makah despite other conservation legislation? Is it morally wrong to hunt whales in the first place? Is "cultural subsistence" a valid argument for taking a whale? What stance should the international community take?

For more information, visit these websites:

www.makah.com/whales.htm
http://www.nwifc.wa.gov/whaling/makah_frank.asp
htpp://www.nwr.noaa.gov/factshet/mak-inf.pdf
http://www.olypen.com/makahwc/
http://www.seashepherd.org/issues/whales/makahhunt1200.html

If that happened, food production would almost certainly decline. Hundreds of millions of people would be driven from their homes by famine, flood, and drought. Billions more would be hard pressed to maintain their current lifestyles, much less aspire to an American standard of living. It has been stated that never mind Botswana and Bangladesh, Cambodia, and Cameroon—in 2050, perhaps even people in the United States may not be able to live the way Americans do today. What are your thoughts? Is there an ethical argument to the question of consumption in the United States? Do we consume too much?

Global Environmental Ethics

In 1990, Noel Brown, the director of the United Nations North American Environmental Programme, stated:

> Suddenly and rather uniquely the world appears to be saying the same thing. We are approaching what I have termed a consensual moment in history, where suddenly from most quarters we get a sense that the world community is now agreeing that the environment has become a matter of global priority and action.

This new sense of urgency and common cause about the environment is leading to unprecedented cooperation in some areas. Despite their political differences, Arab, Israeli, Russian, and American environmental professionals have been working together for several years. Ecological degradation in any nation almost inevitably impinges on the quality of life in others. For years, acid rain has been a major irritant in relations between the United States and Canada. Drought in Africa and deforestation in Haiti have resulted in waves of refugees. From the Nile to the Rio Grande, con-

Table 2.1	Major International Environmental Treaties Since 1965

Convention on International Trade in Endangered Species of Wild Fauna and Flora, 1973

Agreement on an International Energy Programme, 1974

Convention on Long-range Transboundary Air Pollution, 1979; amended, 1991

Convention on the Physical Protection of Nuclear Material, 1980

International Tropical Timber Agreement, 1983

Vienna Convention for the Protection of the Ozone Layer, 1985

Montreal Protocol on Substances that Deplete the Ozone Layer, 1987

UN Framework Convention on Climate Change, 1992

    Council of the Parties 1: Berlin Mandate, 1995

    Council of the Parties 3: Kyoto Protocol, 1997

UN Conference on Environment and Development, 1992

    Rio Declaration on Environment and Development

    Agenda 21: Global Programme of Action on Sustainable Development

    Convention on Biological Diversity

http://sedac.ciesin.org/entri/

http://faolex.fao.org/faolex/

www.greenyearbook.org

In addition to these international conferences, many other treaties and agreements address international environmental concerns. This table identifies several major initiatives since 1965.

flicts flare over water rights. The growing megacities of the Third World are time bombs of civil unrest.

Much of the current environmental crisis is rooted in and exacerbated by the widening gap between rich and poor nations. Industrialized countries contain only 20 percent of the world's population, yet they control 80 percent of the world's goods and create most of its pollution. The developing countries are hardest hit by overpopulation, malnutrition, and disease. As these nations struggle to catch up with the developed world and improve the quality of life for their people, a vicious circle begins: Their efforts at rapid industrialization poison their cities, while their attempts to boost agricultural production often result in the destruction of their forests and the depletion of their soils, which lead to greater poverty. (See figure 2.5.)

Perhaps one of the most important questions for the future is, "Will the nations of the world be able to set aside their political differences to work toward a global environmental course of action?" The United Nations Conference on Human Environment held in Stockholm, Sweden, in 1972 was a step

in the right direction. Out of that international conference was born the UN Environment Programme, a separate department of the United Nations that deals with environmental issues. A second world environmental conference was held in 1992 in Brazil. It followed up the Stockholm conference with many new international initiatives. A major world conference on climate change was held in Kyoto, Japan, in 1997. (See Global Perspective: Earth Summit, and Global Perspective: The Kyoto Protocol in chapter 17.) Through organizations and conferences such as these, nations can work together to solve common environmental problems. Several major international treaties are listed in Table 2.1.

At the individual level, people have begun to respond to increased awareness of global environmental change by altering their values, beliefs, and actions. Changes in individual behavior are necessary but are not enough. As a global species, we are changing the planet. By pooling our knowledge, coordinating our actions, and sharing what the planet has to offer, we can achieve a global environmental ethic.

figure 2.5 **Lifestyle and Environmental Impact** Significant differences in lifestyles and their environmental impact exist between the rich and poor nations of the world. What would be the environmental impact on the Earth if the citizens of China and India and other less-developed countries enjoyed the standard of living in North America? Can we deny them that opportunity?

Antarctica—Resource or Refuge?

Few places on earth have not been exploited by humans. One such place is Antarctica. It is as close to an unpolluted environment as there is on earth, but it is not without its problems.

Seals and whales were the earliest exploited resource in Antarctica. There was money to be made, and this "opportunity" resulted in the near extinction of the southern fur seal, the elephant seal, and the blue whale.

By the 1950s, aboveground nuclear testing had spread radioactive particles around the planet, including Antarctica. Pesticides like DDT were turning up in the tissues and blood of certain Antarctic bird and marine mammal species. A growing hole in the ozone layer above the Antarctic continent developed as a result of the use of chlorofluorocarbons throughout the world. Fossil-fuel combustion contributes to the greenhouse effect, which in turn threatens to melt the ice in Antarctica's Western Peninsula.

During the past several decades, Antarctica has been the site of extensive scientific exploration. Much of this exploration has been economically motivated. For example, government scientists, with the aid of satellites, are advising oil and mineral prospectors. Much of the so-called scientific research is conducted with geopolitical or military objectives in mind.

Antarctica is also being proposed as a tourist attraction. Australia has suggested building a hotel, while Argentina is considering chartering a vessel to transport 600 tourists from South America seven times a year. Several sites are also being viewed as potential ski resorts. Numbers of tourists to Antarctica are increasing yearly. Over 10,000 tourists visited in 1999. This was a 50 percent increase over 1998. Antarctica is being promoted as a tourist destination for those "in search of a new frontier."

From an ecological perspective, Antarctica is fragile. The thin layer on the surface of the ocean, nourished by the sun, supports the tiny shrimplike krill, which sustain fish, whales, seals, and penguins. These short, simple food chains are extremely sensitive to environmental insults.

In the mid-1970s, New Zealand proposed designating the continent an Antarctic World Park. This would turn Antarctica into an international wilderness area, a region on Earth where we recognize that humanity does not belong.

In 1991, 24 countries signed an agreement to ban mineral and oil exploration in Antarctica for 50 years. The agreement, which was hailed as historic by governments and environmental groups, includes new regulations for wildlife protection, waste disposal, marine pollution, and continued monitoring of the Antarctic, which covers nearly one-tenth of the world's land surface. The signing of the agreement in Madrid, Spain, was the result of two years of negotiations. The protocol protects Antarctica's delicate flora and fauna and sets procedures to assess environmental effects of all human activities on the continent.

- Should we turn a continent into a world park?
- Would humanity be better served by developing the natural resources of Antarctica, such as oil and minerals?
- Should the natural beauty of Antarctica be opened up to tourism so it can be enjoyed by many?
- Is it possible to strike a balance between preservation and development in a fragile ecosystem? Can you give examples?

Source: United Nations publications.

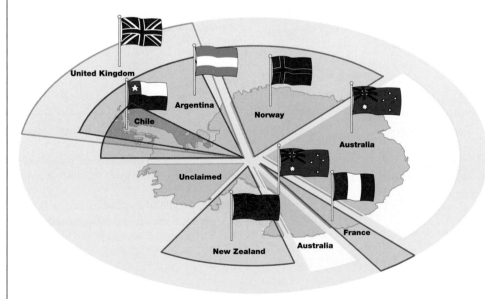

Territorial Claims
No country owns Antarctica, but seven countries have made claims to parts of it. (The United States, which has the largest research population on the continent, has not.) Overlapping claims led to the Antarctic Treaty, signed by 41 countries in 1991, which declares that "Antarctica shall be open to all nations to conduct scientific or other peaceful activities there."

Source: Composite drawing from United Nations publications.

Summary

People of different cultures view their place in the world from different perspectives. Among the things that shape their views are religious understandings, economic pressures, geographic location, and fundamental knowledge of nature. Because of this diversity of backgrounds, different cultures put different values on the natural world and the individual organisms that compose it.

Three prevailing attitudes toward nature are the development ethic, which assumes that nature is for people to use for their own purposes; the preservation ethic, which assumes that nature has value in itself and should not be disturbed; and the conservation ethic, which recognizes that we will use nature but that it should be used in a sustainable manner.

Ethical issues can be examined at several levels. Growth and exploitation have been the prevailing priorities of our society and individual consumers for generations. This does not mean that everyone in society has the same opinions, but the general attitude has been one of development rather than preservation. Most individual environmental decisions have actually been economic decisions, and the rationale has been that if a resource is available for use, it should be used.

Corporate ethics are even more strongly influenced by economics. Corporations exist to make a profit. Any way that they can reduce costs makes them more profitable. Unfortunately, pollution and exploitation of rare resources may be costly to individuals or society while being profitable to corporations. In addition, corporations wield tremendous economic power and can sway public opinion and political will. Many corporations have begun to openly acknowledge their responsibilities to carefully examine their impact on the natural world.

Society and corporations are composed of individuals. An increasing sensitivity of individual citizens to environmental concerns can change the political and economic climate for society and corporations. However, people often do not have a clear idea of what should be done and often do not act in a way that supports their stated beliefs.

Global environmental concerns have become more important. The world is getting "smaller" and more interrelated. As more people are added to the world's population each year, there is increasing competition for the resources needed to live a decent life. An environmental disaster is no longer a local problem but affects us globally. The increasing economic difference between rich and poor nations affects the global environment, since the poor aspire to have what the rich take for granted. All peoples and nations need to work together to solve environmental problems.

Key Terms

anthropocentric *21*	ecocentrism *21*	morals *21*
biocentric *21*	economic growth *24*	preservation ethic *24*
conservation or management ethic *24*	environmental justice *28*	profitability *25*
corporation *25*	ethics *21*	resource exploitation *24*
development ethic *22*	industrial ecology *27*	

Review Questions

1. How does personal wealth relate to ethics? Can you provide personal examples?
2. Why do industries pollute?
3. Why would normal economic forces work against pollution control? Do you feel that this is changing?
4. Is it reasonable to expect a totally unpolluted environment? Why or why not?
5. What has been the dominant societal attitude toward resource use?
6. Describe the differences between development, preservation, and conservation ethics. Must there always be conflict among these ethics?
7. What is a major motivating force of corporate management?
8. Why do decision makers view the actions of corporations differently from the way they view the actions of individuals?

Critical Thinking Questions

1. Using the definitions of moral and ethical judgment as presented in the text, identify at least two moral and ethical responses each to the issue of global climate change. What values, beliefs, and perspectives are at the root of these judgments?

2. What are our responsibilities to future generations regarding the environment? What values, beliefs, and perspectives lead you to think and act the way you do with regard to the environment?

3. Compare and contrast the three approaches to environmental ethics outlined in the text. Which is closest to your own? Why? How does, and how could, your ethical stance influence your actions?

4. The text makes the point that up until recently humans have believed, almost universally, in unchecked growth as a positive good. Now, at the beginning of the twenty-first century, some are beginning to question this belief. What values, beliefs, and perspectives might these critics have? Describe some ways these critics might be received in a developing country. Why?

5. Imagine you are a business executive who wants to pursue an environmental policy for your company that limits pollution and uses fewer raw materials but would cost more. What might be the discussion at your next board of directors meeting? How would you respond to your board of directors and shareholders? Why?

6. In 1997, Ojibwa Indians in northern Wisconsin sat on railroad tracks to block from crossing their reservation a shipment of sulfuric acid that was headed for a controversial injection copper mine in northern Michigan. Try to put yourself in their position. What values, beliefs, and perspectives might have contributed to this action? Now put yourself in the position of the copper miners in northern Michigan. How might these copper miners have responded? What values, beliefs, and perspectives contribute to their action?

7. Read the Environmental Close-Up about environmental philosophy. Do you feel there is too much talk about environmental problems and not enough action? Or too little talk? Or some other problem? Please describe your position on this and your reasons for thinking the way you do.

8. Imagine yourself in the position of a person who lives on a poor Native American reservation that is contemplating building a storage facility for nuclear waste. What preconceptions, values, beliefs, or contextual perspectives might you bring to the issue? What might you propose as a course of action for yourself and for others? Why?

9. Consider environmental ethics issues in the year 2025. At the rate consumers, corporations, and governments are responding to environmental concerns, what quality-of-life consequences do you project for the year 2025? How will your health, lifestyle, income, employment, and community be affected?

Concept Map

Construct a map to show relationships among the following concepts:

ethics

economic growth

environmental justice

morals

anthropocentric theory

corporation

ecocentrism

profitability

biocentrism

Interactive Exploration

Check out the website at **http://www.mhhe.com/environmentalscience** and click on the cover of this textbook for quizzing, career information, case studies, and hot links for the following topics:

Environmental Ethics

Environmental Philosophy

History of Environmental Studies

Environmental and Ecological Organization Sites

Extinction Issues

Environmental Policy, Law and Planning

The Ultimate Ecological Answers

Individual Contributions to Environmental Issues

Environmental Organizations

3 CHAPTER

Risk and Cost:
Elements of Decision Making

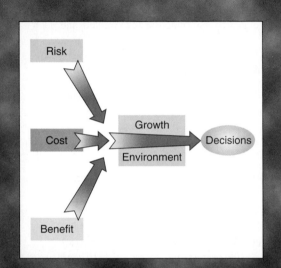

Objectives

After reading this chapter, you should be able to:

- Describe why the analysis of risk has become an important tool in environmental decision making.
- Understand the difference between risk assessment and risk management.
- Describe the issues involved in risk management.
- Understand the difference between true and perceived risks.
- Define what an economic good or service is.
- Understand the relationship between the available supply of a commodity or service and its price.
- Understand how and why cost-benefit analysis is used.
- Understand the concept of sustainable development.
- Understand environmental external costs and the economics of pollution prevention.
- Understand the market approach to curbing pollution.

Chapter Outline

Measuring Risk

Two factors are primary in many decisions in life: risk and cost. We commonly ask such questions as "How likely is it that someone will be hurt?" and "What is the cost of this course of action?" Environmental decision making is no different. If a new air-pollution regulation is contemplated, industry will be sure to point out that it will cost a considerable amount of money to put these controls in place and will reduce profitability. Citizens will point out that their tax money will have to support another governmental bureaucracy. On the other side, advocates will point out the reduced risk of illnesses and the reduced cost of health care for people who live in areas of heavy air pollution.

Risk analysis has become an important decision-making tool at all levels of society. In the area of environmental concerns, assessing and managing risks help us determine what environmental policies are appropriate. The analysis of risk generally involves a probability statement. **Probability** is a mathematical statement about how likely it is that something will happen. Probability is often stated in terms like "The probability of developing a particular illness is 1 in 10,000," or "The likelihood of winning the lottery is 1 in 5 million." It is important to make a distinction between *probability* and *possibility.* When we say something is *possible,* we are just saying that it could occur. It is a very inexact term. *Probability* defines how likely *possible* events are.

Another important consideration is the consequences of an event. If a disease is likely to make 50 percent of the population ill (the probability of becoming ill is 50 percent) but no one dies, that is very different from analyzing the safety of a dam, which if it failed would cause the deaths of thousands of people downstream. We would certainly not accept a 50 percent probability that the dam would fail. Even a 1 percent probability in that case would be unacceptable. The assessment and management

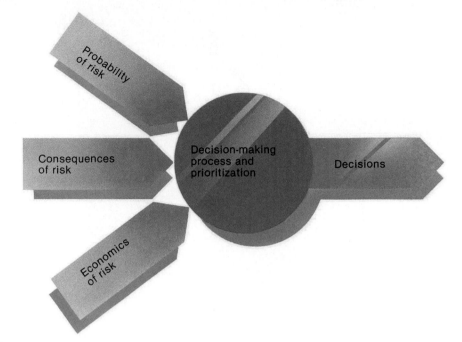

figure 3.1   **Decision-Making Process** The assessment, cost, and consequences of risks are all important to the decision-making process.

of risk involve an understanding of probability and the consequences of decisions. (See figure 3.1.)

Risk assessment involves analyzing a risk to determine the probability of an adverse effect. Risk management is a much broader task that includes assessment and the consequences of risks in the decision-making process. We will look at both these tasks in the next two sections.

Risk Assessment

Environmental **risk assessment** is the use of facts and assumptions to estimate the probability of harm to human health or the environment that may result from exposures to pollutants, toxic agents, or management decisions. What risk assessment provides for environmental decision makers is an orderly, clearly stated, and consistent way to deal with scientific issues when evaluating whether a hazard exists and what the magnitude of the hazard may be.

Calculating the hazardous risk to humans of a particular activity, chemical, or technology is difficult. If a technology is well known, scientists use probabilities based on past experience to

estimate risks. For example, the risk of developing black lung disease from coal dust in mines is well established. To predict the risks associated with new technology, much less accurate statistical probabilities, based on models rather than real-life experiences, must be used.

Risks associated with new chemicals are difficult to quantify. While animal tests are widely accepted in predicting whether or not a chemical will cause cancer in humans, their use in predicting how many cancers will be caused in a group of exposed people is still very controversial. Most risk assessments are *estimates* of the probability that a person will develop cancer or other negative effects. (See figure 3.2.)

Such estimates typically are based on broad assumptions to ensure that a lack of complete knowledge does not result in an underestimation of the risk. For example, people may be more or less sensitive to the effects of certain chemicals than the laboratory animals studied. Also, people vary in their sensitivity to cancer-causing compounds. Thus, what may present no risk to one person may be a high risk to others. Persons with breathing difficulties are more likely to be adversely affected by high levels of

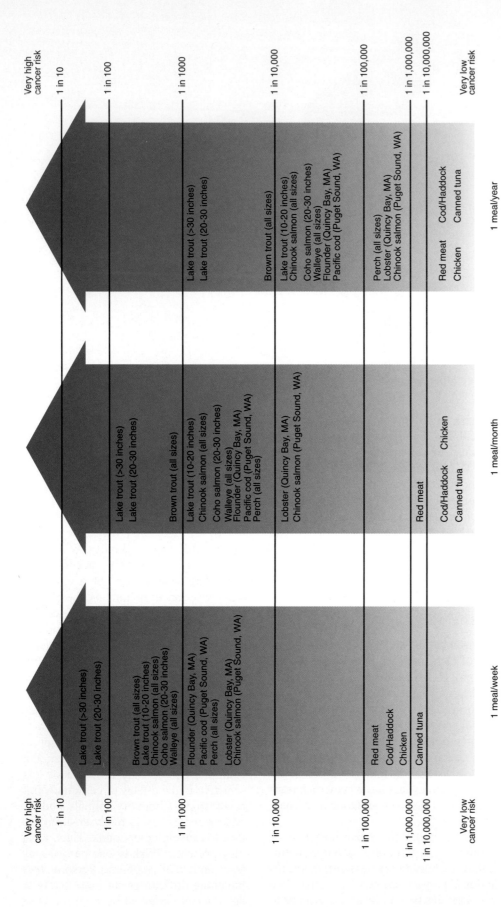

figure 3.2 **Cancer Risk and Fish Consumption** A study by the National Wildlife Federation provides estimates for cancer risks associated with the consumption of sport fish.
Reprinted with permission from "*Great Lakes Fish Consumption Study*," 1989, Great Lakes National Resource Center, Ann Arbor, MI.

environmental
CLOSE-UP

What's in a Number?

Risk values are often stated, shorthand-fashion, as a number. When the risk concern is cancer, the risk number represents a probability of occurrence of additional cancer cases. For example, such an estimate for Pollutant X might be expressed as 1×10^{-6}, or simply 10^{-6}. This number can also be written as 0.000001, or one in a million—meaning one additional case of cancer projected in a population of 1 million people exposed to a certain level of Pollutant X over their lifetimes. Similarly, 5×10^{-7}, or 0.0000005, or five in *10 million,* indicates a potential risk of five additional cancer cases in a population of 10 million people exposed to a certain level of the pollutant. These numbers signify additional cases above the background cancer incidence in the general population. American Cancer Society statistics indicate that the background cancer incidence in the general population is one in three over a lifetime.

If the effect associated with Pollutant X is not cancer but another health effect, perhaps neurotoxicity (nerve damage) or birth defects, then numbers are not typically given as probability of occurrence but rather as levels of exposure estimated to be without harm. This often takes the form of a reference dose (RfD). An RfD is typically expressed in terms of milligrams (of pollutant) per kilogram of body weight per day, e.g., 0.004 mg/kg/day. Simply described, an RfD is a rough estimate of daily exposure to the human population (including sensitive subgroups) that is likely to be without appreciable risk of deleterious noncancerous effects during a lifetime. The uncertainty in an RfD may be one or several orders of magnitude (i.e., multiples of 10).

What's in a number? The important point to remember is that the numbers by themselves don't tell the whole story. For instance, even though the numbers are identical, a cancer risk value of 10^{-6} for the "average exposed person" (perhaps someone exposed through the food supply) is not the same thing as a cancer risk of 10^{-6} for a "most exposed individual" (perhaps someone exposed from living or working in a highly contaminated area). It's important to know the difference. Omitting the qualifier "average" or "most exposed" incompletely describes the risk and would mean a failure in risk communication.

A numerical estimate is only as good as the data it is based on. Just as important as the *quantitative* aspect of risk characterization (the risk numbers), then, are the *qualitative* aspects. How extensive is the database supporting the risk assessment? Does it include human epidemiological data as well as experimental data? Does the laboratory database include less data on more than one species? If multiple species were tested, did they all respond similarly to the test substance? What are the "data gaps," the missing pieces of the puzzle? What are the scientific uncertainties? What science policy decisions were made to address these uncertainties? What working assumptions underlie the risk assessment? What is the overall confidence level in the risk assessment? All of these qualitative considerations are essential to deciding what reliance to place on a number and to characterizing a potential risk.

Source: Data from *EPA Journal.*

---

air pollutants than are healthy individuals. In addition, the estimate of human risk is based on extrapolation from animal tests in which high, chronic doses are used. Human exposure is likely to be lower or infrequent. Because of all these uncertainties, government regulators have decided to err on the side of safety to protect the public health. That approach has been criticized by those who say it carries protection to the extreme, usually at the expense of industry.

Over the past decade, risk assessment has had its largest impact in regulatory practices involving cancer-causing chemicals called **carcinogens.** In the United States, for example, the decisions to continue registration of pesticides, to list substances as hazardous air pollutants under the Clean Air Act, and

to regulate water contaminants under the Safe Drinking Water Act depend to a large degree on the risk assessments for the substances in question.

Risk assessment analysis is also being used to help set regulatory priorities and support regulatory action. Those chemicals or technologies that have the highest potential to cause damage to health or the environment receive attention first, while those perceived as having minor impacts receive less immediate attention. Medical waste is perceived as having high risk, and laws have been enacted to minimize the risk, while the risk associated with the use of fertilizer on lawns is considered minimal and is not regulated.

The science supporting environmental regulatory decisions is complex and

rapidly evolving. Many of the most important threats to human health and the environment are highly uncertain. Risk assessment quantifies risk and states the uncertainty that surrounds many environmental issues. This can help institutions research and plan in a way that is consistent with scientific and public concern for environmental protection.

Risk Management

Risk management is a decision-making process of weighing policy alternatives and selecting the most appropriate regulatory action, integrating the results of risk assessment with engineering data and with social, economic, and political concerns to reach a decision. Risk management includes:

1. Deciding which risks should be given the highest priority
2. Deciding how much money will be needed to reduce each risk to an acceptable level
3. Deciding where the greatest benefit would be realized by spending limited funds
4. Deciding how much risk is acceptable
5. Deciding how the plan will be enforced and monitored

Risk management raises several issues. With environmental concerns such as acid rain, ozone depletion, and hazardous waste, the scientific basis for regulatory decisions is often controversial. As was mentioned in the "Risk Assessment" section, hazardous substances can be tested, but only on animals. Are animal tests appropriate for determining impacts on humans? There is no easy answer to this question. Dealing with global warming, ozone depletion, and acid rain require projecting into the future and estimating the magnitude of future effects. Will the sea level rise? How many lakes will become acidified? How many additional skin cancers will be caused by depletion of the ozone layer? Estimates from equally reputable sources vary widely. Which ones do we believe?

The politics of risk management frequently focus on the adequacy of the scientific evidence. The scientific basis can be thought of as a kind of problem definition. Science determines that some threat or hazard exists, but because scientific facts are open to interpretation, there is controversy. For example, it is a fact that dioxin is a highly toxic material known to cause cancer in laboratory animals. It is also very difficult to prove that human exposure to dioxin has led to the development of cancer, although high exposures have resulted in acne in exposed workers. Acne is a common result of exposure to molecules like dioxin.

This is why problem definition is so important. Defining the problem helps to determine the rest of the policy process (making rules, passing laws, or issuing statements). If a substance poses little or no risk, then policy action is unnecessary. For example, some observers believe

chemicals pose many threats that need to be addressed. Others believe chemicals pose little threat; instead, they see scare tactics and government regulations as unnecessary attacks on businesses. Logging forests poses risks of soil erosion and the loss of resident animal species. The timber industry sees these risks as minimal and as a threat to its economic well-being, while many environmentalists consider the risks unacceptable. These and similar disagreements are often serious public-relations problems for both government and business because most of the public has a poor understanding of the risks they accept daily.

True and Perceived Risks

People often overestimate the frequency and seriousness of dramatic, sensational, well-publicized causes of death and underestimate the risks from more familiar causes that claim lives one by one. Risk estimates by "experts" and by the "public" on many environmental problems differ significantly. This discrepancy and the reasons for it are extremely important because the public generally does not trust experts to make important risk decisions alone.

While public health and environmental risks can be minimized, eliminating all risks is impossible. Almost every daily activity—driving, walking, working—involves some element of risk. (See table 3.1.)

From a risk management standpoint, whether one is dealing with a site-specific situation or a national standard, the deciding question ultimately is:

What degree of risk is acceptable? In general, we are not talking about a "zero risk" standard but rather a concept of **negligible risk:** At what point is there really no significant health or environmental risk? At what point is there an adequate safety margin to protect public health and the environment?

Risk management involves comparing the estimated true risk of harm from a particular technology or product with the risk of harm perceived by the general public. The public generally perceives involuntary risks, such as nuclear power plants or nuclear weapons, as greater than voluntary risks, such as drinking alcohol or smoking. In addition, the public perceives newer technologies, such as genetic engineering or toxic-waste incinerators, as greater risks than more familiar technologies, such as automobiles and dams. Many people are afraid of flying for fear of crashing; however, automobile accidents account for a far greater number of deaths—about 45,000 in the United States each year, compared to less than a thousand annually from plane crashes. (See figure 3.3.)

One of the most profound dilemmas facing decision makers and public health scientists is how to address the apparently growing discrepancy between the scientific and public perceptions of environmental risks. Numerous studies have shown that in the last 20 years, environmental hazards truly affecting health status in the country are not those receiving the highest attention, whether measured by public opinion polls, news coverage, congressional actions, or government expenditures.

Table 3.1 Risks of Death

Deaths per million hours of exposure

Mountain climbing	40,000
Canoeing	10,000
Cigarette smoking	3000
Swimming	2560
Automobile travel	1200
Hunting	1000
Air travel	500
Living beside a nuclear power plant	0.5

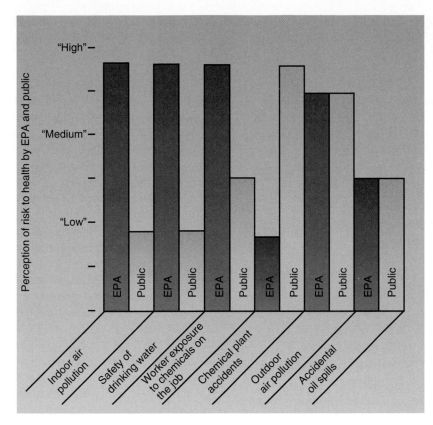

figure 3.3 **Perception of Risk** Professional regulators and the public do not always agree on what risks are.

Indoor air pollution, in its various forms, receives relatively little attention compared with outdoor sources and yet probably accounts for as much, if not more, poor health. Hazardous waste dumps, on the other hand, which are difficult to associate with any measurable ill health, attract much attention and resources. The same chemicals in the form of common consumer products, such as household cleaners, pesticides, and fuel (gasoline), account for much more exposure and ill health and yet are comparatively of little concern to the public.

Several explanations exist for this difference in perception, the major ones relating to the fact that the public uses a number of criteria other than health risk to establish its concerns. In a world of finite resources, however, there is a real health cost in focusing attention on risks that have little measurable health impact and, at least by default, thus result in poorer funding of interventions that address significant risks.

An option for consideration is to explicitly separate these risks by keeping those hazards with significant health risks within the health arena and placing others, perhaps equally as important for public policy but less so for public health (including litter control and wildland preservation), in the category of environmental quality improvement along with many other worthwhile programs of this kind.

It has been stated that environmental programs that should remain in the health arena at present are:

- Controlling household pollution exposures, including radon, environmental tobacco smoke, allergens, lead, consumer products, and wood smoke
- Controlling outdoor air pollution where doing so is more effective than controlling indoor or localized exposure sources of the same pollutants
- Applying stringent criteria in judging any proposed new product or activity that will affect large numbers of people (e.g., new consumer chem-

ical products, household materials, and additives to food or fuel)

- Applying stronger and more uniform restrictions on occupational hazards, which tend to be much larger than those that confront the general public and, in contrast, often have demonstrable impacts on health

Some researchers argue that the public is frequently misled by the politics of public health and environmental safety. This is understandable since many prominent people become involved in such issues and use their public image to encourage people to look at issues from a particular point of view.

Whatever the issue, it is hard to ignore the will of the people, particularly when sentiments are firmly held and not easily changed. A fundamental issue surfaces concerning the proper role of a democratic government and other organizations in a democracy when it comes to matters of risk. Should the government focus available resources and technology where they can have the greatest tangible impact on human and ecological well-being, or should it focus them on problems about which the public is most upset? What is the proper balance? For example, would adequate prenatal health care for all pregnant women have a greater effect on the health of children than removing asbestos from all school buildings?

Obviously, there are no clear answers to these questions. However, experts and the public are both beginning to realize that they each have something to offer concerning how we view risk. Many risk experts who have been accustomed to looking at numbers and probabilities are now conceding that a rationale exists for looking at risk in broader terms. At the same time, the public is being supplied with more data to enable them to make more informed judgments.

Throughout this discussion of risk assessment and management, we have made numerous references to costs and economics. It is not economically possible to eliminate all risk. As risk is eliminated, the cost of the product or service increases. Many environmental issues are difficult to evaluate from a purely economic point of view, but economics is one of the tools useful to analyzing any environmental problems.

Economics and the Environment

Environmental problems are primarily economic problems. While this may be an overstatement, it often is difficult to separate economics from environmental issues or concerns. Basically, economics deals with resource allocation. It is a description of how we value goods and services. We are willing to pay for things or services we value highly and are unwilling to pay for things we think there is plenty of. For example, we will readily pay for a warm, safe place to live but would be offended if someone suggested that we pay for the air we breathe.

Our goal as a society is to seek long-term economic growth that creates jobs while improving and sustaining the environment. Achieving this goal requires an environmental strategy that repairs past environmental damage; helps us shift from waste management to pollution prevention; and uses valuable resources more efficiently.

In the three decades since the first Earth Day, many nations have made considerable progress in responding to threats to public health and the environment. Yet major challenges remain. We can put in place a set of policies and programs that will establish a new course for the development and use of environmental technologies into the twenty-first century. We must, however, broaden our environmental tool kit, replacing those instruments that are no longer effective with a new set of tools designed to meet today's challenges and tomorrow's needs.

Economic Concepts

An economic good or service can be defined as anything that is scarce. Scarcity exists whenever the demand for anything exceeds its supply. We live in a world of general scarcity. **Resources** are anything that contributes to making desired goods and services available for consumption. Resources are limited, relative to the desires of humans to consume. The **supply** is the amount of a good or service available to be purchased. **Demand** is the amount of a product that consumers are willing and able to buy at various prices. In economic terms, supply depends on:

1. The raw materials available to produce a good or service using present technology
2. The amounts of those materials available
3. The costs of extracting, shipping, and processing the raw materials
4. The degree of competition for those materials among users
5. The feasibility and cost of recycling already used material
6. The social and institutional arrangements that might have an impact. (See figure 3.4.)

The relationship between available supply of a commodity or service and its price is known as a **supply/demand curve.** (See figure 3.5.) The price of a product or service reflects the strength of the demand for and the availability of the commodity. When demand exceeds supply, the price rises. Cost increases cause people to seek alternatives or to decide not to use a product or service, which results in lower quantity demanded.

For example, food production depends heavily on petroleum for the energy to plant, harvest, and transport food crops. In addition, petrochemicals are used to make fertilizer and chemical pest-control agents. As petroleum prices rise, farmers reduce their petroleum use. Perhaps they farm less land or use less fertilizer or pesticide. Regardless, the price of food must rise as the price of petroleum rises. As the prices of certain foods rise, consumers seek less costly forms of food.

When the supply of a commodity exceeds the demand, producers must lower their prices to get rid of the product, and eventually, some of the producers go out of business. Ironically, this happens to farmers when they have a series of good years. Production is high, prices fall, and some farmers go out of business.

Market-Based Instruments

With the growing interest in environmental protection during the past three decades, policy makers are examining new methods to reduce harm to the environment. One area of growing interest is market-based instruments (MBIs). MBIs provide an alternative to the common command-and-control legislation because they use economic forces and the ingenuity of entrepreneurs to achieve a high degree of environmental protection at a low cost. Instead of dictating how industry should conduct its activities, MBIs provide incentives by imposing costs on pollution-causing activities. This approach allows companies to decide for themselves how best to achieve the required level of environmental protection. To date, most of these market-based policies have been implemented in developed nations and in some rapidly growing developing nations. In virtually all cases, they have been introduced as supplements to, not substitutes for, traditional government regulations.

Implementing market-based policies is not easy in any nation. To succeed, nations optimally require open, dynamic market systems, sound macroeconomic conditions, political and institutional stability, and full development of human rights.

In developing nations with large informal sectors, other policies—capacity-building, overall policy reform, community participation, investments in education and health care—initially may be more effective than market-based mechanisms.

Nevertheless, market-based policies may offer valuable opportunities to achieve environmental goals more efficiently and at lower cost to governments and entrepreneurs. They also offer opportunities for governments to ride the momentum of economic growth, while leading it to a more sustainable footing.

Market-based instruments acknowledge the fact that environmental resources are often underpriced. First,

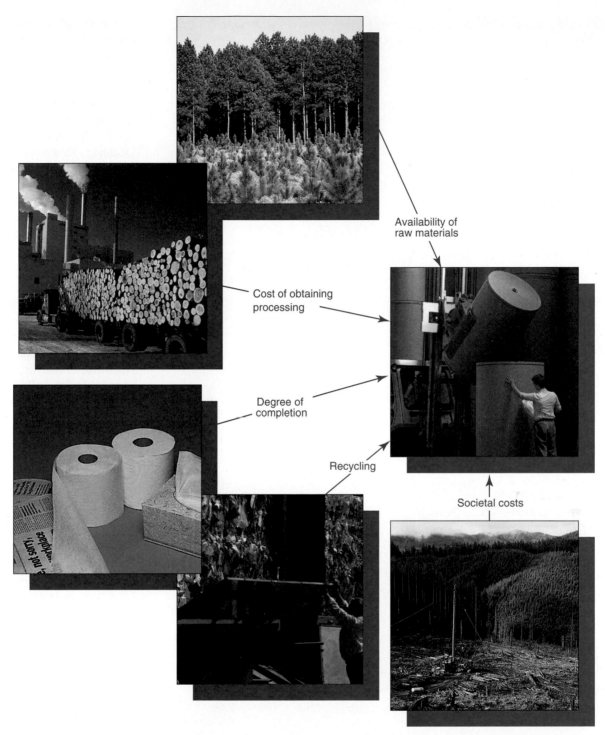

figure 3.4 **Factors That Determine Supply** The supply of a good or service depends on the several factors shown here.

subsidies—on water, electricity, fossil fuels, road transport, and agriculture—are an incentive to overuse a resource; reducing or removing subsidies creates an incentive to use resources more efficiently. Second, market prices reflect only private cost, not the external damages caused by pollution or resource extraction. Instead of inflexible, top-down government directives, market-based policies take advantage of price signals and give entrepreneurs the freedom to choose the solution most economically efficient for them.

MBIs can be grouped into five basic categories:

1. **Information programs**
   These programs rely on informed consumers' market choices to reduce environmental problems.

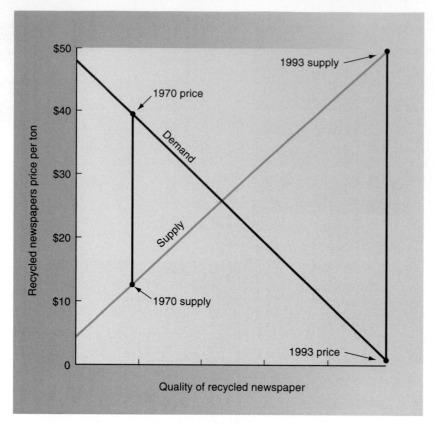

a.

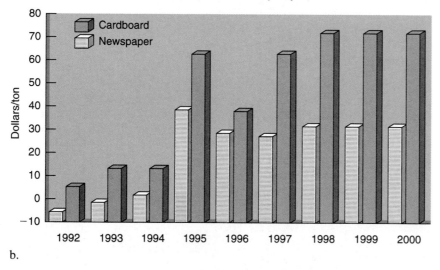

b.

figure 3.5 **Supply and Demand** (*a*) As a result of increased interest in recycling, the supply of recycled newspapers grew substantially between 1970 and 1993. The demand changed very little; therefore, the price in 1993 was extremely low. (*b*) Beginning in 1994, the demand for recycled newspaper and cardboard increased, thus the price rose significantly. Prices were more stable through 2000.

Source: (*b*) Data from *Recycling Times.*

Information about the environmental or risk consequences of choices make clear to consumers that it is in their personal interest to change their decisions or behavior. Examples include radon or lead testing or labeling pesticide products. Another type of program, such as the Toxic Release Inventory in the United States, discloses information on environmental releases by polluters. This provides corporations with incentives to improve their environmental performance to enhance their public image.

2. **Tradable emissions permits**

These permits give companies the right to emit specified quantities of pollutants. Companies that emit less than the specified amounts can sell their permits to other firms or "bank" them for future use. Businesses responsible for pollution have an incentive to internalize the external cost they were previously imposing on society: If they clean up their pollution sources, they can realize a profit by selling their permit to pollute. Once a business recognizes the possibility of selling its permit, it sees that pollution is not costless. The creation of new markets (the permits) to reduce pollution reduces the external aspects to waste disposal and makes them internal costs, just like costs of labor and capital.

3. **Emission fees, taxes, and charges**

These fees provide incentives for environmental improvement by making environmentally damaging activity or products more expensive. Businesses and individuals reduce their level of pollution wherever it is cheaper to abate the pollution than to pay the charges. Emissions fees can be useful when pollution is coming from many small sources, such as vehicular emissions or agricultural runoff, where direct regulation or trading schemes are impractical. Taxes contribute to government revenue; charges are used to fund environmental cleanup programs.

Global Perspective

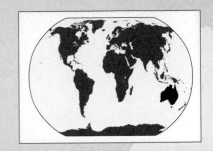

Wombats and the Australian Stock Exchange

The global New Economy can soon claim platypuses, wombats, numbats, and wallabies among its most charismatic recruits.

These native Australian mammals, at risk from the steady loss of their habitat, are working assets of Earth Sanctuaries Ltd., which is taking a controversial new approach to environmental activism. In 2000, Earth Sanctuaries was listed on the Australian Stock Exchange, making it the world's first conservation company to go public.

The venture, which makes money mostly from tourists attracted by access to rare animals at three Australian sanctuaries, is considered an extreme example of a trend intriguing environmentalists and investors around the world. Many believe calls to altruism have failed to reverse the rapid loss of species. It's time, they say, to focus on the bottom line.

Debate is growing, however, over this new approach. On the one side, researchers see this approach as a beautiful alignment of incentives. They would argue that people care first of all about themselves and their families, and it just has not worked to try to make them choose. Do you save to give your daughter a good education or help rescue a marsupial? They argue that this way, you can have both.

Yet some worry about the fund-raising approach. Their argument would be that a tension will always exist between conservation and maximizing profits. If shareholders pressure the company to make higher returns, will the company compromise its values by building megalodges or overstocking the reserves? Some question whether this approach is really conserving natural ecosystems or just creating large zoos.

Is this argument purely idealistic or accurate? Does the marketplace have a role in habitat preservation? Can you identify other applications of this approach?

In China, a pollution tax system is intended to raise revenue for investment in industrial pollution control, help pay for regulatory activities, and encourage enterprises to comply with emission and effluent standards. The system imposes noncompliance fees on discharges that exceed standards and assesses fines and other charges on violations of regulations.

In the Netherlands, a system of effluent charges on industrial wastewater has been viewed as successful. Especially among larger companies, the tax worked as an incentive to reduce pollution. In a survey of 150 larger companies, about two-thirds said the tax was the main factor in their decision to reduce discharges. As the volume of pollution from industrial sources dropped, rates were increased to cover the fixed costs of sewage water treatment plants. Rising rates are providing a further incentive for more companies to start purifying their sewage water.

4. **Performance bond/Deposit-refund programs**

   These programs place a surcharge on the price of a product, which is refunded when the used product is returned for reuse or recycling.

   Some nations—including Indonesia, Malaysia, and Costa Rica—use performance bonds to ensure that reforestation takes place after timber harvesting.

   The United States also has used this kind of approach to ensure that strip-mined lands are reclaimed. Before a mining permit can be granted, a company must post a performance bond sufficient to cover the cost of reclaiming the site in the event the company does not complete reclamation. The bond is not fully released until all performance standards have been met and full reclamation of the site, including permanent revegetation, is successful—a five-year period in the East and Midwest and 10 years in the arid West. The bond can be partially

released as various phases of reclamation are successfully completed.

Deposit/refund schemes have been widely used to encourage recycling. In Japan, deposits are made for the return of bottles. The deposits are passed on from manufacturers to shops and ultimately consumers, who get the deposit refunded when they return the used packages and bottles. Under this system, Japan recycled 92 percent of its beer bottles, 50 percent of waste paper, 43 percent of aluminum cans, and 48 percent of glass bottles by the late 1980s.

5. **Subsidies**

   Subsidies may include consumer rebates for purchases of environmentally friendly goods, loans for businesses planning to implement environmental products, and other monetary incentives designed to reduce the costs of improving environmental performance.

   A **subsidy** is a gift from government to a private enterprise that is considered important to the public interest.

environmental CLOSE-UP

Georgia-Pacific Corporation: Recycled Urban Wood—A Case Study in Extended Product Responsibility

Georgia-Pacific manufactures particleboard from multispecies wood recovered from commercial disposal or general urban solid waste. The company has agreements with five recycling and processing companies that accept or collect wood at various sites. The wood is cleaned of contaminants and sent to a Georgia-Pacific particleboard manufacturing plant in Martell, California, or to other end users.

The project involves five stakeholder groups: (1) wood waste producers (e.g., operations involved with construction and demolition debris, cut-to-size lumber, commercial wood waste from furniture), (2) collection agents, (3) processors of wood waste, which make the waste into a product that can be reused, (4) transportation contractors, shippers, and haulers, and (5) end users (e.g., Georgia-Pacific's Martell plant). The project has a variety of goals, including increasing the availability of the wood supply for particleboard production, contributing to Georgia-Pacific's goals of product stewardship, and contributing to California's mandated reduction in solid waste (e.g., 50 percent reduction by 2002).

Business factors driving the project include the shortage of fiber for the particleboard plant, rising costs of landfilling, and mandated solid waste reductions. Benefits include an expanded fiber supply in the Northwest United States. Contamination is one of the most significant barriers to the wood recovery program. Often the collected wood is mixed in with metal, plastic, and paper and must be cleared of these contaminants to be usable. The captured paper, plastic, and nonferrous metals are sent to a landfill. Wood by-products that cannot be used in particleboard processing are sold for use as animal bedding, playground cover, soil additive, and lawn or garden mulch. Currently virgin fiber, a by-product from sawmills, is often less expensive than recovered fiber for use in particleboard. As wood becomes more scarce, however, the economics will reverse.

Agriculture, transportation, space technology, and communication are frequently subsidized by governments. These gifts, whether loans, favorable tax situations, or direct grants, are all paid for by taxes on the public.

Subsidies are costly in two ways: First, the bureaucracy necessary to administer a subsidy costs money, and the subsidy is an indirect way of keeping the market price of a product low. The actual cost is higher because these subsidy costs must be added to the market price to arrive at the product's true cost. Second, in many cases, subsidies encourage activities that in the long term may be detrimental to the environment. For example, the transportation subsidies for highway construction encourage use of inefficient individual automobiles. Higher taxes on automobile use to cover the cost of building and repairing highways would encourage the use of more energy-efficient public transport. Most nations are moving toward reducing subsidies in many areas, bringing the cost of delivering resources such as water and electricity much closer to market cost.

Once established, resource subsidies are understandably difficult to dis-

lodge. But experience suggests that subsidies can be reduced or removed without disrupting rural economic development. In China, subsidy rates for coal declined from an estimated 61 percent in 1985 to 9 percent in 2000. Private mines now account for about half of all production, and some 80 percent of the coal is now sold at international prices. These reforms have had numerous benefits. Energy intensity in China has fallen by about 50 percent since 1980, operating losses at state-owned mines dropped from $1.4 billion to $230 million over the 1990–95 period, and the government's total subsidy for fossil fuels fell from about $25 billion in 1990–91 to $9 billion in 2000.

Land use subsidies also can create perverse incentives. In France since the mid-nineteenth century, a tax on undeveloped land had encouraged conversion of environmentally sensitive woodlands and wetlands. Under reforms introduced in 1992, this tax has been reduced and the economic incentive to convert less productive natural areas into productive lands decreased.

While research is still needed to determine how MBIs should be structured and used, there has been sufficient experience to justify their further deployment as a means of attaining environmental objectives at least possible cost. Successful programs so far include a sulfur dioxide trading market and hundreds of pay-by-the-bag trash collection programs in many countries.

When used correctly, market-based approaches will allow us to reach a level of environmental protection at lower total cost than would be possible with the traditional means of command and control. Market forces tend to drive decisions toward least-cost solutions. Offer the right incentives, and business will develop and adopt better pollution-control technology, rather than stagnating at "commanded" technology.

Market-oriented policies, however, will not always work. It could also be stated that no single one of these economic incentive approaches will be a panacea for all problems. It could be a mistake to start with a policy instrument, then go in search of applications. But this kind of flexible approach, using several solutions, is gaining favor among the regulated and the regulators.

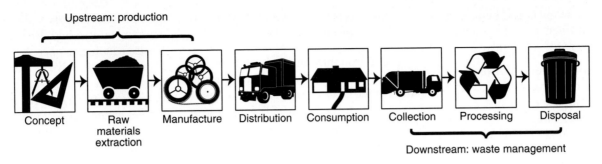

Upstream: production

Concept | Raw materials extraction | Manufacture | Distribution | Consumption | Collection | Processing | Disposal

Downstream: waste management

figure 3.6 **The Life Cycle of a Typical Product**

Source: *Environment,* vol. 39, no. 7, September 1997.

Although the various economic instruments discussed have their own niches, they can be used effectively in combination. For example, trading or pricing approaches work better if supported by information programs: communities that adopted pay-by-the-bag systems of trash disposal had fewer problems if households were given adequate information well in advance. Environmental tax systems can incorporate trading features (e.g., taxes can be levied on net emissions after trades). It will be increasingly important to use the various MBI methods together. The challenge is to design the most appropriate instruments to deal with environmental problems, bearing in mind the relevant policy objectives: steady progress in reducing risks, cost-effectiveness, encouragement of technological innovation, fairness, and administrative simplicity.

Extended Product Responsibility

Extended product responsibility (EPR) is an emerging principle for a new generation of pollution prevention policies that focus on product systems instead of production facilities. It relies for its implementation on life cycle analysis to identify opportunities to prevent pollution and reduce resource and energy use in each stage of the product chain through changes in product design and process technology. All factors along the product chain share responsibility for the life cycle environmental impacts of products, from the upstream impacts inherent in selection of materials and impacts from the manufacturing process itself to

downstream impacts from the use and disposal of the products. (See figure 3.6.)

EPR had its origins in Western Europe. The term *extended product responsibility* is derived from the term *extended producer responsibility,* which is often applied to the German packaging ordinance and similar policies in other Western European countries. The second term is something of a misnomer, however, because most extended responsibility schemes allocate the burden of environmental protection all along the product chain rather than placing it entirely on producers. Under the German packaging ordinance, for instance, consumers, retailers, and packaging manufacturers all share this responsibility, with the financial burden of waste management falling on the last two. In the United States, extended product responsibility has gained greater currency because it highlights systems of shared responsibility.

In most cases, manufacturers have the greatest ability, and therefore the greatest responsibility, to reduce the environmental impacts of their products. Companies that are accepting the challenge are recognizing that product stewardship also represents a substantial business opportunity. By rethinking their products, and their relationships with the supply chain and the ultimate customer, some manufacturers are dramatically increasing their productivity, reducing costs, fostering product and market innovation, and providing customers with more value at less environmental impact. Reducing use of toxic substances, designing for reuse and recyclability, and creating take-back programs are just a few of the many opportunities for companies to become

better environmental stewards of their products. In the twenty-first century, forward-thinking businesses have recognized that demonstrated corporate citizenship and maximum resource productivity are essential components to creating competitive advantage and increasing shareholder wealth.

As the sector with the closest ties to consumers, retailers are one of the gateways to extended product responsibility. Retailers express their support of extended product responsibility when they choose suppliers who offer greater environmental performance, educate the consumer on how to choose environmentally preferable products, and enable consumers to return products for recycling.

All products are designed with a consumer in mind. Ultimately, it is the consumer who makes the choice between competing products and who must use and dispose of products responsibly. Without consumer engagement in extended product responsibility, the loop cannot be closed. Consumers must make responsible buying choices, which consider environmental impacts. They must use products safely and efficiently. Finally, they must take the extra steps to recycle products that they no longer need.

Specific benefits of EPR include:

Cost savings, particularly through the process of producers taking back used products, allow manufacturers to recover valuable materials, reuse them, and save money.

Companies are looking at designing for recycling and disassembly—innovations necessitated by end-of-life management of products, which, in many cases, helps companies realize how to assemble products more efficiently.

There are more efficient environmental protections, since product-based environmental strategies often are a more cost-effective method of complying with environmental regulations and avoiding environmental liabilities than existing facility-based programs.

Despite these benefits, however, obstacles to EPR exist. These include:

Cost of EPR

Lack of information and tools to assess overall product system impacts

Difficulty in building relationships among actors in different life cycle stages

Hazardous waste regulations that require hazardous waste permits for collection and take-back of certain products

Antitrust laws that make it difficult for companies to cooperate

These are the kinds of issues that further explorations of EPR and, most important, real-world experience with EPR programs can help to understand and address. Ultimately, EPR is an opportunity to explore new models of environmental policy that are less costly and more flexible.

Cost-Benefit Analysis

People use **cost-benefit analysis** to determine whether a policy generates more social costs than social benefits and, if benefits outweigh costs, how much activity would obtain optimal results. Steps in cost-benefit analysis include:

1. Identification of the project to be evaluated

2. Determination of all impacts, favorable and unfavorable, present and future, on all of society

3. Determination of the value of those impacts, either directly through market values or indirectly through price estimates

4. Calculation of the net benefit, which is the total value of positive impacts less the total value of negative impacts

For example, the cost of reducing the amount of lead in drinking water in the United States to acceptable limits is estimated to be about $125 million a year. The benefits to the nation's health from such a program are estimated at nearly $1 billion per year. Thus, under a cost-benefit analysis, the program is economically sound. Recycling of solid waste, which was once cost-prohibitive, is now cost-effective in many communities because the costs of landfills have risen dramatically. Table 3.2 gives examples of the kinds of costs and benefits involved in improving air quality. Although not a complete list, the table indicates the kinds of considerations that

Table 3.2 Costs and Benefits of Improving Air Quality

Costs	Benefits
Installation and maintenance of new technology	Reduced deaths and disease
1. Scrubbers on smokestacks	Fewer respiratory problems
2. Automobile emissions control	Reduced plant and animal damage
Redesign of industries and machines	Lower cleaning costs for industry and public
Additional energy costs to industry and public	More clear, sunny days; better visibility
Retraining of employees to use new technology	Less eye irritation
Costs associated with monitoring and enforcement	Fewer odor problems

Cost-Benefit Analysis

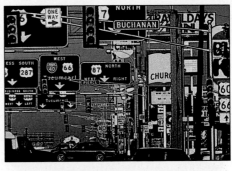

figure 3.7   **Does Everything Have an Economic Value?** The use of water and land is often based on the economic benefits obtained. Anything that humans value creates economic benefits—they just are not all easy to measure.

go into a cost-benefit analysis. Some of these are easy to measure in monetary terms; others are not.

Concerns About the Use of Cost-Benefit Analysis

Does everything have an economic value? Critics of cost-benefit analysis raise this point, among others. Some people argue that if economic thinking pervades society, many simple noneconomic values like beauty or cleanliness can survive only if they prove to be "economic." (See figure 3.7.)

It has long been the case in many developed countries that major projects, especially those undertaken by the government, require some form of cost-benefit analysis with respect to environmental impacts and regulations. In the United States, for example, such requirements were established by the National Environmental Policy Act of 1969, which requires environmental impact statements for major government-supported projects. Increasingly, similar analyses are required for projects supported by national and international lending institutions such as the World Bank.

There are clearly benefits to requiring such analysis. Although environmental issues must be considered at some point during project evaluation, efforts to do so are hampered by the difficulty of assigning specific value to environmental resources. In cases of Third World development projects, these already difficult environmental issues are made more difficult by cultural and socioeconomic differences. A less-developed country, for example, may be less inclined to insist on or be able to afford expensive emissions-treatment technology on a project that will provide jobs and economic development.

One particularly compelling critique of cost-benefit analysis is that for analysis to be applied to a specific policy, the analyst must decide which preferences count—that is, which preferences have "standing" in cost-benefit analysis. In theory, cost-benefit analysis should count all benefits and costs associated with the policy under review, regardless of who benefits or bears the costs. In practice, however, this is not always done. For example, if a cost is spread thinly over a great many people, it may not be recognized as a cost at all. The cost of air pollution in many parts of the

world could fall into such a category. Debates over how to count benefits and costs for future generations, inanimate objects such as rivers, and nonhumans, such as endangered species, are also common.

Economics and Sustainable Development

The most commonly used definition of the term **sustainable development** is one that originated with the 1987 report, *Our Common Future,* by the World Commission on Environment and Development (known as the Bruntland Commission). By that formulation, sustainable development is "development that meets the needs of the present without compromising the ability of future generations to meet their own needs."

Since the release of the Bruntland Commission report, the phrase has been broadened and modified. The term *sustainable* has gained usage because of increasing concern over exploitation of natural resources and economic development at the expense of environmental quality. Although disagreement exists as to the precise meaning of the term beyond respect for the quality of life of

environmental CLOSE-UP

"Green" Advertising Claims—Points to Consider

Like many consumers, you may be interested in buying products that are less harmful to the environment. You have probably seen products with such "green" claims as "environmentally safe," "recyclable," "degradable," or "ozone friendly." But what do these claims really mean? How can you tell which products really are less harmful to the environment? Here are some pointers to help you decide.

1. Look for environmental claims that are specific. Read product labels to determine whether they have specific information about the product or its packaging. For example, if the label says "recycled," check how much of the product or packaging is recycled. Labels with "recyclable" claims mean that these products can be collected and made into useful products. This is relevant to you, however, only if this material is collected for recycling in your community or if you can find a place to send the material for recycling.

2. Be wary of overly broad or vague environmental claims. These claims provide little information to help you make purchasing decisions. Labels with unqualified claims that a product is "environmentally friendly," "eco-safe," or "environmentally safe" have little meaning for two reasons. First, all products have some environmental impact, though some may have less impact than others. Second, these phrases alone do not provide the specific information needed to compare products and packaging on their environmental merits.

3. Some products claim to be "degradable." Degradable materials will not help save landfill space. Biodegradable materials, like

food and leaves, break down and decompose into elements found in nature when exposed to air, moisture, and bacteria or other organisms. Photodegradable materials, usually plastics, disintegrate into smaller pieces when exposed to enough sunlight. Either way, however, degradation of any material occurs very slowly in landfills, where most solid waste is sent. That is because modern landfills are designed to minimize the entry of sunlight, air, and moisture. Even organic materials like paper and food may take decades to decompose in a landfill.

Three types of recycling symbols are commonly used in the United States. *(a)* This symbol simply means that the object is potentially recyclable, not that it has been or will be recycled. *(b)* This symbol indicates that a product contains recycled material, but it does not indicate how much recycled material is in the product (it could be only a very small amount). *(c)* This symbol states explicitly the percentage of recycled content found in the product.

---

future generations, most definitions refer to the viability of natural resources and ecosystems over time and to maintenance of human living standards and economic growth. The popularity of the term stems from the melding of the dual objectives of environmental protection and economic growth. A sustainable agricultural system, for example, can be defined as one that can indefinitely meet the demands for food and fiber at socially acceptable economic costs and environmental impacts.

Finally, as pointed out by Tan Sri Razali, former chairman of the United Nations' Commission on Sustainable Development, the transfer of modern, environmentally sound technology to

developing nations is the "key global action to sustainable development."

The past president of the Japan Economic Research Center, Saburo Okita, once stated that a slowdown of economic growth is needed to prevent further deterioration of the environment. Whether or not a slowdown is necessary provokes sharp differences of opinion.

One school of thought argues that economic growth is essential to finance the investments necessary to prevent pollution and to improve the environment by a better allocation of resources. Another school of thought, which is also progrowth, stresses the great potential of science and technology to solve problems and advocates relying on techno-

logical advances to solve environmental problems. Neither of these schools of thought sees any need for fundamental changes in the nature and foundation of economic policy. Environmental issues are viewed mainly as a matter of setting priorities in the allocation of resources.

A newer school of economic thought believes that economic and environmental well-being are mutually reinforcing goals that must be pursued simultaneously if either one is to be reached. Economic growth will create its own ruin if it continues to undermine the healthy functioning of Earth's natural systems or to exhaust natural resources. It is also true that healthy economies are most likely to provide the necessary financial investments in environmental protection. For this reason, one of the principal objectives of environmental policy must be to ensure a decent standard of living for all. The solution, at least in the broad scope, would be for a society to manage its economic growth in such a way as to do no irreparable damage to its environment. The term *sustainable development* has been criticized as ambiguous and open to a wide range of interpretations, many of which are contradictory. The confusion arises because "sustainable growth" and "sustainable use" have been used interchangeably, as if their meanings were the same. They are not. "Sustainable growth" is a contradiction in terms: Nothing physical can grow indefinitely. "Sustainable use" is applicable only to renewable resources: it means using them at rates within their capacity for renewal.

By balancing economic requirements with ecological concerns, the needs of the people are satisfied without jeopardizing the prospects of future generations. While this concept may seem to be common sense, the history of the world shows that it has not been a common practice. A major obstacle to sustainable development in many countries is a social structure that gives most of the nation's wealth to a tiny minority of its people. It has been said that a person who is worrying about his next meal is not going to listen to lectures on protecting the environment. What to residents in the Northern Hemisphere seem like some of the worst environmental outrages—cutting rainforests to make charcoal for sale as cooking fuel, for example—are often committed by people who have no other source of income.

The disparities that mark individual countries are also reflected in the planet as a whole. Most of the wealth is concentrated in the Northern Hemisphere. From the Southern Hemisphere's point of view, it is the rich world's growing consumption patterns—big cars, refrigerators, and climate-controlled shopping malls—that are the problem. The problem for the long term is that people in developing countries now want those consumer items that make life in the industrial world so comfortable—and these are the items that are environmentally so costly.

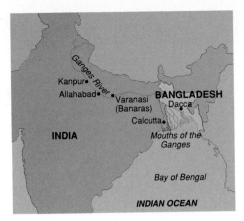

figure 3.8 **Indian Deforestation Causes Floods in Bangladesh** Because the Ganges River drains much of India and the country of Bangladesh is at the mouth of the river, deforestation and poor land use in India can result in devastating floods in Bangladesh.

If the standard of living in China and India were to rise to that of Germany or the United States, the environmental impact on the planet would be significant.

If sustainable development is to become feasible, it will be necessary to transform our approach to economic policy. Steps in that direction would include changing the definition of gross national product (GNP) to include environmental improvement or decline. The concept of sustainable development may seem simple, but implementing it will be a very complex process.

Historically, rapid exploitation of resources has provided only short-term economic growth, and the environmental consequences in some cases have been incurable. For example, 40 years ago, forests covered 30 percent of Ethiopia. Today, forest covers only 1 percent, and deserts are expanding. One-half of India once was covered by trees; today, only 14 percent of the land is in forests. As the Indian trees and topsoil disappear, the citizens of Bangladesh drown in India's runoff. (See figure 3.8.)

Sustainable development requires choices based on values. Both depend on information and education, especially regarding the economics of decisions that affect the environment. A. W. Clausen, in his final address as president of the World Bank, noted the

increasing awareness that environmental precautions are essential for continued economic development over the long run. Conservation, in

its broadest sense, is not a luxury for people rich enough to vacation in scenic parks. It is not just a motherhood issue. Rather, the goal of economic growth itself dictates a serious and abiding concern for resource management.

High-income developed nations, such as the United States, Japan, and much of Europe, are in a position to promote sustainable development. They have the resources to invest in research and the technologies to implement research findings. Some believe that the world should not impose environmental protection standards upon poorer nations without also helping them move into the economic mainstream.

Gaylord Nelson, the founder of the first Earth Day, lists five characteristics that define sustainability:

1. *Renewability:* A community must use renewable resources, like water, topsoil, and energy sources, no faster than they can replace themselves. The rate of consumption of renewable resources cannot exceed the rate of regeneration.

2. *Substitution:* Whenever possible, a community should use renewable resources instead of nonrenewable resources. This can be difficult, because there are barriers to substitution. To be sustainable, a community has to make the transition before the nonrenewable resources become prohibitively scarce.

Table 3.3 Economic Solutions to Pollution and Resource Waste

Solution	Internalizes External Costs	Innovation	International Competitiveness	Administrative Costs	Increases Government Revenue
Regulation	Partially	Can encourage	Decreased*	High	No
Subsidies	No	Can encourage	Increased	Low	No
Withdrawing harmful subsidies	Yes	Can encourage	Decreased*	Low	Yes
Tradable rights	Yes	Encourages	Decreased*	Low	Yes
Green taxes	Yes	Encourages	Decreased*	Low	Yes
User fees	Yes	Can encourage	Decreased*	Low	Yes
Pollution-prevention bonds	Yes	Encourages	Decreased*	Low	No

*Unless more cost-effective and productive technologies are developed.

3. *Interdependence:* A sustainable community recognizes that it is a part of a larger system and that it cannot be sustainable unless the larger system is also sustainable. A sustainable community does not import resources in a way that impoverishes other communities, nor does it export its wastes in a way that pollutes other communities.

4. *Adaptability:* A sustainable community can absorb shocks and can adapt to take advantage of new opportunities. This requires a diversified economy, educated citizens, and a spirit of solidarity. A sustainable community invests in, and uses, research and development.

5. *Institutional commitment:* A sustainable community adopts laws and political processes that mandate sustainability. Its economic system supports sustainable production and consumption. Its educational systems teach people to value and practice sustainable behavior.

External Costs

Many of the important environmental problems facing the world today arise because modern production techniques and consumption patterns transfer waste disposal, pollution, and health costs to society. Such expenses, whether they are measured in monetary terms or in diminished environmental quality, are borne by someone other than the individuals who use a resource. They are referred to as **external costs.** For example, consider an individual who attempts to spend a day of leisure fishing in a lake. Transportation, food, fishing equipment, and bait for the day's activities can be purchased in readily accessible markets. But suppose that upon arrival at the lake, the individual finds a lifeless, polluted body of water. Let us further suppose that a chemical plant situated on the lakeshore is responsible for degrading the water quality on the lake to the point that all or most of the fish are destroyed. In this example, fishers collectively bear the external costs of chemical production in the form of lost recreational opportunities.

Pollution-control costs include pollution-prevention costs and pollution costs. **Pollution-prevention costs** are those incurred either in the private sector or by government to prevent, either entirely or partially, the pollution that would otherwise result from some production or consumption activity. The cost incurred by local government to treat its sewage before dumping it into a river is a pollution-prevention cost; so is the cost incurred by a utility to prevent air pollution by installing new equipment.

Pollution-prevention costs can often be factored into a life cycle analysis. Life cycle analysis can help us understand the full cost, potential, and impact on new products and their associated technologies. As a systems approach, life cycle analysis examines the entire set of environmental consequences of a product, including those that result from its manufacture, use, and disposal. Because the relationships among industrial processes are complex, life cycle analysis requires understanding of material flows, resource reuse, and product substitution. Shifting to an approach that considers all resources, products, and waste as an interdependent system will take time, but governments can facilitate the shift by encouraging the transition to a systems approach.

Pollution costs can be broken down into two categories:

1. The private or public expenditures to avoid pollution damage once pollution has already occurred

2. The increased health costs and loss of the use of public resources because of pollution

The large cost of cleaning up spills, such as that from the *Exxon Valdez* and from the war in the Persian Gulf, is an example of a pollution cost, as is the increased health risk to humans from eating seafood contaminated from the oil. By the middle 1990s, several new economic solutions were being used to both internalize external costs and to prevent pollution. (See table 3.3.)

Common Property Resource Problems

Economists have stated that, when everybody shares ownership of a resource, there is a strong tendency to overexploit and misuse that resource. Thus, common public ownership could be better described as effectively having no owner.

Global Perspective

Pollution Prevention Pays!

While the concept of extended product responsibility (EPR) is relatively new, successful examples of it are in operation. For example, several years ago, 3M's European chemical plant in Belgium switched from a polluting solvent to a safer but more expensive water-based substance to make the adhesive for its Scotch™ Brand Magic™ Tape. The switch was not made to satisfy any environmental law in Belgium or the European Union. 3M managers were complying with company policy to adopt the strictest pollution-control regulations that any of its subsidiaries is subject to—even in countries that have no pollution laws at all.

Part of the policy is founded on corporate public relations, a response to growing customer demand for "green" products and environmentally responsible companies. But as many North American multinationals with similar global environmental policies are discovering, cleaning up waste, whether voluntarily or as required by law, can cut costs dramatically.

Since 1975, 3M's "Pollution Prevention Pays" program—or 3P—has cut the company's air, water, and waste pollution around the world in half and, at the same time, has saved nearly $600 million on changes in its manufacturing process, including $100 million overseas. Less waste has meant less spending to comply with pollution-control laws. But, in many cases, 3M actually has made money selling wastes it formerly hauled away. And, because of recycling prompted by the 3P program, it has saved money by not having to buy as many raw materials.

AT&T followed a similar path. In 1990, it set voluntary goals for the company's 40 manufacturing and 2500 nonmanufacturing sites worldwide. According to its latest estimates, AT&T has (1) reduced toxic air emissions, many caused by solvents used in the manufacture of computer circuit boards, by 73 percent; (2) reduced emissions of chlorofluorocarbons—gases blamed for destroying the ozone layer in the Earth's atmosphere—by 76 percent; and (3) reduced manufacturing waste 39 percent.

Xerox Corporation had focused on recycling materials in its global environment efforts. It provides buyers of its copiers with free United Parcel Service pickup of used copier cartridges, which contain metal-alloy parts that otherwise would wind up in landfills. The cartridges and other parts are now cleaned and used to make new ones.

Other recycled Xerox copier parts include power supplies, motors, paper transport systems, printed wiring boards, and metal rollers. In all,

Control and Prevention Technologies—Some Examples

Control/Treatment/Disposal

- Sewage treatment
- Industrial wastewater treatment
- Refuse collection
- Incineration
- Off-site recovery and recycling of wastes
- Landfilling
- Catalytic conversion and oxidation
- Particulate controls
- Flue-gas desulfurization
- Nitrogen oxides control technology
- Volatile organic compound control and destruction
- Contaminated site remediation

Prevention

- Improved process control to use energy and materials more efficiently
- Improved catalysis or reactor design to reduce by-products, increase yield, and save energy in chemical processes
- Alternative processes (e.g., low or no-chlorine pulping)
- In-process material recovery (e.g., vapor recovery, water reuse, and heavy metals recovery)
- Alternatives to chlorofluorocarbons and other organic solvents
- High-efficiency paint and coating application
- Substitutes for heavy metals and other toxic substances
- Cleaner or alternative fuels and renewable energy
- Energy-efficient motors, lighting, heat exchangers, etc.
- Water conservation
- Improved "housekeeping" and maintenance in industry

Source: Data from *EPA Journal.*

1 million parts per year are remanufactured. The initial design and equipment investment was $10 million. Annual savings total $200 million.

The philosophy of pollution prevention is that pollution should be prevented or reduced at the source whenever feasible. It is increasingly being shown that preventing pollution can cut business costs and thus increase profits. Pollution prevention, then, does make cents!

For example, common ownership of the air makes it virtually costless for any industry or individual to dispose of wastes by burning them. The air-pollution cost is not reflected in the economics of the polluter but becomes an external cost to society. Common ownership of the ocean makes it inexpensive for cities to use the ocean as a dump for their wastes. (See figure 3.9.)

Similarly, nobody owns the right to harvest whales. If any one country delays in getting its share of the available supply of whales, other countries may beat that country to the supply. Thus, there is a strong incentive to overharvest the whale population, which creates a consequent threat to the survival of the

Urban Coastal Population

617 million

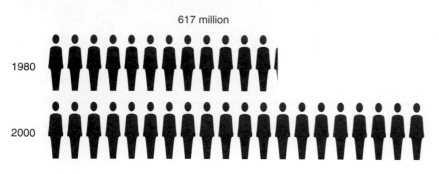

1980

2000

a.

997 million

b.

figure 3.9 **Abuse of the Ocean** Since the oceans of the world are a shared resource that nobody owns, there is a tendency to use the resource unwisely. (*a*) Growing populations in coastal areas lead to more marine pollution and destruction of coastal habitats. (*b*) Many countries use the ocean as a dump for unwanted trash. Some 6.5 million tons of litter find their way into the sea each year.

Source: (*a*) Data from UN Environment Programme; World Resources Institute.

species. What is true on an international scale for whales is true also for other species within countries. Note that endangered species are wild and undomesticated; the survival of privately owned livestock is not a concern.

Finally, common ownership of land resources, such as parks and streets, is the source of other environmental problems. People who litter in public parks do not generally dump trash on their own property. The lack of enforceable property rights to commonly owned resources explains much of what economist John Kenneth Galbraith has termed "public squalor amid private affluence."

Economic Decision Making, and the Biophysical World

For most natural scientists, current crises like biodiversity loss, climate change, and many other environmental problems are symptoms of an imbalance between the socioeconomic system and the natural world. While it is true that humans have always changed the natural world, it is also clear that this imprint is much greater now than anything experienced in the past. One reason for the profound effect of human activity on the natural world is the fact that there are so many of us.

One of the most serious consequences of the growing human impact on the natural world is the loss of biodi-

versity. Biological diversity is thought to affect the stability of ecosystems and the ability to cope with crises. In the 570-million-year history of complex life on Earth, there have been several major extinction events. The loss of biodiversity after major extinction episodes ranged from 20 percent to more than 90 percent. The loss in biodiversity caused by human activity since the Industrial Revolution alone is somewhere between 10 and 20 percent. If current trends continue, losses are likely to reach 50 percent by the end of the twenty-first century. After each extinction event, it took between 20 million and 100 million years for biodiversity to recover to previous levels, a length of time between 100 and 500 times longer than the 200,000-year history of *Homo sapiens*.

The greatest single cause of the loss of biodiversity is habitat destruction, that is, the destruction of the web of organisms and functions that support individual species. The recognition that individual species are supported by others within the ecosystem is foreign to the way markets view the world.

The example of biodiversity illustrates the conflicting frameworks of economics and ecology. Market decisions fail to account for the context of a species or the interconnections between resource quality and ecosystem functions. For example, the value of land used for beef production is measured ac-

cording to its contribution to output. Yet long before output and the use value of land decreases, the diversity of grass varieties, microorganisms in the soil, or groundwater quality may be affected by intensive beef production. As long as yields are maintained, these changes go unnoticed by markets and are unimportant to land-use decisions. It should be pointed out that in Zimbabwe and other African nations, some ranchers now earn more money managing native species of wildlife for ecotourism that they would from raising cattle for beef production.

Another obvious difference between economics and ecology is the relevant time frame in markets and ecosystems. The biophysical world operates in tens of thousands and even millions of years. The time frame for market decisions is short. Particularly where economic policy is concerned, two- to four-year election cycles are the frame of reference; for investors and dividend earners, performance time frames of three months to one year are the rule.

Space or place is another issue. For ecosystems, place is critical. Take groundwater as an example. Soil quality, hydrogeological conditions, regional precipitation rates, plants that live in the region, and losses from evaporation, transpiration, and groundwater flow all contribute to the size and location of groundwater reservoirs. These capacities

Placing a Value on Ecosystem Services

Many services provided by functioning ecosystems are taken for granted. Protection of watersheds by forested land has long been known to be of great value. New York City found that it could provide water to its residents less expensively by protecting the watershed from which the water comes than by building expensive water purification plants to clean water from local rivers. Ducks Unlimited, an organization that supports waterfowl hunting, uses money provided by its members to protect nesting habitat for ducks and geese. Many countries have planted trees to help remove carbon dioxide from the air. All of these services can be converted into monetary terms, since it takes money to purify water, purchase land, and purchase and plant trees.

Since choices between competing uses for ecosystems often are determined by financial values assignable to ecosystems, it is important to have some kind of idea about the value of the "free" services provided by functioning ecosystems. Many environmental thinkers have begun to try to put a value on the many services provided by intact, functioning ecosystems. Obviously, this is not an easy task and many will belittle these initial attempts to put monetary values on ecosystem services, but it is an important first step in forcing people to consider the importance of ecosystem services when making economic decisions about how ecosystems should be used. The following table represents approximate values for ecosystems services assigned by a panel of experts including ecologists, geographers, and economists. The total of $33 trillion per year is an estimate which many consider to be low. The current world GNP is about $18 trillion per year. Therefore, the "free" services of ecosystems must not be overlooked when decisions are made about land use and how natural resources should be managed.

Categories of Services	Examples of Services	Estimated Yearly Value (trillion 1994 U.S. dollars)
Soil formation	Weathering, organic matter	17.1
Recreation	Outdoor recreation, ecotourism, sport fishing	3.0
Nutrient cycling and waste treatment	Nitrogen, phosphorus, organic matter	2.4
Water services	Irrigation, industry, transportation, watersheds, reservoirs, aquifers	2.3
Climate regulation	Greenhouse gas regulation	1.8
Refuges	Nursery areas for animals, stopping places during migration, overwintering areas	1.4
Disturbance regulation and erosion control	Protection from storms and floods, drought recovery	1.2
Food and raw materials	Hunting and gathering, fishing, lumber, fuel wood, food for animals	0.8
Genetic resources	Medicines, reservoir of genes for domesticated plants and animals	0.8
Atmospheric gas balance	Carbon dioxide, oxygen, ozone, sulfur oxides	0.7
Pollination and pest control	Increased production of fruits	0.5
Total		33*

*Disparity is due to rounding.

Source: Data from R. Costanza, et al., "The Value of the World's Ecosystem Services and Natural Capital," *Nature,* 387 (1997).

are not simply transferable from one location to another. For economic activities, place is increasingly irrelevant. Topography, location and function within a bioregion, or local ecological features do not enter into economic calculations except as simple functions of transportation costs or comparative advantage. Production is transferable, and the preferred location is anywhere production costs are the lowest.

Another difference between economics and ecology is that they are measured in different units. The unifying measure of market economics is money. Progress is measured in monetary units that everyone uses and understands to some degree. Biophysical systems are measured in physical units such as calories of energy, carbon dioxide absorption, centimeters of rainfall, or parts per million of nitrate contamination. Focusing only on the economic value of resources while ignoring biophysical health may mask serious changes in environmental quality or function.

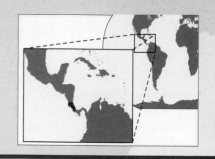

Costa Rican Forests Yield Tourists and Medicines

The Central American republic of Costa Rica has been very successful in protecting a significant amount of its remaining forests. It has a rich variety of different kinds of tropical forests and other natural resources. Mangrove swamps, cloud forests, rainforests, dry tropical forests, volcanoes, and beaches on both the Pacific Ocean and the Caribbean Sea are all part of the mix found in an area about the size of West Virginia.

Nearly 20 percent of the land in Costa Rica is protected as parks, reserves, and refuges. This is the result of several factors. The government is committed to preserving natural areas. A major part of the parks program involves educating the local people about the values of the parks, including the biological value of the large number of species of plants and animals and the economic value of the parks as a tourist attraction. The job of educating the people is made easier by the fact that 93 percent of the people are literate.

Many jobs in Costa Rica are in the developing ecotourism market. People who wish to visit natural areas require guides, transportation, food, and lodging. The jobs created in these industries encourage local people to preserve their natural resources because their livelihood depends on it. Furthermore, when people are employed, they are less likely to try to convert forested land into farmland.

In addition to using the natural resources of forests for tourists, the government of Costa Rica signed a two year agreement in 1991 between its Instituto Nacional de Biodiversidad (INBio) and Merck and Company, Inc., to prospect for possible drugs from many tropical plants and animals found in the forests. In return, Merck paid $1 million to the Costa Rican government. Specially trained local parataxonomists (technicians trained in the identification of plants and animals) were involved in this search, resulting in additional jobs.

- What conditions have contributed to the apparent success of Costa Rica in protecting its forests?
- What are possible sources of failure in other countries?
- Who benefits from ecotourism?

Economics, Environment, and Developing Nations

As previously mentioned, the earth's "natural capital," on which humankind depends for food, security, medicines, and many industrialized products, is its biological diversity. The majority of this diversity is in the developing world, and much of it is threatened by exploitation and development. To pay for development projects, many economically poorer nations are forced to borrow money from banks in the developed world.

So great is the burden of external debt that many developing nations see little option but to overexploit their natural resource base. By 2001, the debt in the developing nations had risen to over US $1,900 billion, a figure almost half their collective gross national product. The debt burdens have led what investment there is in many developing countries to projects with safe, short-term returns and programs absolutely necessary for immediate survival. Environmental impacts are often neglected, the

view being that severely indebted countries cannot afford to pay attention to environmental costs until other problems are resolved. This strategy suggests that environmental problems can be "corrected" once a country has reached a higher income level, but it ignores the growing realization that environmental impacts frequently cause international problems. Many countries under pressure from their debt crisis feel forced to overexploit their natural resources rather than manage them sustainably.

One new method of helping manage a nation's debt crises is referred to as debt-for-nature exchange. Debt-for-nature exchanges are an innovative mechanism for addressing the debt issue while encouraging investment in conservation and sustainable development. The exchanges, or swaps, allow debt to be bought at discount but redeemed at a premium, in local currency, for use in conservation and sustainable development projects. Debt-for-nature originated in 1987, when a nonprofit organization, Conservation International,

bought $650,000 of Bolivia's foreign debt in exchange for Bolivia's promise to establish a national park. By 2000, at least 16 debtor countries—in the Caribbean, Africa, Eastern Europe, and Latin America—had made similar deals with official and nongovernmental organizations. By 2000, nearly US $135 million of debt around the world had been purchased at a cost of some US $28 million but redeemed for the equivalent of US $72 million. This money was used to establish biosphere reserves and national parks, develop watershed protection programs, build inventories of endangered species, and develop environmental education.

In debt-for-nature exchanges, debtor countries benefit from the reduction of their foreign-currency debt obligations and add to their expenditure invested at home. The conservation investor receives a premium on the investment. This can be used for conservation and to establish sustainable development projects. Creditor banks gain by converting their nonpaying debts. Although they receive only part of their initial loan, some return is better than a total loss.

The primary goal of debt-for-nature exchanges has not been debt reduction but the funding of natural-resource-management investment. The contribution made by exchanges could increase, as in the case of the Dominican Republic, where 10 percent of the country's outstanding foreign commercial debt is to be redeemed by exchanges. Although eliminating the debt crisis alone is no guarantee of investment in environmentally sound projects, instruments like debt-for-nature exchanges can, on a small scale, reduce the mismanagement of natural resources and encourage sustainable development.

Attitudes of banks in the industrialized nations also seem to be changing. For example, the World Bank, which lends money for Third World development projects, has long been criticized by environmental groups for backing large, ecologically unsound programs, such as a cattle-raising project in Botswana that led to overgrazing. During the past few years, however, the World Bank has been factoring environmental concerns into its programs. One product of this new approach is an environmental action plan for Madagascar. The 20-year plan, which has been drawn up jointly with the World Wildlife Fund, is aimed at heightening public awareness of environmental issues, setting up and managing protected areas, and encouraging sustainable development.

The Tragedy of the Commons

The problems inherent in common ownership of resources were outlined by biologist Garrett Hardin in a now classic essay entitled "The Tragedy of the Commons" (1968). The original "commons" were areas of pastureland in England that were provided free by the king to anyone who wished to graze cattle.

There are no problems on the commons as long as the number of animals is small in relation to the size of the pasture. From the point of view of each herder, however, the optimal strategy is to enlarge his or her herd as much as possible:

If my animals do not eat the grass, someone else's will. Thus, the size of each herd grows, and the density of stock increases until the commons becomes overgrazed. The result is that everyone eventually loses as the animals die of starvation. The tragedy is that, even though the eventual result should be perfectly clear, no one acts to avert disaster. In a democratic society, there are few remedies to keep the size of herds in line.

The ecosphere is one big commons stocked with air, water, and irreplaceable mineral resources—a "people's pasture," but a pasture with very real limits. Each nation attempts to extract as much from the commons as possible while enough remains to sustain the herd. Thus, the United States and other industrial nations consume far more than their share of the total world resource harvest each year, much of it imported from less-developed nations. The nations of the world compete frantically for all the fish that can be taken from the sea before the fisheries are destroyed. Each nation freely uses the commons to dispose of its wastes, ignoring the dangers inherent in overtaxing the waste-absorbing capacity of rivers, oceans, and the atmosphere.

The tragedy of the commons also operates on an individual level. Most people are aware of air pollution, but they continue to drive their automobiles. Many families claim to need a second or third car. It is not that these people are antisocial; most would be willing to drive smaller or fewer cars if everyone else did, and they could get along with only one small car if public transport were adequate. But people frequently get "locked into" harmful situations, waiting for others to take the first step, and many unwittingly contribute to tragedies of the commons. After all, what harm can be done by the birth of one more child, the careless disposal of one more beer can, or the installation of one more air conditioner?

Lightening the Load

Ship captains pay careful attention to a marking on their vessels called the Plimsoll line. If the water level rises above the Plimsoll line, the boat is too heavy and is in danger of sinking. When the line is submerged, rearranging items on the ship will not help much. The problem is the total weight, which has surpassed the carrying capacity of the ship.

This analogy points out that human activity can reach a scale that the earth's natural systems can no longer support. In 1992, more than 1600 scientists, including 102 Nobel laureates, underscored this point by collectively signing a "Warning to Humanity." Their warning stated in part that "a new ethic is required, a new attitude towards discharging our responsibility for caring for ourselves and for the earth. . . . This ethic must motivate a great movement, convincing reluctant leaders and reluctant governments and reluctant peoples themselves to effect the needed changes."

Such a new successful global effort to lighten humanity's load on the Earth would need to directly address three major driving forces of environmental decline: the inequitable distribution of income, resource consumptive economic growth, and rapid population growth. It would redirect technology and trade to buy time for this great change to occur. Although there is much to say about each of these challenges, some key points bear noting.

Wealth inequality may be the most difficult problem, since it has existed for centuries. The difference today, however, is that the future of rich and poor alike depends on reducing poverty and thereby eliminating this driving force of global environmental decline. In this way, self-interest joins ethics as a motive for redistributing wealth and raises the chances that it might be done.

Important actions to narrow the income gap must include reducing Third World debt. This was talked about a great deal in the 1980s, but little was accomplished. In addition, the developed nations must focus more foreign aid, trade, and international lending policies directly on improving the living standards of the world's poor.

A key description for reducing the kinds of economic growth that harm the environment is the same as that for making technology and trade more sustainable: internalizing environmental costs. If this is done through the adoption of environmental taxes, such as taxing based on pollution emitted, governments could avoid imposing heavier taxes overall by lowering income taxes accordingly. In addition, establishing better measures of economic accounting is critical. Since the calculations used to produce the gross national product do not account for the destruction or depletion of natural resources, this popular economic measure is extremely misleading. It tells us we are making progress even as our ecological foundations are being diminished. A better guide toward a sustainable path is essential. The United Nations and several governments have been working to develop better accounting methods, and while the progress has been slow, there is growing hope in the heightened awareness that a change is necessary.

As our discussion has shown, the economics of environmental problems is complex and difficult. The single most difficult problem to overcome is the assignment of an appropriate economic value to resources that have not previously been examined from an economic perspective. When air, water, scenery, and wildlife are assigned an economic value, they are looked at from an entirely different point of view.

issues–analysis issues–analysis issues–

Shrimp, Turtles, and Turtle Excluder Devices

The nets used for trawling for shrimp unfortunately do not catch only shrimp. One historic victim of such nets has been turtles, including the endangered Kemp's Ridley sea turtle as well as threatened loggerhead and endangered green sea turtles. Until the mid-1990s, shrimp boats traveling the South Atlantic and Gulf of Mexico could accidentally catch an estimated 45,000 sea turtles in their nets, of which some 12,000 would drown.

The U.S. National Marine Fisheries Service (NMFS), looking for a technological innovation to stop the accidental netting and killing of the turtles, developed a device to keep turtles out of shrimp nets. The turtle excluder device, or TED, attaches to standard shrimp nets. A TED is a grid of bars with an opening either at the top or bottom. The grid is fitted into the neck of a shrimp trawl. Small animals like shrimp slip through the bars and are caught in the bag end of the trawl. Large animals such as turtles and sharks, when caught at the mouth of the trawl, strike the grid bars and are ejected through the opening. Data compiled by the NMFS show that TEDs can reduce turtle captures in shrimp nets by 97 percent with minimal loss of shrimp.

TEDs were first introduced to the shrimp industry on a voluntary basis in 1982. TEDs, however, were not widely accepted, and many fishermen claimed that it was not fair for only U.S. fishermen to use them while their counterparts in Mexico did not. The U.S. fishermen argued that they were operating under an unfair economic policy. The shrimpers who opposed the TEDs also claimed that the devices were expensive and dangerous, and cut down on their catches.

TEDs were redesigned with input from the fishermen, and there soon followed legislation replacing the voluntary program with a mandated one. The NMFS ensured that the TED requirements were phased in gradually to minimize the impact on the fishery. By 2000, inspectors from the NMFS uncovered only 76 TED violations in 2724 visits aboard shrimp boats.

In 1993, Mexico also mandated the use of TEDs. By the late 1990s, the use of TEDs had spread to many parts of the world. The NMFS and the U.S. Department of State have worked closely with Mexico and the other shrimp-supplying nations in Latin America to help them develop comparable TED programs. TEDs have also been successfully implemented in over 20 countries, including Thailand, Malaysia, and the Philippines. After actual demonstration and dissemination of results of experiments with TEDs, shrimp fishermen did not resist their use and understood the need to use them.

- Can you think of other examples where the development of a new technology has ensured economic growth without endangering one or more species?
- Is it possible to always have a "technological fix"?

Trawling for shrimp also kills several endangered and threatened species of turtles in the South Atlantic and Gulf of Mexico.

Summary

Risk assessment is the use of facts and assumptions to estimate the probability of harm to human health or the environment that may result from exposures to pollutants, toxic agents, or management decisions. While it is difficult to calculate risks, risk assessment is used in risk management, which analyzes factors in decision making. The politics of risk management focus on the adequacy of scientific evidence, which is often open to divergent interpretations. In assessing risk, people frequently overestimate new and unfamiliar risks, while underestimating familiar ones.

To a large degree, environmental problems can be viewed as economic problems. Economic policies and concepts, such as supply and demand, and subsidies, play important roles in environmental decision making. Another important economic tool is cost-benefit analysis. Cost-benefit analysis is concerned with whether a policy generates more social benefits than social costs. Criticism of cost-benefit analysis is based on the question of whether everything has an economic value. It has been argued that if economic thinking dominates society, then even noneconomic values, like beauty, can survive only if a monetary value is assigned to them.

A newer school of economic thought is referred to as sustainable development. Sustainable development has been defined as actions that address the needs of the present without compromising the ability of future generations to meet their own needs. Sustainable development requires choices based on values.

Pollution is extremely costly. When the costs are imposed on society, they are referred to as external costs. Costs of pollution control include pollution-prevention costs and pollution costs. Prevention costs are less costly, especially from a societal perspective.

Economists have stated that when everyone shares ownership of a resource, there is a strong tendency to overexploit and misuse the resource. This concept was developed by Garrett Hardin in "The Tragedy of the Commons."

Recently, a market approach to curbing pollution has been proposed that would assign a value to not polluting, thereby introducing a profit motive to pollution reduction.

Economic concepts are also being applied to the debt-laden developing countries. One such approach is the debt-for-nature swap. This program, which involves transferring loan payments for land that is later turned into parks and wildlife preserves, is gaining popularity.

Key Terms

carcinogens *41*

cost-benefit analysis *50*

demand *44*

external costs *54*

negligible risk *42*

pollution costs *54*

pollution-prevention costs *54*

probability *39*

resources *44*

risk assessment *39*

risk management *41*

subsidy *47*

supply *44*

supply/demand curve *44*

sustainable development *51*

Review Questions

1. How is risk assessment used in environmental decision making?
2. What is incorporated in a cost-benefit analysis? Develop a cost-benefit analysis for a local issue.
3. What are some of the concerns about the use of cost-benefit analysis in environmental decision making?
4. What concerns are associated with sustainable development?
5. What are some examples of environmental external costs?
6. Define what is meant by pollution-prevention costs.
7. Define the problem in common property resource ownership. Provide some examples.
8. Describe the concept of debt-for-nature.

Critical Thinking Questions

1. If you were a regulatory official, what kind of information would you require in order to make a decision about whether a certain chemical was "safe" or not? What level of risk would you deem acceptable for society? For yourself and your family?

2. Why do you suppose some carcinogenic agents, like those in cigarettes, are so difficult to regulate?

3. Imagine you were assessing the risk of a new chemical plant being built along the Mississippi River in Louisiana. Identify some of the risks that you would want to assess. What kinds of data would you need to assess whether the risk was acceptable, or not? Do you think that some risks are harder to quantify than others? Why?

4. Granting polluting industries or countries the right to buy and sell emissions permits is a controversial idea. Some argue that the market is the best way to limit pollution. Others argue that trade in permits allows polluting industries to continue to pollute and concentrates that pollution. What do you think?

5. Imagine you are an independent economist who is conducting a cost-benefit analysis of a hydroelectric project. What might be the costs of this project? The benefits? How would you quantify the costs of the project? The benefits? What kinds of costs and benefits might be hard to quantify or might be too tangential to the project to figure into the official estimates?

6. Do you think environmentalists should or should not stretch traditional cost-benefit analysis to include how development impacts the environment? What are the benefits to this? The risks?

7. Looking at your own life, what kinds of risks do you take? What kinds are you unwilling to take? What criteria do you use to make a decision about acceptable and unacceptable risk?

8. Is current worldwide growth and development sustainable? If there were less growth, what would be the effect on developing countries? How could we achieve a just distribution of resources and still limit growth?

9. Should our policies reflect an interest in preserving resources for future generations? If so, what level of resources should be preserved? What would you be willing to do without in order to save for the future?

Concept Map

Construct a map to show relationships among the following concepts:

risk assessment	external costs	resources
pollution prevention	probability	supply
cost-benefit analysis	demand	subsidy

Interactive Exploration

Check out the website at **http://www.mhhe.com/environmentalscience** and click on the cover of this textbook for quizzing, career information, case studies, and hot links for the following topics:

Environmental Policy and Decision Making

Environmental Policy, Law and Planning

Environmental Philosophy

Ecological Economics

Individual Contributions to Environmental Issues

Urbanization and Sustainable Cities

Sustainable Agriculture

Ecological
Principles and
Their
Applications

"Fixing" Nature?: Restoration and the Florida Everglades

by Bradley C. Bennett

Florida International University/Florida

"In the Everglades, one is most aware of the superb monotony of saw grass under the world of air. But below that and before it, enclosing and causing it, is the water." The words of Marjory Stoneman Douglas are crucial to understanding the ecosystems of southern Florida. A complex wetland mosaic known as the Everglades dominates the tip of the state. Douglas coined the name "River of Grass" in her 1947 classic book about the Everglades. Extending from the headwaters of the Kissimmee River to Florida Bay, the original Everglades Basin was more than 350 kilometers (217 miles) long and up to 70 kilometers (43.4 miles) wide. South of Lake Okeechobee, the basin dipped gradually toward the southwest. The region's broad marshes are dotted with tree islands and wet prairies. On slightly elevated limestone outcrops, pinelands once thrived and tropical hardwood forests (also known as hammocks) dominated the highest portions of the carbonate Pleistocene and Pliocene reefs that outcrop in southern Florida.

Florida, the second wettest state in the continental United States, receives an average of 600 billion liters (150 billion gallons) of rainwater each day. Few areas in the southern part of the state are more than a meter or two above sea level. Topographic changes of one-half-meter produce dramatic changes in the natural communities. In addition to water, fire shapes the landscape. Frequent lightening sparks fires, which prevent marshes, prairies, and pinelands from becoming forests.

Human intervention in Florida began with the Paleo-Indians who migrated to the state more than 10,000 years ago. The region's indigenous inhabitants harvested fish and wildlife, excavated small canals, built mounds, and used fire to hunt and manage vegetation. These changes, however, were minor compared to those initiated by Hamilton Disston, a Philadelphia toolmaker. In the late 1800s, Disston demonstrated the feasibility of draining the Everglades. Others soon undertook the task of "reclaiming" the Everglades, and by the early 1900s, several drainage canals traversed the Glades. The Tamiaimi Trail, completed in 1928, connected Tampa and Miami and cut through the heart of the Everglades. Devastating hurricanes in 1926 and 1928 fostered major flood control projects.

The final initiative to control water flow in southern Florida began in 1948 when the U.S. Congress authorized the Central and Southern Florida Project for Flood Control and Other Purposes. Hurricanes in 1947 and 1948 produced severe flooding, threatening the post-World War II development boom. In 1949, the Florida Legislature created the Central and Southern Florida Flood Control District, predecessor to the South Florida Water Management District. The U.S. Army Corps of Engineers began an ambitious plan to control flooding by expanding the network of dikes, levees, and canals. The district was charged with operation and maintenance of the structures. Between 1980 and 1989, the district discharged four times more water into the Atlantic Ocean than it allowed to flow into Everglades National Park.

Altered hydrology has devastated both the plant communities and the animals that they support. White ibis, snowy egret, and wood stork populations have plummeted since drainage began. Wood storks are particularly sensitive to water level. A breeding pair requires at least 200 kilograms (441 pounds) of fish to feed themselves and fledge their young. Drainage alters not only the freshwater wetlands of the Everglades, but also the mangroves and sea grass communities in Florida Bay. Some researchers believe that the effects extend even further, harming the reefs along the Florida Keys.

Water quantity is not the only concern; water quality also is an issue. The Everglades is naturally a nutrient-poor system. The original phosphorous levels in the Everglades are thought to have been around 5-to-10 parts per billion (ppb). Runoff from sugar cane fields south of Lake Okeechobee and other agriculture areas often exceeds concentrations of 100 ppb. Cattails favor phosphorous-rich waters at the expense of saw grass and other desirable species. Periphyton, a crucial component of Everglades wetlands, is particularly sensitive to increased levels of phosphorus. Periphyton is composed of an algal matrix along with bacteria, fungi, and other microorganisms. It provides a habitat for macroinvertebrates and contributes as much as 50 percent of the productivity of freshwater wetlands.

In December 2000, President Clinton signed the Comprehensive Everglades Restoration Plan (CERP). The $8 billion project is the largest of its kind. Approximately 1.7 billion gallons of water drains from the Everglades to coastal waters each day. The proposed restoration would store this water until it is needed to replenish the natural system or to supply urban and agricultural needs. Major components of the plan include:

1. Surface water storage reservoirs
2. Water preserve areas
3. Management of Lake Okeechobee as an ecological resource
4. Improved water deliveries to the estuaries
5. Underground water storage
6. Treatment wetlands
7. Improved water deliveries to the Everglades
8. Removal of barriers to sheetflow
9. Storage of water in existing quarries
10. Reuse of wastewater
11. Pilot projects
12. Improved water conservation

Will it work? About half of the original Everglades has been converted to agriculture and housing for southern Florida's 6 million inhabitants. Considering only the ecological objectives, no doubt much of the remaining Everglades could be enhanced. But planners also must cope with the human population. Can enough water make it to the Everglades without flooding residents who now encroach the region's eastern fringe? Will there be enough water for agriculture, industry, and the burgeoning human population? The world's most extensive (and expensive) restoration project will serve as an experiment to test whether urban populations and natural areas can coexist. Marjory was right. The water is everything.

Study Questions

1. Engineering solutions created the crisis in the Everglades. New engineering solutions have been proposed to repair the system. When and to what extent should humans intervene to manage nature?
2. For each of the major components of CERP, discuss both the ecological and social impacts.
3. The human disturbance of one southern Florida ecosystem is often transferred to other ecosystems. How do impacts in natural communities in your region affect surrounding communities?

Interrelated Scientific Principles: Matter, Energy, and Environment

CHAPTER 4

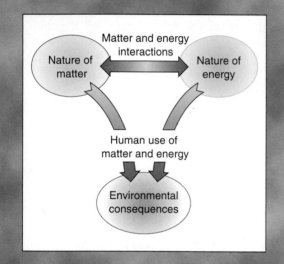

Objectives

After reading this chapter, you should be able to:

- Understand that science is usually reliable because information is gathered in a manner that requires impartial evaluation and continuous revision.
- Understand that matter is made up of atoms that have a specific subatomic structure of protons, neutrons, and electrons.
- Recognize that each element is made of atoms that have a specific number of protons and electrons and that isotopes of the same element may differ in the number of neutrons present.
- Recognize that atoms may be combined and held together by chemical bonds to produce molecules.
- Understand that rearranging chemical bonds results in chemical reactions and that these reactions are associated with energy changes.
- Recognize that matter may be solid, liquid, or gas, depending on the amount of kinetic energy contained by the molecules.
- Realize that energy can be neither created nor destroyed, but when energy is converted from one form to another, some energy is converted into a less useful form.
- Understand that energy can be of different qualities.

Chapter Outline

Scientific Thinking
    The Scientific Method
    Observation
    Questioning and Exploring
    Constructing Hypotheses
    Testing Hypotheses
    The Development of Theories and Laws

Environmental Close-Up: *Typical Household Chemicals*

Limitations of Science

The Structure of Matter
    Atomic Structure
    The Molecular Nature of Matter
    Acids, Bases, and pH
    Inorganic and Organic Matter
    Chemical Reactions
    Chemical Reactions in Living Things

Energy Principles
    Kinds of Energy
    States of Matter
    First and Second Laws of Thermodynamics
    Environmental Implications of Energy
        Flow

Issues—Analysis: *Improvements in Lighting Efficiency*

Scientific Thinking

Since environmental science involves the analysis of data, it is useful to understand how scientists gather and evaluate information. It is also important to understand some chemical and physical principles as a background for evaluating environmental issues. An understanding of these scientific principles will also help you appreciate the ecological concepts in the chapters that follow.

The word *science* creates a variety of images in the mind. Some people feel that it is a powerful word and are threatened by it. Others are baffled by scientific topics and have developed an unrealistic belief that scientists are brilliant individuals who can solve any problem. For example, there are those who believe that the conservation of fossil fuels is unnecessary because scientists will soon "find" a replacement energy source. Similarly, many are convinced that if government really wanted to, it would allocate sufficient funds to allow scientists to find a cure for AIDS. Such images do not accurately portray what science is really like.

The Scientific Method

Science is a process used to solve problems or develop an understanding of nature that involves testing possible answers. Science is distinguished from other fields of study by how knowledge is acquired rather than by what is studied. The process has become known as the *scientific method*. The **scientific method** is a way of gaining information (facts) about the world by forming possible solutions to questions, followed by rigorous testing to determine if the proposed solutions are valid. Furthermore, when using the scientific method, scientists make several fundamental assumptions. They presume that:

1. There are specific causes for events observed in the natural world;
2. That the causes can be identified;
3. That there are general rules or patterns that can be used to describe what happens in nature;
4. That an event that occurs repeatedly probably has the same cause;
5. That what one person perceives can be perceived by others; and
6. That the same fundamental rules of nature apply regardless of where and when they occur.

For example, we have all observed lightning associated with thunderstorms. According to the assumptions just stated, we should expect that there is an explanation that would account for all cases of lightning regardless of where or when they occur and that all people could make the same observations. We know from scientific observations and experiments that lightning is caused by a difference in electrical charge, that the behavior of lightning follows general rules that are the same as those seen with static electricity, and that all lightning that has been measured has the same cause wherever and whenever it occurred.

The scientific method requires a systematic search for information and a continual checking and rechecking to see if previous ideas are still supported by new information. If the new evidence is not supportive, scientists discard or change their original ideas. Scientific ideas undergo constant reevaluation, criticism, and modification. The scientific method involves several important identifiable components, including careful observation, asking questions about observed events, constructing and testing hypotheses, an openness to new information and ideas, and a willingness to submit one's ideas to the scrutiny of others. Underlying all of these activities is constant attention to accuracy and freedom from bias.

The scientific method is not, however, an inflexible series of steps that must be followed in a specific order. Figure 4.1 shows how these steps may be linked.

Observation

Scientific inquiry often begins with an observation that an event has occurred.

An **observation** occurs when we use our senses (smell, sight, hearing, taste, touch) or an extension of our senses (microscope, tape recorder, X-ray machine, thermometer) to record an event. Observation is more than a casual awareness. You may hear a sound or see an image without really observing it. Do you know what music was being played in the shopping mall? You certainly heard it, but if you are unable to tell someone else what it was, you didn't "observe" it. If you had prepared yourself to observe the music being played, you would be able to identify it. When scientists talk about their observations, they are referring to careful, thoughtful recognition of an event—not just casual notice. Scientists train themselves to improve their observational skills, since careful observation is important in all parts of the scientific method.

Because many of the instruments used in scientific investigations are complicated, we might get the feeling that science is incredibly complex, when in reality, these sophisticated tools are being used simply to answer questions that are relatively easy to understand. For example, a microscope has several knobs to turn and a specially designed light source. It requires considerable skill to use properly, but it is essentially a fancy magnifying glass that allows small objects to be seen more clearly. The microscope has enabled scientists to answer some relatively fundamental questions such as: Are there living things in pond water? and Are living things made up of smaller subunits? Similarly, chemical tests allow us to determine the amount of specific materials dissolved in water, and a pH meter allows us to determine how acidic or basic a solution is. Both are simple activities, but if we are not familiar with the procedures, we might consider the processes hard to understand.

Questioning and Exploring

Observations often lead one to ask questions about the observations. Why did this event happen? Will it happen again in the same circumstances? Is it related to something else? Some questions may be simple speculation, but others may

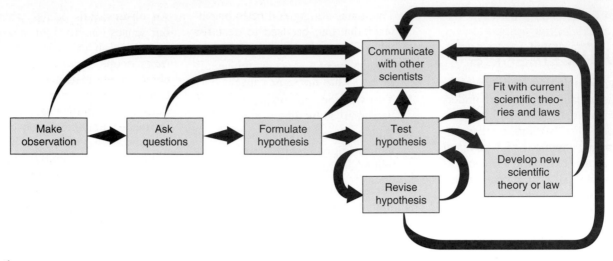

figure 4.1 **Elements of the Scientific Method** The scientific method consists of several kinds of activities. Observation of a natural phenomenon is usually the first step. Observation often leads people to ask questions about the observation they have made or to try to determine why the event occurred. This questioning is typically followed by the construction of a hypothesis that explains why the phenomenon occurred. The hypothesis is then tested to see if it is supported. Often this involves experimentation. If the hypothesis is not substantiated, it is modified and tested in its new form. It is important at all times that others in the scientific community be informed by publishing observations of unusual events, their probably cause, and the results of experiments that test hypotheses. Occasionally, this method of inquiry leads to the development of theories that tie together many bits of information into broad statements that state why things happen in nature and serve to guide future thinking about a specific area of science. Scientific laws are similar broad statements that describe how things happen in nature.

inspire you to further investigation. The formation of the questions is not as simple as it might seem, because the way the questions are asked will determine how you go about answering them. A question that is too broad or too complex may be impossible to answer; therefore, a great deal of effort is put into asking the question in the right way. In some situations, this can be the most time-consuming part of the scientific method; asking the right question is critical to how you look for answers. For example, you observe that robins eat the berries of many plants but avoid others. You could ask the following questions:

1. Do the robins dislike the flavor of some berries?

2. Will robins eat more of one kind of berry if given a choice between two kinds of berries?

The second question is obviously easier to answer.

Once a decision has been made about what question to ask, scientists *explore other sources of knowledge* to gain more information. Perhaps the question already has been answered by someone else or several possible answers already have been rejected. Knowing what others have already done saves one time. This process usually involves reading appropriate science publications, exploring information on the Internet, or contacting fellow scientists interested in the same field of study. Even if the particular question has not already been answered, scientific literature and other scientists can provide insights that may lead to a solution. After exploring the appropriate literature, a decision is made about whether to continue to explore the question. If the scientist is still intrigued by the question, a formal hypothesis is constructed, and the process of inquiry continues at a different level.

Constructing Hypotheses

A **hypothesis** is a statement that provides a possible answer to a question or an explanation for an observation that can be tested. A good hypothesis must be logical, account for all the relevant information currently available, allow one to predict future events relating to the question being asked, and be testable. Furthermore, if one has the choice of several competing hypotheses, one should use the simplest hypothesis with the fewest assumptions. Just as deciding which questions to ask is often difficult, the formation of a hypothesis requires much critical thought and mental exploration. If the hypothesis does not account for all the observed facts in the situation, doubt will be cast on the work and perhaps eventually on the validity of the scientist's work. If a hypothesis is not testable or is not supported by the evidence, the explanation will be only hearsay and no more useful than mere speculation.

Keep in mind that a hypothesis is based on observations and information gained from other knowledgeable sources and predicts how an event will occur under specific circumstances. Scientists test the predictive ability of a hypothesis to see if the hypothesis is supported or is disproved. If you disprove the hypothesis, it is rejected, and a new hypothesis must be constructed. However, if you cannot disprove a hypothesis, it increases your confidence in the hypothesis, but it does not prove it to be true in all cases and for all time.

Science always allows for the questioning of ideas and the substitution of new ones that more completely describe what is known at a particular point in time. It could be that an alternative hypothesis you haven't thought of explains the situation or that you have not made the appropriate observations to indicate that your hypothesis is wrong.

Testing Hypotheses

The test of a hypothesis can take several forms. It may simply involve the collection of pertinent information that already exists from a variety of sources. For example, if you visited a cemetery and observed from reading the tombstones that an unusually large number of people of different ages died in the same year, you could hypothesize that there was an epidemic of disease or a natural disaster that caused the deaths. Consulting historical newspaper accounts would be a good way to test this hypothesis.

In other cases, a hypothesis may be tested by simply making additional observations. For example, if you hypothesized that a certain species of bird used cavities in trees as places to build nests, you could observe many birds of the species and record the kinds of nests they built and where they built them.

Another common method for testing a hypothesis involves devising an experiment. An **experiment** is a recreation of an event or occurrence in a way that enables a scientist to support or disprove a hypothesis. This can be difficult because a particular event may involve a great many separate happenings called **variables.** The best experimental design is a **controlled experiment** in which two groups differ in only one way. For example, tumors of the skin and liver occur in the fish that live in certain rivers *(observation)*. This raises the question: What causes the tumors? Many people feel that the tumors are caused by toxic chemicals that have been released into the rivers by industrial plants *(hypothesis)*. However, it is possible that the tumors are caused by a virus, by exposure to natural substances in the water, or are the result of genes present in the fish. How could an exper-

iment be conducted to test the hypothesis that industrial contaminants cause the tumors? Fish could be collected from the river and placed in one of two groups. One group (the control group) would be raised in a container through which the normal river water passes. The second group (experimental group) would be raised in an identical container through which water from the industrial facility passes. There would need to be large numbers of fish in both groups. This kind of experiment is called a controlled experiment. If the fish in the experimental group have a significantly larger number of tumors than the control group, something in the water from the plant is the probable cause of the tumors. This is particularly true if the chemicals present in the water are already known to cause tumors. After the data have been evaluated, the results of the experiment would be published.

The results of a well-designed experiment should be able to support or disprove a hypothesis. However, this does not always occur. Sometimes the results of an experiment are inconclusive. This means that a new experiment has to be conducted or that more information has to be collected. Often, it is necessary to have large amounts of information before a decision can be made about the validity of a hypothesis. The public often finds it difficult to understand why it is necessary to perform experiments on so many subjects or to repeat experiments again and again.

The concept of **reproducibility** is important to the scientific method. Because it is often not easy for scientists to eliminate unconscious bias, independent investigators must be able to reproduce the experiment to see if they get the same results. To do this, they must have a complete and accurate written document to work from. That means the scientists must publish the methods and results of their experiment. This process of publishing one's work for others to examine and criticize is one of the most important steps in the process of scientific discovery. If a hypothesis is supported by many experiments and by different investigators, it is considered reliable.

The Development of Theories and Laws

When broad consensus exists about an area of science, it is known as a theory or law. A **theory** is a widely accepted, plausible generalization about fundamental concepts in science that explains *why* things happen. An example of a scientific theory is the **kinetic molecular theory,** which states that all matter is made up of tiny, moving particles. As you can see, this is a very broad statement, and it is the result of years of observation, questioning, experimentation, and data analysis. Because we are so confident that the theory explains the nature of matter, we use this concept to explain why materials disperse in water or air, why materials change from solids to liquids, and why different chemicals can interact during chemical reactions.

Theories and hypotheses are different. A hypothesis provides a possible explanation for a specific question; a theory is a broad concept that shapes how scientists look at the world and how they frame their hypotheses. Because they are broad, unifying statements, there are few theories. However, just because a theory exists does not mean that testing stops. As scientists continue to gain new information, they may find exceptions to a theory or even in rare cases disprove a theory.

It is important to recognize that the word *theory* is often used in a much less restrictive sense. Often it is used incorrectly to describe a vague idea or a hunch. This is not a theory in the scientific sense. So when you see or hear the word *theory,* you must look at the context to see if the speaker or writer is referring to a theory in the scientific sense.

A **scientific law** is a uniform or constant fact of nature that describes *what* happens in nature. An example of a scientific law is the **law of conservation of mass,** which states that matter is not gained or lost during a chemical reaction. While laws describe what happens and theories describe why things happen, in a way, laws and theories are similar. They have both been examined repeatedly and are regarded as excellent predictors of how nature behaves.

environmental CLOSE-UP

Typical Household Chemicals

Modern society uses many different kinds of chemicals. A survey of a typical household would probably yield the following inorganic chemicals:

Common Name	Chemical Name	Use
Table salt	Sodium chloride, NaCl	Flavor
Saltpeter	Potassium nitrate, KNO_3	Preservative
Baking soda	Sodium bicarbonate, $NaHCO_3$	Leavening agent
Ammonia	Ammonia, NH_3	Disinfectant
Bleach	Sodium hypochlorite, NaHClO	Bleaching
Lye	Sodium hydroxide, NaOH	Drain cleaner

Other products we use contain mixtures of inorganic chemicals. Fertilizers are good examples. They usually contain a nitrate such as ammonium nitrate (NH_4NO_3), a phosphate compound (PO_4^{3+}), and potash, which is potassium oxide (K_2O).

In addition, we use a vast array of organic chemicals: ethyl alcohol in alcoholic beverages, acetic acid in vinegar, methyl alcohol for fuel, and cream of tartar (tartaric acid) for flavoring. We also use many complex mixtures of organic molecules in flavorings, pesticides, cleaners, and other applications.

Most of us know very little about the activities of the molecules we use. Many of them can be dangerous if used improperly. Fertilizer is poisonous, caustic soda can cause severe burns, and bleach or ammonia in high enough concentrations can damage skin or other tissues. The disposal of unused or unwanted household chemicals is a problem. Many of them should not just be dumped down the sink but should be disposed of in such a way that the material is converted to a harmless product or stored in a secure place. Unfortunately, most people do not know how to dispose of unwanted chemicals. For this reason, many manufacturers of household chemicals that have a potential to cause harm print statements on the containers explaining how to properly dispose of the unused product and the container. In addition, many communities have regular cleanup efforts for household hazardous waste, in which volunteers who know the contents of such products help determine how to dispose of them properly.

CARPET BEETLES—Thoroughly apply as a spot treatment. Spray along baseboards and edges of carpeting, under carpeting, rugs and furniture, in closets and on shelving, or wherever these insects are seen or suspected. **FLEAS, BROWN DOG TICKS**—Remove soiled bed bedding and clean thoroughly or destroy. Spray sleeping quarters of pets, along baseboards, windows, door frames, cracks and crevices, carpets, rugs, floors where these pests may be found. Put fresh bedding in pet quarters after spray has dried. **DO NOT SPRAY ANIMALS.** Pets should be treated with FLEA-B-GON® Flea Killer (aerosol) or ORTHO Pet Flea & Tick Spray Formula II (pump spray).
STORAGE: To store, rotate nozzle to closed position. Keep pesticide in original container. Do not put concentrate or dilute into food or drink containers. Avoid contamination of feed and foodstuffs. Store in a cool, dry place, preferably in a locked storage area.
DISPOSAL: PRODUCT—Partially filled bottle may be disposed of by securely wrapping original container in several layers of newspaper and discard in trash. **CONTAINER**—Do not reuse empty bottle. Rinse thoroughly before discarding in trash.

NOTICE: Buyer assumes all responsibility for safety and use not in accordance with directions.

Chevron Chemical Company © 1984
Ortho Consumer Products Division
P.O. Box 5047   San Ramon CA 94583-0947
Form 10152-N Product 5466   Made in U.S.A.
EPA Reg. No. 239-2490-AA
EPA Est. 239-IA-3

0   8
71549 01980   C

Limitations of Science

Science is a powerful tool for developing an understanding of the natural world, but it cannot analyze international politics, decide if family-planning programs should be instituted, or evaluate the significance of a beautiful landscape. These tasks are beyond the scope of scientific investigation. This does not mean that scientists cannot comment on such issues. They often do. But they should not be regarded as more knowledgeable on these issues just because they are scientists. Scientists may know more about the scientific aspects of these issues, but they struggle with the same moral and ethical questions that face all people, and their judgments on these matters can be just as biased as anyone else's. Consequently, major differences of opinion often exist among lawmakers, regulatory agencies, special interest groups, and members of scientific organizations about the significance or value of specific scientific information.

It is important to differentiate between the scientific data collected and the opinions scientists have about what the data mean. Scientists form and state opinions that may not always be supported by fact, just as other people do. Equally reputable scientists commonly state opinions that are in direct contradiction. This is especially true in environmental science, where predictions about the future must be based on inadequate or fragmentary data. The issue of climate change (covered in chapter 18) is an example of this.

It is important to recognize that some scientific knowledge can be used to support both valid and invalid conclusions. For example, the following statements are all factual.

1. Many of the kinds of chemicals used in modern agriculture are toxic to humans and other animals.

2. Small amounts of agricultural chemicals have been detected in some agricultural products.

3. Low levels of some toxic materials have been strongly linked to a variety of human illnesses.

This does not mean that all foods grown with the use of chemicals are less nutritious or are dangerous to health or that "organically grown" foods are necessarily more nutritious or healthful because they have been grown without agricultural chemicals. The idea that something that is artificial is necessarily bad and something natural is necessarily good is an oversimplification. After all, many plants such as tobacco, poison ivy, and rhubarb leaves naturally contain toxic materials, while the use of chemical fertilizers has contributed to the health and well-being of the human population because fertilizer use accounts for about one-third of the food grown in the world. However, it is appropriate to question if the use of agricultural chemicals is always necessary or if trace amounts of specific agricultural chemicals in food are dangerous. It is often easy to jump to conclusions or confuse fact with hypothesis, particularly when we generalize.

The Structure of Matter

Now that we have an appreciation for the methods of science, it is time to explore some basic information and theories about the structure and function of various kinds of matter. **Matter** is anything that takes up space and has mass. Air, water, trees, cement, and gold are all examples of matter. As stated earlier, the kinetic molecular theory is a central theory that describes the structure and activity of matter. This theory states that all matter is made up of tiny objects that are in constant motion. Although different kinds of matter have different properties, they all are similar in one fundamental way. They are all made up of one or more kinds of smaller subunits called atoms.

Atomic Structure

Atoms are fundamental subunits of matter. They in turn are made up of protons, neutrons, and electrons. There are 92 kinds of atoms found in nature. Each kind forms a specific type of matter known as an **element.** Gold (Au), oxygen (O), and mercury (Hg) are examples of elements. All atoms have a central region known as a **nucleus,** which is composed of two kinds of relatively heavy particles: positively charged particles called **protons** and uncharged particles called **neutrons.** Surrounding the nucleus of the atom is a cloud of relatively lightweight, fast-moving, negatively charged particles called **electrons.** The atoms of each element differ from one another in the number of protons, neutrons, and electrons present. For example, a typical mercury atom contains 80 protons and 80 electrons; gold has 79 of each, and oxygen only eight of each. (See figure 4.2.) (Appendix 3 contains a periodic table of the elements.) All atoms of an element always have the same number of protons and electrons,

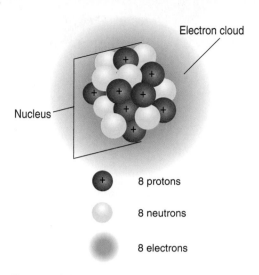

figure 4.2 **Diagrammatic Oxygen Atom** Most oxygen atoms are composed of a nucleus containing eight positively charged protons and eight neutrons without charges. Eight negatively charged electrons move in a cloud around the nucleus.

but the number of neutrons may vary from one atom to the next. Atoms of the same element that differ from one another in the number of neutrons they contain are called **isotopes.** For example, there are three isotopes of the element hydrogen. All atoms have one proton and one electron, but one isotope of hydrogen has no neutrons, one has one neutron, and one has two neutrons. These isotopes behave the same chemically but have different masses since they contain different numbers of neutrons. (See figure 4.3.)

The Molecular Nature of Matter

The kinetic molecular theory states that all matter is made of tiny particles that are in constant motion. However, there are several kinds of these tiny particles. In some instances, atoms act as individual particles. In other instances, atoms bond to one another chemically to form stable units called **molecules.** In still other cases, atoms or molecules may gain or lose electrons and thus become electrically charged particles called **ions.** Atoms or molecules that lose electrons are positively charged because they have more protons $(+)$ than electrons $(-)$. Those that gain electrons have more

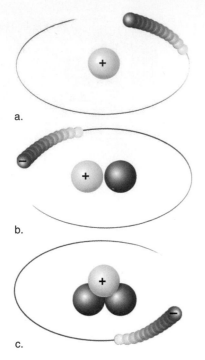

figure 4.3 **Isotopes of Hydrogen**
(*a*) The most common form of hydrogen is the isotope that is 1 AMU. It is composed of one proton and no neutrons. (*b*) The isotope deuterium is 2 AMU and has one proton and one neutron. (*c*) Tritium, 3 AMU, has two neutrons and one proton. Each of these isotopes of hydrogen also has one electron, but because the mass of an electron is so small, the electrons do not contribute significantly to the mass as measured in AMU. All three isotopes of hydrogen are found on Earth, but the most frequently occurring has 1 AMU and is commonly called hydrogen. Most scientists use the term *hydrogen* in a generic sense, that is, the term is not specific but might refer to any or all of these isotopes.

Table 4.1	Relationships Between the Kinds of Subunits Found in Matter	
Category of Matter	**Subunits**	**Characteristics**
Subatomic particles	protons	Positively charged
		Located in nucleus of the atom
	neutrons	Have no charge
		Located in nucleus of the atom
	electrons	Negatively charged
		Located outside the nucleus of the atom
Elements	atoms	Atoms of an element are composed of specific arrangements of protons, neutrons, and electrons.
		Atoms of different elements differ in the number of protons, neutrons, and electrons present.
Compounds	molecules or ions	Compounds are composed of two or more atoms or ions chemically bonded together.
		Different compounds contain specific atoms or ions in specific proportions.
Mixtures	atoms, molecules, or ions	The molecular particles in mixtures are not chemically bonded to each other.
		The number of each kind of molecular particle present is variable.

electrons (−) than protons (+) and are negatively charged. Oppositely charged ions are attracted to one another and may form stable units similar to molecules; however, they typically split into their individual ions when dissolved. For example, table salt (NaCl) is composed of sodium ions (Na$^+$) and chloride ions (Cl$^-$). It is a white, crystalline material when dry but separates into individual ions when placed in water. When two or more atoms or ions are bonded chemically, a new kind of matter called a **compound** is formed.

While only 92 kinds of atoms are commonly found, there are millions of ways atoms can be combined to form compounds. Water (H$_2$O), table sugar (C$_6$H$_{12}$O$_6$), table salt (NaCl), and methane gas (CH$_4$) are examples of compounds.

Many other kinds of matter are **mixtures,** variable combinations of atoms, ions, or molecules. Honey is a mixture of several sugars and water; concrete is a mixture of cement, sand, gravel, and reinforcing rods; and air is a mixture of several gases of which the most common are nitrogen and oxygen. Table 4.1 summarizes the various kinds of matter and the subunits of which they are composed.

Acids, Bases, and pH

Acids and bases are two classes of compounds that are of special interest. Their characteristics are determined by the nature of their chemical bonds. When acids are dissolved in water, hydrogen ions (H$^+$) are set free. A *hydrogen ion* is positive because it has lost its electron and now has only the positive charge of its proton. Therefore, a hydrogen ion is a proton. An **acid** is any compound that releases hydrogen ions (protons) in a solution. Some familiar examples of com-

mon acids are sulfuric acid (H$_2$SO$_4$) in automobile batteries and acetic acid (HCH$_2$COOH) in vinegar.

A **base** is the opposite of an acid in that it accepts hydrogen ions in solution. Many common bases release **hydroxide ions** (OH$^-$). This ion is composed of an oxygen atom and a hydrogen atom bonded together but with an additional electron. The hydroxide ion is negatively charged. It is a base because it is able to accept hydrogen ions in solution to form water (H$^+$ + OH$^-$ → H$_2$O). A very strong base often used in oven cleaners is sodium hydroxide (NaOH). Sometimes people refer to a base as an alkali and solutions that are basic as alkaline solutions.

The concentration of an acid or base solution is given by a number called its **pH.** The pH scale is a measure of hydrogen ion concentration. However, the pH scale is different from what you might expect. First, it is a reciprocal scale, which means that the lower the pH, the greater the number of hydrogen ions present. Second, the scale is logarithmic, which means that a difference between two consecutive pHs is really a difference of 10 times. For example, a pH of 7 indicates that the solution is neutral and has an equal number of H⁺ ions and OH⁻ ions, but a pH of 6 means that the solution has 10 times more hydrogen ions than it would at a pH of 7. As the number of hydrogen ions in the solution increases, the pH gets smaller. A number higher than seven indicates that the solution has more OH⁻ than H⁺. As the number of hydroxide ions increases, the pH gets larger. The higher the pH, the more concentrated the hydroxide ions. (See figure 4.4.)

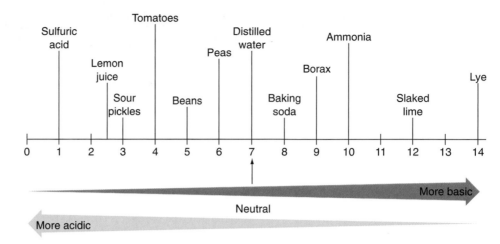

figure 4.4 **The pH Scale** The concentration of acid is greatest when the pH is lowest. As the pH increases, the concentration of base increases. At a pH of 7.0, the concentrations of H⁺ and OH⁻ are equal. We usually say as the pH gets smaller, the solution becomes more acid. As the pH gets larger, the solution becomes more basic or alkaline.

Inorganic and Organic Matter

Inorganic and organic matter are usually distinguished from one another by one fact: organic matter consists of molecules that contain carbon atoms that are usually bonded to form chains or rings. Consequently, organic molecules can be very large. Many different kinds of organic compounds exist. Inorganic compounds generally consist of small molecules and combinations of ions, and relatively few kinds exist. All living things contain molecules of organic compounds. They must either be able to manufacture organic compounds from inorganic compounds or to modify organic compounds they obtain from eating organic material. Typically, chemical bonds in organic molecules contain a large amount of chemical energy that can be released when the bonds are broken and new inorganic compounds are produced. Salt, water, metals, sand, and oxygen are examples of inorganic matter. Sugars, proteins, and fats are examples of organic compounds that are produced and used by living things. Natural gas, oil, and coal are all examples of organic substances that were originally produced by living things but have been modified by geologic processes.

Chemical Reactions

When atoms or ions combine to form compounds, they are held together by chemical bonds. (See figure 4.5.) **Chemical bonds** are attractive forces between atoms resulting from the interaction of their electrons. Each chemical

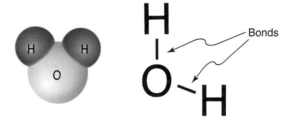

figure 4.5 **Water Molecule** A water molecule consists of an atom of oxygen bonded to two atoms of hydrogen. The molecule is not symmetrical. The hydrogen atoms are not on opposite sides of the oxygen atom.

bond contains a certain amount of energy. When chemical bonds are broken or formed, a chemical reaction occurs. During chemical reactions, the amount of energy within the chemical bonds changes. If the chemical bonds in the new compounds have less chemical energy than the previous compounds, some of the energy may be released as heat and light. These kinds of reactions are called **exothermic reactions.** In other cases, the newly formed chemical bonds contain more energy than was present in the compounds from which they were formed. Such reactions are called **endothermic reactions.** For such a reaction to occur, the additional energy must come from an external source.

A common example of an exothermic reaction is the burning of natural gas. The primary ingredient in natural gas is the compound methane. When methane and oxygen are mixed together

and a small amount of energy is used to start the reaction, the chemical bonds in the methane and oxygen (reactants) are rearranged to form two different compounds, carbon dioxide and water (products). In this kind of reaction, the products have less energy than the reactants. The leftover energy is released as light and heat. (See figure 4.6.) In every reaction, the amount of energy in the reactants and in the products can be compared and the differences accounted for by energy loss or gain. Even energy-yielding reactions usually need an input of energy to get the reaction started. This initial input of energy is called **activation energy.** In certain cases, the amount of activation energy required to start the reaction can be reduced by the use of a catalyst. A **catalyst** is a substance that alters the rate of a reaction, but the catalyst itself is not consumed or altered in the process. Catalysts are used in catalytic converters, which are attached to automobile exhaust systems. The purpose of the catalytic converter is to bring about more complete burning of the fuel, thus resulting in less air pollution. Most of the materials that are not completely burned by the engine require high temperatures to react further; with the presence of catalysts, these reactions can occur at lower temperatures.

Endothermic reactions require an input of energy in order to occur. For example, nitrogen gas and oxygen gas can be combined to form nitrous oxide ($O_2 + N_2 + heat \rightarrow 2NO$). Another important endothermic reaction is the process of photosynthesis, which is discussed in the section on "Chemical Reactions in Living Things."

Chemical Reactions in Living Things

Living things are constructed of cells that are themselves made up of both inorganic and organic matter in very specific arrangements. The chemical reactions that occur in living things are regulated by protein molecules called **enzymes** that function to reduce the activation energy needed to start the reactions. Enzymes are important since the high temperatures required to start these reactions without enzymes would destroy living organisms. Many enzymes are arranged in such a way that they cooperate in controlling a chain of reactions, as in photosynthesis and respiration.

Photosynthesis is the process plants use to convert inorganic material into organic matter, with the assistance of light energy. Light energy enables the smaller inorganic molecules (water and carbon dioxide) to be converted into organic sugar molecules. In the process, molecular oxygen is released.

$$6CO_2 + 6H_2O \xrightarrow[\text{in chloroplasts}]{\text{light energy}} C_6H_{12}O_6 + 6O_2$$

$$\text{Carbon dioxide} + \text{Water} \rightarrow \text{Sugar} + \text{Oxygen}$$

Molecules of the green pigment, chlorophyll, are found in cellular structures called chloroplasts. Chlorophyll is responsible for trapping the sunlight en-

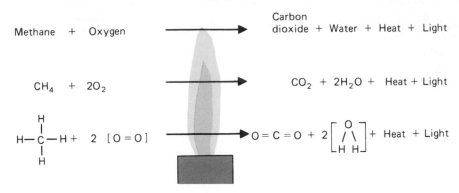

figure 4.6 **A Chemical Reaction** When methane is burned, chemical bonds are rearranged, and the excess chemical bond energy is released as light and heat. The same atoms are present in the reactants as in the products, but they are bonded in different ways, resulting in molecules of new substances.

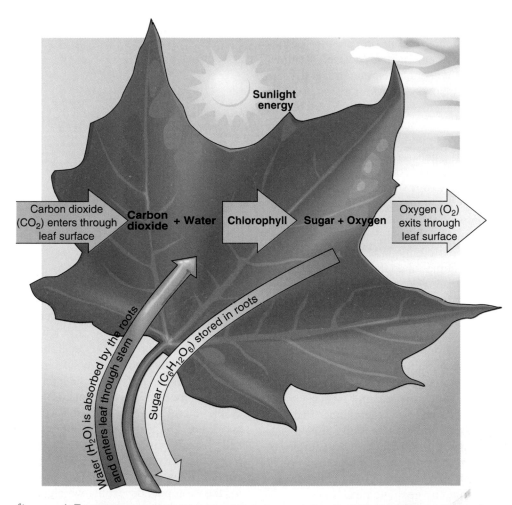

figure 4.7 **Photosynthesis** This reaction is an example of one that requires an input of energy (sunlight) to combine low-energy molecules (CO_2 and H_2O) to form sugar ($C_6H_{12}O_6$) with a greater amount of chemical bond energy. Molecular oxygen (O_2) is also produced.

ergy needed in the process of photosynthesis. Therefore, photosynthesis takes place in the green portions of the plant, usually the leaves. (See figure 4.7.) The

organic molecules produced as a result of photosynthesis can be used as an energy source by the plants and by organisms that eat the plants.

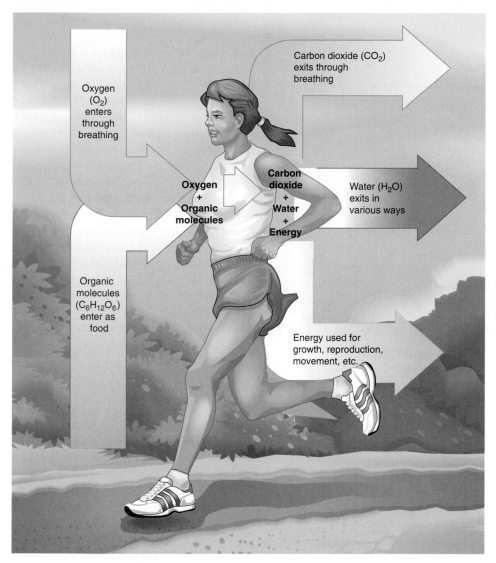

figure 4.8 **Respiration** Respiration involves the release of energy from organic molecules when they react with oxygen. In addition to providing energy in a usable form, respiration produces carbon dioxide and water.

Respiration involves the use of atmospheric oxygen to break down large, organic molecules (sugars, fats, and proteins) into smaller, inorganic molecules (carbon dioxide and water). This process releases energy the organisms can use.

$$C_6H_{12}O_6 + 6O_2 \longrightarrow 6CO_2 + 6H_2O + Energy$$

$$Sugar + Oxygen \longrightarrow Carbon\ dioxide + Water + Energy$$

(See figure 4.8.) All organisms must carry on some form of respiration, since all organisms need a source of energy to maintain life.

Energy Principles

The "Chemical Reactions in Living Things" section started out with a description of matter, yet it used the concept of energy to describe chemical bonds, chemical reactions, and molecular motion. That is because energy and matter are inseparable. It is difficult to describe one without the other. **Energy** is the ability to do work. Work is done when an object is moved over a distance. This occurs even at the molecular level.

Kinds of Energy

There are several kinds of energy. Heat, light, electricity, and chemical energy are common forms. The energy contained by moving objects is called **kinetic energy.** The moving molecules in air have kinetic energy, as does water running downhill or a dog chasing a ball. In contrast, **potential energy** is the energy matter has because of its position. The water behind a dam has potential energy by virtue of its elevated position. (See figure 4.9.) An electron moved to a position farther from the nucleus has increased potential energy due to the increased distance between the electron and the nucleus.

States of Matter

Depending on the amount of energy present, matter can occur in three common states: solid, liquid, or gas. The physical nature of matter changes when a change occurs in the amount of kinetic energy its molecular particles contain, but the chemical nature of matter and the kinds of chemical reactions it will undergo remain the same. For example, water vapor, liquid water, and ice all have the same chemical composition but differ in the arrangement and activity of their molecules. The amount of kinetic energy molecules have determines how rapidly they move. (See figure 4.10.) In solids, the molecular particles have comparatively little energy, and they vibrate in place very close to one another. In liquids, the particles have more energy, are farther apart from one another, and will roll, tumble, and flow over each other. In gases, the molecular particles move very rapidly and are very far apart. All that is necessary to change the physical nature of a substance is an energy change. Heat energy must be added or removed.

When two forms of matter have different temperatures, heat energy will flow from the one with the higher temperature to the one with the lower temperature. The temperature of the cooler matter increases while that of the warmer matter decreases. You experience this whenever you touch a cold or hot object.

figure 4.9 **Kinetic and Potential Energy** Kinetic and potential energy are interconvertible. The potential energy possessed by the water behind a dam is converted to kinetic energy as the water flows to a lower level.

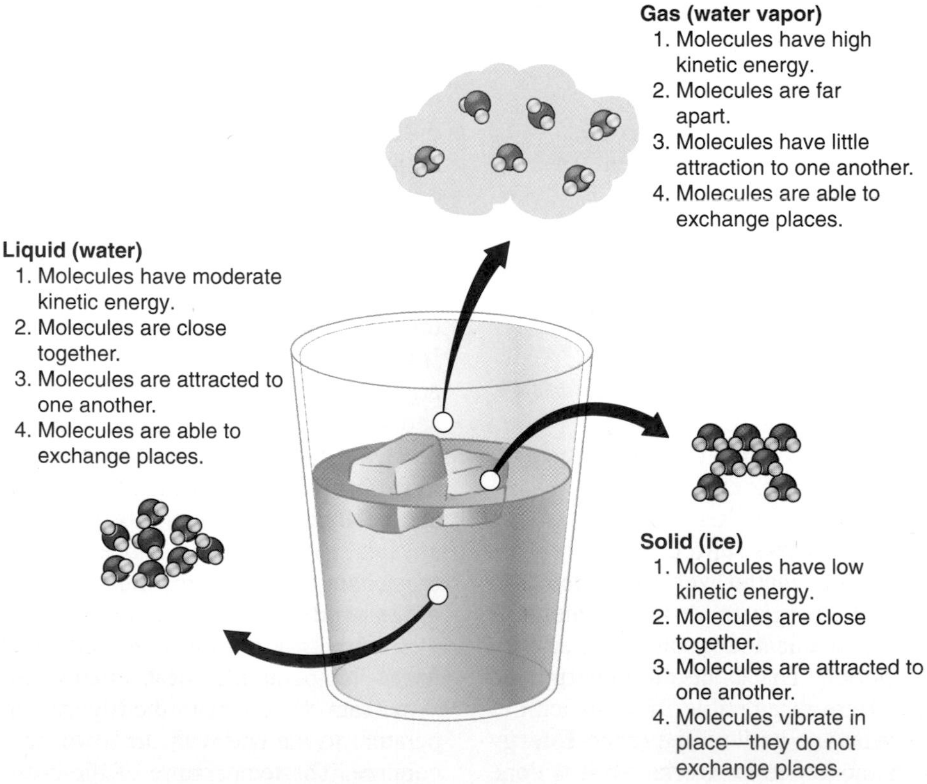

Gas (water vapor)
1. Molecules have high kinetic energy.
2. Molecules are far apart.
3. Molecules have little attraction to one another.
4. Molecules are able to exchange places.

Liquid (water)
1. Molecules have moderate kinetic energy.
2. Molecules are close together.
3. Molecules are attracted to one another.
4. Molecules are able to exchange places.

Solid (ice)
1. Molecules have low kinetic energy.
2. Molecules are close together.
3. Molecules are attracted to one another.
4. Molecules vibrate in place—they do not exchange places.

figure 4.10 **States of Matter** Matter exists in one of three states, depending on the amount of kinetic energy the molecules have. The higher the amount of energy, the greater the distance between molecules and the greater their degree of freedom of movement.

This is referred to as a **sensible heat** transfer. When heat energy is used to change the state of matter from solid to liquid at its melting point or liquid to gas at its boiling point, heat is transferred, but the temperature of the matter does not change. This is called a **latent heat** transfer. You have experienced this effect when water evaporates from your skin. Your body supplies the heat necessary to convert liquid water to water vapor. While the temperature of the water does not change, the physical state of the water does. The heat that was transferred to the water caused it to evaporate, and your body cooled. When substances change from gas to liquid at the boiling point or liquid to solid at the freezing point, there is a corresponding release of heat energy without a change in temperature.

First and Second Laws of Thermodynamics

Energy can exist in several different forms, and it is possible to convert one form of energy into another. However, the total amount of energy remains constant. The **first law of thermodynamics** states that energy can neither be created nor destroyed; it can only be changed from one form into another. From a human perspective, some forms of energy are more useful than others. We tend to make extensive use of electrical energy for a variety of purposes, but there is very little electrical energy present in nature. Therefore, we convert other forms of energy into electrical energy. When converting energy from one form to another, some of the useful energy is lost. This is the **second law of thermodynamics.** The energy that cannot be used to do useful work is called **entropy.** Therefore, another way to state the second law of thermodynamics is to say that when energy is converted from one form to another, entropy increases. An alternative way to look at the idea of entropy is to say that entropy is a measure of disorder and that the amount of disorder (entropy) typically increases when energy conversions take place. Obviously, it is possible to generate greater order in a system (living things are a good example of highly ordered things), but when things become more highly ordered, the disorder of the surroundings must increase. For example, all living things release heat to their surroundings. It is important to understand that when energy is converted from one form to another, there is no loss of *total* energy, but there is a loss of *useful* energy. For example, coal, which contains chemical energy, can be burned in a power plant to produce electrical energy. The heat from the burning coal is used to heat water to form steam, which turns turbines that generate electricity. At each step in the process, some heat energy is lost from the system. Therefore, the amount of useful energy (electricity) coming from the plant is much less than the total amount of chemical energy present in the coal that was burned. (See figure 4.11.)

Within the universe, energy is being converted from one form to another continuously. Stars are converting nuclear energy into heat and light. Animals are converting the chemical potential energy found in food into kinetic energy that allows them to move. Plants are converting sunlight energy into the chemical bond energy of sugar molecules. In each of these cases, some energy is produced that is not able to do useful work. This is generally in the form of heat lost to the surroundings.

Environmental Implications of Energy Flow

The heat produced when energy conversions occur is dissipated throughout the universe. This is a common experience. All machines and living things that manipulate energy release heat. It is also true that organized matter tends to become more disordered unless an external source of energy is available to maintain the ordered arrangement. Houses fall into ruin, automobiles rust, and appliances wear out unless work is done to maintain them. In reality, all of these phenomena involve the loss of heat. The organisms that decompose the wood in our houses release heat. The chemical reaction that causes rust releases heat. Friction, caused by the

figure 4.11 **Second Law of Thermodynamics** Whenever energy is converted from one form to another, some of the useful energy is lost, usually in the form of heat. The conversion of fuel to electricity produces heat, which is lost to the atmosphere. As the electricity moves through the wire, resistance generates some additional heat. When the electricity is converted to light in a light bulb, heat is produced as well. All of these steps produce low-quality heat in accordance with the second law of thermodynamics.

movement of parts of a machine against each other, generates heat and causes the parts to wear.

Ultimately, orderly arrangements of matter, such as clothing, automobiles, or living organisms, become disordered. There is an increase in entropy. Eventually, nonliving objects wear out and living things die and decompose. This process of becoming more disordered coincides with the constant flow of energy toward a dilute form of heat. This dissipated, low-quality heat has little value to us, since we are unable to use it.

It is important to understand that some forms of energy are more useful to us than others. Some forms, such as electrical energy, are of high quality because they can be easily used to perform a variety of useful actions. Other forms, such as the heat in the water of the ocean, are of low quality because we are not able to use them for useful purposes.

Although the total *quantity* of heat energy in the ocean is much greater than the total amount of electrical energy in the world, little useful work can be done with the heat energy in the ocean because it is of *low quality*. Therefore, it is not as valuable as other forms of energy that can be used to do work for us.

The reason the heat of the ocean is of little value is related to the small temperature difference between two sources of heat. When two objects differ in temperature, heat will flow from the warmer to the cooler object. The greater the temperature difference, the more useful the work that can be done. For example, fossil-fuel power plants burn fuel to heat water and convert it to steam. High temperature steam enters the turbine, while cold cooling-water condenses the steam as it leaves the turbine. This steep temperature gradient also provides a steep pressure gradient as heat energy flows from the steam to the cold water, which causes a turbine to turn, which generates electricity. Because the average temperature of the ocean is not high, and it is difficult to find another object that has a greatly lower temperature than the ocean, it is difficult to use the huge heat content of the ocean to do useful work for us.

These quantitative and qualitative factors are also evident in the energy expended by a stream as the water runs downhill. The steeper the slope, the greater the amount of energy expended per kilometer of its length. If there is no point along the stream where the slope is very steep, the stream has low-quality energy, because the energy is dissipated along the entire length of the stream. To make this a high-quality (concentrated) source of energy, the water must be dammed so that it will drop a long distance at one point. This means that it will give up much of its energy over a short distance. With damming, the *quantity* of energy has not changed but the *quality* has. Whenever low-quality energy is converted to high-quality energy, work must be done.

It is important to understand that energy that is of low quality from our point of view may still have significance to the world in which we live. For example, the distribution of heat energy in the ocean tends to moderate the temperature of coastal climates, contributes to weather patterns, and causes ocean currents that are extremely important in many ways. It is also important to recognize that we can sometimes figure new ways to convert low-quality energy to high-quality energy. For example, it is possible to use the waste heat from power plants to heat cities if the power plants are located in the cities. Scientists have recently made major improvements in wind turbines and photovoltaic cells that allow us to economically convert to low-quality light or wind to high-quality electricity.

From the point of view of ecological systems, organisms such as plants do photosynthesis and are able to convert low-quality light energy to high-quality chemical energy in the organic molecules they produce. Eventually, they will use this stored energy for their needs, or it will be used by some other organism that has eaten the plant. In accordance with the second law of thermodynamics, all organisms, including humans, are in the process of converting high-quality energy into low-quality energy. Waste heat is produced when the chemical-bond energy in food is converted into the energy needed to move, grow, or respond. The process of releasing chemical-bond energy from food by organisms is known as cellular respiration. From an energy point of view, it is comparable to the process of **combustion,** which is the burning of fuel to obtain heat, light, or some other form of useful energy. The efficiency of cellular respiration is relatively high. About 40 percent of the energy contained in food is released in a useful form. The rest is dissipated as low-quality heat. Table 4.2 lists the efficiencies of many common energy conversion systems.

An unfortunate consequence of energy conversion is pollution. The heat lost from most energy conversions is a pollutant. The wear of the brakes used to stop cars results in pollution. The emissions from power plants pollute. All of these are examples of the effect of the second law of thermodynamics. If each individual on earth used less energy, there would be less waste heat and other forms of pollution that result from energy conversion. The amount of energy in the universe is limited. Only a small portion of that energy is of high quality. The use of high-quality energy decreases the amount of useful energy available, as more low-quality heat is generated. All life and all activities are subject to these important physical principles described by the first and second laws of thermodynamics.

Table 4.2 Approximate Efficiency of Some Conversion Systems

Energy Conversion System	% Efficiency*
Electric motor	95
Hydroelectric power plant	85
Home oil furnace	65
Jet engine	55
Steam-power plant	47
Diesel engine	45
Fluorescent lamp	40
Automobile engine (gasoline)	25
Incandescent lamp	10

*Efficiency with which the energy of the power or fuel source is converted to a useful form.

Source: Data from several sources.

PART TWO Ecological Principles and Their Application

Improvements in Lighting Efficiency

In the United States, about 25 percent of the electrical energy consumed is used for lighting. Improvements in the efficiency of lighting would significantly reduce the demand for electrical energy. All forms of lighting involve the conversion of electricity into light, but some systems are much more efficient than others. The incandescent light bulb is extremely inefficient yet is still used in many homes and commercial buildings. Standard incandescent light bulbs are 5 to 10 percent efficient. That means that 90 to 95 percent of the energy entering an incandescent light bulb is being released as heat. Compact fluorescent lights are four times more efficient. They use about 25 percent of the energy of an incandescent bulb to produce the same amount of light and can be used in the standard incandescent light bulb socket. Modern compact fluorescent bulbs have color qualities similar to incandescent bulbs. In many situations such as commercial buildings, newer high-efficiency fluorescent lights with electronic ballasts can reduce energy consumption by about 15 percent over standard fluorescent bulbs, which are already four times more efficient than incandescent bulbs.

In some cases, fluorescent lighting is not practical. It does not work well in the cold and in most situations cannot be used with dimmer switches. Other kinds of higher efficiency lighting are available, however. Halogen lights are incandescent lights with about 10 percent better efficiency than standard incandescent lights. Sodium vapor, mercury vapor, and metal halide lights are very efficient but produce a light of a different color than normal daylight. These lights also require several minutes to come up to full lighting power. They are used in places where color is not important and where they are not turned on and off repeatedly, such as exterior lighting in parking lots. The U.S. Department of Energy has helped develop a new sulfur lamp that is even more efficient than fluorescent lighting and has better color than other high-efficiency lamps.

Any improvement in the efficiency with which electricity is converted to light also reduces the amount of waste heat produced, which reduces the amount of electricity needed to cool buildings. Often the cost of replacing inefficient incandescent light bulbs is offset by subsidies from local electric utilities, since increased lighting efficiency reduces the demand for electricity and allows utilities to put off building expensive new power plants.

- How many incandescent light bulbs do you have in your home?
- Why haven't they been replaced?

Summary

Science is a method of gathering and organizing information. It involves observation, asking questions, exploring alternative sources of information, hypothesis formation, the testing of hypotheses, and publication of the results for others to evaluate. A hypothesis is a logical prediction about how things work that must account for all the known information and be testable. The process of science attempts to be careful, unbiased, and reliable in the way information is collected and evaluated. This often involves conducting experiments to test the validity of a hypothesis. If a hypothesis is continually supported by the addition of new facts, it may be incorporated into a theory. A theory is a broadly written, widely accepted generalization that ties together large bodies of information and explains why things happen. Similarly, a law is a broad statement that describes what happens in nature.

The fundamental unit of matter is the atom, which is made up of protons and neutrons in the nucleus surrounded by a cloud of moving electrons. The number of protons for any one type of atom is constant, but the number of neutrons in different atoms of the same type of atom may vary. The number of electrons is equal to the number of protons. Protons have a positive charge, neutrons lack a charge, and electrons have a negative charge.

Molecules are units made of a combination of two or more atoms bonded to one another. Chemical bonds are physical attractions between atoms resulting from the interaction of their electrons. When chemical bonds are broken or formed, a chemical reaction occurs, and the amount of energy within the chemical bonds is changed. Chemical reactions require activation energy to get the reaction started.

Matter that is composed of only one kind of atom is known as an element. Matter that is composed of small units containing different kinds of atoms bonded in specific ratios is known as a compound. An atom or molecule that has gained or lost electrons so that it has an electric charge is known as an ion.

Matter can occur in three states: solid, liquid, and gas. These three differ in the amount of energy the molecular units contain and the distance between the units. Kinetic energy is the energy contained by moving objects. Potential energy is the energy an object has because of its position.

The first law of thermodynamics states that the amount of energy in the universe is constant, that energy can neither be created nor destroyed. The second law of thermodynamics states that when energy is converted from one form to another, some of the useful energy is lost (entropy increases). Some forms of energy are more useful than others. The quality of the energy determines how much useful work can be accomplished by expending the energy. Low-temperature heat sources are of poor quality, since they cannot be used to do useful work.

Key Terms

acid *72*
activation energy *73*
atom *71*
base *72*
catalyst *73*
chemical bond *73*
combustion *78*
compound *72*
controlled experiment *69*
electron *71*
element *71*
endothermic reaction *73*
energy *75*
entropy *76*
enzyme *74*

exothermic reaction *73*
experiment *69*
first law of thermodynamics *76*
hydroxide ion *72*
hypothesis *68*
ion *71*
isotope *71*
kinetic energy *75*
kinetic molecular theory *69*
latent heat *76*
law of conservation of mass *69*
matter *71*
mixture *72*
molecule *71*
neutron *71*

nucleus *71*
observation *67*
pH *72*
photosynthesis *74*
potential energy *75*
proton *71*
reproducibility *69*
respiration *75*
science *67*
scientific law *69*
scientific method *67*
second law of thermodynamics *76*
sensible heat *76*
theory *69*
variable *69*

Review Questions

1. How do scientific disciplines differ from nonscientific disciplines?
2. What is a hypothesis? Why is it an important part of the way scientists think?
3. Why are events that happen only once difficult to analyze from a scientific point of view?
4. What is the scientific method, and what processes does it involve?
5. How are the second law of thermodynamics and pollution related?

6. Diagram an atom of oxygen and label its parts.
7. What happens to atoms during a chemical reaction?
8. State the first and second laws of thermodynamics.
9. How do solids, liquids, and gases differ from one another at the molecular level?
10. List five kinds of energy.
11. Are all kinds of energy equal in their capacity to bring about changes? Why or why not?

Critical Thinking Questions

1. You observe that a high percentage of frogs, especially sensitive to environmental poisons, in small ponds in your agricultural region have birth defects. Suspecting agricultural chemicals present in runoff to be the culprit, state the hypothesis in your own words. Next devise an experiment that might help you support or reject your hypothesis.
2. Given the experiment you proposed in Critical Thinking Question 1, imagine some results that would support that hypothesis. Now imagine you are a different scientist, one who is very skeptical of the initial hypothesis. How convincing do you find these data? What other possible explanations (hypotheses) might there be to explain the results? Devise a different experiment to test this new hypothesis.
3. Increasingly, environmental issues like global climate change are moving to the forefront of world concern. What role should science play in public policy decisions? How should we decide between competing scientific explanations about an environmental concern like global climate change? What might be some of the criteria for deciding what is "good science" and what is "bad science"?

4. How important are the first and second laws of thermodynamics to explaining environmental issues? Using the concepts in these laws of thermodynamics, try to explain a particular environmental issue. How does an understanding of thermodynamics change your conceptual framework regarding this issue?
5. The text points out that incandescent light bulbs are only 5–10 percent efficient at using energy to accomplish their task, while new, initially more expensive, compact fluorescent lighting uses significantly less electricity to provide the same quantity of light. Examine the contextual framework of those who advocate for new lighting methods and the contextual framework of those who continue to design and build using the old methods. What are the major differences in perspective? What could you suggest be done to help bring these different perspectives closer together?
6. Some scientists argue that living organisms constantly battle against the principles of the second law of thermodynamics using the principles of the first law of thermodynamics. What might they mean by this? Do you think this is accurate? What might be some of the implications of this for living organisms?

Concept Map

Construct a map to show relationships among the following concepts:

scientific method

law

hypothesis

atom

molecule

observation

isotope

compound

ion

element

matter

energy

kinetic molecular theory

Interactive Exploration

Check out the website at **http://www.mhhe.com/environmentalscience** and click on the cover of this textbook for quizzing, career information, case studies, and hot links for the following topics:

Scientific Method

Chemistry of Biology

Inorganic Chemistry

Atoms

Molecules

Energy and Laws of Thermodynamics

Conventional Energy Sources

Sustainable Energy

Interactions: Environments and Organisms

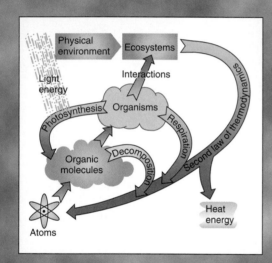

Objectives

After reading this chapter, you should be able to:

- Identify and list abiotic and biotic factors in an ecosystem.
- Define *niche*.
- Describe the process of natural selection as it operates to refine the fit between organism, habitat, and niche.
- Describe predator-prey, parasite-host, competitive, mutualistic, and commensalistic relationships.
- Differentiate between a community and an ecosystem.
- Define the roles of producer, herbivore, carnivore, omnivore, scavenger, parasite, and decomposer.
- Describe energy flow through an ecosystem.
- Relate the concepts of food webs and food chains to trophic levels.
- Explain the cycling of nutrients such as nitrogen, carbon, and phosphorus through an ecosystem.

Chapter Outline

Ecological Concepts

The science of **ecology** is the study of the ways organisms interact with each other and with their nonliving surroundings. Ecology deals with the ways in which organisms are adapted to their surroundings, how they make use of these surroundings, and how an area is altered by the presence and activities of organisms. These interactions involve energy and matter. Living things require a constant flow of energy and matter to assure their survival. If the flow of energy and matter ceases, the organisms die.

All organisms are dependent on other organisms in some way. One organism may eat another and use it for energy and raw materials. One organism may temporarily use another without harming it. One organism may provide a service for another, such as when animals distribute plant seeds or bacteria break down dead organic matter for reuse. The study of ecology can be divided into many specialties and be looked at from several levels of organization. (See figure 5.1.) Before we can explore the field of ecology in greater depth, we must become familiar with some of the standard vocabulary of this field.

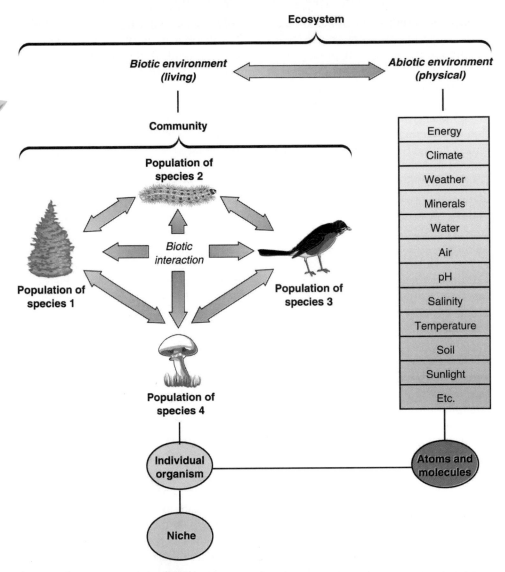

figure 5.1 **Levels of Organization in Ecology** Ecology is the science that deals with the interactions between organisms and their environment. This study can take place at several different levels, from the broad ecosystem level through community interactions to populations studies and the study of the niche of individual organisms. Ecology also involves study of the physical environment and the atoms and molecules that make up both the living and nonliving parts of an ecosystem.

Environment

Everything that affects an organism during its lifetime is collectively known as its **environment.** Environment is a very broad concept. For example, during its lifetime, an animal such as a raccoon is likely to interact with millions of other organisms (bacteria, food organisms, parasites, mates, predators), drink copious amounts of water, breathe huge quantities of air, and respond to daily changes in temperature and humidity. This list only begins to describe the various components that make up the raccoon's environment. Because of this complexity, it is useful to subdivide the concept of environment into **abiotic** (nonliving) and **biotic** (living) **factors.**

Abiotic factors can be organized into several broad categories: energy, nonliving matter, and processes that involve the interactions of nonliving matter and energy. All organisms require a source of energy to maintain themselves. The ultimate source of energy for almost all organisms is the sun; in the case of plants, the sun directly supplies the energy necessary for them to maintain themselves. Animals obtain their energy by eating plants or other animals that eat plants. Ultimately, the amount of living material that can exist in an area is determined by the amount of energy plants, algae, and bacteria can trap.

All forms of life require atoms of elements such as carbon, nitrogen, and phosphorus, and molecules such as water to construct and maintain themselves. Organisms constantly obtain these materials from their environment. The atoms become part of an organism's body structure for a short time period, and eventually all of them are returned to the environment through respiration, excretion, or death and decay.

The structure and location of the space organisms inhabit is also an important abiotic aspect of their environment. Some are at sea level; others are at high elevations. Some spaces are homogeneous and flat; others are a jumble of rocks of different sizes. Some are close to the equator; others are near the poles.

Important ecological processes involve interactions of matter and energy. The climate (average weather patterns over a number of years) of an area involves energy in the form of solar radiation interacting with the matter that makes up the Earth. The kind of climate present is determined by many different factors, including the amount of solar radiation, the proximity to the equator, prevailing wind patterns, and closeness to water. The intensity and duration of sunlight in an area causes daily and seasonal changes in temperature. Differences in temperature generate wind. Solar radiation is also responsible for generating ocean currents and the evaporation of water into the atmosphere that subsequently falls as precipitation. Depending on the climate, precipitation may be of several forms: rain, snow, hail, or fog. Furthermore, there may be seasonal precipitation patterns. Soil building processes are influenced by prevailing weather patterns, local topography, and the geologic history of the region. These factors interact to produce soils that range from sandy, dry, and infertile to fertile and moist with fine particles.

The biotic factors of an organism's environment include all forms of life with which it interacts. Some broad categories are: plants that carry on photosynthesis; animals that eat other organisms; bacteria and fungi that cause decay; bacteria, viruses, and other parasitic organisms that cause disease; and other individuals of the same species.

Limiting Factors

Although organisms interact with their surroundings in many ways, certain factors may be critical to a particular species' success. A shortage or absence of this factor restricts the success of the species; thus, it is known as a **limiting factor.** Limiting factors may be either abiotic or biotic and can be quite different from one species to another. Many plants are limited by scarcity of water, light, or specific soil nutrients. Animals may be limited by climate or the availability of a specific food. For example,

many snakes and lizards are limited to the warmer parts of the world because they have difficulty maintaining their body temperature in cold climates and cannot survive long periods of cold. Monarch butterflies are limited by the number of available milkweed plants since their developing caterpillars use this plant as their only food source.

The limiting factor for many species of fishes is the amount of dissolved oxygen in the water. In a swiftly flowing, tree-lined mountain stream, the level of dissolved oxygen is high and so provides a favorable environment for trout. (See figure 5.2.) As the stream continues down the mountain, the steepness of the slope decreases, which results in fewer rapids where the water tumbles over rocks and becomes oxygenated. In addition, as the stream becomes wider, the canopy of trees over the stream usually is thinner, allowing more sunlight to reach the stream and warm the water. Warm water cannot hold as much dissolved oxygen as cool water. Therefore, slower-flowing, warm-water streams contain less oxygen than rapidly moving, cool streams. Fishes such as black bass and walleye are adapted to such areas, since they are able to tolerate lower oxygen concentrations and higher water temperatures. Trout are not able to survive under such conditions and are not found in warm, less well-oxygenated water. Each of these species has a specific **range of tolerance** to oxygen concentration and water temperature. Thus, low levels of oxygen and high water temperatures are limiting factors for the distribution of trout.

Other factors, such as the abundance of silt, may influence the ability of water to support certain species of fishes. Silt reduces visibility, making it difficult for fish to find food, and covers gravel beds needed for spawning. Reduced light also lessens the amount of photosynthesis that takes place, which has an impact on the amount of oxygen in the water. Since silt causes the water to darken, the particles in the water absorb sunlight and cause the water to become warmer. Under these conditions, the bass and walleye may be replaced by

such species as carp and catfish, which have an even greater ability to withstand high temperatures and low oxygen concentrations and are, therefore, better able to survive in water with a high amount of silt.

Habitat and Niche

As we have just seen, it is impossible to understand an organism apart from its environment. The environment influences the organism, and organisms affect the environment. To focus attention on specific elements of this interaction, ecologists have developed two concepts that need to be clearly understood: habitat and niche.

The **habitat** of an organism is the space that the organism inhabits, the place where it lives (its address). We tend to characterize an organism's habitat by highlighting some prominent physical or biological feature of their environment such as soil type, availability of water, climatic conditions, or predominant plant species that exist in the area. For example, mosses are small plants that must be covered by a thin film of water in order to reproduce. In addition, many kinds dry out and die if they are exposed to sunlight, wind, and drought. Therefore, the typical habitat of moss is likely to be cool, moist, and shady. (See figure 5.3.) Likewise, a rapidly flowing, cool, well-oxygenated stream with many bottom-dwelling insects is good trout habitat, while open prairie with lots of grass is preferred by bison, prairie dogs, and many kinds of hawks and falcons. Elm bark beetles will reside only in areas where elm trees are found. The particular biological requirements of an organism determine the kind of habitat in which it is likely to be found.

The **niche** of an organism is the functional role it has in its surroundings (its profession). A description of an organism's niche includes all the ways it affects the organisms with which it interacts as well as how it modifies its physical surroundings. In addition, the description of a niche includes all of the things that happen to the organism. For example, beavers frequently flood areas

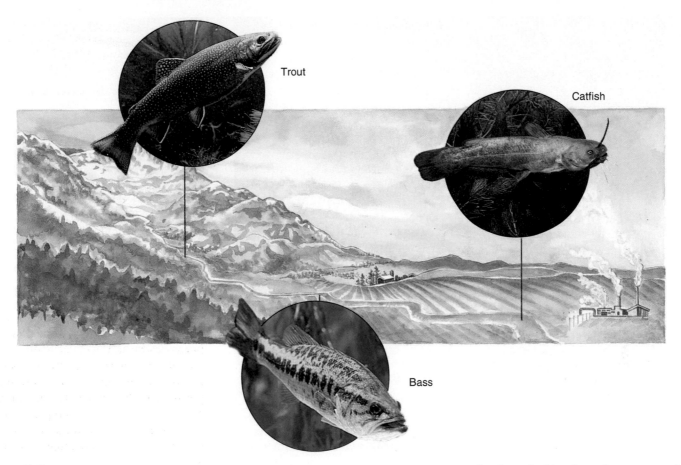

figure 5.2 **Limiting Factors** In aquatic habitats, the amount of oxygen dissolved in the water is often a limiting factor for many species of fish. Cool, highly oxygenated water, which is typical of the rapidly flowing upper sections of a river system, supports trout, but warmer, less oxygenated water is unsuited for trout. Other fish, which are more tolerant of low levels of oxygen, such as bass, catfish, bullheads, and carp, occupy the lower sections of the river, where the water is warmer, there is less oxygen, and the river contains much silt and other soil particles.

by building dams of mud and sticks across streams. (See figure 5.4.) The flooding has several effects. It provides beavers with a larger area of deep water, which they need for protection; it provides a pond habitat for many other species of animals like ducks and fish; and it kills trees that cannot live in saturated soil. The animals attracted to the pond and the beavers often fall to predators. After the beavers have eaten all the suitable food, such as aspen, they abandon the pond and migrate to other areas along the stream and begin the whole process over again.

In this recitation of beaver characteristics, we have listed several effects that the animal has on its local environment. It changes the physical environment by flooding, it kills trees, it enhances the environment for other animals, and it is a food source for predators. This is only a superficial glimpse

of the many aspects of the beaver's interaction with its environment. A complete catalog of all aspects of its niche would make up a separate book.

Another familiar organism is the dandelion. (See figure 5.5.) It is an opportunistic plant that rapidly becomes established in sunny, disturbed sites. In a few days, it can produce thousands of parachutelike seeds that are easily carried by the wind over long distances. (You have probably helped this process by blowing on the fluffy, white collections of seeds of a mature dandelion fruit.) Furthermore, it often produces several sets of flowers per year. Since there are so many seeds and they are so easily distributed, the plant can easily establish itself in any sunny, disturbed site, including lawns. Since it is a plant, one major aspect of its niche is the ability to carry on photosynthesis and grow. Dandelions need direct sunlight to grow

figure 5.3 **Moss Habitat** The habitat of mosses is typically cool, moist, and shady, since many mosses die if they are subjected to drying. In addition, mosses must have a thin layer of water present in order to reproduce sexually.

figure 5.4 **Ecological Niche** The ecological niche of an organism is a complex set of interactions between an organism and its surroundings, which includes all of the ways an organism influences its surroundings as well as all of the ways the organism is affected by its environment. A beaver's ecological niche includes building dams and flooding forested areas, killing trees, providing habitat for ducks and other animals, serving as food for predators, and many other effects.

figure 5.5 **The Niche of a Dandelion** A dandelion is a familiar plant that commonly invades disturbed sites because it produces many seeds that are blown easily to new areas. It serves as food to various herbivores, supplies nectar to bees, and can regrow quickly from its root if its leaves are removed.

develop an understanding of the processes that led to this situation. Furthermore, since the mechanisms that result in adaptation occur within a species, we need to understand the nature of a species.

Genes, Populations, and Species

We can look at an organism from several points of view. We can consider an individual, groups of individuals of the same kind, or groups that are distinct from other groups. This leads us to discuss three interrelated concepts. **Genes** are distinct pieces of DNA that determine the characteristics an individual displays. There are genes for structures like leaf shape or feather color, behaviors like cricket chirps or migratory activity, physiological processes like photosynthesis or muscular contractions. Each individual has a particular set of genes. A **population** is considered to be all the organisms of the same kind found within a specific geographic region. The individuals of a population will have very similar sets of genes, although there will be some individual variation. Because some genetic difference exists among individuals in a population, a population contains more kinds of genes than any individual within the population. Reproduction also takes place among individuals in a population so that genes are passed from

successfully. Mowing lawns helps provide just the right conditions for dandelions, since the vegetation is never allowed to get so tall that dandelions are shaded. Many kinds of animals, including some humans, use the plant for food. The young leaves may be eaten in a salad, and the blossoms can be used to make dandelion wine. Bees visit the flowers regularly to obtain nectar and pollen.

The Role of Natural Selection and Evolution

Since organisms generally are well adapted to their surroundings and fill a particular niche, it is important that we

Habitat Conservation Plans: Tool or Token?

The U.S. Endangered Species Act places strict regulations on the destruction of the habitat of an endangered species. Since endangered species typically have narrow niches, they are restricted to specific habitats and often have very local distribution. Therefore, preservation of specific patches of habitat is critical to their survival. The Endangered Species Act states that persons cannot "incidentally take" (accidentally kill) members of the endangered species. Since many economically important land uses (farming, development, mining) would alter the habitat and result in the incidental taking of members of the endangered species, many landowners feel that the presence of an endangered species on their land unfairly deprives them of the use of their land. Many landowners have argued that, since they have lost the use of their land due to the presence of an endangered species, they should be compensated by the government for the value they have lost. The cost to the government would be enormous. However, the Endangered Species Act allows for habitat conservation plans to be put in place that would allow the landowner some limited use of the land while ensuring the protection of the endangered species.

The process of developing a habitat conservation plan results in a negotiated settlement between the federal government and the landowner. Often these plans allow the landowner to use part of the land while setting other portions aside as protected areas. Sometimes travel corridors must be maintained, or critical nesting sites must be protected. Many conservationists argue that the process is totally inadequate because it is impossible to determine all the critical habitat features of poorly understood endangered species. Therefore, habitat conservation plans are very likely to be inadequate to protect the species from extinction. An additional problem is that once a habitat conservation plan is established, it becomes a binding document and allows landowners to continue to operate under the habitat conservation plan even if new scientific information becomes available that demonstrates that the plan is inadequate. Landowners feel that a binding plan is necessary. Without a binding plan, they could be held responsible for unforeseen consequences that would put their investments at risk. Critics feel that most plans are political compromises that do not protect the endangered species and that the plans make no provisions for the recovery of the species. And since modifications to the habitat are often allowed, the ultimate fate of most species the plans are supposed to protect is likely to be extinction.

From a purely scientific point of view, we know that each species has a specific niche and has critical habitat requirements necessary for its survival. If a species is endangered and it is to be preserved, its habitat must be protected. However, the economic and political forces of human populations are also important. Habitat conservation plans are compromises between total protection and total conversion of habitat to human use. Some well-constructed plans will be successful, while others will temporarily delay the inevitable extinction of vulnerable species.

one generation to the next. The concept of a species is an extension of this thinking about genes, groups, and reproduction. A **species** is a population of all the organisms potentially capable of reproducing naturally among themselves and having offspring that also reproduce. Therefore, the concept of a species is a population concept. *An individual organism is not a species but a member of a species.* It is also a genetic concept, since individuals that are of different species are not capable of exchanging genes through reproduction. This working definition of species contains three points that require further discussion. First, it is obvious that there are individuals in any population that never reproduce, and many pairs of individuals that will never meet one another. However, they still have the potential to interbreed and are considered members of the species. The second point involves the ability to produce fertile offspring. In some instances, two kinds of organisms may interbreed and produce offspring, but the offspring are sterile and never reproduce. For example, horses and donkeys can breed and produce offspring called mules, but since mules are sterile, horses and donkies are considered separate species.

There is a third issue that is more difficult to accommodate within this species definition. Some organisms reproduce primarily by asexual reproduction; that is, they do not mate but simply produce copies of themselves. Those that only reproduce asexually do not fit this definition. However, most organisms that reproduce asexually also reproduce sexually at certain times and can be assigned to a species based on those times when they do mate.

Some species are easy to recognize. We easily recognize humans as a distinct species. Most people recognize a dandelion when they see it and do not confuse it with other kinds of plants that have yellow flowers. Other species are not as easy to recognize. Most of us cannot tell one species of mosquito from another or identify different species of grasses. Because of this, we tend to lump organisms into large categories and do not recognize the many subtle niche differences that exist among the similar-appearing species. However, different species of mosquitoes are quite distinct from one another genetically and also occupy different niches. Only certain ones carry and transmit the human disease malaria. Other species transmit the dog heartworm parasite.

Each mosquito species is active during certain portions of the day or night. And each species requires specific conditions to reproduce.

Natural Selection

As we have seen, each species of organism is specifically adapted to a particular habitat in which it has a very specific role (niche). But how is it that each species of plant, animal, fungus, or bacterium fits into its environment in such a precise way? Since most of the structural, physiological, and behavioral characteristics organisms display are determined by the genes they possess, these characteristics are passed from one generation to the next when individuals reproduce. The process that leads to this close fit between the characteristics organisms display and the demands of their environment is known as natural selection. **Natural selection** is the process that determines which individuals within a species will reproduce and pass their genes to the next generation. The changes that we see in the genes and the characteristics displayed by successive generations of a population of organisms over time is known as **evolution.** Thus, natural selection is the mechanism that causes evolution to occur. Several conditions and steps are involved in the process of natural selection.

1. *Individuals within a species show genetically determined variation; some of the variations are useful and others are not.* For example, individual animals that are part of the same species show color variations. Some colors make the animal more conspicuous while others make it less conspicuous.

2. *Organisms within a species typically produce many more offspring than are needed to replace the parents when they die. Most of the offspring die.* One apple tree may produce hundreds of apples with several seeds in each apple, or a pair of rabbits may have three to four litters of offspring each summer, with several young in each litter. Few of the seeds or baby rabbits become reproducing adults.

3. *The excess number of individuals results in a shortage of specific resources.* Individuals within a species must compete with each other for food, space, mates, or other requirements that are in limited supply. If you plant 100 bean seeds in a pot, many of them will begin to grow, but eventually, some will become taller and get the majority of the sunlight while the remaining plants are shaded. Great horned owls typically produce two young at a time, but if food is in short supply, the larger of the two young will get the majority of the food.

4. *Because of variation among individuals, some have a greater chance of obtaining needed resources and, therefore, have a greater likelihood of surviving and reproducing than others.* Individuals that have genes that allow them to obtain needed resources and avoid threats to their survival will be more likely to survive and reproduce. Even if less well-adapted individuals survive, they may mature more slowly and not be able to reproduce as many times as the more well-adapted members of the species. The degree to which organisms are adapted to their environment influences their reproductive success and is referred to as fitness. It is important to recognize that fitness does not necessarily mean the condition of being strong or vigorous. In this context, it means how well the organism fits in with all the aspects of its surroundings so that it successfully passes its genes to the next generation. For example, in a lodgepole pine forest, many lodgepole pine seedlings become established following a fire. Some grow more rapidly and obtain more sunlight and nutrients. Those that grow more rapidly are more likely to survive. These are also likely to reproduce for longer periods and pass more genes to future generations than those that die or grow more slowly.

5. *As time passes and each generation is subjected to the same process of natural selection, the percentage of individuals showing favorable variations will increase and those having unfavorable variations will decrease.* Those that reproduce more successfully pass on to the next generation the genes for the characteristics that made them successful in their environment, and the genes that made them successful become more common in future generations. Thus, each species of organism is continually refined to be adapted to the environment in which it exists. One modern example of genetic change resulting from natural selection involves the development of pest populations that are resistant to the pesticides previously used to control them. Figure 5.6 shows a graph of the number of species of weeds that have populations that are resistant to commonly used herbicides. When an herbicide is first used against weed pests, it kills most of them. However, in many cases, some individual weed plants within the species happen to have genes that allow them to resist the effects of the herbicide. These individuals are better adapted to survive in the presence of the herbicide and have a higher likelihood of surviving. When they reproduce, they pass on to their offspring the same genes that contributed to their survival. After several generations of such selection, a majority of the individuals in the species will contain genes that allow them to resist the herbicide, and the herbicide is no longer effective against the weed.

Evolutionary Patterns

When we look at the effects of natural selection over time, we can see considerable change in the characteristics of a species and kinds of species present. Some changes take thousands or millions of years to occur. Others, like resistance to pesticides, can occur in a few years. (See Figure 5.6.) Natural selection involves the processes that bring about change in species, and the end result of

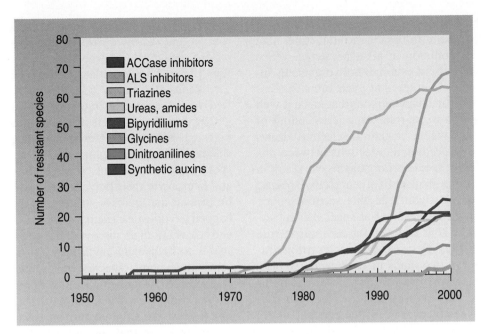

figure 5.6 **Evolutionary Change** Populations of weed plants that have been subjected repeatedly to herbicides often develop resistant populations. Those individual weed plants that were able to resist the affects of the herbicide lived to reproduce and pass their genes for resistance on to their offspring; thus, resistant populations of weeds develop.

Source: Data from Ian Heap, "The International Survey of Herbicide Resistant Weeds." [cited 5 November 2001]. Internet.

the natural selection process observable in organisms is called evolution.

Scientists have continuously shown that this theory of natural selection can explain the development of most aspects of the structure, function, and behavior of organisms. It is the central idea that helps explain how species adapt to their surroundings. When we discuss environmental problems, it is helpful to understand that species change and that as the environment is changed, either naturally or by human action, some species will adapt to the new conditions while others will not.

There are many examples that demonstrate the validity of the process of natural selection and the evolutionary changes that result from natural selection. In recent times, we have become aware that many species of insects, weeds, and bacteria have become resistant to the insecticides, herbicides, and antibiotics that formerly were effective against them.

When we look at the evolutionary history of organisms in the fossil record over long time periods, it becomes obvious that new species come into being while other species disappear. The production of new species from previously existing species is known as **speciation** and is thought to occur as a result of a species dividing into two isolated subpopulations. If the two subpopulations contain some genetic differences and their environments are somewhat different, natural selection will work on the two groups differently and they will begin to diverge from each other. Eventually, the differences may be so great that the two subpopulations are not able to interbreed. At this point, they are two different species.

Among plants is another common mechanism known as polyploidy that results in new species. Polyploidy occurs when the number of sets of chromosomes in the cells of the plant is increased. Many organisms are diploid; that is, they have two sets of chromosomes. They got a set from each parent, one set in the egg and one set in the sperm. Polyloid organisms may have several sets of chromosomes. The details of how polyploidy comes about are not important for this discussion. It is sufficient to recognize that many species

of plants appear to have extra sets of chromosomes, and they are not able to reproduce with closely related species that have a different number of sets of chromosomes.

The environment in which organisms exist does not remain constant over long time periods. Those species that lack the genetic resources to cope with a changing environment go extinct. **Extinction** is the loss of an entire species and is a common feature of the evolution of organisms. Of the estimated 500 million species of organisms that are believed to have ever existed on Earth since life began, perhaps 5 million to 10 million are currently active. This represents an extinction rate of 98 to 99 percent. Obviously, these numbers are estimates, but the fact remains that extinction has been the fate of most species of organisms. In fact, studies of recent fossils and other geologic features show that only thousands of years ago, huge glaciers covered much of Europe and the northern parts of North America. Humans coexisted with mammoths, saber-toothed tigers, and giant cave bears. As the climate became warmer and the glaciers receded, and humans continued to prey on these animals, new pressures affected the organisms in the area. Some, including the mammoths, saber-toothed tigers, and giant cave bears, did not adapt and became extinct. Others, such as humans, horses, and many kinds of plants, adapted to the new conditions and so survive to the present.

It is also possible to have the extinction of specific populations of a species. Most species consist of many different populations that may differ from one another in significant ways. Often some of these populations have small local populations that can easily be driven to extinction. While these local extinctions are not the same as the extinction of an entire species, local extinctions often result in the loss of specific gene combinations. Many of the organisms listed on the endangered species list are really local populations of a more widely distributed species.

Natural selection is constantly at work shaping organisms to fit a changing environment. It is clear that humans have had a significant impact on the extinction of many kinds of species. Wherever humans have modified the environment for their purposes (farming, forestry, cities, hunting, and introducing exotic organisms), species are typically displaced from the area. If large areas are modified, entire species may be displaced. Ultimately, humans are also subject to evolution and the possibility of extinction as well.

Coevolution is the concept that two or more species of organisms can reciprocally influence the evolutionary direction of the other. In other words, organisms affect the evolution of other organisms. Since all organisms are influenced by other organisms, this is a common pattern. For example, grazing animals and the grasses they consume have coevolved. Grasses that are eaten by grazing animals grow from the base of the plant near the ground rather than from the tips of the branches as many plants do. Furthermore, grasses have hard materials in their cell walls that make it difficult for animals to crush the cell walls and digest them. Grazing animals have different kinds of adaptations that overcome these deterrents. Many grazers have teeth that are very long or grow continuously to compensate for the wear associated with grinding hard cell walls. Others, such as cattle, have complicated digestive tracts that allow microorganisms to do most of the work of digestion. Similarly, the red color and production of nectar by many kinds of flowers is attractive to hummingbirds, which pollinate the flowers at the same time as they consume nectar from the flower. The "Kinds of Organism Interactions" section will explore in more detail the ways that organisms interact and the results of long periods of coevolution.

Kinds of Organism Interactions

Ecologists look at organisms and how they interact with their surroundings. Perhaps the most important interactions occur between organisms. Ecologists

have identified several general types of organism-to-organism interactions that are common in all ecosystems. When we closely examine how organisms interact, we see that each organism has specific characteristics that make it well suited to its role. An understanding of the concept of natural selection allows us to see how interactions between different species of organisms can result in species that are finely tuned to a specific role. As you read this section, notice how each species has special characteristics that equip it for its specific role (niche). Because these interactions involve two kinds of organisms interacting, we should expect to see examples of coevolution. If the interaction between two species is the result of a long period of interaction, we should expect to see that each species has characteristics that specifically adapt it to be successful in its role.

Predation

One common kind of interaction called **predation** occurs when one organism, known as a **predator,** kills and eats another, known as the **prey.** (See figure 5.7.) The predator benefits from killing and eating the prey and the prey is harmed. Some examples of predator-

prey relationships are lions and zebras, robins and earthworms, wolves and moose, and toads and beetles. Even a few plants show predatory behavior. The Venus flytrap has specially modified leaves that can quickly fold together and trap insects that are then digested. To succeed, predators employ several strategies. Some strong and speedy predators (lions, sharks) chase and overpower their prey; other species lie in wait and quickly strike prey that happen to come near them (many lizards and hawks); and some (spiders) use snares to help them catch prey. At the same time, prey species have many characteristics that help them avoid predation. Many have keen senses that allow them to detect predators, others are camouflaged so they are not conspicuous, and many can avoid detection by remaining motionless when predators are in the area. An adaptation common to many prey species is a high reproductive rate. For example, field mice may have 10 to 20 offspring per year, while hawks typically have two to three. Because of this high reproductive rate, prey species can endure a high mortality rate and still maintain a viable population. Certainly, the *individual* organism that is killed and eaten is harmed, but the prey *species* is not, since the prey

figure 5.7 **Predator-Prey Relationship** Lions are predators on zebras. The quicker lions are more likely to get food, and the slower, sickly, or weaker zebras are more likely to become prey.

PART TWO Ecological Principles and Their Application

individuals that die are likely to be the old, the slow, the sick, and the less well-adapted members of the population. The healthier, quicker, and better-adapted individuals are more likely to survive. When these survivors reproduce, their offspring are more likely to have characteristics that help them survive; they are better adapted to their environment. At the same time, a similar process is taking place in the predator population. Since poorly adapted individuals are less likely to capture prey, they are less likely to survive and reproduce. The predator and prey species are both participants in the natural selection process. This dynamic relationship between predator and prey species is a complex one that continues to intrigue ecologists.

Competition

A second type of interaction between species is **competition,** in which two organisms strive to obtain the same limited resource. In the process, both organisms are harmed to some extent. (See figure 5.8.) However, this does not mean that there is no winner. If a large number of lodgepole pine trees begin growing close to one another, they will compete for water, minerals, and sunlight. None of the trees grows as rapidly as it could because their access to resources is restricted by the presence of the other trees. Eventually, some of the pines will grow faster and will get a greater share of the resources. The taller trees will get more sunlight and the shorter trees will receive less. In time, some of the smaller trees die. Similarly, when two robins are competing for the same worm, only one gets it. Both organisms were harmed because they had to expend energy in fighting for the worm, but one got some food and was harmed less than the one that fought and got nothing. These examples of competition, in which members of the same species compete for resources, is known as **intraspecific competition.** Other examples of intraspecific competition include corn plants in a field competing for water and nutrients, male elk competing with one another for the right to mate with the females, and certain species of woodpeckers competing for the holes in dead trees to use for nesting sites.

Competition among members of the same species is a major force in shaping the evolution of a species. When resources are limited, less well-adapted individuals are more likely to die or be denied mating privileges. Because the most successful organisms are likely to have larger numbers of offspring, each succeeding generation will contain more of the genetic characteristics that are favorable for survival of the species in that particular environment. Since individuals of the same species have similar needs, competition among them is usually very intense. A slight advantage on the part of one individual may mean the difference between survival and death.

Competition between organisms of different species is called **interspecific competition.** Many species of predators (hawks, owls, foxes, coyotes) may use the same prey species (mice, rabbits) as a food source. If the supply of food is inadequate, intense competition for food will occur and certain predator species may be more successful than others. In grasslands, the same kind of competition for limited resources occurs. Rapidly growing, taller grasses get more of the water, minerals, and sunlight, while shorter species are less successful. Often the shorter species are found to be more abundant when the taller species are removed by grazers, fire, or other activities.

As with intraspecific competition, one of the effects of interspecific competition is that the species that has the larger number of successful individuals emerges from the interaction better adapted to its environment than its less successful rivals. The more similar two species are, the more intense will be the competition between them. If one of the two competing species is better adapted to live in the area than the other, the less-fit species must evolve into a slightly different niche, migrate to a different geographic area, or become extinct. This concept is often formally called the **competitive exclusion principle,** which states that no two species can occupy the same ecological niche in the same place at the same time. When the niche requirements of two similar species are examined closely, we usually find significant differences between the niches of the two species. The difference in niche requirements reduces the intensity of the competition between the two species. For example, many small forest birds eat insects. However, they may obtain them in different ways; a flycatcher sits on a branch and makes short flights to snatch insects from the air, a woodpecker excavates openings to obtain insects in rotting wood, and many warblers flit about in the foliage capturing insects. Even among these categories there are specialists. Different

figure 5.8 **Competition** Whenever a needed resource is in limited supply, organisms compete for it. This competition may be between members of the same species and is called intraspecific competition, or it may be between different species and is called interspecific competition. This photograph shows several vultures competing for a food source.

species of warblers look in different parts of trees for their insect food.

Symbiotic Relationships

Symbiosis is a close, long-lasting, physical relationship between two different species. In other words, the two species are usually in physical contact and at least one of them derives some sort of benefit from this contact. There are three different categories of symbiotic relationships: parasitism, commensalism, and mutualism.

Parasitism

Parasitism is a relationship in which one organism, known as the **parasite,** lives in or on another organism, known as the **host,** from which it derives nourishment. Generally, the parasite is much smaller than the host. Although the host is harmed by the interaction, it is generally not killed immediately by the parasite, and some host individuals may live a long time and be relatively little affected by their parasites. Some parasites are much more destructive than others, however. Newly established parasite-host relationships are likely to be more destructive than those that have a long evolutionary history. With a long-standing interaction between the parasite and the host, the two species generally evolve in such a way that they can accommodate one another. It is not in the parasite's best interest to kill its host. If it does, it must find another. Likewise, the host evolves defenses against the parasite, often reducing the harm done by the parasite to a level the host can tolerate.

Many parasites have complex life histories that involve two or more host species for different stages in the parasite's life cycle. Many worm parasites have their adult, reproductive stage in a carnivore (the definitive host), but they have an immature stage that reproduces asexually in another animal (the intermediate host) that the carnivore uses as food. Thus, a common dog tapeworm is found in its immature form in certain internal organs of rabbits. Other parasite life cycles involve animals that carry the parasite from one host to another. These carriers are known as **vectors.** For example, many biting insects and mites can transmit parasites when they obtain a blood meal. Malaria, lyme disease, and sleeping sickness are transmitted by vectors.

Parasites that live on the surface of their hosts are known as **ectoparasites.** Fleas, lice, and some molds and mildews are examples of ectoparasites. (See figure 5.9.) Many other parasites, like tapeworms, malaria parasites, many kinds of bacteria, and some fungi, are called **endoparasites,** because they live inside the bodies of their hosts. A tapeworm lives in the intestines of its host, where it is able to resist being digested and makes use of the nutrients in the intestine. If a host has only one or two tapeworms, it can live for some time with little discomfort, supporting itself and its parasites. If the number of parasites is large, the host may die.

Even plants can be parasites. Mistletoe is a flowering plant that is parasitic on trees. It establishes itself on the surface of a tree when a bird transfers the seed to the tree. It then grows down into the water-conducting tissues of the tree and uses the water and minerals it obtains from these tissues to support its own growth.

Parasitism is a very common life strategy. If we were to categorize all the organisms in the world, we would find many more parasitic species than nonparasitic species. Each organism, including you, has many others that use it as a host.

Commensalism

If the relationship between organisms is one in which one organism benefits while the other is not affected, it is called **commensalism.** It is possible to visualize a parasitic relationship evolving into a commensal one. Since parasites generally evolve to do as little harm to their host as possible and the host is combating the negative effects of the parasite, they might eventually evolve to the point where the host is not harmed at all. Many examples of commensal relationships exist. Many orchids use trees as a surface upon which to grow. The tree is not harmed or helped, but the orchid needs a surface

Flea (external parasite)

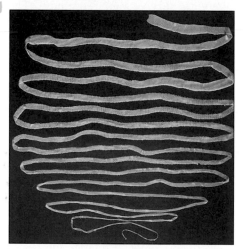

Tapeworm (internal parasite)

figure 5.9 **Parasitism** Fleas are small insects that live in the feathers of birds or the fur of mammals, where they bite their hosts to obtain blood. Since they live on the outside of their hosts, they are called ectoparasites. Tapeworms live inside the intestines of their hosts, where they absorb food from their hosts' intestines. Since they live inside their hosts, they are called endoparasites.

upon which to establish itself and also benefits by being close to the top of the tree, where it can get more sunlight and rain. Some mosses, ferns, and many vines also make use of the surfaces of trees in this way.

In the ocean, many sharks have a smaller fish known as a remora attached to them. Remoras have a sucker on the top of their heads that they can use to attach to the shark. In this way, they can hitchhike a ride as the shark swims along. When the shark feeds, the remora frees itself and obtains small bits of food that the shark misses. Then, the remora reattaches. The shark does not appear to

figure 5.10 **Commensalism** Remoras hitchhike a ride on sharks and feed on the scraps of food lost by the sharks. This is a benefit to the remoras. The sharks do not appear to be affected by the presence of the remoras.

be positively or negatively affected by remoras. (See figure 5.10.) Many commensal relationships are rather opportunistic and may not involve long-term physical contact. For example, many birds rely on trees of many different species for places to build their nests but do not use the same tree year after year. Similarly, in the spring, bumble bees typically build nests in underground mouse nests that are no longer in use.

Mutualism

Mutualism is another kind of symbiotic relationship and is actually beneficial to both species involved. In many mutualistic relationships, the relationship is obligatory; the species cannot live without each other. In others, the species can exist separately but are more successful when they are involved in a mutualistic relationship. Some species of *Acacia*, a thorny tree, provide food in the form of sugar solutions in little structures on their stems. Certain species of ants feed on the solutions and live in the tree, which they will protect from other animals by attacking any animal that begins to feed on the tree. Both organisms benefit; the ants receive food and a place to live, and the tree is protected from animals that would use it as food.

One soil nutrient that is usually a limiting factor for plant growth is nitrogen. Many kinds of plants, such as legumes (beans, clover, and acacia trees) and alder trees, have bacteria that

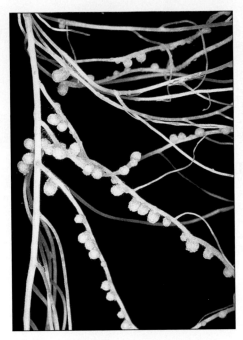

figure 5.11 **Mutualism** The growths on the roots of this plant contain beneficial bacteria that make nitrogen available to the plant. The relationship is also beneficial to the bacteria, since the bacteria obtain necessary raw materials from the plant. It is a mutually beneficial relationship.

live in their roots in little nodules. The roots form these nodules when they are infected with certain kinds of bacteria. The bacteria do not cause disease but provide the plants with nitrogen-containing molecules that the plants can use for growth. The nitrogen-fixing bacteria benefit from the living site and nutrients that the plants provide, and the plants benefit from the nitrogen they receive. (See figure 5.11.) Similarly, many kinds of fungi form an association with the roots of plants. The root-fungus associations are called **mycorrhizae.** The fungus obtains organic molecules from the roots of the plant, and the branched nature of the fungus assists the plant in obtaining nutrients such as phosphates and nitrates. In many cases, it is clear that the relationship is obligatory.

Some Relationships Are Difficult to Categorize

Sometimes it is not easy to categorize the relationships that organisms have with each other. For example, it is not always easy to say whether a relation-

figure 5.12 **Nest (Brood) Parasitism** This red-eyed vireo is feeding the nestling of a brown-headed cowbird. A female cowbird laid its egg in the vireo's nest. The vireo is harmed because it is not raising its own young, and the cowbird benefits because it did not need to expend energy to build and defend a nest or collect food for its own young.

ship is a predator-prey relationship or a host-parasite relationship. How would you classify a mosquito or a tick? Both of these animals require blood meals to live and reproduce. They don't kill and eat their prey. Neither do they live in or on a host for a long period of time. This question points out the difficulty encountered when we try to place all kinds of organism interactions into a few categories. However, we can eliminate this problem if we call them temporary parasites or blood predators.

Another relationship that doesn't fit well is the relationship that certain birds like cowbirds and European cuckoos have with other birds. Cowbirds and European cuckoos do not build nests but lay their eggs in the nests of other species of birds, who are left to care for a foster nestling at the expense of their own nestlings, who generally die. This situation is usually called nest parasitism or brood parasitism. (See figure 5.12.)

What about grazing animals? Are they predators or parasites on the plants that they eat? Sometimes they kill the

environmental CLOSE-UP

Human Interaction—A Different Look

In terms of our ability to modify ecosystems, humans are the dominant organisms on Earth. Our niche is very broad and we interact in many ways with the organisms with which we share the planet. If we examine our activities, we can see that we have complicated interactions with other organisms, and these interactions can be placed into the same categories we use to describe relationships between nonhuman organisms.

Predator—Humans throughout the world use animals as food. Some actually kill animals themselves, while others rely on employees of slaughterhouses to do the killing for them.

Herbivore—Humans rely on many kinds of plants as their primary source of food.

Grazing and browsing involve consuming part of a living plant without killing the entire plant. We graze or browse parts of plants such as asparagus, rhubarb, lettuce, broccoli, and many other kinds of plants.

Foraging involves searching for food that is available from nature. Hunter-gatherer peoples spend a great deal of time foraging for edible plant materials such as roots, fruits, and seeds. Even individuals from sophisticated cultures engage in foraging activities when they pick wild berries, mushrooms, or asparagus.

Scavenger—Scavenging involves finding and consuming animals that are already dead. Our distant ancestors were probably actively engaged in scavenging by seizing the kills of more efficient carnivores. Even today in places where protein-rich food is in short supply, any recently dead animal is a valued food source. And many states in the United States have laws that allow people to take animals killed by collisions with automobiles.

Commensalism—Humans find themselves on both sides of commensal relationships. Many kinds of organisms use our homes as places to live without affecting us. Birds may nest on our buildings, spiders build webs in our windows, and rats may live under our decks. Other animals benefit from the animals we accidentally kill along our highways. We also derive benefit from organisms without affecting them, such as when we are able to get out of the hot sun by sitting under the shade of a tree or when we rely on decomposers to decay wastes.

Parasitism—Although humans do not live in or on other living things, we do engage in relationships that are parasitic in nature. In some African cultures, blood is drawn from cows and mixed with milk to serve as food. Maple syrup is made by "bleeding" maple trees. Similarly, humans tap rubber trees to obtain sap that is used to make rubber. Many human activities regularly rob animals. Honeybees are robbed of their honey and chickens are robbed of their eggs.

Mutualism—Humans have many mutualistic relationships with plants and animals. Our domesticated plants and animals rely on us for support and nutrition, and we extract from them payment in the form of companionship, food, or other valuable resources.

Competition—Humans are in competition with all other organisms on Earth. As we convert land and aquatic resources to our uses, we deprive other organisms of what they need to survive. When humans hunt and kill large grazing animals such as bison or gazelles, we are in direct competition with other predators such as wolves or leopards. Because of our technological superiority and our huge population, we usually win in the game of competition.

---

plant they eat, while at other times they simply remove part of the plant and the rest continues to grow. In either case, the plant has been harmed by the interaction and the grazer has benefited.

There are also mutualistic relationships that do not require permanent contact between the participants in the relationship. Bees and the flowering plants they pollinate both benefit from their interactions. The bees obtain pollen and nectar for food and the plants are pollinated. But the active part of the relationship involves only a part of the life of any plant, and the bees are not restricted to any one species of plant for the food. They must actually switch to different flowers at different times of the year.

Community and Ecosystem Interactions

Thus far, we have discussed specific ways in which individual organisms interact with one another and with their physical surroundings. However, often it is useful to look at ecological relationships from a broader perspective. Two concepts that focus on relationships that involve many different kinds of interactions are community and ecosystem. A **community** is an assemblage of all the interacting populations of different species of organisms in an area. Some species play minor roles, while others play major roles, but all are part of the community. For example, the grasses of the prairie have a major role since they carry on photosynthesis and provide food and shelter for the animals that live in the area. Grasshoppers, prairie dogs, and bison are important consumers of grass. Meadowlarks consume many kinds of insects, and though they are a conspicuous and colorful part of the prairie scene, they have a relatively minor role and little to do with maintaining a prairie community. Bacteria and fungi in the soil break down the bodies of dead plants and animals and provide nutrients to plants.

Communities consist of interacting populations of different species, but these species interact with their physical world as well. An ecosystem is a defined space in which interactions take place between a community, with all its complex interrelationships, and the physical environment. The physical world has a major impact on what kinds of plants and animals can live in an area. We do not expect to see a banana tree in the Arctic or a walrus in the Mississippi River. Banana trees are adapted to warm, moist, tropical areas, and walruses require cold ocean waters. Some ecosystems, such as grasslands and certain kinds of forests, are shaped by periodic fires. The kind of soil and the amount of moisture also influence the kinds of organisms found in an area.

While it is easy to see that the physical environment places limitations on the kinds of organisms that can live in an area, it is also important to recognize that organisms impact their physical surroundings. Trees break the force of the wind, grazing animals form paths, and earthworms create holes that aerate the soil. While the concepts of community and ecosystem are closely related, an ecosystem is a broader concept because it involves physical as well as biological processes.

Every system has parts that are related to one another in specific ways. A bicycle has wheels, a frame, handlebars, brakes, pedals, and a seat. These parts must be organized in a certain way or the system known as a bicycle will not function. Similarly, ecosystems have parts that must be organized in specific ways or the systems will not operate. To more fully develop the concept of ecosystem, we will look at ecosystems from three points of view: the major

roles played by organisms, the way energy is utilized within ecosystems, and the way atoms are cycled from one organism to another.

Major Roles of Organisms in Ecosystems

Several categories of organisms are found in any ecosystem. Producers are organisms that are able to use sources of energy to make complex, organic molecules from the simple inorganic substances in their environment. In nearly all ecosystems, energy is supplied by the sun, and organisms such as, plants, algae, and tiny aquatic organisms called phytoplankton use light energy to carry on photosynthesis. Since producers are the only organisms in an ecosystem that can trap energy and make new organic material from inorganic material, all other organisms rely on producers as a source of food, either directly or indirectly. These other organisms are called consumers because they consume organic matter to provide themselves with energy and the organic molecules necessary to build their own bodies. An important part of their role is the process of respiration in which they break down organic matter to inorganic matter.

However, some consumers have significantly different roles from others. Primary consumers, also known as herbivores, are animals that eat producers (plants or phytoplankton) as a source of food. Herbivores, such as leaf-eating

insects and seed-eating birds, are usually quite numerous in ecosystems, where they serve as food for the next organisms in the chain. Secondary consumers or carnivores are animals that eat other animals. Secondary consumers can be further subdivided into categories based on what kind of prey they capture and eat. Some carnivores, like ladybird beetles, primarily eat herbivores, like aphids; others, such as eagles, primarily eat fish that are themselves carnivores. While these are interesting conceptual distinctions, most carnivores will eat any animal they can capture and kill. In addition, many animals, called omnivores, include both plants and animals in their diet. Even animals that are considered to be carnivores (foxes, bears) regularly include large amounts of plant material in their diets. Conversely, animals often thought of as herbivores (mice, squirrels, seed-eating birds) regularly consume animals as a source of food.

A final category of consumer is the decomposer. Decomposers are organisms that use nonliving organic matter as a source of energy and raw materials to build their bodies. Whenever an organism sheds a part of itself, excretes waste products, or dies, it provides a source of food for decomposers. Since decomposers carry on respiration, they are extremely important in recycling matter by converting organic matter to inorganic material. Many small animals, fungi, and bacteria fill this niche. (See table 5.1.)

Table 5.1 Roles in an Ecosystem

Category	Major Role or Action	Examples
Producer	Converts simple inorganic molecules into organic molecules by the process of photosynthesis	Trees, flowers, grasses, ferns, mosses, algae
Consumer	Uses organic matter as a source of food	Animals, fungi, bacteria
Herbivore	Eats plants directly	Grasshopper, elk, human vegetarian
Carnivore	Kills and eats animals	Wolf, pike, dragonfly
Omnivore	Eats both plants and animals	Rats, raccoons, most humans
Scavenger	Eats meat but often gets it from animals that died by accident or illness, or were killed by other animals	Coyote, vulture, blowflies
Parasite	Lives in or on another living organism and gets food from it	Tapeworm, many bacteria, some insects
Decomposer	Returns organic material to inorganic material; completes recycling of atoms	Fungi, bacteria, some insects and worms

Keystone Species

Ecosystems typically consist of many different species interacting with each other and their physical surroundings. However, some species have more central roles than others. In recognition of this idea, ecologists have developed the concept of keystone species. A keystone species is one that has a critical role to play in the maintenance of specific ecosystems. In prairie ecosystems, grazing animals are extremely important in maintaining the mix of species typical of a grassland. Without the many influences of the grazers, the nature of the prairie changes. A study of the American tallgrass prairie indicated that when bison are present, they increase the biodiversity of the site. Bison typically eat grasses and, therefore, allow smaller plant species that would normally be shaded by tall grasses to be successful. In ungrazed plots, the tall grasses become the dominant vegetation and biodiversity decreases. Bison dig depressions in the soil, called wallows, to provide themselves with dust or mud with which they can coat themselves. These wallows retain many species of plants that typically live in disturbed areas. Bison urine has also been shown to be an important source of nitrogen for the plants.

The activities of bison even affect the extent and impact of fire, another important feature of grassland ecosystems. Since bison prefer to feed on recently burned sites and revisit these sites several times throughout the year, they tend to create a patchwork of grazed and ungrazed areas. The grazed areas are less likely to be able to sustain a fire, and fires likely will be more prevalent in ungrazed patches.

The concept of keystone species has also been studied in marine ecosystems. The relationship between sea urchins, sea otters, and kelp forests suggests that sea otters are a keystone species. Sea otters eat sea urchins, which eat kelp. A reduction in the number of otters results in an increase in the number of sea urchins. Increased numbers of sea urchins lead to heavy grazing of the kelp by sea urchins. When the amount of kelp is severely reduced, fish and many other animals that live within the kelp beds lose their habitat and biodiversity is significantly reduced.

The concept of keystone species is useful to ecologists and resource managers because it helps them to realize that all species cannot be treated equally. Some species have pivotal roles and their elimination or severe reduction can significantly alter ecosystems. In some cases, the loss of a keystone species can result in the permanent modification of an ecosystem into something considerably different from the original mix of species.

Energy Flow Through Ecosystems

An ecosystem is a stable, self-regulating unit. This does not mean that an ecosystem is unchanging. The organisms within it are growing, reproducing, dying, and decaying. In addition, an ecosystem must have a continuous input of energy to remain its stability. The only significant source of energy for most ecosystems is sunlight. Producers are the only organisms that are capable of trapping solar energy through the process of photosynthesis and making it available to the ecosystem. The energy is stored in the form of chemical bonds in large organic molecules such as carbohydrates (sugars, starches), fats, and proteins. The energy stored in the molecules of producers is transferred to other organisms when the producers are eaten. Each step in the flow of energy through an ecosystem is known as a **trophic level.** Producers (plants, algae, phytoplankton) constitute the first trophic level, and herbivores constitute the second trophic level. Carnivores that eat herbivores are the third trophic level, and carnivores that eat other carnivores are the fourth trophic level. Omnivores, parasites, and scavengers occupy different trophic levels, depending on what they happen to be eating at the time. If we eat a piece of steak, we are at the third trophic level; if we eat celery, we are at the second trophic level. (See figure 5.13.)

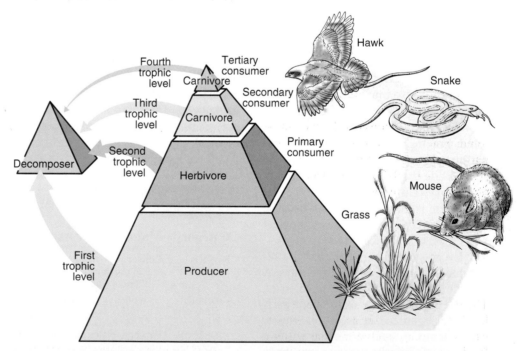

figure 5.13 **Energy Flow Through an Ecosystem** As energy flows through an ecosystem, it passes through several levels known as trophic levels. Each trophic level contains a certain amount of energy. Each time energy flows to another trophic level, approximately 90 percent of the useful energy is lost, usually as heat to the surroundings. Therefore, in most ecosystems, higher trophic levels contain less energy and fewer organisms.

The second law of thermodynamics states that whenever energy is converted from one form to another, some of the energy is converted to a nonuseful form (typically, low-quality heat). Thus, there is always less useful energy following an energy conversion. When energy passes from one trophic level to the next, there is less useful energy left with each successive trophic level. This loss of low-quality heat is dissipated to the surroundings and warms the air, water, or soil. In addition to this loss of heat, organisms must expend energy to maintain their own life processes. It takes energy to chew food, defend nests, walk to waterholes, or produce and raise offspring. Therefore, the amount of energy contained in higher trophic levels is considerably less than that at lower levels. Approximately 90 percent of the useful energy is lost with each transfer to the next highest trophic level. So in any ecosystem, the amount of energy contained in the herbivore trophic level is only about 10 percent of the energy contained in the producer trophic level. The amount of energy at the third trophic level is approximately 1 percent of that found in the first trophic level.

Because it is difficult to actually measure the amount of energy contained in each trophic level, ecologists often use other measures to approximate the relationship between the amounts of energy at each level. One of these is the biomass. The **biomass is the weight of living material in a trophic level. It is often possible in a simple ecosystem to collect and weigh all the producers, herbivores, and carnivores**. The weights often show the same 90 percent loss from one trophic level to the next as happens with the amount of energy.

Food Chains and Food Webs

The passage of energy from one trophic level to the next as a result of one organism consuming another is known as a **food chain.** For example, willow trees grow well in very moist soil, perhaps near a pond. The trees' leaves capture sunlight and convert carbon dioxide and water into sugars and other organic molecules. The leaves serve as a food source

for insects, such as caterpillars and leaf beetles, that have chewing mouth parts and a digestive system adapted to plant food. Some of these insects are eaten by spiders, which fall from the trees into the pond below, where they are consumed by a frog. As the frog swims from one lily pad to another, a large bass consumes the frog. A human may use an artificial frog as a lure to entice the bass from its hiding place. A fish dinner is the final step in this chain of events that began with the leaves of a willow tree. (See figure 5.14.) This food chain has six trophic levels. Each organism occupies a specific niche and has special abilities that fit it for its niche, and each organism in the food chain is involved in converting energy and matter from one form to another.

Some food chains rely on a constant supply of small pieces of dead organic material being supplied from situations where photosynthesis is taking place. The small bits of nonliving organic material are called **detritus.** Detritus food chains are found in a variety of situations. The bottoms of the deep lakes and oceans are too dark for photosynthesis. The animals and decomposers that live there rely on a steady rain of small bits of organic matter from the upper layers of the water where photosynthesis does take place. Similarly, in most streams, leaves and other organic debris serve as the major source of organic material and energy. A sewage treatment plant is also a detritus food chain in which particles and dissolved organic matter are constantly supplied to a series of bacteria and protozoa that use this material for food.

In another example, the soil on a forest floor receives leaves, which fuel a detritus food chain. In detritus food chains, a mixture of insects, crustaceans, worms, bacteria, and fungi cooperate in the breakdown of the large pieces of organic matter, while at the same time feeding on one another. When a leaf dies and falls to the forest floor, it is colonized by bacteria and fungi, which begin the breakdown process. An earthworm will also feed on the leaf and at the same time consume the bacteria and fungi. If that earthworm is eaten by a bird, it becomes part of a larger food chain that in-

figure 5.14 **Food Chain** As one organism feeds on another organism, energy flows through the series. This is called a food chain.

cludes material from both a detritus food chain and a photosynthesis-driven food chain. When several food chains overlap and intersect, they make up a **food web.** (See figure 5.15.) This diagram is typical of the kinds of interactions that take place in a community. Each organism is likely to be a food source for several other kinds of organisms. Even the simplest food webs are complex.

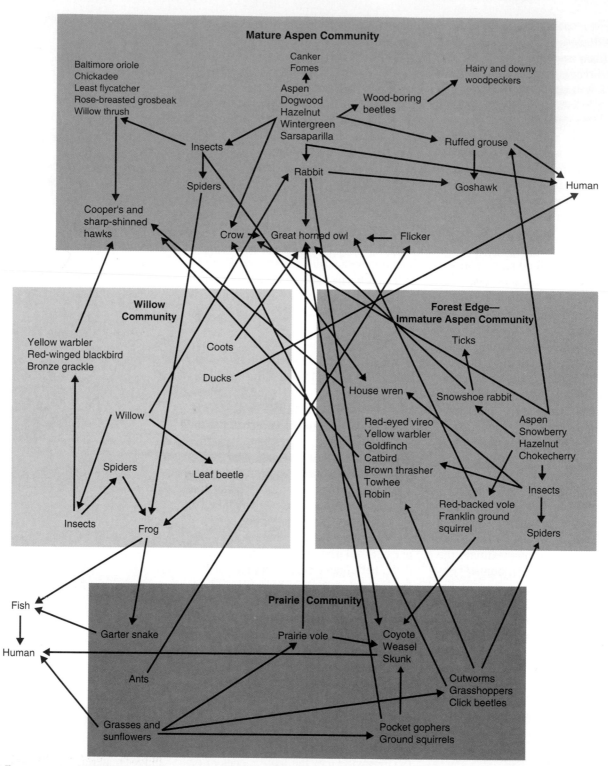

figure 5.15 **Food Web** The many kinds of interactions among organisms in an ecosystem constitute a food web. In this network of interactions, several organisms would be affected if one key organism were reduced in number. Look at the rabbit in the mature aspen community and note how many organisms use it as food.

Source: R.D. Bird, "Biotic Communities of the Aspen Parkland of Central Canada," *Ecology*, 11 (April 1930.): 410.

Contaminants in the Food Chain of Fish from the Great Lakes

As organisms eat, molecules move through the food chain from one organism to another. Most molecules are simply recycled through nutrient pathways because they can be broken down by decay organisms. However, humans have invented a large number of organic molecules that are not easily decomposed, and these tend to remain in food chains and accumulate in higher concentrations in organisms that are at higher trophic levels.

The Great Lakes area developed as an industrial center because the lakes provided an efficient way to move raw materials and products and because water was important for manufacturing processes. In the past, many of these industries released heavy metals and organic molecules into the water as an accidental by-product of the manufacturing process or because it was a cheap way to get rid of unwanted material. Many of the organic molecules are products of modern organic synthesis and, therefore, are not something that bacteria and fungi are able to decay. Abnormally high concentrations of inorganic materials in the environment can also lead to abnormally high concentrations in organisms.

Approximately 500 different organic compounds that scientists think are contaminants have been identified in the bodies of fish from the Great Lakes. Most are present in extremely small amounts and probably do not represent a serious hazard, but others are present in high enough concentrations to cause public health officials to be concerned. Since these materials do not break down, fish tend to accumulate more of these toxic materials in their bodies as they get older. Furthermore, carnivorous fish that are feeding at higher trophic levels accumulate more of these compounds. It just so happens that most of the fish that people catch and eat are carnivores, which tend to have the highest concentrations of contaminants.

If people place themselves in this food chain by eating the contaminated fish, they will tend to accumulate the contaminants in their bodies, which could have health effects. It is not possible to check every fish caught to see if it is fit to eat. The cost of doing so would be on the order of several hundred to several thousands of dollars per individual fish, depending on which contaminants are looked for. Therefore, the states and provinces surrounding the Great Lakes

have developed advisory statements to help people avoid unsafe fish. While there are some minor differences among state and provincial governments, the following advisories are typical.

1. Some species of fish, such as carp and catfish, feed on the bottom and tend to accumulate contaminants that are in the bottom sediments. In many areas of the Great Lakes, people are advised not to eat these fish.

2. Larger fish have generally consumed more food and have had an opportunity to accumulate more contaminants. Therefore, people are advised to eat only smaller fish.

3. Since many of the organic contaminants are fat soluble, removal of the fat or cooking in such a way that the fat is allowed to separate from the flesh is also advised in many areas.

4. Because the amount of contamination a person is exposed to is directly related to the number of fish a person eats, people are advised to limit the number of fish consumed, and women of childbearing age and young children are often advised to not eat the fish at all.

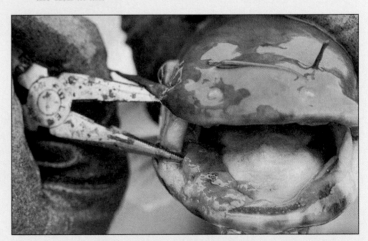

The lip cancer on this bullhead indicates carcinogens in Wisconsin's Fox River. In some tributaries of the Great Lakes, fish cancer rates may reach 84 percent.

Notice in the upper-left-hand corner of figure 5.15 that the Cooper's and sharp-shinned hawks use many different kinds of birds as a source of food. These hawks fit into several food chains. If one source of prey is in short supply, they can switch to something else without too much trouble. These kinds of complex food webs tend to be more stable than simple food chains with few cross-links.

Nutrient Cycles in Ecosystems—Biogeochemical Cycles

All matter is made up of atoms. These atoms are cycled between the living and nonliving portions of an ecosystem. The activities involved in the cycling of atoms include biological, geological, and chemical processes. Therefore, these nutrient cycles are often called **biogeochemical cycles.** Some atoms are more common in living things than are others. Carbon, nitrogen, oxygen, hydrogen, and phosphorus are found in important organic molecules like proteins, DNA, carbohydrates, and fats, which are found in all kinds of living things. Organic molecules contain large numbers of carbon atoms attached to one another.

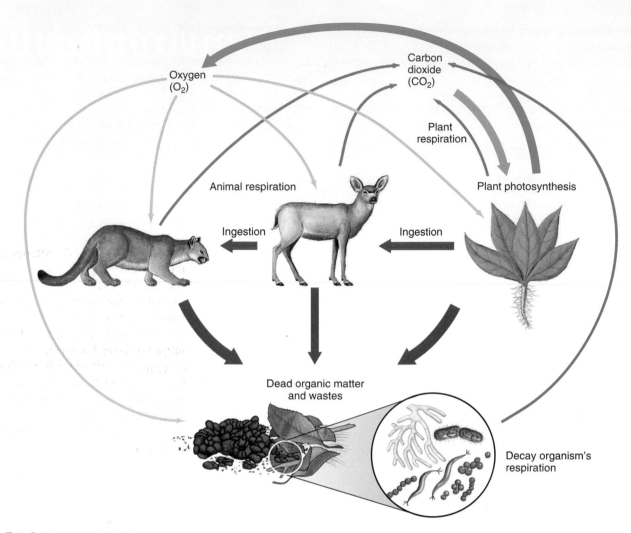

figure 5.16 **Carbon Cycle** Carbon atoms are cycled through ecosystems. Plants can incorporate carbon atoms from carbon dioxide into organic molecules when they carry on photosynthesis. The carbon-containing organic molecules are passed to animals when they eat plants or other animals. Organic wastes or dead organisms are consumed by decay organisms. All organisms, plants, animals, and decomposers return carbon atoms to the atmosphere when they carry on respiration. Oxygen atoms are being cycled at the same time that carbon atoms are being cycled.

These organic molecules are initially manufactured from inorganic molecules by the activities of producers and are transferred from one living organism to another in food chains. The processes of respiration and decay ultimately break down the complex organic molecules of organisms and convert them to simpler, inorganic constituents which are returned to the abiotic environment. In this section, we will look at the flow of three kinds of atoms within communities and between the abiotic and abiotic portions of an ecosystem: carbon, nitrogen, and phosphorus.

Carbon Cycle

All living things are composed of organic molecules that contain atoms of the element carbon. The **carbon cycle** includes the processes and pathways involved in capturing inorganic carbon-containing molecules, converting them into organic molecules that are used by organisms, and the ultimate release of inorganic carbon molecules back to the abiotic environment. (See figure 5.16.) Carbon and oxygen combine to form the molecule carbon dioxide (CO_2), which is present in small quantities as a gas in the

atmosphere and dissolved in water. During photosynthesis, carbon dioxide from the atmosphere is taken into the leaves of plants where it is combined with hydrogen from water molecules (H_2O), which are absorbed from the soil by the roots and transported to the leaves. Many kinds of aquatic organisms such as algae and some bacteria also perform photosynthesis but absorb carbon dioxide and water molecules from the water in which they live. The energy needed to perform photosynthesis is provided by sunlight. As a result of photosynthesis, complex organic molecules such as carbohydrates (sugars) are formed. Oxygen molecules (O_2) are released into the atmosphere or water during the process of photosynthesis because water molecules are split to provide hydrogen atoms necessary to manufacture carbohydrate molecules. The remaining oxygen is released as a waste product of photosynthesis. In this process, light energy is converted to chemical-bond energy in organic molecules, such as sugar. Plants and other producer organisms use these sugars for growth and to provide energy for other necessary processes.

Herbivores can use these complex organic molecules as food. When an herbivore eats plants or algae, it breaks down the complex organic molecules into simpler organic molecular building blocks, which can be reassembled into the specific organic molecules that are part of its chemical structure. The carbon atom, which was once part of an organic molecule in a producer, is now part of an organic molecule in an herbivore. Nearly all organisms also carry on the process of respiration, in which oxygen from the atmosphere is used to break down large organic molecules into carbon dioxide and water. Much of the chemical-bond energy is released by respiration and is lost as heat, but the remainder is used by the herbivore for movement, growth, and other activities.

In similar fashion, when an herbivore is eaten by a carnivore, some of the carbon-containing molecules of the herbivore become incorporated into the body of the carnivore. The remaining organic molecules are broken down in the process of respiration to obtain energy and carbon dioxide and water are released.

The organic molecules contained in animal waste products and dead organisms are acted upon by decomposers that use these organic materials as a source of food. The decay process of decomposers involves respiration and releases carbon dioxide and water so that naturally occurring organic molecules are typically recycled.

In the carbon cycle, all organisms require organic molecules for their survival and must either manufacture them or consume them. Photosynthetic organisms (producers) capture inorganic carbon in the form of carbon dioxide molecules and manufacture organic molecules. Nearly all organisms, including plants, carry on respiration, in which organic molecules are broken down to provide energy and inorganic carbon dioxide is released. The same carbon atoms are used over and over again. In fact, you are not exactly the same person today that you were yesterday. Some of your carbon atoms are different. Furthermore, those carbon atoms have been involved in many other kinds of living things over the past several billion years. Some of them were temporary residents in dinosaurs, extinct trees, or insects, but at this instant, they are part of you.

Although this is a terrestrial example, it is important to recognize that this same cycle operates in aquatic systems as well. Carbon dioxide dissolves in water and, therefore, is available for aquatic plants and algae to use in the process of photosynthesis. When the consumers in the food web use molecules of oxygen that are dissolved in water to carry on respiration, carbon dioxide is released back into the water.

Fossil fuels (coal, oil, and natural gas) are part of the carbon cycle as well. At one time, these materials were organic molecules in the bodies of living organisms. The organisms were buried and the organic compounds in their bodies were modified by geologic forces. Thus, the carbon atoms present in fossil fuels were removed temporarily from the active, short-term carbon cycle. When we burn fossil fuels, the carbon reenters the active carbon cycle.

Nitrogen Cycle

Another very important nutrient cycle, the **nitrogen cycle,** involves the cycling of nitrogen atoms between the abiotic and biotic components and among the organisms in an ecosystem. Seventy-eight percent of the gas in the air we breathe is made up of molecules of nitrogen gas (N_2). However, the two nitrogen atoms are bound very tightly to each other, and very few organisms are able to use nitrogen in this form. Since plants and other producers are at the base of nearly all food chains, they must make new nitrogen-containing molecules, such as proteins and DNA. Plants and other producers are unable to use the nitrogen in the atmosphere and must get it in the form of nitrate (NO_3^-) or ammonia (NH_3). Because atmospheric nitrogen is not usable by plants, nitrogen-containing compounds are often in short supply and the availability of nitrogen is often a factor that limits the growth of plants. The primary way in which plants obtain nitrogen compounds they can use is with the help of bacteria that live in the soil.

Bacteria, called **nitrogen-fixing bacteria,** are able to convert the nitrogen gas (N_2) that enters the soil into ammonia that plants can use. Certain kinds of these bacteria live freely in the soil and are called **free-living nitrogen-fixing bacteria.** Others, known as **symbiotic nitrogen-fixing bacteria,** have a mutualistic relationship with certain plants and live in nodules in the roots of plants known as legumes (peas, beans, and clover) and certain trees such as alders. Some grasses and evergreen trees appear to have a similar relationship with certain root fungi that seem to improve the nitrogen-fixing capacity of the plant.

Once plants and other producers have nitrogen available in a form they can use, they can construct proteins, DNA, and other important nitrogen-containing organic molecules. When herbivores eat plants, the plant protein molecules are broken down to smaller building blocks called amino acids. These amino acids are then reassembled to form proteins typical for the herbivore. This same process is repeated throughout the food chain.

Bacteria and other types of decay organisms are involved in the nitrogen cycle also. Dead organisms and their waste products contain molecules, such as proteins, urea, and uric acid, that contain nitrogen. Decomposers break down these nitrogen-containing organic molecules, releasing ammonia, which can be used directly by many kinds of plants. Still other kinds of soil bacteria called **nitrifying bacteria** are able to convert ammonia to nitrite, which can be converted to nitrate. Plants can use nitrate

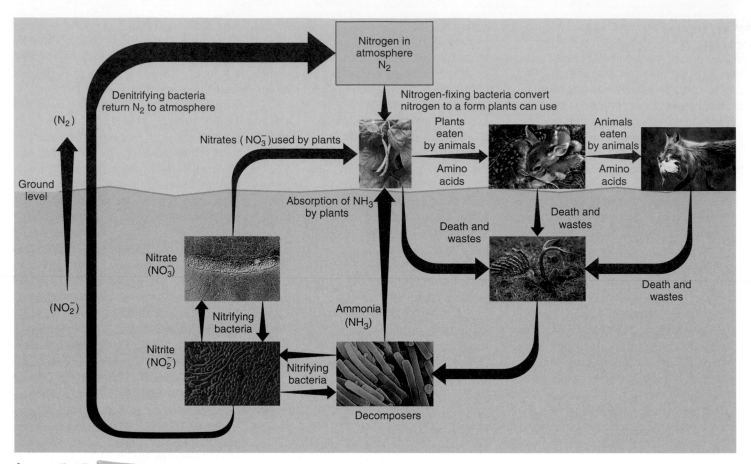

figure 5.17 **Nitrogen Cycle** Nitrogen atoms are cycled in ecosystems. Atmospheric nitrogen is converted by nitrogen-fixing bacteria to a form that plants can use to make protein and other compounds. Proteins are passed to other organisms when one organism is eaten by another. Dead organisms and waste products are acted on by decay organisms to form ammonia, which may be reused by plants or converted to other nitrogen compounds by other kinds of bacteria. Denitrifying bacteria are able to convert inorganic nitrogen compounds into atmospheric nitrogen.

as a source of nitrogen for synthesis of nitrogen-containing organic molecules.

Finally, bacteria, known as **denitrifying bacteria,** are, under conditions where oxygen is absent, able to convert nitrite to nitrogen gas (N_2), which is ultimately released into the atmosphere. These nitrogen atoms can reenter the cycle with the aid of nitrogen-fixing bacteria. (See figure 5.17.)

Although a cyclic pattern is present in both the carbon cycle and the nitrogen cycle, the nitrogen cycle shows two significant differences. First, most of the difficult chemical conversions are made by bacteria and other microorganisms. Without the activities of bacteria, little nitrogen would be available and the world would be a very different place. Second, although nitrogen enters organisms by way of nitrogen-fixing bacteria and returns to the atmosphere through

the actions of denitrifying bacteria, there is a secondary loop in the cycle that recycles nitrogen compounds directly from dead organisms and wastes directly back to producers.

In naturally occurring soil, nitrogen is often a limiting factor of plant growth. To increase yields, farmers provide extra sources of nitrogen in several ways. Inorganic fertilizers are a primary method of increasing the nitrogen available. These fertilizers may contain ammonia, nitrate, or both.

Since the manufacture of nitrogen fertilizer requires a large amount of energy, fertilizer is expensive. Therefore, farmers use alternative methods to supply nitrogen and reduce their cost of production. Several different techniques are effective. Farmers can alternate nitrogen-yielding crops like soybeans with nitrogen-demanding crops like

corn. Since soybeans are legumes that have symbiotic nitrogen-fixing bacteria in their roots, if soybeans are planted one year, the excess nitrogen left in the soil can be used by the corn plants grown the next year. Some farmers even plant alternating strips of soybeans and corn in the same field. A slightly different technique involves growing a nitrogen-fixing crop for a short period of time and then plowing the crop into the soil and letting the organic matter decompose. The ammonia released by decomposition serves as fertilizer to the crop that follows. This is often referred to as green manure. Farmers can also add nitrogen to the soil by spreading manure from animal production operations or dairy farms on the field and relying on the soil bacteria to decompose the organic matter and release the nitrogen for plant use.

Phosphorus Cycle

Phosphorus is another kind of element common in the structure of living things. It is present in many important biological molecules such as DNA and in the membrane structure of cells. In addition, the bones and teeth of animals contain significant quantities of phosphorus. The phosphorus cycle differs from the carbon and nitrogen cycles in one important respect. Phosphorus is not present in the atmosphere as a gas. The ultimate source of phosphorus atoms is rock. In nature, new phosphorus compounds are released by the erosion of rock and become dissolved in water. Plants use the dissolved phosphorus compounds to construct the molecules they need. Animals obtain the phosphorus they need when they consume plants or other animals. When an organism dies or excretes waste products, decomposer organisms recycle the phosphorus compounds back into the soil. Phosphorus compounds that are dissolved in water are ultimately precipitated as deposits. Geologic processes elevate these deposits and expose them to erosion, thus making these deposits available to organisms. Waste products of animals often have significant amounts of phosphorus. In places where large numbers of seabirds or bats congregate for hundreds of years, the thickness of their droppings (called guano) can be a significant source of phosphorus for fertilizer. (See figure 5.18.) In many soils, phosphorus is in short supply and must be provided to crop plants to get maximum yields. Phosphorus is also in short supply in aquatic ecosystems.

Fertilizers usually contain nitrogen, phosphorus, and potassium compounds. The numbers on a fertilizer bag indicate the percentage of each in the fertilizer. For example, a 6–24–24 fertilizer has 6 percent nitrogen, 24 percent phosphorus, and 24 percent potassium compounds. In addition to carbon, nitrogen, and phosphorus, potassium and other elements are cycled within ecosystems. In an agriculture ecosystem, these elements are removed when the crop is harvested. Therefore, farmers must not only return the nitrogen, phosphorus, and potassium, they must also analyze for other less

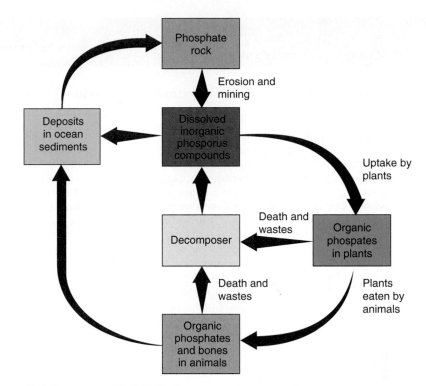

figure 5.18 **Phosphorus Cycle** The source of phosphorus is rock that, when dissolved, provides the phosphate used by plants and animals.

prominent elements and add them to their fertilizer mixture as well. Aquatic ecosystems are also sensitive to nutrient levels. High levels of nitrates or phosphorus compounds often result in rapid growth of aquatic producers. In aquaculture, such as that used to raise catfish, fertilizer is added to the body of water to stimulate the production of algae which is the base of many aquatic food chains.

Human Impact on Nutrient Cycles

To appreciate how ecosystems function, it is important to have an understanding of how elements such as carbon, nitrogen, and phosphorus flow through them. When we look at these cycles from a global perspective, it is apparent that humans have significantly altered them in many ways. Two activities have caused significant changes in the carbon cycle: burning fossil fuels and converting forests to agricultural land. Fossil fuels are carbon-containing molecules produced when organisms were fossilized. Burning fossil fuels (coal, oil, and natural gas) releases large amounts of car-

bon dioxide into the atmosphere. The conversion of forest ecosystems that tend to store carbon for long periods to agricultural ecosystems that store carbon only temporarily has also disrupted the natural carbon cycle. Less carbon is being stored in the bodies of large, long-lived plants such as trees. One consequence of these actions is that the amount of carbon dioxide in the atmosphere has been increasing steadily since humans began to use fossil fuels extensively. It has become clear that increasing carbon dioxide is causing changes in the climate of the world, and many nations are seeking to reduce energy use and prevent deforestation. This topic will be discussed in more detail in chapter 17, which deals with air pollution.

The burning of fossil fuels has also altered the nitrogen cycle. When fossil fuels are burned, the oxygen and nitrogen in the air are heated to high temperatures and a variety of nitrogen-containing compounds are produced. These compounds are used by plants as nutrients for growth. Many people suggest that these sources of nitrogen, along with that provided by fertilizers, have doubled the

Reintroducing Wolves to the Yellowstone Ecosystem

At one time, large predators such as wolves, mountain lions, and grizzly bears were an important part of the community of organisms in the region of Yellowstone National Park. For generations, ranchers in the area killed these large predators because they were a threat to livestock. It was even government policy to kill wolves in national parks. The last wolves in Yellowstone National Park were killed in 1924. By the early 1900s, the wolf was extinct in the U.S. Rocky Mountains and, in 1967, was listed as an endangered species. Without large predators, the populations of large grazing animals in the park (elk, moose, and bison) increased to levels that were not natural. In 1995, the U.S. Fish and Wildlife Service began to reintroduce wolves into the park to restore a more natural mix of species and to control the burgeoning grazer populations.

By 2002, about 24 packs of wolves and several lone wolves were active in and around Yellowstone National Park. The total population was about 216. The wolves are successful in reducing the elk population. During the winter, when elk are weaker, a typical pack of wolves will kill an elk every two to three days. Another effect of the wolf introduction has been the reduction in the coyote population, since wolves kill coyotes. Historically, coyotes were not important in Yellowstone, but when wolves were eliminated, the coyote population rose significantly. The reintroduction of wolves has restored a more natural balance between these two predators.

However, many of the ranchers in the area were not pleased with the reintroduction of wolves and fought the plan in the courts. It is highly likely that as the population of wolves increases, the wolves

amount of nitrogen available today as compared to pre-industrial times.

Fertilizer is used in agriculture to increase the growth of crops. These nutrients are intended to become incorporated into the bodies of the plants and animals that we raise for food. However, if too much nitrogen or phosphorus is applied as fertilizer or if they are applied at the wrong time, much of this fertilizer is carried into aquatic ecosystems. In addition, raising large numbers of animals for food in concentrated settings results in huge amounts of animal waste that contain nitrogen and phosphorus compounds that often enter local water sources. This addition of nitrogen and phosphorus to aquatic ecosystems is particularly significant since aquatic ecosystems normally are starved for these nutrients. The presence of large amounts of these nutrients in either freshwater or saltwater results in increased rates of growth of bacteria, algae, and aquatic plants. Increases in the number of these organisms can have many different effects. Many algae are toxic, and when their numbers increase significantly, fish are killed and incidents of human poisoning occur. An increase in the number of plants and algae in aquatic ecosystems also can lead to low oxygen concentrations in the water. When these organisms die, decomposers use oxygen

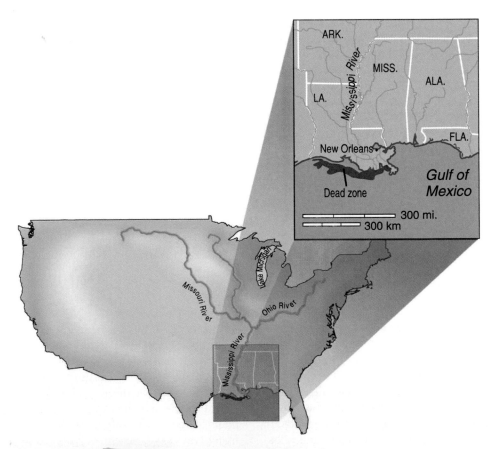

figure 5.19 **Nutrient Impact on Aquatic Organisms** The Mississippi River drainage system carries water from the agricultural center of the United States to the Gulf of Mexico. Extensive use of fertilizer for agriculture results in nitrogen and phosphorus compounds being carried to the Gulf of Mexico. When the organisms that grew as a result of the fertilizer die, a major region of the Gulf of Mexico has oxygen concentrations too low to support most kinds of life.

will migrate from the park and prey on livestock. This has already happened in a few instances. The losses of livestock to wolves was anticipated, and a fund was set up by a private organization—the Defenders of Wildlife—to reimburse ranchers for livestock they lost to wolf predation. Annual payments vary but are generally on the order of $20,000 to $60,000. Some ranchers even return their payments. However, many ranchers were not satisfied with this mechanism and filed a lawsuit to have the wolves removed from the park. The lawsuit was based on a technicality of the Endangered Species Act. The wolves in Yellowstone received a "nonessential experimental" designation under the Endangered Species Act. Because the wolves were migrating from Canada into the area, the reintroduction program simply accelerated the process. The special designation would allow ranchers to kill reintroduced wolves that attack their livestock; however, wolves that migrate naturally from Canada were not to be killed. The ranchers argued that they would not be able to tell reintroduced "nonessential experimental" wolves from those that arrived by migration from other areas.

In December 1997, a U.S. District judge ordered the removal of the wolves from Yellowstone. The ruling was appealed. Finally, in April 2000, the American Farm Bureau, which had been challenging the reintroduction program on behalf of local ranchers, announced that it would not pursue further legal action. Therefore, wolves are again a legal as well as an important biological component of the Yellowstone ecosystem.

- Is it important to return species to areas that were part of their former range?
- Should the economic interests of ranchers be more important than the biological interests of managing Yellowstone National Park as an example of a natural ecosystem?
- Should special interests be allowed to use "technicalities" to circumvent the generally accepted policy?

---

from the water as they break down the dead organic matter. This lowers the oxygen concentration and many organisms die. For example, each summer a major "dead zone" develops in the Gulf of Mexico off the mouth of the Mississippi River. In 2002 it covered 22,000 square kilometers (8,600 square miles). This "dead zone" contains few fish and bottom-dwelling organisms. It is caused by low oxygen levels brought about by the rapid growth of algae and bacteria in the nutrient-rich waters. The nutrients can be traced to the extensive use of fertilizer in the major farming areas of the central United States, farming areas drained by the Mississippi River and its tributaries. Thus, fertilizer use in the agricultural center of the United States results in the death of fish in the Gulf of Mexico. (See figure 5.19.)

Summary

Everything that affects an organism during its lifetime is collectively known as its environment. The environment of an organism can be divided into biotic (living) and abiotic (non-living) components.

The space an organism occupies is known as its habitat, and the role it plays in its environment is known as its niche. The niche of a species is the result of natural selection directing the adaptation of the species to a specific set of environmental conditions.

Organisms interact with one another in a variety of ways. Predators kill and eat prey. Organisms that have the same needs compete with one another and do mutual harm, but one is usually harmed less and survives. Symbiotic relationships are those in which organisms live in physical contact with one another. Parasites live in or on another organism and derive benefit from the relationship, harming the host in the process. Commensal organisms derive benefit from another organism but do not harm the host. Mutualistic organisms both derive benefit from their relationship.

A community is the biotic portion of an ecosystem that is a set of interacting populations of organisms. Those organisms and their abiotic environment constitute an ecosystem. In an ecosystem, energy is trapped by producers and flows from producers through various trophic levels of consumers (herbivores, carnivores, omnivores, and decomposers). About 90 percent of the energy is lost as it passes from one trophic level to the next. This means that the amount of biomass at higher trophic levels is usually much less than that at lower trophic levels. The sequence of organisms through which energy flows is known as a food chain. Several interconnecting food chains constitute a food web.

The flow of atoms through an ecosystem involves all the organisms in the community. The carbon, nitrogen, and phosphorus cycles are examples of how these materials are cycled in ecosystems.

Key Terms

abiotic factors *83*	environment *83*	nitrogen cycle *101*
biogeochemical cycles *99*	evolution *88*	nitrogen-fixing bacteria *101*
biomass *97*	extinction *89*	omnivore *95*
biotic factors *83*	food chain *97*	parasite *92*
carbon cycle *100*	food web *97*	parasitism *92*
carnivore *95*	free-living nitrogen-fixing bacteria *101*	population *86*
coevolution *90*	genes *86*	predation *90*
commensalism *92*	habitat *84*	predator *90*
community *94*	herbivore *95*	prey *90*
competition *91*	host *92*	primary consumer *95*
competitive exclusion principle *91*	interspecific competition *91*	producer *95*
consumer *95*	intraspecific competition *91*	range of tolerance *84*
decomposer *95*	keystone species *96*	secondary consumer *95*
denitrifying bacteria *102*	limiting factor *84*	speciation *89*
detritus *97*	mutualism *93*	species *87*
ecology *83*	mycorrhizae *93*	symbiosis *92*
ecosystem *95*	natural selection *88*	symbiotic nitrogen-fixing bacteria *101*
ectoparasite *92*	niche *84*	trophic level *96*
endoparasite *92*	nitrifying bacteria *101*	vector *92*

Review Questions

1. Define *environment*.
2. Describe, in detail, the niche of a human.
3. How is natural selection related to the concept of niche?
4. List five predators and their prey organisms.
5. How is an ecosystem different from a community?
6. Humans raising cattle for food is what kind of relationship?
7. Give examples of organisms that are herbivores, carnivores, and omnivores.
8. What are some different trophic levels in an ecosystem?
9. Describe the carbon cycle, the nitrogen cycle, and the phosphorus cycle.
10. Analyze an aquarium as an ecosystem. Identify the major abiotic and biotic factors. List members of the producer, primary consumer, secondary consumer, and decomposer trophic levels.

Critical Thinking Questions

1. Ecologists and political scientists look at habitat destruction differently. Consider the Environmental Close-Up about habitat conservation plans and the political, economic, and scientific issues that surround conservation plans. Identify some perspectives each discipline has to contribute to our understanding of habitat destruction. What values does each place on the ideas of the other discipline? What do you think about the issue of creating conservation plans? Protecting habitat from destruction? Why?
2. Even before humans entered the scene, many species of plants and animals were extinct and new ones had developed. Why are we even concerned about endangered species, given the fact that species have always come and gone?
3. Humans have had a major impact on many ecosystems. Consider the questions in the Issues and Analysis section about reintroducing wolves to Yellowstone Park. Is this the appropriate way for humans to try to repair the environmental damage they have caused? Identify some of the values and beliefs that make up these different perspectives: ranchers and conservationists. Which do you find the most compelling? Why?
4. Concentrations of industrial chemicals are high in some species of fish, high enough to call for an advisory to limit the number of fish a person should eat within a given period of time. Many of these chemicals are thought to cause cancer, but cancer is an effect that is often not felt for decades after exposure. How do scientists decide how many fish can be safely eaten? Is there any "safe" level? What evidence would convince you that there is danger? How could you tell?
5. You notice that after using pesticides on your farm field that the number of insects declines for a year. The next year, though, they come back and you need to reapply the pesticide. This time, though, there is less of an effect on the insect population. A third application in another year has even less of an effect. What is your hypothesis about what is happening here? Design an experiment that tests your hypothesis.

Concept Map

Construct a map to show relationships among the following concepts:

producer

trophic level

consumer

carnivore

herbivore

food chain

ecosystem

nitrogen cycle

decomposer

nitrogen-fixing bacteria

Interactive Exploration

Check out the website at **http://www.mhhe.com/environmentalscience** and click on the cover of this textbook for quizzing, career information, case studies, and hot links for the following topics:

General Ecology Sites

Animal Population Ecology

Evolution

Natural Selection

Community Ecology

Competition

Parasitism, Predation, Herbivory

Mutualism

Field Methods for Studies of Populations

Field Methods for Studies of Ecosystems

Species Abundance, Diversity, and Complexity

Food Webs

Nutrient Cycling

Extinction Issues

Biodiversity

Trees of Life

6 CHAPTER

Kinds of Ecosystems and Communities

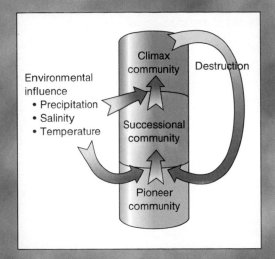

Objectives

After reading this chapter, you should be able to:

- Recognize the difference between primary and secondary succession.
- Describe the process of succession from pioneer to climax community in both terrestrial and aquatic situations.
- Associate typical plants and animals with the various terrestrial biomes.
- Recognize the physical environmental factors that determine the kind of climax community that will develop.
- Differentiate the forest biomes that develop based on temperature and rainfall.
- Describe the various kinds of aquatic ecosystems and the factors that determine their characteristics.

Chapter Outline

Succession

Ecosystems are dynamic, changing units. On a daily basis, plants grow and die, animals feed on plants and on one another, and decomposers recycle the chemical elements that make up the biotic portion of any ecosystem. Abiotic factors (such as temperature, rainfall, intensity of sunlight, and seasonality) also have a major influence on the kind of community that will be established. Since all organisms are linked together in a community, any change in the community affects many organisms within it. Certain conditions within a community are keys to the kinds of organisms that are found associated with one another. Grasshoppers need grass for food, robins need trees to build nests, and herons need shallow water to find food. Each organism has specific requirements that must be met in the community, or it will not survive.

Over long time periods, it is possible to see trends in the way the structure of a community changes and to recognize that climate greatly influences the kind of community that becomes established in an area. Generally, this series of changes eventually results in a relatively long-lasting, stable combination of species that is self-perpetuating. The concept that communities proceed through a series of recognizable, predictable changes in structure over time is called **succession.** The relatively stable, long-lasting community that is the result of succession is called a **climax community.** As ecologists have studied the process of succession, they have come to recognize that the process is not always as predictable as once was thought. However, to begin to understand the process of succession, it often is valuable to look at general models before looking at the exceptions. In the traditional view of succession, the kind of climax community that develops is primarily determined by climate. Some communities will be forests, while others will be grassland or deserts. Succession occurs because the activities of organisms cause changes in their surroundings that make the environment suitable for other kinds of organisms. When new species become established, they compete with the original inhabitants. In some cases, original species may be replaced completely. In other cases, early species may not be replaced but may become less numerous as invading species take a dominant role. Slowly, over time, it is recognized that a significantly different community has become established. Several factors determine the pace and direction of the successional process. For example, climate, locally available seed sources, frequent disturbance, and invasions of organisms from outside the area all greatly affect the course of succession. Ecologists have traditionally recognized two kinds of succession. **Primary succession** is a successional progression that begins with a total lack of organisms and bare mineral surfaces or water. Such conditions can occur when volcanic activity causes lava flows or glaciers scrape away the organisms and soil. Similarly, a lowering of sea level exposes new surfaces for colonization by terrestrial organisms. Primary succession often takes an extremely long time, since there is no soil and few readily available nutrients for plants to use for growth. **Secondary succession** is much more commonly observed and generally proceeds more rapidly, because it begins with the destruction or disturbance of an existing ecosystem. Fire, flood, windstorm, or human activity can destroy or disturb a community of organisms. Usually, however, there is at least some soil and often there are seeds or roots from which plants can begin growing almost immediately.

Primary Succession

Primary succession can begin on a bare rock surface, pure sand, or standing water. Since succession on rock and sand is somewhat different from that which occurs with watery situations, we deal with them separately. We discuss terrestrial succession first.

Terrestrial Primary Succession

Several factors determine the rate of succession and the kind of climax community that will develop in an area. The kind of substrate (rock, sand, clay) will

figure 6.1 **Pioneer Organism** The lichen growing on this rock is able to accumulate bits of debris, carry on photosynthesis, and aid in breaking down the rock. All of these activities contribute to the formation of a thin layer of soil, which is necessary for plant growth in the early stages of succession.

greatly affect the kind of soil that will develop. The kinds of spores, seeds, or other reproductive structures will determine the species available to colonize the area. The climate will determine the species that will be able to live in an area and how rapidly they will grow. The rate of growth will determine how quickly organic matter will accumulate in the soil. The kind of substrate, climate, and amount of organic matter will influence the amount of water available for plant growth. Let's look at a specific example of how these factors are interrelated in an example of primary succession from bare mineral surfaces.

Bare rock or sand is a very inhospitable place for organisms to live. The temperature changes drastically, there is no soil, there is little moisture, the organisms are exposed to the damaging effect of the wind, few nutrients are available, and few places are available for organisms to attach themselves or hide. However, wind-blown spores or other tiny reproductive units of a few kinds of organisms can become established and survive in even this inhospitable environment. This collection of organisms is known as the **pioneer community** because it is the first to colonize bare rock. (See figure 6.1.)

A common kind of organism in this initial community is something called a lichen. Lichens are actually mutualistic

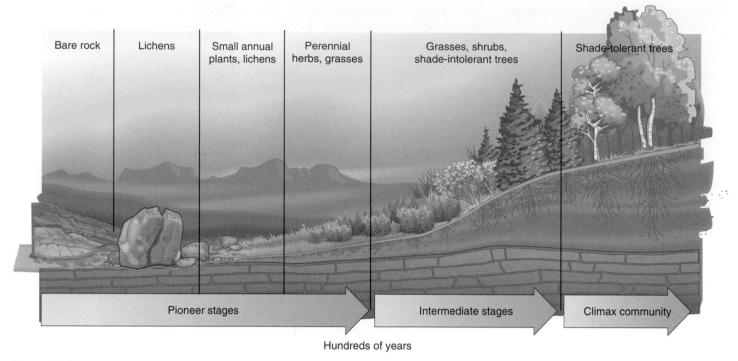

Bare rock | Lichens | Small annual plants, lichens | Perennial herbs, grasses | Grasses, shrubs, shade-intolerant trees | Shade-tolerant trees

Pioneer stages → Intermediate stages → Climax community →

Hundreds of years

figure 6.2 **Primary Succession on Land** The formation of soil is a major step in primary succession. Until soil is formed, the area is unable to support large amounts of vegetation, which modify the harsh environment. Once soil formation begins, the site proceeds through an orderly series of stages toward a climax community.

relationships between two kinds of organisms: algae or bacteria that carry on photosynthesis and fungi that attach to the rock surface and retain water. The growth and development of lichens is often a slow process. It may take lichens 100 years to grow as large as a dinner plate. Lichens are the producers in this simple ecosystem, and many tiny consumer organisms may be found associated with lichens. Some feed on the lichen and many use it as a place of shelter, since even a drizzle is like a torrential rain for a microscopic animal. Since lichens are firmly attached to rock surfaces, they also tend to accumulate bits of airborne debris and store small amounts of water that would otherwise blow away or run off the rock surface. Acids produced by the lichen tend to cause the breakdown of the rock substrate into smaller particles. This fragmentation of rock, aided by physical and chemical weathering processes, along with the trapping of debris and the contribution of organic matter by the death of lichens and other organisms, ulti-

mately leads to the accumulation of a very thin layer of soil.

This thin layer of soil is the key to the next stage in the successional process. The layer can retain some water and support some fungi, certain small worms, insects, bacteria, protozoa, and perhaps a few tiny annual plants that live for only one year but produce flowers and seeds that fall to the soil and germinate the following growing season. Many of these initial organisms or their reproductive structures are very tiny and will arrive as a result of wind and rain. As these organisms grow, reproduce, and die, they contribute additional organic material for the soil-building process, and the soil layer increases in thickness and is better able to retain water. This stage, which is dominated by annual plants, eliminates the lichen community because the plants are taller and shade the lichens, depriving them of sunlight. This annual plant stage is itself replaced by a community of small perennial grasses and herbs. The perennial grasses and herbs are often replaced by larger

perennial woody shrubs, which are often replaced by larger trees that require lots of sunlight, which often are replaced in turn by trees that can tolerate shade. Sun-loving (shade-intolerant) trees are replaced by shade-tolerant trees because the seedlings of shade-intolerant trees cannot grow in the shade of their parents, while seedlings of shade-tolerant trees can. Eventually, a relatively stable, long-lasting, complex, and interrelated climax community of plants, animals, fungi, and bacteria is produced. Each step in this process from pioneer community to climax community is called a **successional stage,** or **seral stage,** and the entire sequence of stages—from pioneer community to climax community—is called a **sere.** (See figure 6.2.)

Although in this example we have described a successional process that began with a lichen pioneer community and ended with a climax forest, it is important to recognize that the process of succession can stop at any point along this continuum. In certain extreme climates, lichen communities may last for

PART TWO Ecological Principles and Their Application

Tens to hundreds of years

figure 6.3 **Primary Succession from a Pond to a Wet Meadow** A shallow pond will fill slowly with organic matter from producers in the pond. Eventually, a wet soil will form and grasses will become established. In many areas, this will be succeeded by a climax forest.

hundreds of years and must be considered climax communities. Others reach a grass-herb stage and proceed no further. The specific kind of climax community produced depends on such things as climate and soil type, which are discussed in greater detail later in this chapter. However, when successional communities are compared to climax communities, climax communities show certain characteristics.

1. Climax communities maintain their mix of species for a long time, while successional communities are temporary.

2. Climax communities tend to have many specialized ecological niches, while successional communities tend to have more generalized niches.

3. Climax communities tend to have many more kinds of organisms and kinds of interactions among organisms than do successional communities.

4. Climax communities tend to recycle nutrients and maintain a relatively constant biomass, while successional communities tend to accumulate large amounts of new material.

The general trend in succession is toward increasing complexity and more efficient use of matter and energy compared to the successional communities that preceded them.

Aquatic Primary Succession

The principal concepts of land succession can be applied to aquatic ecosystems. Except for the oceans, most aquatic ecosystems are considered temporary. Certainly, some are going to be around for thousands of years, but eventually they will disappear and be replaced by terrestrial ecosystems as a result of normal successional processes. All aquatic ecosystems receive a continuous input of soil particles and organic matter from surrounding land, which results in the gradual filling in of shallow bodies of water like ponds and lakes.

In deep portions of lakes and ponds, only floating plants and algae can exist. However, as sediment accumulates, the water depth becomes less, and it becomes possible for certain species of submerged plants to establish their roots in the sediments of the bottom of shallow bodies of water. They carry on photosynthesis, which results in a further accumulation of

organic matter. These plants also tend to trap sediments that flow into the pond or lake from streams or rivers, resulting in a further decrease in water depth. Eventually, as the water becomes shallower, emergent plants become established. They have leaves that float on the surface of the water or project into the air. The network of roots and stems below the surface of the water results in the accumulation of more material, and the water depth decreases as material accumulates on the bottom. As the process continues, a wet soil is formed and grasses and other plants that can live in wet soil become established. This successional stage is often called a wet meadow. The activities of plants tend to draw moisture from the soil, and, as more organic matter is added to the top layer of the soil, it becomes somewhat drier. Once this occurs, the stage is set for a typical terrestrial successional series of changes, eventually resulting in a climax community typical for the climate of the area. (See figure 6.3.)

Since the shallower portions of most lakes and ponds are at the shore, it is often possible to see the various stages in aquatic succession from the shore. In the central, deeper portions of the lake, there are only floating plants and algae. As we approach the shore, we first find submerged plants like *Elodea* and algal mats, then emergent vegetation like water lilies and cattails, then grasses and sedges that can tolerate wet soil, and on the shore, the beginnings of a typical terrestrial succession resulting in the climax community typical for the area.

In many northern ponds and lakes, sphagnum moss forms thick, floating mats. These mats may allow certain plants that can tolerate wet soil conditions to become established. The roots of the plants bind the mat together and establish a floating bog, which may contain small trees and shrubs as well as many other smaller, flowering plants. (See figure 6.4.) Someone walking on such a

figure 6.4 **Floating Bog** In many northern regions, sphagnum moss forms a floating mat that can be colonized by plants that tolerate wet soils. A network of roots ties the mat together to form a floating community.

mat would recognize that the entire system is floating when they noticed the trees sway or when they stepped through a weak zone in the mat and sunk to their hips in water. Eventually, these bogs will become increasingly dry and the normal climax vegetation for the area will succeed the more temporary bog stage.

Secondary Succession

The same processes and activities drive both primary and secondary succession. The major difference is that secondary succession occurs when an existing community is destroyed but much of the soil and some of the organisms remain. A forest fire, a flood, or the conversion of a natural ecosystem to agriculture may be the cause. Since much of the soil remains and many of the nutrients necessary for plant growth are still available, the process of succession can advance more rapidly than primary succession. In addition, because some plants and other organisms may survive the disturbance and continue to grow and others will survive as roots or seeds, they can quickly reestablish themselves in the area. Furthermore, undamaged communities adjacent to the disturbed area can serve as sources of seeds and animals that migrate into the disturbed area. Thus, the new climax community is likely to resemble the one that was destroyed. Figure 6.5 shows old field succession—the typical secondary

succession found on abandoned farmland in the southeastern United States.

Similarly, when beavers flood an area, the existing terrestrial community is replaced by an aquatic ecosystem. As the area behind the dam fills in with sediment and organic matter, it goes through a series of changes that may include floating plants, submerged plants, emergent plants, and wet meadow stages, but it eventually returns to the typical climax community for the area. (See figure 6.6.)

Many kinds of communities exist only as successional stages and are continually reestablished following disturbances. Many kinds of woodlands along rivers exist only where floods remove vegetation, allowing specific species to become established on the disturbed flood plain. Some kinds of forest and shrub communities exist only if fire occasionally destroys the mature forest. Windstorms such as hurricanes are also important in causing openings in forests that allow the establishment of certain kinds of plant communities.

Modern Concepts of Succession and Climax

The discussion of the nature of succession and climax communities in the "Succession" section is an oversimplification of the true nature of the process. Some historical perspective will help to

Mature oak/hickory forest destroyed	Farmland abandoned	Annual plants	Grasses and biennial herbs	Perennial herbs and shrubs begin to replace grasses and biennials	Pines begin to replace shrubs	Young oak and hickory trees begin to grow	Pines die and are replaced by mature oak and hickory trees	Mature oak/hickory forest
		1–2 years	**3–4 years**	**4–15 years**	**5-15 years**	**10–30 years**	**50–75 years**	

figure 6.5 **Secondary Succession on Land** A plowed field in the southeastern United States shows a parade of changes over time, involving plant and animal associations. The general pattern is for annual weeds to be replaced by grasses and other perennial herbs, which are replaced by shrubs, which are replaced by trees. As the plant species change, so do the animal species.

PART TWO Ecological Principles and Their Application

clarify how ecologists have changed their concept of successional change. When European explorers traveled across the North American continent, they saw huge expanses of land dominated by specific types of communities: hardwood forests in the East, evergreen forests in the North, grasslands in central North America, and deserts in the Southwest. These regional communities came to be considered the steady-state or normal situation for those parts of the world. When ecologists began to explore the way in which ecosystems developed over time, they began to think of these ecosystems as the end point or climax of a long journey, beginning with the formation of soil and its colonization by a variety of plants and other organisms.

As settlers removed the original forests or grasslands and converted the land to farming, the original "climax" community was destroyed. Eventually, as poor farming practices destroyed the soil, many farms were abandoned and the land was allowed to return to its

"original" condition. This secondary succession often resulted in forests that resembled those that had been destroyed. However, in most cases, these successional forests contained fewer species and in some cases were entirely different kinds of communities from the originals. These new stable communities were also called climax communities, but they were not the same as the original climax communities.

In addition, the introduction of species from Europe and other parts of the world changed the mix of organisms that might colonize an area. Many grasses and herbs that were introduced either on purpose or accidentally have become well established. Today, some communities are dominated by these introduced species. Even diseases have altered the nature of climax communities. Chestnut blight and Dutch elm disease have removed tree species that were at one time dominant species in certain plant communities.

Ecologists began to recognize that there was no fixed, predetermined com-

munity for each part of the world and began to modify the way they looked at the concept of climax communities. The concept today is a more plastic one. It is still used to talk about a stable stage following a period of change, but ecologists no longer feel that land will eventually return to a "preordained" climax condition. They have also recognized in recent years that the type of climax community that develops depends on many factors other than simply climate. One of these is the availability of seeds to colonize new areas. Some seeds may lie dormant in the soil for a decade or more, while others may be carried to an area by wind, water, or animals. Two areas with very similar climate and soil characteristics may develop very different successional and "climax" communities because of the seeds that were present in the area when the lands were released from agriculture. Furthermore, we need to recognize that the only thing that differentiates a "climax" community from a successional one is the time scale over which change occurs. "Climax" communities do not

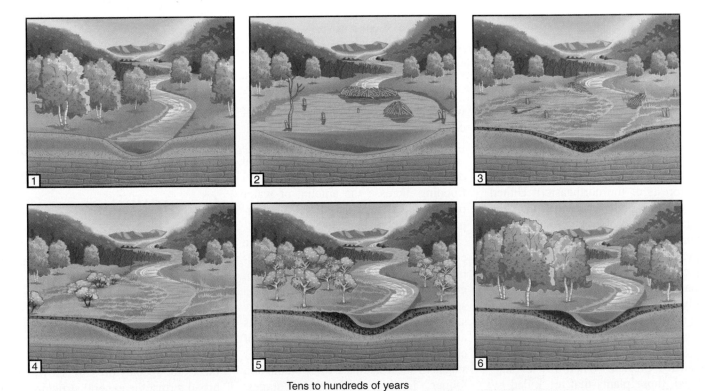

Tens to hundreds of years

figure 6.6 **Secondary Succession from a Beaver Pond** A colony of beavers can dam up streams and kill trees by the flooding that occurs and by using trees for food. Once the site is abandoned, it will slowly return to the original forest community by a process of succession.

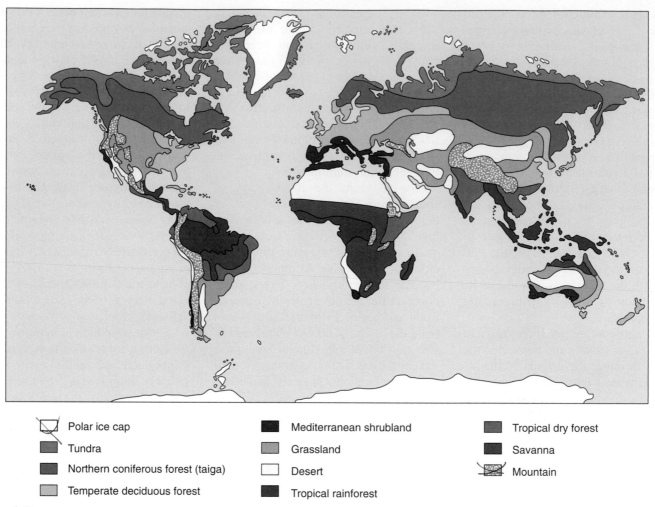

Polar ice cap	Mediterranean shrubland	Tropical dry forest
Tundra	Grassland	Savanna
Northern coniferous forest (taiga)	Desert	Mountain
Temperate deciduous forest	Tropical rainforest	

figure 6.7 **Biomes of the World** Although most biomes are named for a major type of vegetation, each includes a specialized group of animals adapted to the plants and the biome's climatic conditions.

change as rapidly as successional ones. However, all communities are eventually replaced, as were the swamps that produced coal deposits, the preglacial forests of Europe and North America, and the pine forests of the northeastern United States. Many human activities alter the nature of the successional process. Agricultural practices obviously modify the original community to allow for the raising of crops. However, several other management practices have also significantly altered communities. Regular logging returns a forest to an earlier stage of succession. The suppression of fire in many forests has also changed the mix of organisms present. When fire is suppressed, those plants that are killed by regular fires become more common and those that are able to resist fire become less common. Changing the amount of

water present will also change the kind of community. Draining an area makes it less suitable for the original inhabitants and more suitable for those that live in drier settings. Similarly, irrigation and flooding increase the amount of water present and change the kinds of organisms that can live in an area.

So what should we do with these concepts of succession and climax communities? Although the climax concept embraces a false notion that there is a specific end point to succession, it is still important to recognize that there is an identifiable, predictable pattern of change during succession and that later stages in succession are more stable and longer lasting than early stages. Whether we call a specific community of organisms a climax community is not really important.

Biomes: Major Types of Terrestrial Climax Communities

Biomes are terrestrial climax communities with wide geographic distribution. (See figure 6.7.) Although the concept of biomes is useful for discussing general patterns and processes, it is important to recognize that when different communities within a particular biome are examined, there will be variations in the exact species present. However, in broad terms, the general structure of the ecosystem and the kinds of niches and habitats present are similar. Two primary nonbiological factors have major

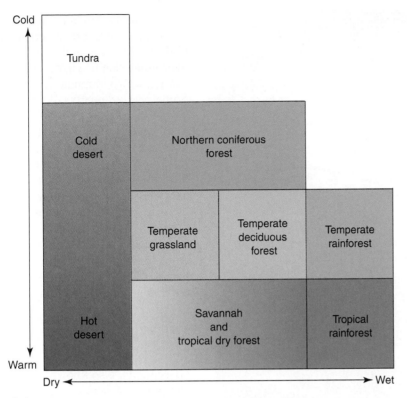

figure 6.8 **Influence of Precipitation and Temperature on Vegetation** Temperature and moisture are two major factors that influence the kind of vegetation that can occur in an area. Areas with low moisture and low temperatures produce tundra; areas with high moisture and freezing temperatures during part of the year produce deciduous or coniferous forests; dry areas produce deserts; moderate amounts of rainfall or seasonal rainfall support grasslands or savannas; and areas with high rainfall and high temperatures support tropical rainforests.

impacts on the kind of climax community that develops in any part of the world: precipitation and temperature. Several aspects of precipitation are important: the total amount of precipitation per year, the form in which it arrives (rain, snow, sleet), and its seasonal distribution. Precipitation may be evenly spaced throughout the year, or it may be concentrated at particular times so that there are wet and dry seasons.

The temperature patterns are also important and can vary considerably in different parts of the world. Tropical areas have warm, relatively unchanging temperatures throughout the year. Areas near the poles have long winters with extremely cold temperatures and relatively short, cool summers. Other areas are more evenly divided between cold and warm periods of the year. (See figure 6.8.)

Although temperature and precipitation are of primary importance, several other factors may influence the kind of

climax community present. Periodic fires are important in maintaining some grassland and shrub climax communities because the fires prevent the establishment of larger, woody species. Some parts of the world have frequent, strong winds that prevent the establishment of trees and cause rapid drying of the soil. The type of soil present is also very important. Sandy soils tend to dry out quickly and may not allow the establishment of more water-demanding species like trees, while extremely wet soils may allow only certain species of trees to grow. Obviously, the kinds of organisms currently living in the area are also important, since their offspring will be the ones available to colonize a new area.

The Effect of Elevation on Climate and Vegetation

The distribution of terrestrial ecosystems is primarily related to precipitation and temperature. The temperature is

warmest near the equator and becomes cooler toward the poles. Similarly, as the height above sea level increases, the average temperature decreases. This means that even at the equator, it is possible to have cold temperatures on the peaks of tall mountains. As one proceeds from sea level to the tops of mountains, it is possible to pass through a series of biomes that are similar to what would be encountered as one traveled from the equator to the North Pole. (See figure 6.9.)

In the next sections, we will look at the major biomes of the world and highlight the abiotic and biotic features typical of each biome.

Desert

A lack of water is the primary factor that determines that an area will be a desert. **Deserts** are areas that generally average less than 25 centimeters (10 inches) of precipitation per year. (See figure 6.10.) When and how precipitation arrives is quite variable in different deserts. Some deserts receive most of the moisture as snow or rain in the winter months, while in others rain comes in the form of thundershowers at infrequent intervals. If rain comes as heavy thundershowers, much of the water does not sink into the ground but runs off into gullies. Also, since the rate of evaporation is high, plant growth and flowering usually coincide with the periods when moisture is available. Deserts are also likely to be windy. We often think of deserts as hot, dry wastelands devoid of life. However, many deserts are quite cool during a major part of the year. Certainly, the Sahara Desert and the deserts of the southwestern United States and Mexico are hot during much of the year, but the desert areas of the northwestern United States and the Gobi Desert in Central Asia can be extremely cold during winter months and have relatively cool summers. Furthermore, the temperature can vary greatly during a 24-hour period. Since deserts receive little rainfall, it is logical that most will have infrequent cloud cover. With no clouds to block out the sun, during the day the soil surface and the air above it tend to heat up rapidly.

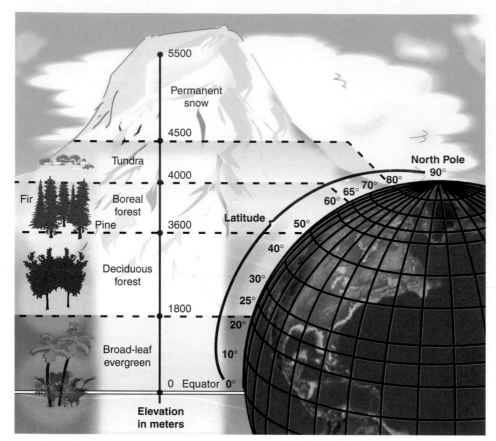

figure 6.9 **Relationship Between Height above Sea Level, Latitude, and Vegetation**
As one travels up a mountain, the climate changes. The higher the elevation, the cooler the climate. Even in the tropics, tall mountains can have snow on the top. Thus, it is possible to experience the same change in vegetation by traveling up a mountain as one would experience traveling from the equator to the North Pole.

After the sun has set, the absence of an insulating layer of clouds allows heat energy to be reradiated from the Earth, and the area cools off rapidly. Cool to cold nights are typical even in "hot" deserts, especially during the winter months.

Another misconception about deserts is that few species of organisms live in the desert. There are many species, but they typically have low numbers of individuals. For example, a conspicuous feature of deserts is the dispersed nature of the plants. There is a significant amount of space between them. Similarly, animals do not have large, dense populations. However, those species that are present are specially adapted to survive in dry, often hot environments. For example, water evaporates from the surfaces of leaves. As an adaptation to this condition, many desert plants have very small leaves that

allow them to conserve water. Some even lose their leaves entirely during the driest part of the year. Some, like cactus, have the ability to store water in their spongy bodies or their roots for use during drier periods. Other plants have parts or seeds that lie dormant until the rains come. Then they grow rapidly, reproduce, and die, or become dormant until the next rains. Even the perennial plants are tied to the infrequent rains. During these times, the plants are most likely to produce flowers and reproduce. Many desert plants are spiny. The spines discourage large animals from eating the leaves and young twigs.

The desert has many kinds of animals. However, they are often overlooked because their populations are low, numerous species are of small size, and many are inactive during the hot part of the day. They also aren't seen in large, conspicuous groups. Many insects,

lizards, snakes, small mammals, grazing mammals, carnivorous mammals, and birds are common in desert areas. All of the animals that live in deserts are able to survive with a minimal amount of water. Some receive nearly all of their water from the moisture in the food they eat. They generally have an outer skin or cuticle that resists water loss, so they lose little water by evaporation. They also have physiological adaptations, such as extremely efficient kidneys, that allow them to retain water. They often limit their activities to the cooler part of the day (the evening) and may spend considerable amounts of time in underground burrows during the day, which allows them to avoid extreme temperatures and to conserve water.

Grassland

Grasslands, also known as **prairies** or **steppes,** are widely distributed over temperate parts of the world. As with deserts, the major factor that contributes to the establishment of a grassland is the amount of available moisture. Grasslands generally receive between 25 and 75 centimeters (10 to 30 inches) of precipitation per year. These areas are windy with hot summers and cold-to-mild winters. In many grasslands, fire is an important force in preventing the invasion of trees and releasing nutrients from dead plants to the soil. Grasses make up 60 to 90 percent of the vegetation. Many other kinds of flowering plants are interspersed with the grasses. (See figure 6.11.) Typically, the grasses and other plants are very close together and their roots form a network that binds the soil together. Trees, which generally require greater amounts of water, are rare in these areas except along watercourses.

The primary consumers are animals that eat the grasses, such as large herds of migratory, grazing mammals like bison, wildebeests, wild horses, and various kinds of sheep, cattle, and goats. While the grazers are important as consumers of the grasses, they also supply fertilizer from their dung and discourage invasion by woody species of plants because they eat the young shoots.

City: Cairo, Egypt
Latitude: 29° 52′N
Altitude: 116 m (381 ft.)
Yearly precipitation: 1.8 cm (0.7 in.)

Climate name: Hot desert
Other cities with similar climates:
Mecca, Karachi

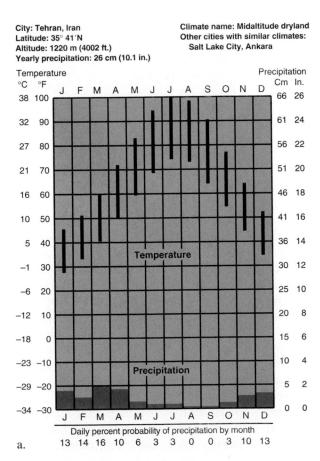

Daily percent probability of precipitation by month

J	F	M	A	M	J	J	A	S	O	N	D
3	3	3	0	0	0	0	0	0	0	3	3

a.

b.

figure 6.10 **Desert** (a) Climagraph for Cairo, Egypt. (b) The desert receives less than 25 centimeters (10 inches) of precipitation per year, yet it teems with life. Cactus, sagebrush, lichens, snakes, small mammals, birds, and insects inhabit the desert. Because daytime temperatures are often high, most animals are active only at night, when the air temperature drops significantly. Cool deserts also exist in many parts of the world, where rainfall is low but temperatures are not high.

City: Tehran, Iran
Latitude: 35° 41′N
Altitude: 1220 m (4002 ft.)
Yearly precipitation: 26 cm (10.1 in.)

Climate name: Midaltitude dryland
Other cities with similar climates:
Salt Lake City, Ankara

Daily percent probability of precipitation by month

J	F	M	A	M	J	J	A	S	O	N	D
13	14	16	10	6	3	3	0	0	3	10	13

a.

b.

figure 6.11 **Grassland** (a) Climagraph for Tehran, Iran. (b) Grasses are better able to withstand low water levels than are trees. Therefore, in areas that have moderate rainfall, grasses are the dominant plants.

environmental CLOSE-UP

Grassland Succession

Because there are many kinds of grasslands, it is difficult to generalize about how succession takes place in these areas. Most grasslands in North America have been heavily influenced by agriculture and the grazing of domesticated animals. The grasslands reestablished in these areas may be quite different from the original ecosystem. However, there appear to be several stages typically involved in grassland succession.

After land is abandoned from cultivation, a short period of one to three years elapses in which the field is dominated by annual broad-leaf weeds. In this respect, grassland succession is like deciduous forest succession. The next stage varies in length (10 or more years) and is dominated by annual grasses. Usually, in these early stages, the soil is in poor condition, lacking organic matter and nutrients. After several years, the soil fertility increases as organic material accumulates from the death and decay of annual grasses. This leads to the next stage in development, perennial grasses. Eventually, a mature grassland develops as prairie flowers invade the area and become interspersed with the grasses. In general, throughout this sequence, the soil becomes more fertile and of higher quality.

Because so much of the original North American grassland has been used for agriculture, when the land is allowed to return to a prairie, there may not be seeds of all of the original plants native to the area. Thus, the grassland that results from secondary succession may not be exactly like the original; some species may be missing. Consequently, in many managed restorations of prairies, seeds that are no longer available in the local soils are introduced from other sources.

The low amount of rainfall and the fires typical of grasslands generally cause the successional process to stop at this point. However, if more water becomes available or if fire is prevented, woody trees may invade moist sites.

Actively farmed

Recently abandoned

Several years of succession

In addition to grazing mammals, many kinds of insects, including grasshoppers and other herbivorous insects, dung beetles (which feed on the dung of grazing animals), and several kinds of flies are common. Some of these flies bite to obtain blood. Others lay their eggs in the dung of large mammals. Some feed on dead animals and lay their eggs in carcasses. Small herbivorous mammals, such as mice and ground squirrels, are also common. Birds are often associated with grazing mammals. They eat the insects stirred up by the mammals or feed on the insects that bite them. Other birds feed on seeds and other plant parts. Reptiles (snakes and lizards) and other carnivores like coyotes, foxes, and hawks feed on small mammals and insects.

Most of the moist grasslands of the world have been converted to agriculture, since the rich, deep soil that developed as a result of the activities of centuries of soil building is useful for growing cultivated grasses like corn (maize) and wheat. The drier grasslands have been converted to the raising of domesticated grazers like cattle, sheep, and goats. Therefore, little undisturbed grassland is left, and those fragments that remain need to be preserved as refuges for the grassland species that once occupied huge portions of the globe.

Savanna

Tropical parts of Africa, South America, and Australia have extensive grasslands spotted with occasional trees or patches of trees. (See figure 6.12.) This kind of a biome is often called a **savanna.** Although savannas receive 50 to 150 centimeters (20 to 60 inches) of rain per year, the rain is not distributed evenly throughout the year. Typically, a period of heavy rainfall is followed by a prolonged drought. This results in a very seasonally structured ecosystem. The plants and animals time their reproductive activities to coincide with the rainy period, when limiting factors are least

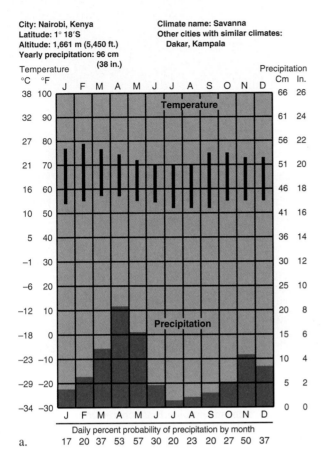

City: Nairobi, Kenya
Latitude: 1° 18′S
Altitude: 1,661 m (5,450 ft.)
Yearly precipitation: 96 cm
(38 in.)

Climate name: Savanna
Other cities with similar climates:
Dakar, Kampala

Daily percent probability of precipitation by month
17 20 37 53 57 30 20 23 20 27 50 37

a.

b.

figure 6.12 **Savanna** (a) Climagraph for Nairobi, Kenya. (b) Savannas develop in tropical areas that have seasonal rainfall. They typically have grasses as the dominant vegetation with drought- and fire-resistant trees scattered through the area.

confining. The predominant plants are grasses, but many drought-resistant, flat-topped, thorny trees are common. As with grasslands, fire is a common feature of the savanna and the trees present are resistant to fire damage. Many of these trees are particularly important because they are legumes that are involved in nitrogen fixation. They also provide shade and nesting sites for animals. As with grasslands, the predominant mammals are the grazers. Wallabies in Australia, wildebeests, zebras, elephants, and various species of antelope in Africa, and capybaras (rodents) in South America are examples. In Africa, the large herds of grazing animals provide food for many different kinds of large carnivores (lions, hyenas, leopards). Many kinds of rodents, birds, insects, and reptiles are associated with this biome. Among the insects, mound-building termites are particularly common.

Mediterranean Shrublands (Chaparral)

The **Mediterranean shrublands** are located near an ocean and have wet, cool winters and hot, dry summers. Rainfall is 40 to 100 centimeters (15 to 40 inches) per year. As the name implies, this biome is typical of the Mediterranean coast and is also found in coastal southern California, the southern tip of Africa, a portion of the west coast of Chile, and southern Australia. The vegetation is dominated by woody shrubs that are adapted to withstand the hot, dry summer. (See figure 6.13.) Often the plants are dormant during the summer. Fire is a common feature of this biome, and the shrubs are adapted to withstand occasional fires. The kinds of animals vary widely in the different regions of the world with this biome. Many kinds of insects, reptiles, birds, and mammals

are found in these areas. In the chaparral of California, rattlesnakes, spiders, coyotes, lizards, and rodents are typical inhabitants.

Tropical Dry Forest

Another biome that is heavily influenced by seasonal rainfall is known as the **tropical dry forest.** Many of the tropical dry forests have a monsoon climate in which several months of heavy rainfall are followed by extensive dry periods ranging from a few to as many as eight months. (See figure 6.14.) The rainfall may be as low as 50 centimeters (20 inches) or as high as 200 centimeters (80 inches), but since the rainfall is highly seasonal, many of the plants have special adaptations for enduring drought. In many of the regions that have extensive dry periods, many of the trees drop their leaves during the dry period. Tropical dry

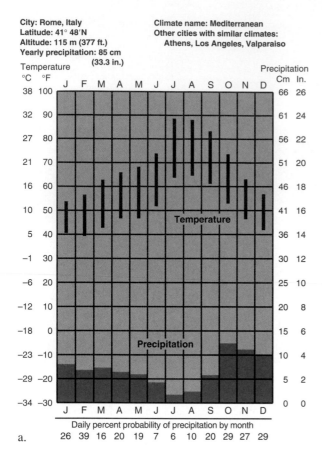

City: Rome, Italy
Latitude: 41° 48′N
Altitude: 115 m (377 ft.)
Yearly precipitation: 85 cm
(33.3 in.)

Climate name: Mediterranean
Other cities with similar climates:
Athens, Los Angeles, Valparaiso

Daily percent probability of precipitation by month
26 39 16 20 19 7 6 10 20 29 27 29

a.

b.

figure 6.13 Mediterranean Shrubland (a) Climagraph for Rome, Italy. (b) Mediterranean shrublands are characterized by a period of winter rains and a dry, hot summer. The dominant plants are drought-resistant, woody shrubs.

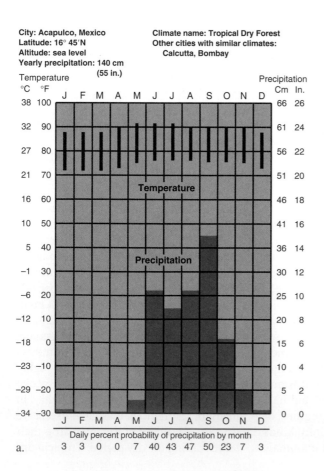

City: Acapulco, Mexico
Latitude: 16° 45′N
Altitude: sea level
Yearly precipitation: 140 cm
(55 in.)

Climate name: Tropical Dry Forest
Other cities with similar climates:
Calcutta, Bombay

Daily percent probability of precipitation by month
3 3 0 0 7 40 43 47 50 23 7 3

a.

b.

figure 6.14 Tropical Dry Forest (a) Climagraph for Acapulco, Mexico. (b) Tropical dry forests typically have a period of several months with no rain. In places where the drought is long, many of the larger trees lose their leaves.

PART TWO   Ecological Principles and Their Application

Tropical Rainforests—A Special Case?

Today, there is considerable political and economic interest in how tropical rainforests are used. Some would preserve them in their current state, while others would use the trees and other forest resources for economic gain. However, since tropical rainforests are located in countries in which there are large numbers of poor people, there are strong pressures to exploit forests for short-term economic benefit. Most economic uses of the rainforest result in its destruction or reduce its biodiversity.

Two hundred years ago, tropical rainforests covered about 1500 million hectares (3700 million acres), an area the size of Europe, but today, only 900 million hectares (2200 million acres) remain. Modern technology makes short work of clearing the rainforest, taking less than an hour to clear 1 hectare (2.47 acres). Exactly how much rainforest is disappearing is not known, but it seems likely that over 20 million hectares (50 million acres) are destroyed each year. At this rate, there will be no rainforests remaining in 50 years. The causes of deforestation are easy to identify.

Logging

Rainforests were spared from exploitation in earlier years because of their inaccessibility, the relative low value of most of the trees for timber purposes, and the limited world demand. Recently, this situation has changed and a wide variety of tree species previously considered worthless are now used for pulp or as cellulose for the production of plastics. With new machines and better transportation, it has become profitable to remove trees from previously remote areas. Often logging companies are interested only in one or two hardwood species, such as teak or mahogany. As there may be only three suitable trees in a hectare, to remove them may not seem a threat. However, heavy machinery is needed to clear a path, and when the tree is felled and dragged through the forest, the amount of damage to other trees can be enormous. Faced with a high demand for their forest products, most countries with rainforests have been willing to sign over timber rights to foreign companies, hoping thereby to increase their national incomes. Unfortunately, most of these timber contracts contain few or no provisions for conservation.

Farming

Poor people seeking farmland make use of roads built by logging and mining companies to gain access to previously remote areas. Government policies often encourage people to settle in logged areas, even though these areas are usually unsuitable for growing crops. The settlers establish a shifting form of agriculture in which the trees are cut down and burned. They grow vegetables in the clearing for two to three years until the soil is depleted of its nutrients, then abandon the clearing and move on to another place in the rainforest. The activities of settlers prevent succession from occurring, so rainforest is permanently lost.

Ranching

Large areas of rainforest are being burned down and converted into ranchland, even though this is the worst possible use of the land. Central America has already lost two-thirds of its rainforest to cattle ranching. Where tropical rainforests have been cleared for pasture and cattle ranches established, the production of meat hardly reaches 50 kilograms of meat/hectare/year (45 lb/acre/year) whereas North American farms produce more than 600 kilograms of meat/hectare/year (535 lb/acre/year). The soil quickly loses its scarce nutrients and becomes useless. When this happens, the land is abandoned and more rainforest is cleared.

Mining

Many rainforests are rich in oil deposits and mineral reserves such as bauxite, coal, copper, diamonds, gold, iron ore, nickel, tin, and uranium. While mining is a minor cause of deforestation, still land is cleared for access and the mines are often in areas that are the only habitats of certain plant and animal species. The small amount of deforestation can cause a disproportionate amount of damage, and the mining often releases toxic wastes into rivers, destroying the animal and plant life.

Biodiversity Resources

Tropical rainforests contain an amazing array of different kinds of organisms. Many kinds of foods and other useful products are derived from rainforest plants. Every time we drink coffee, eat chocolate, bananas, or nuts, or use anything made of rubber, we are using products originally discovered in the rainforests.

The following list points out the value of the diverse kinds of plants that live in the tropics.

Fruit: pineapples, bananas, oranges, lemons, limes, grapefruit, and tangerines

Vegetables: tomatoes, avocados, and many types of beans

Nuts: Brazil nuts, cashew nuts, peanuts, sesame seeds, and coconuts

Spices: chilies, pepper, cloves, nutmeg, vanilla, cinnamon, and turmeric

Drinks: tea, coffee, cocoa, chocolate, kola nut extract used in Coca-Cola

Gums and resins: rubber for household and industrial goods, copals for paints and varnishes, chicle gum used for chewing gum, balata used in golf balls

One of the greatest potential benefits of rainforests is the possibility of discovering new drugs. Plants contain many substances known as phytochemicals that help them deter insects from eating their leaves or have other value to the plant. These same chemicals often have use as drugs. There is a one-in-four chance that the next time you enter a pharmacy for a prescription, you will leave with a product derived from the rainforest. Seventy percent of plants identified

continued on next page

—continued

as having anticancer properties come from the rainforests. In fact, almost 50 percent of our medications are derived from plants, yet less than 1 percent of tropical rainforest species have been examined for their possible value in medicine.

Global Rainforest Services

Many people value rainforests for the services they provide. Large expanses of rainforest alter weather conditions, protect soil from erosion, and store large amounts of carbon. This carbon-storing function is particularly important as we recognize we are adding large amounts of carbon to the atmosphere as we burn fossil fuels. While these may be valuable services, it is hard to put a monetary value on them. Therefore, in most cases, the short-term economic benefits of logging and agriculture tend to outweigh the long-term biodiversity and service values provided by rainforests.

forests are found in parts of Central and South America, Australia, Africa, and Asia (particularly India and Myranmar). Many of the species of animals found here are also found in more moist tropical forests of the region. However, there are fewer kinds in dry forests than in rainforests.

Tropical Rainforest

Tropical rainforests are located near the equator in Central and South America, Africa, Southeast Asia, and some islands in the Caribbean Sea and Pacific Ocean. (See figure 6.15.) The temperature is normally warm and relatively constant. There is no frost, and it rains nearly every day. Most areas receive in excess of 200 centimeters (80 inches) of rain per year. Some receive 500 centimeters (200 inches) or more. Because of the warm temperatures and abundant rainfall, most plants grow very rapidly; however, soils are usually poor in nutrients because water tends to carry away any nutrients not immediately taken up by plants. Many of the trees have extensive root networks, associated with fungi (mycorrhiza), near the surface of the soil that allow them to capture nutrients from decaying vegetation before the nutrients can be carried away. Because most of the nutrients in a tropical rainforest are tied up in the biomass, not in the soil, these areas do not make good farmland. Rainfall is a source of new nutrients since atmospheric particles and gases dissolve as the rain falls. The canopy contains many kinds of epiphytic plants (plants that live on the surface of other plants) that trap many of these nutrients in the canopy before they can reach the soil. Many understory species are vines that attach themselves to the tall trees as they grow toward the sun. When the vines reach the canopy, they can compete effectively with their supporting tree for available sunlight. In addition to supporting various vines, each tree serves as a surface for the growth of ferns, mosses, and orchids.

Tropical rainforests have a greater diversity of species than any other

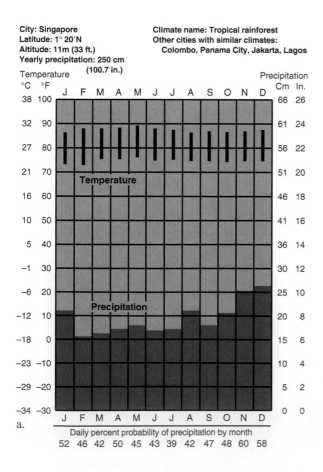

City: Singapore
Latitude: 1° 20′N
Altitude: 11m (33 ft.)
Yearly precipitation: 250 cm (100.7 in.)

Climate name: Tropical rainforest
Other cities with similar climates:
Colombo, Panama City, Jakarta, Lagos

Daily percent probability of precipitation by month
52 46 42 50 45 43 39 42 47 48 60 58

a.

b.

figure 6.15 **Tropical Rainforest** (a) Climagraph for Singapore. (b) Tropical rainforests develop in areas with high rainfall and warm temperatures. They have an extremely diverse mixture of plants and animals.

PART TWO Ecological Principles and Their Application

Forest Canopy Studies

In the past few years, a new frontier of ecological study has developed. Scientists have traditionally looked at forest ecosystems at ground level. Trees were identified, understory species were categorized, and the animals that live on or near the forest floor were studied. Gradually, scientists began to realize that many kinds of plants and animals that are important parts of forest ecosystems rarely descend from the tops of trees to the forest floor. They began to devise methods for studying these canopy-dwelling organisms. Initially, they relied on techniques and gear used by mountain climbers, but that approach was labor intensive and dangerous. Recently, ecologists have established several sites where large construction cranes have been built in forests. These cranes allow researchers to study the chemistry, climate, and organisms found in the canopy.

Several surprising discoveries have been made, including the discovery of new species of insects. A canopy study in an old-growth temperate rainforest on Vancouver Island, British Columbia, identified more than 60 new species of insects. Researchers estimate that hundreds of new species will eventually be identified. Similar studies in tropical forests have identified hundreds more new species of insects. Studies in Panama recognized that vines (lianas) can constitute up to 70 percent of the forest canopy and that these vines compete for sunlight with the trees that support them.

The way in which animals use the canopy is another interesting area of research. Birds, bats, monkeys, squirrels, and insects use specific parts of the forest canopy for food, nesting, and travel routes. Researchers are learning much as they spend time in the forest canopy with the animals.

biome. More species are found in the tropical rainforests of the world than in the rest of the world combined. A small area of a few square kilometers is likely to have hundreds of species of trees. Furthermore, it is typical to have distances of a kilometer or more between two individuals of the same species. Balsa, teakwood, and many other ornamental woods are from tropical trees. Each of those trees is home to a set of animals and plants that use it as food, shelter, or support. The canopy, which forms a solid wall of leaves between the sun and the forest floor, consists of two or three levels. A few trees, called emergent trees, protrude above the canopy. Below the canopy is a layer of understory tree species. Recently, biologists discovered a whole new community of organisms that live in the canopy of these forests. Since most of the sunlight is captured by the trees, only shade-tolerant plants live beneath the trees' canopy.

Associated with this variety of plants is an equally large variety of animals. Insects, such as ants, termites, moths, butterflies, and beetles, are particularly abundant. Birds also are extremely common, as are many climbing mammals, lizards, and tree frogs. The insects are food for many of these species. Since flowers and fruits are available throughout the year, there are many kinds of nectar- and fruit-feeding birds and mammals. Their activities are important in pollination and spreading seeds throughout the forest. Because of the low light levels and the difficulty of maintaining visual contact with one another, many of the animals communicate by making noise.

Tropical rainforests are under intense pressure from logging and agriculture. Many of the countries where tropical rainforests are present are poor and seek to obtain jobs and money by exploiting this resource. Generally, agriculture has not been successful because the soils have few nutrients and cannot withstand constant agricultural activity. Forestry can be a sustainable activity, but in many cases, it is not. The forests are being cut down with no effort to protect them for long-term productivity.

Temperate Deciduous Forest

Forests in temperate areas of the world that have a winter–summer change of seasons typically have trees that lose their leaves during the winter and replace them the following spring. This kind of forest is called a **temperate deciduous forest** and is typical of the eastern half of the United States, parts of south central and southeastern Canada,

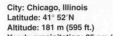

City: Chicago, Illinois
Latitude: 41° 52'N
Altitude: 181 m (595 ft.)
Yearly precipitation: 85 cm (33.3 in.)

Climate name: Humid continental (warm summer)
Other cities with similar climates:
    New York, Berlin, Warsaw

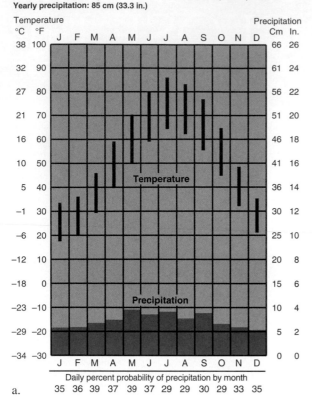

Daily percent probability of precipitation by month
35 36 39 37 39 37 29 29 30 29 33 35

a.

b.

figure 6.16 **Temperate Deciduous Forest** (a) Climagraph for Chicago, Illinois. (b) A temperate deciduous forest develops in areas that have significant amounts of moisture throughout the year but where the temperature falls below freezing for parts of the year. During this time, the trees lose their leaves. This kind of forest once dominated the eastern half of the United States and southeastern Canada.

southern Africa, and many areas of Europe and Asia.

These areas generally receive 75 to 100 centimeters (30 to 60 inches) of relatively evenly distributed precipitation per year. The winters are relatively mild, and plants are actively growing for about half the year. Each area of the world has certain species of trees that are the major producers for the biome. (See figure 6.16.) In contrast to tropical rainforests, where individuals of a tree species are scattered throughout the forest, temperate deciduous forests generally have many fewer species, and many forests may consist of two or three dominant tree species. In deciduous forests of North America and Europe, common species are maples, aspen, birch, beech, oaks, and hickories. These tall trees shade the forest floor, where many small flowering plants bloom in the spring. These spring wildflowers store food in underground structures. In the spring, before the leaves come out on the trees, the wildflowers can capture sunlight and reproduce before they are shaded. Many

smaller shrubs also are found in the understory of these forests.

These forests are home to a great variety of insects, many of which use the leaves and wood of trees as food. Beetles, moth larvae, wasps, and ants are examples. The birds that live in these forests are primarily migrants that arrive in the spring of the year, raise their young during the summer, and leave in the fall. Many of these birds rely on the large summer insect population for their food. Others use the fruits and seeds that are produced during the summer months. A few kinds of birds, including woodpeckers, grouse, turkeys, and some finches, are year-round residents. Amphibians (frogs, toads, salamanders) and reptiles (snakes and lizards) prey on insects and other small animals. Several kinds of small and large mammals inhabit these areas. Mice, squirrels, deer, shrews, moles, and opossums are common examples. Major predators on these mammals are foxes, badgers, weasels, coyotes, and birds of prey.

Taiga, Northern Coniferous Forest, or Boreal Forest

Throughout the southern half of Canada, parts of northern Europe, and much of Russia, there is an evergreen coniferous forest known as the **taiga, northern coniferous forest,** or **boreal forest.** (See figure 6.17.) The climate is one of short, cool summers and long winters with abundant snowfall. The winters are extremely harsh and can last as long as six months. Typically, the soil freezes during the winter. Precipitation ranges between 25 and 100 centimeters (10 to 40 inches) per year. However, the climate is typically humid because there is a great deal of snow melt in the spring and generally low temperatures reduce evaporation. The landscape is typically dotted with lakes, ponds, and bogs.

Conifers such as spruces, firs, and larches are the most common trees in these areas. These trees are specifically adapted to winter conditions. Winter is relatively dry as far as the trees are concerned because the moisture falls as

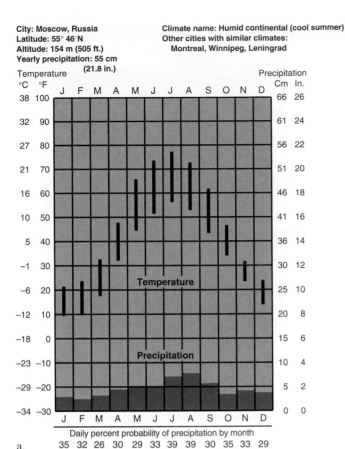

City: Moscow, Russia
Latitude: 55° 46′N
Altitude: 154 m (505 ft.)
Yearly precipitation: 55 cm
 (21.8 in.)

Climate name: Humid continental (cool summer)
Other cities with similar climates:
 Montreal, Winnipeg, Leningrad

Temperature
°C °F

Precipitation
Cm In.

Temperature

Precipitation

Daily percent probability of precipitation by month
35 32 26 30 29 33 39 39 30 35 33 29

a.

b.

figure 6.17   **Taiga, Northern Coniferous Forest, or Boreal Forest** (a) Climagraph for Moscow. (b) The taiga, northern coniferous forest, or boreal forest occurs in areas with long winters and heavy snowfall. The trees have adapted to these conditions and provide food and shelter for the animals that live there.

snow and stays above the soil until it melts in the spring. The needle-shaped leaves are adapted to prevent water loss; in addition, the larches lose their needles in the fall. The branches of these trees are flexible, allowing them to bend under a load of snow so that the snow slides off the pyramid-shaped trees without greatly damaging them. As with the temperate deciduous forest, many of the inhabitants of this biome are temporarily active during the summer. Most birds are migratory and feed on the abundant summer insect population, which is not available during the long, cold winter. A few birds, such as woodpeckers, owls, and grouse, are permanent residents. Typical mammals are deer, caribou, moose, wolves, weasels, mice, snowshoe hares, and squirrels. Because of the cold, few reptiles and amphibians live in this biome.

Tundra

North of the taiga is the **tundra,** a biome that lacks trees and has a permanently frozen subsurface soil. This frozen soil layer is known as **permafrost.** (See figure 6.18.) Because of the permanently frozen soil and extremely cold, windy climate (up to 10 months of winter), no trees can live in the area. Although the amount of precipitation is similar to that in some deserts—less than 25 centimeters (10 inches) per year—the short summer is generally wet because the winter snows melt in the spring and summer temperatures are usually less than 10°C (50°F), which reduces the evaporation rate. Since the permafrost does not let the water sink into the soil, waterlogged soils and many shallow ponds and pools are present. Many waterfowl like ducks and geese migrate to the tundra in the spring; there, they mate and raise their young during the summer before migrating south in the fall. When the top few centimeters (inches) of the soil thaw, many plants (grasses, dwarf birch, dwarf willow) and lichens, such as reindeer moss, grow. The plants are short, usually less than 20 centimeters (8 inches) tall.

Clouds of insects are common during the summer and serve as food for migratory birds. Permanent resident birds are the ptarmigan and snowy owl. No reptiles or amphibians survive in this extreme climate. A few hardy mammals like musk oxen, caribou (reindeer), arctic hare, and lemmings can survive by feeding on the grasses and other plants that grow during the short, cool summer. Arctic foxes, wolves, and owls are the primary predators in this region. Because of the very short growing season, damage to this kind of ecosystem is slow to heal, so the land must be handled with great care.

Scattered patches of tundralike communities also are found on mountaintops throughout the world. These are known as **alpine tundra.** Although the general appearance of the alpine tundra is similar to true tundra, many of the species of plants and animals are different. Many of the birds and large mammals migrate up to the alpine tundra during the summer and return to lower elevations as the weather turns cold.

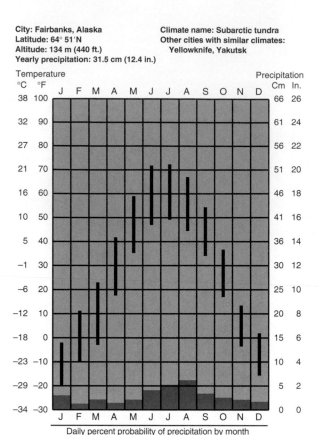

City: Fairbanks, Alaska
Latitude: 64° 51'N
Altitude: 134 m (440 ft.)
Yearly precipitation: 31.5 cm (12.4 in.)

Climate name: Subarctic tundra
Other cities with similar climates:
Yellowknife, Yakutsk

Temperature

Precipitation

a.

Daily percent probability of precipitation by month
32 21 19 13 29 33 42 48 33 35 33 29

b.

figure 6.18 **Tundra** (a) Climagraph for Fairbanks, Alaska. (b) In the northern latitudes and on the tops of some mountains, the growing season is short and plants grow very slowly. Trees are unable to live in these extremely cold areas, in part because there is a permanently frozen layer of soil beneath the surface, known as the permafrost. Because growth is so slow, damage to the tundra can still be seen generations later.

Major Aquatic Ecosystems

Terrestrial biomes are determined by the amount and kind of precipitation and by temperatures. Other factors, such as soil type and wind, also play a part. Aquatic ecosystems also are shaped by key environmental factors. Several important factors are the ability of the sun's rays to penetrate the water, the depth of the water, the nature of the bottom substrate, the water temperature, and the amount of dissolved salts.

Marine Ecosystems

An important determiner of the nature of aquatic ecosystems is the amount of salt dissolved in the water. Those that have little dissolved salt are called **freshwater ecosystems,** and those that have a high salt content are called **marine ecosystems.**

Pelagic Marine Ecosystems

In the open ocean, many kinds of organisms float or swim actively. Crustaceans, fish, and whales swim actively as they pursue food. Organisms that are not attached to the bottom are called **pelagic** organisms, and the ecosystem they are a part of is called a **pelagic ecosystem.** The term **plankton** is used to describe aquatic organisms that are so small and weakly swimming that they are simply carried by currents. As with all ecosystems, the organisms at the bottom of the energy pyramid carry on photosynthesis. The planktonic organisms that carry on photosynthesis are called **phytoplankton.** In the open ocean, a majority of these organisms are small, microscopic, floating algae and bacteria. The upper layer of the ocean, where the sun's rays penetrate, is known as the **euphotic zone.** It is in this euphotic zone where phytoplankton are most common. The thickness of the euphotic zone varies with the degree of clarity of the water but in clear water can be up to 150 meters

(500 feet) in depth. Small, weakly swimming animals of many kinds, known as **zooplankton,** feed on the phytoplankton. Zooplankton are often located at a greater depth in the ocean than the phytoplankton but migrate upward at night and feed on the large population of phytoplankton. The zooplankton are in turn eaten by larger animals like fish and larger shrimp, which are eaten by larger fish like salmon, tuna, sharks, and mackerel. (See figure 6.19.)

A major factor that influences the nature of a marine community is the kind and amount of material dissolved in the water. Of particular importance is the amount of dissolved, inorganic nutrients available to the organisms carrying on photosynthesis. Phosphorus, nitrogen, and carbon are all required for the construction of new living material. In water, these are often in short supply. Therefore, the most productive aquatic ecosystems are those in which these essential nutrients are most common. These areas include places in oceans where currents bring up nutrients that

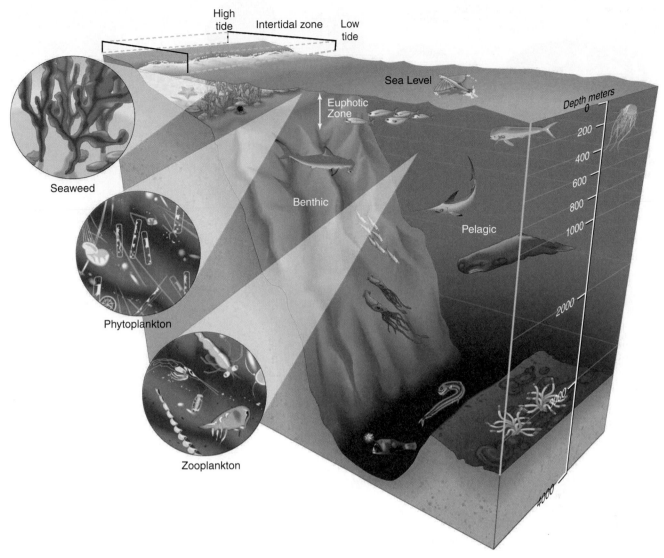

figure 6.19 **Marine Ecosystems** All of the photosynthetic activity of the ocean occurs in the shallow water called the euphotic zone, either by attached algae near the shore or by minute phytoplankton in the upper levels of the open ocean. Consumers are either free-swimming pelagic organisms or benthic organisms that live on the bottom. Small animals that feed on phytoplankton are known as zooplankton.

have settled to the bottom and places where rivers deposit their load of suspended and dissolved materials.

Benthic Marine Ecosystems

Organisms that live on the ocean bottom, whether attached or not, are known as **benthic** organisms, and the ecosystem of which they are a part is called a **benthic ecosystem.** Some fish, clams, oysters, various crustaceans, sponges, sea anemones, and many other kinds of organisms live on the bottom. In shallow water, sunlight can penetrate to the bottom, and a variety of attached photosynthetic organisms commonly called

seaweeds are common. Since they are attached and some, like kelp, can grow to very large size, many other bottom-dwelling organisms, such as sea urchins, worms, and fish, are associated with them.

The substrate is very important in determining the kind of benthic community that develops. Sand tends to shift and move, making it difficult for large plants or algae to become established, although some clams, burrowing worms, and small crustaceans find sand to be a suitable habitat. Clams filter water and obtain plankton and detritus or burrow through the sand, feeding on other inhabitants. Mud may provide

suitable habitats for some kinds of rooted plants, such as mangrove trees or sea grasses. Although mud usually contains little oxygen, it still may be inhabited by a variety of burrowing organisms that feed by filtering the water above them or that feed on other animals in the mud. Rocky surfaces in the ocean provide a good substrate for many kinds of large algae. Associated with this profuse growth of algae is a large variety of animals. (See figure 6.20.)

Temperature also has an impact on the kind of benthic community established. Some communities, such as coral reefs or mangrove swamps, are found only in areas where the water is warm.

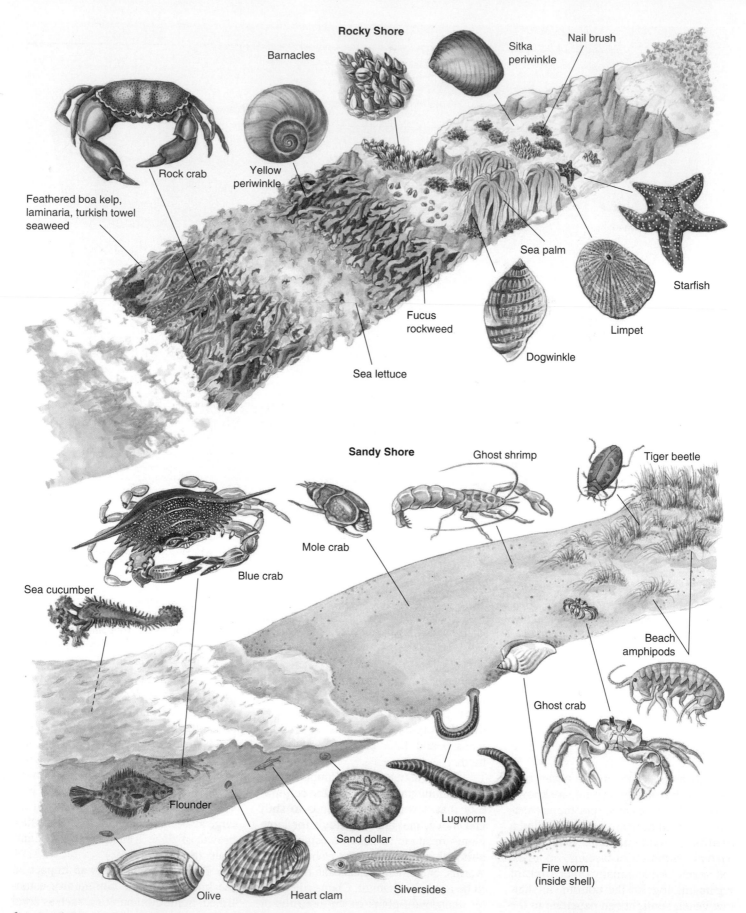

Rocky Shore

Barnacles

Sitka periwinkle

Nail brush

Rock crab

Yellow periwinkle

Feathered boa kelp, laminaria, turkish towel seaweed

Sea palm

Starfish

Fucus rockweed

Limpet

Dogwinkle

Sea lettuce

Sandy Shore

Ghost shrimp

Tiger beetle

Mole crab

Blue crab

Sea cucumber

Beach amphipods

Ghost crab

Flounder

Sand dollar

Lugworm

Fire worm (inside shell)

Olive

Heart clam

Silversides

figure 6.20 **Types of Shores** The kind of substrate determines the kind of organisms that can live near the shore. Rocks provide areas for attachment that sands do not, since sands are constantly shifting. Muds usually have little oxygen in them; therefore, the organisms that live there must be adapted to those kinds of conditions.

PART TWO   Ecological Principles and Their Application

figure 6.21 **Coral Reef** Corals are small sea animals that secrete external skeletons. They have a mutualistic relationship with certain algae, which allows both kinds of organisms to be very successful. The skeletal material serves as a substrate upon which many other kinds of organisms live.

figure 6.22 **Mangrove Swamp** Mangroves are tropical trees that are able to live in very wet, salty muds found along the ocean shore. Since they are able to trap additional sediment, they tend to extend farther seaward as they reproduce.

Coral reef ecosystems are produced by coral animals that build cup-shaped external skeletons around themselves. Corals protrude from their skeletons to capture food and expose themselves to the sun. Exposure to sunlight is important because corals contain single-celled algae within their bodies. These algae carry on photosynthesis and provide both themselves and the coral animals with the nutrients necessary for growth. This mutualistic relationship between algae and coral is the basis for a very productive community of organisms. The skeletons of the corals provide a surface upon which many other kinds of animals live. Some of these animals feed on corals directly, while others feed on small plankton and bits of algae that establish themselves among the coral organisms. Many kinds of fish, crustaceans, sponges, clams, and snails are members of coral reef ecosystems.

Because they require warm water, coral ecosystems are found only near the equator. Coral ecosystems also require shallow, clear water since the algae must have ample sunlight to carry on photosynthesis. Coral reefs are considered one of the most productive ecosystems on Earth. (See figure 6.21.)

Mangrove swamp ecosystems occupy a region near the shore. The dominant organisms are special kinds of trees that are able to tolerate the high salt content of the ocean. In areas where the water is shallow and wave action is not too great, the trees can become established. They have long seeds that float in the water. When the seeds become trapped in mud, they take root. The trees can excrete salt from their leaves. They also have extensively developed roots that extend above the water, where they can obtain oxygen and prop up the plant. The trees trap sediment and provide places for oysters, crabs, jellyfish, sponges, and fish to live. The trapping of sediment and the continual extension of mangroves into shallow areas result in the development of a terrestrial ecosystem in what was once shallow ocean. Mangroves are found in south Florida, the Caribbean, Southeast Asia, Africa, and other parts of the world where tropical mudflats occur. (See figure 6.22.)

At great depths in the ocean is a benthic ecosystem that must rely on a continuous rain of organic matter from the euphotic zone. These areas are known as abyssal areas, and the ecosystem is known as an **abyssal ecosystem.** No light penetrates to this region, and the amount of food available is limited. Essentially, all of the organisms in this environment are scavengers that feed on whatever drifts their way. Many of the animals are small and generate light that they use for finding or attracting food.

Estuaries

An **estuary** is a special category of aquatic ecosystem, that consists of shallow, partially enclosed areas where freshwater enters the ocean. The saltiness of the water in the estuary changes with tides and the flow of water from rivers. The organisms that live here are specially adapted to this set of physical conditions, and the number of species is less than in the ocean or in freshwater. Estuaries are particularly productive ecosystems because of the large amounts of nutrients introduced into the basin from the rivers that run into them. This is further enhanced by the fact that the shallow water allows light to penetrate to most of the water in the basin. Phytoplankton and attached algae and plants are able to use the sunlight and the nutrients for rapid growth. This photosynthetic activity supports many kinds of organisms in the

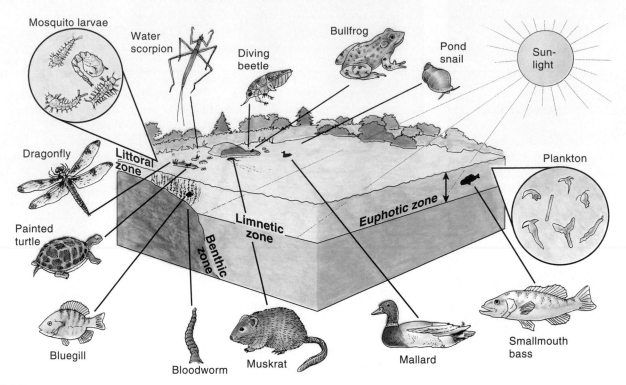

figure 6.23 **Lake Ecosystem** Lakes are similar in structure to oceans except that the species are different because most marine organisms cannot live in freshwater. Insects are common organisms in freshwater lakes, as are many kinds of fish, zooplankton, and phytoplankton.

estuary. Estuaries are especially important as nursery sites for fish and crustaceans like flounder and shrimp. The adults enter these productive, sheltered areas to reproduce and then return to the ocean. The young spend their early life in the estuary and eventually leave as they get larger and are more able to survive in the ocean. Estuaries also trap sediment. This activity tends to prevent many kinds of pollutants from reaching the ocean and also results in the gradual filling in of the estuary, which may eventually become a salt marsh and then part of a terrestrial ecosystem.

Freshwater Ecosystems

Freshwater ecosystems differ from marine ecosystems in several ways. The amount of salt present is much less, the temperature of the water can change greatly, the water is in the process of moving to the ocean, oxygen can often be in short supply, and the organisms that inhabit freshwater systems are different.

Freshwater ecosystems can be divided into two categories: those in which the water is relatively stationary, such as lakes, ponds, and reservoirs, and those in which the water is running downhill, such as streams and rivers.

Lakes and Ponds

Large lakes have many of the same characteristics as the ocean. If the lake is deep, there is a euphotic zone at the top, with many kinds of phytoplankton, and zooplankton that feed on the phytoplankton. Small fish feed on the zooplankton and are in turn eaten by larger fish. The species of organisms found in freshwater lakes are different from those found in the ocean, but the roles played are similar, so the same terminology is used.

Along the shore and in the shallower parts of lakes, many kinds of flowering plants are rooted in the bottom. Some have leaves that float on the surface or protrude above the water and are called **emergent plants.** Cattails, bulrushes, arrowhead plants, and water lilies are examples. Rooted plants that stay submerged below the surface of the water are called **submerged plants.** *Elodea* and *Chara* are examples.

Many kinds of freshwater algae also grow in the shallow water, where they may appear as mats on the bottom or attached to vegetation and other objects in the water. Associated with the plants and algae are a large number of different kinds of animals. Fish, crayfish, clams, and many kinds of aquatic insects are common inhabitants of this mixture of plants and algae. This region, with rooted vegetation, is known as the **littoral zone,** and the portion of the lake that does not have rooted vegetation is called the **limnetic zone.** (See figure 6.23.)

The productivity of the lake is determined by several factors. Temperature is important, since cold temperatures tend to reduce the amount of photosynthesis. Water depth is important because shallow lakes will have light penetrating to the lake bottom, and therefore, photosynthesis can occur throughout the entire water column. Shallow lakes also tend to be warmer as a result of the warming effects of the sun's rays. A third factor that influences the productivity of lakes is the amount of nutrients present. This is primarily determined by the rivers and streams that carry nutrients to the lake. River systems that run through areas that donate many nutrients will carry the nutrients to the lakes. Farming and construction expose soil and release nutrients, as do other human activities such as depositing sewage into streams and lakes. Deep, clear, cold, nutrient-poor lakes are low in productivity and are called **oligotrophic lakes.** Shallow, murky, warm, nutrient-rich lakes are called **eutrophic lakes.**

Although the water molecule (H_2O) has oxygen as part of its structure, this oxygen is not available to organisms. The oxygen that they need is dissolved molecular oxygen (O_2), which enters water from the air or when it is released as a result of photosynthesis by aquatic plants. When water tumbles over rocks in a stream or crashes on the shore as a result of wave action, air and water mix, which allows more oxygen to dissolve in the water.

The dissolved oxygen content of the water is important since the quantity of oxygen determines the kinds of organisms that can inhabit the lake. When organic molecules enter water, they are broken down by bacteria and fungi.

These decomposer organisms use oxygen from the water as they perform respiration. The amount of oxygen used by decomposers to break down a specific amount of organic matter is called the **biochemical oxygen demand** or **BOD.** Organic materials enter aquatic ecosystems in several ways. The organisms that live in the water produce the metabolic wastes. When organisms that live in or near water die or shed parts, their organic matter is contributed to the water. The amount of nutrients entering the water is also important, since the algae and plants whose growth is stimulated will eventually die and their decomposition will reduce oxygen concentration. Many bodies of water experience a reduced oxygen level during the winter when producers die. The amount and kinds of organic matter determine, in part, how much oxygen is left to be used by other organisms, such as fish, crustaceans, and snails. Many lakes may experience periods when oxygen is low, resulting in the death of fish and other organisms. Human activity often influences the health of bodies of water, because we tend to introduce nutrients from agriculture and organic wastes from a variety of industrial, agricultural, and municipal sources. These topics are discussed in greater depth in chapter 16.

Streams and Rivers

Streams and rivers are a second category of freshwater ecosystem. Since the water is moving, planktonic organisms are less important than are attached organisms. Most algae grow attached to rocks and other objects on the bottom. This collection of attached algae, animals, and fungi is called the **periphyton.** Since the water is shallow, light can penetrate easily to the bottom (except for large or extremely muddy rivers). Even so, it is difficult for photosynthetic organisms to accumulate the nutrients necessary for growth, and most streams are not very productive. As

a matter of fact, the major input of nutrients is from organic matter that falls into the stream from terrestrial sources. These are primarily the leaves from trees and other vegetation, as well as the bodies of living and dead insects.

Within the stream is a community of organisms that are specifically adapted to use the debris as a source of food. Bacteria and fungi colonize the organic matter, and many kinds of insects shred and eat this organic matter along with the fungi and bacteria living on it. The feces (intestinal wastes) of these insects and the tiny particles produced during the eating process become food for other insects that build nets to capture the tiny bits of organic matter that drift their way. These insects are in turn eaten by carnivorous insects and fish.

Organisms in larger rivers and muddy streams, which have less light penetration, rely in large part on the food that drifts their way from the many streams that empty into the river. These larger rivers tend to be warmer and to have slower-moving water. Consequently, the amount of oxygen is usually less, and the species of plants and animals change. Any additional organic matter added to the river system adds to the BOD, further reducing the oxygen in the water. Plants may become established along the river bank and contribute to the ecosystem by carrying on photosynthesis and providing hiding places for animals.

Just as estuaries are a bridge between freshwater and marine ecosystems, swamps and marshes are a transition between aquatic and terrestrial ecosystems. **Swamps** are wetlands that contain trees that are able to live in places that are either permanently flooded or flooded for a major part of the year. **Marshes** are wetlands that are dominated by grasses and reeds. Many swamps and marshes are successional states that eventually become totally terrestrial communities.

Protecting Old-Growth Temperate Rainforests of the Pacific Northwest

The coastal areas of northern California, Oregon, Washington, British Columbia, and southern Alaska have an unusual set of environmental conditions that support a special kind of forest, a temperate rainforest. The prevailing winds from the west bring moisture-laden air to the coast. As this air meets the coastal mountains and is forced to rise, it cools and the moisture falls as rain or snow. Most of these areas receive 200 or more centimeters (80 or more inches) of precipitation per year. This abundance of water, along with fertile soil and mild temperatures, results in a luxuriant growth of plants.

Sitka spruce, Douglas fir, and western hemlock are typical evergreen coniferous trees. Undisturbed (old-growth) forests of this region have trees as old as 800 years that are almost as tall as the length of a football field. Deciduous trees of various kinds (red alder, bigleaf maple, black cottonwood) grow in places where they can get enough light. All trees are covered with mosses, ferns, and other plants that grow on their surface. The dominant color is green, since most surfaces have something growing on them.

When a tree dies and falls to the ground, it rots in place and often serves as a site for the establishment of new trees. This is such a common feature of the forest that the fallen, rotting trees are called nurse trees. The fallen trees also serve as a food source for a variety of insects, which are food for a variety of other animals. Several endangered or threatened animals, such as the northern spotted owl, the marbled murrelet (a seabird), and the Roosevelt elk, are dependent on undisturbed forest for their survival.

Because of the rich resource of trees, 90 percent of the original temperate rainforest has been logged. What remains has become a source of controversy. Some maintain that it should be protected as a remnant of the original forest of the region, just as small patches of prairie and eastern woodland have been preserved in other parts of North America. Others point out that the intact old-growth forest must be preserved to protect endangered species. Some maintain that since the trees are old and will die in the near future, they should be harvested for their timber value and the jobs the logging will provide. Others counter that dying and dead trees are important to the maintenance of this unique ecosystem. The fate of this unusual ecosystem is not likely to be resolved without legislation or legal action.

- List three components you would include in a law that would protect this ecosystem.
- What compromises can you offer to those who feel the forest should be used for timber?
- If there were no endangered species in the region, would you feel differently?

Summary

Ecosystems change as one kind of organism replaces another in a process called succession. Ultimately, a relatively stable stage is reached, called the climax community. Succession may begin with bare rock or water, in which case it is called primary succession, or may occur when the original ecosystem is destroyed, in which case it is called secondary succession. The stages that lead to the climax are called successional stages.

Major regional terrestrial climax communities are called biomes. The primary determiners of the kinds of biomes that develop are the amount and yearly distribution of rainfall and the yearly temperature cycle. Major biomes are desert, grassland, savanna, Mediterranean shrublands, tropical dry forest, tropical rainforest, temperate deciduous forest, taiga, and tundra. Each has a particular set of organisms that is adapted to the climatic conditions typical for the area. As one proceeds up a mountainside, it is possible to witness the same kind of change in biomes that occurs if one were to travel from the equator to the North Pole.

Aquatic ecosystems can be divided into marine (saltwater) and freshwater ecosystems. In the ocean, some organisms live in open water and are called pelagic organisms. Light penetrates only the upper layer of water; therefore, this region is called the euphotic zone. Tiny photosynthetic organisms that float near the surface are called phytoplankton. They are eaten by small animals known as zooplankton, which in turn are eaten by fish and other larger organisms.

The kind of material that makes up the shore determines the mixture of organisms that live there. Rocky shores provide surfaces for organisms to attach; sandy shores do not. Muddy shores are often poor in oxygen, but marshes and swamps may develop in these areas. Coral reefs are tropical marine ecosystems dominated by coral animals. Mangrove swamps are tropical marine shoreline ecosystems dominated by trees. Estuaries occur where freshwater streams and rivers enter the ocean. They are usually shallow, very productive areas. Many marine organisms use estuaries for reproduction.

Insects are common in freshwater and absent in marine systems. Lakes show a structure similar to that of the ocean, but the species are different. Deep, cold-water lakes with poor productivity are called oligotrophic, while shallow, warm-water, highly productive lakes are called eutrophic. Streams differ from lakes in that most of the organic matter present in them falls into them from the surrounding land. Thus, organisms in streams are highly sensitive to the land uses that occur near the streams.

Key Terms

abyssal ecosystem *129*	limnetic zone *130*	primary succession *109*
alpine tundra *125*	littoral zone *130*	savanna *118*
benthic *127*	mangrove swamp ecosystem *129*	secondary succession *109*
benthic ecosystem *127*	marine ecosystem *126*	seral stage *110*
biochemical oxygen demand (BOD) *131*	marsh *131*	sere *110*
biome *114*	Mediterranean shrublands *119*	steppe *116*
boreal forest *124*	northern coniferous forest *124*	submerged plants *130*
climax community *109*	oligotrophic lake *131*	succession *109*
coral reef ecosystem *129*	pelagic *126*	successional stage *110*
desert *115*	pelagic ecosystem *126*	swamp *131*
emergent plants *130*	periphyton *131*	taiga *124*
estuary *129*	permafrost *125*	temperate deciduous forest *123*
euphotic zone *126*	phytoplankton *126*	tropical dry forest *119*
eutrophic lake *131*	pioneer community *109*	tropical rainforest *122*
freshwater ecosystem *126*	plankton *126*	tundra *125*
grassland *116*	prairie *116*	zooplankton *126*

Review Questions

1. Describe the process of succession. How does primary succession differ from secondary succession?
2. How does a climax community differ from a successional community?
3. List three characteristics typical of each of the following biomes: tropical rainforest, desert, tundra, taiga, savanna, Mediterranean shrublands, tropical dry forest, grassland, and temperate deciduous forest.
4. What two primary factors determine the kind of terrestrial biome that will develop in an area?
5. How does height above sea level affect the kind of biome present?
6. What areas of the ocean are the most productive?
7. How does the nature of the substrate affect the kinds of organisms found at the shore?
8. What is the role of each of the following organisms in a marine ecosystem: phytoplankton, zooplankton, algae, coral animals, and fish?
9. List three differences between freshwater and marine ecosystems.
10. What is an estuary? Why are estuaries important?

Critical Thinking Questions

1. Does the concept of a "climax community" make sense? Why or why not?
2. What do you think about restoring ecosystems that have been degraded by human activity? Should it be done or not? Why? Who should pay for this reconstruction?
3. Identify the biome in which you live. What environmental factors are instrumental in maintaining this biome? What is the current health of your biome? What are the current threats to its health? How might your biome have looked 100, 1000, 10,000 years ago?
4. Imagine you are a conservation biologist who is being asked by local residents what the likely environmental outcomes of development would be in the tropical rainforest in which they live. What would you tell them? Why do you give them this evaluation? What evidence can you cite for your claims?
5. The text says that 90 percent of the old-growth temperate rainforest in the Pacific Northwest has been logged. What to do with the remaining 10 percent is still a question. Some say it should be logged, and others say it should be preserved. What values, beliefs, and perspectives are held by each side? What is your ethic regarding logging old-growth in this area? What values, beliefs, and perspectives do you hold regarding this issue?
6. Much of the old-growth forest in the United States has been logged, economic gains have been realized, and second-growth forests have become established. This is not the case in the tropical rainforests, although they are being lost at alarming rates. Should developed countries, which have already "cashed in" on their resources, have anything to say about what is happening in developing countries? Why do you think the way you do?

Concept Map

Construct a map to show relationships among the following concepts:

primary succession	pioneer community	northern coniferous forest
secondary succession	prairie	tundra
biome	temperate deciduous forest	successional stage
climax community		

Interactive Exploration

Check out the website at **http://www.mhhe.com/environmentalscience** and click on the cover of this textbook for quizzing, career information, case studies, and hot links for the following topics:

Field Methods for Studies of Ecosystems	Taiga
Succession and Stability	Tundra
Biomes and Environmental Habitats	Marine Habitats
Deserts	Marine Ecology
Savannah	The Marine Pelagic Zone
Grasslands	The Marine Deep Sea Zone
Rangelands and Land Use Issues	Shallow Subtidal Communities
Tropical Rainforests	Estuaries
Tropical Rainforests and Land Use Issues	Mangroves
Tropical Forests and Extinction	Coral Reefs
Temperate Forests	Freshwater Habitats
Nontropical Forests and Land Use Issues	

Population Principles

Objectives

After reading this chapter, you should be able to:

- Understand that birthrate and death rate are both important in determining the population growth rate.
- Define the following characteristics of a population: natality, mortality, sex ratio, age distribution, biotic potential, and spatial distribution.
- Explain the significance of biotic potential to the rate of population growth.
- Describe the lag, exponential growth, and stable equilibrium phases of a population growth curve. Explain why each of these stages occurs.
- Describe how the limiting factors determine the carrying capacity for a population.
- List the four categories of limiting factors.
- Describe a death phase that is typical of some kinds of populations.
- Recognize that humans are subject to the same forces of environmental resistance as are other organisms.
- Understand the implications of overreproduction.
- Understand that the human population is still growing rapidly.
- Explain how human population growth is influenced by social, theological, philosophical, and political thinking.

Chapter Outline

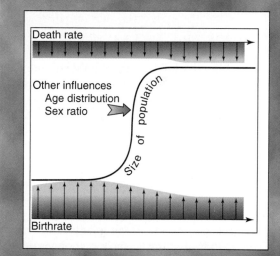

Population Characteristics

A **population** can be defined as a group of individuals of the same species inhabiting an area. Just as individuals within a population are recognizable, different populations of the same species have specific characteristics that distinguish them from one another. Some important ways in which populations differ include natality (birthrate), mortality (death rate), sex ratio, age distribution, growth rates, density, and spatial distribution.

Natality and Mortality

Natality refers to the number of individuals added to the population through reproduction over a particular time period. There are two ways in which new individual organisms are produced—asexual reproduction and sexual reproduction. Bacteria and other tiny organisms reproduce primarily asexually when they divide to form new individuals that are identical to the original parent organism. Even plants and many kinds of animals, such as sponges, jellyfish, and many kinds of worms, reproduce asexually by dividing into two parts or by budding off small portions of themselves that become independent individuals. Even some insects and lizards have a special kind of asexual reproduction in which the females lay unfertilized eggs that are genetically identical to the female. However, most species have some stage in their life cycle in which they reproduce sexually. In plant populations, sexual reproduction results in the production of numerous seeds, but the seeds must land in appropriate soil conditions before they will germinate to produce a new individual. Animal species also typically produce large numbers of offspring as a result of sexual reproduction. In human populations, natality is usually described in terms of the **birthrate,** the number of individuals born per 1000 individuals per year. For example, if a population of 2000 indi-

viduals produced 20 offspring during one year, the birthrate would be 10 per thousand per year. The natality for most species is typically quite high. Most species produce many more offspring than are needed to replace the parents.

It is important to recognize that the growth of a population is not determined by the birthrate (natality) alone. **Mortality,** the number of deaths in a population over a particular time period, is also important. For most species, mortality rates are very high, particularly among the younger individuals. For example, of all the seeds that plants produce, very few will result in a mature plant that itself will produce offspring. Many seeds are eaten by animals, some do not germinate because they never find proper soil conditions, and those that germinate must compete with other organisms for nutrients and sunlight.

In human population studies, mortality is usually discussed in terms of the **death rate,** the number of people who die per 1000 individuals per year. Compared to the high mortality of the young of most species, the infant death rate of long-lived animals like humans is relatively low. In order for the size of a population to grow, the number of individuals added by reproduction must be greater than the number leaving it by dying. (See figure 7.1)

The **population growth rate** is the birthrate minus the death rate. In human population studies, the population growth rate is usually expressed as a percentage of the total population. For example, in the United States, the birthrate is 15 births per thousand individuals in the population. The death rate is 9 per thousand. The difference between the two is 6 per thousand, which is equal to an annual population increase of 0.6 percent (6/1000).

Another way to look at mortality is to look at how likely it is that an offspring will survive to a specific age. One way of visualizing this is with a survivorship curve. A **survivorship curve** shows the proportion of individuals likely to survive to each age. While each species is different, three general types of survivorship curves can be recognized:

species that have high mortality among their young, species in which mortality is evenly spread over all age groups, and species in which survival is high until old age, when mortality is high. Figure 7.2 gives examples of species that fit these three general categories.

Sex Ratio and Age Distribution

The population growth rate is greatly influenced by sex ratio and age distribution of the population. The **sex ratio** refers to the relative numbers of males and females. (Many kinds of organisms, such as earthworms and most plants, have both kinds of sex organs in the same body; sex ratio has no meaning for these species.) The number of females is very important, since they ultimately determine the number of offspring produced in the population. In polygamous species, one male may mate with many females. Therefore, the number of males is less important to the population growth rate than the number of females. In monogamous species, a male and female pair up, mate, and raise their young together. Unpaired females are not likely to be fertilized and raise young. Even if an unpaired female is fertilized, she will be less successful in raising young.

It is typical in most species that the sex ratio is about 1:1 (one female to one male). However, there are populations in which this is not true. In populations of many species of game animals, the males are shot (have a higher mortality) and the females are not. This results in an uneven sex ratio in which the females outnumber the males. In many social insect populations (bees, ants, and wasps), the number of females greatly exceeds the number of males at all times, though most of the females are sterile. In humans, about 106 males are born for every 100 females. However, in the United States, by the time people reach their mid-twenties, a higher death rate for males has equalized the sex ratio. The higher male death rate continues into old age, when women outnumber men.

The **age distribution,** the number of individuals of each age in the population,

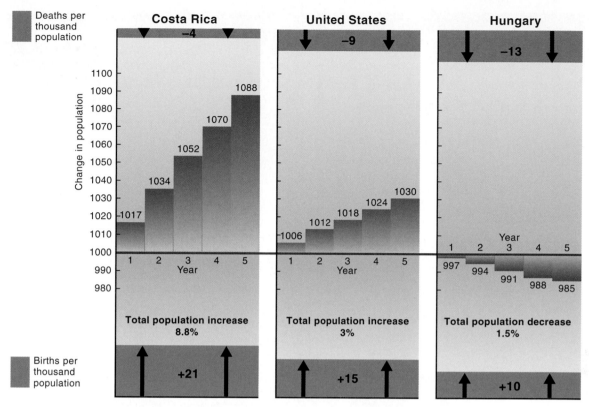

figure 7.1 **Effect of Birthrate and Death Rate on Population Size** For a population to grow, the birthrate must exceed the death rate for a period of time. These three human populations illustrate how the combined effects of births and deaths would change population size if birthrates and death rates were maintained for a five-year period.

Source: Data from World Population Data Sheet 2002, Population Reference Bureau, Inc., Washington, D.C.

greatly influences the population growth rate. As you can see in figure 7.3, some are prereproductive juveniles, some are reproducing adults, and some are postreproductive adults. The age distribution greatly influences the population growth rate. If the population has a large number of prereproductive juveniles, it would be expected to grow in the future as the young become sexually mature. If the majority of a population is made up of reproducing adults, the population should be growing. If the population is made up of old individuals whose reproductive success is low, the population is likely to fall.

Many species, particularly those that have short life spans, have age distributions that change significantly during the course of a year. Species typically produce their young at particular parts of the year. The seeds of many annual plants germinate in the spring or following a rainy period of the year. Therefore, during one part of the year, most of the individuals are prereproductive. As time passes, nearly all of those that survive become reproducing adults and produce seeds. Later in the year, they all die. A similar pattern is seen in many insects that emerge from eggs as larvae, transform into adults, mate and lay eggs, and die. Other animals typically produce their young at a time when food is available. For many birds, this is during the spring and summer. Many herbivores produce their young in the spring or following rains, when plants begin to grow.

In species that live a long time, it is possible for a population to have an age distribution in which the proportion of individuals in these three categories is relatively constant. Since mortality is generally higher among young individuals, such populations typically have more prereproductive individuals than reproductive individuals and more reproductive individuals than postreproductive individuals.

Human populations exhibit several types of age distribution. (See figure 7.3.) Kenya's population has a large prereproductive and reproductive component. This means that it will continue to increase rapidly for some time. The United States has a very large reproductive component with a declining number of prereproductive individuals. Eventually, if we were to have no immigration, the U.S. population would begin to decline if current trends in birthrates and death rates continue. Germany has an age distribution with high postreproductive and low prereproductive portions of the population. With low numbers of prereproductive individuals entering their reproductive years, the population of Germany has begun to decline.

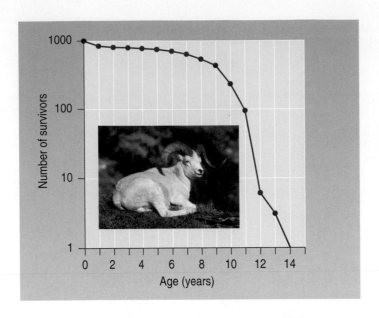

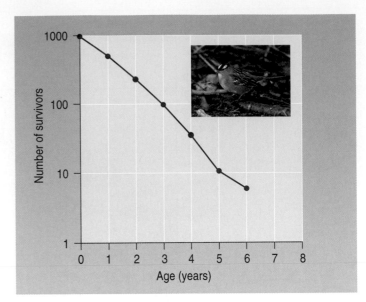

figure 7.2 **Types of Survivorship Curves** (a) The Dall sheep is a large mammal that produces relatively few young. Most of the young survive, and survival is high until individuals reach old age, when they are more susceptible to predation and disease. (b) The curve shown for the white-crowned sparrow is typical of that for many kinds of birds. After a period of high mortality among the young, the mortality rate is about equal for all ages of adult birds. (c) Many small animals and plants produce enormous numbers of offspring. Mortality is very high in the younger individuals, and few individuals reach old age.

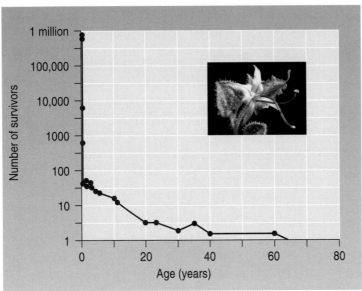

Population Density and Spatial Distribution

Because of such factors as soil type, quality of habitat, and availability of water, organisms normally are distributed unevenly. Some populations have many individuals clustered into a small space, while other populations of the same species may be widely dispersed. **Population density** is the number of organisms per unit area. For example, fruit-fly populations are very dense around a source of rotting fruit, while they are rare in other places. Similarly, humans are often clustered into dense concentrations we call cities, with lower densities in rural areas. When the population density is too great, all individuals within the population are injured because they compete severely with each other for necessary resources. Plants may compete for water, soil nutrients, or sunlight. Animals may compete for food, shelter, or nesting sites. In animal populations, overcrowding might cause some individuals to ex-plore and migrate into new areas. This movement from densely populated locations to new areas is called **dispersal.** It relieves the overcrowded conditions in the home area and, at the same time, increases the population in the places to which they migrate. Often, it is juvenile individuals that relieve overcrowding by leaving. The pressure to migrate from a population (**emigration**) may be a result of seasonal reproduction leading to a rapid increase in population size or environmental changes that intensify

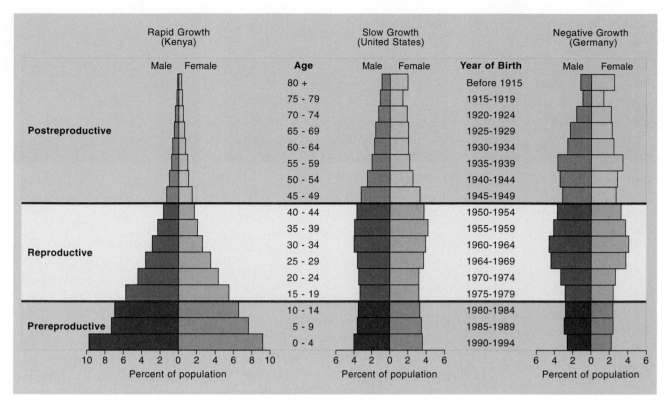

figure 7.3 **Age Distribution in Human Populations** The relative numbers of individuals in each of the three categories (prereproductive, reproductive, and postreproductive) are good clues to the future growth of a population. Kenya has a large number of young individuals who will become reproducing adults. Therefore, this population is likely to grow rapidly. The United States has a large proportion of reproductive individuals and a moderate number of prereproductive individuals. Therefore, this population is likely to grow slowly. Germany has a declining number of reproductive individuals and a very small number of prereproductive individuals. Therefore, its population has begun to decline.

Source: Data from Population Reference Bureau.

competition among members of the same species. For example, as water holes dry up, competition for water increases, and many desert birds emigrate to areas where water is still available.

The organisms that leave one population often become members of a different population. This migration into an area (**immigration**) may introduce characteristics that were not in the population originally. When Europeans immigrated to North America, they brought genetic and cultural characteristics that had a tremendous impact on the existing Native American population. Among other things, Europeans brought diseases that were foreign to the Native Americans. These diseases increased the death rate and lowered the birthrate of Native Americans, resulting in a sharp decrease in the size of their populations.

Droughts, wars, and political persecution have caused people to emigrate from their native lands to other countries. In 2002, the United Nations High Commissioner for Refugees estimated that there were about 20 million refugees worldwide. Often, the receiving countries are unable to cope with the large influx of new inhabitants.

Summary of Factors That Influence Population Growth Rates

Populations have an inherent tendency to increase in size. However, as we have just seen, many factors influence the rate at which a population can grow. At the simplest level, the rate of increase is determined by subtracting the number of individuals leaving the population from

the number entering. Individuals leave the population either by death or emigration. Individuals enter the population by birth or immigration. Birthrates and death rates are influenced by several factors, including the number of females in the population and their age. In addition, the density of a population may encourage individuals to leave because of intense competition for a limited supply of resources.

A Population Growth Curve

Sex ratios and age distributions directly influence the rate of reproduction within a population. However, each species

Apples

Geese

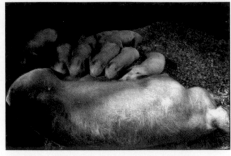
Pigs

figure 7.4 **Biotic Potential** The ability of a species to reproduce greatly exceeds the number necessary to replace those who die. Here are some examples of the prodigious reproductive abilities of some species.

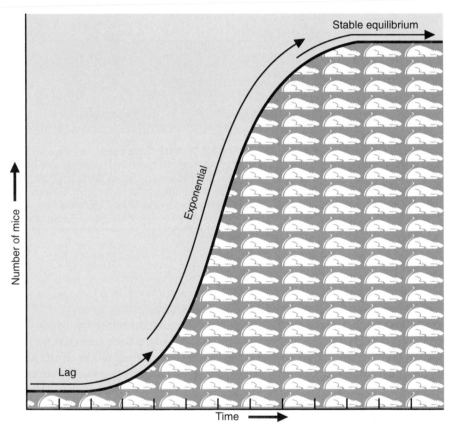

figure 7.5 **A Typical Population Growth Curve** In this mouse population, there is little growth during the lag phase. During the exponential growth phase, the population rises rapidly as increasing numbers of individuals reach reproductive age. Eventually, the population reaches a stable equilibrium phase, during which the birthrate equals the death rate.

Because most species have a high biotic potential, there is a natural tendency for populations to increase. If we consider a hypothetical situation in which mortality is not a factor, we could have the following situation. If two mice produced four offspring and they all lived, eventually they would produce offspring of their own, while their parents continued to reproduce as well. Under these conditions, the population will grow exponentially. (Exponential growth results in the number of individuals in each succeeding generation being a multiple of the previous generation; for example, 2, 4, 8, 16, 32, etc.). While populations cannot grow exponentially forever, they often have a exponential period of growth.

Population growth often follows a particular pattern, consisting of a lag phase, an exponential growth phase, and a stable equilibrium phase. Figure 7.5 shows a typical population growth curve. During the first portion of the curve, known as the **lag phase,** the population grows very slowly because there are few births, since the process of reproduction and growth of offspring takes time. Organisms must mature into adults before they can reproduce. When the offspring begin to mate and have young, the parents may be producing a second set of offspring. Since more organisms now are reproducing, the population begins to increase at an accelerating rate. This stage is known as the **exponential growth phase (lag phase).** This growth will continue for as long as the birthrate exceeds the death rate. Eventually, however, the death rate and the birthrate will come to

also has an inherent reproductive capacity, or **biotic potential,** which is its biological ability to produce offspring. Some species, like apple trees, may produce thousands of offspring (seeds) per year, while others, like pigs or geese, may produce 10 to 12 young per year. (See figure 7.4.) Some large animals, like bears or elephants, may produce

one young every two to three years. Although there are large differences among species, generally, adults produce many more offspring during their lifetimes than are needed to replace themselves when they die. However, most of the young die, so only a few survive to become reproductive adults themselves.

equal one another, and the population will stop growing and reach a relatively stable population size. This stage is known as the **stable equilibrium phase.**

It is important to recognize that although the size of the population may not be changing, the individuals are changing. As new individuals enter by birth or immigration, others leave by death or emigration. For most organisms, the first indication that a population is entering a stable equilibrium phase is an increase in the death rate. A decline in the birthrate may also contribute to the stabilizing of population size. Usually, this occurs after an increase in the death rate. To understand why populations cannot grow continuously, it is necessary to discuss the concept of carrying capacity.

Carrying Capacity

The **carrying capacity** of an area is the number of individuals of a species that can be maintained in an area, over time, without harming the habitat. The concept of carrying capacity is usually applied to relatively long-lasting habitats and is helpful when we examine why populations stabilize. However, nothing is permanent, and as the habitats change, because of disturbance or succession, the carrying capacity for a species changes also. Seasonal changes also influence the numbers of individuals that can be supported in an area. The number of individuals that can be supported during the summer may be much larger than can be supported in winter. Indeed, some animals (such as birds) use certain habitats only during the summer and migrate in the fall when living conditions become difficult. The combination of factors that sets the carrying capacity for an area is called **environmental resistance.**

When a particular condition or factor can be identified as a key component that limits the size of a population, it is identified as a **limiting factor.** For most populations, four categories of limiting factors are recognized as components of envi-

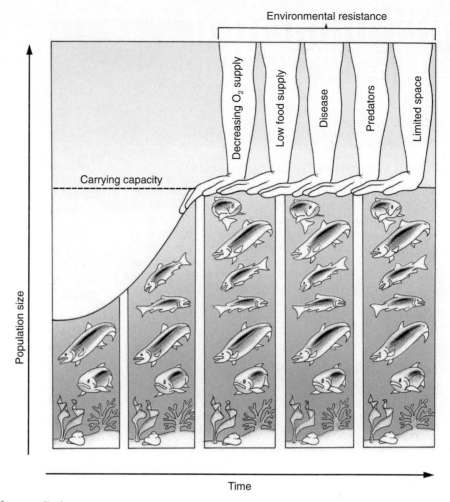

figure 7.6 **Carrying Capacity** A number of factors in the environment, such as oxygen supply, food supply, diseases, predators, and space, determine the number of organisms that can survive in a given area—the carrying capacity of that area. The environmental factors that limit populations are known collectively as environmental resistance.

ronmental resistance that set the carrying capacity. These are: (1) the availability of raw materials, (2) the availability of energy, (3) the accumulation of waste products and their means of disposal, and (4) interactions among organisms.

In some cases, limiting factors are easy to identify. Lack of food, lack of oxygen, competition with other species, or disease are examples. In other cases, the limiting factors may be less obvious. (See figure 7.6.) For example, plants need nitrogen and magnesium from the soil as raw materials for the manufacture of chlorophyll. If these minerals are not present in sufficient quantities, the plant population cannot increase. The application of fertilizers containing these minerals removes this limiting factor, and the individual plants grow and reproduce, re-

sulting in a larger population. In effect, the carrying capacity has been increased because this limiting factor has been removed. A carrying capacity still exists, but it is set at a new level and some new primary limiting factor will emerge. Perhaps it will be the amount of water, the number of insects that feed on the plants, or competition for sunlight. Because plants require energy in the form of sunlight for photosynthesis, the amount of light can be a limiting factor for many plants. When small plants are in the shade of trees, they often do not grow well and have small populations because they do not receive enough sunlight.

Accumulation of waste products is not normally a limiting factor for plants, since they produce few wastes, but it can be for other kinds of organisms.

environmental CLOSE-UP

Population Growth of Invading Species

When a new species is introduced into an area suitable for its survival, it often has great potential to increase its population size, since it may not have natural enemies to keep mortality high. New species may also be able to compete favorably for available resources and lower the populations of native species. A typical population growth curve starts with a few individuals being released. Once established, the population will increase in number and expand its range. The zebra mussel, for example, is thought to have entered the Great Lakes about 1985. Today, it is found throughout all the Great Lakes and has been introduced into the Mississippi River and its tributaries, where it has been discovered as far south as New Orleans. Similarly, the gypsy moth has spread from its place of original release to much of the forested land of the Midwest and is still expanding its range. Another invading species is the kudzu vine, which has become a pest in many areas of the southern United States.

Eventually, the invading species occupies all the habitat suitable to it and the population stabilizes. Dandelions and starlings are introduced species that are no longer expanding their range. They have simply become a normal part of the biology of North America.

Kudzu vine

Gypsy moths

Dandelion

Zebra mussels

Starlings

Bacteria, other tiny organisms, and many kinds of aquatic organisms that live in small ecosystems like puddles, pools, or aquariums may be limited by wastes. When a small number of a species of bacterium are placed on a petri plate with nutrient agar (a jellylike material containing food substances), the population growth follows a curve shown in figure 7.7. As expected, it begins with a lag phase, continues through an exponential growth phase, and eventually levels off in a stable equilibrium phase. However, in this small, enclosed space, there is no way to get rid of the toxic waste products, which accumulate, eventually killing the bacteria. This decline in population size is known as the **death phase.** When a population decreases rapidly, it is said to crash.

Interactions among organisms are also important in determining population size. Since white-tailed deer and cottontail rabbits eat the twigs of many species of small trees, they have a limiting effect on the size of some tree populations. Many aquatic organisms may produce waste products that build up to toxic levels. Decomposer organisms that break down these toxic waste materials may allow for increased populations because they prevent the buildup of toxic wastes. Parasites and predators may cause the premature death of individuals, thus limiting the size of the population. A good example of predator-prey interaction is the relationship between the cat known as the Canada lynx and a member of the rabbit family known as the varying hare. The varying hare produces large numbers of young. In peak reproductive years a female varying hare can produce 16 to 18 young. As with many animals a primary cause of death is predation. The varying hare population is a good food source for a variety of predators including the lynx. When the population of varying hares increases, it provides an abundant source of food for the lynx and the size of the lynx population rises, and when the population of hares decreases, so does that of the lynx. This pattern repeats itself in a 10-year cycle. (See figure 7.8.)

Recent studies indicate that one of the causes of the decline in varying hare populations is a reduction in their birth-

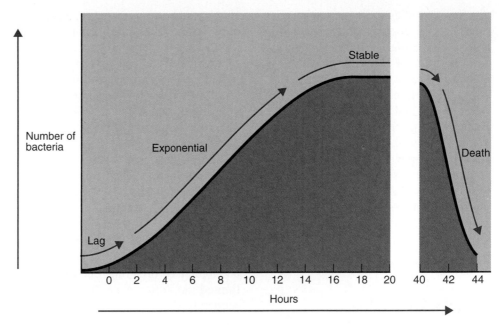

figure 7.7 **A Bacterial Growth Curve** The initial change in population size follows a typical population growth curve until waste products become lethal. The buildup of waste products lowers the carrying capacity. When a population begins to decline, it enters the death phase.

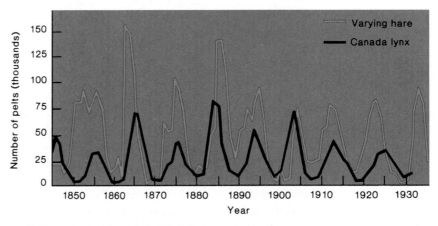

figure 7.8 **Interaction of Predator–Prey Populations** Interaction between predator and prey species is complex and often difficult to interpret. These data were collected from the records of the number of pelts purchased by the Hudson Bay Company. They show that the two populations fluctuate with about 10 years between successive high populations. The change in the lynx population usually followed changes in the varying hare population.

Source: Data from D.A. MacLulich, "Fluctuations in the Numbers of the Varying Hare" (*Lepus americanus*), in *University of Toronto Studies, Biology Series #43 1937* (reprinted 1974), University of Toronto Press.

rate. The causes of this reduction may be related to a variety of factors, including reduced quality of food and higher levels of stress resulting from greater difficulty in finding food and avoiding predators. With reduced reproduction and continued high predation, the varying hare population drops. With reduced numbers of hares, lynx populations

drop. Eventually, the birthrate of hares increases and the population rebounds, followed by a rebound in the lynx population as well. It appears that both food availability and predation are important limiting factors that determine the size of the varying hare population, and the number of varying hares is a primary limiting factor for the lynx.

Some studies indicate that populations can be controlled by interaction among individuals within the population. A study of laboratory rats shows that crowding causes a breakdown in normal social behavior, which leads to fewer births and increased deaths. The changes observed include abnormal mating behavior, decreased litter size, fewer litters per year, lack of maternal care, and increased aggression in some rats or withdrawal in others. Thus, limiting factors can reduce birthrates as well as increase death rates.

Reproductive Strategies and Population Fluctuations

So far, we have talked about population growth as if all organisms reach a stable population when they reach the carrying capacity. That is an appropriate way to begin to understand population changes, but the real world is much more complicated. Species can be divided into two broad categories based on their reproductive strategies. **K-strategists** are usually large organisms that have relatively long lives, produce few offspring, and provide care for their offspring. Their populations typically stabilize at the carrying capacity for the area. Their reproductive strategy is to invest a great deal of energy in producing a few offspring that have a good chance of living to reproduce. Deer, lions, and swans are examples of this kind of organism. Humans generally produce single offspring, and even in countries with high infant mortality, 80 percent of the children survive beyond one year of age, and the majority of these will reach adulthood. Generally, populations of K-strategists are controlled by density-dependent limiting factors. **Density-dependent limiting factors** are those that become more severe as the size of the population increases. For example, there is a carrying capacity that limits the size of a hawk population.

Hawks feed on mice, snakes, and small birds. As the size of the hawk population increases, the competition among hawks for available food becomes more severe. Increased competition for food leads to less for the young in the nest. Many of the young die, and population growth rate slows as the carrying capacity for the area is reached.

The **r-strategist** is typically a small organism that has a short life, produces many offspring, and does not reach a carrying capacity. Examples are grasshoppers, gypsy moths, and some mice. The reproductive strategy of r-strategists is to expend large amounts of energy producing many offspring but to provide limited care (often none) for them. Consequently, there is high mortality among the young. For example, one female oyster may produce a million eggs, but few of them ever find suitable places to attach themselves and grow. Typically, these populations are limited by **density-independent limiting factors** in which the size of the population has nothing to do with the limiting factor. Typical density-independent limiting factors include changing weather conditions that kill large numbers of organisms, habitat loss when a pond dries up or fire destroys a forest, or events such as a deep snow or flood that buries sources of food and leads to the death of entire populations. The population size of r-strategists is likely to fluctuate wildly. They reproduce rapidly, and the size of the population increases until some density-independent factor causes the population to crash; then they begin the cycle all over again. (The letters *K* and *r* in *K-strategist* and *r-strategist* come from a mathematical equation in which *K* represents the carrying capacity of the environment and *r* represents the biotic potential of the species.)

Since humans are K-strategists, it may be difficult for us to appreciate that the r-strategy can be viable from an evolutionary point of view. There is no carrying capacity for temporary resources. Resources that are present only for a short time can be exploited most effectively if many individuals of one species monopolize the resource, while denying other species access to it. Rapid reproduction can place a species in a position

to compete against other species that are not able to increase numbers as rapidly. Obviously, most of the individuals will die, but not before they have left some offspring or resistant stages that will be capable of exploiting the resource should it become available again.

Even K-strategists, however, have population fluctuations for a variety of reasons. One reason is that the environment is not constant from year to year. Floods, droughts, fires, extreme cold, and similar events may affect the carrying capacity of an area, thus causing fluctuations in population size. Epidemic disease or increased predation may also lead to populations that vary from year to year. Figure 7.8 shows rather substantial changes in the populations of lynx and varying hare. The size of the lynx population seems to be tied to the size of the hare population, which is logical since lynx eat hares. However, the causes of fluctuations in the varying hare population are unclear. One possibility is that periodic epidemic disease may cause dense populations of organisms like hares to crash, leading to the crash of their predators' populations.

Although local human populations often show fluctuations, the worldwide human population has increased continually for the past several hundred years. Humans have been able to reduce environmental resistance by eliminating competing organisms, increasing food production, and controlling disease organisms.

Human Population Growth

The human population growth curve has a long lag phase followed by a sharply rising exponential growth phase that is still rapidly increasing. (See figure 7.9.) A major reason for the continuing increase is that the human species has lowered its death rate. When various countries reduce environmental resistance by increasing food production or controlling disease, they share this technology

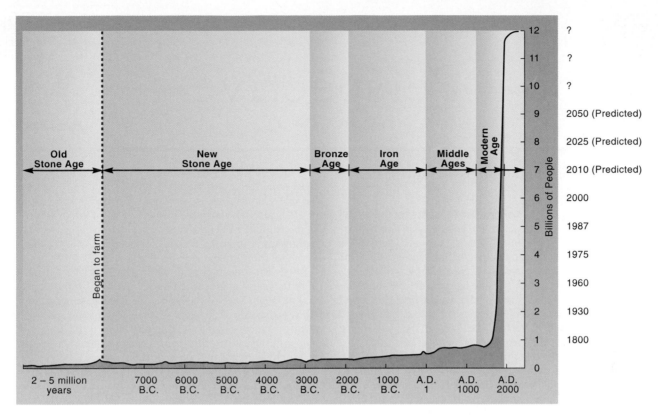

figure 7.9 **The Historical Human Population Curve** From A.D. 1800 to A.D. 1930, the number of humans doubled (from 1 billion to 2 billion) and then doubled again by 1975 (4 billion) and could double again (8 billion) by the year 2025. How long can this pattern continue before the Earth's ultimate carrying capacity is reached?

Source: Data from Jean Van Det Tak, et al., "Our Population Predicament: A New Look," *Population Bulletin*, vol. 34, no. 5 (December 1979), Population Reference Bureau, Washington, D.C.; and more recent data from the Population Reference Bureau.

throughout the world. Developed countries send health care personnel to all parts of the globe to improve the quality of life for people in less-developed countries. Physicians offer advice on nutrition, and engineers develop wastewater treatment systems. Improved sanitary facilities in India and Indonesia, for example, decreased deaths caused by cholera. These advancements tend to reduce death rates while birthrates remain high. Thus, the size of the human population increases rapidly.

Let us examine the human population situation from a different perspective. The world population is currently increasing at an annual rate of 1.3 percent. That may not seem like much, but even at 1.3 percent, the population is growing rapidly. It can be difficult to comprehend the impact of a 1 or 2 percent annual increase. Remember that a growth rate in any population compounds itself, since many of the additional individuals eventually reproduce, thus adding more individuals. One way

to look at this growth is to determine how much time is needed to double the population. This is a valuable method because most of us can appreciate what life would be like if the number of people in our locality were doubled, particularly if the doubling were to occur within our lifetime.

Figure 7.10 shows the relationship between the rate of annual increase for the human population and the number of years it would take to double the population if that rate were to continue. The doubling time for the human population is easily calculated by dividing the number 70 by the annual rate of increase. Thus, at a 1 percent rate of annual increase, the population will double in 70 years (70/1). At a 2 percent rate of annual increase, the human population will double in 35 years (70/2). The current worldwide rate of annual increase of about 1.3 percent will double the world human population in about 54 years.

What does this very rapid rate of growth mean to the human species? As a

species, humans are subject to the same limiting factors as all other species. We cannot increase beyond our ability to acquire raw materials and energy and safely dispose of our wastes. We also must remember that interactions with other species and with other humans will help determine our carrying capacity. Let us look at these four factors in more detail.

Available Raw Materials

Many of us think of raw materials simply as the amount of food available. However, we have become increasingly dependent on technology, and our lifestyles are directly tied to our use of other kinds of resources, such as irrigation water, genetic research, and antibiotics. Food production is becoming a limiting factor for some segments of the world's human population. Malnutrition is a serious problem in many parts of the world because sufficient food is not available. Currently, about 1 billion people (one-sixth of the world's population)

Managing Elephant Populations—Harvest or Birth Control?

During the 1980s, uncontrolled hunting of elephants for ivory and food had reduced the African elephant, *Loxodonta africana,* population from 1.3 million to 650,000. In addition, expansion of agriculture into areas that had been excellent elephant habitat compounded the problem because it reduced the amount of land available to elephants. Many conservation organizations became alarmed by the rapid decline in the numbers of elephants and supported a ban on the export of ivory. In 1989, the Convention on International Trade in Endangered Species of Wild Fauna and Flora (CITES) banned all international trade in elephant products, of which ivory is the most important. The ban worked. The price of ivory fell and poaching became unprofitable. Although the total population of elephants in Africa continued to fall to about 500,000 by the year 2000, the decline is not true for all parts of Africa. Some southern African countries actually have experienced an elephant population explosion that threatens to destroy the limited habitat available to elephants and increasingly results in conflicts between farmers and elephants.

The African elephant requires huge amounts of food (150 to 250 kilograms per day [330 to 550 pounds per day]) to sustain its large body. The animals strip bark from trees, uproot the trees, and eat large quantities of grass. Where humans and elephants share the same habitat, elephants can do great damage to crops. Because elephants have been increasingly confined to national parks and nature preserves by agricultural expansion, two natural options for relieving population pressure (migration or starvation) have not been available. Migration from the park results in increased agricultural damage and risk of injury to farmers, and increasing populations can cause irreparable damage to the protected habitat they occupy as they seek food. If they are allowed to starve, there will be enormous public pressure from wildlife groups, and the park managers will be condemned. If the populations are to remain healthy, not destroy their habitat, and not come into conflict with farming communities, the size of the population must be controlled. There are only two ways to do that—increase the death rate (culling) or decrease the birthrate (sterilization or birth control).

Because they have growing elephant populations, several southern African countries, such as Botswana, Namibia, Zimbabwe, and South Africa, have argued that they should be allowed to manage their elephant populations as a natural resource in the same way that other large game animals (deer, elk, caribou) are managed. The sustainable harvest of surplus elephants would be wise use of a natural resource that would provide income to the local people as well as a much needed source of protein in this region. Furthermore, the income from harvesting elephants could be used to provide additional funding for park management. Those who oppose harvesting maintain that killing the animals for trophies or ivory will take many older individuals from the herd. It is well known that older elephants serve as leaders. Selectively killing older animals, therefore, could remove important memories that could disrupt traditional migration patterns and other important behaviors necessary to their survival.

In 1997, CITES approved the sale of 60 tonnes (66 U.S. tons) of ivory from Botswana, Namibia, and Zimbabwe to Japan. Following the sale, conservation organizations reported an increase in poaching of elephants and unlawful sale of ivory. In 2000, several countries again petitioned CITES to approve additional sales of ivory. Their request was not approved because of concerns about poaching and monitoring of illegal sales.

The Humane Society of the United States has advocated the use of birth control to reduce the birthrate and solve the elephant population problem. Several methods have been tested and they can work. However, critics of this approach suggest that preventing female elephants from conceiving for long periods of time may disrupt the normal social structure of the herd. Under most circumstances, a female that is pregnant will not come into heat again for two years. If females do not conceive, they will come into heat approximately every 15 weeks, and there will be more females in heat than is normal throughout the year. Since the mating activity of elephants is disruptive of their usual pattern of activity, some people are concerned that the normal social structure of the herd will be disrupted. Furthermore, a herd usually includes various ages of immature individuals. The younger individuals learn from older siblings or cousins. The use of birth control could disrupt the typical spacing of pregnancies and interfere with the normal process of "educating" young elephants. What looks like a simple problem to solve has resulted in two diametrically opposed camps: those who advocate increased deaths by harvesting and those who advocate decreased births by manipulating births. Neither is "natural," but something must be done or the survival of elephants could be threatened.

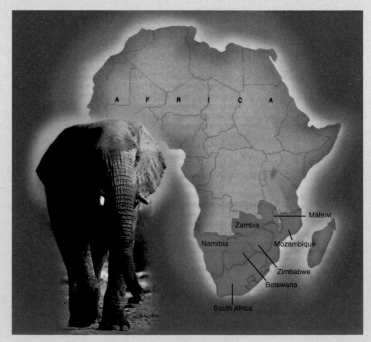

suffer from a lack of adequate food. Chapter 8 deals in greater detail with the problems of food production and distribution and their relationship to human population growth.

Available Energy

The second factor, available energy, involves problems similar to those of raw materials. Essentially, all species on earth are ultimately dependent on sunlight for their energy. New, less disruptive methods of harnessing this energy must be developed to support an increasing population. Currently, the world population depends on fossil fuels to raise food, modify the environment, and move from place to place. When energy prices increase, much of the world's population is placed in jeopardy because incomes are not sufficient to pay the increased costs for energy and other essentials.

Waste Disposal

Waste disposal is the third factor determining the carrying capacity for humans. Most pollution is, in reality, the waste product of human activity. Lack of adequate sewage treatment and safe drinking water causes large numbers of deaths each year. Some people are convinced that disregard for the quality of our environment will be a major limiting factor. In any case, it makes good sense to control pollution and to work toward cleaning our environment.

Interaction with Other Organisms

The fourth factor that determines the carrying capacity of a species is interaction with other organisms. We need to become aware that we are not the only species of importance. When we convert land to meet our needs, we displace other species from their habitats. Many of these displaced organisms are not able to compete with us successfully and must migrate or become extinct. Unfortunately, as humans expand their domain, the areas available to these displaced organisms become more rare.

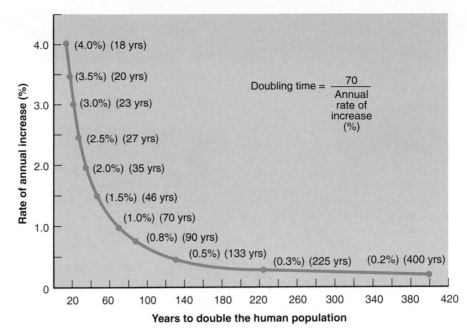

figure 7.10 **Doubling Time for the Human Population** This graph shows the relationship between the rate of annual increase in percent and doubling time. A population growth rate of 1 percent per year would result in the doubling of the population in about 70 years. A population growth rate of 3 percent per year would result in a population doubling in about 23 years.

Parks and natural areas have become tiny refuges for the plants and animals that once occupied vast expanses of land. If these refuges fall to the developer's bulldozer or are converted to agricultural use, many organisms will become extinct. What today seems like an unimportant organism, one that we could easily do without, may someday be seen as an important link to our very survival. It is also important to recognize that many organisms provide services that we enjoy without thinking about them. Forest trees release water and moderate temperature changes, bees and other insects pollinate crops, insect predators eat pests, and decomposers recycle dead organisms. All of these activities illustrate how we rely on other organisms. Eliminating the services of these valuable organisms would be detrimental to our way of life.

Social Factors Influence Human Population

Human survival depends on interaction and cooperation with other humans.

Current technology and medical knowledge are available to control human population growth and to improve the health of the people of the world. Why, then, does the population continue to increase, and why do large numbers of people continue to live in poverty, suffer from preventable diseases, and endure malnutrition? Humans are social animals who have freedom of choice and frequently do not do what is considered "best" from an unemotional, uninvolved, biological point of view. People make decisions based on history, social situations, ethical and religious considerations, and personal desires. The biggest obstacles to controlling human population are not biological but are the province of philosophers, theologians, politicians, and sociologists. People in all fields need to understand that the cause of the population problem has both biological and social components if they are to successfully develop strategies for addressing it. Chapter 8 will cover many of these social, political, and ethical topics in greater detail.

Wolves and Moose on Isle Royale

Isle Royale in Lake Superior has been the site of a long-term study of the relationship between moose and wolf populations. Wolves (probably a single pair) reached Isle Royale by crossing the ice from Canada during the winter of 1947–48 and had reached a population of about 20 by 1958, when a succession of biologists began a long-term study of the wolves and their relationships with moose and other organisms on the island. In over 50 years, they have learned a great deal about the food habits of wolves, pack behavior, the relationships among various wolf packs, and the relationship between the predator (wolf) and prey (moose) populations. Although several factors affect population size for wolves and moose, an examination of the fluctuations in moose and wolf populations suggests that they have an effect on one another. When wolf populations decline, moose populations rise, and when wolf populations rise, moose populations fall. The major changes in moose and wolf populations since 1980 appear to have had triggering events that resulted in major

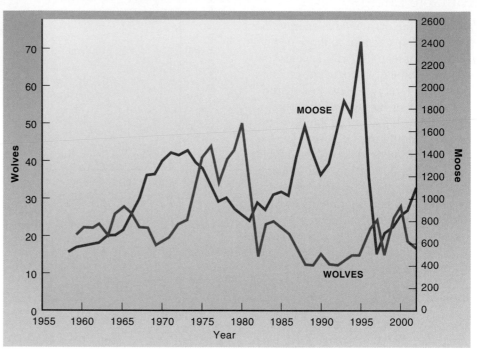

Source: Data from Rolf O. Peterson, "Ecological Studies of Wolves on Isle Royale," *Annual Report 1995–96* and subsequent reports.

Ultimate Size Limitation

The human population is subject to the same biological constraints as other species of organisms. We can say with certainty that our population will ultimately reach its carrying capacity and stabilize. There is disagreement about how many people can exist when the carrying capacity is reached. Some people suggest we are already approaching the carrying capacity, while others maintain that we could more than double the population before the carrying capacity is reached. Furthermore, uncertainty exists about what the primary limiting factors will be and about the quality of life the inhabitants of a more populous world would have. If the human population continues to reproduce at its current rate, the population will double from its current 6.2 billion to 12 billion people by 2055.

If the reproduction rate falls so that each woman produces only two children during her lifetime, the world will contain 9 billion people by the year 2050. As with all K-strategist species, when the population increases, density-dependent limiting factors will become more forceful. Some people suggest that a lack of food, a lack of water, or increased waste will ultimately control the size of the human population. Still others suggest that, in the future, social controls will limit population growth. These social controls could be either voluntary or involuntary. In the economically developed portions of the world, families have voluntarily lowered their birthrates to fewer than two children per woman. Most of the poorer countries of the world have higher birthrates. What kinds of measures are needed to encourage them to limit their populations? Will voluntary compliance with stated national goals be enough, or will enforced sterilization and economic

penalties become the norm? Others are concerned that countries will launch wars to gain control of limited resources or to simply eliminate people who compete for the use of those resources.

It is also important to consider the age structure of the world population. In most of the world, there are many reproductive and prereproductive individuals. Since most of these individuals are currently reproducing or will reproduce in the near future, even if they reduce their rate of reproduction, there will be sharp increase in the number of people in the world in the next few years.

No one knows what the ultimate human population size will be or what the most potent limiting factors will be, but most agree that we are approaching the maximum sustainable human population. If the human population continues to increase, eventually the amount of agricultural land available will not be able to satisfy the demand for food.

reductions in one of the two populations that had a subsequent effect on the other. For example, the major decline in wolves in the early 1980s from about 50 to 13 wolves was probably initiated by the introduction of a viral disease of domesticated dogs (parvovirus). The reduction in the number of wolves allowed a greater number of young moose to survive, and the moose population increased rapidly.

The rapid decline in the moose population during the mid-1990s was probably initiated by a combination of factors. The rapid growth in the years prior to 1995 had reduced the amount of food available and resulted in malnutrition, slow growth, and poor reproduction. A very severe winter and a tick infestation probably contributed to the die-off as well. During the winter of 1995–96, the moose population suffered a 50 percent reduction as it fell to about 1200. This decline continued as the population fell to about 500 individuals in 1997, an 80 percent reduction from its maximum. However, it rebounded to about 700 individuals in 1998. What was a bad time for moose was a good time for wolves, since weak and dying moose provided an abundant food supply, and the wolf population rose to 24 individuals, its highest population since the early 1980s. However, this abundant food supply was short-lived, and the remaining moose were younger, healthy individuals that are harder to take as prey. In 1998, the wolf population fell to 14, raising concern that the population may not be able to sustain itself. On the positive side, a smaller wolf population reduces competition for food. In addition, most of the remaining wolves are young and have the potential to produce many sets of pups over the next several years. On the negative side, recent survival of pups has been low. This could be due to several factors. Perhaps the canine parvovirus is affecting the survival of the pups. Lack of genetic variety may also be contributing to the low survival, since the entire population is thought to be descended from a single mated pair.

Since 1998, the moose population has steadily increased and was about 1100 individuals in 2002. However, the wolf population has fluctuated. It rose from 14 in 1998 to 29 in 2000 but fell to 19 in 2001 and declined further to 17 in 2002. The decline between 2001 and 2002 appears to be related to a mild winter with low snowfall, which made it more difficult for wolves to kill moose. Observations over the next few years may identify causes for the fluctuations in the wolf population. For example, blood samples taken from captured wolves could clarify questions about the presence of parvovirus and concerns about genetic variety. It is also possible that the wide fluctuations seen in these two populations may be normal for isolated, island populations in which prey and predator populations significantly influence one another.

- When populations of animals become endangered, efforts are often made to preserve them. Should special efforts be made to preserve this wolf population?
- If lack of genetic variety is shown to be a contributing factor to the decline of the wolf population, should efforts be made to introduce genetic variety?
- In what ways is this wolf population different from that of populations of California condors or whooping cranes?

Summary

A population is a group of organisms of the same species that inhabits an area. The birthrate (natality) is the number of individuals entering the population by reproduction during a certain period. The death rate (mortality) measures the number of individuals that die in a population during a certain period. Population growth is determined by the combined effects of the birthrate and death rate.

The sex ratio of a population is a way of stating the relative number of males and females. Age distribution and the sex ratio have a profound impact on population growth. Most organisms have a biotic potential much greater than that needed to replace dying organisms.

Interactions among individuals in a population, such as competition, predation, and parasitism, are also important in determining population size. Organisms may migrate into (immigrate) or migrate out of (emigrate) an area as a result of competitive pressure.

A typical population growth curve shows a lag phase followed by an exponential growth phase and a stable equilibrium phase at the carrying capacity. The carrying capacity is determined by many limiting factors that are collectively known as environmental resistance. The four major categories of environmental resistance are available raw materials, available energy, disposal of wastes, and interactions among organisms. Some populations experience a death phase following the stable equilibrium phase.

K-strategists typically are large, long-lived organisms that reach a stable population at the carrying capacity. Their population size is usually controlled by density-dependent limiting factors. Organisms that are r-strategists are generally small, short-lived organisms that reproduce very quickly. Their populations do not generally reach a carrying capacity but crash because of some density-independent limiting factor.

The human population is increasing at a rapid rate. The Earth's ultimate carrying capacity for humans is not known. The causes for human population growth are not just biological but also social, political, philosophical, and theological.

Key Terms

age distribution *136*
biotic potential *140*
birthrate *136*
carrying capacity *141*
death phase *143*
death rate *136*
density-dependent limiting factors *144*
density-independent limiting factors *144*
dispersal *138*

emigration *138*
environmental resistance *141*
exponential growth phase (lag phase) *140*
immigration *139*
K-strategists *144*
lag phase *140*
limiting factor *141*
lag phase *140*
mortality *136*

natality *136*
population *136*
population density *138*
population growth rate *136*
r-strategists *144*
sex ratio *136*
stable equilibrium phase *141*
survivorship curve *136*

Review Questions

1. How is biotic potential related to the rate at which a population will grow?
2. List three characteristics populations might have.
3. Why do some populations grow? What factors help to determine the rate of this growth?
4. Under what conditions might a death phase occur?
5. List four factors that could determine the carrying capacity of an animal species.
6. How do the concepts of birthrate and population growth differ?
7. How does the population growth curve of humans compare with that of bacteria on a petri dish?
8. How do r-strategists and K-strategists differ?
9. As the human population continues to increase, what might happen to other species?
10. All successful organisms overproduce. What advantage does this provide for the species? What disadvantages may occur?

Critical Thinking Questions

1. Why do you suppose some organisms display high natality and others display lower natality? For example, why do cottontail rabbits show high natality and wolves relatively low natality? Why wouldn't all organisms display high natality?
2. Do you think African elephants should be managed by southern African countries like deer are managed in the United States? What values, beliefs, and perspectives lead you to your conclusion?
3. Where do you stand on the issue of birth control for elephants or culling of the herd? Why do you think the way you do? What is the position of the Humane Society in the United States? Why do you think they hold the position they do? Do you think conservationists in the United States should have a say in what is done with the elephants in Africa? Should Africans have a say in what is done in the United States? Why?
4. Consider the differences between r-strategists and K-strategists. What costs are incurred by adopting either strategy? What evolutionary benefits does each strategy enjoy?
5. Why do invading species, which can survive in a new environment, often show exponential growth rates soon after they are introduced to that environment?
6. If a population of organisms, like the population of wolves on Isle Royale, is declining for "natural" reasons, should humans intervene and try to preserve these populations? Why?

Concept Map

Construct a map to show relationships among the following concepts:

carrying capacity
limiting factors
death rate

birthrate
immigration
dispersal

r-strategist
K-strategist

Interactive Exploration

Check out the website at **http://www.mhhe.com/environmentalscience** and click on the cover of this textbook for quizzing, career information, case studies, and hot links for the following topics:

Animal Population Ecology

Field Methods for Studies of Populations

Life Histories

Field Methods for Studies of Ecosystems

Species Abundance, Diversity, and Complexity

Population Density of Animals

Population Growth

Human Population Growth

Invasive and Introduced Species

Endangered Species

8

Human Population Issues

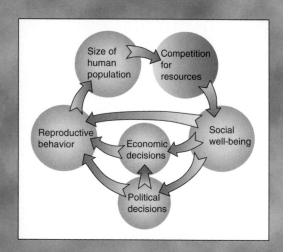

Objectives

After reading this chapter, you should be able to:

- Apply some of the principles discussed in chapter 7 to the human population.
- Differentiate between birthrate and population growth rate.
- Describe the current population situation in the United States.
- Explain why the age distribution and the status and role of women affect population growth projections.
- Recognize that countries in the developed world are experiencing an increase in the average age of their populations.
- Recognize that most countries of the world have a rapidly growing population.
- Describe the implications of the demographic transition concept.
- Understand how an increasing world population will alter the worldwide ecosystem.
- Recognize that rapid population growth and poverty are linked.
- Explain why less-developed nations have high birthrates and why they will continue to have a low standard of living.
- Recognize that the developed nations of the world will be under greater pressure to share their abundance.

Chapter Outline

Human Population Trends and Implications

Global Perspective: *Thomas Malthus and His Essay on Population*

Factors That Influence Population Growth

  Biological Factors

  Social Factors

Environmental Close-Up: *Control of Births*

  Political Factors

Population Growth and Standard of Living

Population and Poverty—A Vicious Cycle?

Hunger, Food Production, and Environmental Degradation

The Demographic Transition Concept

Global Perspective: *The Urbanization of the World's Population*

The U.S. Population Picture

Anticipated Changes with Continued Population Growth

Global Perspective: *North America— Population Comparisons*

Issues—Analysis: *The Impact of AIDS on Populations*

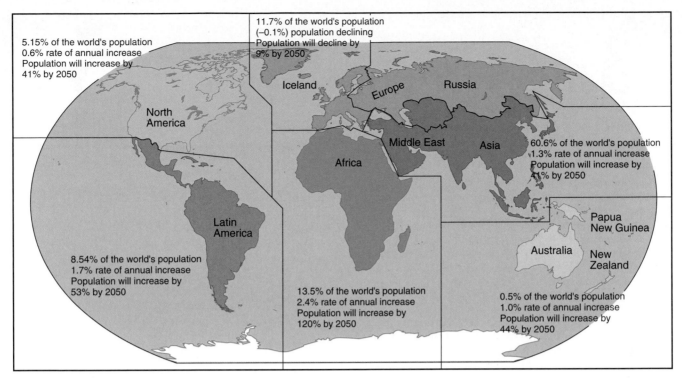

figure 8.1 **Population Growth in the World (2002)** The population of the world is not evenly distributed. Currently, over 82 percent of the world's population is in Latin America, Africa, and Asia. These areas also have the highest rates of increase and are generally considered less developed. Because of the high birthrates, they are likely to remain less developed and will constitute over 87 percent of the world's population by the year 2050.

Map labels:

5.15% of the world's population
0.6% rate of annual increase
Population will increase by 41% by 2050

11.7% of the world's population (−0.1%) population declining
Population will decline by 9% by 2050

60.6% of the world's population
1.3% rate of annual increase
Population will increase by 41% by 2050

8.54% of the world's population
1.7% rate of annual increase
Population will increase by 53% by 2050

13.5% of the world's population
2.4% rate of annual increase
Population will increase by 120% by 2050

0.5% of the world's population
1.0% rate of annual increase
Population will increase by 44% by 2050

Region labels: North America, Iceland, Europe, Russia, Middle East, Asia, Africa, Latin America, Papua New Guinea, Australia, New Zealand

Human Population Trends and Implications

The human population dilemma is very complex. To appreciate it, we must understand current population trends and how they are related to social, political, and economic conditions. Currently, the world population is over 6 billion people. By the year 2025, this is expected to increase to over 7.8 billion. Much of this increase is expected to occur in Africa, Asia, and Latin America, which already have over 82 percent of the world's population. (See figure 8.1.) If these trends continue, the total population of Africa, Asia, and Latin America will increase from the current 5.1 billion to over 6.7 billion by 2025, when these continents will contain over 85 percent of the world's people.

Human population growth is a contributing factor in nearly all environmental problems. Current population growth has led to famine in areas where food production cannot keep pace with increasing numbers of people; political unrest in areas with great disparities in availability of resources (jobs, goods, food); environmental degradation (erosion, desertification) by poor agricultural practices; water pollution by human and industrial waste; air pollution caused by the human need to use energy for personal and industrial applications; extinctions caused by people converting natural ecosystems to managed agricultural ecosystems; and destructive effects of exploitation of natural resources (strip mining, oil spills, groundwater mining).

Several factors interact to determine the impact of a society on the resources of its country. These include the size of the population, the land area the people occupy, and their degree of technological development. The larger the size of a population, the greater the demand on the resources of the country. However, population size is not the only important factor. **Population density,** the number of people per unit of land area, is also important. A million people spread out over the huge area of the Amazon have much less impact on resources than that same million people in a small island country, because the impact is distributed over a greater land surface.

The degree of technological development and affluence is also important. The environmental impact of the developed world is often underestimated because the population in these countries is relatively stable and local environmental conditions are good. However, the people in highly developed countries consume huge amounts of resources. Citizens of these countries eat more food, particularly animal protein, which requires larger agricultural inputs than does a vegetarian diet. They have more material possessions and consume vast amounts of energy. Developed countries purchase goods and services from other parts of the world, often degrading environmental conditions in less-developed countries. Thus, the environmental impact of highly developed regions like North America, Japan, and Europe is often felt in distant places, while the

Global Perspective

Thomas Malthus and His Essay on Population

In 1798, Thomas Robert Malthus, an Englishman, published an essay on human population. In it, he presented an idea that was contrary to popular opinion. His basic thesis was that human population increased in a geometric or exponential manner (2, 4, 8, 16, 32, 64, etc.), while the ability to produce food increased only in an arithmetic manner (1, 2, 3, 4, 5, 6, etc.). The ultimate outcome of these different rates would be that population would outgrow the ability of the land to produce food. He concluded that wars, famines, plagues, and natural disasters would be the means of controlling the size of the human population. His predictions were hotly debated by the intellectual community of his day. His assumptions and conclusions were attacked as erroneous and against the best interest of society. At the time he wrote the essay, the popular opinion was that human knowledge and "moral constraint" would be able to create a world that would supply all human needs in abundance. One of Malthus's basic postulates was that "commerce between the sexes" (sexual intercourse) would continue unchanged, while other philosophers of the day believed that sexual behavior would take less procreative forms and human population would be limited. Only within the past 50 years, however, have really effective conception-control mechanisms become widely accepted and used, and they are used primarily in developed countries.

Malthus did not foresee the use of contraception, major changes in agricultural production techniques, or the exporting of excess people to colonies in the Americas, Australia, and other parts of the world. These factors, as well as high death rates, prevented the most devastating of his predictions from coming true. However, in many parts of the world today, people are experiencing the forms of population control (famine, epidemic disease, wars, and natural disasters) predicted by Malthus in 1798. Many people feel that his original predictions were valid—only his time scale was not correct—and that we are seeing his predictions come true today.

---

impact on resources in the developed region may be minimal.

While controlling world population growth would not eliminate all environmental problems, it could reduce the rate at which environmental degradation is occurring. It is also generally believed that the quality of life for many people in the world would improve if their populations grew less rapidly. Why, then, does the human population continue to grow at such a rapid rate?

Factors That Influence Population Growth

In chapter 7, we examined populations from a biological point of view. We looked at their characteristics, the causes of growth, and the forces that cause populations to stabilize. All of these biological factors apply to human as well as nonhuman populations. There is an ultimate carrying capacity for the human population. Eventually, limiting factors will cause human populations to stabilize. However, unlike other kinds of organisms, humans are also influenced by social, political, economic, and ethical factors. We have accumulated knowledge that allows us to predict the future. We can make conscious decisions based on the likely course of events and adjust our lives accordingly. Part of our knowledge is the certainty that as populations continue to increase, death rates and birthrates will become equal. This can happen by allowing the death rate to rise or by choosing to limit the birthrate. Controlling human population would seem to be a simple process. Once people understand that lowering the birthrate is more humane than allowing the death rate to rise, they should make the "correct" decision and control their birthrates; however, it is not quite that simple.

Biological Factors

The scientific study of human populations, their characteristics, how these characteristics affect growth, and the consequences of that growth is known as **demography.** Demographers can predict the future growth of a population by looking at several biological indicators. Currently, in almost all countries of the world, the birthrate exceeds the death rate. Therefore, the size of the population must increase. (See table 8.1.) Some countries that have high birthrates and high death rates—with birthrates greatly exceeding the death rates—will grow rapidly (Nigeria and Ethiopia). Such countries usually have an extremely high mortality rate among children because of disease and malnutrition; but because the birthrate still greatly exceeds the death rate, the populations will grow rapidly.

Some countries have high birthrates and low death rates and will grow extremely rapidly (Mexico and Indonesia). Infant mortality rates are moderately high in these countries. Other countries have low birthrates and death rates that closely match the birthrates; they will grow slowly (Japan and the United Kingdom). These and other more-developed countries typically have very low infant mortality rates. The disruption caused by the political upheaval in the former Soviet Union and Eastern

Table 8.1 Population Characteristics of the 20 Most Populous Countries (2002)

Country	Current Population (Millions)	Births per 1000 Individuals	Deaths per 1000 Individuals	Infant Mortality Rate (Deaths per 1000 Live Births)	Total Fertility Rate (Children per Woman per Lifetime)	% Married Women Using Birth Control	Rate of Natural Increase (Annual %)	Projected Population Change 2002–2050 (%)
World	*6215*	*21*	*9*	*54*	*2.8*	*61*	*1.3*	*+46%*
Russia	143.5	9	16	16	1.3	67	(−0.7)	(−29%)
Germany	82.4	9	10	4.4	1.3	75	(−0.1)	(−18%)
United Kingdom	60.2	11	10	5.6	1.6	72	0.1	+9%
Japan	127.4	9	8	3.2	1.3	56	0.2	(−21%)
United States	287.4	15	9	7.1	2.1	76	0.6	+44%
China	1,280.7	13	6	31	1.8	84	0.7	+9%
Thailand	62.6	14	6	20	1.8	72	0.8	+15%
Iran	65.6	18	6	32	2.5	74	1.2	+47%
Brazil	173.8	20	7	33	2.2	76	1.3	+42%
Vietnam	79.7	19	5	30	2.3	74	1.4	+47%
Turkey	67.3	22	7	35	2.5	64	1.5	+44%
Indonesia	217.0	22	6	46	2.6	57	1.6	+46%
India	1,049.5	26	9	68	3.2	48	1.7	+55%
Egypt	71.2	27	7	44	3.5	56	2.0	+62%
Pakistan	143.5	30	9	86	4.8	28	2.1	+131%
Mexico	101.7	26	5	25	2.9	69	2.1	+48%
Philippines	80.0	28	6	26	3.5	47	2.2	+82%
Bangladesh	133.6	30	8	66	3.3	54	2.2	+56%
Ethiopia	67.7	40	15	97	5.9	8	2.5	+155%
Nigeria	129.9	41	14	75	5.8	15	2.7	+134%

Source: Data from *World Population Data Sheet 2002,* Population Reference Bureau, Washington, D.C.

Europe has resulted in several countries (e.g., Russia and Germany) having death rates that are equal to or exceed birthrates, causing their populations to decline. Because of these countries and the generally low rates of growth in the rest of Europe, the European region as a whole has a declining population.

The most important determinant of the rate at which human populations grow is related to how many women in the population are having children and the number of children each woman will have. The **total fertility rate** of a population is the number of children born per woman in her lifetime. A total fertility rate of 2.1 is known as **replacement fertility,** since parents produce 2 children who will replace the parents when they die. Eventually, if the total fertility rate is maintained at 2.1, population growth will stabilize. A rate of 2.1 is used rather than 2.0 because some children do not live very long after birth and

therefore will not contribute to the population for very long. When population is not growing, and the number of births equals the number of deaths, it is said to exhibit **zero population growth.**

A total fertility rate of 2.1 will not necessarily immediately result in a stable population with zero growth for several reasons. First, the death rate may fall as living conditions improve and people live longer. If the death rate falls faster than the birthrate, there will still be an increase in the population even though it is reproducing at the replacement rate.

The **age distribution,** the number of people of each age in the population, also has a great deal to do with the rate of population growth. If a population has many young people who are raising families or who will be raising families in the near future, the population will continue to increase even if the families limit themselves to two children. Depending on the number of young people

in a population, it may take 20 years to a century for the population of a country to stabilize so that there is no net growth.

Social Factors

It is clear that populations in economically developed countries of the world have low fertility rates and low rates of population growth and that the less-developed countries have high fertility rates and high population growth rates. It also appears obvious that reducing fertility rates would be to everyone's advantage; however, not everyone in the world feels that way. Several factors influence the number of children a couple would like to have. Some are religious, some are traditional, some are social, and some are economic.

The major social factors that determine family size are the status and desires of women in the culture. In many male-dominated cultures, the traditional

environmental CLOSE-UP

Control of Births

The use of technology to control disease and famine has greatly reduced the death rate of the human population. Technological developments can also be used to control the birthrate. Birth control refers to anything that reduces the number of births. Some processes prevent conception and can be called contraceptives. A variety of contraceptive methods are available to help people regulate their fertility. Research is continuing to develop more effective, more acceptable, and less expensive methods of controlling conception. Because of cultural and religious differences, some forms of contraception may be more acceptable to one segment of the world's population than to another.

The most common methods of contraception are oral contraceptive pills, diaphragms and spermicidal jelly, spermicidal vaginal foam, condoms, vasectomy, and tubal ligation. The range of effectiveness of these methods, shown in the table, is the result of individual fertility differences and the degree of care employed in the use of each method.

Some birth control devices such as intrauterine contraceptive devices (IUDs) are thought to prevent embryos from attaching to the uterine wall and completing development and, therefore, control births but do not prevent conception. Abortion is the physical intervention in the development of the embryo in the uterus that is used to terminate unwanted pregnancies. Most countries with low birthrates, such as the United States, Japan, and many European countries, have ready access to contraceptive methods and allow abortions.

Effectiveness of Various Methods of Contraception

Method	Percent of Women Experiencing an Unintended Pregnancy Within the First Year of Use	
	Typical Use	Perfect Use
No contraceptive method used	85	85
Spermicidal foams, creams, gels, suppositories, and vaginal films	26	6
Cervical cap		
Women who have had children	40	26
Women who have not had children	20	9
Sponge		
Women who have had children	40	20
Women who have not had children	20	9
Female condom	21	5
Diaphragm with spermicide	20	6
Withdrawal	19	4
Male condom	14	3
Periodic abstinence (natural family planning)		
Calendar method		9
Ovulation method		3
Temperature method		2
Postovulation method		1
Intrauterine device (IUD)	2	1.5
Female sterilization (tubal ligation)	0.5	0.5
Contraceptive pill		0.5
Contraceptive injection (Depo-Provera)	0.3	0.3
Male sterilization (vasectomy)	0.15	0.10
Contraceptive implant (Norplant)	0.05	0.05

Source: Data from J. Trussel, "Contraceptive Efficacy," in R. A. Hatcher, et al., *Contraceptive Technology*, 17th rev. ed., 1998, Irvington Publishers.

role of women is to marry and raise children. Often this role is coupled with strong religious input as well. Typically, little value is placed on educating women, and early marriage is encouraged. In these cultures, women are totally dependent on their husbands and children in old age. Because early marriage is encouraged, fertility rates are high, since women are exposed to the probability of pregnancy for more of their fertile years. Lack of education reduces options for women in these cultures. They do not have the option to not marry or to delay marriage and thus reduce the number of children they will bear. By contrast, in much of the developed world, women are educated, delay marriage, and have fewer children. It has been said that the single most important activity needed to reduce the world population growth rate is to educate women. Whenever the educational level of women increases, fertility rates fall. Figure 8.2 compares total fertility rates and educational levels of women in the 20 most populous countries of the world. Educational level of women is strongly correlated with the total fertility rate and economic well-being of a population.

Data on the age of mothers giving birth indicate that 17 percent of births in Africa are to women in the 15- to 19-year-old range. This is true of 16 percent of births in Latin America and 9 percent of the births in Asia. The total

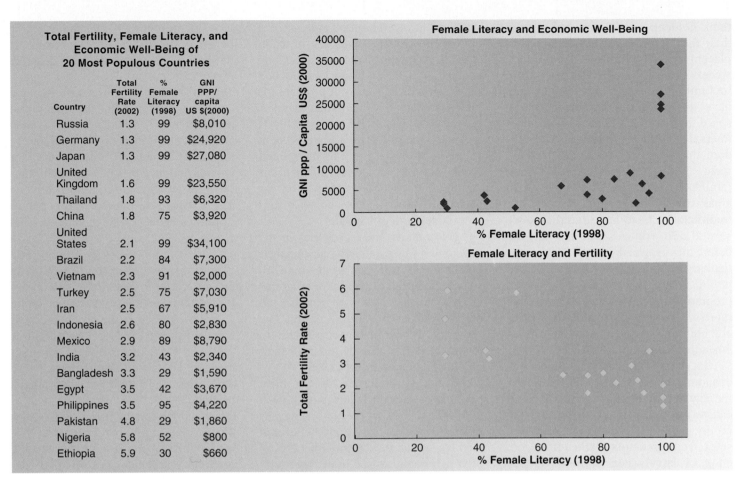

Total Fertility, Female Literacy, and Economic Well-Being of 20 Most Populous Countries			
Country	Total Fertility Rate (2002)	% Female Literacy (1998)	GNI PPP/ capita US $(2000)
Russia	1.3	99	$8,010
Germany	1.3	99	$24,920
Japan	1.3	99	$27,080
United Kingdom	1.6	99	$23,550
Thailand	1.8	93	$6,320
China	1.8	75	$3,920
United States	2.1	99	$34,100
Brazil	2.2	84	$7,300
Vietnam	2.3	91	$2,000
Turkey	2.5	75	$7,030
Iran	2.5	67	$5,910
Indonesia	2.6	80	$2,830
Mexico	2.9	89	$8,790
India	3.2	43	$2,340
Bangladesh	3.3	29	$1,590
Egypt	3.5	42	$3,670
Philippines	3.5	95	$4,220
Pakistan	4.8	29	$1,860
Nigeria	5.8	52	$800
Ethiopia	5.9	30	$660

figure 8.2   **Relationship Between Literacy, Fertility, and Economic Well-Being**   The literacy rate of a population is directly correlated with the fertility rate and economic well-being.

Data from Population Reference Bureau 2002 Data Sheet, and *Human Development Report 2000*, United Nations Development Program.

fertility rates of these areas are 5.2, 2.7, and 2.6 children per woman per lifetime, respectively. In the developed world, the average age of first marriage is much higher, between ages 25 and 27; early marriages are rare; about 3 percent of births are to mothers who are between 15 to 19 years of age, and the total fertility rate is less than replacement fertility at about 1.6 children per woman per lifetime.

Even childrearing practices have an influence on population growth rates. In countries where breast feeding is practiced, several benefits accrue. Breast milk is an excellent source of nutrients for the infant as well as a source of antibodies against some diseases. Furthermore, since many women do not return to a normal reproductive cycle until after they have stopped nursing, during the months a woman is breast feeding her child, she is less likely to become preg-

nant again. Since in many cultures, breast feeding may continue for one to two years, it serves to increase the time between successive births. Increased time between births results in a lower mortality among women of childbearing age.

As women become better educated and obtain higher paying jobs, they become financially independent and can afford to marry later and consequently have fewer children. Better-educated women are also more likely to have access to and use birth control. In economically advanced countries, a high proportion of women typically use contraception. In the less-developed countries, contraceptive use is much lower—about 26 percent in Africa, about 70 percent in Latin America, and about 64 percent in Asia (about 52 percent if China is excluded).

It is important to recognize that access to birth control alone will not solve

the population problem. What is most important is the desire of women to limit the size of their families. In developed countries, use of birth control is extremely important in regulating the birthrate. This is true regardless of religion and previous historical birthrates. For example, Italy and Spain are both traditionally Catholic countries, and have low total fertility rates: 1.2 for Spain and 1.3 for Italy. The average for the developed countries of the world is 1.6. Obviously, women in these countries make use of birth control to help them regulate the size of their families. (See Environmental Close-Up: Control of Births.) By contrast, Mexico, which is also a traditionally Catholic country, has a total fertility rate of 2.9, which is typical of birthrates in the less-developed world regardless of religious tradition.

Women in the less-developed world typically have more children than they think is ideal, and the number of children they have is higher than the replacement fertility rate of 2.1 children. Access to birth control will allow them to limit the number of children they have to their desired number and to space their children at more convenient intervals, but they still desire more children than the 2.1 needed for replacement. Why do they desire large families? There are several reasons. In areas where infant mortality is high, it is traditional to have large families since several of a woman's children may die before they reach adulthood. This is particularly important in the less-developed world, where there is no government program of social security. Parents are more secure in old age if they have several children to contribute to their needs when they can no longer work.

In less-developed countries, the economic benefits of children are extremely important. Even young children can be given jobs that contribute to the family economy. They can protect livestock from predators, gather firewood, cook, or carry water. In the developed world, large numbers of children are an economic drain. They are prevented by law from working, they must be sent to school at great expense, and they consume large amounts of the family income. Many parents in the developed world make an economic decision about having children in the same way they buy a house or car: "We are not having children right away. We are going to wait until we are better off financially."

Political Factors

Two other factors that influence the population growth rate of a country are government policies on population growth and immigration. Many countries in Europe have official policies that state that their population growth rates are too low. As their populations age and there are few births, they are concerned about a lack of working-age people in the future and have instituted programs that are meant to encourage people to have children. For example, Hungary, Sweden, and several other European countries provide paid maternity leave for

mothers during the early months of a child's life and the guarantee of a job when the mother returns to work. Many countries provide childcare facilities and other services that make it possible for both parents to work. This removes some of the economic barriers that tend to reduce the birthrate. The tax system in many countries, such as the United States, provides an indirect payment for children by allowing a deduction for each child. Canada pays a bonus to couples on the birth of a child.

By contrast, most countries in the developing world publicly state that their population growth rates are too high. To reduce the birthrate, they have programs that provide information on maternal and child health and on birth control. The provision of free or low-cost access to contraceptives is usually a part of their population-control effort as well.

China and India are the two most populous countries in the world, each with over a billion people. China has taken steps to control its population and now has a total fertility rate of 1.8 children per woman while India has a total fertility rate of 3.2. This difference between these two countries is the result of different policy decisions over the last 50 years. The history of China's population policy is an interesting study of how government policy affects reproductive activity among its citizens. When the People's Republic of China was established in 1949, the official policy of the government was to encourage births, because more Chinese would be able to produce more goods and services, and production was the key to economic prosperity. The population grew from 540 million to 614 million between 1949 and 1955, while economic progress was slow. Consequently, the government changed its policy and began to promote population control.

The first family-planning program in China began in 1955 as a means of improving maternal and child health. Birthrates fell. (See figure 8.3.) In addition, other social changes resulted in widespread famine, increased death rates, and low birthrates in the late 1950s and early 1960s.

The present family-planning policy began in 1971 with the launching of the *wan xi shao* campaign. Translated, the phrase means "later" (marriages), "longer" (intervals between births), and "fewer" (children). This program raised the legal ages for marriage. For women and men in rural areas, the ages were raised to 23 and 25, respectively; for women and men in urban areas, the ages were raised to 25 and 28, respectively. These policies resulted in a reduction of birthrates by nearly 50 percent between 1970 and 1979.

An even more restrictive, one-child campaign was begun in 1978–79. The program offered incentives for couples to restrict their family size to one child. Couples enrolled in the program would receive free medical care, cash bonuses for their work, special housing treatment, and extra old-age benefits. Those who broke their pledge were penalized by the loss of these benefits as well as other economic penalties. By the mid-1980s, less than 20 percent of the eligible couples were signing up for the program. Rural couples particularly desired more than one child. In fact, in a country where about 70 percent of the population is rural, the rural total fertility rate was 2.5 children per woman. In 1988, a second child was sanctioned for rural couples if their first child was a girl, which legalized what had been happening anyway.

The current total fertility rate is 1.8 children per woman. Over 80 percent of couples use contraception; the most commonly used forms are male and female sterilization and the intrauterine device. Abortion is also an important aspect of this program, with a ratio of over 600 abortions per 1000 live births.

By contrast, during the same 50 years, India has had little success in controlling its population. In 2000, a new plan was unveiled that had the goal of bringing the total fertility rate from its current 3.2 children per woman to 2 (replacement rate) by 2010. In the past, the emphasis of government programs was on meeting goals of sterilization and contraceptive use, but this has not been successful. Today, about 48 percent of couples use contraceptives. This new

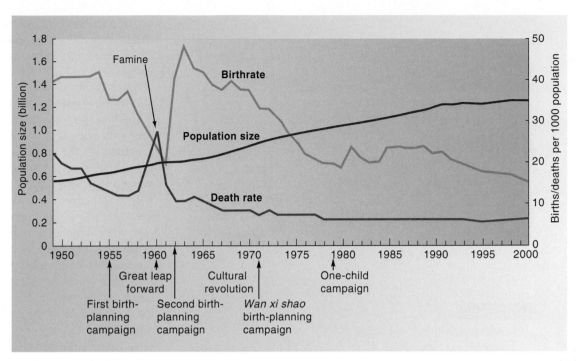

figure 8.3   **Population Changes in China** China has had a long history of actively promoting population control. This graph shows the changes that have occurred in birthrates, death rates, and total population as a result of significant policy initiatives.

Source: Data from H. Yuan Tien, "China's Demographic Dilemmas," *Population Bulletin, 1992*, Population Reference Bureau, Inc., Washington, D.C., and National Family Planning Commission of China; and more recent data from the Population Reference Bureau.

plan emphasizes improvements in the quality of life of the people. The major thrusts are to reduce infant and maternal death, immunize children against preventable disease, and encourage girls to attend school. It is hoped that improved health will remove the perceived need for large numbers of births. Currently, less than 50 percent of the women in India can read and write. The emphasis on improving the educational status of women is related to the experiences of other developing countries. In many other countries, it has been shown that an increase in the education level of women is linked to lower fertility rates.

The immigration policies of a country also have a significant impact on the rate at which the population grows. Birthrates are currently so low in several European countries, Japan, and China that these countries will likely have a shortage of those of working age in the near future. One way to solve this problem is to encourage immigration from other parts of the world.

The developed countries are under tremendous pressure to accept immigrants. The standard of living in these countries is a tremendous magnet for refugees or people who seek a better life than is possible where they currently live. The continuing economic and political reorganization in central Europe is causing significant increases in the number of immigrants and placing considerable strain on the social systems of Germany, Austria, and other countries of the region. In the United States, approximately one-third of the population increase experienced each year is the result of immigration. Canada encourages immigrants and has set a goal of accepting 300,000 new immigrants each year. This is 1 percent of its current population.

Population Growth and Standard of Living

There appears to be an inverse relationship between the rate at which the popu-

lation of a country is growing and its standard of living. The **standard of living** is an abstract concept that attempts to quantify the quality of life of people. Standard of living is a difficult concept to quantify since various cultures have different attitudes and feelings about what is desirable. However, several factors can be included in an analysis of standard of living: economic well-being, health conditions, and the ability to changes one's status in the society. Figure 8.4 lists several factors that are important in determining standard of living and contrasts three countries with very different standards of living (the United States, Argentina, and Kenya). One important economic measure of standard of living is the average purchasing power per person. One index of purchasing power is the **gross national income (GNI).** The GNI is an index that measures the total goods and services generated within a country as well as income earned by citizens of the country who are living in other countries. Since the prices of goods and services vary from one country to another, a true

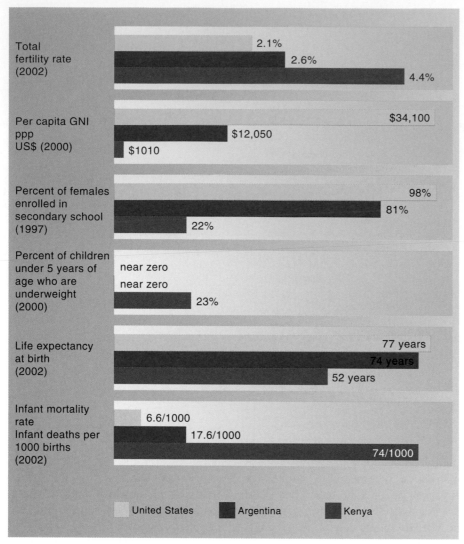

quate nutrition. Kenya has a low life expectancy (52 years), many undernourished children (23 percent are underweight), and a high infant mortality rate (74 per 1000). The United States has a low infant mortality rate (6.6 per 1000). Argentina has an intermediate infant mortality rate (17.6 per 1000).

Finally, the educational status of people determines the kinds of jobs that are available and the likelihood of being able to improve one's status. In general, men are more likely to receive an education than women, but the educational status of women has a direct bearing on the number of children they will have and, therefore, on the economic well-being of the family, Nearly all girls in the United States attend high school, while 81 percent do so in Argentina, but only 22 percent do so in Kenya. Obviously, tremendous differences exist in the standard of living among these three countries. What the average U.S. citizen would consider poverty level would be considered a luxurious life for the average person in Kenya.

figure 8.4   **Standard of Living and Population Growth in Three Countries** Standard of living is a measure of how well one lives. It is not possible to get a precise definition, but when we compare the United States, Argentina, and Kenya, it is obvious that there are great differences in how the people in these countries live. Kenya has a high population growth rate, a low life expectancy, a high infant mortality rate, and many people without adequate food. Furthermore, their incomes are low and they are poorly educated. The United States has a low population growth rate, a high life expectancy, a low infant mortality rate, and many people who eat too much. People in the United States have high educational levels and high incomes. Argentina is intermediate in all of these characteristics.

Population and Poverty—A Vicious Cycle?

It is clear that the areas of the world where the human population is growing most rapidly are those that have the lowest standard of living. The developed regions of the world (Europe, North America, Japan, Australia, and New Zealand) have abundant wealth and have relatively slowly growing populations. The less-developed countries of Africa, Latin America, and Asia are generally poor and have high population growth rates. Although not all cases are the same, poverty, high birthrates, poor health, and lack of education seem to be interrelated.

comparison of purchasing power requires some adjustments. Therefore, a technique used to compare economic well-being across countries is a measure called the GNI PPP (gross national income purchasing power parity). Finally, the GNI PPP can be divided by the number of people in the country to get a per capita (per person) GNI PPP. As you can see from figure 8.4, a wide economic gap exists between economically advanced countries and those that are less developed. Yet the people of less-developed countries aspire to the same standard of living enjoyed by people in the developed world.

Health criteria reflect many aspects of standard of living. Access to such things as health care, safe drinking water, and adequate food are reflected in life expectancy, infant mortality, and growth rates of children. The United States and Argentina have similar life expectancies (over 70 years) and ade-

1. Poor people cannot afford birth control and, since they are often poorly educated, may not be able to read the directions on how to use various birth control mechanisms correctly.

Therefore, they have more children than they may wish to have.

2. Poor people need to obtain income in many ways. Often this includes taking children out of school so that they are able to work on the farm or in other jobs to provide income for the family. Poorly educated people cannot get high paying jobs and so remain in poverty.

3. Poor people have little access to health care and are therefore more likely to suffer preventable illnesses that lower their ability to earn an income.

4. Women in poor countries are usually poorly educated and do not have disposable income. Therefore, they are dependent on their husbands or the family unit for their livelihood. Women who do not have independent income are more likely to have children they do not want because they cannot afford birth control.

5. High infant mortality rates result from poor health, but children (particularly sons) are desired by parents because the sons will provide for the parents when the parents are old. High infant mortality rates make it likely that parents will desire larger numbers of offspring, since some offspring will die as children.

At the United Nations International Conference on Population and Development held in Cairo, Egypt, in September 1994, attention was focused on breaking this cycle of poverty and high population growth rates. Several important conclusions were reached that have the potential to break this cycle.

1. There was recognition that economic well-being is tied to solving the population problem. However, the huge, rapidly growing population of the poor countries of the world cannot hope to consume at the rate the rich countries do. Furthermore, the rich countries of the world need to reduce their rate of consumption.

2. Improving the educational status of women was promoted. This would lead to improved financial standing for women, which could allow them to have fewer children.

3. Access to birth control and health care would reduce infant and maternal deaths.

In fact, several countries have instituted population control programs that focus on improving the health of women and children, increasing the educational levels of women, and making birth control universally available.

Hunger, Food Production, and Environmental Degradation

As the human population increases, the demand for food also rises. People must either grow food themselves or purchase it. Most people in the developed world purchase what they need and have more than enough food to eat. Most people in the less-developed world must grow their own food and have very little money to purchase additional food. Typically, these farmers have very little surplus. If crops fail, people starve. Even in countries with the highest population (China and India), the majority of the people live on the land and farm. (Sixty-two percent of Chinese and 72 percent of Indians live in rural areas.)

The human population can increase only if the populations of other kinds of plants and animals decrease. Each ecosystem has a maximum biomass that can exist within it. There can be shifts within ecosystems to allow an increase in the population of one species, but this always adversely affects certain other populations because they are competing for the same basic resources.

When humans need food, they convert natural ecosystems to artificially maintained agricultural ecosystems. The natural mix of plants and animals is destroyed and replaced with species useful to humans. If these agricultural ecosystems are mismanaged, the region's total productivity may fall below that of the original ecosystem. The dust bowl of North America, desertification in Africa, and destruction of tropical rainforests are well-known examples. In countries where food is in short supply and the population is growing, pressure is intense to convert remaining natural ecosystems to agriculture. Typically, these areas are the least desirable for agriculture and will not be productive. However, to a starving population, the short-term gain is all that matters. The long-term health of the environment is sacrificed for the immediate needs of the population.

A consequence of the basic need for food is that people in less-developed countries generally feed at lower trophic levels than do those in the developed world. (See figure 8.5.) Converting the less-concentrated carbohydrates of plants into more nutritionally valuable animal protein and fat is an expensive process. During the process of feeding plants to animals and harvesting animal products, approximately 90 percent of the energy in the original plants is lost. (See Chapter 5.) Although many modern agricultural practices in the developed world obtain better efficiencies than this, most of the people in the developing world are not able to use such sophisticated systems. Thus, these people approach the 90 percent loss characteristic of natural ecosystems. Therefore, in terms of economics and energy, people in less-developed countries must consume the plants themselves rather than feed the plants to animals and then consume the animals. In most cases, if the plants were fed to animals, many people would starve to death. On the other hand, a lack of protein in diets that consist primarily of plants can lead to malnutrition. Many people in the less-developed world suffer from a lack of adequate protein, which stunts their physical and mental development.

In contrast, in most of the developed world, meat and other animal protein sources are important parts of the diet. Many people suffer from overnutrition (they eat too much); they are "malnourished" in a different sense. About fifty percent of North Americans are overweight and 25 percent are obese.

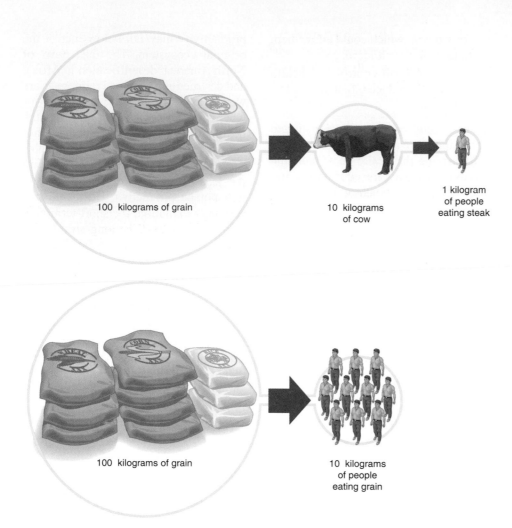

100 kilograms of grain

10 kilograms of cow

1 kilogram of people eating steak

100 kilograms of grain

10 kilograms of people eating grain

figure 8.5 **Population and Trophic Levels** The larger a population, the more energy it takes to sustain the population. Every time one organism is eaten by another organism, approximately 90 percent of the energy is lost. Therefore, when countries are densely populated, they usually feed at the herbivore trophic level because they cannot afford the 90 percent energy loss that occurs when plants are fed to animals. The same amount of grain can support 10 times more people at the herbivore level than at the carnivore level.

The ecological impact of one person eating at the carnivore level is about 10 times that of a person eating at the herbivore level. If people in the developed world were to reduce their animal protein intake, they would significantly reduce their demands on world resources. Almost all of the corn and soybeans grown in the United States is used as animal feed. If these grains were used to feed people rather than animals, less grain would have to be grown and the impact on farmland would be less.

In countries where food is in short supply, agricultural land is already being exploited to its limit, and there is still a need for more food. This makes the United States, Canada, Australia, Argentina, New Zealand, and the European Union net food exporters. Many countries, like India and China, are able to grow enough food for their people but do not have any left for export. Others, including many nations of the former Soviet Union, are not able to grow enough to meet their own needs and, therefore, must import food.

A country that is a net food importer is not necessarily destitute. Japan and some European countries are net food importers but have enough economic assets to purchase what they need. Hunger

occurs when countries do not produce enough food to feed their people and cannot obtain food through purchase or humanitarian aid.

The current situation with respect to world food production and hunger is very complicated. It involves the resources needed to produce food, such as arable land, labor, and machines; appropriate crop selection; and economic incentives. It also involves the maldistribution of food within countries. This is often an economic problem, since the poorest in most countries have difficulty finding the basic necessities of life, while the rich have an excess of food and other resources. In addition, political activities often determine food availability. War, payment of foreign debt, corruption, and poor management often contribute to hunger and malnutrition.

Improved plant varieties, irrigation, and improved agricultural methods have dramatically increased food production in some parts of the world. In recent years, India, China, and much of southern Asia have moved from being food importers to being self-sufficient and, in some cases, food exporters.

The areas of greatest need are in sub-Saharan Africa. Africa is the only major region of the world where per capita grain production has decreased over the past few decades. People in these regions are trying to use marginal lands for food production, as forests, scrubland, and grasslands are converted to agriculture. Often, this land is not able to support continued agricultural production. This leads to erosion and desertification.

What should be done about countries that are unable to raise enough food for their people and are unable to buy the food they need? This is not an easy question. A simple humanitarian solution to the problem is for the developed countries to supply food. Many religious and humanitarian organizations do an excellent service by taking food to those who need it and save many lives. However, the aim should always be to provide temporary help and insist that the people of the country develop mechanisms for solving their own problem.

Often, emergency food programs result in large numbers of people migrating from their rural (agricultural) areas to cities, where they are unable to support themselves. They become dependent on the food aid and stop working to raise their own food, not because they do not want to work but because they need to leave their fields to go to the food distribution centers. Many humanitarian organizations now recognize the futility of trying to feed people with gifts from the developed world. The emphasis must be on self-sufficiency.

The Demographic Transition Concept

The relationship between the standard of living and the population growth rate seems to be that countries with the highest standard of living have the lowest population growth rate, and those with the lowest standard of living have the highest population growth rate. This has led many people to suggest that countries naturally go through a series of stages called **demographic transition.** This model is based on the historical, social, and economic development of Europe and North America. In a demographic transition, the following four stages occur (See figure 8.6):

1. Initially, countries have a stable population with a high birthrate and a high death rate. Death rates often vary because of famine and epidemic disease.

2. Improved economic and social conditions (control of disease and increased food availability) bring about a period of rapid population growth as death rates fall. Birthrates remain high.

3. As countries become industrialized, the birthrates begin to drop because people desire smaller families and use contraceptives.

4. Eventually, birthrates and death rates again become balanced, with low birthrates and low death rates.

This is a very comfortable model because it suggests that if a country can become industrialized, then social, political, and economic processes will naturally cause its population to stabilize.

However, the model leads to some serious questions. Can the historical pattern exhibited by Europe and North America be repeated in the less-developed countries of today? Europe, North America, Japan, and Australia passed through this transition period when world population was lower and when energy and natural resources were still abundant. It is doubtful whether these supplies are adequate to allow for the industrialization of the major portion of the world currently classified as less developed.

A second concern is the time element. With the world population increasing as rapidly as it is, industrialization probably cannot occur fast enough to have a significant impact on population growth. As long as people in less-developed countries are poor, there is a strong incentive to have large numbers of children. Children are a form of social security because they take care of their elderly parents. Only people in developed countries can save money for their old age. They can choose to have children, who are expensive to raise, or to invest money in some other way.

When the countries of Europe and North America passed through the demographic transition, they had access to large expanses of unexploited lands, either within their boundaries or in their colonies. This provided a safety valve for expanding populations during the early stages of the transition. Without this safety valve, it would have been impossible to deal adequately with the population while simultaneously encouraging economic development. Today, less-developed countries may be unable to accumulate the necessary capital to develop economically, since they do not have uninhabited places to which their people can migrate and an ever-increasing population is a severe economic drain.

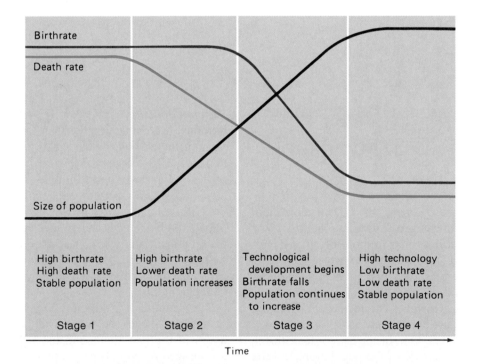

figure 8.6 **Demographic Transition** The demographic transition model suggests that as a country develops technologically, it automatically experiences a drop in the birthrate. This certainly has been the experience of the developed countries of the world. However, the developed countries make up about 20 percent of the world's population. It is doubtful whether the less-developed countries can achieve the kind of technological advances experienced in the developed world.

Global Perspective

The Urbanization of the World's Population

The United Nations estimates that currently about 47 percent of the world's population lives in cities. This is expected to increase to over 50 percent by 2005 and reach 60 percent by 2025. The economically developed world is currently about 75 percent urban, and the less-developed world is about 40 percent urban. Therefore, the increase in the urban population will occur primarily in the less-developed world, where resources to deal with urban problems are least available.

Urbanization is not necessarily bad. The economic activity of the developed world takes place primarily in cities. Cities can be planned and managed to be healthy, interesting places to live. Cities offer jobs, health care, schools, and other services that are usually lacking in rural areas. Often the rural economy of a country is considered to be of low status, and young people wish to move to cities, where higher status jobs and desirable cultural amenities are available.

Dense populations of people have a concentrated impact on local resources. Often water must be transported long distances, wastes are difficult to get rid of, and air quality drops as each additional person places a burden on the local environment. Unfortunately, in the less-developed world, it is impossible to plan for growth and provide basic resources such as drinking water, sewers, transportation, and housing as fast as the population is growing. New migrants to the cities often lack education or training for the high wage jobs available, have lost the opportunity to raise their own food, and lack the support of their family units. Consequently, when these poor people migrate to the cities, they often live in shantytowns on the edge of the city or form squatter settlements on hillsides and other unused land. These centers of poverty lack basic services since they are often outside the bureaucratic structure of the city. The rich in such cities live a short distance from the poor and may be relatively isolated, but the negative environmental impact of the entire urban region (air and water pollution, transportation problems, etc.) affects all the residents. Even the long-established urban centers of the developed world have major problems. Many contain pockets of poverty that have led to problems of social unrest and crime.

The U.S. Population Picture

In many ways, the U.S. population, which has a total fertility rate that is low (2.1), is similar to those of other developed countries of the world with low birthrates. One might expect the population to be stabilizing under these conditions. However, two factors are operating to cause significant change over the next 50 years. One factor has to do with the age structure of the population, and the other has to do with immigration policy. The U.S. population includes a **postwar baby boom** component, which has significantly affected population trends. These baby boomers were born during an approximately 15-year period (1947–1961) following World War II, when birthrates were much higher than today, and constitute a bulge in the age distribution profile. (See figure 8.7.) As members of this group have raised families, they have had a significant influence on how the U.S. population has grown. As this population bulge ages and younger people limit their family size, the population will gradually age. By 2030, about 20 percent of the population will be 65 years of age or older.

A changing age structure will lead to social changes as well. The baby boom of the late 1940s and the 1950s encouraged growth in service industries needed by young families. Maternity wards had to be expanded, schools could not be built fast enough, baby-care companies saw unprecedented sales, and the toy industry flourished. Today, these "babies" are in their forties and fifties. They are buying homes, cars, and appliances but are raising fewer children than did their parents. However, because baby boomers are such a large segment of the population, they have contributed a large number of children to the population. The children of the baby boomers are now having children, and in many parts of the country, schools are seeing increased numbers of children in the early grades. At the same time, there are more elderly, because people are living longer. That creates a need for additional services for the elderly. This trend toward an aging population will be accentuated as the baby boomers retire. What will the social needs be in the year 2020 when many of these baby boomers will have retired?

Both legal and illegal immigration significantly influence future population growth trends. Even with the current

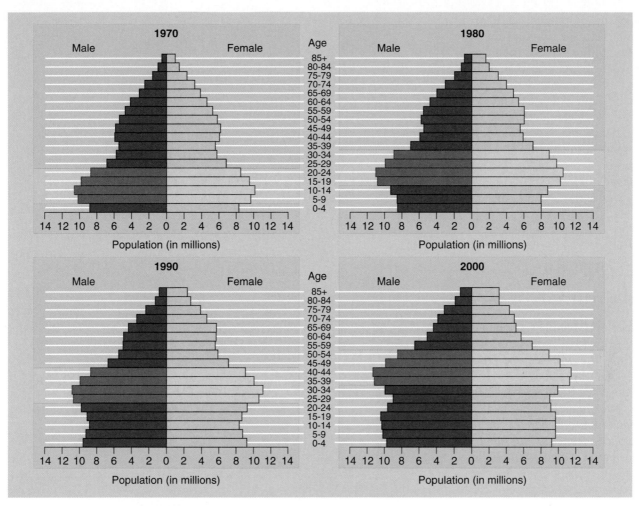

figure 8.7   **Changing Age Distribution of U.S. Population (1970–2000)** These graphs show the number of people in the United States at each age level. Notice that in the year 2000, a bulge begins to form at age 35 to 39 and ends at about age 50 to 54. These people represent the "baby boom" that followed World War II. As you compare the age distribution for 1970, 1980, and 1990 with 2000, you can see that this group of people moves through the population. As this portion of the population has aged, it has had a large impact on the nature of the U.S. population. In the 1970s and 1980s, baby boomers were in school. In the 1990s, they were in their middle working years. In 2000, some of them were beginning to retire, and many more will do so throughout the decade.

Source: Data from the U.S. Department of Commerce, Bureau of the Census.

total fertility rate of 2.1 children per woman, the population is still growing by about 1.1 percent per year. About 0.6 percent is the result of natural increases due to the difference between birthrates and death rates. The remainder is the result of immigration into the United States. The U.S. Census Bureau projects that immigration will increase significantly and account for 50 percent of population growth by the year 2050.

Current immigration policy in the United States is difficult to characterize. Strong measures are being taken to reduce illegal immigration across the southern border. This is in part due to pressures placed on Congress by states

that receive large numbers of illegal immigrants. Illegal immigrants add to the education and health care that states must fund. At the same time, some segments of the U.S. economy (agriculture, tourism) maintain that they are unable to find workers to do certain kinds of work. Consequently, special guest workers are allowed to enter the country for limited periods to serve the needs of these segments of the economy. There is also a consistent policy of allowing immigration that reunites families of U.S. residents. Obviously, the families that fall into this category are likely to be U.S. citizens who were recent immigrants themselves. It is obvious that

most immigration policy is the result of political decisions rather than decisions that relate to population policy or a concern about the rate at which the U.S. population is growing.

Projections based on the 2000 census indicate that the population will continue to grow and does not seem to be moving toward zero growth. This would result in an increase from about 275 million people in 2000 to about 404 million by 2050. (See figure 8.8.)

Differences in family size exist between different segments of the U.S. population. The Hispanic and Asian-American portions of the population tend to have larger families, so these portions of the population will grow rapidly, and the caucasian portion will decline. In addition, recent immigrants tend to have larger families than nonimmigrants. Since immigrants tend to come from Latin America and Asia, these portions of the population will increase. Thus, the U.S. population will be much more ethnically diverse in the future.

Anticipated Changes with Continued Population Growth

As the world human population continues to increase, pressure for the necessities of life will become greater. Differences in the standard of living between developed and less-developed countries will remain significant because population will increase most in less-developed countries. The supply of fuel and other resources is dwindling. Pressure for these resources will intensify as industrialized countries seek to maintain their current standard of living. People in less-developed countries will seek more land to raise crops and feed themselves unless major increases in food production per hectare occur. Since most of these people live in tropical areas, tropical forests will be cleared for farmland. The resulting erosion or alteration of the soil will make it no longer suitable for forest or crops. This conversion of natural ecosystems to agricultural ecosystems could cause profound changes in the world ecosystem.

Developed countries may have to choose between several alternatives: helping the less-developed countries, thus maintaining their friendship; allowing increased immigration from the developing world; or isolating themselves from the problems of the less-developed nations. None of these policies will be able to prevent a change in lifestyle as world population increases. The resources of the world are finite.

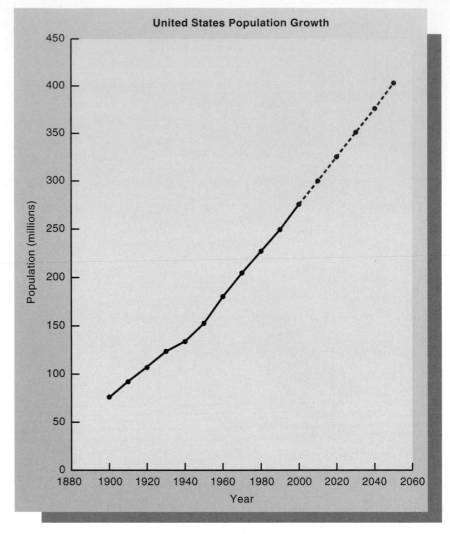

figure 8.8 **U.S. Population Growth** The population of the United States has grown continuously since colonization. The graph indicates that the size of the population was about 275 million people in 2000. Current U.S. Census Bureau projections are based on the current total fertility rate of 2.1 and immigration of about 1 million people per year. If these projections are correct, the U.S. population will reach about 404 million by 2050. This is an increase of nearly 50 percent in 50 years.

Source: Data from the U.S. Department of Commerce, Bureau of the Census.

Even if industrialized countries continue to use a disproportionate share of the world's resources, the amount available per person will decline as population rises. As world population increases, the less-developed areas will probably maintain their low standard of living. It is difficult to see how that standard of living could get much lower, since some people in these countries are already starving to death. Lifestyles in developed nations will probably change very little but may become less consumption-oriented. What some people currently view as necessities

(meals in restaurants, vacations to remote sites, and two cars per family) will probably become luxuries. Many people who enjoy the freedom of mobility associated with the automobile will have their travel limited as public transportation replaces private transportation. Recreation may cease to involve expensive, energy-demanding machines (powerboats, motorcycles, and electric-powered toys) and emphasize instead such activities as hiking, bicycling, and reading. These changes will not come quickly, unless some catastrophic political or economic

 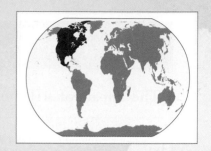

Global Perspective

North America—Population Comparisons

The three countries that make up North America (Canada, the United States, and Mexico) interact politically, socially, and economically. The characteristics of their populations determine how they interact. Canada and the United States are both wealthy countries with similar age structures and low total fertility rates. The United States has a total fertility rate of 2.1 and Canada has a total fertility rate of 1.5. Both countries have a relatively small number of young people (about 20 percent of the population is under 15 years of age) and a relatively large number of older people (about 13 percent are 65 or older). Without immigration, these countries would have stable or falling populations. Therefore, they must rely on immigration to supply additional people to fill the workplace. This is particularly true for jobs with low pay. The United States currently receives about 1 million immigrants per year. Canada currently receives about 175,000 immigrants per year but is planning to increase that to about 300,000 in the future.

Mexico, on the other hand, has a young, rapidly growing population. About 33 percent of the population is under 15 years of age, and the total fertility rate is 2.9. At that rate, the population will double in about 33 years. Furthermore, the average Mexican has a purchasing power about 40 percent of that in Canada and about 25 percent of that in the United States. These conditions create a strong incentive for individuals to migrate from Mexico to other parts of the world. The United States is the usual country of entry. The number is hard to evaluate, since many enter the United States illegally or as seasonal workers and eventually plan to return to Mexico. Current estimates are that about 250,000 people enter the United States from Mexico each year.

In response to the large number of illegal immigrants from Mexico, the United States has erected fences and increased surveillance. In 2000, over a million people were apprehended attempting to cross the border from Mexico to the United States. Thus, instead of crossing at normal points of entry, illegal immigrants are likely to cross the border in remote desert areas, which has resulted in numerous deaths due to exposure.

The economic interplay between Mexico and the United States has several components. Working in the United States allows Mexican immigrants to improve their economic status. Furthermore, their presence in the United States has a significant effect on the economy of Mexico, since many immigrants send much of their income to Mexico to support family members. The North American Free Trade Agreement (NAFTA) allows for relatively free exchange of goods and services between Canada, the United States, and Mexico. Consequently, many Canadian, U.S., and European businesses have built assembly plants in Mexico to make use of the abundant inexpensive labor, particularly along the U.S.-Mexican border. Labor leaders in Canada and the United States are concerned about the effect access to low-cost labor will have on their membership and complain that many high-paying jobs have been moved to Mexico, where labor costs are lower.

	Population Size 2002 (Millions)	Birthrate per 1000	Death Rate per 1000	Rate of Natural Increase	Total Fertility Rate (Children per Woman)	Life Expectancy (Years)	Infant Mortality Rate (Deaths per 1000 Births)	Per Capita GNI PPP 2000 (U.S. $)	Percent Under 15 Years of Age	Percent Over 65 Years of Age	Annual Immigration/ Emigration (Estimate)*
Canada	31.3	11	7	0.3	1.5	79	5.5	$22,120	19	13	+175,000
United States	287.4	15	9	0.6	2.1	77	6.6	$34,100	21	13	+1,000,000
Mexico	101.7	26	5	2.1	2.9	75	25.0	$8,770	33	5	−250,000

Source: Population Reference Bureau, 2002 Population Data Sheet.

*Immigration/emigration estimates from United States and Canadian government sources.

force causes major worldwide adjustments. Most likely, changes will occur only as economic pressures affect families. Many economists and political thinkers feel that as the economics of the world become more linked and jobs flow more freely from country to country, wealth will be redistributed from the former rich countries to emerging economies in developing countries. The possibility of such a redistribution has become a major political issue in many of the developed countries of the world.

The Impact of AIDS on Populations

The human immunodeficiency virus (HIV), which causes AIDS (acquired immunodeficiency syndrome), continues to spread throughout the world. Because the epidemic has affected all parts of the world, it has been called a pandemic. The disease is spread through direct transfer of body fluids containing the virus into the bloodstream of another person. Sharing of contaminated needles among

Adults and children estimated to be living with HIV/AIDS as of the end of 2001

North America 940,000

Western Europe 560,000

Eastern Europe & Central Asia 1 million

East Asia & Pacific 1 million

Caribbean 420,000

North Africa & Middle East 440,000

South & Southeast Asia 6.1 million

Latin America 1.4 million

sub-Saharan Africa 28.1 million

Australia & New Zealand 15,000

Total: 40 million

Summary

Many of the problems of the world are caused by or made worse by an increasing human population. Currently, the world's population is growing very rapidly. Most of the growth is occurring in the less-developed areas of the world (Africa, Asia, and Latin America), where people have a low standard of living. The developed regions of the world, with their high standard of living, have relatively slow population growth and, in some instances, declining populations.

Demography is the study of human populations and the things that affect them. Demographers study the sex ratio and age distribution within a population to predict future growth. Population growth rates are determined by biological factors such as birthrate, which is determined by the number of women in the population and the age of the women, and death rate. Sociological and economic conditions are also important, since they affect the number of children desired by women, which helps set the population growth rate. In developed countries, women usually have access to jobs. Couples marry later, and they make decisions about the number of children they will have based on the economic cost of raising children. In the less-developed world, women marry earlier, and children have economic value as additional workers, as future caregivers for the parents, and as status for either or both parents.

The current U.S. population will continue to grow. The average age of the population will increase, and the racial mix of the population will change significantly.

The demographic transition model suggests that as a country becomes industrialized, its population begins to stabilize. However, there is little hope that the Earth can support the entire world in the style of the industrialized nations. It is

intravenous drug users and sexual contact are the most likely methods of passage. In the United States, the disease was once considered a problem only for the homosexual community and those who use intravenous drugs. This perception is rapidly changing. Many of the new cases of AIDS are being found in women infected by male sex partners and in children born to infected mothers.

The World Health Organization (WHO) estimated that by 2001, about 60 million people had been infected with HIV worldwide. WHO estimates that 5 million people died of AIDS in 2001 and that over 20 million have died since the pandemic began. About 40 million people are currently living with HIV. The distribution of the virus is lowest in the economically developed countries and highest in the developing countries. The figure shows estimates by the World Health Organization of the numbers of HIV-infected people in various parts of the world at the end of 2001. In the less-developed world, there is little medical care to treat AIDS and a lack of resources to identify those who have HIV. Many people do not know they are infected and will continue to pass the disease to others.

Sub-Saharan Africa has been hit hardest by this disease. In Africa, AIDS has always been primarily a sexually transmitted disease spread through heterosexual contact. Many people believe that in the poor countries of central Africa, permissive sexual behavior and prostitution have created conditions for a rapid spread of the disease. This is particularly evident along major transportation routes. The World Health Organization estimates that nearly 70 percent of all persons infected with HIV (28.1 million) live in this region of the world. This is about 8.4 percent of the population of the region. In some large urban hospitals in this part of Africa, over 50 percent of the hospital beds are taken up by AIDS patients, and AIDS is the leading cause of death from disease. Since those who

are dying are young, there is a change in the age structure of the population of heavily infected countries. This has resulted in large numbers of AIDS orphans—about 12.1 million AIDS orphans in sub-Saharan Africa by the end of 2001. With the death of young infected adults, villages are composed primarily of older people and children.

The economic burden on these countries is tremendous. Those with AIDS symptoms are unable to work and need medical care and medication. Because of poverty, there is often little medical care available. Also, the millions of orphaned children have resulted in an additional economic burden on relatives.

In the developed world, drugs are available that can extend life and perhaps prevent transmission between infected mothers and their infants. These drugs are expensive and therefore are unavailable to most of the people in the world who need them. In 2001, several drug companies agreed to sell some of these drugs to impoverished people at reduced prices (about $700 per person per year), but even at these prices, the drugs are still beyond the economic reach of most people, since the average per capita income is less than $1500.

- Should the developed countries of the world provide funding and drugs to combat the AIDS epidemic in Africa?
- What actions should the governments of these countries institute to combat AIDS?
- Should economic aid be given to countries that have large numbers of AIDS patients?

Source: Data from World Health Organization of the United Nations.

doubtful whether there are enough energy resources and other natural resources to develop the less-developed countries or whether there is enough time to change trends of population

growth. Highly developed nations should anticipate increased pressure in the future to share their wealth with less-developed countries.

Key Terms

age distribution *155*
demographic transition *163*
demography *154*
gross national income (GNI) *159*

population density *153*
postwar baby boom *164*
replacement fertility *155*
standard of living *159*

total fertility rate *155*
zero population growth *155*

Review Questions

1. What is demography?
2. What is demographic transition? What is it based on?
3. What is a baby boom?
4. What does age distribution of a population mean?
5. List 10 differences between your standard of living and that of someone in a less-developed country.
6. Why do people who live in overpopulated countries use plants as their main source of food?
7. Although predicting the future is difficult, describe what you think your life will be like in 10 years. Why?
8. List five changes you might anticipate if world population were to double in the next 50 years.
9. Which three areas of the world have the highest population growth rate? Which three areas of the world have the lowest standard of living?
10. How many children per woman would lead to a stable U.S. population?
11. What role does the status of women play in determining population growth rates?
12. Describe three reasons why women in the less-developed world might desire more than two children.

Critical Thinking Questions

1. Do you think it is appropriate for developed countries to persuade less-developed countries to limit their population growth? What would be appropriate and inappropriate interventions, according to your ethics? Now imagine you are a citizen of a less-developed country. What might be your reply to those who live in more-developed countries? Why?
2. China's government has been very involved in regulating that country's population growth. What do you think about this kind of government intervention in China's population problem? What values, beliefs, and perspectives do you hold that lead you to think the way you do?
3. Population growth causes many environmental problems. Identify some of these problems. What role do you think technology will play in solving these problems? Are you optimistic or pessimistic about these problems being solved through technology? Why?
4. Do you think that demographic transition will be a viable option for world development? What evidence leads you to your conclusions? What role should the developed countries play in the current demographic transition of developing countries? Why?
5. The U.S. Census Bureau projects that by 2050, immigration will account for 50 percent of the population growth in the United States. What values and perspectives should guide our immigration policy? Why?
6. Imagine a debate between an American and a Sudanese about human population and the scarcity of resources. What perspectives do you think the American might bring to the debate? What perspectives do you think the Sudanese would bring? What might be their points of common ground? On what might they differ?
7. Many people in developing countries hope to achieve the standard of living of those in the developed world. What might be the effect of this pressure on the environment in developing countries? On the political relationship between developing countries and already developed countries? What ethical perspective do you think should guide this changing relationship?
8. The demographic changes occurring in Mexico have an influence on the United States. What problems does Mexico face regarding its demographics? Should the United States be involved in Mexican population policy?

Concept Map

Construct a map to show relationships among the following concepts:

demographic transition
gross national income

standard of living
replacement fertility

total fertility rate

Interactive Exploration

Check out the website at **http://www.mhhe.com/environmentalscience** and click on the cover of this textbook for quizzing, career information, case studies, and hot links for the following topics:

Human Population Growth
Ecological Economics
Human Interactions Between Populations

Global Ecology
Urbanization and Sustainable Cities

P A R T

iii

Energy

Skull Valley: A Reservation for Disaster?

by Barbara Wachocki

Weber State University/Utah

Currently, about 20 percent of our nation's electricity is generated by 103 nuclear reactors. Approximately 44,000 tonnes (48,501 U.S. tons) of spent nuclear fuel is being stored, and by 2035, the amount could be 105,000 tonnes (115,742 U.S. tons). It takes 10,000 years for spent fuel rods to reach safe levels of radioactivity, resulting in a need for a permanent, high-level nuclear waste storage facility.

For decades, the federal government has promised that a permanent facility would be established. In 1987, Congress instructed the Department of Energy to study Yucca Mountain, Nevada, as a permanent underground storage site. In July 2002, Congress approved the measure.

Yucca Mountain could accept waste within 10 years. Until then, waste must be temporarily stored. All reactor facilities have the capacity to store spent fuel on-site; however, local communities don't want it. States have begun setting on-site storage limits, which could force the premature shutdown of some reactors. Hence, a centralized temporary storage facility would be politically and financially desirable.

With a permanent solution delayed, eight public utilities formed a limited liability company, Private Fuel Storage (PFS), to develop a temporary national waste storage site. In 1992, the Skull Valley Band of the Goshute Tribe volunteered its reservation for this purpose and received a federal grant to investigate the establishment of a facility at Skull Valley, Utah.

The Goshutes have lived in the area for thousands of years, and their homeland spanned several hundred square miles. The Goshute Tribe, once numbering 20,000, now has fewer than 500 members, 124 of whom belong to the Skull Valley Band. Only about 25 members currently live on the 18,000-acre reservation in Tooele County, Utah (129 kilometers, or 80 miles, from Salt Lake City).

An 1863 treaty recognized tribes as domestic dependent nations. The federal government has a responsibility to tribes, their interests, and assets. The treaty gave the government the right to establish free routes of travel, telegraph lines, and railways through tribal lands as well as agricultural and mining settlements. Currently, Dugway Proving Grounds, a chemical and biological weapons development and testing site, lies just south of the Skull Valley Reservation. In 1968, a chemical agent escaped, killing 6,000 sheep. Nerve gas is now stored there, and a nerve gas incinerator is nearby. A coal-fired electric plant, the Envirocare Low-Level Radioactive Disposal Site, and the MagCorp plant, the nation's worst air polluter, also surround the reservation. The tribe was never consulted in the siting of these facilities.

The Skull Valley Goshutes signed a lease with PFS allowing the company to use some of its land for high-level nuclear waste storage. PFS is seeking a site license from the Nuclear Regulatory Commission, with a decision expected by year's end. The facility could store 40,000 tonnes (44,092 U.S. tons) of waste in specially designed, above ground casks for 40 years and would cost approximately $3.1 billion. The amount guaranteed to the tribe has not been disclosed. The project would create about 60 local jobs, allowing some tribal members to move back to the reservation. It would also improve conditions on the reservation.

Opponents of the Skull Valley facility include the state of Utah, its governor, its Department of Environmental Quality, legislators, and groups such as HEAL Utah, a citizens' alliance formed to protect against the risks of nuclear and toxic wastes. The Governing Body of the Confederated Tribes of the Goshute Reservation expressed concern that the Skull Valley Tribe should consider the impacts on neighboring property owners, adjacent states, and Indian tribes in general.

Concerns regarding the PFS site are numerous. Current regulations prohibit private, off-site storage of spent nuclear fuel except at federally owned facilities. Reactors could store waste on-site, eliminating the risk of cross-country railway transport. Skull Valley would house 4,000 175 U.S. tons storage casks, equivalent to all of the nation's current commercial high-level nuclear waste. Uncovered, concrete casks would be at risk from sabotage and earthquakes. The facility design lacks seismic monitoring and earthquake proofing to meet current NRC requirements. Once tipped, the huge casks must be upright within 48 hours to avoid overheating. In addition, the site's airspace is used for training of F-16 pilots from nearby Hill Air Force Base and Cruise missile testing. Curtailing military operations could put the base, a major employer, at a disadvantage in future base closure reviews. The stigma associated with nuclear waste could also impact Utah's economy. Furthermore, PFS has not adequately addressed either its ability to fund the 40-year project or provide for emergency response training and equipment, or its extent of liability. Finally, it is questionable whether the facility is temporary, since the Yucca Mountain site's capacity (77,000 tonnes; 84,800 U.S. tons) will not likely meet the nation's needs.

This project involves critical issues: tribal sovereignty, our nation's treatment of Native Americans, state sovereignty, the federal government's failure to provide permanent waste storage for an industry it encouraged, and local communities' responsibility for their own waste.

Study Questions:

1. To what extent should any individual or group be allowed to exercise their rights before the impacts of their actions on others in society outweigh those rights?

2. As a nation, we have put some communities at higher environmental risk than others. What factors do you think are considered when establishing hazardous waste facilities near a community? (e.g., area surrounding the Skull Valley Reservation) ?

3. Should we as a nation continue to do things "because we can," without fully considering the long-term consequences of those actions? (e.g., nuclear power production, our foreign policy decisions, etc.)?

Energy and Civilization: Patterns of Consumption

9
CHAPTER

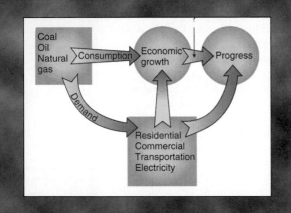

Objectives

After reading this chapter, you should be able to:

- Explain why all organisms require a constant input of energy.
- Describe how per capita energy consumption increased as civilization developed from hunting and gathering to primitive agriculture to advanced cultures.
- Describe how advanced modern civilizations developed as new fuels were used to run machines.
- Correlate the Industrial Revolution with social and economic changes.
- Explain how cheap oil and natural gas led to a consumption-oriented society.
- Explain how the automobile changed people's lifestyles.
- Explain why overall energy use in the United States declined during the 1970s and 1980s.

Chapter Outline

History of Energy Consumption
    Biological Energy Sources
    Increased Use of Wood
    Fossil Fuels and the Industrial Revolution

Energy and Economics
    Economic Growth and Energy Consumption
    The Role of the Automobile
    Gasoline Prices and Government Policy

Global Perspective: *Five Ways to Curb Traffic*

How Energy Is Used
    Residential and Commercial Energy Use
    Industrial Energy Use
    Transportation Energy Use

Environmental Close-Up: *Hybrid Electric Vehicles*
    The Variability of Gasoline Prices

Electrical Energy

Energy Consumption Trends

Environmental Close-Up: *Alternative-Fuel Vehicles*

Global Perspective: *Energy Development in China*

Global Perspective: *Potential World Petroleum Resources*
    OPEC

History of Energy Consumption

Every form of life and all societies require a constant input of energy. If the flow of energy through organisms or societies ceases, they stop functioning and begin to disintegrate. Some organisms and societies are more energy efficient than others. In general, history shows that complex industrial societies use the most energy. If societies are to survive, they must continue to expend energy. However, they may need to change their pattern of energy consumption as traditional sources become limited.

Biological Energy Sources

Energy is essential to maintain life. In every ecosystem, the sun provides that energy. (See chapter 5.) The first transfer of energy occurs during photosynthesis, when plants convert light energy into chemical energy in the production of food. Herbivorous animals utilize the food energy in the plants. The herbivores, in turn, are a source of energy for carnivores. Because nearly all of their energy requirements were supplied by food, primitive humans were no different from other animals in their ecosystems. In such hunter-gatherer cultures, nearly all human energy needs were met by using plants and animals as food, tools, and fuel. (See figure 9.1.)

Early in human history, people began to use additional sources of energy to make their lives more comfortable. They domesticated plants and animals to provide a more dependable supply of food. They no longer needed to depend solely on gathering wild plants and hunting wild animals for sustenance. Domesticated animals also furnished a source of energy for transportation, farming, and other tasks. (See figure 9.2.) Wood provided a source of fuel for heating and cooking. Eventually, this biomass energy was used in simple technologies, such as shaping tools and extracting metals.

figure 9.1 **Hunter-Gatherer Society** In this type of society, people obtain nearly all of their energy from the collection of wild plants and the hunting of animals. These societies do not make large demands on fossil fuels.

figure 9.2 **Animal Power** This bas-relief panel from an Egyptian tomb depicts an important accomplishment in the development of human civilization. With the use of domesticated animals, people had a source of power other than their own muscles.

Increased Use of Wood

Early civilizations, such as the Aztecs, Chinese, Indians, Greeks, Egyptians, and Romans were culturally advanced, but their societies used human muscle, animal muscle, and fire as sources of energy. Except for limited use of some wind-powered and water-powered devices such as ships and canoes, the controlled use of fire was the first use of energy in a form other than food. Wood was the primary fuel. (Wood was also used for building materials and other cultural uses.) The energy provided by wood enabled people to cook their food, heat their dwellings, and develop a primitive form of metallurgy. Such advances separated humans from other animals. When dense populations of humans made heavy use of wood for fuel and building materials, they eventually used up the readily available sources and had to import wood or seek alternative forms of fuel.

Because of a long history of high population density, India and some other parts of the world experienced a wood shortage hundreds of years before Europe and North America did. In many of these areas, animal dung replaced wood as a fuel source. It is still used today in some parts of the world.

Western Europe and North America were able to use wood as a fuel for a longer period of time. The forests of Europe supplied sufficient fuel until the thirteenth century. In North America, vast expanses of virgin forests supplied adequate fuel until the late nineteenth century. Fortunately, when local supplies of wood declined in Europe and North America, coal, formed from fossilized plant remains, was available as an alternative energy source. By 1880, coal had replaced wood as the primary energy source.

Fossil Fuels and the Industrial Revolution

Fossil fuels are the remains of plants, animals, and microorganisms that lived millions of years ago. (The energy in these fuels is stored sunlight, just as the biomass of wood represents stored sunlight.) During the Carboniferous period, 286 to 362 million years ago, when the Earth's climate was warmer and wetter than it is now, conditions were conducive to the formation of large deposits of coal. (See figure 9.3.) Oil and natural gas formed primarily from one-celled marine organisms whose bodies accumulated in large quantities on the seafloor and became compressed over millions of years. Heat and pressure from overlying sediment layers eventually converted the organic matter into oil and gas by distillation. Ever since machines replaced muscle power, the major energy sources for the world have been fossil remains from the distant past.

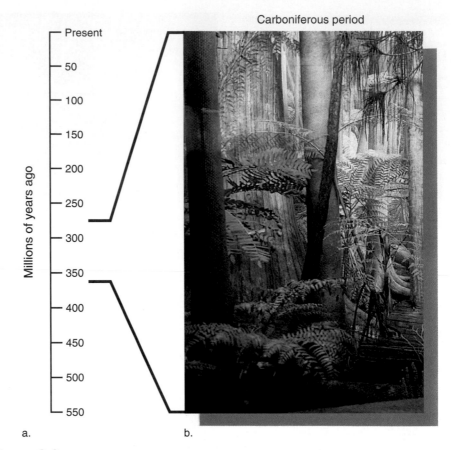

Carboniferous period

a. b.

figure 9.3   **Carboniferous Period** Approximately 300 million years ago, this kind of ecosystem was common throughout the world. Plant material accumulated in these swamps and was ultimately converted to coal.

Historically, the first fossil fuel to be used extensively was coal. In the early eighteenth century, regions of the world that had readily available coal deposits were able to switch to this new fuel and participate in a major cultural change known as the **Industrial Revolution.** The Industrial Revolution began in England and spread to much of Europe and North America. It involved the invention of machines that replaced human and animal labor in manufacturing and transporting goods. Central to this change was the invention of the steam engine, which could convert heat energy into the energy of motion. The steam engine made possible the large-scale mining of coal. Before steam engines, coal mines flooded and thus were not economical. The source of energy for steam engines was either wood or coal; wood was quickly replaced by coal in most cases. Nations without a source of coal or those possessing coal reserves that were not easily exploited did not participate in the Industrial Revolution.

Prior to the Industrial Revolution, Europe and North America were predominately rural. Goods were manufactured on a small scale in the home. As machines and the coal to power them became increasingly available, the factory system of manufacturing products replaced the small home-based operation. Because expanding factories required a constantly increasing labor supply, people left the farms and congregated in areas surrounding the factories. Villages became towns, and towns became cities. Widespread use of coal in cities resulted in increased air pollution. In spite of these changes, the Industrial Revolution was viewed as progress. Energy consumption increased, economies grew, and people prospered. Within a span of 200 years, the daily per capita energy consumption of industrialized nations increased eightfold. This energy

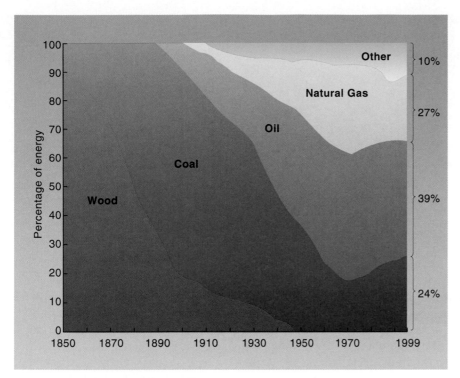

figure 9.4 **Oil Replaces Coal** Just as wood was replaced by coal, coal was later replaced by oil. This graph represents the production of energy by various sources in the United States. Oil has remained a dominant energy source for the past 40 years; coal has decreased slightly; natural gas has increased; and other sources such as nuclear, hydroelectric, wind, and solar power have been increasing.

Source: Data from *Annual Energy Review,* 1999, Energy Information Administration, and *BP Statistical Review of World Energy,* 1999 and previous editions.

was furnished primarily by coal, but a new source of energy was about to be discovered: oil.

The Chinese used some gas and oil as early as 1000 B.C., yet these resources remained virtually untapped until fairly recently. The oil well that Edwin L. Drake, an early oil prospector, drilled in Pennsylvania in 1859 was not the world's first oil well, but it was the beginning of the modern petroleum era. By 1870, oil production in the United States had reached over 4 million barrels a year and supplied 1 percent of the nation's energy requirements. It grew to nearly 50 percent by 1970 and currently contributes just over 40 percent. (See figure 9.4.)

For the first 60 years of production, the principal use of oil was to make kerosene, a fuel for lamps. The gasoline produced was discarded as a waste product. During this time, oil was abundant relative to its demand, and thus, it had a low price. The low price led the

oil industry to have the federal government control the price of oil. However, the automobile dramatically increased the demand. In 1900, the United States had only 8000 automobiles. By 1920, it had 8 million cars, and by 2001, 164 million. More oil was needed to make automobile fuel and lubricants.

The use of natural gas did not increase as rapidly as the use of oil. It was used primarily for home heating. In the early 1900s, 90 percent of the natural gas was "flared," that is, burned as a waste product at oil wells.

A series of events involving the U.S. government and the need for more oil was ultimately responsible for increased use of natural gas in the United States. World War II greatly increased the energy demand for manufacturing and transportation. In 1943, a federally financed pipeline was constructed to transport oil within the United States. This 2000-kilometer (1200-mile) pipeline transported oil more efficiently

from wells in Texas, Louisiana, and Oklahoma to refineries and factories in the eastern section of the country. In 1944, a longer (2400 kilometers [1500 miles]) federally financed pipeline was built to increase the flow of oil to the country's eastern and midwestern regions.

After the war, the federal government sold these pipelines to private corporations. The corporations converted the pipelines to transport natural gas. Thus, a direct link was established between the natural gas fields in the Southwest and the markets in the Midwest and East. By 1971, there were 400,000 kilometers (250,000 miles) of long-range transmission pipelines and 986,000 kilometers (612,000 miles) of distribution pipelines in the United States. Approximately 1,613,000 kilometers (1 million miles) of natural gas distribution pipelines are used today. Natural gas is used in many parts of the developed world for both home heating and industrial purposes.

Energy and Economics

Most industrial societies want to ensure a continuous supply of affordable energy. The higher the price of energy, the more expensive goods and services become. To keep costs down, many countries have subsidized their energy industries and maintained energy prices at artificially low levels. International trade in fossil fuels has a major influence on the world economy and politics. The emphasis on low-priced fuels has encouraged high rates of consumption.

Economic Growth and Energy Consumption

A direct link exists between economic growth and the availability of inexpensive energy. The replacement of human and animal energy with fossil fuels began with the Industrial Revolution and was greatly accelerated by the supply of cheap, easy-to-handle, and highly efficient fuels. Because the use of inexpensive fossil fuels allows each worker to produce more goods and services, productivity increased. The result was unprecedented economic growth in Europe, North America, and the rest of the industrialized world.

In North America, World War II was a prime factor in ending the economic depression of the 1930s. Military activities created millions of defense jobs. Almost everyone was employed, but there was a scarcity of consumer goods. After World War II, consumer goods that had been unavailable during the war were in great demand. Industries set up to produce military goods turned to the production of consumer goods. High employment, a rapidly expanding population, and a supply of inexpensive energy encouraged a period of rapid economic growth.

In Europe and Japan, where much of the industrial base was destroyed by the war, the recovery was slow. However, foreign aid and an intense rebuilding effort resulted in a reindustrialization of these countries. Because they had to build new facilities, often their technology was better than that of North America, where few factories were affected by the war.

The Role of the Automobile

The cheap, abundant energy that fueled industries produced an ever-increasing amount and array of consumer goods. One product was the automobile. The growth of the automobile industry, first in the United States and then in other industrialized countries, led to roadway construction, which required energy. Thus, the energy costs of driving a car were greater than just the fuel consumed in travel. As roads improved, higher speeds were possible. People demanded faster cars, and automobile companies were quick to build them. Bigger and faster cars required more fuel and even better roads. So roads were continually being improved, and better cars were being produced. A cycle of *more chasing more* had begun. In North America and much of Europe, the convenience of the automobile encouraged two-car

figure 9.5 **Energy-Demanding Lifestyle** Building private homes on large individual lots some distance from shopping areas and places of employment is directly related to the heavy use of the automobile as a mode of transportation. Heating and cooling a large enclosed shopping mall, along with the gasoline consumed in driving to the shopping center, increase the demand for energy. This has also been referred to as urban sprawl.

families, which created a demand for more energy. It requires energy to mine ore, process it into metals, form the metals into automobile components, and transport all the materials. As the economy grew, so did energy needs.

More cars meant more jobs in the automobile industry, the steel industry, the glass industry, and hundreds of other industries. Constructing thousands of kilometers of roads created additional jobs. From their beginnings as suppliers of lamp oil, the oil companies grew into one of the largest industries in the world. Thus, the automobile industry played a major role in the economic development of the industrialized world. All this wealth gave people more money for cars and other necessities of life. The car, originally a luxury, was now considered a necessity.

The car not only created new jobs but also altered people's lifestyles. Vacationers could travel greater distances. New resorts and chains of motels, restaurants, and other businesses developed to serve the motoring public, creating thousands of new jobs. Because people could live farther from work, they began to move to the suburbs. (See figure 9.5.) Large shopping centers in suburban areas hastened the decline of central business districts. Today, fewer than 50 percent of retail sales are made in central business districts of North American cities, which has resulted in a loss of jobs in these areas. In Philadelphia, 79 percent of all retail jobs were

located in the city in 1930; by 1970, they had declined to 43 percent.

As people moved to the suburbs, they also changed their buying habits. Labor-saving, energy-consuming devices became essential in the home. The vacuum cleaner, dishwasher, garbage disposal, and automatic garage door opener are only a few of the ways human power has been replaced with electrical power. Eleven percent of the electrical energy in North America is used to operate home appliances. Other aspects of our lifestyles illustrate our energy dependence. The small, horse-powered, family-owned farm of yesterday has grown into the huge, diesel-powered, corporate farm of today. Regardless of where we live, we expect Central American bananas, Florida oranges, California lettuce, Texas beef, Hawaiian pineapples, Ontario fruit, and Nova Scotia lobsters to be readily available at all seasons. What we often fail to consider is the amount of energy required to process, refrigerate, and transport these items. The car, the modern home, the farm, and the variety of items on our grocery shelves are only a few indications of how our lifestyles are based on a continuing supply of cheap, abundant energy.

Gasoline Prices and Government Policy

The price of a liter of gasoline is determined by two major factors: (1) the cost

Global Perspective

Five Ways to Curb Traffic

Here is how six cities around the world are discouraging traffic in already congested areas.

Hong Kong

Electronic sensors on cars record highway travel and time of day. Drivers are issued a monthly bill (commuter hours are the most expensive).

Singapore

Automobiles entering downtown Singapore during rush hour are required to display a $30-a-month sticker, but cars carrying four or more passengers may pass without charge.

Gothenburg, Sweden

To encourage pedestrian traffic, the central business district has been divided like a pie into zones, with cars prohibited from moving directly from one zone to another. Autos move from zone to zone by way of a peripheral ring road.

Rome and Florence

All traffic except buses, taxis, delivery vehicles, and cars belonging to area residents have been banned between 7:30 A.M. and 7:30 P.M.

Tokyo

Before closing the sale, the buyer of a standard-size vehicle must show evidence that a permanent parking space is available for the car. To comply with the law, some drivers have constructed home garages with lifts to permit double parking!

of purchasing and processing crude oil into gasoline and (2) various taxes. Most of the differences in gasoline prices among countries are a result of taxes and reflect differences in government policy toward motor vehicle transportation. (See figure 9.6.)

A major objective of governments is to collect money to build and repair roads. Governments often charge road users by taxing the fuel their cars or trucks run on. Governments can also discourage the use of automobiles by raising fuel taxes. An increase in fuel costs also creates a demand for increased fuel efficiency in all forms of motor transport.

Many European countries raise more money from fuel taxes than they spend on building and repairing roads. The United States, on the other hand, raises approximately 60 percent of the monies needed for roads from fuel taxes. The relatively low cost of fuel in the United States encourages more travel, which increases road repair costs. The cost of taxes to the U.S. consumer is about one-third of the retail gasoline price, while in Japan and many European countries, the cost is one-half to three-quarters.

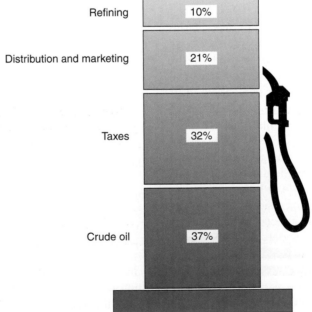

What We Pay for in a Gallon of Regular Gasoline

Retail price: $1.42 per gallon

(October 2002)

Refining	10%
Distribution and marketing	21%
Taxes	32%
Crude oil	37%

figure 9.6 **What We Pay for in a Gallon of Regular Gasoline** The price at the pump reflects several costs in addition to the cost of the crude oil.

Source: U.S. Dept. of Energy

How Energy Is Used

The amount of energy consumed by countries of the world varies widely. (See table 9.1.) The highly industrialized countries use most of the world's energy; less-developed countries use much less. Even countries with the same level of development vary in the amount of energy they use as well as in how they use it. To maintain their style of living, individuals in Canada and the United States use about twice as much energy as people in France or Japan and about 25 times as much energy as people in Africa.

Countries also use energy in different ways. Industrialized nations use energy about equally for three purposes: (1) residential and commercial uses, (2) industrial uses, and (3) transportation. Less-developed nations with little industry use most of their energy for residential purposes (cooking and heating). Developing countries use much of their energy to develop their industrial base.

Residential and Commercial Energy Use

The amount of energy required for residential and commercial use varies greatly throughout the world. Although a country with a high gross domestic product (GDP) uses a large amount of energy, it uses a lower percentage of its energy per capita for residential and commercial needs than does a less-developed country. For example, about 30 percent of the energy used in North America is for residential and commercial purposes, while in India, 90 percent of the energy is for residential uses. The ways residential and commercial energy is used also vary widely. In North America, 75 percent is used for air conditioning, refrigeration, water heating, and space heating. New demands for energy also have emerged. The Internet, for example, consumes an estimated 300 billion kilowatt hours of electrical power per year. Studies show that computers and the Internet now consume 13 percent of the available electrical power in the United States, up from only 1 percent seven years ago. That figure is expected to reach a range from 20 percent to 30 percent in just a few years as access to computers and the Internet expands. In India, however, almost all of the energy is used in the home for cooking since the scarcity and high cost of fuel preclude other uses.

As countries industrialize, their consumption of energy increases. Energy consumption and industrialization are interrelated.

The current pattern of residential and commercial energy use in each region of the world determines what conservation methods will be effective. In Canada, which has a cold climate, 40 percent of the residential energy is used for heating. Proper conservation practices could reduce this by 50 percent. In Africa, almost half of the energy used in the home is for cooking. (See figure 9.7.) Using fuel-efficient stoves instead of open fires could reduce these energy requirements by 50 percent.

Industrial Energy Use

The amount of energy countries use for industrial processes varies considerably. Nonindustrial countries use little energy for industry. Countries that are developing new industries dedicate a high percentage of their energy use to them. They divert energy to the developing industries at the expense of other sectors of their economy. Highly industrialized countries use a significant amount of their energy in industry, but their energy use is high in other sectors as well. In the United States, industry claims about 30 percent of the energy used.

The amount of energy required in a country's industrial sector depends on the types of industrial processes used. Many countries use inefficient processes and could reduce their energy consumption by converting to more energy-efficient ones. However, they need capital investment to upgrade their industries and reduce energy consumption. Some countries cannot afford the upgrade. For example, India, a nation with few coal deposits, still uses outdated open-hearth furnaces to produce steel. These furnaces require nearly double the worldwide energy average to produce a metric ton of steel. The high cost of converting forces India to continue to use this energy-expensive method rather than convert to the more efficient electric-hearth method.

Transportation Energy Use

As with residential, commercial, and industrial uses, the amount of energy used for transportation varies widely throughout the world. In some of the less-developed nations, transportation uses are very small. Per capita energy use for transportation is larger in developing countries and highest in highly developed countries. (See table 9.2.)

Table 9.1 Energy Consumption, 2000

Region	Energy Consumption per Capita per Year (Metric Tons of Oil Equivalent)
Africa	0.32
Latin America	0.67
Japan	3.70
France	4.05
Germany	4.11
Canada	7.63
United States	7.85

Source: Data from *BP Statistical Review of World Energy,* June 2001, and Population Reference Bureau, *2001 World Population Data Sheet.*

figure 9.7 **Open-Fire Cooking** About half of the energy demand in Africa is for cooking. Using a solar stove instead of an open fire could greatly reduce this energy need.

Table 9.2 Per Capita Energy Use for Transportation

Per Capita Use in Various Countries (2000)

Country	Energy Use in Gigajoules per Capita
India	2
Zimbabwe	4
Mexico	17
Argentina	18
Russia	26
Japan	28
Netherlands	41
Denmark	43
Australia	86
United States	105

U.S. Driving Habits

Average travel time to work in the United States in minutes:	24.3
In New York:	31.2
In North Dakota:	15.4
Percent of Americans who get to work by:	
Driving alone:	76.3
Carpooling:	11.2
Taking public transportation:	5.2
Staying home:	3.2
Walking:	2.7
Riding a bicycle:	0.4
Riding a motorcycle:	0.1
Other:	0.9

Source: U.S. Census Bureau, 2000.

Once a country's state of development has been taken into account, the specific combination of bus, rail, waterways, and private automobiles is the main factor in determining a country's energy use for transportation. In Europe, Latin America, and many other parts of the world, rail and bus transport are widely used because they are more efficient than private automobile travel, governments support these transportation methods, or a large part of the populace is unable to afford an automobile. In countries with high population densities, rail and bus transport is particularly efficient. (See figure 9.8.) In these countries, automobiles require about four times more energy per passenger kilometer than does bus or rail transport. In addition, most of these countries have high taxes on fuel, which raise the cost to the consumer and encourage the use of public transport. In North America, the situation is different. Government policy has kept the cost of energy artificially low and supported the automobile industry while removing support for bus and rail transport. Consequently, the automobile plays a dominant role, and public transport is primarily used only in metropolitan areas. Rail and bus transport are about twice as energy efficient as private automobiles. Private automobiles in North America consume about 40 percent of the gasoline produced in the world. Air travel is relatively expensive in terms of energy, although it is slightly more efficient than private automobiles. Passengers, however, are paying for the convenience of rapid travel over long distances.

environmental CLOSE-UP

Hybrid Electric Vehicles

In many metropolitan areas, air pollution is a significant problem. Emissions from automobiles are a major contributor to this air-quality problem. An increase in the efficiency with which the chemical energy of fuel is converted to the motion of automobiles would greatly reduce air pollution. Several conceptually simple modifications to automobiles can significantly reduce the energy needed to move an automobile. Reducing the weight by using lighter-weight materials to build automobiles and streamlining the shape are techniques that have been used for many years. To further reduce energy consumption (increase energy efficiency), more innovative techniques are required.

A great deal of energy is needed to get a vehicle to begin moving from a stop (accelerate), and it takes an equal amount of energy to stop a vehicle once it is moving (decelerate). Internal combustion engines operate most efficiently when they are running at a specific speed (rpm) and work at less than peak efficiency when the vehicle is accelerating or decelerating. Thus, using an internal combustion engine to accelerate contributes significantly to air pollution. When the brakes are applied to stop the vehicle, the kinetic energy possessed by the moving vehicle is converted to heat in the braking system. So the energy that has just been used to accelerate the vehicle is lost as heat when the vehicle is brought to a stop. Furthermore, in metropolitan areas, an automobile is sitting in traffic with its engine running a significant amount of the time. Thus, the stop-and-go traffic common in city driving provides conditions that significantly reduce the efficient transfer of the chemical energy of fuel to the kinetic energy of turning wheels.

Hybrid electric vehicles (HEVs) combine the internal combustion engine of a conventional vehicle with the battery and electric motor of an electric vehicle. This combination offers the extended range and rapid refueling that consumers expect from a conventional vehicle, with a significant portion of the energy and environmental benefits of an electric vehicle. The practical benefits of HEVs include improved fuel economy and lower emissions compared to conventional vehicles. The flexibility of HEVs will allow them to be used in a wide range of applications, from personal transportation to commercial hauling.

HEVs have several advantages over conventional vehicles:

- Regenerative braking capability helps minimize energy loss and recover the energy used to slow down or stop a vehicle.
- Engines can be sized to accommodate average load, not peak load, which reduces the engine's weight.

- Fuel efficiency is greatly increased (hybrids consume significantly less fuel than vehicles powered by gasoline alone).
- Emissions are greatly decreased.
- HEVs can reduce dependency on fossil fuels because they can run on alternative fuels.

Hybrids are growing in importance to automakers and government agencies because they offer greater fuel efficiency than conventional vehicles and are more appealing to consumers than pure electric vehicles. A Toyota official stated that Toyota's single year 1999 hybrid sales exceeded the sales of all pure electric vehicles from 1970 through 1999.

The GM Precept, introduced at the 2000 North American International Auto Show, is an example of the new type of HEV. The Precept is an aerodynamic five-passenger sedan designed to achieve 80 miles per gallon (gasoline equivalent). The Precept was developed by GM as its contribution to the Partnership for a New Generation of Vehicles (PNGV). The PNGV is a collaboration that began in 1993 between the U.S. government and the domestic auto industry. The specific aims of this partnership are lower emissions and up to three times the fuel efficiency of conventional cars without compromising safety, performance affordability, or utility. GM's Precept has nearly 130 technology innovations and 44 records of invention.

In October 2000, southern California was among the first regions in the United States to receive hybrid electric transit buses, from New Flyer of America, that feature series hybrid propulsion systems developed by GM's Allison Transmission Division. The hybrid technology, which uses a combination of electric motors, batteries, and an internal combustion engine, significantly reduces emissions, increases fuel economy by up to 50 percent, and increases acceleration by up to 50 percent. The system will cut nitrous oxide emissions by up to 50 percent, while cutting particulate, hydrocarbon, and carbon monoxide emissions by up to 90 percent when fueled with low-sulfur fuel.

General Motors will produce a full-size pickup truck featuring a hybrid power train beginning in 2004. GM's full-size hybrid pickup trucks, versions of the Chevrolet Silverado and the GMC Sierra, will deliver nearly 15 percent better fuel economy. GM will test its hybrid pickups in demonstration fleets in several U.S. cities. Each fleet of 10 trucks will feature a conventional power train and driveline with an electric motor integrated between the engine and transmission. Low-voltage lead-acid batteries will power the motor.

figure 9.8 **Public Transportation** In regions of the world where energy is expensive, such as in Germany, people make maximum use of cheaper public transportation.

The Variability of Gasoline Prices

By the first quarter of 2002, the retail price of gasoline in the United States was at its lowest since 1999. This follows a year in which gasoline prices were at their highest recorded level in many parts of the country. Although U.S. consumers are pleased at this reduction in driving costs, many are puzzled at the speed and magnitude of the decrease and wonder when gasoline prices will rise again—and whether they will increase as quickly as they fell. While there is no certainty what the future may hold, a number of factors behind gasoline prices can help to explain the recent rapid price changes and may give us some indications about which way they will go next.

The first step is to understand what drives gasoline prices. Three main factors influence price fluctuations for gasoline and most other petroleum products.

Changes in crude oil prices—Generally speaking, any change in the underlying price of crude oil is passed through to the prices of refined products, though these changes are sometimes offset by other factors, as described later. After peaking in 2001 at their highest levels since the Gulf War,

world crude oil prices have declined as lower global demand has allowed inventories to increase from the relatively low levels of 2000.

Seasonality in gasoline supply/ demand balance—Gasoline demand is significantly (about a million barrels per day in the United States) higher during the summer than at its low point in the winter, usually in January. Production of gasoline by refineries can also be adjusted but not enough to meet the increased summer demand, leaving a larger portion of summer demand to be satisfied by imports and withdrawals from inventories. Typically, after subtracting out crude oil prices, gasoline prices average about 10 cents per gallon higher at their summer high than at their winter low.

Unusual events or trends affecting the supply/demand balance—Events such as refinery or pipeline problems can affect supply, as can events such as changes in the specifications of products that can be sold in a particular region or season. This latter type of influence occurred in 2000–01 in California and the midwestern United States, where more stringent air quality requirements were instituted. Beginning in the second half of 2001, demand was negatively impacted by a weakening economy, while

potential refinery gasoline production increased due to a drop in jet fuel demand following the September 11 terrorist attacks. (See figure 9.9).

The market forces that help determine the price of gasoline vary from the state of the local economy to the marketing strategies of the companies that produce the gasoline. Customer-buying preferences and the number of stations in any given area are also factors. In addition to the above, the wholesale price of gasoline at any given time is also a factor. These factors can all affect the price of the gasoline you buy.

You have also no doubt witnessed what are referred to as price wars over gasoline at your local stations. Price wars can begin when a single dealer may use gasoline as the "loss leader" to attract you to the station to purchase other items such as car parts or groceries. The evolution from traditional gas stations to "mini-marts" has increased price wars and "loss leader" marketing.

Gasoline is a very competitive market, and the visible pricing of the cost of a gallon or liter will attract customers. Pricing, however, in the long run will always stabilize to provide the station owner a reasonable profit after accounting for wholesale delivery and operating costs.

Governmental bodies at both the federal and state level closely monitor gasoline price swings. Such swings are indicative of market competition and often benefit the consumer by offering lower prices.

Electrical Energy

Electrical energy is such a large proportion of energy consumed in most countries that it deserves special comment. Electricity is both a way that energy is consumed and a way that it is supplied. Almost all electric energy is produced as a result of burning fossil fuels. Thus, we can look at electrical energy as a use to which fossil fuel energy is put. In the same way we use natural gas to heat

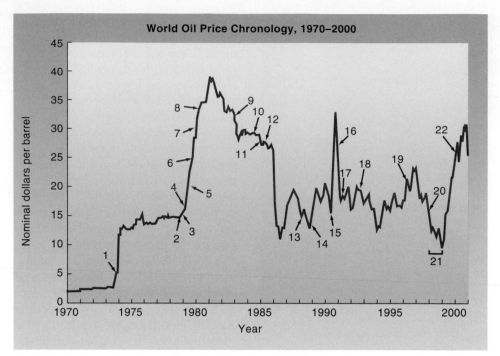

World Oil Price Chronology, 1970–2000

U.S. Oil Imports		
Source	Barrels per day 2001 (in millions)	
*Saudi Arabia	1.6	18%
Mexico	1.4	15%
Canada	1.3	15%
*Venezuela	1.3	14%
*Nigeria	0.8	9%
*Iraq	0.7	8%
Angola	0.3	3%
Norway	0.3	3%
Colombia	0.2	3%
*Kuwait	0.2	3%
Other nations	1.1	9%

*These countries are members of OPEC

Source: U.S. Energy Information Administration.

1. Organization of Petroleum Exporting Countries (OPEC) oil embargo begins (October 19–20, 1973).
2. OPEC decides on 14.5 percent price increase for 1979.
3. Iranian revolution; Shah deposed.
4. OPEC raises prices 14.5 percent on April 1, 1979.
5. OPEC raises prices 15 percent.
6. Iran takes hostages; President Carter halts imports from Iran; Iran cancels U.S. contracts; non-OPEC output hits 17.0 million barrels per day.
7. Saudis raise market crude price from $19 to $26 per barrel.
8. Kuwait, Iran, and Libya production cuts drop OPEC oil production to 27 million barrels per day.
9. OPEC cuts prices by $5 per barrel and agrees to 17.5 million barrels per day output.
10. Norway, United Kingdom, and Nigeria cut prices.
11. OPEC accord cuts Saudi Light price to $28 per barrel.
12. OPEC output falls to 13.7 million barrels per day.
13. OPEC/non-OPEC meeting fails.
14. Exxon's *Valdez* tanker spills 11 million gallons of crude oil.
15. Iraq invades Kuwait.
16. Operation Desert Storm begins; 17.3 million barrels of Strategic Petroleum Reserves crude oil sales are awarded.
17. Persian Gulf war ends.
18. OPEC production reaches 25.3 million barrels per day, the highest in over a decade.
19. Extremely cold weather in the United States and Europe.
20. OPEC raises its production ceiling by 2.5 million barrels per day to 27.5 million barrels per day. This is the first increase in four years.
21. Oil prices continue to plummet as increased production from Iraq coincides with no growth in Asian oil demand due to the Asian economic crisis and increases in world oil inventories following two unusually warm winters.
22. Oil prices triple between January 1999 and September 2000 due to strong world oil demand, OPEC oil production cutbacks, and other factors, including weather and low oil stock levels.

figure 9.9 **The Impact of Events on Oil Prices** This chronology shows how various events can affect the price of crude oil.

Source: U.S. Department of Energy, Office of Strategic Petroleum Reserve, 2000 chronology.

homes, we can use natural gas to produce electricity. Because the transportation of electrical energy is so simple and the uses to which it can be put are so varied, electricity is a major energy source for many people of the world.

As with other forms of energy use, electrical consumption in different regions of the world varies widely. The amount of electricity used by all of the less-developed nations of the world, which have about 80 percent of the world's population, is less than that used by the United States alone. The per capita use of electricity in North America is 25 times greater than average per capita use in the less-developed countries. In Nepal, the annual per capita use of electricity is 23 kilowatt hours, which is enough to light a 100-watt lightbulb for one week. The per capita consumption of electricity in North America is 270 times greater than in Nepal. There is also a wide variation in electrical consumption among developed countries. For example, the per capita use in Europe is about half that in North America.

The production and distribution of electricity is a major step in the economic development of a country. In developed nations, about a quarter of the electricity is used by industry. The remainder is used primarily for residential and commercial purposes. In nations that are developing their industrial base, over half of the electricity is used by industry. For example, industries consume 55 percent of the electricity used in Mexico and 70 percent of that used in South Korea.

Energy Consumption Trends

From a historical point of view, it is possible to plot changes in energy consumption. Economics, politics, public attitudes, and many other factors must be incorporated into an analysis of energy use trends. (See tables 9.3 and 9.4)

Table 9.3	World Total Energy Consumption by Region and Fuel, 1990–2010		
(Quadrillion Btu)			
	History	**Projections**	
Region/Fuel	**2000**	**2005**	**2010**
OECD	214.4	227.4	239.0
Oil	89.6	94.7	98.2
Natural gas	45.6	49.5	53.9
Coal	40.5	42.4	44.0
Nuclear	18.8	19.2	18.7
Renewables	19.8	21.5	24.0
EE/FSU	63.7	69.4	74.7
Oil	12.8	17.5	21.0
Natural gas	29.1	32.5	34.0
Coal	15.4	15.4	15.0
Nuclear	3.4	3.3	3.3
Renewables	3.1	3.4	3.8
Non-OECD Asia	79.4	92.1	104.2
Oil	29.5	32.7	34.6
Natural gas	4.7	5.9	7.2
Coal	39.1	45.7	52.8
Nuclear	1.3	1.8	1.9
Renewables	4.8	6.0	7.5
Middle East	15.2	16.9	18.4
Oil	9.4	10.4	11.4
Natural gas	5.4	6.0	6.3
Coal	0.2	0.2	0.3
Nuclear	0.0	0.0	0.0
Renewables	0.3	0.4	0.4
Africa	12.6	13.6	14.6
Oil	5.8	6.3	6.9
Natural gas	1.9	2.1	2.4
Coal	4.1	4.2	4.4
Nuclear	0.1	0.1	0.1
Renewables	0.7	0.8	0.9
Central and South America	17.2	19.1	20.8
Oil	9.4	10.2	10.9
Natural gas	2.5	3.1	3.7
Coal	1.1	1.4	1.6
Nuclear	0.2	0.3	0.3
Renewables	4.0	4.2	4.4
World Total	402.6	438.6	471.7
Oil	156.5	170.0	181.3
Natural gas	89.2	98.1	106.8
Coal	100.5	109.4	118.0
Nuclear	23.7	24.6	24.4
Renewables	28.1	36.4	41.1

Notes: OECD = Organization for Economic Cooperation and Development. EE/FSU = Eastern Europe/Former Soviet Union.

Sources: History data from Energy Information Administration, *International Energy Annual,* DOE/EIA-0219(92), Washington, D.C.; projections from EIA, *World Energy Projections,* 2000.

Table 9.4 — International Petroleum Supply and Demand

(Million Barrels per Day)

	Year				Year		
	1999	2000	2001		1999	2000	2001
Demand				**Supply**			
OECD				OECD			
United States (50 states)	19.5	19.6	20.0	United States (50 states)	9.0	9.0	9.0
U.S. territories	0.3	0.4	0.4	Canada	2.6	2.7	2.8
Canada	1.9	1.9	1.9	North Sea	6.3	6.6	6.7
Europe	14.5	14.7	14.9	Other OECD	1.5	1.7	1.7
Japan	5.6	5.5	5.6				
Australia and New Zealand	1.0	1.0	1.0				
Total OECD	42.8	43.2	43.8	Total OECD	19.5	20.1	20.1
Non-OECD				Non-OECD			
Former Soviet Union	3.6	3.7	3.7	OPEC	29.3	30.4	31.9
Europe	1.6	1.6	1.7	Former Soviet Union	7.4	7.6	7.7
China	4.3	4.5	4.8	China	3.2	3.3	3.3
Other Asia	8.8	9.2	9.7	Mexico	3.4	3.6	3.6
Other Non-OECD	13.5	13.9	14.3	Other Non-OECD	11.2	11.3	11.5
Total Non-OECD	31.9	32.9	34.2	Total Non-OECD	54.5	56.2	58.0
Total World Demand	74.7	76.1	78.0	Total World Supply	74.0	76.3	78.1

OECD: Organization for Economic Cooperation and Development: Australia, Austria, Belgium, Canada, Denmark, Finland, France, Germany, Greece, Iceland, Ireland, Italy, Japan, Luxembourg, the Netherlands, New Zealand, Norway, Portugal, Spain, Sweden, Switzerland, Turkey, the United Kingdom, and the United States. The Czech Republic, Hungary, Mexico, Poland, and South Korea are all members of OECD but are not yet included in our OECD estimates.

OPEC: Organization of Petroleum Exporting Countries: Algeria, Indonesia, Iran, Iraq, Kuwait, Libya, Nigeria, Qatar, Saudi Arabia, the United Arab Emirates, Gabon, and Venezuela.

Former Soviet Union: Armenia, Azerbaijan, Belarus, Estonia, Georgia, Kazakhstan, Kyrgyzstan, Latvia, Lithuania, Moldova, Russia, Tajikistan, Turkmenistan, Ukraine, and Uzbekistan.

Sources: Energy Information Administration: latest data available from EIA databases supporting the following reports: *International Petroleum Statistics Report,* DOE/EIA-0520; Organization for Economic Cooperation and Development, *Annual and Monthly Oil.*

In 2001, world energy consumption was around 26 million metric tons of oil equivalent per day, an increase of 19 percent since 1985. Of this total, conventional fossil fuels—oil, natural gas, and coal—accounted for about 90 percent.

Over half of world energy is consumed by the 25 countries that are members of the Organization for Economic Cooperation and Development (OECD). These countries (Australia, New Zealand, Japan, Canada, Mexico, the United States, and the countries of Europe) are the developed nations of the world. In the last decade, per capita energy consumption in OECD countries has risen moderately while economic growth has continued. There has also been a shift toward service-based economies, with energy-intensive industries moving to non-OECD countries. In contrast, in countries that are becoming more economically advanced, energy consumption is increasing at a faster rate.

Oil remains the world's major source of energy, accounting for about 40 percent of primary energy demand. Coal accounts for 24 percent, and natural gas for 27 percent; the remainder is supplied mainly by nuclear energy and hydropower.

Since 1973, the year of the first "oil shock," demand for natural gas has increased faster than the demand for other fossil fuels. From 1973 to 2000, consumption of natural gas increased by about 100 percent, coal consumption by about 40 percent, and oil consumption by about 30 percent.

Since 1950, North American energy consumption has steadily increased. However, there were two periods of decline, one beginning in 1973 and the other in 1979. Both of these episodes were the result of political turmoil in the Middle East and the increasing influence of the Organization of Petroleum Exporting Countries (OPEC). (See figure 9.10.) The same trend occurred in Western Europe, Japan, and Australia, resulting in a slowing of the world's growth in energy consumption. (See figure 9.11.) Several factors contributed to these declines in energy consumption. Increased prices for oil and all other forms of energy forced businesses and individuals to become more energy conscious and to expand efforts to conserve energy. From 1970 to 1983, the amount of energy used

Alternative-Fuel Vehicles

Electricity is the only zero-emission option now available for vehicles, but many experts believe that other fuels will prove viable in *reduced*-emission vehicles, although emissions are still produced in the generation of electricity at the power plant. Here are some of the options.

Compressed Natural Gas (CNG)

Status:
Fuels thousands of vehicles already on U.S. highways. Mass production is still a couple of years away, at the earliest.

Advantages:
- Substantial U.S. supply of natural gas
- Easy—though expensive—to convert conventional gasoline cars to CNG
- Cheaper: the equivalent of buying gasoline at 70 cents a gallon

Disadvantages:
- Requires heavy, bulky fuel tanks
- Public refueling stations expensive to build
- Lower performance and shorter range than gasoline vehicles

Emissions Benefits:
- Burns 80 percent cleaner than gasoline

Methanol/Flex-Fuel

Status:
Some heavy vehicles—including 354 buses operated by Southern California's Metropolitan Transportation Authority—run on pure methanol. More than 10,000 flex-fuel vehicles—running on methanol-gasoline combinations—are now on U.S. roads.

Advantages:
- Can be made from natural gas (98 percent of current production) and a range of renewable sources
- Requires only modest changes in gasoline engines and fueling infrastructure
- Higher octane than gasoline, giving 5 percent better performance

Disadvantages:
- Highly corrosive
- Lower energy content than gasoline, so requires bigger fuel tanks
- Fuel cost slightly higher than gasoline

Emissions Benefits:
- An 85 percent methanol mix burns 30 to 50 percent cleaner than pure gasoline; pure methanol vehicles potentially could be much cleaner

Hydrogen

Status:
Still largely experimental. Vehicles could be powered by hydrogen fuel cells or by engines that burn the gas. Practical, wide-scale use is at least a decade away.

Advantages:
- A high-energy fuel, boosting vehicle range
- Potentially unlimited fuel supply if produced from water; currently made from natural gas
- Produces no carbon dioxide, the gas blamed for global warming

Disadvantages:
- Highly explosive, though ultimately could be less dangerous than gasoline
- Costly and difficult to produce
- Costly and difficult to store

Emissions Benefits:
- If produced from water using solar energy and then used in a fuel cell, it would be a zero-emissions technology.

Propane

Status:
Broadly used in transportation worldwide

Advantages:
- A proven, low-cost fuel
- Public refueling infrastructure in place
- Nontoxic, so storage tanks are exempt from environmental regulations

Disadvantages:
- Supplies are limited, compared with other alternatives. At most, it could replace 10 percent of gasoline
- Highly flammable

Emissions Benefits:
- At least 50 percent cleaner than conventional gasoline

Sources: Data from California Air Resources Board, California Energy Commission, South Coast Air Quality Management District, Ward's Communications, Ford Motor Co., General Motors Corp., Chrysler Corp., and LP Gas Coalition.

for heat per dwelling in the United States declined by 20 percent. Comparable reductions were made in Denmark, West Germany, and Sweden.

Since 1970, however, energy consumption has increased in North America and Europe. Like the stabilization of energy use in 1973 and 1979, the increase since 1980 is due to the price of oil. In 1979, oil was selling for about $40 a barrel. Beginning in 1980, the price began to drop, and in 1998, oil was selling for less than $15 a barrel. When Iraq invaded Kuwait in August 1990, prices rose dramatically but decreased at the end of hostilities.

During the 1980s, with energy costs declining, people in North America and Europe became less concerned about their energy consumption. They used more energy to heat and cool their homes and buildings, used more home appliances, and bought bigger cars. Obviously, the two primary factors that determine energy use are political stability in parts of the world that supply oil and the price of that oil. Insecurity leads to a price rise, but governments can also manipulate prices by changing taxes, granting subsidies, and using other means. The energy consumption behavior of most people is motivated by economics rather than by a desire to wisely use energy resources.

Over the past several years, world oil prices have been extremely volatile. In 1999, consumers benefited from oil prices that fell to $10 per barrel—a result of oversupply caused by lower demand for oil in both southeast Asia, which was suffering from an economic recession,

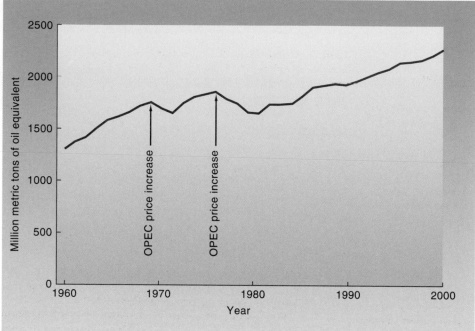

figure 9.10 **Changes in U.S. Energy Consumption** Energy consumption in the United States rose gradually until 1973, when the OPEC countries increased the price of oil. The rise in price reduced consumption. A similar action in 1979 resulted in another decrease. Since 1983, however, consumption has steadily increased.

Source: Data from *BP Statistical Review of World Energy*, 2001.

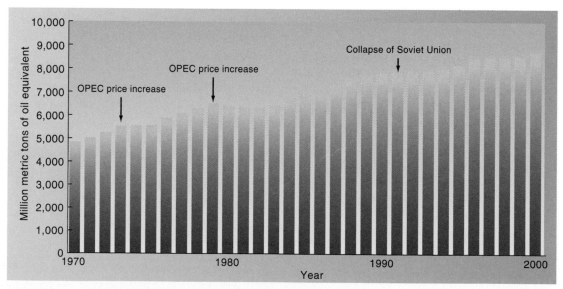

figure 9.11 **World Energy Consumption** Energy consumption has increased steadily for the past 30 years. The slowdown in growth beginning in 1973 and the decline in the early 1980s is the result of price increases instituted by OPEC. The slight decline noted in 1991 and 1992 is the result of decreased energy consumption related to economic problems in the former Soviet Union and Eastern Europe.

Source: Data from *BP Statistical Review of World Energy*, 2001.

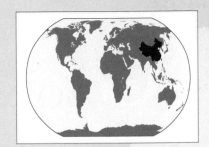

Global Perspective

Energy Development in China

China is the second largest energy consumer in the world, following the United States (about 39 quadrillion Btu in 1999 versus 90 quadrillion Btu in the United States). China's rapidly growing economy will drive energy demand growth of about 4 to 5 percent annually through 2015 (compared with growth of about 1 percent in the industrialized countries). China consumes about 10 percent of the world's energy and accounts for about 10 percent of the world's energy production.

China became a net importer of energy in 1995 and is expected to become increasingly dependent on imports; however, it is expected to remain a net exporter of coal. China has been a net importer of oil since 1993. Most production is from onshore fields. Offshore and some onshore areas are now open to foreign investment. At current rates, China's oil imports could exceed 2 million barrels daily by 2003.

In 2000, coal accounted for 70 percent of China's primary energy production. In the same year, petroleum accounted for 19 percent, hydroelectricity 11 percent, natural gas 2 percent, and nuclear power 0.5 percent. China's electricity is generated overwhelmingly by coal (about 70 percent), followed by hydroelectricity at about 20 percent. Because of China's extensive domestic coal resources and its wish to minimize dependence on foreign energy sources, it is expected that coal will remain the main energy source for electricity generation in China for the foreseeable future. Since 1979, the government has pursued a policy of replacing oil consumption with coal in the generation of electric power. The proportion of coal-fired electric-power generation has grown from 60 percent in 1980 to 70 percent in 1999.

High coal use has serious environmental repercussions. China plans to diversify by doubling its 1990 level of hydropower production by 2003 and increasing nuclear capacity. The Three Gorges dam, when completed, will contribute significantly to the country's hydropower production.

One of the biggest concerns regarding energy consumption in China is that of carbon emissions and the threat of increasing global warming. China's carbon emissions are expected to increase 4 percent annually through 2015, driven by fast economic growth and a rapid increase in coal use. The country's total carbon emissions should exceed 2 billion metric tons (2.2 tons) shortly after 2015, over three times the 1986 emissions and more than double the 1995 emissions.

The growing demand for power has driven reforms in China's energy-sector process and structure. The traditional, centralized regulatory regime is moving toward an increasingly decentralized and market-oriented regime.

Until recently, the Ministry of Electric Power (MOEP) regulated the electric power industry of China. The Ninth People's Congress held in 1998 introduced reform measures to streamline the central government. This resulted in the abolition of a number of industrial ministries, including MOEP. Much of the former MOEP's function has now been taken over by the State Power Corporation of China (SPCC). SPCC's main duties in this respect are:

- Formulating China's electric-power development strategies, legislation, and policies, including investment policy, technical policy, and major energy production and consumption policies.
- Formulating unified energy-industry planning in collaboration with the State Development Planning Commission and other governmental agencies.
- Supervising the implementation of related national policies decrees and plans.
- Providing services to regional and provincial electric power enterprises.

Furthermore, SPCC owns all of China's state-owned power-generating assets, other than those owned by companies not directly managed by SPCC and a few smaller units directly owned by local governments.

Source: Data from Energy Information Administration, U.S. Department of Energy, *International Energy Outlook,* 2001.

Source: History data from Energy Information Administration, *International Petroleum Statistics Report,* DOE/EIA–520 (92/08), August 1992, Washington, D.C. Projections from EIA, National Energy Modeling System, International Energy Module, 2000.

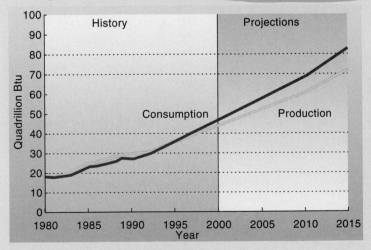

a. China's energy production and consumption

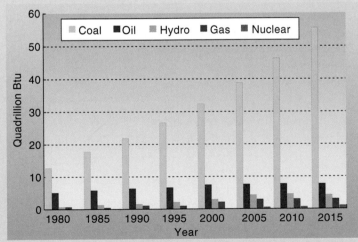

b. Energy production by fuel type, 1980–2015

Global Perspective

Potential World Petroleum Resources

Oil is a finite, nonrenewable resource. It will eventually be exhausted. The primary questions about when we will run out of oil revolve around price, rates of use, and rates at which new discoveries are made. In 2000, the U.S. Geological Survey (USGS) issued an assessment of potential world petroleum resources. The report stated that undiscovered oil and gas resources of the world had increased, with a 20 percent increase in undiscovered oil and a slight decrease in undiscovered natural gas. This assessment estimates the volume of oil and gas, exclusive of the United States, that may be added to the world's reserves in the next 30 years.

The assessment indicates that there is more oil and gas in the Middle East and in the offshore areas of western Africa and eastern South America than previously reported, less oil and gas in Canada and Mexico, and significantly lower volumes of natural gas in the former Soviet Union.

With the evolution of technology and new understandings of petroleum systems, the USGS *World Petroleum Assessment 2000* is the first of its kind to provide a rigorous geologic foundation for estimating undiscovered energy resources for the world. The results have important implications for energy prices, policy, security, and the global resource balance.

The assessment provides a snapshot of current information about the location and abundance of undiscovered oil and gas resources. Such an overview provides exploration geologists, economists, and investors with a general picture of where oil and gas resources are likely to be developed in the future. The USGS periodically estimates the amount of oil and gas remaining to be found. Since 1981, the last three of these studies have shown a slight increase in the combined volume of identified reserves and undiscovered resources.

In the USGS 2000 report, the world was divided into approximately 1000 petroleum provinces, based primarily on geologic factors, and grouped into eight regions. Estimates of reserve growth were based on the following:

- As drilling and production within discovered fields progresses, new pools or reservoirs are found that were not previously known.
- Advances in exploration technology make it possible to identify new targets within existing fields.
- Advances in drilling technology make it possible to recover oil and gas not previously considered recoverable in the initial reserve estimates.

To view the entire report, go to the USGS home page at www.usgs.gov.

Estimation of Undiscovered Oil and Natural Gas

Region	Undiscovered Oil (Billion Barrels)	Percent of World Total	Undiscovered Natural Gas (Trillion Cubic Feet)	Percent of World Total
1. Former Soviet Union	116	17.9%	1611	34.5%
2. Middle East and North Africa	230	35.4%	1370	29.3%
3. Asia-Pacific	30	4.6%	379	8.1%
4. Europe	22	3.4%	312	6.7%
5. North America*	70	10.9%	154	3.3%
6. Central and South America	105	16.2%	487	10.4%
7. Sub-Saharan Africa and Antarctica	72	11.0%	235	5.0%
8. South Asia	4	0.6%	120	2.6%
WORLD TOTALS*	649		4669	

*Exclusive of the United States.

Source: USGS *World Petroleum Assessment* 2001.

and North America and Western Europe, which had warmer than expected winters. In 2000, however, world oil prices rebounded strongly, reaching a daily peak of $37 per barrel, rates not seen since the Persian Gulf War or 1990–1991. The high prices can be traced to a tightening of production by OPEC and several key non-OPEC countries, such as Russia, Mexico, and Norway. In addition, oil companies did not want to commit capital to major development efforts for fear of a return to low prices. Unrest in the Middle East has also affected the price volatility.

The industrialized world was impacted by the high world oil prices of 2000. Concerns in the United States about a recurrence of the previous winter's shortage of home heating fuel oil for the Northeast led the Clinton administration to allow industry access to as much as 30 million barrels of crude oil from the nation's Strategic Petroleum Reserve. By 2001, however, the price of oil was again declining. The slowing of the world's economy and mild winter temperatures led to the price of oil approaching five-year lows.

Oil provides a larger share of world energy consumption than any other energy source, and it is expected to remain in that position through 2020. The share of total world energy consumption attributed to oil is projected to remain unchanged over the 2002–20 time period at 40 percent. Oil's market share does not increase because countries in many parts of the world are expected to switch to natural gas and other fuel, particularly for electricity generation. World oil consumption is projected to increase by 2.3 percent annually over the 20-year projection period, from 75 million barrels per day in 1999 to 120 million barrels per day in 2020.

OPEC

OPEC began in September 1960, when the governments of five of the world's leading oil-exporting countries agreed to form a cartel. Three of the original members—Saudi Arabia, Iraq, and Kuwait—were Arab countries, while Venezuela and Iran were not. Today, 12 countries belong to OPEC. These include seven Arab states—Saudi Arabia, Kuwait, Libya, Algeria, Iraq, Qatar, and United Arab Emirates—and five non-Arab members—Iran, Indonesia, Nigeria, Gabon, and Venezuela. OPEC nations control over 75 percent of the world's estimated oil reserves of 1200 billion barrels of oil. Middle Eastern OPEC countries control over 60 percent of this total, which makes OPEC and the Middle East important world influences.

During the Arab-Israeli War of 1973, OPEC was supplying 55 percent of the world's oil. To protest the war, seven Arab members of OPEC reduced their production of oil. This resulted in a worldwide oil shortage, which caused all oil companies, in OPEC and non-OPEC countries, to increase their prices. In the United States, the price of oil skyrocketed from $3.39 to $13.93 per barrel, resulting in oil shortages, long lines at gas stations, and increased domestic exploration. The unity of OPEC, however, was not to last into the 1980s.

During the 1980s, OPEC members had important differences concerning oil pricing and production rules. These differences weakened the cartel. Then, the 1990 invasion of Kuwait by Iraq deeply divided the OPEC countries.

Owing in part to friction caused by the Kuwait conflict and the decline in world oil prices during the middle and late 1990s, OPEC's power has continued to slide. The price of oil began to rise in 1999 and, by the middle of 2000, had reached over $31 per barrel. Gasoline prices in the United States began to exceed $2 per gallon. The United States exerted diplomatic pressure on OPEC to increase production levels to lower the cost of gasoline. OPEC stated that gasoline production in the United States, rather than a shortage of OPEC crude oil output, was to blame for high prices. Iran, OPEC's second largest producer, remained strongly against higher oil production, stating that there was no shortage of crude.

While OPEC includes 11 of the world's top 20 net petroleum exporters, the second and third largest exporters, Russia and Norway, are not OPEC members. Russian total petroleum production and exports have risen more than most analysts anticipated over the past few years, and the country is expected to raise production and exports into the foreseeable future. Non-OPEC Mexico, the tenth largest net petroleum exporter, has also been increasing production. The overall effect of such non-OPEC production will have an impact on the influence of OPEC in the global oil market.

Summary

A constant supply of energy is required by all living things. Energy has a major influence on society. A direct correlation exists between the amount of energy used and the complexity of civilizations.

Wood furnished most of the energy and construction materials for early civilizations. Heavy use of wood in densely populated areas eventually resulted in shortages, so fossil fuels replaced wood as a prime source of energy. Fossil fuels were formed from the remains of plants, animals, and microorganisms that lived about 300 million years ago. Fossil-fuel consumption in conjunction with the invention of labor-saving machines resulted in the Industrial Revolution, which led to the development of technology-oriented societies today in the developed world.

Throughout the world, residential and commercial uses, industries, transportation, and electrical utilities require energy. Because of financial, political, and other factors, nations vary in the amount of energy they use as well as in how they use it. Analysts expect the worldwide demand for energy to increase steadily.

Key Terms

fossil fuels *176*

Industrial Revolution *176*

Review Questions

1. Why was the sun able to provide all energy requirements for human needs before the Industrial Revolution?
2. In addition to food, what energy requirements does a civilization have?
3. Why were some countries unable to use the technologies developed during the Industrial Revolution?
4. What factors caused a shift from wood to coal as a source of energy?
5. How were energy needs in World War II responsible for the subsequent increased consumption of natural gas?
6. What part does government regulation play in changing the consumption of natural gas and oil?
7. Why was much of the natural gas that was first produced wasted?
8. What was the initial use of oil? What single factor was responsible for a rapid increase in oil consumption?
9. List the three purposes for which a civilization uses energy.
10. Why is OPEC important in the world's economy?

Critical Thinking Questions

1. Imagine you are an historian writing about the Industrial Revolution. Imagine that you also have your new knowledge of environmental science and its perspective. What kind of a story would you tell about the development of industry in Europe and the United States? Would it be a story of triumph or tragedy, or some other story? Why?
2. What might be some of the effects of raising gasoline taxes in the United States to the rate that most Europeans pay for gasoline? Why? What do you think about this possibility?
3. Some argue that the price of gasoline in the United States is artificially low because it does not take into account all of the costs of producing and using gasoline. If you were to figure out the "true" cost of gasoline, what kinds of factors would you want to take into account?
4. How has the ubiquitous nature of automobiles changed the United States? Do you feel these changes are, in balance, positive or negative? What should the future look like regarding automobile use in the United States? How can this be accomplished?
5. Some energy experts predict that the Organization of Petroleum Exporting Countries (OPEC) will control about 65 percent of the known oil reserves by the year 2010. What political and economic effects do you think this will this have? Will this have any effect on energy use?
6. How do you think projected energy consumption will affect world politics and economics, given current concerns about global warming?

Concept Map

Construct a map to show relationships among the following concepts:

energy consumption

economic growth

fossil fuels

industrial revolution

energy demand

Interactive Exploration

Check out the website at **http://www.mhhe.com/environmentalscience** and click on the cover of this textbook for quizzing, career information, case studies, and hot links for the following topics:

Conventional Energy Sources

Oil, Coal, and Gas as Energy Sources

Nuclear Energy

Sustainable Energy

Air Pollution

Energy Sources

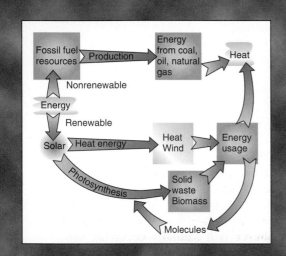

Objectives

After reading this chapter, you should be able to:

- Differentiate between resources and reserves.
- Identify peat, lignite, bituminous coal, and anthracite coal as steps in the process of coal formation.
- Recognize that natural gas and oil are formed from ancient marine deposits.
- Explain how various methods of coal mining can have negative environmental impacts.
- Explain why surface mining of coal is used in some areas and underground mining in other areas.
- Explain why it is more expensive to find and produce oil today than it was in the past.
- Recognize that secondary recovery methods have been developed to increase the proportion of oil and natural gas obtained from deposits.
- Recognize that transport of natural gas is still a problem in some areas of the world.
- Explain why the amount of energy supplied by hydroelectric power is limited.
- Describe how wind, geothermal, and tidal energy are used to produce electricity.
- Recognize that wind, geothermal, and tidal energy can be developed only in areas with the proper geologic or geographical features.
- Describe the use of solar energy in passive heating systems, active heating systems, and the generation of electricity.
- Recognize that fuelwood is a major source of energy in many parts of the less-developed world and that fuelwood shortages are common.
- Describe the potential and limitations of biomass conversion and waste incineration as sources of energy.
- Recognize that energy conservation can significantly reduce our need for additional energy sources.

Chapter Outline

Energy Sources

Chapter 9 outlined the historical development of energy consumption and how advances in civilizations were closely linked to the availability and exploitation of energy. New manufacturing processes relied on dependable sources of energy. Technology accelerated in the twentieth century. Between 1900 and 2000, world energy consumption increased by a factor of 14, the quantity of products manufactured increased by nearly fortyfold, but population increased only threefold. (See table 10.1.)

The energy sources most commonly used by industrialized nations are the fossil fuels: oil, coal, and natural gas, which supply about 90 percent of the world's commercially traded energy. Fossil fuels were formed hundreds of millions of years ago. They are the accumulation of energy-rich organic molecules produced by organisms as a result of photosynthesis over millions of years. We can think of fossil fuels as concentrated, stored solar energy. The rate of formation of fossil fuels is so slow that no significant amount of fossil fuels will be formed over the course of human history. Since we are using these resources much faster than they can be produced and the amount of these materials is finite, they are known as **nonrenewable energy sources.** (See figure 10.1.) Eventually, human demands will exhaust the supplies of coal, oil, and natural gas.

Table 10.1	World Population, Economic Output, and Fossil-Fuel Consumption		
	Population (Billions)	Gross World Product (Trillion 1995 Dollars)	Fossil-Fuel Consumption (Billion Tons Coal Equivalent)
1900	1.6	0.6	1
1950	2.5	2.9	3
2000	6.0	34.0	15

Source: Data from *CIA World Fact Book.*

In addition to nonrenewable fossil fuels, there are several renewable energy sources. **Renewable energy sources** replenish themselves or are continuously present as a feature of the solar system. For example, in plants, photosynthesis converts light energy into chemical energy.

$$\text{Carbon Dioxide} + \text{Water} + \text{Sunlight Energy} \rightarrow \text{Biomass (Chemical Energy)} + \text{Oxygen}$$

This energy is stored in the organic molecules of the plant as wood, starch, oils, or other compounds. Any form of biomass—plant, animal, alga, or fungus—can be traced back to the energy of the sun. Since biomass is constantly being produced, it is a form of renewable energy. Solar, geothermal, and tidal energy are renewable energy sources because they are continuously available. Anyone who has ever lain in the sun, seen a geyser or hot springs, or been swimming in the surf has experienced these forms of energy. However, many technical problems must be solved before these renewable energy forms can contribute significantly to meeting humans' energy demands.

Resources and Reserves

When discussing deposits of nonrenewable resources, such as fossil fuels, we must differentiate between deposits that can be extracted and those that cannot. From a technical point of view, a **resource** is a naturally occurring substance of use to humans that can *potentially* be extracted using current technology. **Reserves** are known deposits from which materials *can* be extracted profitably with existing technology *under certain economic conditions.* It is important to recognize that *reserves* is an economic concept and is

Predicting Energy Needs

Rate of Average Energy Use

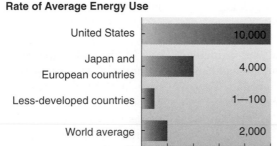

	Watts per person 24 hours per day
United States	10,000
Japan and European countries	4,000
Less-developed countries	1—100
World average	2,000

Annual Total World Energy Consumption

12 TWyr — 6 billion people x 2000 watts per person = 12 trillion watts/year (TWyr)

Future Energy Needs in the Lifetime of Today's Children

30 TWyr

Assuming that there will be 10 billion people and that each person can reach a satisfactory living standard with less than one-third the energy an average American uses now, the world will consume

(10 billion x 3000 watts per person) = 30 trillion watts/year.

figure 10.1 **Predicting Energy Needs**

Source: Princeton Plasma Physics Laboratory, U.S. Department of Energy, Washington, D.C.

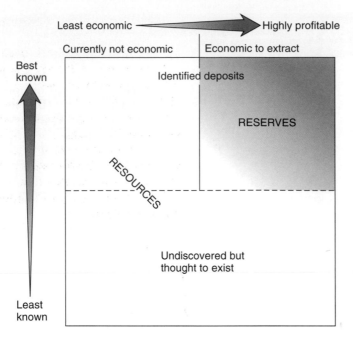

figure 10.2 **Resources and Reserves** Each term describes the amount of a natural resource present. Reserves are those known deposits that can be profitably obtained using current technology under current economic conditions. Reserves are shown in the box in the upper right-hand corner in this diagram. The darker the color, the more valuable the reserve. Resources are much larger quantities that include undiscovered deposits and deposits that currently cannot be profitably used, although it might be feasible to do so if technology or market conditions change.

Source: Adapted from the U.S. Bureau of Mines.

only loosely tied to the total quantity of a material present in the world. Therefore, reserves are smaller than resources. (See figure 10.2.) Both terms are used when discussing the amount of mineral or fossil-fuel deposits a country has at its disposal. This can cause considerable confusion if the difference between these concepts is not understood. The total amount of a resource such as coal or oil changes only by the amount used each year. The amount of a reserve changes as technology advances, new deposits are discovered, and economic conditions vary. There can be large changes in the amount of reserves, while the resource remains almost constant.

When we read about the availability of fossil fuels, we must remember that energy is needed to extract the wanted material from the Earth. If the material is dispersed or hard to reach, it is expensive to extract. If the cost of removing and processing a fuel is greater than the fuel's market value, no one is going to produce it. Also, if the amount of energy

used to produce, refine, and transport a fuel is greater than its potential energy, the fuel will not be produced. A net useful energy yield is necessary to exploit the resource. However, in the future, new technology or changing prices may permit the profitable removal of some fossil fuels that currently are not profitable. If so, those resources will be reclassified as reserves.

To further illustrate the concept of reserves and how technology and economics influence their magnitude, let us look at the history of oil. The ancient Chinese are said to have been the first to use oil as a fuel. The only oil available to them was the small amount that naturally seeped out of the ground. These seepages represented the known oil reserves as well as the known oil resources at that time.

Nearly 2000 years passed before oil reserves increased significantly. When the first oil well in North America was drilled in Pennsylvania in 1859, it greatly expanded the estimate of the

amount of oil in the Earth. There was a sudden increase in the known oil reserves. In the years that followed, new deposits were discovered. Better drilling techniques led to the discovery of deeper oil deposits, and offshore drilling established the location of oil under the ocean floor. At the time of their initial discovery, these deep deposits and the offshore deposits added to the estimated size of the world's oil resources. But they did not necessarily add to the reserves because it was not always profitable to extract the oil. With advances in drilling and pumping methods and increases in oil prices, it eventually became profitable to obtain oil from many of these deposits. As it became economical to extract them, they were reclassified as reserves.

The amount of fossil fuel reserves is in a constant state of flux. For example, prior to 1973, many oil wells were capped because it was not profitable to remove the oil. The oil still in the ground at these wells was not economic to produce and was not included in the reserves category. During the oil embargoes of 1973–74 and 1979–80, when some members of the Organization of Oil Exporting Countries (OPEC) reduced oil production, the price of oil increased. With the increase in oil prices, these wells became profitable, and the oil was reclassified as reserves. Then, in the 1980s, an oil surplus developed because the high oil prices resulting from the embargo caused increased production worldwide. Prices fell. Many of these same wells were capped again, and the oil was no longer classified as part of the reserve. With the exception of 1990 and 1991, when Iraq invaded Kuwait, the price of oil in the 1990s was relatively stable. (See figure 10.3.)

By the end of the decade, however, oil prices were once again rising. Beginning in 1999, the price of oil began to climb, and, by the first half of 2001, oil prices had risen to over $40 per barrel, up from $12 a barrel only two years earlier. The United States met frequently with OPEC leaders in the hope of having OPEC increase its production of oil and thus reduce the price at the pump in the United States. By the summer of 2000,

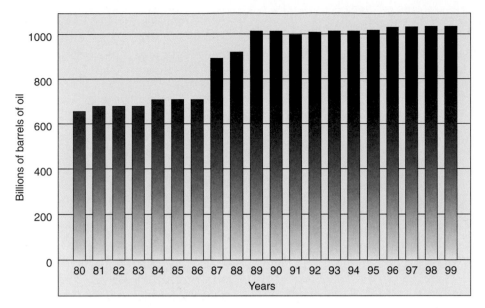

figure 10.3 **Changes in Oil Reserves** The figure shows the changes in oil reserves over a 20-year period. The changes that occurred in 1987 and 1989 are the result of changes in reporting, not the result of new discoveries. Since 1989, new discoveries and revisions have matched consumption. Thus, reserves have remained nearly constant.

Source: Data from *BP Statistical Review of World Energy,* June 2001.

the price of a gallon of gasoline exceeded $2 in many parts of the United States—double what it was a year earlier.

Many reasons were given for this sharp increase. Gasoline and petroleum stocks were low. With little crude oil or gasoline in inventory, refiners began to purchase crude oil in a market short on supplies, and that translated to higher prices. At the same time, refineries, squeezed by the high crude prices, were not operating at maximum capacity. By the spring of 2000, refineries on average were running at about 85 percent capacity, compared with nearly 92 percent at the same time a year earlier. In addition to the supply issues, the oil companies claimed that new environmental rules governing the manufacture of gasoline contributed to the increased cost of gasoline at the pump. By the summer of 2002, the price of a gallon of gasoline in the United States had declined to the 1999 range. This decline was a result of a global economic recession, OPEC disunity, and the September 11, 2001, terrorist attacks. Perhaps the only certainty of gasoline pricing is its continuing uncertainty.

Fossil-Fuel Formation

Fossil fuels are the remains of once living organisms that were preserved and altered as a result of geologic forces. Significant differences exist in the formation of coal from that of oil and natural gas.

Coal

Tropical freshwater swamps covered many regions of the Earth 300 million years ago. Conditions in these swamps favored extremely rapid plant growth, resulting in large accumulations of plant material. Because this plant material collected under water, decay was inhibited, and a spongy mass of organic material formed, called peat. Peat moss (peat) deposits are 90 percent water, 5 percent carbon, and 5 percent volatile materials. In some parts of the world such as Ireland, Latvia, and parts of Russia, peat is cut, dried, and used as fuel. However, because of its high water content, it is regarded as a low-grade fuel.

Due to geological changes in the Earth, some of the swamps containing peat were submerged by seas. The plant material that had collected in the swamps was then covered by sediment. The weight of the plant material plus the weight of the sediment on top of it compressed it into a harder form of low-grade coal known as lignite, which contains less water and a higher proportion of burnable materials.

Given sufficient thickness of overlying sediment, heat from the Earth, and the passing of geologic time, the lignite would have changed into bituminous (soft) coal. The major change from lignite to bituminous coal is a reduction in the water content from over 40 percent to about 3 percent. If the heat and pressure continued over time, some of the bituminous coal could have changed to a hard-grade coal known as anthracite, which is about 96 percent carbon. Through this combination of events, which occurred over hundreds of millions of years, present-day coal deposits were created. Most parts of the world have coal deposits. (See figure 10.4.)

Oil and Natural Gas

Oil and natural gas, like coal, are products from the past. They probably originated from microscopic marine organisms. When these organisms died and accumulated on the ocean bottom and were buried by sediments, their breakdown released oil droplets. Gradually, the muddy sediment formed rock called shale, which contained dispersed oil droplets. Although shale is common and contains a great deal of oil, extraction from shale is difficult because the oil is not concentrated. If a layer of porous rock called sandstone formed on the top of the oil-containing shale, and an impermeable layer of rock formed on top of the sandstone, conditions might be suitable for oil pools to form. Usually, the trapped oil does not exist as a liquid mass but rather as a concentration of oil within sandstone pores, where it accumulates because water and gas pressure force it out of the shale. (See

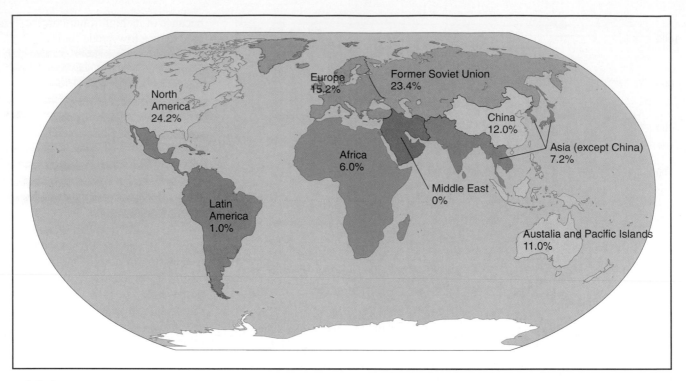

figure 10.4 **Recoverable Coal Reserves of the World 2000** The percentage indicates the coal reserves in different parts of the world. This coal can be recovered under present local economic conditions using available technology.

Source: Data from *BP Statistical Review of the World Energy*, June 2001.

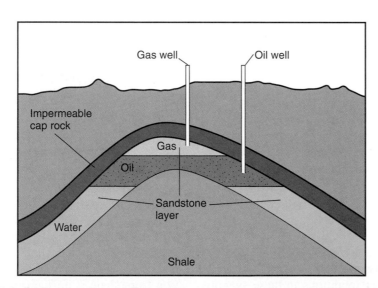

figure 10.5 **Crude Oil and Natural Gas Pool** Water and gas pressure force oil and gas out of the shale and into sandstone capped by impermeable rock.

Adapted with permission from Arthur N. Strahler, *Planet Earth*. Copyright ©1972 by Arthur N. Strahler.

figure 10.5.) These accumulations of oil are more likely to occur if the rock layers were folded by geological forces.

Natural gas, like coal and oil, forms from fossil remains. If the heat generated within the Earth reached high enough temperatures, natural gas could have formed along with or instead of oil. This would have happened as the organic material changed to lighter, more volatile (easily evaporated) hydrocarbons than those found in oil. The most common hydrocarbon in natural gas is the gas methane (CH_4). Water, liquid hy-drocarbons, and other gases may be present in natural gas as it is pumped from a well.

The conditions that led to the formation of oil and gas deposits were not evenly distributed throughout the world. Figure 10.6 illustrates the geographic distribution of oil reserves, and figure 10.7 illustrates the geographic distribution of natural gas reserves. Some of these deposits are easy to extract, while others are not.

Issues Related to the Use of Fossil Fuels

As previously mentioned, of the world's commercial energy, over 90 percent is furnished by the three nonrenewable fossil-fuel resources: coal, oil, and natural gas. Coal supplies about 24 percent, oil supplies about 40 percent, and natural gas supplies about 27 percent. Each fuel has advantages and disadvantages

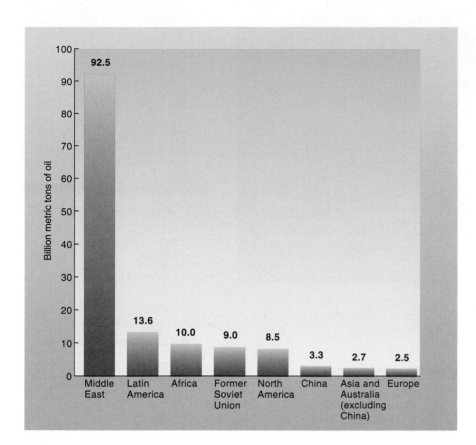

figure 10.6 **World's Oil Reserves 2000** The world's supply of oil is not distributed equally. Certain areas of the world enjoy an economic advantage because they control vast amounts of oil. Reserves are given in billions of metric tons of oil.

Source: Data from *BP Statistical Review of World Energy,* June 2001.

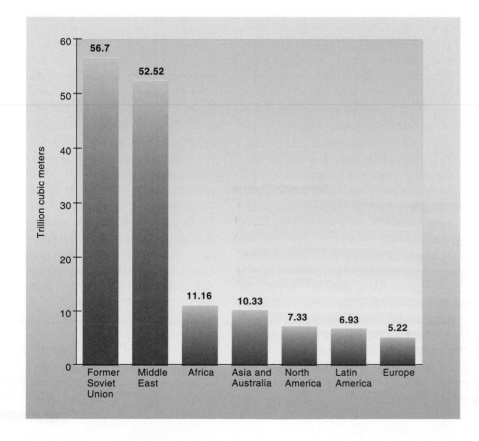

figure 10.7 **Natural Gas Reserves 2000** Natural gas reserves, like oil and coal, are concentrated in certain regions of the world. Figures are in trillion cubic meters.

Source: Data from *BP Statistical Review of World Energy,* June 2001.

figure 10.8 **Surface Mining** Large power draglines in Kentucky are used to remove the overburden, which is piled to the side. The coal can then be loaded into trucks. When the coal has been removed, the overburden is placed back in the trench.

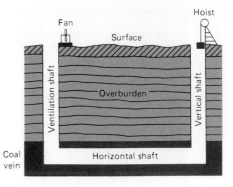

a. Vertical shaft

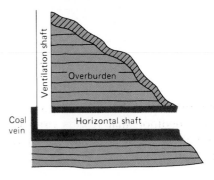

b. Drift mine

figure 10.9 **Underground Mining** If the overburden is too thick to allow surface mining, underground mining must be used. (*a*) If the coal vein is not exposed, a vertical shaft is sunk to reach the coal. (*b*) In hilly areas, if the vein is exposed, a drift mine is used in which miners enter from the side of the hill.

and requires special techniques for its production and use.

Coal Use

Coal is the world's most abundant fossil fuel, but it supplies only about 24 percent of the energy used in the world. It varies in quality and is generally classified in three categories: lignite, bituminous, and anthracite. Lignite coal has a high moisture content and is crumbly in nature, which makes it the least desirable form. Bituminous coal is the most widely used because it is the easiest to mine and the most abundant. It supplies about 20 percent of the world's energy requirements. Coal is primarily used for electric power generation and other industrial uses. For most uses, anthracite coal is the most desirable because it furnishes more energy than the other grades and is the cleanest burning. Anthracite is not as common, however, and is usually more expensive because it is found at great depths and is difficult to obtain.

Because coal was formed as a result of plant material being buried under layers of sediment, it must be mined. There are two methods of extracting coal: surface mining and underground mining. **Surface mining** (strip mining) involves removing the material located on top of a vein of coal, called **overburden,** to get

at the coal beneath. (See figure 10.8.) Coal is usually surface mined when the overburden is less than 100 meters (328 feet) thick. This type of mining operation is efficient because it removes most of the coal in a vein and can be profitably used for a seam of coal as thin as half a meter. For these reasons, surface mining results in the best utilization of coal reserves. Advances in the methods of surface mining and the development of better equipment have increased surface mining activity in the United States from 30 percent of the coal production in 1970 to more than 60 percent today. This trend toward increased surface mining has also occurred in Canada, Australia, and the former Soviet Union.

If the overburden is thick, surface mining becomes too expensive, and the coal is extracted through **underground mining.** The deeply buried coal seam can be reached in two ways. In the first, in flat country where the vein of coal lies buried beneath a thick overburden, the coal is reached by a vertical shaft. (See figure 10.9a.) In the second, in hilly areas where the coal seam often comes to the surface along the side of a hill, the coal is reached from a drift-mine opening. (See figure 10.9b.)

The mining, transportation, and use of coal as an energy source present several significant problems. Surface mining disrupts the landscape, as the topsoil

and overburden are moved to access the coal. It is possible to minimize this disturbance by reclaiming the area after mining operations are completed. (See figure 10.10.) However, reclamation rarely, if ever, returns the land to its previous level of productivity. The cost of reclamation is passed on to the consumer in the form of higher coal prices. Underground mining methods do not disrupt the surface environment as much as surface mining does, but subsidence (sinking of the land) occurs if the mine collapses. In addition, large waste heaps are produced around the mine entrance from the debris that must be removed and separated from the coal.

Health and safety are important concerns related to coal mining, which is one of the most dangerous jobs in the world. This is particularly true with underground mining. Many miners

a.   b.

figure 10.10   **Surface-Mine Reclamation**   (*a*) This photograph shows a large area that has been surface mined with little effort to reclaim the land. The windrows created by past mining activity are clearly evident, and little effort has been made to reforest the land. By contrast, (*b*) is an example of effective surface-mining reclamation. The sides of the cut have been graded and planted with trees. The topsoil has been returned, and the level land is now productive pasture.

suffer from **black lung disease,** a respiratory condition that results from the accumulation of fine coal-dust particles in the miners' lungs. The coal particles inhibit the exchange of gases between the lungs and the blood. The health care costs and death benefits related to black lung disease are an indirect cost of coal mining. Since these costs are partially paid by the federal government, their full price is not reflected in the price of coal but is paid by taxpayers in the form of federal taxes and higher health premiums.

Because coal is bulky, shipping presents a problem. Generally, the coal can be used most economically near where it is produced. Rail shipment is the most economic way of transporting coal from the mine. Rail shipment costs include the expense of constructing and maintaining the tracks, as well as the cost of the energy required to move the long strings of railroad cars. In some areas, the coal is transferred from trains to ships.

Coal mining and transport generate a great deal of dust. The large amounts of coal dust released into the atmosphere at the loading and unloading sites can cause local air-pollution problems. If a boat or railroad car is used to transport coal, there is the expense of cleaning it before other types of goods can be shipped. In some cases, the coal can be ground and mixed with water to form a slurry that can be pumped through pipelines. This helps to alleviate some of the air-pollution problems without causing significant water-pollution problems.

Air pollution from coal burning releases millions of metric tons of material into the atmosphere and is responsible for millions of dollars of damage to the environment. The burning of coal for electric generation is the prime source of this type of pollution.

Since coal is a fossil fuel formed from plant remains, it contains sulfur, which was present in the proteins of the original plants. Sulfur is associated with **acid mine drainage** and air pollution. Acid mine drainage occurs when the combined action of oxygen, water, and certain bacteria causes the sulfur in coal to form sulfuric acid. Sulfuric acid can seep out of a vein of coal even before the coal is mined. However, the problem becomes worse when the coal is mined and the overburden is disturbed, allowing rains to wash the sulfuric acid into streams. Streams may become so acidic that they can support only certain species of bacteria and algae. Today, many countries regulate the amount of runoff allowed from mines, but underground and surface mines abandoned before these regulations were enacted continue to contaminate the water.

Currently, a form of acid pollution called acid deposition is becoming a serious problem. Acid deposition occurs when coal is burned and sulfur oxides are released into the atmosphere, causing acid-forming particles to accumulate. Each year, over 150 million metric tons of sulfur dioxide are released into the atmosphere worldwide. This problem is discussed in greater detail in chapter 17.

The release of carbon dioxide from the burning of coal has become a major issue in recent years. Increasing amounts of carbon dioxide in the atmosphere are strongly implicated in global warming. Environmentalists have suggested that the use of coal be decreased, since the other fossil fuels (oil and natural gas) produce less carbon dioxide for an equivalent amount of energy.

Because coal is difficult to transport and often has a high sulfur content resulting in air pollution, people seek alternative sources of fuel. The most common alternatives for coal are oil and natural gas.

Oil Use

Oil has several characteristics that make it superior to coal as a source of energy. Its extraction causes less environmental damage than does coal mining. It is a more concentrated source of energy than coal, it burns with less pollution, and it can be moved easily through pipes. These characteristics make it an ideal fuel for automobiles. However, it is often difficult to find. Today, geologists use a series of tests to locate underground formations that may contain oil. When a likely area is identified, a test well is

a.

figure 10.11 **Offshore Drilling** Once the drilling platform is secured to the ocean floor, a number of wells can be sunk to obtain the gas or oil.

Source: (*b*) American Petroleum Institute.

Drilling derrick

Crew quarters

Ocean

Seabed

Platform legs imbedded in seabed

Directionally drilled wells

Impermeable cap rock

Oil and gas reservoir

Reservoir rock

Impermeable cap rock

Oil and gas reservoir

Reservoir rock

b.

drilled to determine if oil is actually present. Since the many easy-to-reach oil fields have already been tapped, drilling now focuses on smaller amounts of oil in less accessible sites, which means that the cost of oil from most recent discoveries is higher than that from the large, easy-to-locate sources of the past. As oil deposits on land have become more difficult to find, geologists have widened the search to include the ocean floor. Building an offshore drilling platform can cost millions of dollars. To reduce the cost, as many as 70 wells may be sunk from a single platform. (See figure 10.11.) Total proven world reserves of oil in 2000 were estimated at about 1000 billion barrels. Of this, more than three-quarters are in OPEC countries and more

than half lie in four Middle Eastern countries—Saudi Arabia, Iraq, Kuwait, and Iran.

One of the problems of extracting oil is removing it from the ground. If the water or gas pressure associated with a pool of oil is great enough, the oil is forced to the surface when a well is drilled. When the natural pressure is not great enough, the oil must be pumped to the surface. Present technology allows only about one-third of the oil in the ground to be removed. This means that two barrels of oil are left in the ground for every barrel produced. In most oil fields, **secondary recovery** is used to recover more of the oil. Secondary recovery methods include pumping water or gas into the well to drive the oil out or even

starting a fire in the oil-soaked rock to liquefy thick oil. As oil prices increase, more expensive and aggressive secondary recovery methods will need to be used.

Processing crude oil to provide useful products generates a variety of problems. Oil, as it comes from the ground, is not in a form suitable for use. It must be refined. The various components of crude oil can be separated and collected by heating the oil in a distillation tower. (See figure 10.12.) After distillation, the products may be further refined by "cracking." In this process, heat, pressure, and catalysts are used to produce a higher percentage of volatile chemicals, such as gasoline, from less volatile liquids, such as diesel fuel and furnace oils. It is possible, within limits, to obtain

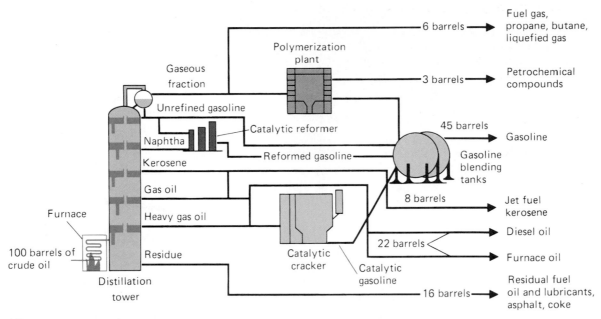

figure 10.12 **Uses of Crude Oil** A great variety of products can be obtained from distilling and refining crude oil. A barrel of crude oil produces slightly less than half a barrel of gasoline. This figure shows the many steps in the refining process and the variety of products that can be obtained from crude oil.

Source: *Man, Energy, Society* by Earl Cook. W.H. Freeman and Company. Copyright © 1976. Reprinted with permission.

many products from one barrel of oil. In addition, petrochemicals from oil serve as raw materials for a variety of synthetic compounds. (See figure 10.13.) All of these processing activities are opportunities for accidental or routine releases that may cause air or water pollution. The petrochemical industry is a major contributor to air pollution.

The environmental impacts of producing, transporting, and using oil are somewhat different from those of coal. Oil spills in the oceans have been widely reported by the news media. (See figure 10.14.) However, these accidental spills are responsible for only about one-third of the oil pollution resulting from shipping. Nearly 60 percent of the oil pollution in the oceans is the result of routine shipping operations. Oil spills on land can contaminate soil and underground water. The evaporation of oil products and the incomplete burning of oil fuels contribute to air pollution. These problems are discussed in chapter 17.

Natural Gas Use

Natural gas, the third major source of fossil-fuel energy, supplies 27 percent of the world's energy. The drilling opera-

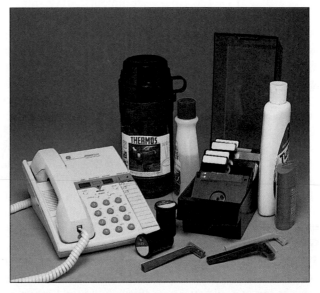

figure 10.13 **Oil-Based Synthetic Materials** These common household items are produced from chemicals derived from oil. Although petrochemicals represent only about 3 percent of each barrel of oil, they are extremely profitable for the oil companies.

tions to obtain natural gas are similar to those used for oil. In fact, a well may yield both oil and natural gas. As with oil, secondary recovery methods that pump air or water into a well are used to obtain the maximum amount of natural gas from a deposit. After processing, the gas is piped to the consumer for use.

Transport of natural gas still presents a problem in some parts of the world. In the Middle East, Mexico, Venezuela, and Nigeria, wells are too far from consumers to make pipelines practical, so much of the natural gas is burned as a waste product at the wells. However, new methods of transporting natural gas

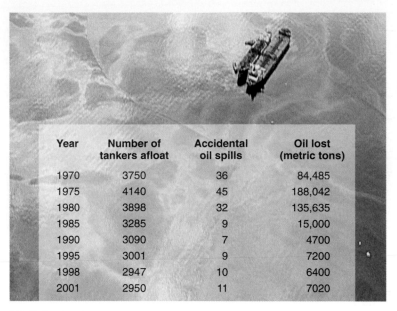

Year	Number of tankers afloat	Accidental oil spills	Oil lost (metric tons)
1970	3750	36	84,485
1975	4140	45	188,042
1980	3898	32	135,635
1985	3285	9	15,000
1990	3090	7	4700
1995	3001	9	7200
1998	2947	10	6400
2001	2950	11	7020

figure 10.14 **Oil Spills** Accidents involving oil tankers are sources of water pollution.

Source: Data from Tanker Advisory Center Reference Data for 1973–2001.

and converting it into other products are being explored. At −162°C (−126°F), natural gas becomes a liquid and has only 1/600 of the volume of its gaseous form. Tankers have been designed to transport **liquefied natural gas** from the area of production to an area of demand. In 1999, over 75 billion cubic meters (2600 billion cubic feet) of natural gas were shipped between countries as liquefied natural gas. This is nearly 4.5 percent of the natural gas consumed in the world. Of that amount, Japan alone imported 58 billion cubic meters (2000 billion cubic feet). As the demand for natural gas increases, the amount of it wasted will decrease and new methods of transportation will be employed. Higher prices will make it profitable to transport natural gas greater distances between the wells and the consumers.

One concern about transporting liquefied natural gas is accidents that might cause tankers to explode. Another, safer process converts natural gas to methanol, a liquid alcohol, and transports it in that form.

Of the three fossil fuels, natural gas is the least disruptive to the environment. A natural gas well does not produce any unsightly waste, although there may be local odor problems. Except for the danger of an explosion or fire, natural gas poses no harm to the environment during transport. Since it is clean burning, it causes almost no air pollution. The products of its combustion are carbon dioxide and water. However, the burning of natural gas does produce carbon dioxide, which contributes to global warming. Global warming is discussed in chapter 17.

Although natural gas is used primarily for heat energy, it does have other uses, such as the manufacture of petrochemicals and fertilizer. Methane contains hydrogen atoms that are combined with nitrogen from the air to form ammonia, which can be used as fertilizer.

More than two-thirds of the world's natural gas reserves are located in Russia and the Middle East. In fact, more than a third of the total reserves are located in 10 giant fields; six are located in Russia and the remainder are in Qatar, Iran, Algeria, and the Netherlands.

From 1985 to 1999, world production of natural gas rose by more than 25 percent. In Russia, which has the largest natural gas reserves, production increased by 40 percent from 1985 through 1991 but has since fallen as a result of the economic disruption associated with the collapse of the Soviet Union.

Because oil reserves have been located beneath the North Sea, several European countries (the United Kingdom, Norway, and the Netherlands) have sizable reserves of natural gas. During oil shortages in the 1970s, they increased their use of natural gas to compensate for the increased cost of oil. This trend has continued; natural gas consumption increased by 45 percent from 1985 to 2000.

Renewable Sources of Energy

The three nonrenewable fossil-fuel sources of energy—coal, oil, and natural gas—furnish about 90 percent of the world's commercially traded energy. Nuclear power provides over 7 percent of the world's energy. The remainder is supplied by renewable energy sources, including hydroelectric power, tidal power, geothermal power, wind power, solar energy, fuelwood, biomass conversion, and solid waste. Hydroelectric power accounts for nearly 98 percent of electric utilities generated by renewables. (See figure 10.15.) These data do not include the use of wood, animal dung, and other locally produced energy sources typically used in the less-developed world.

Hydroelectric Power

People have long used water to power a variety of machines. Some early uses of water power were to mill grain, saw wood, and run machinery for the textile industry. Hydroelectric power can be produced in three ways:

Impoundment: An impoundment facility, typically a large hydropower system, uses a dam to store river water in a reservoir. The water may be released either to meet changing electricity needs or to maintain a constant reservoir level.

Diversion: A diversion, sometimes called run-of-river, facility channels a portion of a river through a canal or channel. It may not require the use of a dam.

Pumped storage: When the demand for electricity is low, a

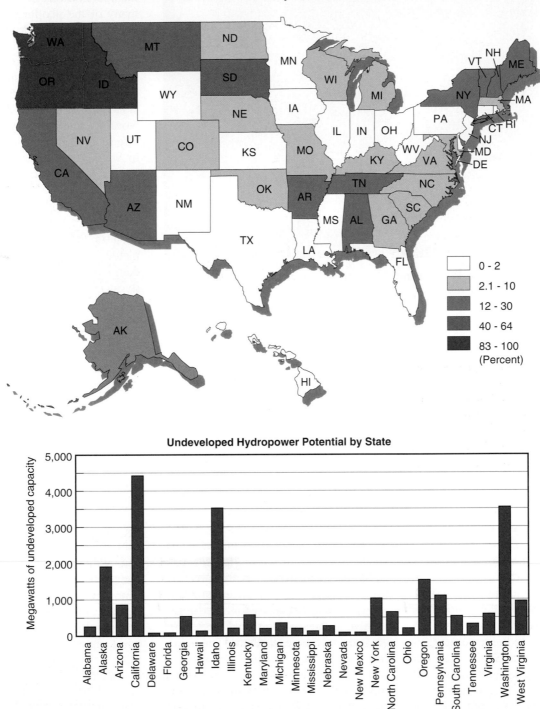

figure 10.15 **Hydroelectric Power and Potential** The top map indicates the percent of hydroelectric power in the United States. The bottom graph indicates the current undeveloped hydropower in the United States. (For comparison, many large coal-fired power plants have about 1000 megawatts of generating capacity.)

Source: U.S. Geological Survey and the Foundation for Water and Energy Education, 2000.

pumped storage facility stores energy by pumping water from a lower reservoir to an upper reservoir. During periods of high electrical demand, the water is released back to the lower reservoir to generate electricity.

Today, water power is used almost exclusively to generate electricity. As the water flows from higher to lower levels, it supplies the energy to turn a generator

figure 10.16 **Hydroelectric Power Plant** The water impounded in this reservoir is used to produce electricity. In addition, this reservoir serves as a means of flood control and provides an area for recreation.

and produce electricity. Hydroelectric power plants are commonly located on human-made reservoirs. (See figure 10.16.) The impounded water represents a potential energy source. In some areas of the world where the streams have steep gradients and a constant flow of water, hydroelectricity may be generated without a reservoir. Such sites are usually found in mountainous regions and can support only small power-generating stations. At present, hydroelectricity produces about 2.5 percent of the world's commercially traded energy.

Hydroelectric power potential is distributed among the continents in rough proportion to land area; China alone possesses one-tenth of the world's potential. Mountainous regions and large river valleys are the most promising. Besides the United States, the eastern area of the former Soviet Union, and southern Canada, the regions that have done the most to harness hydroelectric energy are Europe and Japan. Europe has exploited almost 60 percent of its potential. Although it has only one-fourth of Asia's resources, it generates nearly twice as much hydroelectric power. In contrast, Africa has developed only 5 percent of its potential, half of which comes from only three dams: Kariba in East Africa, Aswan on the Nile, and Akosombo in Ghana.

In some areas of the world, hydroelectric power is the main source of electricity. More than 35 nations already obtain more than two-thirds of their electricity from falling water. In South America, 73 percent of the electricity used comes from hydroelectric power, compared to 44 percent in the developing world as a whole. Norway gets 99 percent of its electricity and 50 percent of all its energy from falling water.

It is important to recognize that the construction of a reservoir for a hydroelectric plant causes environmental and social problems, including loss of fertile farmland, destruction of the natural aquatic ecosystem, relocation of entire communities, and a reduction in the amount of nutrient-rich silt deposited on downriver agricultural lands. The building of the Tellico Dam in Tennessee was delayed for several years because it might have caused the extinction of the snail darter, a fish that lived only in streams that would be flooded. The construction of the Aswan Dam in Egypt resulted in the displacement of 80,000 people. It created an environment that increased schistosomiasis, a waterborne disease caused by flatworm parasites that spend part of their life cycle in snails that live in slowly moving water. The irrigation canals built to distribute the water from the Aswan Dam provide

ideal conditions for the snails. Now many of the people who use the canal water for cooking, drinking and bathing are infected with the flatworm parasites.

Even though hydroelectric projects cause problems, new sites continue to be developed. From 1985 to 2000, the energy furnished by hydroelectricity for world use increased by over 30 percent. The most significant change took place in South America, where hydroelectric use increased about 40 percent over that time.

China is constructing a huge hydroelectric dam, known as the Three Gorges Dam, on the Yangtze River. If it is completed, it will be the largest hydroelectric dam in the world. (See Global Perspectives: The Three Gorges Dam.)

Today's large dams rank among humanity's greatest engineering feats. The Itaipu hydroelectric power plant is the largest development of its kind in operation in the world (until the completion of the Three Gorges Dam in China). Built from 1975 to 1991 in a development on the Parana River, Itaipu is a joint project of Brazil and Paraguay. The power plant's 18 generating units add up to a total production capacity of 12,600 megawatts. The magnitude of the project also can be demonstrated by the fact that in 1997, Itaipu provided 25 percent of the energy supply in Brazil and 80 percent in Paraguay. By means of comparison, Itaipu produces enough power to meet most of California's annual need. The lakes created by major dams also number among the planet's largest freshwater bodies. Ghana's Lake Volta covers 8500 square kilometers (3280 square miles), an area the size of Lebanon. Hydroelectric dam projects figure prominently in the economic and investment plans of many developing countries. Egypt electrified virtually all of its villages with power from Aswan. Since the 1950s, large-scale hydropower development in these countries has occurred primarily because energy-intensive industries need cheap electricity and because global lending institutions have been willing to advance multibillion-dollar loans for these projects. According to the World Bank, developing countries

Global Perspective

Hydroelectric Sites

Approximately 17 percent of the potential hydroelectric sites of the world have been developed. (See the tables.) The World Energy Conference estimates that the electricity produced by hydropower will increase six times by the year 2020. The less-developed countries, which have developed about 10 percent of their hydropower, will experience most of this growth.

The projected increase will come mainly from the development of plants on large reservoirs. However, construction of "mini-hydro" (less than 10 megawatts) and "micro-hydro" (less than 1 megawatt) plants is also increasing. Such plants can be built in remote places and supply electricity to small areas. China has over 80,000 such small stations, and the United States has nearly 1500.

About 50 percent of the U.S. hydroelectric capacity has been developed. However, this statistic is somewhat misleading. The Wild and Scenic Rivers Act (1968) prevents the construction of dams on designated streams. Presently, 37 potential hydroelectric sites are on streams protected by this act. The hydroelectric generating potential often quoted for the United States includes these areas, even though, at present, construction of plants at these sites is not possible. In fact, some dams on rivers in states such as Maine and Washington are scheduled to be destroyed in order to improve native fisheries. The U.S. political climate, which now favors the protection of certain rivers, might change if the demand for more energy becomes acute.

Developed Hydroelectric Sites, 2001

Region	Percent of Hydropower Developed
Asia	19
South America	18
Africa	10
North America	62
Russia	10
Europe	90
Oceania	34

Source: Data from *Survey of Energy Resources,* World Energy Conference.

World Consumption of Hydroelectricity and Other Renewable Energy by Region (Quadrillion BTU)

Region/Country	Historical to 2000	Projections 2005	Projections 2010
OECD	19.8	21.5	24.0
United States*	7.8	8.3	9.4
Canada	4.1	4.7	5.3
Mexico	0.3	0.4	0.5
Japan	1.8	1.9	2.0
OECD Europe	5.3	5.7	6.2
United Kingdom	0.2	0.2	0.2
France	0.7	0.7	0.7
Germany	0.3	0.4	0.4
Italy	0.6	0.7	0.7
Netherlands	0.1	0.1	0.2
Other Europe	3.5	3.6	3.9
Other OECD	0.5	0.5	0.7
EE/FSU	3.1	3.4	3.8
FSU	2.6	2.8	2.9
Eastern Europe	0.5	0.6	1.0
Non-OECD Asia	4.8	6.0	7.5
China	2.6	3.5	4.5
Other Asia	2.1	2.6	3.0
Middle East	0.3	0.4	0.4
Africa	0.7	0.8	0.9
Central and South America	4.0	4.2	4.4
Total World	32.6	36.4	41.1

*Includes the 50 states, the District of Columbia, and U.S. territories.

Notes: OECD = Organization for Economic Cooperation and Development. EE/FSU = Eastern Europe/Former Soviet Union.

Sources: History data from Energy Information Administration, *International Energy Annual,* DOE/EIA–0219(92), Washington, D.C.; Projections from EIA, *Annual Energy Outlook,* DOE/EIA–0383(95), Table B1, January 1995, Washington, D.C.; and World Energy Projections 2000.

will need to raise an estimated $100 billion (U.S.) by the year 2002 for hydroelectric plants currently in the planning stage. Given the environmental concerns of large dam construction, the trend for the future will probably be to build small-scale hydro plants that can generate electricity for a single community.

Tidal Power

Tidal flow is another source of energy related to local geological conditions.

Global Perspective

The Three Gorges Dam

A project that began in late 1997 is already being compared to the building of the pyramids and the Great Wall of China. China's Three Gorges Dam across the world's third longest river, the Yangtze, is a project that when completed will be visible from the moon and will be the largest dam in the world. The Three Gorges Dam will stretch 2 kilometers (1.3 miles) across the Yangtze River, tower 185 meters (610 feet) into the air, and create a 600-kilometer (385-mile) reservoir behind it. The dam is expected to cost in excess of $40 billion before the estimated date of completion in 2009. The government of China contends that the dam is necessary for several reasons. The primary reason is the 18,000 megawatts of electric power that the dam will generate. China's rapidly developing economy needs the energy for its expanding industrial centers. It is also argued that the project will transform the Yangtze River, especially the upper regions, into a more navigable and, hence, economic waterway. The final major argument for the dam relates to flood control in the middle and lower reaches of the river. These parts of the river have been prone to frequent and disastrous floods. The economic arguments, however, are not the only issues being debated in relation to the Three Gorges Dam.

Scientists in China and from other nations have warned that the project could threaten migratory fish, concentrate water pollution, and endanger to the point of extinction the Chinese alligator, river dolphins, the Siberian white crane, and the Chinese sturgeon. Environmental concerns relating to the project were instrumental in the 1993 decision of the United States to withdraw technical assistance. The impact on the regional Chinese population is also a critical issue. Upon completion, the lake formed by the dam will inundate 153 towns, 4500 villages, numerous archeological sites, and the scenic canyons of the Three Gorges that have inspired poets and painters for centuries.

- Do you feel that the benefits of this massive project exceed its economic and environmental costs?
- Would you have supported the project in its developmental phase?

Source: Magellan Geographix/ABCNEWS.com. Data from International Rivers Network.

The gravitational pull of the moon and the sun, along with the Earth's rotation, causes tides. The tidal movement of water represents a great deal of energy. For years, engineers have suggested that this moving water could be used to produce electricity. The principle is the same as that employed in a hydroelectric plant. As water flows from a higher level to a lower one, it can be used to generate electricity. The greater the difference between high and low tides, the more energy can be extracted.

Since tidal changes are greatest near the poles and are accentuated in narrow bays and estuaries, suitable sites of constructing power plants are limited. In the 1930s, the United States explored the possibility of constructing a tidal electrical generating facility at Passamaquoddy, Maine, on the Bay of Fundy. After spending considerable time and money on a feasibility study, it abandoned the idea.

In 1966, France constructed a commercial tidal generating station. (See figure 10.17.) Located on the Rance River Estuary on the Brittany coast, it is the world's only large tidal generating station. It was built to generate 240 megawatts, but because of the cyclic

figure 10.17 **Tidal Generating Station** The Rance River Estuary Power Plant in France is the world's largest tidal electrical generating station.

nature of the tides, it usually produces only 62 megawatts. This satisfies the electric power needs of about 100,000 people (less than 0.2 percent of the population of France). There is also a 16-megawatt plant in Nova Scotia, Canada.

The British government studied the feasibility of constructing a 7200-megawatt tidal station in the Severn Estuary in southwestern England. It would have 30 times the capacity of the French station. The facility could be built for a price comparable to that of a coal-fired plant of similar generating capacity, but it would disrupt the normal estuary flow and would concentrate pollutants in the area. The plant has not been built.

A new style of tidal power generator is being developed in the Dalupiri Ocean Power Plant in the Philippines. This new system is a submerged turbine, referred to as a hydro turbine. Similar to an underwater windmill, the hydro turbine will not obstruct fish migration, silt transport, or water flow, problems often associated with traditional tidal power generation facilities.

Changing tides in the area generate ocean currents that can turn hydro turbines. The hydro turbines will have slow-moving blades, pose little danger to marine life, and not impede the flow of silt and water. A protective fence prevents large marine mammals from approaching the turbines. It is estimated that a 1-kilometer (0.62-mile) "tidal bridge" made up of a series of these turbines could generate more electricity than a large nuclear plant. The completed project in the Philippines will span 4 kilometers (2.5 miles) and will include 274 turbines with a generating capacity of 2.3 gigawatts during peak tidal flow.

Geothermal Power

Geothermal energy flow from the Earth is not uniformly distributed throughout the Earth's surface area but is linked to geologically active areas where heat from the Earth can reach the surface through thinner crust. In particular, countries around the Pacific plate (in a region called the circum-Pacific "Ring of Fire"), mid-ocean ridges (such as Iceland), continental rift zones, and other hot spots experience significantly higher geothermal energy fluxes from the Earth's interior.

In areas where steam is trapped underground, **geothermal energy** is tapped by drilling wells to obtain the steam. The steam is then used to power electrical generators. At present, geothermal energy is practical only in areas where this hot mass is near the surface. (See figure 10.18.) Geothermal energy is not considered a true renewable energy source but rather an alternative one.

The United States has about half of the world's geothermal electrical generating capacity. More than 130 generating plants are operating in 12 other countries. The Philippines, Italy, Mexico, Japan, New Zealand, and Iceland produce sizable amounts of electricity by geothermal methods.

California alone produces 40 percent of the world's geothermal electricity, about 1900 megawatts. One megawatt provides energy for about 1000 households. The Pacific Gas and Electric Company (PG&E) has been producing electricity from geothermal energy since 1960. PG&E's complex of generating units located north of San Francisco is the largest in the world and provides 700 megawatts of power, enough for 700,000 households, or 2.9 million people.

In addition to producing electricity, geothermal energy is used directly for heating. In Iceland, half of the geothermal energy is used to produce electricity and half is used for heating. In the capital, Reykjavik, all of the buildings are heated with geothermal energy at a cost that is less than 25 percent of what it would be if oil were used.

Geothermal energy creates some environmental problems. The steam contains hydrogen sulfide gas, which has the odor of rotten eggs and is an unpleasant form of air pollution. (The sulfides from geothermal sources can, however, be removed.) The minerals in the steam corrode pipes and equipment, causing maintenance problems. The minerals are also toxic to fish.

All objects contain heat energy, which can be extracted from and transferred to other locations. A refrigerator extracts heat from the interior and exports it to the coils on the back of the unit. Similarly, pipes placed within the

Earth can extract heat and transfer it to a home. It is important to recognize that such an application of geothermal energy requires the expenditure of energy through an electric-powered device. In some instances, obtaining heat in this manner is less expensive than other traditional sources, such as oil or natural gas, but it is not a "free" source of heat. Geothermal energy use, however, does conserve fossil fuels and does not contribute to global warming.

Electricity from the Ground Up

The naturally occurring heat beneath the planet's surface powers volcanoes, hot springs, and geysers such as Yellowstone National Park's Old Faithful. In a few places in the world, such as Iceland and parts of California, natural heat and natural water come together in sufficient quantities to provide energy. In most places where there is abundant subsurface water, useful heat is 5 kilometers

(3 miles) or more below the surface, beyond the reach of economical drilling technology. In places where heat is within reach, there is often no subsurface water. (See figure 10.18b.)

Searching for cheap, nonpolluting energy, scientists at the Los Alamos National Laboratory in New Mexico have begun a process to "mine heat." Mining heat is a new effort to use geothermal energy. In this process, water is pumped at high pressure 3 kilometers (1.9 miles)

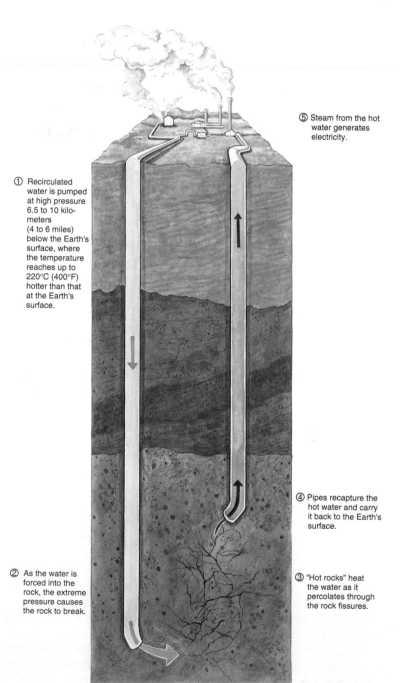

① Recirculated water is pumped at high pressure 6.5 to 10 kilometers (4 to 6 miles) below the Earth's surface, where the temperature reaches up to 220°C (400°F) hotter than that at the Earth's surface.

② As the water is forced into the rock, the extreme pressure causes the rock to break.

③ "Hot rocks" heat the water as it percolates through the rock fissures.

④ Pipes recapture the hot water and carry it back to the Earth's surface.

⑤ Steam from the hot water generates electricity.

figure 10.18 **Geothermal Power Plant** Steam obtained from geothermal wells is used in the production of electricity.

a.

b.

down into the Earth through a well. When the water comes back up through a parallel well, it has been heated beyond the boiling point by the Earth's natural heat. The Hot Dry Rock Project is an attempt to bring heat and water together as a geothermal energy source.

On average, temperatures below the Earth's surface increase about 27°C per kilometer (80°F per mile). In much of the western United States, however, residual heat from ancient volcanoes increases temperatures by more than 69°C per kilometer (200°F per mile). At the Los Alamos site, the temperature at the base of the well is about 240°C (430°F).

According to researchers at Los Alamos, a fairly conservative estimate is that there are at least 500,000 quads (quadrillion British thermal units, or Btus) of useful heat in hot dry rock at accessible drilling depths beneath the United States. This is about 6000 times the total amount of energy used in the country in one year.

The fundamental question now being asked is if the energy can be developed economically for commercial use. At this point, the answer is no; however, if new technology is developed, the answer could change. Other questions also must be addressed: Will the heat in a particular well last long enough to justify the cost? Can the system be sealed to minimize water leakage? Can a constant flow of water be sustained?

Even if it proves commercially competitive, hot dry rock mining might encounter obstacles involving site access, hookups to distant power grids, and legal disputes over water rights. If such problems can be overcome, maybe one day the energy used to run your computer will come from water heated kilometers beneath the surface of the Earth.

Wind Power

As the sun's radiant energy strikes the Earth, that energy is converted into heat, which warms the atmosphere. The Earth is unequally heated because various portions receive different amounts of sunlight. Since warm air is less dense and rises, cooler, denser air flows in to take

figure 10.19 **Wind Energy** Fields of wind-powered generators such as these in California can produce large amounts of electricity.

its place. This flow of air is wind. For centuries, wind has been used to move ships, grind grains, pump water, and do other forms of work. In more recent times, wind has been used to generate electricity. (See figure 10.19.)

In 2001, California was using 15,100 wind turbines to produce 370 megawatts of electricity at a cost competitive with newly constructed coal or nuclear plants. By 2010, according to the American Wind Energy Association, wind could power 10 million U.S. households. The U.S. Department of Energy has stated that the Great Plains could supply 48 states with 75 percent of their electricity. In 1999, wind facilities produced enough power for 425,000 homes. That figure is expected to increase with new wind "farms" in Minnesota, Iowa, Wyoming, and Texas coming online. Wind power is now considered competitive with new coal and natural gas plants and cheaper than nuclear plants. Since the 1980s, costs associated with wind power have decreased considerably due in part to evolving technology and government actions. In addition, many consumers are willing to pay extra for clean energy. In Colorado, through a program called Windsource, 10,000 residents and businesses pay an additional $2.50 a month for every 100-kilowatt block of wind power.

According to the U.S. Department of Energy, wind power was the world's

fastest-growing energy source in the 1990s. Technological improvements in the last 20 years cut the cost from 40 cents per kilowatt hour to between 4 and 5 cents per kilowatt hour. The cost of electricity produced by wind is expected to fall more as the technology advances. Meanwhile, a push for energy deregulation and concerns about smog, acid rain, and global warming are driving policy makers to require utilities to sell electricity from renewable sources. Eight states—Texas, Wisconsin, Massachusetts, Connecticut, New Jersey, Minnesota, Nevada, and Pennsylvania—require utilities to provide some "green" electricity. At least 36 utilities include wind energy as a component of their green power programs.

Most U.S. wind sources remain untapped. Texas and the Dakotas alone have enough wind to power the nation, but that's not likely to happen. Variable wind speeds make it unreliable as a primary energy source; energy companies and regulators view it as supplementary to fossil fuels. While places like the Dakotas have the strongest winds, they are remote from energy-using population centers and lack suitable transmission grids.

India, China, Germany, and Spain have plans for major increases in wind-generated electricity through the year 2005. Since 1998, the world's wind-energy capacity has grown more than

35 percent, topping 10,000 megawatts. Germany contributed a third of the recent growth, largely by guaranteeing wind farms access to the power grid at a competitive price for the power they generate. As U.S. companies begin building 1- or 2-megawatt wind turbines, European firms are exploring 5-megawatt machines. In inner Mongolia, nomadic herders carry with them small, portable, wind-driven generators that provide electricity for light, television, and movies in their tents as well as for electric fences to contain their animals.

A steady and dependable source of wind makes the use of wind power more productive in some regions than others. Wide, open areas, such as the Great Plains in North America, are better suited for wind power than are heavily wooded areas. Wind-generated electricity is usually used in conjunction with other sources of electricity that take over when the wind does not blow. Wind generators do have some negative effects. The moving blades are a hazard to birds and produce a noise that some find annoying. Newer windmills, however, have slower moving rotors that many birds like the golden eagle find easier to avoid. Vibrations from the generators can also cause structural problems. In addition, some people consider the sight of a large number of wind generators, such as are found in California, to be visual pollution.

Since 1998, wind power has been the fastest-growing new source of electricity in the world, expanding an average of 30 percent a year. The value of the world's electrical power generated from renewable sources such as wind and solar is about $7 billion—up from less than $1 billion a decade ago but still a small fraction of the total electricity market. The market for renewable energy is expected to exceed $80 billion by 2010 as technological advances lower the price and make renewables easier to use.

Solar Energy

The sun is often mentioned as the ultimate answer to the world's energy problems. It provides a continuous supply of energy that far exceeds the world's demands. In fact, the amount of energy re-

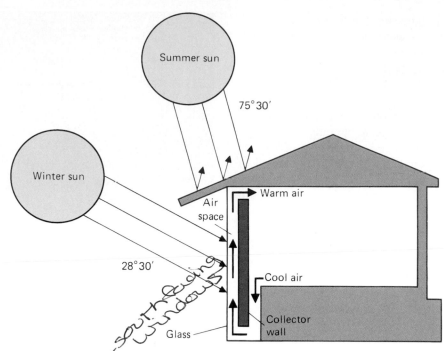

figure 10.20 **Passive Solar Heating** The length of overhang in this home is designed for solar heating at the latitude of St. Louis, Missouri (38°N). In this design, a wall 30 to 40 centimeters thick is used to collect and store heat. The collector wall is located behind a glass wall and faces south. During a midwinter day, when the sun's angle is 28 degrees, light energy is collected by the wall and stored as heat. At night, the heat stored in the wall is used to warm the house. Natural convection causes the air to circulate past the wall, and the house is heated. During a midsummer day, when the sun's angle is 75 degrees, the overhang shades the collector wall from the sun.

ceived from the sun each day is 600 times greater than the amount of energy produced each day by all other energy sources combined. The major problem with solar energy is its intermittent nature. It is available only during the day and when it is sunny. All systems that use solar energy must store energy or use supplementary sources of energy when sunlight is not available. Because of differences in the availability of sunlight, some parts of the world are more suited to the use of solar energy than others. Solar energy is also very diffuse, which is another disadvantage of its use.

Solar energy is utilized in three ways:

1. In a passive heating system, the sun's energy is converted directly into heat for use at the site where it is collected.
2. In an active heating system, the sun's energy is converted into heat, but the heat must be transferred from the collection area to the place of use.

3. The sun's energy also can be used to generate electricity, which may be used to operate solar batteries or may be transmitted along normal transmission lines.

Passive Solar Systems

Anyone who has walked barefoot on a sidewalk or a blacktopped surface on a sunny day has experienced the effects of passive solar heating. In a **passive solar system,** light energy is transformed to heat energy when it is absorbed by a surface. Some of the earliest uses of passive solar energy were to dry food and clothes and to evaporate seawater to produce salt. Homes and buildings may now be designed to use passive solar energy for heating. (See figure 10.20.) Such systems require a large window through which sunlight may enter and a large mass that collects and stores the heat. This large mass may be a thick wall or floor. The type of construction material, the color of the roofing, and

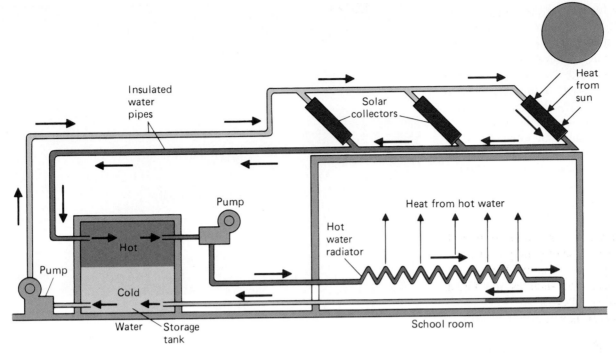

the landscaping all are important factors in passive solar heating. Since there are no moving parts, a passive solar system is maintenance free. None of the energy is used to transfer heat within the system, and there are no operating costs. However, passive solar design is usually practical only in new construction.

Active Solar Systems

Sun's energy converted to heat, put then transported somewhere else

An **active solar system** requires a solar collector, a pump, and a system of pipes to transfer the heat from the site of production to the area to be heated. (See figure 10.21.) Active solar systems are most easily installed in new buildings but in some cases can be installed in existing structures. A major consideration in the use of an active solar system is the initial cost of installation. An active system requires a specially designed collector, consisting of a series of liquid-filled tubes; a pump; and pipes to transfer warm liquid from the collector to the space to be heated. Because an active system has moving parts, it also has operation and maintenance costs like any other heating or cooling system.

Rock, water, or specially produced products are used to store heat. The hot liquid in the pipes heats the storage

b.

figure 10.21 **Active Solar System** Collectors provide 65 percent of the hot water in this dental school building in California.

medium, which releases its heat when the sun is not shining.

Solar-Generated Electricity

When the first **photovoltaic (PV) cell,** a bimetallic unit that allows the direct conversion of sunlight to electricity, was developed by Bell Laboratories in 1954, it was regarded as an expensive novelty.

However, as more efficient batteries were developed and production costs were reduced, practical uses were found for photovoltaic cells. By the mid-1980s, more than 60 million solar calculators were being produced annually. These calculators used over 10 percent of the photovoltaic cells manufactured.

Photovoltaic cells appear to be emerging as sources of small amounts of electricity for special uses like running equipment in remote regions. The normal system of generating electricity in large, centrally located plants and distributing it by high tension lines is costly and is practical only in highly populated areas. Photovoltaic cells are more practical in remote regions of the world.

Although sales of solar cells for items like highway signs, roofs, and radios increased 20 percent from 1998 to 1999, most of that growth was outside the United States. Over 70 percent of solar cells made in the United States are sent abroad, often to remote spots, like rural India, that are not connected to the grid. Since 1997, the U.S. share of the global market for solar cell products has dropped from 44 percent to 35 percent. It is expected that Japan, where electricity is relatively expensive, will move ahead to lead solar PV cell sales worldwide. Not coincidentally, Japan spends $240 million annually on solar power versus $72 million in 1999 for the United States. Over 10,000 solar-electric homes have been built in parts of Alaska and the Australian outback. (See figure 10.22.) The French government has subsidized the installation of over 2000 solar electric units on 18 islands in the Pacific. These units provide electricity for a thousand homes and five hospitals. Many of the developing countries of the world will introduce electricity to villages through the use of photovoltaic cells rather than the use of generators that require fuel and distribution lines.

It has to be argued that to compete with fossil fuels, solar engineers will have to think bigger. A novel Dutch effort is setting out to do just that. Near Amersfoort, the Netherlands, the NV REMU power company is leading a $13 million project to build 500 houses with roofs covered with PV cell panels. By the time the homes are finished, they should be drawing 1.3 megawatts of energy from the sun, enough to supply about 60 percent of the community's energy needs. The goal of the Amersfoort project is to demonstrate the construction of a solar energy system at the level of an entire community.

The price of photovoltaic cells has been falling as better technology is de-

figure 10.22 **Solar Energy** In some remote areas, solar energy is an economical method of electricity production.

figure 10.23 **Solar Generation of Electricity** This solar-powered electricity generating plant is capable of generating electricity at a cost that is competitive with other methods of generating electricity.

veloped. Eventually, the cells may become competitive with other energy sources, particularly as the cost of fossil fuels rises.

Solar energy is also being used to generate electricity in a more conventional way. A company named Luz International in California has built solar collector troughs that can heat oil in pipes to 390°C (734°F). (See figure 10.23.) This heat can be transferred to water, which is turned into steam that is used to run conventional electricity-generating turbines. As with photovoltaic cells, the cost of producing electricity in this manner is falling and is becoming competitive with conventional sources.

Limitations of Solar Energy

Solar energy provides less than 1 percent of the world's energy for several reasons. The most obvious is that it works only during the day, which means

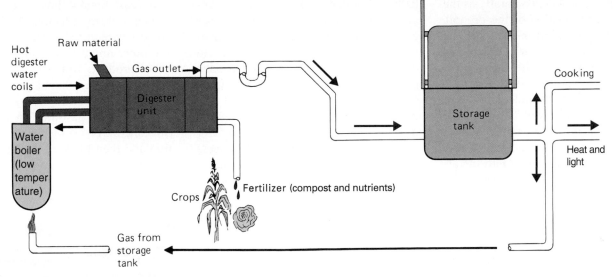

figure 10.24 **Methane Digester** In the digester unit, anaerobic bacteria convert animal waste into methane gas. This gas is then used as a source of fuel. The sludge from this process serves as a fertilizer. In many less-developed countries, this type of digester has the advantages of providing a source of energy and a supply of fertilizer and managing animal wastes, which helps reduce disease.

that some type of heat or electrical storage mechanism is needed for night use, which adds to the expense of relying on solar energy. The fact that solar heating is most practical in new construction also limits its use. In colder climates, solar heat is inadequate as the sole source of heat, and some type of a conventional heating system is required for backup. Climate is also a problem, since many areas have extensive cloudy periods, which reduce the amount of energy that can be collected. Although the price of collectors and related equipment has decreased in recent years, many collector systems are still expensive. For example, electricity from photovoltaic cells is still more expensive than conventional electric generating systems. Prices of photovoltaic cells continue to fall, however, and the cost of generating electricity from conventional fuels is rising.

Biomass Conversion

Biomass is any accumulation of organic material produced by living things. The most commonly used biomass sources are fuelwood, agricultural residue from the harvesting of crops, crops grown for their energy content, and animal waste. (Because of its major impact on world energy resources, fuelwood will be discussed separately.) These traditional,

often noncommercial sources of fuel provide more than 10 percent of the world's energy but are not reported in most statistics about global energy. In many developing countries, these sources of fuel are a large proportion of the energy available.

Biomass conversion is the process of obtaining energy from the chemical energy stored in biomass. It is not a new idea; burning wood is a form of biomass conversion that has been used for thousands of years. Biomass can be burned directly as a source of heat for cooking, burned to produce electricity, converted to alcohol, or used to generate methane. (See figure 10.24.) China has 500,000 small methane digesters in homes and on farms; India has 100,000; and Korea has 50,000. Brazil is the largest producer of alcohol from biomass. The low price of sugar coupled with the high price of oil have prompted Brazil to use its large crop of sugar cane as a source of energy. Alcohol provides 50 percent of Brazil's automobile fuel.

Biomass conversion raises some environmental and economic concerns. Countries that use large amounts of biomass for energy are usually those that have food shortages. Biomass conversion means that fewer nutrients are being returned to the soil, and this compounds the food shortage. If the price of food

rises or the price of oil falls, there could be less biomass conversion.

The energy required to produce usable energy stocks from biomass must be taken into account. Growing corn to produce alcohol requires large energy inputs. The amount of energy present in the alcohol produced from the corn is actually less than the amount of energy that went into producing the alcohol. Obviously, this makes no sense from an energy point of view. However, the convenience of a liquid fuel may be worth paying for in economic terms.

Biomass Conversion Technologies

There exist several technologies capable of converting biomass into energy. These include anaerobic digestion, pelletising, direct combustion and cogeneration, pyrolysis, gasification, and ethanol production.

Anaerobic Digestion

Anaerobic digestion is the decomposition of wet and green biomass, through bacterial action in the absence of oxygen, to produce a mixed gas output of methane and carbon dioxide known as biogas. The anaerobic digestion of municipal solid waste buried in landfill sites produces a gas known as landfill gas. This process occurs naturally as the bacterial decomposition of the organic matter continues over time. The methane gas produced in landfill sites eventually escapes into the atmosphere. However, the landfill gas can be extracted by inserting perforated pipes into the landfill. In this way, the gas will travel through the pipes, under natural pressure, to be used as an energy source, rather than simply escaping into the atmosphere to contribute to greenhouse gas emissions.

Pelletising

Pelletising involves the compaction of biomass at high temperatures and very high pressures. The biomass particles are compressed in a die to produce briquettes or pellets. These products have significantly smaller volume than the original biomass and thus have a higher volumetric energy density (VED), making them a more compact source of energy. They are also easier to transport and store than natural biomass. The pellets can be used directly on a large scale as direct combustion feed or on a small scale in domestic stoves or wood heaters.

Direct Combustion and Cogeneration

Direct combustion is the main process adopted for utilizing biomass energy. The energy produced can be used to provide heat or steam for cooking, space heating, and industrial processes, or for electricity generation. Large biomass power-generation systems can have comparable efficiencies to fossil-fuel systems, but this comes at a higher cost due to the design of the burner to handle the higher moisture content of biomass. However, by using the biomass in a combined heat- and electricity-production system (or cogeneration system), the economics are significantly improved.

Pyrolysis

Pyrolysis is the basic thermochemical process for converting solid biomass to a more useful liquid fuel. Biomass is heated in the absence of oxygen, or par-

tially combusted in a limited oxygen supply, to produce a hydrocarbon-rich gas mixture, an oil-like liquid, and a carbon-rich solid residue. Traditionally, in developing countries, the solid residue produced is charcoal, which has a higher energy density than the original fuel. The traditional charcoal kilns are simply mounds of wood covered with earth or pits in the ground. However, the process of carbonization is very slow and inefficient in these kilns, and more sophisticated kilns are replacing the traditional ones. The pyrolitic residue or "bio-oil" produced can be easily transported and refined into a series of products. The process is similar to refining crude oil.

Gasification

Gasification is a form of pyrolysis, carried out with more air and at high temperatures, to optimize the gas production. The resulting gas, known as producer gas, is a mixture of carbon monoxide, hydrogen and methane, together with carbon dioxide and nitrogen. The gas is more versatile than the original solid biomass, and it can be used as a source of heat or used in internal combustion engines or gas turbines to produce electricity. During the Second World War, countries such as Australia and Germany even used it to power vehicles.

Ethanol Production

Ethanol can be produced from certain biomass materials that contain sugars, starch, or cellulose. The best-known feedstock for ethanol production is sugar cane, but other materials can be used, including wheat, corn, other cereals, and sugar beets. Starch-based biomass is usually cheaper than sugar-based materials but requires additional processing. Similarly, cellulose materials, such as wood and straw, are readily available but require expensive preparation.

Ethanol is produced by a process known as fermentation. Typically, sugar is extracted from the biomass crop by crushing and mixing with water and yeast, and then keeping the mixture warm in large tanks called fermenters.

The yeast breaks down the sugar and converts it to ethanol. A distillation process is required to remove the water and other impurities from the dilute alcohol product. Brazil has a successful industrial-scale ethanol project, which produces ethanol from sugar cane for blending with gasoline. In the United States, corn is used for ethanol production and then blended with gasoline to produce "gasahol."

Fuelwood

In less-developed countries, wood has been the major source of fuel for centuries. In fact, wood is the primary source of energy for nearly half of the world's population. In these regions, the primary use of wood is for cooking.

The use of wood as a prime energy source, a rapid population increase, and the high cost of other types of fuel have combined to create some serious environmental problems in many areas of the world. It is estimated that 1.3 billion people are not able to obtain enough wood or must harvest wood at a rate that exceeds its growth. This has resulted in the destruction of much forest land in Asia and Africa and has hastened the rate of desertification in these regions. (See figure 10.25.)

Because of its bulk and low level of energy compared to equal amounts of coal or oil, wood is not practical to transport over a long distance, so most of it is used locally. In the United States, Norway, and Sweden, wood furnishes 10 percent of the energy for home heating. Canada obtains 3 percent of its total energy, not just home-heating energy, from wood. Most of this energy is used in forest product industries, such as lumbering and paper mills.

Burning wood is also a source of air pollution. Studies indicate that more than 75 organic compounds are released when wood is burned, 22 of which are hydrocarbons known or suspected to be carcinogens. Often, woodstoves are not operated in the most efficient manner, and high amounts of particulate matter and other products of incomplete combustion, such as carbon monoxide, are released, contributing to ill health and death.

In areas with a high population density, the heavy use of wood releases large amounts of fly ash into the air. In Missoula, Montana, in recent years, 55 percent of the particles in the air during the summer were from burning wood. In the winter, wood was responsible for 75 percent of the particles. A number of steps have been taken to reduce air pollution resulting from burning wood. Some cities, such as London, England, have a total ban on burning wood. Vail, Colorado, permits only one wood-burning stove per dwelling. Many areas require woodstoves to have special pollution controls that reduce the amount of particulates and other pollutants released.

Solid Waste

Residents of New York City discard in excess of 25,000 metric tons (27,500 tons) of waste each day, which is 1.8 kilograms (4 pounds) per person. In fact, New Yorkers lead the world in the production of municipal trash. High-income cities such as New York usually produce more waste than low-income cities. (See table 10.2.) About 80 percent of this waste is combustible and, therefore, represents a potential energy source. (See figure 10.26.)

"Trash power," the use of municipal waste as a source of energy, requires several steps. First, the waste must be sorted so that the burnable organic material is separated from the inorganic material. The sorting is accomplished most economically by the person who produces the waste. That means the producer must separate trash before putting it out for pickup. It must be separated into garbage, burnable materials, glass, and metals, and picked up by compartmentalized collection trucks. Second, the efficient use of waste as a source of fuel requires a large volume and a dependable supply. If a community constructs a facility to burn 200 metric tons (220 tons) of waste a day, the community must generate and collect 200 metric tons (220 tons) of waste each day.

Communities have been burning trash as a means of reducing the volume of waste for a number of years. The first

figure 10.25 **Desertification** The demand for fuelwood in many regions has resulted in the destruction of forests. This is a major cause of desertification.

Table 10.2	Amount of Solid Waste Produced per Capita	
High-Income Cities	**Kilograms per Day**
New York	1.8
Singapore	1.6
Tokyo	0.94
Rome	0.72
Low-Income Cities	
Tunis, Tunisia	0.56
Medellin, Colombia	0.54
Calcutta, India	0.51
Kano, Nigeria	0.46

figure 10.26 **Waste to Energy** Municipal trash can be burned to produce heat and electricity. This refuse pit is used to feed hoppers of high-temperature furnaces.

Global Perspective

Are Fuel Cells the Future?

With growing concern regarding global climate change and fossil-fuel depletion, attention is increasingly turning to renewable energy sources. At the forefront of this development are hydrogen fuel cells, which, although first conceived over a century ago, are now beginning to rise in popularity due to their high efficiency and low emissions.

Fuel cells are often likened to batteries. The primary difference between fuel cells and batteries, however, is that fuel cells do not need to be recharged; while batteries require a long recharging period and release toxic heavy metals when disposed of, fuel cells run continuously if provided with adequate fuel input. This fuel may take the form of natural gas, methanol, gasoline, or pure hydrogen—almost anything from which hydrogen may be extracted.

Although numerous types of fuel cells are available, the most common is the proton exchange membrane (PEM) fuel cell. Although the specific reactions within the fuel cell involve several steps, they can be summarized as follows. Pressurized hydrogen gas enters the fuel cell and comes in contact with a platinum catalyst that causes the hydrogen molecules (H_2) to split into hydrogen ions (H^+), also known as protons, and electrons. A proton exchange membrane allows protons (H^+)—but not the electrons—to flow through it. The electrons instead flow through an electric circuit to do work such as generating light or running a motor. The hydrogen ions flow through the proton exchange membrane and recombine with the electrons that have passed through the circuit and with oxygen to form water.

Numerous obstacles are delaying the process of fuel cell development. Most prominently, hydrogen used to power the cell is difficult to obtain in a pure state. While pure hydrogen gas produces

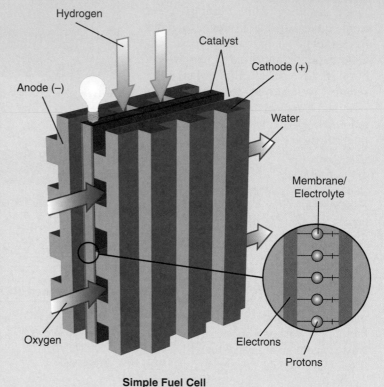

Simple Fuel Cell

A fuel cell is an electrochemical energy conversion device.

consolidated incineration of waste was done in Nottingham, England, in 1874. Burning the municipal waste of Munich, Germany, not only reduces the volume of the waste but supplies the energy for 12 percent of the city's electricity. Rotterdam, the Netherlands, operates a 55-megawatt power plant from its garbage. In the United States, only 3 percent of household waste is burned. This is low compared to the 26 percent burned in Japan, 51 percent in Sweden, and 75 percent in Switzerland.

Burning trash is not profitable. It is a way to decrease the cost of trash disposal because it reduces the need for landfill sites. Baltimore produces gas from 1000 metric tons (1100 tons) of waste per day. This gas generates steam for the Baltimore Gas and Electric Com-

pany. The daily waste also yields 80 metric tons (88 tons) of a charcoal-like material (char), 70 metric tons (77 tons) of ferrous metals, and 170 metric tons (190 tons) of glass. These products all have potential uses. The char can be burned as an additional source of energy. The ferrous metals can be sold, reducing the cost of operating the plant, as well as conserving mineral resources. The glass can be sold for recycling, further reducing the costs of the plant.

Although the burning of trash reduces the trash volume and furnishes energy, it poses environmental concerns, one of which is air pollution. Many of the older incinerating plants do not comply with today's air-quality standards. Also, much of the waste material, such as bleached paper and plastics, have

chlorine-containing organic compounds. When burned, these compounds can form dioxins, which are highly toxic and suspected carcinogens. Another problem associated with waste-to-energy systems is the popularity of recycling. Many of the items that are now recycled, such as plastics and wood, have high heat content. The reduction in the amount of these items in the waste stream reduces its value as an energy source.

Energy Conservation

Many observers have pointed out that demanding more energy while failing to conserve is like demanding more water to fill a bathtub while leaving the drain

218

PART THREE   Energy

the highest efficiency and lowest emissions, it is highly flammable, difficult to store, and not available to consumers as readily as oil-derived fuels. The hydrogen may be extracted from hydrocarbons through a process known as fuel reformation, but this results in slightly impure hydrogen, decreasing the efficiency of the fuel cell. Fuel reformation also releases carbon dioxide, nitrous oxide, and particulate matter, although in vastly lesser quantities than are released by traditional electricity production techniques. Manufacturing costs are currently high because the economies of scale experienced with mass production have not yet developed.

Fuel cells have been developed for specialized use in numerous fields. The National Aeronautics and Space Administration (NASA) developed cells for use aboard spacecrafts in the 1960s due to their size, weight, and nontoxic emissions, which, as an added bonus, provided pure drinking water for the crew. Some schools, banks, and offices have installed fuel cells to provide electricity. While large-scale power generation and industrial applications have not yet been developed, fuel cells may soon be adapted for these purposes, and residential applications may soon follow.

The advantages of fuel cells over other power sources are numerous. With a low operating temperature and minimal noise generation, fuel cells are safe to install in semiexposed areas. Cells are also self-sustaining, making them ideal for locations far removed from traditional electricity production facilities. Because they can operate separately from the power lines associated with most electricity distribution systems, the risk of weather-related power outages is also diminished. Their most important benefit, however, is that they do not pollute. When fueled by pure hydrogen, water, heat, and electricity are the only by-products. Although currently powered by modified fossil fuels, fuel cells are capable of extracting far greater quantities of energy from each unit of fuel than other methods of power generation, thereby reducing the need for fossil fuel extraction.

One of the most significant prospects for future development of fuel cells is in the automotive industry. Although there are some technical difficulties arising from adapting fuel cells for vehicular rather than stationary use, many leading automotive companies agree that fuel cells will be the successor to the internal combustion engine.

Fuel cells can be both a stationary and mobile power source.

Experimental buses are already on the road in selected European and American cities, and similar cars are expected to follow.

Some renewable energy enthusiasts foresee the development of a "hydrogen economy" in which thousands of jobs would be created for the development of fuel cell technology. The manufacture of fuel cells, the construction of hydrogen fuel storage and transport systems, and the development of associated products could potentially employ more people than currently work in fossil-fuel operations.

What do you think? Should fuel cells be pursued as a viable alternative to fossil-fuel combustion? Do the costs of research and development outweigh the benefits of fossil-fuel conservation? Are there other alternative energy sources that should be pursued with more vigor? Is the "hydrogen economy" a feasible vision for our energy future?

For more information, check out:

www.fuelcellstore.com
www.fuelcelltoday.com
www.howstuffworks.com/fuel-cell.htm

open. To be sure, conservation and efficiency strategies by themselves will not eliminate demands for energy, but they can make the demands much easier to meet, regardless of what options are chosen to provide the primary energy.

Much of the energy we consume is wasted. This statement is not meant as a reminder to simply turn off lights and lower furnace thermostats; it is a technological challenge. Our use of energy is so inefficient that most potential energy in fuel is lost as waste heat, becoming a form of environmental pollution.

Many conservation techniques are relatively simple and highly cost effective. More efficient and less energy-intensive industry, and domestic practices could save large amounts of energy. Improved automobile efficiency, better mass transit, and increased railroad use for passenger and freight traffic are sim-

ple and readily available means of conserving transportation energy. In response to the 1970s' oil price shocks, automobile gas-mileage averages in the United States more than doubled, from 5.55 kilometers per liter (13 miles per gallon) in 1975 to 12.3 kilometers per liter (28.8 miles per gallon) in 1988. Unfortunately, the oil glut and falling fuel prices of the late 1980s discouraged further conservation. Between 1990 and 1997, the average slipped to only 11.8 kilometers per liter (27.6 miles per gallon). It remains to be seen if the sharp increase of gasoline early in 2000 will translate into increased miles per gallon in new car design.

Conservation is not a way of generating energy, but it is a way of reducing the need for additional energy consumption and saves money for the consumer. Some conservation technologies are so-

phisticated, while others are quite simple. For example, if a small, inexpensive wood-burning stove were developed and used to replace open fires in the less-developed world, energy consumption in these regions could be reduced by 50 percent.

Several technologies that reduce energy consumption are now available. (See figure 10.27.) Highly efficient fluorescent light bulbs that can be used in regular incandescent fixtures give the same amount of light for 25 percent of the energy, and they produce less heat. Since lighting and air conditioning (which removes the heat from inefficient incandescent lighting) account for 25 percent of U.S. electricity consumption, widespread use of these lights could significantly reduce energy consumption. Low-emissive glass for windows can reduce the amount of heat entering a

The Arctic National Wildlife Refuge and Oil

The Arctic National Wildlife Refuge (ANWR) has been a source of controversy for many years. The major players are environmentalists who seek to preserve this region as wilderness; the state of Alaska, which funds a major portion of its activities with dividends from oil production; Alaska residents, who receive a dividend payment from oil revenues; oil companies that want to drill in the refuge; and members of Congress who see the oil reserves in the region as important economic and political issues.

In 1960, 3.6 million hectares (8.9 million acres) were set aside as the Arctic National Wildlife Range. Passage of the Alaskan National Interest Lands Conservation Act in 1980 expanded the range to 8 million hectares (19.8 million acres) and established 3.5 million hectares (8.6 million acres) as wilderness. The act also renamed the area the Arctic National Wildlife Refuge. There are international implications to this act. The refuge borders Canada's Northern Yukon National Park. Many animals, particularly members of the Porcupine caribou herd, travel across the border on a regular yearly migration. The United States is obligated by treaty to protect these migration routes.

The act requires specific authorization from Congress before oil drilling or other development activities can take place on the coastal plain in the refuge. The coastal plain has the greatest concentration of wildlife, is the calving ground for the Porcupine caribou, and has the greatest potential for oil production. In each case, the potential authorization caused a collision of three forces: environmental protection, economic development, and political benefit. Furthermore, great differences of opinion exist within each of the competing interest groups. Some Alaskan citizens support

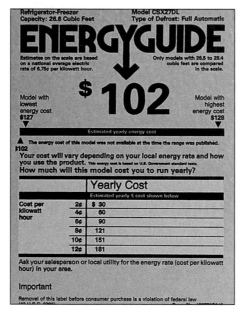

figure 10.27 **Energy Conservation** The use of fluorescent light bulbs, energy-efficient appliances, and low-emissive glass could reduce energy consumption significantly.

building while allowing light to enter. The use of this glass in new construction and replacement windows could have a major impact on the energy picture. Many other technologies, such as automatic dimming devices or automatic light-shutoff devices, are being used in new construction.

The shift to more efficient use of energy needs encouragement. Often, poorly designed, energy-inefficient buildings and machines can be produced inexpensively. The short-term cost is low, but the long-term cost is high. The public needs to be educated to look at the long-term economic and energy costs of purchasing poorly designed buildings and appliances.

Electric utilities have recently become part of the energy conservation picture. In some states, they have been allowed to make money on conservation efforts; previously, they could make money only by building more power plants. This encourages them to become involved in energy conservation education, because teaching their customers how to use energy more efficiently allows them to serve more people without building new power plants.

Conservation is cheaper than building more power plants to meet increased demands

drilling; others oppose it. The Inupiat Eskimos who live along the north Alaskan coast mostly are in favor of drilling in ANWR. The Inupiat believe oil revenues and land-rental fees from oil companies will raise their living standards. The other Native American tribe in the region, the Gwich'in, who live on the southern fringe of the refuge, oppose drilling. They argue that the drilling will impact the caribou migration through the area every fall and thus affect their ability to provide food for their families. Members of Congress are similarly split. Even members of the Department of the Interior have provided conflicting testimony about the risks and benefits of drilling for oil in the refuge.

In 1998, the secretary of the Interior, acting under the recommendation of President Bill Clinton, cleared the way for oil development on Alaska's North Slope. Under the plan, about a third of a 4.6-million-acre study area in the northeastern corner of the federal reserve would be off limits to drilling. Oil leases would be sold on 4 million acres of the government's National Petroleum Reserve west of the Prudhoe Bay oil fields. Some of the leases would allow only slant drilling because the surface is to be protected. In 2000, the Energy Information Administration (EIA) released a report on the potential oil production from the coastal plain of ANWR. The coastal plain region, which comprises approximately 8 percent of the 7.7 million hectares (19 million acres) ANWR, is the largest un-

explored, potentially productive geologic onshore basin in the United States. A decision on permitting the exploration and development is up to the U.S. Congress. In 2002, President George W. Bush reconfirmed his support for drilling. The EIA report estimated a 95 percent probability that at least 5.7 billion barrels of technically recoverable undiscovered oil are in the ANWR coastal plain. There is a 5 percent probability that at least 16 billion barrels of oil are recoverable. The report states that once oil has been discovered, more than 80 percent of the technically recoverable oil is commercially developable at a price of $25 per barrel (oil was $25 per barrel in July 2002). The value of the oil in 2002 dollars could be between $125 and $350 billion. Oil companies have repeatedly stated that the oil can be recovered without endangering wildlife or the fragile Arctic ecosystem. Conservationists have argued that none of the reserve should be developed when improvements in energy could reduce demand. They argue that drilling in the reserve will harm the habitat of millions of migratory birds, caribou, and polar bears. Only time will tell which side, if either, is correct.

• What do you think of the question of drilling in ANWR?
• Is a compromise position between the two sides possible?
• If you had to make this decision, what would it be? Why?

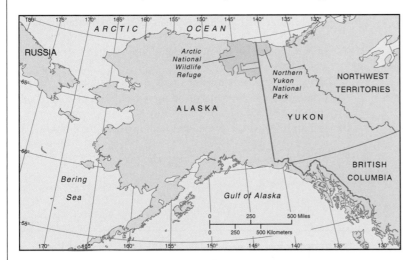

Migrating caribou in the Arctic National Wildlife Refuge.

Summary

A resource is a naturally occurring substance of use to humans, a substance that can potentially be extracted using current technology. Reserves are known deposits from which materials can be extracted profitably with existing technology under present economic conditions.

Coal is the world's most abundant fossil fuel. Coal is obtained by either surface mining or underground mining. Problems associated with coal extractions are disruption of the landscape due to surface mining and subsidence due to under-

ground mining. Black lung disease, waste heaps, water and air pollution, and acid mine drainage are additional problems. Oil was originally chosen as an alternative to coal because it was more convenient and less expensive. However, the supply of oil is limited. As oil becomes less readily available, multiple offshore wells, secondary recovery methods, and increased oil exploration will become more common. Natural gas is another major source of fossil-fuel energy. The primary problem associated with natural gas is transport of the gas to consumers.

Fossil fuels are nonrenewable: The amounts of these fuels are finite. When the fossil fuels are exhausted, they will have to be replaced with other forms of energy, probably renewable forms. Hydroelectric power can be increased significantly, but its development must flood areas and in so doing may require the displacement of people. The use of geothermal and tidal energy is limited by geographic locations. Wind power may be used to generate electricity but may require wide, open areas and a large number of wind generators. Solar energy can be collected and used in either passive or active systems and can also be used to generate electricity. Lack of a constant supply of sunlight is solar energy's primary limitation. Fuelwood is a minor source of energy in industrialized countries but is the major source of fuel in many less-developed nations. Biomass can be burned to provide heat for cooking or to produce electricity, or it can be converted to alcohol or used to generate methane. In some communities, solid waste is burned to reduce the volume of the waste and also to supply energy.

Energy conservation can reduce energy demands without noticeably changing standards of living.

Key Terms

acid mine drainage *201*
active solar system *213*
biomass *215*
black lung disease *201*
geothermal energy *209*
liquefied natural gas *204*

nonrenewable energy sources *195*
overburden *200*
passive solar system *212*
photovoltaic cell *213*
renewable energy sources *195*

reserves *195*
resources *195*
secondary recovery *202*
surface mining *200*
underground mining *200*

Review Questions

1. Why are fossil fuels important?
2. Distinguish between reserves and resources.
3. What are the advantages of surface mining of coal compared to underground mining? What are the disadvantages of surface mining?
4. Compare the environmental impacts of the use of coal and the use of oil.
5. What are some limiting factors in the development of new hydroelectric generating sites?
6. What factors limit the development of tidal power as a source of electricity?
7. In what parts of the world and why is geothermal energy available?
8. Why can wind be considered a form of solar energy?
9. Compare a passive solar-heating system with an active solar-heating system.
10. What problems are associated with the use of solid waste as a source of energy?
11. List three energy conservation techniques.

Critical Thinking Questions

1. Given what you know about the economic and environmental costs of different energy sources, would you recommend that your local utility company use hydroelectricity or coal to supplement electric production? What criteria would you use to make your recommendation?
2. Coal-burning electric power plants in the Midwest have contributed to acid rain in the eastern United States. Other energy sources would most likely be costlier than coal, thereby raising electricity rates. Should citizens of another state be able to pressure these utility companies to change the method of generating electricity? What mechanisms might be available to make these changes? How effective are these mechanisms?
3. Imagine you are an official with the Department of Energy and are in the budgeting process for alternative energy research. Where would you put the money? Why?
4. Given your choices from question 3, what do you think the political repercussions of your decision would be? Why?
5. Do you believe that large dam projects like the Three Gorges Dam project in China are, on the whole, beneficial or not? What alternatives would you recommend? Why?
6. Energy conservation is one way to decrease dependence on fossil fuels. What are some things you can do at home, work, or school that would reduce fossil-fuel use and save money?
7. What alternative energy resources that the text has outlined are most useful in your area? How might these be implemented?

Concept Map

Construct a map to show relationships among the following concepts:

renewable energy sources

reserves

nonrenewable energy sources

fossil fuels

hydroelectric power

energy conservation

solar energy

oil and natural gas

Interactive Exploration

Check out the website at **http://www.mhhe.com/environmentalscience** and click on the cover of this textbook for quizzing, career information, case studies, and hot links for the following topics:

Conventional Energy Sources

Oil, Coal, and Gas as Energy Sources

Nuclear Energy

Sustainable Energy

Solar Energy

Hydropower

Wind Energy

Nuclear Energy:
Benefits and Risks

CHAPTER 11

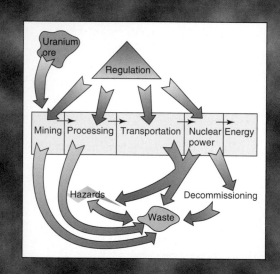

Objectives

After reading this chapter, you should be able to:

- Explain how nuclear fission has the potential to provide large amounts of energy.
- Describe how a nuclear reactor produces electricity.
- Describe the basic types of nuclear reactors.
- Explain the steps involved in the nuclear fuel cycle.
- List concerns regarding the use of nuclear power.
- Explain the problem of decommissioning a nuclear plant.
- Describe how high-level radiation waste is stored.
- Describe the accident at Chernobyl.
- Explain how a breeder reactor differs from other nuclear reactors.
- List the technical problems associated with the design and operation of a liquid metal fast-breeder reactor.
- Explain the process of fusion.

Chapter Outline

The Nature of Nuclear Energy

The History of Nuclear Energy Development

Nuclear Reactors
  Plans for New Reactors Worldwide
  Plant Life Extension

Breeder Reactors

Nuclear Fusion

The Nuclear Fuel Cycle

Nuclear Material and Weapons Production

Nuclear Power Concerns
  Reactor Safety: The Effects of Three Mile Island and Chernobyl
  Exposure to Radiation
  Thermal Pollution
  Decommissioning Costs
  Radioactive Waste Disposal

Global Perspective: *The Nuclear Legacy of the Soviet Union*

Environmental Close-up: *The Hanford Facility: A Legacy of Contamination*

The Nature of Nuclear Energy

Energy from disintegrating atomic nuclei has a tremendous potential to do good for the people of the world. We routinely use X rays to examine bones for fractures, treat cancer with radiation, and diagnose disease with the use of radioactive isotopes. About 17 percent of the electrical energy generated in the world comes from nuclear power plants. Engineers in the former Soviet Union used nuclear explosions to move large amounts of earth and rock to construct

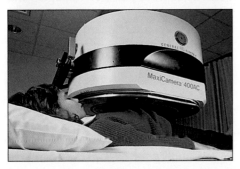

figure 11.1 **Uses of Nuclear Energy** Each of these photographs illustrates the uses of the energy available from the splitting of atoms.

dams, canals, and underground storage facilities. The U.S. government briefly considered the possibility of using nuclear explosions to build a new Panama Canal. (See figure 11.1.)

On the other hand, nuclear energy has the potential to do great harm. The Japanese cities of Hiroshima and Nagasaki were destroyed by nuclear bombs. The manufacture of nuclear weapons and their storage have left a legacy of radioactive wastes. In many cases, these wastes have been mismanaged or carelessly disposed of.

To understand where nuclear energy comes from, it is necessary to review some of the aspects of atomic structure presented in chapter 4. All atoms are composed of a central region called the nucleus, which contains positively charged protons, and neutrons that have no charge. Moving around the nucleus are smaller, negatively charged electrons. Since the positively charged particles in the nucleus repel one another, energy is needed to hold the protons and neutrons together. However, some isotopes of atoms are **radioactive;** that is, the nuclei of these atoms are unstable and spontaneously decompose. Neutrons, electrons, protons, and other larger particles are released during nuclear disintegration, along with a great deal of energy. The rate of decomposition is consistent for any given isotope. It is measured and expressed as **radioactive**

half-life, which is the time it takes for one-half of the radioactive material to spontaneously decompose. Table 11.1 lists the half-lives of several radioactive isotopes.

Nuclear disintegration releases energy from the nucleus as **radiation,** of which there are three major types:

Alpha radiation consists of a moving particle composed of two neutrons and two protons. Alpha radiation usually travels through air for less than a meter and can be stopped by a sheet of paper or the outer layer of the skin.

Beta radiation consists of electrons released from nuclei. Beta particles travel more rapidly than alpha particles and will travel through air for a couple of meters. They are stopped by a layer of clothing, glass, or aluminum.

Gamma radiation is a type of electromagnetic radiation, like X rays, light, and radio waves. It can pass through several centimeters of concrete.

If the radiation reaches living tissue, equivalent doses of beta and gamma radiation cause equal amounts of biological damage. Alpha particles are more massive than beta particles (electrons) or gamma rays (photons); therefore, they cause more damage to biological tissues.

Table 11.1 The Half-Life of Some Radioactive Isotopes

Radioactive Isotope	Half-Life
Iodine 132	2.4 hours
Technetium 99	6.0 hours
Rhodium 105	36.0 hours
Xenon 133	5.3 days
Barium 140	12.8 days
Cerium 144	284.0 days
Cesium 137	30 years
Carbon 14	5730 years
Uranium 234	250,000 years
Chlorine 36	300,000 years
Beryllium 10	4.5 million years
Potassium 40	1.3 billion years
Helium 4	12.5 billion years

The ingestion or inhalation of radioactive materials places the source of the radiation in direct contact with cells; thus, this kind of exposure is much more dangerous than exposure from distant sources.

The release of neutrons is particularly important in obtaining energy from nuclear disintegration. In addition to releasing alpha, beta, and gamma radiation when they disintegrate, the nuclei of a few kinds of atoms release neutrons. When moving neutrons hit the nuclei of certain other atoms, they can cause those nuclei to split as well. This process is known as **nuclear fission.** If these splitting nuclei also release neutrons, they can strike the nuclei of other atoms, which also disintegrate, resulting in a continuous process called a **nuclear chain reaction.** Only certain kinds of atoms are suitable for the development of a nuclear chain reaction. The two materials commonly used in nuclear reactions are uranium-235 and plutonium-239. In addition, there must be a certain quantity of nuclear fuel (a critical mass) in order for a nuclear chain reaction to occur. It is this process that results in the large amounts of energy released from bombs or nuclear reactors.

The History of Nuclear Energy Development

The first controlled fission of an atom occurred in Germany in 1938, but the United States was the first country to develop an atomic bomb. In 1945, the U.S. military dropped atomic bombs on the Japanese cities of Hiroshima and Nagasaki. The incredible devastation of these two cities demonstrated the potential of nuclear energy for destruction. For many years, most atomic research involved military applications of nuclear energy as bombs and as power sources for ships. During the 50 years following World War II, the two major military powers of the world—the United States and the former Soviet Union—conducted secret nuclear research projects related to

the building and testing of bombs. This continued to be a primary focus of nuclear research until the recent changes in the former Soviet Union, which led to a world in which nuclear war is much less of a concern—although the explosion of nuclear devices in June of 1998 by India and Pakistan has heightened concern about nuclear war somewhat. A legacy of this military research is a great deal of soil, water, and air contaminated with radioactive material. Many of these contaminated sites have come to light recently and require major cleanup efforts. The U.S. Department of Energy has begun to clean up the pollution created by its weapons production activities and is shipping transuranic waste to a storage site in New Mexico.

After World War II, people began to see the potential for using nuclear energy for peaceful purposes rather than as weapons. The world's first electricity-generating reactor was constructed in the United States in 1951, and the Soviet Union built its first reactor in 1954. In December 1953, President Dwight D. Eisenhower, in his "Atoms for Peace" speech, made the following prediction:

> Nuclear reactors will produce electricity so cheaply that it will not be necessary to meter it. The users will pay an annual fee and use as much electricity as they want. Atoms will provide a safe, clean, and dependable source of electricity.

Fifty years have passed since Eisenhower's predictions. Although nuclear power is being used throughout the world as a reliable source of electricity, it has not fulfilled such overly optimistic promises. Several serious accidents have caused worldwide concern about safety, and construction of most new nuclear power projects has stopped. At the same time, many energy experts predict a rebirth of the nuclear power industry as energy demands increase and a new generation of safer nuclear power plants is designed. These experts believe the public will favor nuclear power plants because they do not generate carbon dioxide, which contributes to global warming.

Nuclear Reactors

A **nuclear reactor** is a device that permits a controlled fission chain reaction. In the reactor, neutrons are used to cause a controlled fission of heavy atoms, such as uranium. **Uranium-235 (U-235)** is a uranium isotope used to fuel nuclear fission reactors. When the nucleus of a U-235 atom is struck by a slowly moving neutron from another atom, the nucleus of the atom of U-235 is split into several smaller particles. Because the nucleus will split, it is said to be **fissionable.** When the nucleus is split, two to three rapidly moving neutrons are released, along with large amounts of energy. This energy is an important product of nuclear fission reactions. The neutrons released strike the nuclei of other atoms of U-235 and cause them to undergo fission, which, in turn, releases more energy and more neutrons, thus resulting in a chain reaction. (See figure 11.2.) Once begun, this chain reaction continues to release energy until the fuel is spent or the neutrons are prevented from striking other nuclei.

In addition to fuel rods containing uranium, reactors contain control rods of cadmium, boron, graphite, or other nonfissionable materials used to control the rate of fission by absorbing neutrons. When control rods are lowered into a reactor, they absorb the neutrons produced by fissioning uranium. There are fewer neutrons to continue the chain reaction, and the rate of fission decreases. If the control rods are withdrawn, more fission occurs, and more particles, radiation, and heat are produced.

The fuel rods housed in a reactor are surrounded by water or some other type of moderator. A **moderator** absorbs energy, which slows neutrons, enabling them to split the nuclei of other atoms more effectively. Fast-moving neutrons are less effective at splitting atoms than slow-moving neutrons. As U-235 undergoes fission, the energy of the fast-moving neutrons is transferred to water; the neutrons slow down, and the water is heated.

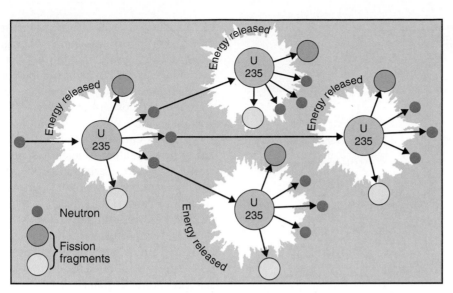

figure 11.2 **Nuclear Fission Chain Reaction** When a neutron strikes a nucleus of U-235, energy is released, and several fission fragments and neutrons are produced. These new neutrons may strike other atoms of U-235, causing their nuclei to split. This series of events is called a chain reaction.

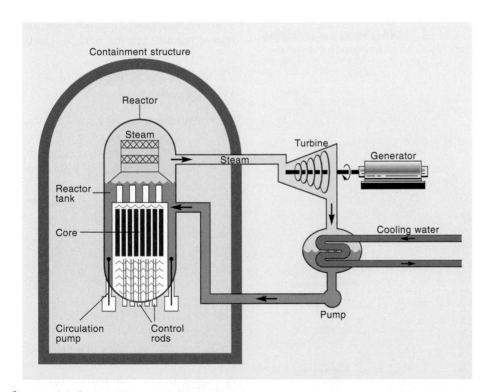

figure 11.3 **Boiling-Water Reactor** A boiling-water reactor is a type of light-water reactor that produces steam to directly power the turbine and produce electricity. Water is used as a moderator and as a reactor-core coolant.

In the production of electricity, a nuclear-powered reactor serves the same function as any fossil-fueled boiler. It produces heat, which converts water to steam to operate a turbine that generates electricity. After passing through the tur-bine, the steam must be cooled, and the water is returned to the reactor to be heated again. Various types of reactors have been constructed to furnish heat for the production of steam. They differ in the moderator used, in how the reactor core is cooled, and in how the heat from the core is used to generate steam. Water is the most commonly used reactor-core coolant and also serves as a neutron mod-erator. **Light-water reactors (LWR),** which make up 90 percent of reactors op-erating today, use ordinary water, which contains the lightest, most common isotope of hydrogen, having an atomic mass of one. The two types of LWRs are **boiling-water reactors (BWR)** and **pressurized-water reactors (PWR).**

In a BWR (the type of construction used in about 20 percent of the nuclear reactors in the world), the water func-tions as both a moderator and reactor-core coolant. (See figure 11.3.) Steam is formed within the reactor and transferred directly to the turbine, which generates electricity. A disadvantage of the BWR is that the steam passing to the turbine must be treated to remove any radiation. Even then, some radioactive material is left in the steam; therefore, the generat-ing building must be shielded.

In a PWR (the type of construction used in over 70 percent of the nuclear re-actors in the world), the water is kept un-der high pressure so that steam is not allowed to form in the reactor. (See fig-ure 11.4.) A secondary loop transfers the heat from the pressurized water in the re-actor to a steam generator. The steam is used to turn the turbine and generate electricity. Such an arrangement reduces the risk of radiation in the steam but adds to the cost of construction by requiring a secondary loop for the steam generator.

A third type of reactor that uses water as a coolant is a **heavy-water reactor (HWR),** developed by Canadi-ans. It uses water that contains the hydro-gen isotope deuterium in its molecular structure as the reactor-core coolant and moderator. Since the deuterium atom is twice as heavy as the more common hy-drogen isotope, the water that contains deuterium weighs slightly more than or-dinary water. An HWR also uses a steam generator to convert regular water to steam in a secondary loop. Thus, it is similar in structure to a PWR. The major advantage of an HWR is that naturally occurring uranium isotopic mixtures serve as a suitable fuel. This is possible

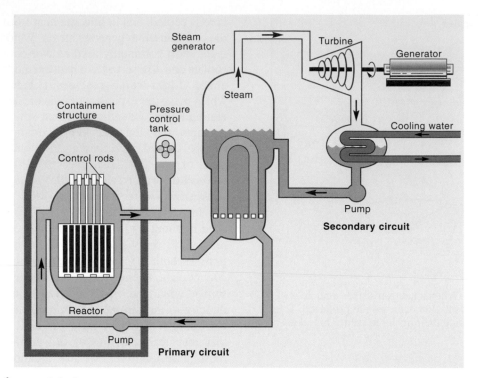

figure 11.4 **Pressurized-Water Reactor** A pressurized-water reactor is a type of light-water reactor that uses a steam generator to form steam and a secondary loop to transfer this steam to the turbine.

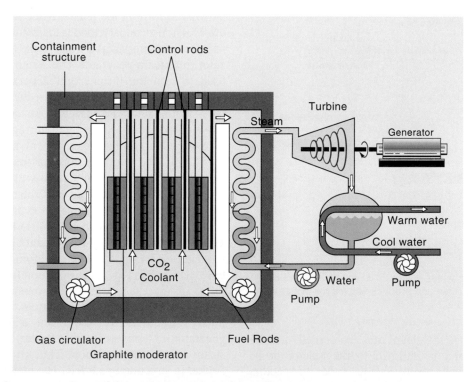

figure 11.5 **Advanced Gas-Cooled Reactor** This type of reactor uses graphite as a moderator and the gas carbon dioxide as the reactor-core coolant. A steam generator forms steam and a secondary loop is used to transfer steam to the turbine.

because heavy water is a better neutron moderator than is regular water, while other reactors require that the amount of U-235 be enriched to get a suitable fuel. Since it does not require enriched fuel, the operating costs of an HWR are less than that of an LWR.

The **gas-cooled reactor (GCR)** was developed by atomic scientists in the United Kingdom. Carbon dioxide serves as a coolant for a graphite-moderated core. As in the HWR, natural isotopic mixtures of uranium are used as a fuel. (See figure 11.5.)

The various types of reactors represent differing approaches to building safe, economical plants. Each method has its advantages and disadvantages, and its supporters and critics. Modifications will continue to be made to these basic types. Figure 11.6 shows the distribution of nuclear power plants in North America, and table 11.2 shows the numbers of reactors present in each country.

Plans for New Reactors Worldwide

Currently, there are 439 nuclear power reactors in 31 countries, with a combined capacity of 354 gigawatts. In 2000, these provided 2447 billion kilowatt hours, over 16 percent of the world's electricity. Although some countries, notably Japan, China, Taiwan, India, Russia, Slovak Republic, and the Republic of Korea, intend to continue major nuclear power construction programs, the rate of growth of installed nuclear-generating capacity over the next 10 years is expected to be low.

Some 32 power reactors are currently being constructed in 10 countries. Construction is well advanced on many of them and, based on reported progress and allowing for delays in countries, 13 with a total net capacity of over 9000 megawatts are expected to be in operation before 2004. This excludes two Russian and two Ukrainian reactors for which funding is uncertain.

After about 2005, forecasts of installed nuclear capacity become much less certain. When the present construction programs are completed, most

figure 11.6 **Distribution of Nuclear Power Plants in North America** The United States has the largest number of nuclear power plants of any country in the world—over 100. Canada has about 20 nuclear power plants, and Mexico has two.

significant nuclear power growth is expected to continue only in Asia and parts of the former Soviet Union. The International Atomic Energy Agency (IAEA) forecasts that the total installed nuclear capacity in 2015 will be little more than that in 2000, with the nuclear share of world electricity output decreased from 17 percent in 1997 to 13 percent in 2015.

Eight countries have plans to build new power reactors beyond those now under construction. In all, 32 power reactors with a total net capacity of over 34,460 megawatts are planned. Only one of these is in Western Europe or North or South America; it is in Argentina.

In France, the only Western European country that has had an active nuclear power construction program, the national utility announced in 1994 that it would order no new generating capacity, nuclear or conventional, until about 2002. Sites have, however, been designated for new power reactors, and new reactor construction is expected to resume by 2004.

In 2002, Germany announced that it would close all of its 19 nuclear power plants by 2021, making it the world's largest industrialized nation to willingly forgo the technology.

Germany, Europe's largest economy, currently receives one-third of its electri-cal power from nuclear sources. The decision reached in Germany limits nuclear plants to an average 32 years of operation. That means that the most modern plants likely will close around 2021, and Germany will join nations such as Italy and Austria in abandoning nuclear power.

In Eastern Europe, the Russian government in 1997 approved a nuclear power construction program. In addition to the three reactors under construction, another three, taking total capacity to about 26,400 megawatts, are planned to be operating by 2010. Several of the oldest Russian reactors are expected to be retired by 2010, and it is Russia's announced intention to replace retired nuclear capacity with new construction at the same site, to optimize the use of established infrastructure and personnel.

Most planned reactors are in the Asian region, which has fast-growing economies and rapidly rising electricity demand. Nuclear power will continue to play a major role in the future electricity supply mix in both South Korea and Japan. South Korea plans to bring a further 12 reactors, with a total capacity of 13,100 megawatts, into operation by the year 2015. Japan has plans and, in most cases, designated sites and announced timetables for a further 15 power reactors, totaling about 20,000 megawatts, and some of these are negotiating the governmental approval process. Fulfilling the necessary conditions for approval can take over a decade.

China, with four operating reactors, has begun the next phase of its nuclear power program. Construction has started on seven reactors. These are expected to start up from 2002 to 2005 and to add some 5770 megawatts to China's existing 2767 megawatts nuclear capacity.

Plant Life Extension

Most nuclear power plants originally had a nominal design lifetime of up to 40 years, but engineering assessments of many plants over the last decade has established that many can operate longer. In the United States, most reactors now have confirmed life spans of 40 to 60 years, and in Japan, 40 to 70 years. In the

Table 11.2 World Nuclear Power Reactors, 2000–2001, and Uranium Requirements

Country	Reactors Operating March 2002		Reactors Building March 2002		On Order or Planned March 2002		Uranium Required 2001
	No.	MWe	No.	MWe	No.	MWe	tonnes U
Argentina	2	935	0	0	1	692	133
Armenia	1	376	0	0	0	0	68
Belgium	7	5728	0	0	0	0	1109
Brazil	2	1855	0	0	0	0	296
Bulgaria	6	3538	0	0	0	0	618
Canada	14	9998	6	3598	0	0	1343
China	4	2767	7	5770	0	0	572
Taiwan	6	4884	2	2600	0	0	966
Czech Republic	5	2560	1	912	0	0	519
Finland	4	2656	0	0	0	0	553
France	59	63203	0	0	0	0	10159
Germany	19	21141	0	0	0	0	3712
Hungary	4	1755	0	0	0	0	425
India	14	2548	3	1916	3	1340	312
Iran	0	0	1	950	0	0	0
Japan	54	444301	3	3696	12	15858	7393
Korea DPR (North)	0	0	0	0	2	1900	0
Korea RO (South)	16	12970	4	3800	8	9200	2466
Lithuania	2	2370	0	0	0	0	359
Mexico	2	1310	0	0	0	0	232
Netherlands	1	452	0	0	0	0	115
Pakistan	2	425	0	0	0	0	56
Romania	1	655	0	0	1	620	90
Russia	30	20793	3	2625	3	2950	3411
Slovak Rep.	6	2472	2	840	0	0	528
Slovenia	1	679	0	0	0	0	131
South Africa	2	1842	0	0	0	0	363
Spain	9	7345	0	0	0	0	1613
Sweden	11	9460	0	0	0	0	1533
Switzerland	5	3170	0	0	0	0	599
Ukraine	13	11195	0	0	2	1900	1893
United Kingdom	33	12528	0	0	0	0	2588
USA	104	98248	0	0	0	0	20801
WORLD	**439**	**354,159**	**32**	**26,707**	**32**	**34,460**	**64,956**

Source: Reprinted by permission of Uranium Information Centre Ltd., Melbourne, Australia.

United States, the first two reactors have been granted license renewals, which extends their operating lives to 60 years.

When the oldest commercial nuclear power stations in the world, Calder Hall and Chapelcross in the United Kingdom, were built in the 1950s, it was assumed that they would have a useful lifetime of 20 years. As of 2001, they were still authorized to operate.

Sweden's oldest reactor, which started up in 1971, has been fully rebuilt at a cost equivalent to 8 percent of a replacement unit, and all Sweden's reactors are maintained so that a further 20 years of life is in prospect.

It should be noted, however, that economic, regulatory, and political considerations have led to the premature closure of some power reactors. In the United States, reactor numbers have fallen from 110 to 104. Germany, as previously mentioned, plans to close all of its plants by 2021.

Breeder Reactors

During the early stages of the development of nuclear power plants, breeder reactor construction was seen as the logical step after nuclear fission development. Although a regular fission reactor produces some additional radioactive material that can serve as nuclear fuel, it is smaller than the quantity of fuel consumed. A **nuclear breeder reactor** is a nuclear fission reactor that produces heat to be converted to steam to generate electricity and also forms a new supply of radioactive isotopes. If a fast-moving neutron hits a uranium-238 (U-238) nucleus and is absorbed, an atom of fissionable **plutonium-239 (Pu-239)** is produced. (See figure 11.7.) In a breeder reactor, water is not used as a moderator because water slows the neutrons too much and Pu-239 is not produced. Breeder reactors need a moderator that allows the neutrons to move more rapidly and has good heat transfer properties.

The **liquid metal fast-breeder reactor (LMFBR)** appears to be the most promising model. In this type of reactor, the fuel rods in the core are surrounded by rods of U-238 and liquid sodium. The energy of the neutrons released from U-235 in the fuel rods heats the sodium to a temperature of 620°C. In addition to furnishing heat, these fast-moving neutrons are absorbed by the rods containing U-238, and some of these atoms are converted into Pu-239. After approximately 10 years of operation, during which electricity is produced, the LMFBR will have also produced enough radioactive material to operate a second reactor.

There are some serious drawbacks to the LMFBR. Sodium reacts violently if it comes into contact with water or air. Therefore, costly, highly specialized equipment is required to contain and pump the sodium. If these systems fail, the sodium boils, which allows the chain reaction to proceed at a faster rate and could damage the reactor, leading to a nuclear accident.

There are also problems in the startup of an LMFBR. When a breeder reactor is starting up or going on-line

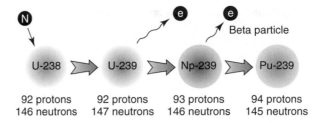

figure 11.7 **Formation of Pu-239 in a Breeder Reactor** When a fast-moving neutron (N) is absorbed by the nucleus of a U-238 atom, a series of reactions results in the formation of Pu-239 from U-238. Two intermediate atoms are U-239 and Np-239, which release beta particles (electrons) from their nuclei. (A neutron in the nucleus can release a beta particle and become a proton.) This is an important reaction because, while the U-238 does not disintegrate readily and therefore is not a nuclear fuel, the Pu-239 is fissionable and can serve as a nuclear fuel.

after a shutdown, the solid sodium cannot be moved by the pumps until it becomes a liquid. This presents a technical problem in developing LMFBRs.

Another problem is that reaction rates are extremely rapid and very difficult to regulate. The instrumentation needed to monitor an LMFBR must be of extremely high quality because control of the reaction involves precise adjustments over very short periods of time.

Finally, the product of a breeder reactor, plutonium-239, is extremely hazardous to humans who come in contact with it. And, because plutonium-239 can be made into nuclear weapons, it must be transported, processed, or produced under very close security. The more breeder reactors in use, the more difficult the security problems and the more likely the chance that the small amount of plutonium needed to manufacture a bomb could be stolen.

Because of these problems, no breeder reactors are scheduled for commercial use in the United States. In fact, the only experimental U.S. breeder reactor is scheduled to be shut down. The Clinch River Fast-Breeder Reactor Plant near Knoxville, Tennessee, which was to be operational in the 1970s, was never completed.

In Europe, as in the United States, the development of breeder reactors has slowed. In 1981, after running smoothly for eight years, a 250-megawatt French prototype plant developed a leak in the cooling system that resulted in a sodium fire. This raised some serious questions about the safety of breeder reactors. In 1982, the French government announced

that it was scaling down the planned construction of breeder reactors from five to one. Today, it is doubtful that even that one will be built. A total of five LMFBRs are in operation in the world today, in France, Japan, Kazakhstan, and Russia.

Nuclear Fusion

Another aspect of nuclear power that may have promise for the future involves the even more advanced technology of nuclear fusion. When two lightweight atomic nuclei combine to form a heavier nucleus, a large amount of energy is released. This process is known as **nuclear fusion.** The energy produced by the sun is the result of fusion. Most studies of fusion have involved small atoms like hydrogen. Most hydrogen atoms have one proton and no neutrons in the nucleus. The hydrogen isotope deuterium (H^2) has a neutron and a proton. Tritium (H^3) has a proton and two neutrons. When deuterium and tritium isotopes combine to form heavier atoms, large amounts of energy are released. (See figure 11.8.) The energy that would be released by combining the deuterium in 1 cubic kilometer of ocean water would be greater than that contained in the world's entire supply of fossil fuels.

Although fusion could solve the world's energy problems, technology must answer several questions before we can actually use fusion power. Three conditions must be met simultaneously for fusion to occur: high temperature, adequate density, and confinement. If heat is

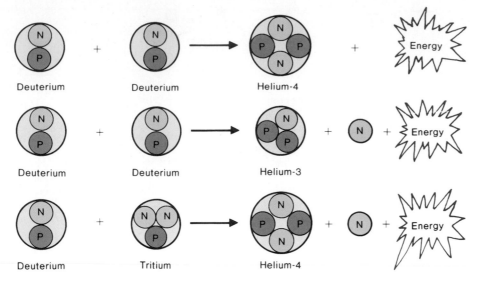

figure 11.8 **Nuclear Fusion** In nuclear fusion, small atomic nuclei are combined to form heavier nuclei. Large amounts of energy are released when this occurs. Different isotopes of hydrogen can be used in the process of fusion. There are three possible types of fusion: (1) two deuterium isotopes can combine to form helium-4 and energy; (2) two deuterium isotopes can combine to form helium-3, a free neutron, and energy; and (3) a deuterium and a tritium isotope can combine to form helium-4, a free neutron, and energy.

used to provide the energy necessary for fusion, the temperature must approach that of the center of the sun. At the same time, the walls of the vessel confining the atoms must be protected from the heat, or they will vaporize. However, the main problem is containment of the nuclei. Because they have a positive electric charge, the nuclei repel one another.

Even though, in theory, fusion promises to furnish a large amount of energy, technical difficulties appear to prevent its commercial use in the near future. Even the governments of nuclear nations are budgeting only modest amounts of money for fusion research. And, as with nuclear fission and the breeder reactor, economic costs and fear of accidents may continue to delay the development of fusion reactors.

The Nuclear Fuel Cycle

To appreciate the consequences of using nuclear fuels to generate energy, it is important to understand how the fuel is processed. The nuclear fuel cycle begins with the mining operation. (See figure 11.9.) Low-grade uranium ore is obtained by underground or surface mining. The ore contains about 0.2 percent uranium by weight. After it is mined, the ore goes through a milling process. It is crushed and treated with a solvent to concentrate the uranium. Milling produces yellowcake, a material containing 70 to 90 percent uranium oxide.

Naturally occurring uranium contains about 99.3 percent nonfissionable U-238 and only 0.7 percent fissionable U-235. This concentration of U-235 is not high enough for most types of reactors, so the amount of U-235 must be increased by enrichment. Since the masses of the isotopes U-235 and U-238 vary only slightly, and the chemical differences are very slight, enrichment is a difficult and expensive process. However, it increases the U-235 content from 0.7 percent to 3 percent.

Fuel fabrication converts the enriched material into a powder, which is then compacted into pellets about the size of a pencil eraser. These pellets are sealed in metal fuel rods about 4 meters in length, which are then loaded into the reactor.

As fission occurs, the concentration of U-235 atoms decreases. After about three years, a fuel rod does not have enough radioactive material to sustain a chain reaction, and the spent fuel rods must be replaced by new ones. The spent rods are still very radioactive, containing about 1 percent U-235 and 1 percent plutonium. These rods are the major source of radioactive waste material produced by a nuclear reactor.

When nuclear reactors were first being built, scientists proposed that spent fuel rods could be reprocessed. The remaining U-235 could be enriched and used to manufacture new fuel rods. Since plutonium is fissionable, it could also be fabricated into fuel rods. Besides providing new fuel, reprocessing would reduce the amount of nuclear waste. At present, India, Japan, Russia, France, and the United Kingdom operate reprocessing plants that reprocess spent fuel rods as an alternative to storing them as a nuclear waste.

Each step in the nuclear fuel cycle involves the transport of radioactive materials. The uranium mines are some distance from the processing plants. The fuel rods must be transported to the power plants, and the spent rods must be moved to a reprocessing plant or storage area. Each of these links in the fuel cycle presents the possibility of an accident or mishandling that could release radioactive material. Therefore, the methods of transport are extremely carefully designed and tested before they are used. Many people are convinced that the transport of radioactive materials is hazardous, while others are satisfied that the utmost care is being taken and that the risks are extremely small.

Ultimately, both high-level and low-level radioactive wastes must be stored. Thus, each step in the nuclear fuel cycle, from the mining of uranium to the storage of nuclear waste, poses health and environmental concerns.

Nuclear Material and Weapons Production

Producing nuclear materials for weapons and other military uses involves many of the same steps used to produce nuclear

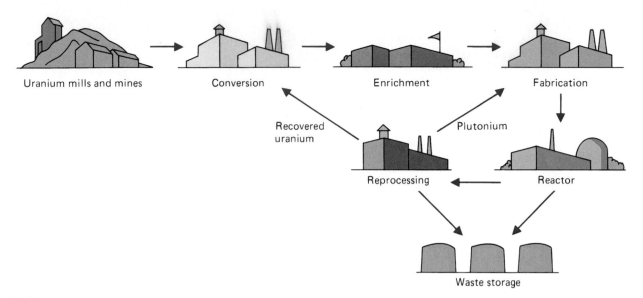

figure 11.9 **Steps in the Nuclear Fuel Cycle** The process of obtaining nuclear fuel involves mining, extracting the uranium from the ore, concentrating the U-235, fabricating the fuel rods, installing and using the fuel in a reactor, and disposing of the waste. Some countries reprocess the spent fuel as a way of reducing the amount of waste they must deal with.

Source: U.S. Department of Energy.

fuel for power reactors. In fact, the nuclear power industry is an outgrowth of the weapons industry. In the United States, the Department of Energy is responsible for nuclear research for both weapons and peaceful uses and stewardship of the facilities used for research and weapons production. Some facilities are used for both processing nuclear fuel and providing materials for weapons. In both cases, uranium must be mined, concentrated, and transported to sites of use. Furthermore, military uses involve the production and concentration of plutonium. The production and storage facilities invariably become contaminated, as does the surrounding land.

Research and production facilities have typically dealt with hazardous chemicals and low-level radioactive wastes by burying them, pumping them into the ground, storing them in ponds, or releasing them into rivers. Despite environmental regulations that prevented such activities, the Department of Energy (DOE) (formerly the Atomic Energy Commission) maintained that it was exempt from such federal environmental legislation. As a result, the DOE has become the steward of a large number of sites that are contaminated with both hazardous chemicals and radio-

active materials. The magnitude of the problem is huge. There are 8600 square kilometers (3365 square miles) of DOE properties that are or have been involved in weapons development or production. These include:

- 3700 contaminated sites,
- 330 underground storage tanks with high-level radioactive waste,
- more than a million 55-gallon drums of radioactive, hazardous, or mixed waste in storage,
- 5700 sites where wastes are moving through the soil, and
- millions of cubic meters of low-level and high-level radioactive wastes.

The Department of Energy has pledged to clean up these sites; however, the process could take decades and cost billions of dollars. Several U.S. sites are being cleaned up, but things are not going smoothly. Local residents and the states that are hosts to these facilities distrust the DOE and are insisting that the sites be cleaned completely. This may not be technologically or economically possible. Furthermore, environmental cleanup is a new mission for the DOE, and the department is having difficulty adjusting. Figure 11.10 shows

the location of waste sites the U.S. Department of Energy has responsibility for cleaning up.

An additional problem has arisen as a result of the reduced importance of nuclear weapons. The political disintegration of the Soviet Union and Eastern Europe has made large numbers of nuclear weapons, in both the East and West, unnecessary. This has probably made the world a safer place, but how are nations of the world to dispose of their nuclear weapons? Some nuclear material can be diverted to fuel use in nuclear reactors, and some reactors can be modified to accept enriched uranium or plutonium. The security of these materials, particularly in the former Soviet Union, is a concern.

Nuclear Power Concerns

Nuclear power has provided a significant amount of electricity for the people of the world. Currently, over 7 percent of the energy consumed worldwide and 17 percent of the electricity consumed worldwide comes from nuclear power. However, several accidents have raised

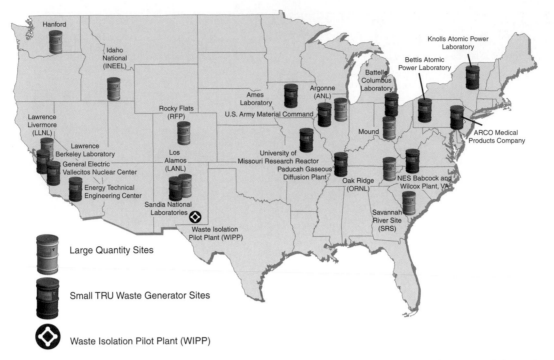

Large Quantity Sites

Small TRU Waste Generator Sites

Waste Isolation Pilot Plant (WIPP)

figure 11.10 **The U.S. Department of Energy Waste Sites** As a result of many activities related to research in nuclear energy and to the production of nuclear weapons, the U.S. Department of Energy has responsibility for cleaning up many contaminated sites.

questions about safety, radiation has been released into the air and water, the production and transportation of fuel have caused contamination, and the disposal of wastes is a continuing problem.

After September 11, 2001, fear arose regarding nuclear plants as potential targets for terrorist attacks. The image of a hijacked airliner crashing into the core of a nuclear reactor brought memories of Chernobyl to the minds of many, and security measures were heightened at power plants nationwide. But just how vulnerable are nuclear plants to terrorist attack?

The concensus is that there is not a great cause for concern. An airliner flying at high speed would not be able to do significant damage to a reactor or containment building. The reactor itself is very small and is encased in steel and concrete in order to survive any natural disaster; this enclosure could also withstand the impact of a small aircraft. The segments of the plant that may be vulnerable to attack, such as cooling and auxiliary equipment, are supplied with emergency cooling systems and power supplies. If all of these should fail, the reactor core could overheat and melt, but its materials would remain contained.

The areas of most concern within a nuclear plant are the storage facilities for spent fuel. These depositories contain more total radioactivity than the reactor itself, but many of the more dangerous radioactive isotopes have already decayed. Also, while a small attack could drain the water out of cooling ponds for spent fuel rods, causing the fuel to overheat and melt, little radioactive material would be dispersed. Similarly, the crash of an airliner into a dry storage facility would release little radioactive particulate matter. If any of these facilities were the target of an attack, residents would have at least eight hours to evacuate the region before unsafe levels of radioactivity were reached; the worst contamination would occur through dispersal of airborne radioactive material into the natural environment.

Reactor Safety: The Effects of Three Mile Island and Chernobyl

Although there have been accidents at nuclear power plants since they were first used, two relatively recent accidents have had the greatest effect on

people's attitudes toward nuclear power plant safety: Three Mile Island in 1979 and Chernobyl in 1986.

On March 14, 1979, the valves closed to three auxiliary pumps at one of the reactors at the Three Mile Island nuclear plant in Pennsylvania. On March 28, 1979, the main pump to the reactor broke down. The auxiliary pumps failed to operate because of the closed valves, and the electrical-generating turbine stopped. At this point, an emergency coolant should have flooded the reactor and stabilized the temperature. The coolant did start to flow, but a faulty gauge indicated that the reactor was already flooded. Relying on this faulty reading, an operator overrode the automatic emergency cooling system and stopped it. Without the emergency coolant, the reactor temperature rose rapidly. The control rods eventually stopped fission, but a partial core meltdown had occurred. Later that day, radioactive steam was vented into the atmosphere.

The crippled reactor was eventually defueled in 1990 at a cost of about $1 billion. It has been placed in monitored storage until the companion reactor, which is still operating, reaches the end

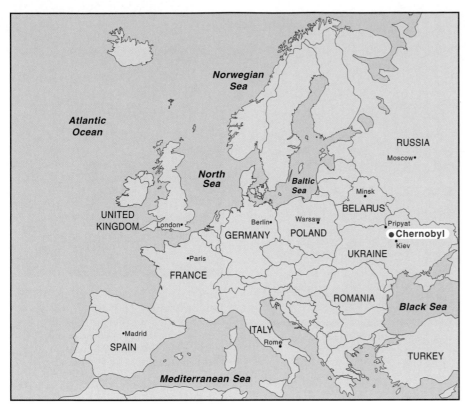

figure 11.11 **Chernobyl** The town of Chernobyl became infamous as the location of the world's worst nuclear power plant accident.

of its useful life. At that time, both reactors will be decommissioned.

Chernobyl is a small city in Ukraine near the border with Belarus, north of Kiev. As is true of many small cities in the world, most people had never heard of it. (See figure 11.11.) However, in the spring of 1986, the world's largest nuclear accident catapulted Chernobyl into the news.

At 1 A.M. on April 25, 1986, at Chernobyl Nuclear Power Station-4, a test was begun to measure the amount of electricity that the still-spinning turbine would produce if the steam were shut off. This was important information since the emergency core cooling system required energy for its operation and the coasting turbine could provide some of that energy until another source became available. The amount of steam being produced was reduced by lowering the control rods into the reactor. But the test was delayed because of a demand for electricity, and a new shift of workers came on duty. The operators failed to program the computer to main-

tain power at 700 megawatts, and output dropped to 30 megawatts. This presented an immediate need to rapidly increase the power, and many of the control rods were withdrawn. Meanwhile, an inert gas (xenon) had accumulated on the fuel rods. The gas absorbed the neutrons and slowed the rate of power increase. In an attempt to obtain more power, operators withdrew all the control rods. This was a second serious safety violation.

At 1 A.M. on April 26, the operators shut off most emergency warning signals and turned on all eight pumps to provide adequate cooling for the reactor following the completion of the test. Just as final stages of the test were beginning, a signal indicated excessive reaction in the reactor. In spite of the warning, the operators blocked the automatic reactor shutdown and began the test.

As the test continued, the power output of the reactor rose beyond its normal level and continued to rise. The operators activated the emergency system designed to put the control rods

back into the reactor and stop the fission. But it was too late. The core had already been deformed, and the rods would not fit properly; the reaction could not be stopped. In 4.5 seconds, the energy level of the reactor increased 2000 times. The fuel rods ruptured, the cooling water turned into steam, and a steam explosion occurred. The lack of cooling water allowed the reactor to explode. The explosion blew the 1000-metric ton (1102 ton) concrete roof from the reactor, and the reactor caught fire.

In less than 10 seconds, Chernobyl became the scene of the world's worst nuclear accident. A core meltdown had occurred. (See figure 11.12.) It took 10 days to bring the runaway reaction under control. By November, the damaged reactor was entombed in a concrete covering, but the hastily built structure, known as the sarcophagus, may have structural flaws. The Ukrainian government is planning to have a second containment structure built around the current structure. The immediate consequences were 31 fatalities; 500 persons hospitalized, including 237 with acute radiation sickness; and 116,000 people evacuated. Of the evacuees, 24,000 received high doses of radiation. The delayed effects are more difficult to assess. Many people suffer from illnesses they feel are related to their exposure to the fallout from Chernobyl. In 1996, it became clear that at least one delayed effect was strongly correlated with exposure. Children or fetuses exposed to fallout are showing increased frequency of thyroid cancer. The thyroid gland accumulates iodine, and radioactive iodine 131 was released from Chernobyl. It is still too early to tell if there are other delayed health effects.

More than a year after the disaster at Chernobyl, the decontamination of 27 cities and villages within 40 kilometers of Chernobyl was considered finished. This does not mean that the area was cleaned up, only that all practical measures were completed. Some areas were simply abandoned. The largest city to be affected was Pripyat, which had a population of 50,000 and was only 4 kilometers from the reactor. A new town was built to accommodate those displaced by the accident, and Pripyat remains a

figure 11.12 **The Accident at Chernobyl** An uncontrolled chain of reactions in the reactor of unit four resulted in a series of explosions and fires. See the circled area in the photograph.

figure 11.13 **Anti-nuclear Demonstration** This anti-nuclear demonstration was held in 2000 in Germany as members of the public protested the transport of nuclear waste across the country to a storage site.

ghost town. Sixteen other communities within the 40-kilometer radius have been cleaned up, and the original inhabitants have returned, although there is still controversy about the safety of these towns. Seventeen years after the Chernobyl explosion, some scientists fear that the worst is yet to come. Compared to the general population, rates of some noncancer diseases—endocrine disorders and stroke, for instance—appear to be rising disproportionately among the roughly 600,000 "liquidators" who cleaned up the heaviest contamination in the plant's vicinity and entombed unit four's lethal remnants in a concrete sarcophagus. Whether people living in the shadow of Chernobyl remain at risk for health problems is a question now under intense scrutiny.

One important impact of Chernobyl is that it deepened public concern about the safety of nuclear reactors. Even before Chernobyl, between 1980 and 1986, the governments of Australia, Denmark, Greece, Luxembourg, and New Zealand had officially adopted a "no nuclear" policy. Since 1980, 10 countries have canceled nuclear plant orders or mothballed plants under construction. Argentina canceled 4 plants;

Brazil, 8; Mexico, 18; and the United States, 54. There have been no orders for new plants in the United States since 1974. Sweden, Austria, Germany, and the Philippines have decided to phase out and dismantle their nuclear power plants.

Before Chernobyl, 65 percent of the people in the United Kingdom were opposed to nuclear power plants; after Chernobyl, 83 percent were against them. In Germany, opposition increased from 46 percent to 83 percent. (See figure 11.13.) Opposition in the United States rose from 67 to 78 percent. Even the French, who have the greatest commitment to nuclear power, were against it by 52 percent. The number of new nuclear plants being constructed has been reduced. In addition, many plants have been shut down prematurely, resulting in the prediction that the amount of energy furnished by nuclear fission reactors will actually decline in the future.

At the same time some people are predicting the death of nuclear power as a viable energy source, others are examining ways to make nuclear power facilities safer. They expect nuclear power to be an important energy source in the future. At both Three Mile Island and

Chernobyl, operator error caused or contributed to the accidents. Operators manually stopped normal safety actions from taking place. However, a contributing factor was the design: active mechanical processes had to work properly to shut the reactors down. Many of the new designs for reactors include passive mechanisms that will shut down the reactor, special catching basins for the reactor core should it melt down, and better containment buildings.

Nuclear power will continue to be a part of the energy mix, particularly in countries that lack fossil-fuel reserves. Furthermore, as fossil fuels are used up, the pressure for energy will probably create a market for nuclear power technology.

Exposure to Radiation

Although nuclear accidents are spectacular, frightening, and can cause immediate death because of the incredible amount of energy involved, the long-term problems resulting from exposure to radiation are even more worrisome. Radiation is converted to other forms of energy when it is absorbed by matter. When organisms are irradiated, this energy conversion causes damage at a cellular, tissue, organ, or organism level. The degree and kind of damage vary with the kind of radiation, the amount of radiation, the duration of the exposure, and the types of cells irradiated.

Radiation can also cause mutations, which are changes in the genetic messages within cells. Mutations can cause two quite different kinds of problems. Mutations that occur in the ovaries or testes can form mutated eggs or sperm, which can lead to abnormal offspring. Care is usually taken to shield these organs from unnecessary radiation. Mutations that occur in other tissues of the body may manifest themselves as abnormal tissue growths known as cancer. Two common cancers that are strongly linked to increased radiation exposure are leukemia and breast cancer. Because mutations are essentially permanent, they may accumulate over time. Therefore, the accumulated effects of radiation over many years may result in the development of cancer later in life.

Human exposure to radiation is usually measured in **rems** (*r*oentgen *e*quivalent *m*an), a measure of the biological damage to tissue. The effects of large doses (1000 to 1 million rems) are easily seen and can be quantified, because there is a high incidence of death at these levels, but demonstrating known harmful biological effects from smaller doses is much more difficult. (See table 11.3.) Moderate doses (10 to 100 rems) are known to increase the likelihood of cancer and birth defects. The higher the dose, the higher the incidence of abnormality. Lower doses may cause temporary cellular changes, but it is difficult to demonstrate long-term effects. Thus, the effects of low-level, chronic radiation generate much controversy. Some people feel that all radiation is harmful, that there is no safe level, and that special care must be taken to prevent exposure. (See figure 11.14.) It should be pointed out, however, that no human is subject to zero exposure and that the average person is exposed to 0.2 to 0.3 rems per year from natural and medical sources. Current research is trying to assess the risks associated with repeated exposure to low-level radiation.

Table 11.3 Radiation Effects

Source	Dose	Biological Effect
Nuclear bomb blast or exposure in a nuclear facility	100,000 rems/incident	Immediate death
	10,000 rems/incident	Coma, death within one to two days
X rays for cancer patients (It is important to distinguish between a whole-body dose of 1000 rems, which is lethal, and a similar dose targeted at a cancerous organ, which often saves lives.)	1000 rem/incident	Nausea, lining of intestine damaged, death in one to two weeks
	100 rems/incident	Increased probability of leukemia
	10 rems/incident	Early embryos may show abnormalities
Upper limit for occupationally exposed people	5 rems/year	Effects difficult to demonstrate
X ray of the intestine	1 rem/procedure	Effects difficult to demonstrate
Upper limit for release from nuclear installations (except nuclear power plants)	0.5 rem/year	Effects difficult to demonstrate
Natural background radiation	0.2–0.3 rem/year	Effects difficult to demonstrate
Upper limit for release by nuclear power plants	0.005 rem/year	Effects difficult to demonstrate

figure 11.14 **Protective Equipment** Persons working in an area subjected to radiation must take steps to protect themselves. These workers are wearing protective clothing and filtering the air they breathe.

Each step in the nuclear fuel cycle poses a radiation exposure problem, beginning with the mining of uranium. Although the radiation level in the ore is low, miners' prolonged exposure to low-level radiation increases their rates of certain cancers, such as lung cancer. Uranium miners who smoke have an even higher lung cancer rate. After mining, the ore must be crushed in a milling process. This releases radioactive dust into the atmosphere, so workers are chronically exposed to low levels of radioactivity. The crushed rock is left on the surface of the ground as mine tailings. There are over 150 million metric tons of low-level radioactive mine tailings in the United States and at least as much in the rest of the world. These tailings constitute a hazard because they are dispersed into the environment. (See figure 11.15.)

In the enrichment and fabrication processes, the main dangers are exposure to radiation and accidental release of radioactive material into the environment. Transport also involves exposure risks. Most radioactive material is transported

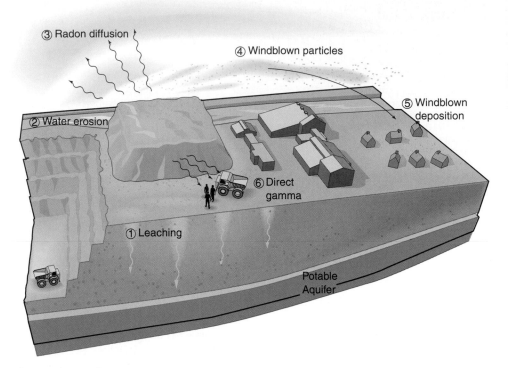

figure 11.15 **Uranium Mine Tailings** Even though the amount of radiation in the tailings is low, the radiation still represents a threat to human health. Radioactivity may be dispersed throughout the environment and come in contact with humans in several ways: (1) radioactive materials may leach into groundwater; (2) radioactive materials may enter surface water through erosion; (3) radon gas may diffuse from the tailings and enter the air; (4) persons living near the tailings may receive particles in the air, (5) come in contact with objects that are coated with particulate, or (6) receive direct gamma radiation.

Source: U.S. Environmental Protection Agency.

figure 11.16 **Cooling Towers** Cooling towers draw air over wet surfaces. The evaporation of the water cools the surface, removing heat from the process and releasing the heat into the air.

by highways or railroads. If a transporting vehicle were involved in an accident, radioactive material could be released into the environment. Even after fuel rods have been loaded into the reactor, people who work in the area of the reactor risk radiation exposure. Countries that reprocess spent fuel rods must be concerned about the exposure of workers and the possible theft of rods or reprocessed material by terrorist organizations that would use the radioactive material to construct an atomic bomb.

Thermal Pollution

Thermal pollution is the addition of waste heat to the environment. This is a problem particularly in aquatic environments, since many aquatic organisms are very sensitive to changes in temperature. When temperatures increase, oxygen solubility decreases while metabolic rates of aquatic organisms increase, raising the demand for oxygen. All industrial processes release waste heat, so the problem of thermal pollution is not unique to nuclear power plants. In both fossil-fuel and nuclear plants, generating steam to produce electricity results in a great deal of waste heat. In a fossil-fuel plant, half of the heat energy produces electricity, and half is lost as waste heat. In a nuclear power plant, only one-third of the heat generates electricity, and two-thirds is waste heat. Therefore, the less efficient nuclear power plant increases the amount of thermal pollution more than does an equivalent fossil-fuel power plant. To reduce the effects of this waste heat, utilities build costly cooling facilities. Cooling processes involve water, so nuclear power plants often are constructed next to a water source. In some cases, water is drawn directly from lakes, rivers, or oceans and returned. In other cases, it is supplied by giant cooling towers. (See figure 11.16.)

Decommissioning Costs

All industrial facilities have a life expectancy, that is, the number of years they can be profitably operated. The life expectancy for an electrical generating plant, whether fossil-fuel or nuclear, is about 30 to 40 years, after which time the plant is demolished. With a fossil-fuel plant, the demolition is relatively simple and quick. A wrecking ball and bulldozers reduce the plant to rubble, which is trucked off to a landfill. The only harm to the environment is usually the dust raised by the demolition.

Demolition of a nuclear plant is not so simple. In fact, nuclear plants are not demolished, they are decommissioned. **Decommissioning** involves removing the fuel, cleaning surfaces, and permanently preventing people from coming into contact with the contaminated buildings or equipment. Today, over 70 nuclear plants in the world are shut down and waiting to be decommissioned, and the number of plants scheduled for decommissioning will grow. By 2005, 68 of the 104 nuclear plants in the United States will be 20 years old or older. The Nuclear Regulatory Commission authorizes their operation for 40 years but is considering extending authorization for an additional 20 years. This would put off the need for decommissioning for a few years, but eventually, it will be necessary. In the United States, many small experimental facilities have been decommissioned, and fourteen plants have been removed from service and are already awaiting decommissioning. Canada has decommissioned three plants and has one scheduled for decommissioning. Dozens of nuclear reactors have been shut down or are expected to close soon in Europe, Canada, the former Soviet republics, and Japan. The great majority of near-term decommissioning work will take place in Europe: France, Germany, and Britain each are decommissioning 20 or more reactors, while nations using less nuclear power, such as Italy, Belgium, Spain, and Sweden, are decommissioning one or more reactors.

The European Commission (EC) is working to standardize national decommissioning policies. Currently, policies vary widely on key questions such as how long sites will be put in safe storage to allow radiation to decay (from a maximum of 30 years in Finland to 135 years in Britain). Waste management also poses a serious challenge. The EC estimates that by 2060, decommissioning will produce more than 2 million metric tons (more than 2.2 million tons) of metallic and concrete waste, but many member states currently have, at best, limited storage and disposal options to handle such amounts of waste.

The ongoing international effort to close unsafe Soviet-designed reactors in central and Eastern Europe has important decommissioning implications. The European Commission has made closure of these reactors a condition for membership in the European Union and has pledged financial support for decommissioning eight reactors in Lithuania, Bulgaria, and the Slovak Republic that are expected to be closed by 2008. The European Bank for Reconstruction and Development is collecting international contributions for these projects and supports predecommissioning work at Ukraine's Chernobyl plant.

Because most Asian reactors are not as far into their licensed operating lives, decommissioning there is a less urgent issue—although it will become a concern as reactors age in Japan, South

Korea, and Taiwan, all of which derive major shares of their electricity from nuclear power. However, none of these countries currently has a viable plan for long-term management of radioactive waste, so decommissioning may loom as a larger problem in coming decades. In September 2000, the Taiwanese economics minister raised concerns by arguing that Taiwan should scrap its partially constructed fourth nuclear power plant and phase out nuclear power by 2025 because it lacked a way to safely dispose of nuclear waste.

The decommissioning of a plant is a two-step process. In stage 1, the plant is shut down and all the fuel rods are removed; all of the water used as a reactor-core coolant, as a moderator, or to produce steam is drained; and the reactor and generator pipes are cleaned and flushed. The spent fuel rods, the drained water, and the material used to clean the pipes are all radioactive and must be safely stored or disposed of. This removes 99 percent of the radioactivity. Stage 2 consists of dismantling all the parts of the plant except the reactor and safely containing the reactor.

Utilities have three decommissioning options: (1) decontaminate and dismantle the plant as soon as it is shut down; (2) shut the plant down for 20 to 100 years to allow radioactive materials that have a short half-life to disintegrate and then dismantle the plant; (3) entomb the plant by covering the reactor with

reinforced concrete and placing a barrier around the plant.

Originally, entombment was thought to be the best method. But because of the long half-life of some of the radioactive material and the danger of groundwater contamination, this method now appears to be the least favorable. Most U.S. utilities plan to allow about 60 years to elapse between stage 1 and stage 2. Japan, which has greater needs for land, plans to mothball its plants for 5 to 10 years before dismantling them. The costs associated with decommissioning include dismantling the plant and then packaging, transporting, and burying the wastes. Because of contamination and activated material, ordinary dismantling methods cannot be used to demolish the buildings. Special remote-control machinery must be developed. Techniques that do not release dust into the atmosphere must be used, and workers must wear protective clothing. Waste disposal is estimated to be 40 percent of the decommissioning costs. Table 11.4 shows the volume of material that would be handled from an 1100-megawatt power plant.

Recent experience with the cost of decommissioning indicates that the cost for decommissioning a large plant will be between $200 and $400 million, about 5 percent of the cost of generating electricity. Although the mechanisms vary among countries, the money for decommissioning is generally collected over the useful life of the plant.

Table 11.4 Low-Level Radioactive Contaminants from Decommissioning an 1100-Megawatt Pressurized-Water Reactor

Material	Volume (Cubic Meters)
Radioactive	618
Activated	
Metal	484
Concrete	707
Contaminated	
Metal	5465
Concrete	10,613
Total	17,887

Source: Data from *Worldwatch Paper 69*, "Decommissioning: Nuclear Power's Missing Link," and U.S. Nuclear Regulatory Commission.

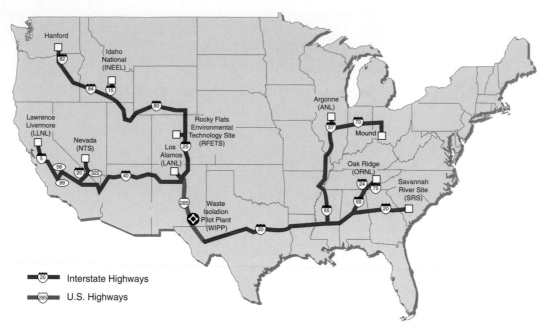

| | Interstate Highways |
| | U.S. Highways |

figure 11.17 **Department of Energy High-Level Radioactive Transuranic Waste Sites** The high-level transuranic radioactive waste generated by the U.S. Department of Energy will be shipped to the Waste Isolation Pilot Plant near Carlsbad, New Mexico, for storage.

Source: Nuclear Regulatory Commission.

Radioactive Waste Disposal

When the world entered the atomic age, the problem of the disposal of nuclear waste was not fully appreciated. Low-level radioactive waste is generated by nuclear power plants, military facilities, hospitals, and research institutions. High-level radioactive waste results from spent fuel rods, obsolete nuclear weapons, and the wastes generated by their manufacture.

High-Level Radioactive Waste

In the United States, 380,000 cubic meters of highly radioactive military waste are temporarily stored at several sites. (See figure 11.17.) A high-level waste site has been constructed near Carlsbad, New Mexico. Known as the Waste Isolation Pilot Plant, it was built to house high-level radioactive waste. The wastes are commonly referred to as **transuranic wastes,** which consist primarily of various isotopes of plutonium. This facility began accepting waste in March 1999. In addition to the high-level waste from weapons programs, 2 million cubic meters of low-level radioactive military and commercial waste are buried at various sites. In addition, about 30,000 metric tons (33,060 tons) of highly radioactive

spent fuel rods are stored in special storage ponds at nuclear reactor sites. Many plants are running out of storage space and have been authorized by the Nuclear Regulatory Commission to store radioactive waste in aboveground casks, because the storage ponds will not accommodate additional waste.

In other countries, spent fuel rods are either stored in special ponds or sent to reprocessing plants. Even though reprocessing is more expensive than manufacturing fuel rods from ore, some countries reprocess as an alternative to waste storage. The reprocessing plants in France and the United Kingdom accept domestic and foreign fuel rods, and any fuel rods fabricated by the former Soviet Union can be returned to reprocessing plants in Russia.

At this time, no country has a permanent storage solution for the disposal of high-level radioactive waste. Most experts feel the best solution is to bury it in a stable geologic formation. Several countries are investigating storage in salt deposits. The waste is mixed with silica, borax, and other chemicals, melted, and poured into metallic (stainless steel) canisters to form borosilicate glass. Each container would then be sealed in a thick-walled, stainless steel canister. Due to the high temperature of radioactive waste,

the containers would be stored above-ground for 10 years. At the end of this time, the temperature of the container would have decreased, and the material could be buried in a salt deposit 600 meters (1968 feet) below the surface.

Sweden intends to store its waste 500 meters (1640 feet) underground in granite. Site construction is not planned until 2010, when Sweden's 12 reactors are scheduled to be decommissioned. (Sweden plans to discontinue nuclear power generation of electricity, but some now doubt that this will occur.) France plans to use its reprocessing plants and "temporary" storage indefinitely.

The politics concerning the disposal of high-level radioactive waste are probably as critical as developing a suitable method. No communities want a radioactive disposal site in their area. In December 1982, the U.S. Congress passed legislation calling for a high-level radioactive disposal site for the storage of spent fuel rods from nuclear power plants to be selected by March 1987 and to be completed by 1998. In 1984, the Department of Energy stated that it was already three years behind schedule. In 2002, the secretary of Energy indicated that the choice of a site at Yucca Mountain in Nevada was based on "scientifically sound and suitable science." The

Global Perspective

The Nuclear Legacy of the Soviet Union

The tremendous political changes that have occurred in the former Soviet Union and Eastern Europe have brought to light the magnitude of nuclear contamination caused by those nations' unwise use of nuclear energy. The Soviet Union used nuclear energy for several purposes: generating power for electricity and nuclear-powered ships, testing and producing weapons, and blasting to move earth. The accident at Chernobyl was only the most recent and most public of many problems associated with the use of nuclear energy in the Soviet Union. About 13 nuclear reactors of the Chernobyl type are still operating in the former Soviet Union. These are considered unsafe by most experts, including scientists from the former Soviet Union.

The production of nuclear fuels and weapons and the reprocessing of nuclear fuels, all of which produced nuclear wastes, occurred at three major sites: Chelyabinsk, Tomsk, and Krasnoyarsk. These facilities were secret until the recent breakup of the Soviet Union. At Chelyabinsk, radioactive waste was dumped into the Techa River and Lake Karachai. Other nuclear wastes were secretly dumped into the ocean or were buried in shallow dumps. It is estimated that there are 600 secret nuclear dump sites in and around Moscow alone.

More than 100 nuclear bombs were exploded to move earth for mining purposes, to create underground storage caverns, or to "dig" canals. In addition, 467 nuclear blasts were conducted to test weapons at a site near Semipalatinsk in Kazakhstan. As a result, nuclear fallout contaminated farmland to the northeast.

The Barents Sea and the Kara Sea have been polluted by nuclear tests and the dumping of nuclear waste. At least 15 nuclear reactors from obsolete ships were dumped into the Kara Sea. Seals that live in the area have high rates of cancer.

None of these problems is unique to the former Soviet Union. Similar secrecy surrounded the early development of nuclear power in the United States and the rest of the world. What makes the situation in the former Soviet Union different is its magnitude, caused by decades of secrecy and a commitment to nuclear weapons as the primary deterrent against enemies. It can be argued that the threat posed to the Soviet Union by its former enemies contributed to the reckless development of nuclear power.

site was chosen for several reasons. It is in an unpopulated area near the Nevada Test Site, where several nuclear devices were exploded. It is a very dry area and the water table is about 600 meters (1968 feet) below the mountain. It was also considered to be geologically stable; however, the site has witnessed seismic activity since being chosen as a site.

Work has begun on the series of tunnels that would serve as storage places for the waste. (See figure 11.18.) However, current work is primarily exploratory and is seeking to characterize the likelihood of earthquake damage and the movement of water through sediments. If completed, the facility would hold about 70,000 metric tons of spent fuel rods and other highly radioactive material. It will not be completed before 2015, and by that time, the total amount of waste produced by nuclear power plants will exceed the storage capacity of the site. The volume of reprocessing waste from nuclear weapons production is larger than the volume of spent fuel from nuclear power plants, but the total activity of the spent fuel is much greater than that of the defense wastes.

A new milestone in the history of Yucca Mountain was passed in February 2002 when President Bush notified congress that he considered the Yucca Mountain site to be qualified to receive a construction permit. This resulted in the governor of Nevada, Kenny Guinn, filing an official notice of disapproval. Subsequently both the House of Representatives and the Senate voted to override Governor Guinn's disapproval and in July 2002 President Bush signed a joint resolution, which allows construction to proceed. However, this is not the final step, since some members of congress have vowed to continue to fight this designation and the state of Nevada has several lawsuits pending in federal courts.

Low-Level Radioactive Waste

Low-level radioactive waste includes the cooling water from nuclear reactors, material from decommissioned reactors, radioactive materials used in the medical field, protective clothing worn by persons working with radioactive materials, and materials from many other modern uses of radioactive isotopes. Disposal of this type of radioactive waste is very difficult to control. Estimates indicate that much of it is not disposed of properly.

In 1970, a U.S. moratorium halted the dumping of radioactive waste in the oceans. Before this, the United States placed some 90,000 barrels of radioactive waste on the ocean floor. European countries also have dumped both high-level and low-level radioactive material into the Atlantic Ocean. Before 1983, when ocean dumping was halted, these countries disposed of 90,000 metric tons of radioactive waste in the ocean.

The United States produces about 800,000 cubic meters of low-level radioactive waste per year. This is presently being buried in disposal sites in South Carolina and Washington. Temporary storage sites have also been opened in Utah and Texas. However, these states balked at accepting all the country's low-level waste. In 1980, Congress set a deadline of 1986 (later extended to 1993) for each state to provide for its own low-level radioactive waste storage site. Later, a change allowed several states to cooperate and form regional coalitions, called compacts. Under this arrangement, one state would provide a disposal site for the entire compact. Several states did not join compacts, and several others have since withdrawn from the compacts they were originally a part of. Figure 11.19 shows the current compacts and the status of

figure 11.19 **Low-Level Radioactive Waste Sites** Each state is responsible for the disposal of its low-level radioactive waste. Many states have formed compacts and selected a state to host the disposal site. None of the proposed sites is currently accepting waste. Many other states do not have an acceptable site and are temporarily relying on a site in South Carolina to accept their waste. Eventually, they will need to find other acceptable sites. The map shows the current compacts and proposed disposal sites.

Source: Nuclear Regulatory Commission.

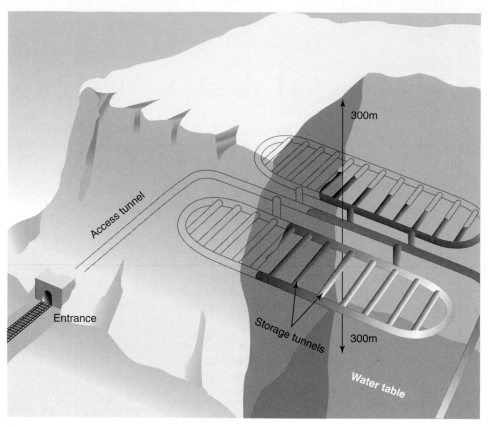

figure 11.18 **High-Level Nuclear Waste Disposal** Current plans for the disposal of high-level nuclear waste involve placing materials in tunnels underground at Yucca Mountain, Nevada. The material could be stored about 300 meters (984 feet) below the surface and about 300 meters above the water table. Because of the dry climate, it is considered unlikely that water infiltrating the soil would move nuclear material downward to the water table.

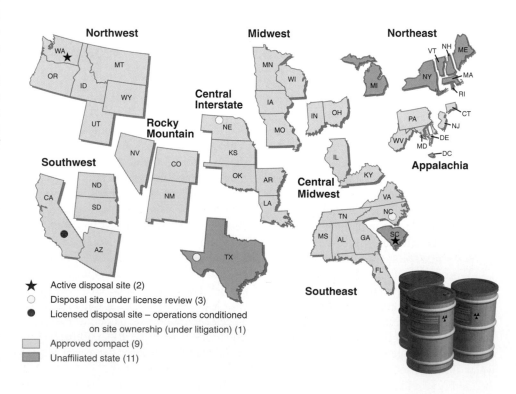

The Hanford Facility:
A Legacy of Contamination

In March 1943, the U.S. government obtained 1650 square kilometers (640 square miles) in southeastern Washington to house a large-scale plutonium production facility. The Hanford site, located on the Columbia River, offered a mild climate, isolation, and abundant water, all of which were necessary for the operation of reactors and reprocessing plants.

Plutonium production on this scale had never before been attempted. However, due to the urgent pressures to develop nuclear weapons during World War II, the normal process of using small-scale research facilities before initiating large production facilities was sidestepped. General Leslie Groves, the Army's head of Hanford's development, expressed concern regarding the safety of industrial-scale plutonium production but accepted the risks because of the urgent need of the war effort. Residents were given three weeks' notice to vacate the area, and construction began immediately.

The first nuclear reactor and chemical processing plants were completed in late 1944. Over the course of the following two decades, eight additional reactors and several processing plants were constructed.

During the height of its production, the Hanford facility was the primary source of plutonium for the United States: the world's first nuclear explosion in 1945 and the atomic bomb dropped on Nagasaki at the end of World War II each contained plutonium isolated at Hanford.

Hanford's plutonium and uranium production facility is no longer in operation. However, it has left a legacy of contamination and waste that has aroused the concern of scientists, citizens, and policy makers alike; this reservation is one of the largest and most complex environmental cleanup sites in the United States.

Nuclear waste at the Hanford reservation exists in three forms: contained liquid and sludge, buried and stored solid waste, and released liquids. Of the 2 million cubic meters (525 million gallons) of high-level waste produced at Hanford over the four decades of its operation, 202,000 cubic meters (53 million gallons) remain onsite, contained within underground tanks, buildings, or concrete basins. This comprises approximately 60 percent of all high-level waste existing in the United States today.

For the first 20 years of operation, Hanford's liquid waste was poured into single-shell carbon-steel tanks for storage. Of the 149 single-shell tanks on the reservation, 67 have cumulatively leaked over 3811 cubic meters (1 million gallons) of waste into the surrounding soil. Following the initial discoveries of these leaks, 28 double-shell tanks were constructed to house the waste. Most of the liquids present in the single-shell tanks have been pumped into the newer tanks, and no leaks have been discovered. However, the design life of the oldest double-shell tanks has been reached.

The vast majority of the remaining irradiated uranium fuel, which never underwent reprocessing, is stored in two water-filled concrete basins. Some of this fuel has corroded, and radionuclides have been detected in the basin water, walls, and sludge, as well as surrounding soil and groundwater. Approximately 11 metric tons (12 tons) of plutonium are also present onsite and are contained within three separate facilities.

Buried and stored solid waste is comprised of boxes, crates, and drums holding materials such as clothing, rags, and tools that were contaminated with chemicals or radioactivity. About 97 percent of this waste is buried in one of 75 solid waste burial grounds on the Hanford reservation, containing nearly 600 metric tons (650 tons) of uranium and 360 kilograms (800 pounds) of plutonium. In 2000, groundwater near one of the burial sites was found to contain 400 times more tritium than drinking water standards allow.

Release of waste has contributed nearly 1 million curies of radioactivity and 90,000 to 270,000 metric tons (100,000 to 300,000 tons) of chemicals to the soil and groundwater beneath Hanford. In the first months of Hanford's operation, low-level liquid waste was discharged directly onto the ground, from whence contamination leached into the soil, sediment, and groundwater. Evaporated waste also left surface residues that were dispersed by wind to surrounding wetlands.

When this contamination became evident, liquid wastes were instead pumped into wells. However, this only accelerated the process of contamination by injecting the waste closer to underground aquifers. Cribs (underground boxlike structures), backfilled trenches, vertical buried pipes, and gravel-filled fields were then constructed to receive the liquid waste. Non- or slightly contaminated liquids were released into open ponds and ditches.

Today, plumes of groundwater contaminated with metals, chemicals, and radionuclides cover 400 square kilometers (150 square miles). Intentionally discharged tank waste has contributed additional radioactive material, and sections of the underlying aquifer have been contaminated with uranium, plutonium, and other metallic isotopes.

Radionuclides were also released into the atmosphere from Hanford's reactors and reprocessing plants throughout the course of their production, exposing residents downwind of Hanford to radioactive argon and iodine. Radioactive material was also released into the Columbia River, where it now remains buried in sediment.

Today, Hanford contains 440 million curies of radioactivity, or 40 percent of all human-made radioactivity in the nuclear weapons complex. The Department of Energy, the Environmental Protection Agency, and the Washington Department of Energy signed an agreement, known as the Tri-Party Agreement, in 1989, calling for cleanup and better management of hazardous materials. However, the continued dispersal of radionuclides and sheer immensity of the project present a formidable challenge to the effective removal of radioactive material.

their disposal sites. Several of the proposed sites are under severe pressure from activists who do not want such sites in their locality. As a result, most states still do not have a permanent place to deposit low-level radioactive wastes and are relying on temporary storage facilities in Richland, Washington, or Barnwell, South Carolina. (See Environmental Close-up: The Hanford Facility: A Legacy of Contamination.) As a further example of the creative politics created by the need to have low-level radioactive waste disposal sites and the unwillingness to host such a site, the states of Texas, Maine, and Vermont have applied to be a compact with Texas serving as the host for the site.

Summary

Nuclear fission is the splitting of the nucleus of the atom. The resulting energy can be used for a variety of purposes. The splitting of U-235 in a nuclear reactor can be used to heat water to produce steam that generates electricity. Various kinds of nuclear reactors have been constructed, including boiling-water reactors, pressurized-water reactors, heavy-water reactors, gas-cooled reactors, and experimental breeder reactors. Scientists are also conducting research on the possibilities of using fusion to generate electricity. All reactors contain a core with fuel, a moderator to control the rate of the reaction, and a cooling mechanism to prevent the reactor from overheating.

The nuclear fuel cycle involves mining and enriching the original uranium ore, fabricating it into fuel rods, using the fuel in reactors, and reprocessing or storing the spent fuel rods. The fuel and wastes must also be transported. At each step in the cycle, there is danger of exposure. During the entire cycle, great care must be taken to prevent accidental releases of nuclear material.

The use of nuclear materials for military purposes has resulted in the same kinds of problems caused by nuclear power generation. The reduced risk of nuclear warfare has resulted in a need to destroy nuclear weapons and clean up the sites contaminated by their construction.

Public acceptance of nuclear power plants has been declining. Initial promises of cheap electricity have not been fulfilled because of expensive construction, cleanup, and decommissioning costs. The accidents at Three Mile Island in the United States and Chernobyl in Ukraine have accelerated concerns about the safety of nuclear power plants.

Other concerns are the unknown dangers associated with exposure to low-level radiation, the difficulty of agreeing to proper long-term storage of high-level and low-level radioactive waste, and thermal pollution.

Key Terms

alpha radiation 225
beta radiation 225
boiling-water reactor (BWR) 227
decommissioning 238
fissionable 226
gamma radiation 225
gas-cooled reactor (GCR) 228
heavy-water reactor (HWR) 227
light-water reactor (LWR) 227

liquid metal fast-breeder reactor (LMFBR) 231
moderator 226
nuclear breeder reactor 231
nuclear chain reaction 226
nuclear fission 226
nuclear fusion 231
nuclear reactor 226
plutonium-239 (Pu-239) 231

pressurized-water reactor (PWR) 227
radiation 225
radioactive 225
radioactive half-life 225
rems 237
thermal pollution 238
transuranic waste 240
uranium-235 (U-235) 226

Review Questions

1. How does a nuclear power plant generate electricity?
2. Name the steps in the nuclear fuel cycle.
3. What is a rem?
4. What is a nuclear chain reaction?
5. How will nuclear fuel supplies, the cost of decommissioning facilities, and the storage and ultimate disposal of nuclear wastes influence the nuclear power industry?
6. What happened at Chernobyl, and why did it happen?
7. Describe a boiling-water reactor, and explain how it works.
8. How is plutonium-239 produced in a breeder reactor?
9. Why is plutonium-239 considered dangerous?
10. Why is fusion not being used as a source of energy?
11. What are the major environmental problems associated with the use of nuclear power?
12. List three environmental problems associated with the construction and subsequent de-emphasis of nuclear weapons.

Critical Thinking Questions

1. Recent concerns about global warming have begun to revive the nuclear industry in the United States. Do you think nuclear power should be used instead of coal for generating electricity? Why?

2. Why has nuclear power not become the power source that is "too cheap to meter," as foreseen by President Eisenhower?

3. Disposal of radioactive wastes is a big problem for the nuclear energy industry. What are some of the things that need to be evaluated when considering nuclear waste disposal? What criteria would you use to judge whether a storage proposal is adequate or not?

4. Nuclear weapons testing has released nuclear radiation into the environment. These tests have always been justified as necessary for national security. Do you agree or not? What are the risks? What are the benefits?

5. Some states allow consumers to choose an electric supplier. Would you choose an alternative to nuclear or coal even if it cost more?

Concept Map

Construct a map to show relationships among the following concepts:

nuclear reactor

thermal pollution

decommission nuclear breeder reactor

fissionable

radioactive half-life

nuclear fission

nuclear fusion

radioactive waste disposal

breeder reactor

Interactive Exploration

Check out the website at **http://www.mhhe.com/environmentalscience** and click on the cover of this textbook for quizzing, career information, case studies, and hot links for the following topic:

<u>Nuclear Energy</u>

PART

iv

Human Influences on Ecosystems

Salton, A Sea of Controversy

Kristin B. Vessey

Bowling Green State University/Ohio

Suppose you live in San Diego: You turn on the tap because it's hot outside, you're thirsty, and you want to keep your pool filled and your lawn watered. Suppose you're a farmer in the Imperial Valley: You turn on the tap to irrigate the lettuce and strawberries you sell all over the United States. Suppose you're a bird migrating along the traditional Pacific flyway to breeding grounds in the far north: You expect hospitable stopover sites like your ancestors always found. Is there a problem? After all, California still has about 9 percent of its wetlands left, the mighty Colorado River runs along its southeastern border, and the Pacific Ocean forms its western boundary. Why worry about water?

Let's back up to 1905, when control structures on irrigation canals, which were dug in 1901 to use the Colorado for irrigating farmland in the Imperial Valley, were overwhelmed by floods. For 18 months, water from the Colorado poured into one of the lowest, driest places in North America, a desert basin that now contains this accidental lake called the Salton Sea—56 kilometers (35 miles) north of the Mexican border and 160 kilometers (130 miles) east of San Diego. It covers 987 square kilometers (381 square miles) and averages 9.5 meters (31 feet) deep; its surface is 69 meters (227 feet) below sea level. With no outlet, its water is lost only through evaporation: this sea is now 25 percent saltier than the ocean.

Both cities and farms get water from the Colorado via a network of canals. Water used for irrigation ends up in the Salton Sea, bringing along pesticides, salts, fertilizers, and nutrients. The New River enters it with industrial and human wastes from Mexicali, Mexico. Nonetheless, the Salton Sea is a state recreation area stocked for sport fishing, and visited by over a million people annually. Even more important, birds use the sea. With few other wetlands available, an estimated 4 million are present on a typical winter day. Designated a national wildlife refuge in 1930, the Salton Sea provides crucial habitat for other species, too, several of which are endangered. Its importance has grown as the Colorado River delta has shrunk due to water diversion; the river is now down to a trickle when it reaches the Gulf of Mexico.

By the 1980s, several problems became apparent. Both the lake level and salinity were rising, nutrients from agriculture supported algal blooms, and parasites and pathogens resulted in die-offs of fish and birds. Concentrations of selenium and other chemicals increased in bird eggs. Fish embryos showed developmental defects. Contaminants accumulated up the food chain. Some fish species stopped reproducing because of the high salt content. Excess nutrients caused eutrophication, leading to decreased oxygen content. In one die-off on August 4, 1999, an estimated 7.6 million fish died. In sum, the ecosystem was rapidly degrading. While the Salton Sea provides a crowded substitute for the once-extensive wetlands, for many species, it has become a death trap instead of a safe haven.

The demand for water by cities that draw from the Colorado keeps increasing. California has long used more than its share, based on an agreement among western states, but that must change: the federal government has required a 15 percent reduction in the amount withdrawn from the river by the end of 2002. Will conservation make up the deficit? The Nevada-Arizona-California region, one of the fastest growing in the country, is characterized by sprawling development. Anxiety about water is growing, too; as of January 2002, a new California law requires developers to provide detailed proof of adequate water supplies that will last at least 20 years.

Our relentless human population growth demands more resources such as water and land, reducing habitat for other species. Technological advances have let us invade drier areas and so-called marginal lands. We must eat, yet at what cost? Some environmental effects of food production are visible in the dilemma posed by the Salton Sea. Natural vegetation was destroyed, wetland storage and purification of water were lost, soil was eroded, water was diverted and became polluted, and native species were displaced.

If less water goes to irrigation so more may be diverted to the cities, the Salton Sea will shrink and get saltier, able to support less life. Some scientists believe it has only five to 25 years left unless serious restoration efforts occur. Do you think allowing it to die is a wise decision? Can we afford to try to remedy this human-caused ecological disaster? Remember that people vote; birds don't.

What Do You Think?

1. How should water, this vital but finite resource, be allotted among cities, farms, and the Salton Sea?
2. Should California's high growth rate be restricted?
3. Should farming occur where irrigation is necessary?
4. Should water transfers from water-rich areas be permitted or required?
5. How should decisions about water use be made?

This case was originally published by the National Center for Case Study Teaching in Science at the University at Buffalo. Used with permission.

12 CHAPTER

Human Impact on Resources and Ecosystems

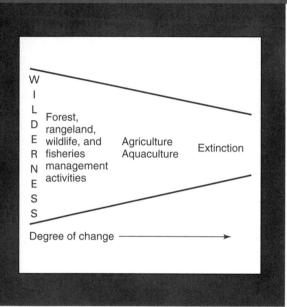

Objectives

After reading this chapter, you should be able to:

- Recognize that humans significantly modify natural ecosystems.
- Define pollution.
- Differentiate between renewable and nonrenewable resources.
- Recognize that since mineral resources are unevenly distributed, there is international trade in these commodities.
- List three types of costs associated with the use of natural resources.
- Understand that some wilderness areas still have minimal human influence.
- Appreciate the ways humans modify forests.
- Identify causes of desertification.

- Recognize that aquatic systems are modified by terrestrial changes.
- Identify changes that occur to aquatic systems as a result of human activity.
- Recognize that wildlife management focuses on specific species.
- Appreciate that waterfowl management is an international problem.
- Recognize that extinction is a natural process.
- Recognize that humans increase the rate of extinction.
- Identify ways that humans cause extinctions.
- Recognize that many extinctions can be prevented if societies are willing to preserve crucial habitats and prevent the hunting of endangered species.

Chapter Outline

The Changing Role of Human Impact

At one time, a human was just another consumer somewhere in the food chain. Humans fell prey to predators and died as a result of disease and accident just like other animals. The simple tools they used would not allow major changes in their surroundings, so these people did not have a long-term effect on their surroundings. They only minimally exploited mineral and energy resources. They mined chert, obsidian, and raw copper for the manufacture of tools and used certain other mineral materials, as well. Salt was used as a nutrient, clay for pottery, and ocher as a pigment. Aside from occasional use of surface seams of coal or surface oil seeps, early humans met most of their energy needs by muscle power or the burning of biomass.

As human populations grew, and as their tools and methods of using them became more advanced, the impact that a single human could have on his or her surroundings increased tremendously. The purposeful use of fire was one of the first events that marked the capability of humans to change ecosystems. Although in many ecosystems fires were natural events, the use of fire by humans to capture game and to clear land for gardens could destroy climax communities and return them to earlier successional stages more frequently than normal.

As technology advanced, wood was needed for fuel and building materials, land was cleared for farming, streams were dammed to provide water power, and various mineral resources were exploited to provide energy and build machines. These modifications allowed larger human populations to survive, but always at the expense of previously existing ecosystems. (See figure 12.1.)

Today, with over 6 billion people on the Earth, nearly all of the Earth's surface has been affected in some way by human activity. Even the polar ice caps show the effects of human activity. Various kinds of organic pollutants and lead residues from the burning of leaded gasoline can be identified in the layers of ice that build up from the continuous accumulation of snow in these areas. The amount of lead is currently decreasing because many countries have switched from burning leaded fuel to burning unleaded fuel in automobiles. Currently, over 80 percent of the gasoline sold in the world is unleaded.

Historical Basis of Pollution

Pollution is any addition of matter or energy that degrades the environment for humans and other organisms. However, when we think about pollution, we usually mean something that people produce in large enough quantities that it interferes with our health or well-being. Two primary factors that affect the amount of damage done by pollution are the size of the population and the development of technology that "invents" new forms of pollution.

When the human population was small and people lived in a simple manner, the wastes produced were biological and so dilute that they usually did not constitute a pollution problem. People used what was naturally available and did not manufacture many products. Humans, like any other animal, fit into their natural ecosystems. Their waste products were **biodegradable** materials that were broken down into simpler chemicals, such as water and carbon dioxide, by the action of decomposer organisms.

Human-initiated pollution became a problem when human populations became so concentrated that their waste materials could not be broken down as fast as they were produced. As the population increased, people began to congregate and establish villages, towns, and cities. The release of large amounts of smoke, biological waste, and trash faster than they could be absorbed and dispersed resulted in pollution, which led to unhealthy living conditions.

Throughout history, humans have sought to improve living conditions and eliminate the misery caused by hunger and disease. In general, we rely on science and technology to improve our quality of life. However, while technological progress can improve the quality of life, it can also generate new sources of pollution.

a.

b.

c.

figure 12.1 **Changes in the Ability of Humans to Modify Their World** As technology has advanced, the ability of people to modify their surroundings has increased significantly. (a) When humans lacked technology, they had only minor impacts on the natural world. (b) The agricultural revolution resulted in many of the suitable parts of the Earth being converted to agriculture. (c) Modern agricultural technology allows major portions of the Earth to be transformed into agricultural land.

Water pollution This sign indicates it is unsafe to swim in this area because of high levels of bacteria.

Smoke Smoke contains small particles that can cause lung problems.

Odors Feedlots create an odor problem that many people find offensive.

Visual pollution Unsightly surroundings are annoying but not hazardous to your health.

Health hazard

Smog Smog that develops when air pollution is trapped is a serious health hazard.

Solvents Solvents evaporate and cause localized pollution.

Thermal pollution Cooling towers release heat into the atmosphere and can cause localized fog.

Litter The presence of litter is unsightly but constitutes only a minor safety hazard.

Annoyance

figure 12.2 **Examples of Pollution** There are many kinds of polution. Some are major health concerns, whereas others merely annoy.

The development of the steam engine allowed machines to replace animal power and human labor but increased the amount of smoke and other pollutants in the air as well as the need for fuel. The modern chemical industry has produced many extremely valuable synthetic materials (plastics, pesticides, medicines), but it has also produced toxic pollutants.

Even identifying pollution is not always easy. To some, the smell of a little wood smoke in the air is pleasant; others do not like the odor. A business may consider advertising signs valuable and necessary; others consider them to be visual pollution. (See figure 12.2.) Even the presence of chemicals in drinking water can be difficult to classify as a clear-cut example of pollution. Toxic heavy metals such as arsenic certainly should be considered hazardous, but we do not know how much arsenic contamination is allowable before harm occurs. Are the small quantities of arsenic in groundwater great enough that, over a period of time, they will accumulate and cause human health problems? Is the arsenic a normal part of the groundwater, or is it the result of past heavy spraying of arsenic-containing pesticides on apple trees? Certainly, if the arsenic is the result of human activity and is causing health problems, it is a pollutant. If it is a natural part of the groundwater, it may still be a health hazard, but technically, many would not consider it to be a pollutant.

Renewable and Nonrenewable Resources

Modern technologies have allowed us to exploit natural resources to a much greater extent than our ancestors were able to achieve. **Natural resources** are structures and processes that humans can use for their own purposes but cannot create. If the supply of a resource is very large and the demand for it is low, the resource may be thought of as free. Sunlight, oceans, and air are often not even thought of as natural resources because their supply is so large. If a resource has always been rare or has been severely reduced by consumption, it is expensive. Pearls and precious metals are expensive because they have always been rare. In the past, land and its covering of soil was considered a limitless natural resource, but as the population grew and the demand for food, lodging, and transportation increased, we began to realize that land is a finite, nonrenewable resource. The economic value of land is highest in

figure 12.3 **Mismanagement of a Renewable Resource** Although soil is a renewable resource, extensive use can permanently damage it. Many of the world's deserts were formed or extended by unwise use of farmland. This photograph shows a once-productive farm, now abandoned to the wind and sand because the soil was mistreated and allowed to erode.

metropolitan areas, where open land is unavailable. Unplanned, unwise, or inappropriate use can result in severe damage to the land and its soil. (See figure 12.3.) The landscape is also a natural resource, as we see in countries with a combination of mountainous terrain and high rainfall that can be used to generate hydroelectric power. Rivers, forests, scenery, climates, and wildlife populations are additional examples of natural resources.

Natural resources are usually categorized as either renewable or nonrenewable. **Renewable resources** can be formed or regenerated by natural processes. Soil, vegetation, animal life, air, and water are renewable primarily because they naturally undergo processes that repair, regenerate, or cleanse them when their quality or quantity is reduced. However, just because a resource is renewable does not mean that it is inexhaustible. Overuse of renewable resources can result in their irreversible degradation. **Nonrenewable resources** are not replaced by natural processes, or the rate of replacement is so slow as to be ineffective. For example, iron ore, fossil fuels, and mountainous landscapes are nonrenewable on human timescales. Therefore, when

nonrenewable resources are used up, they are gone, and a substitute must be found or we must do without.

Costs Associated with Resource Utilization

There are always costs associated with the exploitation of any natural resource. These costs fall into three categories. The first category is the **economic cost** necessary to exploit the resource. Money is needed to lease or buy land, modify the land, construct roads, pay for labor, build equipment, and buy energy to run equipment. A second category is the **energy cost** of exploiting the resource. It takes energy to modify the landscape (build dams, construct roads, level the land) to allow for the use of most resources. It also takes energy to locate, extract, and transport materials such as minerals, forest products, or fish to processing sites. Since energy requires the expenditure of money, energy costs are ultimately converted to economic costs. When energy is inexpensive, inefficient processes may be profitable; however, when the cost of energy rises, energy-

inefficient processes will be eliminated. For example, when fuel for machines is inexpensive, it may be economically viable to transport raw materials long distances to manufacturing sites. When the price of fuel rises, distant resources will not be exploited as heavily.

A third way to look at costs is in terms of environmental effects. Air pollution, water pollution, plant and animal extinctions, and loss of scenic quality are all examples of the **environmental costs** of resource exploitation. Often environmental costs are difficult to assess, since they are not easily converted to monetary values. In addition, since they may not be recognized immediately, environmental costs are often deferred costs. For example, when dams were built on the Colorado River to provide electric power and irrigation water, planners did not anticipate that the changes in the flow of the river would reduce habitat for endangered bird species, lead to the loss of fish species because the water is colder, and result in increased salinity in lower regions of the river. Environmental costs also may include lost opportunities or values because the resource could not be used for another purpose. For example, if houses are built in a forested region, its possible use as a hiking or hunting resource is significantly reduced. Recently, in many parts of the developed world, people are recognizing the importance of environmental costs, and are converting them to economic costs as more strict controls on pollution and environmental degradation are enforced. It takes money to clean up polluted water and air or to reclaim land that has been degraded by extracting or using resources.

Mineral Resources

Mineral resources are one of the major kinds of nonrenewable resources. The distribution of these resources is not uniform throughout the world. The geological and biological processes that formed mineral resources in certain places on the Earth occurred many millions of years ago. Since that time, landmasses have been divided into political entities. Some have a greater wealth of mineral resources than others. Because no country has within its territorial limits all of the mineral resources it needs, an international exchange has developed. In particular, the industrially developed countries import many of the minerals they need from countries that have the resources but no economic ability to develop them.

Not only are mineral resources unevenly distributed, but those that are easiest to use and the least costly to extract have been exploited. As we continue to use mineral resources, they will be harder to find and more costly to develop. As with energy, North America is one of the primary consumers of the world's mineral resources. Reasonable estimates are that each year North America consumes over 30 percent of the minerals produced in the world, which is a disproportionate share, given that the combined population of the United States and Canada is about 5 percent of the world's population.

Steps in Mineral Utilization

The extraction and use of minerals involves several steps: exploration, mining, refining, transportation, and manufacturing. Each of these activities has costs (economic, energy, and environmental) associated with it. The costs associated with locating new sources of minerals (exploration) are primarily economic because exploration takes time and new technology. There are also some energy costs and some very small environmental costs. As the better sources of mineral resources are used up, it will be necessary to look for minerals in areas that are more difficult to explore, such as under the oceans. Therefore, both economic and energy costs will increase. Some areas, such as national parks and preserves, have been off-limits to mineral exploration. As current reserves of mineral resources are depleted, pressures will build to explore in these protected areas, resulting in increased environmental costs. Conversely, as a mineral becomes scarce, its price rises and there is economic incentive to extract the mineral from less-

concentrated ores at greater economic and environmental costs.

Once a mineral resource has been located and the decision made to exploit it, the resource must be taken from the Earth and concentrated before it is transported to where it will be used. These activities involve large expenditures of money for labor and the construction of machines and equipment. The energy costs of mining and refining are large because they basically involve extracting a small amount of desired material from a large amount of unwanted material. Table 12.1 shows the percentage of desired material present in several ores. As energy costs increase and the more concentrated ores are depleted, the economics of producing some materials may change so drastically that seeking substitute materials that require less energy expenditure is the best alternative.

In addition to the economic and energy costs, there are significant environmental costs. Mining and refining affect the environment in several ways. All mining operations involve the separation of the valuable mineral from the surrounding rock. The surrounding rock must then be disposed of in some way. These pieces of rock are usually piled on the surface of the Earth, where they are known as mine tailings and present an eyesore. It is also difficult to get vegetation to grow on these deposits. Some mine tailings contain materials (such as asbestos, arsenic, lead, and radioactive materials) that can be harmful to humans and other living things.

Many types of mining operations require vast quantities of water for the

extraction process. The quality of this water is degraded, so it is unsuitable for drinking, irrigation, or recreation. Since mining disturbs the natural vegetation in an area, water may carry soil particles into streams and cause erosion and siltation. Some mining operations, such as strip mining, rearrange the top layers of the soil, which lessens or eliminates its productivity for a long time. (See figure 12.4.) Strip mining has disturbed approximately 75,000 square kilometers (30,000 square miles) of U.S. land, an area equivalent to the state of Maine.

Transportation and manufacturing are the final components of the overall cost of extracting minerals from the Earth. Transportation is involved in the actual mining process, getting the ore to the refinery, moving the concentrated mineral to the site where it will be made into a finished product, and distributing the product to where it will be sold. The costs of transportation and manufacturing are primarily in the form of money and energy. Table 12.2 summarizes the steps involved in the use of minerals and the principal costs involved.

Recycling of Mineral Materials

To fully appreciate mineral resource use, we must consider the concept of recycling. Many minerals are not actually consumed or used up; they are just temporarily within a structure or a process. When the material is reclaimed and used again for another structure or process, it is recycled. (See chapter 18.) For recycling to be economic and effective,

Table 12.1 Metals in Ores

Metal	Percent Metal Needed in the Ore for Profitable Extraction	Price Range per Kilogram
Iron	30.0	
Chromium	30.0	Less than a dollar
Aluminum	20.0	
Nickel	1.5	
Tin	1.0	Several dollars
Copper	0.5	
Uranium	0.1	Tens of dollars

several conditions must be met. The material must be able to be collected relatively easily, and the material must have a high economic value.

Empty aluminum beverage cans and waste oil are no longer useful to the consumer. However, the aluminum atoms can be reprocessed into new cans or other aluminum products, and the oil can be reprocessed or burned as fuel. For recycling to be economically practical, the obsolete items must be recaptured before they have dispersed into the environment. For example, waste oil is relatively easy to recapture at the gasoline station where engine oil is replaced but difficult to recapture if the oil is dumped on the ground or into a sewer.

In many industries, the cost of purchasing recycled raw materials is higher than the cost of purchasing virgin materials; therefore, it is more costly to make products from recycled material than from virgin materials. For example, disassembling a building and reusing its parts is more costly than using new construction materials. Conversely, it is less costly to remanufacture recycled aluminum, iron, and copper into other products than to produce the metal from ores. Therefore, there is a strong incentive to recycle these metals.

One reason recycling is not more common is that, historically, the monetary costs for energy have been extremely low. As the cost of energy rises, more attention will be given to recycling minerals than to mining new ones. Manufacturing products in an environmentally harmonious way will become economically advantageous in the future.

Utilization and Modification of Terrestrial Ecosystems

Whenever humans make intensive use of natural ecosystems, the ecosystems are modified. Natural ecosystems provide many valuable goods and services, such as pollination, removal of carbon dioxide, genetic resources, medicines, foods, and other useful materials. These values are the result of the activities of the organisms that are present. **Biodiversity** is

figure 12.4 **A Strip-Mining Operation** It is easy to see the important impact a mine of this type has on the local environment. Unfortunately, many mining operations are located in areas that are also known for their scenic beauty.

Table 12.2 Costs Associated with Mineral Exploitation

Steps in Exploitation	Economic Costs	Energy Costs	Environmental Costs
Exploration	High because personnel and technology are expensive	Small	Small
Mining	High because of construction of plants and the labor to run them	High because of the need to move large amounts of material to separate valuable ores from other material	High because of changes in the landscape and large amounts of wastes produced Many mining operations use large amounts of water.
Refining	High because of the labor, equipment, and energy needed to concentrate ores	High because of the need to concentrate ores	Air and water pollution from refining facilities can be a problem.
Transportation	Needed at all stages of the process to move products from the extraction site to the manufacturing site	Energy is involved in all aspects of transportation.	Air pollution can be a problem. The construction of new transportation routes may influence scenic and other ecosystem values.
Manufacturing	Manufacturing plants and labor are needed.	Large amounts of energy are used in all manufacturing processes.	Air pollution and water pollution can be a problem, but most manufacturing sites are strictly regulated.

a measure of the variety of kinds of organisms present in an ecosystem. Natural ecosystems have much greater biodiversity than human-managed ecosystems. Therefore, when we look at intensive human use of ecosystems, we need to recognize that there will be loss of biodiversity.

Impact of Agriculture on Natural Ecosystems

About 40 percent of the world's land surface has been converted to cropland and permanent pasture. Typically, the most productive natural ecosystems (forests and grasslands) are the first to be modified by human use and the most intensely managed. Since nearly all agricultural practices involve the removal of the original vegetation and substitution of exotic domesticated crops and animals for the original inhabitants, the loss of biodiversity is significant. As the human population grows, it needs more space to grow food. The pressures to modify the environment are greatest in areas that have high population density. Often the changes brought about by intense agricultural use can degrade the ecosystem and permanently alter the biotic nature of the area. For example, much of the Mediterranean and Middle East once supported extensive forests that were converted to agriculture but now consist of dry scrubland. Today, agricultural land is being pushed to feed more people, and its wise use is essential to the health and welfare of the people of the world. Chapters 14 and 15 will take a close look at patterns of use of agricultural land.

Managing Forest Ecosystems

The history and current status of the world's forests are well known. The economic worth of the standing timber can be assessed, and the importance of forests for wildlife and watershed protection can be given a value. Originally, almost half of the United States, three-fourths of Canada, almost all of Europe, and significant portions of the rest of the world were forested. The forests were removed for fuel, for building materials, to clear land for farming, and just because they were in the way. This activity returned the forests to an earlier successional stage, which resulted in the loss of certain animal and plant species (biodiversity) that required mature forests for their habitat. In general, all continents still contain significant amounts of forested land. Most of these forest ecosystems have been extensively modified by human activity. Figure 12.5 shows that temperate parts of the world (particularly North America and Europe, which have large forested areas) are maintaining their forests, while tropical forests are being lost.

Economic and Energy Costs of Utilizing Forest Ecosystems

The major economic costs of utilizing forests involve the purchase or leasing of land, paying for equipment and labor, and building roads to allow for transportation of forest products to where they will be converted to lumber, paper, or other uses. The major energy costs are involved in harvesting activities and transportation. Therefore, efficient methods of harvest and transportation are important to reduce the economic cost of using forest resources. Many harvest methods that may have reduced environmental impact are rejected because they have higher energy costs associated with harvest and transportation.

Environmental Costs of Utilizing Forest Ecosystems

It is often difficult to put into monetary terms the environmental value of a resource. Therefore, environmentalists must rely on ethical or biological arguments to make their points. Modern forest-management practices in many parts of the world involve a compromise that allows economic exploitation while maintaining some of the environmental value of the forest. However, the very act of harvesting trees must alter the original "wilderness" character of the forest. Logging removes the trees and, therefore, the habitat for many kinds of animals that require mature stands of timber. The pine martin, grizzly bear, and cougar all require forested habitat that is relatively untouched by human activity. The removal of trees alters plant and animal biodiversity. Today in Australia, a small squirrel-sized marsupial known as a numbat is totally dependent on old forests that have termite- and ant-infested trees. Termites are numbats' major source of food, and the hollow trees and limbs caused by the insects' activities provide numbats with places to hide. Clearly, tree harvesting will have a negative impact on this species, which is already in danger of extinction. (See figure 12.6.)

In addition to serving as habitats for many species of plants and animals, forests provide many other environmental services. Forested areas modify the climate, reduce the rate of water runoff, protect soil from erosion, and provide recreational opportunities. Because trees transpire large quantities of water and shade the soil, their removal often leads to a hotter, dryer climate. Trees and other plants hold water on their surfaces, thus reducing the rate of runoff. Slowing runoff also allows more water to sink into the soil and recharge groundwater resources. Therefore, removal of trees results in more rapid runoff, and flooding and soil erosion are more common. Soil particles can wash into streams, where they cause siltation. The loss of soil particles reduces the soil's fertility. The particles that enter streams may cover spawning sites and eliminate fish populations. If the trees along the stream are removed, the water in the stream will be warmed by increased exposure to sunlight. This may also negatively affect the fish population.

The process of cutting the trees and transporting the logs also disturbs the wildlife in the area. The roads necessary for moving equipment and removing logs are a special problem. Constant travel over these roads removes vegetation and exposes the bare soil to more rapid erosion. When the roads are not properly located and constructed, they eventually

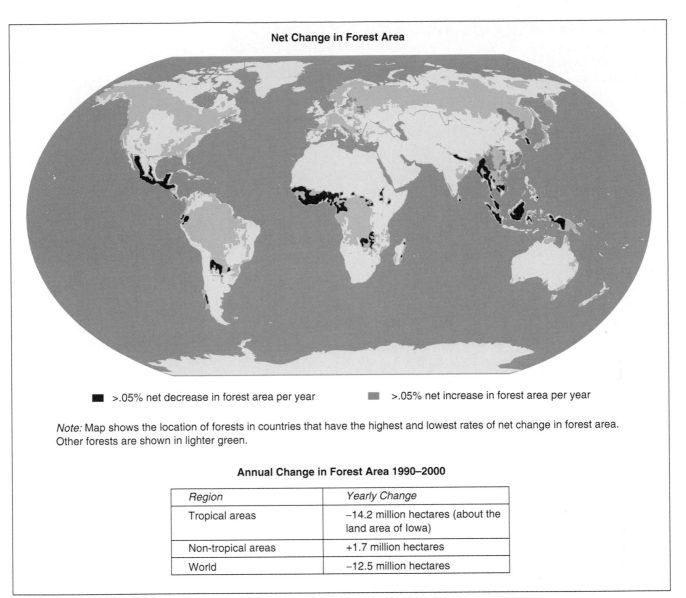

Net Change in Forest Area

■ >.05% net decrease in forest area per year   ■ >.05% net increase in forest area per year

Note: Map shows the location of forests in countries that have the highest and lowest rates of net change in forest area. Other forests are shown in lighter green.

Annual Change in Forest Area 1990–2000

Region	Yearly Change
Tropical areas	−14.2 million hectares (about the land area of Iowa)
Non-tropical areas	+1.7 million hectares
World	−12.5 million hectares

figure 12.5 **Changes in Forest Area** Areas shown in red are losing forests at a rate of greater than 0.5 percent per year. Areas shown in dark green are increasing at a rate of greater than 0.5 percent per year. It is obvious that most tropical forests in Central America, Africa, and much of Asia are being lost, while nontropical areas are holding their own or growing. The table shows actual land area affected.

Source: Food and Agriculture Organization of the United Nations.

become gullies and serve as channels for the flow of water. Most of these environmental problems can be minimized by using properly engineered roads and appropriate harvesting methods.

In many parts of the world, the construction of logging roads increases access to the forest and results in colonization by peasant "squatters" who seek to clear the forest for agriculture. The roads also permit poachers to have greater access to wildlife in the forest. Finally, the "wilderness" nature of the area is destroyed, which results in a loss

of value for many who like to visit mature forests for recreation. An area that has recently been logged is not very scenic, and the roads and other changes often irreversibly alter the area's wilderness character. Obviously, wilderness and logging cannot coexist. Therefore, it becomes necessary to designate specific forests as wilderness or harvestable resources. In 2000, President Bill Clinton officially recognized the role of roads in altering the wilderness nature of forests by issuing an executive order that no new roads will be constructed in feder-

ally held forests. Although the Bush administration initially indicated support for the concept, in December 2001, it issued directives that put implementation of the order on hold for 18 months.

Environmental Implications of Various Harvesting Methods

One of the most controversial logging practices is **clear-cutting.** (See figure 12.7.) As the name implies, all of the

figure 12.6 **A Specialized Marsupial—the Numbat** This small marsupial mammal requires termite- and ant-infested trees for its survival. Termites serve as food, and the hollow limbs and logs provide hiding places. Loss of old-growth forests with diseased trees will lead to the numbat's extinction.

figure 12.7 **Extensive Clear-cutting** Large, clear-cut plots of land can lead to a loss of species and accelerated soil erosion.

trees in a large area are removed. This is a very economical method of harvesting, but it exposes the soil to significant erosive forces. If large blocks of land are cut at one time, it may slow reestablishment of forest and have significant effects on wildlife. On some sites with gentle slopes, clear-cutting is a reasonable method of harvesting trees, and environmental damage is limited. This is especially true if a border of undisturbed forest is left along the banks of any streams in the area. The roots of the trees help to stabilize the stream banks and retard siltation. Also, the shade provided by the trees helps to prevent warming of the water, which might be detrimental to some fish species.

Clear-cutting can be very destructive on sites with steep slopes or where regrowth is slow. Under these circumstances, it may be possible to use **patchwork clear-cutting.** With this method, smaller areas are clear-cut among patches of untouched forest. This reduces many of the problems associated with clear-cutting and can also improve conditions for species of game animals that flourish in successional

forests but not in mature forests. For example, deer, grouse, and rabbits benefit from a mixture of mature forest and early-stage successional forest.

Clear-cut sites where natural reseeding or regrowth is slow may need to be replanted with trees, a process called **reforestation.** Reforestation is especially important for many of the conifer species, which often require bare soil to become established. Many of the deciduous trees will resprout from stumps or grow quickly from the seeds that litter the forest floor, so reforestation is not as important in deciduous forests.

Selective harvesting of some species of trees is also possible but is not as efficient or as economical as other methods, from the point of view of the harvesters. It allows them, however, to take individual, mature, high-value trees without completely disrupting the forest ecosystem. In many tropical forests, high-value trees, such as mahogany, are often harvested selectively. However, there may still be extensive damage to the forest by the construction of roads and to noncommercial trees by the felling of the selected species.

Plantation Forestry

Many forest products companies manage forest plantations in the same way farmers manage crops. They plant single-species, even-aged forests of fast-growing hybrid trees that have been developed in the same way as high-yielding agricultural crops. Fertilizer is applied if needed, and weeds and pests are controlled. Often controlled fires and aerial application of pesticides are used to control competing species and pests. Such forests have low species diversity and are not as valuable for wildlife and other uses as are more natural, mixed-species forests. Furthermore, the trees planted in many managed forests may be exotic species. *Eucalyptus* trees from Australia have been planted in South America, Africa, and other parts of the world; and most of the forests in northern England and Scotland have a mixture of native pines and imported species from the European mainland and North America. In these intensively managed forests, some single-species plantations mature to harvestable size in 20 years, rather than in the approximately 100

The Northern Spotted Owl

In June 1990, the northern spotted owl (*Strix occidentalis caurina*) was listed as a threatened species under the Endangered Species Act. Threatened species are those that are likely to become endangered if current conditions do not change. The northern spotted owl lives in old-growth coniferous forests in the Pacific Northwest. A primary reason for the listing is that the owl requires mature forests as its habitat. Several characteristics of the owl's biology make old-growth forests important. It prefers a relatively closed canopy with rather open spaces under the canopy. These conditions allow it to fly freely as it searches for food. One of its primary food sources is the northern flying squirrel, which feeds primarily on fungi that are common in old-growth forests that have many dead trees and much down timber on the forest floor. Dead and diseased trees are also important because they provide cavities and platforms for the owls to use as nesting sites. As with most carnivores at the top of the food chain, the northern spotted owl requires large areas of forest for hunting. Therefore, relatively large areas of forest are needed for nesting pairs to be successful.

Logging in the Pacific Northwest has already used up most of the trees on private land. Most of the remaining old-growth forest, which is about 10 percent of its original area, is on land managed by the U.S. Forest Service. Many of the trees in these forests are several centuries old, and it takes 150 to 200 years for regenerating forests to become suitable habitat for the northern spotted owl. This timescale would not allow for the continued existence of the northern spotted owl if major sections of the remaining old-growth forest were logged. The reduction in logging has had significant economic impact on companies that relied on harvesting timber from federally owned lands, and as the companies are hurt economically, so too are the communities where these companies function.

Listing the owl as threatened required that the U.S. Forest Service develop and implement a plan to protect it. Since the owl is found only in mature old-growth forests and since it requires large areas for hunting, the plan set aside large tracts of forested land and protected them from logging. The plan to protect the owl resulted in several actions. Most federally owned lands that

were suitable habitat for the owl have indeed been protected from logging, although some salvage logging has been permitted when forests are damaged by fire or other events. Furthermore, private land that is known to have nesting owls was protected as well. As a result of this, individuals and companies have filed lawsuits claiming that since they cannot cut the trees on land they own, they have been deprived of the use of their land. Therefore, they have asked for compensation from the government. One lumber company was awarded $1.8 million in compensation for 22.7 hectares (56 acres) of forest that was "taken" for use as owl habitat. Court cases continue in the area. In 2000, a federal district judge ruled that cutting down trees does not necessarily harm northern spotted owls and that a logging company could proceed to cut a stand of timber. He ruled that to prevent logging of privately held land, the federal government must prove that cutting the trees would harm the owl. In 2001, a coalition of environmental groups filed a lawsuit charging that the U.S. Fish and Wildlife Service had not enforced the Endangered Species Act and had not adequately protected the habitat of the northern spotted owl. It is clear that, at this point, the current arguments about the northern spotted owl are legal maneuvers rather than serious discussions about the protection of a threatened species.

years typical for naturally reproducing mixed forests. However, the quality of the lumber products is reduced. In many of these forests, clear-cutting is the typical method of harvest, and the cut-over area is immediately replanted.

A single-species, even-aged forest is ideal for producing wood but does not support as wide a variety of wildlife as does a mixed-age, mixed-species forest. However, with proper planning, trees can be harvested in small enough patches to encourage some wildlife but in large enough sections to be economi-

cal. Since some species require old trees as part of their niche, some older trees or patches of old-stand timber can be left as refuges for such species. For example, in the southern pine forests of the United States, the endangered red-cockaded woodpecker requires old, diseased pine trees in which to build its nests. (See figure 12.8.) A well-managed forest plantation does not provide these sites, but forest products companies in the southeastern United States have agreed to modify forestry practices on their lands to allow for the retention of

some old, diseased trees needed by the woodpeckers. This is important because not enough suitable woodpecker habitat exists on state and federally owned lands to protect the woodpecker from extinction.

Special Concerns About Tropical Deforestation

Tropical forests have greater species diversity than any other terrestrial ecosystem. The diverse mixture of tree species requires harvesting techniques different

figure 12.8 **Habitat Needs of the Red-Cockaded Woodpecker** As the place for its nest, the red-cockaded woodpecker requires old living pines that have a disease known as red heart. Old diseased trees are rare in intensely managed forests. Therefore, the birds are endangered.

from those traditionally used in northern temperate forests. In addition, because tropical soils have low fertility and are highly erodible, tropical forests are not as likely to regenerate after logging as are temperate forests. Currently, few of these forests are being managed for long-term productivity; they are being harvested on a short-term economic basis only, as if they were nonrenewable resources. Worldwide tropical forests are being lost at a rate of about 0.6 percent per year.

Several concerns are raised by tropical deforestation. First, the deforestation of large tracts of tropical forest is significantly reducing the species diversity of the world. Second, because tropical forests very effectively trap rainfall and prevent rapid runoff and the large amount of water transpired from the leaves of trees tends to increase the humidity of the air, the destruction of these forests also can significantly alter climate, generally resulting in a hotter, more arid climate. Third, high rainfall coupled with the nature of the soils results in the deforested lands being easily eroded. Finally, people have become

concerned about preserving the potential of forests to trap carbon dioxide. As they carry on photosynthesis, trees trap large amounts of carbon dioxide. This may help to prevent increased carbon dioxide levels that contribute to global warming. (See chapter 17 for a discussion of global warming and climate change.)

Another complicating factor is that human population is growing rapidly in tropical regions of the world. More people need more food, which means that forestland will be converted to agriculture and the value of forest for timber, fuel, watershed protection, wildlife habitat, biodiversity, and carbon dioxide storage will be lost.

Managing Rangeland Ecosystems

Rangelands consist of the many arid and semiarid lands of the world that support grasses or a mixture of grasses and drought-resistant shrubs. These lands are too dry to support crops but are often used to raise low-density populations of domesticated or semidomesticated animals. In some cases, the animals are maintained on permanent open ranges, while in others, nomadic herds are moved from place to place in search of suitable grazing. Usually, the animals are introduced species not native to the region. Sheep, cattle, and goats that are native to Europe and Asia have been introduced into the Americas, Australia, New Zealand, and many areas of Africa.

Few of these lands are privately owned, so they may be used free, or a small fee may be payable to the government. Consequently, there are few economic costs of using these lands. Furthermore, the low-intensity use of the lands does not require much energy expenditure. The primary costs of using these lands are environmental.

Environmental Costs of Utilizing Rangelands

The grazing of domesticated animals on rangelands has major impacts on biodi-

versity. In an effort to increase the productivity of rangelands, management techniques may specifically eliminate certain species of plants that are poisonous or not useful as food for the grazing animals, or specific grasses may be planted that are not native to the area. In some cases, native animals are reduced if they are a threat to livestock because they are predators or because they may spread disease to the livestock. In addition, the selective eating habits of livestock tend to reduce certain species of native plants and encourage others.

Because rainfall is low and often unpredictable, it is important to regulate the number of livestock on the range. In areas where the animals are on permanent pastures, the numbers of animals can be adjusted to meet the capacity of the range to provide forage. In many parts of the world, nomadic herders simply move their animals from areas where forage is poor to areas that have better forage. This is often a seasonal activity that involves movement of animals to higher elevations in the summer or to areas where rain has fallen recently. (See figure 12.9.)

In many parts of the world where human population pressures are great, overgrazing is a severe problem. As populations increase, desperate people attempt to graze too many animals on the land. They also cut down the trees for firewood. If overgrazed, many plants die, and the loss of plant cover subjects the soil to wind erosion, resulting in a loss of fertility, which further reduces the land's ability to support vegetation. The cutting of trees for firewood has a similar effect, but it is especially damaging because many of these trees are legumes, which are important in nitrogen fixation. Their removal further reduces soil fertility. This severe overuse of the land results in conversion of the land to a more desertlike ecosystem. This process of converting arid and semiarid land to desert because of improper use by humans is called **desertification.** Desertification can be found throughout the world but is particularly prevalent in northern Africa and parts of Asia, where rainfall is irregular and unpredictable, and where many people are subsistence farmers or nomadic herders who are under considerable pressure to provide food for their families. (See figure 12.10.)

Areas with Minimal Human Impact— Wilderness and Remote Areas

There are still many areas of the world that have had minimal human impact. Some of these are remote areas with harsh environmental conditions, such as the continent of Antarctica, northern Arctic areas, the tops of some tall mountains, or extremely arid areas, such as desert areas of Africa, central Australia, and central Asia. Most of these have been explored and found to be too harsh to sustain agriculture; therefore, dense human habitation is impossible.

Many other areas of the world, such as tropical rainforests, currently support ecosystems that have small human populations but are under threat because these ecosystems are capable of supporting agriculture or other human uses. The primary factor that will determine the survival of these natural areas is population pressure as people seek additional land for farming.

Many countries have established parks and other special designations of land use to protect areas of natural beauty or communities of organisms thought worthy of protection. This has been most noticeable in Africa, Central and South America, North America, and Australia. Until recently, these continents had large amounts of land that were relatively untouched. Parks and protected lands allow a variety of uses, depending on the country and its laws. Some are used primarily as tourist attractions, which generate funds for the country. Others are set up primarily to protect a species or community of organisms, while others simply seek to restrict use so that certain environmental standards are met. For example, harvesting in a forest might be restricted to a certain number of trees

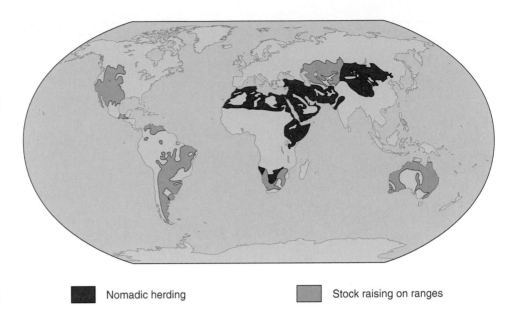

■ Nomadic herding   ▨ Stock raising on ranges

figure 12.9   **Use of Rangelands** The arid and semiarid regions of the world will not support farming without irrigation. In many of these areas, livestock can be raised. Permanent ranges occur where rainfall is low but regular. Nomadic herders can utilize areas that have irregular, sparse rainfall.

each year, and the number of people who visit a particular scenic site might be limited. A particularly sensitive issue is the designation of certain areas as wilderness. Although definitions of wilderness vary from country to country, the U.S. Congress, in the Wilderness Act of 1964, defined **wilderness** as "an area where the Earth and its community of life are untrammeled by man, where man himself is a visitor who does not remain."

Recently, there has been intense pressure on members of Congress from oil-producing states such as Alaska to allow oil exploration in areas currently designated as wilderness, such as the Arctic National Wildlife Refuge. (See Issues—Analysis: The Arctic National Wildlife Refuge and Oil in chapter 10.)

As population increases and resources become more scarce, this pressure will become greater throughout the world.

The oceans of the world represent a major area that has been modified very little by human activities. Some areas, particularly those near the shore, receive intense use, but many areas of the open ocean are poorly understood and little affected by humans.

Managing Aquatic Ecosystems

Aquatic ecosystems are divided into three broad categories: freshwater,

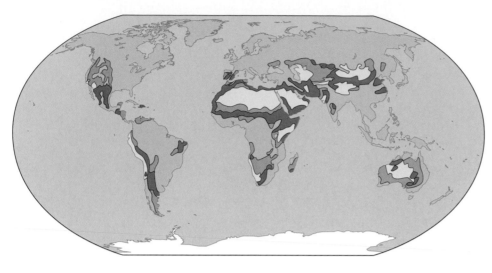

Degree and Risk of Desertification

■ High   ■ Moderate   □ Existing desert

figure 12.10 **Desertification** Arid and semiarid areas can be converted to deserts by overgrazing or unsuccessful farming practices. The loss of vegetation increases erosion by wind and water, increases the evaporation rate, and reduces the amount of water that infiltrates the soil. All of these conditions encourage the development of desertlike areas.

Sources: *United Nations Map of World Desertification,* United Nations Food and Agriculture Organization, United Nations Educational, Scientific, and Cultural Organization, and the World Meteorological Organization for the United Nations Conference on Desertification, 1977, Nairobi, Kenya.

brackish, and saltwater (marine). Freshwater has very low salt concentrations (less than 0.5 grams per liter), brackish water contains moderate amounts of salt (0.5–30 grams per liter), and saltwater has significant amounts of salt (30–40 grams per liter). The primary renewable resources obtained from these ecosystems are fish, shellfish, and crustaceans (lobsters, shrimp). Each of the three kinds of aquatic ecosystems presents a distinct set of management issues. These are typically related to ownership, harvesting methods, and allowable harvest quotas. This section will deal primarily with the food production ability of aquatic ecosystems. Other issues related to water resources will be covered in chapter 16.

The economic and energy costs of utilizing aquatic ecosystems for food production are related to the techniques used to search for and capture aquatic organisms. Since the oceans and most freshwater systems are not owned by individuals, the cost of acquiring the right to fish is usually small. Although many countries may exercise control over specific areas along their coasts, they generally do not charge fees for the right to fish. Fishers may be required to purchase licenses, but their cost is small compared to the cost of boats, equipment, and the energy needed to search for and capture fish and other aquatic organisms.

Environmental Costs Associated with Utilizing Marine Ecosystems

Environmental costs related to utilizing marine ecosystems fall into two broad categories: overfishing and the damaging effect of harvesting practices.

The United Nations estimates that 70 percent of the world's marine fisheries are being overexploited or are being fully exploited and are in danger of overexploitation as the number of fishers increases. To preserve the fisheries, total fishing capacity should be reduced by about 30 percent. Figure 12.11 shows both the amount of fish captured and the amount produced by aquaculture. While

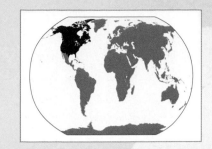

Global Perspective

The History of the Bison

The original ecosystem in the central portions of North America was a prairie dominated by a few species of grasses. The eastern prairie, where moisture was greater, had grasses up to 2 meters (6 feet) tall, while the dryer western grasslands were populated with shorter grasses. Many kinds of animals lived in this area, including prairie dogs, grasshoppers, many kinds of birds, and bison. The bison was the dominant organism. Millions of these animals roamed the prairies of North America with few predators other than the Native Americans, who used the bison for food, their hides for shelter, and their horns for tools and ornaments. The relationship between the bison and Native Americans was a predator-prey relationship in which the humans did not significantly reduce bison numbers.

When Europeans came to North America, they changed this relationship drastically. European-born Americans sought to convert the prairie to agriculture and ranching. However, two things stood in their way: the Native Americans, who resented the intrusion of the "white man" into their territory, and the bison. Since many of the Native American tribes had horses and a history of warlike encounters with other tribes, they attempted to protect their land from this intrusion. It was impossible to use the land for agriculture or ranching with the millions of bison occupying so much area because the groups of migrating bison would damage farmers' fields and compete with livestock for grasses. To facilitate the settling of the West, the U.S. government established a policy of controlling the bison and the Native Americans: Since bison were the primary food source of Native

Americans in many areas, eliminating them would result in the starvation of many Native Americans, which would eliminate them as a problem for the frontier settler. In 1874, the secretary of the Interior stated that "the civilization of the Indian was impossible while the buffalo remained on the plains." Another example of this kind of thinking was expressed by Colonel Dodge, who was quoted as saying, "Kill every buffalo you can; every buffalo dead is an Indian gone." Bison were killed by the millions. Often, only their hides and tongues were taken; the rest of the animal was left to rot. Years later, the bones from these animals were collected and ground up to provide fertilizer. By 1888, the bison was virtually eliminated.

A few bison were left in the Canadian wild and in remote mountain areas of the United States, while others survived in small captive herds. Eventually, in the early 1900s, the U.S. government established the national Bison Range near Missoula, Montana. The Canadian government established a bison reserve in Alberta. These animals have proven to be useful, since many people now desire meat with a lower fat content, which bison have. Crossbreeding cattle with bison has led to a new breed of cattle, with reduced fat content, known as a beefalo. The animal that was barely saved from extinction by a few thoughtful individuals may contribute to better health for all.

the amount of fish captured by fishers has remained essentially constant since 1989, the amount produced by fish farming has increased. It is important to note that the increase in the amount of fish captured is the result of more effective capture methods and increased fishing effort. Since most areas of the oceans are being exploited to their capacity, increased production of marine fishes should not be anticipated.

Another indication that fishery resources are being overexploited is the change in the kinds of fish being caught. Figure 12.12 shows the change in species composition for both bottom-dwelling species and surface-dwelling open-ocean fish. The commercial fishing industry has been attempting to market fish species that previously were re-

garded as unacceptable to the consumer. These activities are the result of reduced catches of desired species. Examples of "newly discovered" fish in this category are monkfish and orange roughy.

For several reasons, the most productive areas of the ocean are those close to land. In shallow water, the entire depth (water column) is exposed to sunlight. Plants and algae can carry on photosynthesis, and biological productivity is high. The nutrients washed from the land also tend to make these waters more fertile. Furthermore, land masses modify currents that bring nutrients up from the ocean bottom. Many of the commercially important fish and other seafood species are bottom dwellers, but fishing for them at great depths is not practical. Therefore, fishing pressure is

concentrated on areas where the water is shallow and relatively nutrient rich. Countries have taken control over the waters near them by establishing a 200-nautical-mile limit (approximately 300 kilometers) within which they control the fisheries. This has not solved conflicts, however, as neighboring countries dispute the fishing practices used in waters they both claim. Canada and the United States have had continuing arguments over the management of the Pacific salmon fishery and argued as well over the North Atlantic cod fishery until its collapse in the mid 1990s.

Open-ocean species of fish are also subject to overexploitation. During the 1960s, anchovy fishing off the coast of Peru was a major industry. From 1971 to 1972, the catch dropped dramatically.

World Capture Fisheries and Aquaculture Production

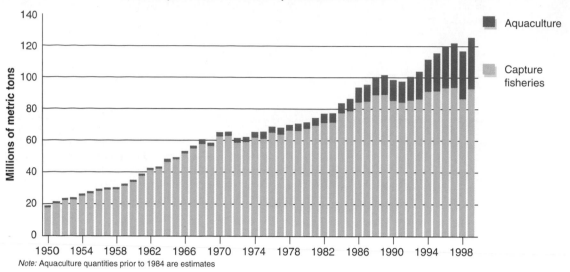

Note: Aquaculture quantities prior to 1984 are estimates

figure 12.11 **Trends in World Fish Production** The amount of fish captured increased steadily until about 1989. Since then, the amount has remained essentially constant. This indicates that the world fisheries are being exploited to their capacity. Aquaculture has continued to increase so the total production continues to increase. These data have been called into question by many because it is thought that China has grossly overreported its fish production statistics. This would mean that the capture fisheries may actually have been declining in recent years.

Source: Food and Agricultural Organization of the United Nations.

Overfishing was believed to be one of the major contributing factors. This was aggravated by an increase in the area's water temperature, which prevented nutrient-rich layers from rising to the euphotic zone. Thus, productivity at all trophic levels decreased. Currently, the Peruvian anchovy fishery has rebounded but is considered to be overexploited and not sustainable. (See figure 12.12b.)

One of the major problems associated with the management of marine fisheries resources is the difficulty in achieving agreement on limits to the harvest. The oceans do not belong to any country; therefore, each country feels it has the right to exploit the resource wherever it wants. Since each country seeks to exploit the resource for its advantage without regard for the sustainable use of the resource, international agreements are required to manage the resource. Such agreements are difficult to achieve because of political differences and disregard for the nature of the resource. (See The Tragedy of the Commons, page 59.)

In recent years, several countries have designated areas within their territorial waters where fishing is not al-lowed. Several studies suggest that these reserves are effective ways to protect a portion of the fishery resource and serve to repopulate surrounding areas that are open to fishing. This approach is particularly effective for fish species that live on the bottom or are associated with specific structures on the bottom.

Another environmental problem associated with these shallow-water, near-shore fisheries is the method used to harvest the fish. The bottom-dwelling fish are generally harvested with trawls—nets that are dragged along the bottom. These nets capture many different species, many of which are not commercially valuable. Typically, 25 percent of the catch consists of species that have no commercial value. These are discarded by being thrown overboard. However, they are usually dead, and their removal further alters the ecological nature of the seafloor. In addition, the trawls disturb the seafloor and create conditions that make it more difficult for the fish populations to recover. Some people have even advocated that the trawl should be banned as a fishing technique because of the damage done to the ocean bottom.

Environmental Costs Associated with Utilizing Freshwater Ecosystems

Because freshwater ecosystems are small and more intimately associated with humanity, few have not been considerably altered by human activity. Changes in water quality and the introduction of exotic species are two primary alterations. The warming of water due to thermal pollution and the addition of nutrients have made it impossible for some native species to survive. The effects of water pollution are discussed in chapter 16.

Although people in many places in the world harvest freshwater fish for food, the quantities are relatively small compared to those produced by marine fisheries. Furthermore, the management of the freshwater fish resource is much more intense since the bodies of water are smaller and human populations have greater access to the resource. Freshwater fish management in much of the world includes managing for both recreational and commercial food purposes. The fisheries resource manager must try to satisfy two interest groups. Both

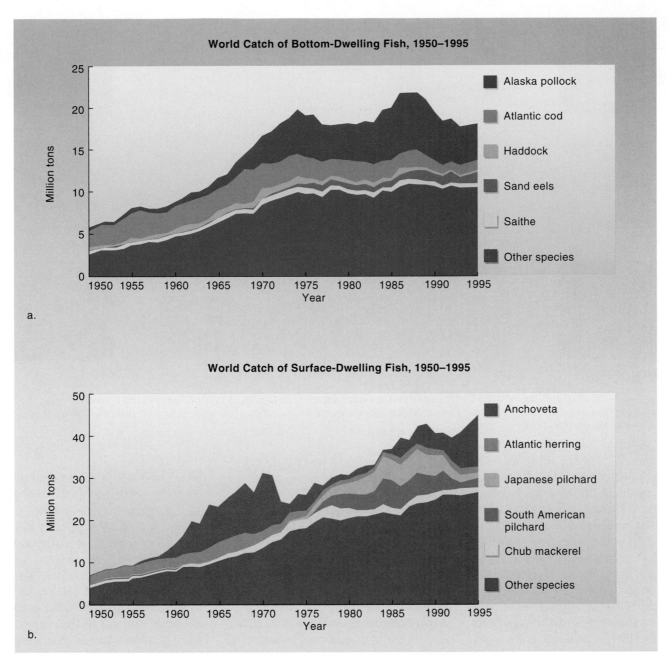

figure 12.12 Changes in World Marine Fish Harvests 1950–1995 Increased fishing pressure has resulted in the significant reduction in many of the traditional fishing stocks. As traditional species have been overfished, other, less desirable species have been replacing them. This is true for both the bottom-dwelling species (*a*) and those that live in the open ocean (*b*).

Source: Food and Agriculture Organization of the United Nations.

sports fishers and those who harvest for commercial purposes must adhere to regulations. However, the regulations usually allow the commercial fisher to use different harvesting methods, such as nets. In much of North America, freshwater fisheries are primarily managed for sport fishing. The management of freshwater fish populations is similar to that of other wild animals. Fish require

cover, such as logs, stumps, rocks, and weed beds, so they can escape from predators. They also need special areas for spawning and raising young. These might be a gravel bed in a stream for salmon, a sandy area in a lake for bluegills or bass, or a marshy area for pike. A freshwater fisheries biologist tries to manipulate some of the features of the habitat to enhance them for the de-

sired species of game fish. This might take the form of providing artificial spawning areas or cover. Regulation of the fishing season so that the fish have an opportunity to breed is also important.

In addition to these basic concerns, the fisheries biologist pays special attention to water quality. Whenever people use water or disturb land near the water, water quality is affected. For example,

Lake trout

Brown trout

Native	Introduced
Brook lamprey	Sea lamprey
Lake trout	Brown trout
Brook trout	Rainbow trout
Whitefish	Pink salmon
Herring	Coho salmon
Ciscoes	Chinook salmon
	Atlantic salmon
Suckers	
Chubs	Carp
Shiners	
Catfish	
Bullheads	
	Smelt
	Alewife
White bass	White perch
Smallmouth bass	Crappies
Rock bass	Sunfish
Yellow perch	
Walleye	

Yellow perch

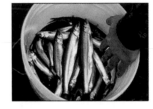
Smelt

figure 12.13 **Native and Introduced Fish Species in the Great Lakes** The Great Lakes have been altered considerably by the introduction of many non-native fish species. Some were introduced accidentally (lamprey, alewife, and carp), and others were introduced on purpose (salmon, brown trout, rainbow trout).

toxic substances kill fish directly, and organic matter in the water may reduce the oxygen. The use of water by industry or the removal of trees lining a stream warms the water and makes it unsuitable for certain species. Poor watershed management results in siltation, which covers spawning areas, clogs the gills of young fish, and changes the bottom so food organisms cannot live there. The fisheries biologist is probably as concerned about what happens outside the lake or stream as what happens within it.

The introduction of exotic fish species has greatly affected naturally occurring freshwater ecosystems. The Great Lakes, for example, have been altered considerably by the accidental and purposeful introduction of fish species. The sea lamprey, smelt, carp, alewife, brown trout, and several species of salmon are all new to this ecosystem. (See figure 12.13.) The sea lamprey is parasitic on lake trout and other species and nearly eliminated the native lake trout population. Controlling the lamprey problem requires use of a very spe-

cific larvicide that kills the immature lamprey in the streams. This technique works because mature lamprey migrate upstream to spawn, and the larvae spend several years in the stream before migrating downstream into the lake. With the partial control of lamprey, lake trout populations have been increasing. (See figure 12.14.) Recovery of the lake trout is particularly desirable, since at one time it was an important commercial species.

Another accidental introduction to the Great Lakes, the alewife, a small fish of little commercial or sport value, became a problem during the 1960s, when alewife populations were so great that they died in large numbers and littered beaches. Various species of salmon were introduced about this time in an attempt to control the alewife and replace the lake trout population, which had been depressed by the lamprey. While this salmon introduction has been an economic success and has generated millions of dollars for the sport fishing industry, it may have had a negative impact on some native fish such as the lake

trout, with which the salmon probably compete. Salmon also migrate upstream, where they disrupt the spawning of native fish. Most species of salmon die after spawning, which causes a local odor problem.

Other more recent arrivals in the Great Lakes include a clam, the zebra mussel, and two fishes, the round gobi and river ruff. All of these probably arrived in ballast water from fresh or brackish water bodies in Europe. These exotic introductions are increasing rapidly and changing the mix of species present in Great Lakes' waters. (See Environmental Close-up: Population Growth of Invading Species, page 142.)

Streams and rivers are modified for navigation, irrigation, flood control, or power production purposes, all of which may alter the natural ecosystem and change the numbers or kinds of fish present. These topics are discussed in greater detail in chapter 16. In the Pacific Northwest, the extensive development of dams to provide power and aid navigation has made it nearly impossible for adult salmon to migrate upstream

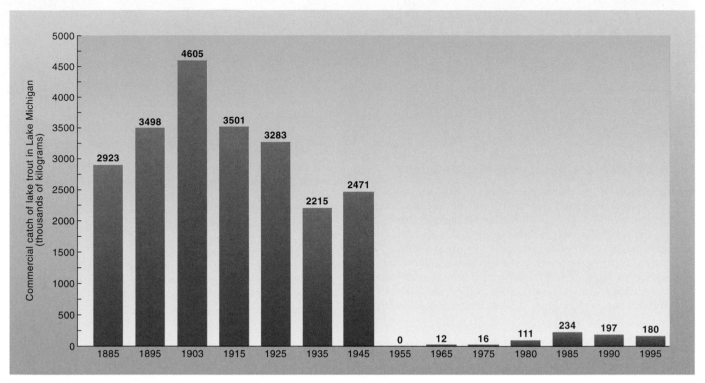

figure 12.14 **The Impact of the Lamprey on Commercial Fishing** The lamprey entered the Great Lakes in 1932. Because it is an external parasite on lake trout, it had a drastic effect on the population of lake trout in the Great Lakes. As a result of programs to prevent the lamprey from reproducing, the number of lamprey has been reduced somewhat, and the lake trout population is recovering with the aid of stocking programs.

to spawn and difficult for young fish to migrate downstream to the ocean. Fish ladders and other techniques have not been successful in allowing the fish to pass the dams. As a result, many populations of Pacific salmon are nearly extinct. The only solution to the problem is to remove or greatly modify several of the dams.

Aquaculture

Fish farming (aquaculture) is becoming increasingly important as a source of fish production. Salmon farming has been particularly successful. This involves raising the various species of salmon in "pens" in the ocean, which allows the introduction of food and other management techniques to achieve rapid growth of the fish. Norway, Chile, Canada, and Scotland are the leading countries in salmon production. The production of salmon from fish farms has increased rapidly. In 1988, less than 20 percent of the salmon sold were from fish farms, compared to 50 percent to-

day. During the same period of time, the production of wild-caught salmon has been relatively constant. However, the farming method of producing fish is not without its environmental effects. Raising fish in such concentrated settings results in increased nutrients in the surrounding water from uneaten food and the wastes the fish release. This can cause local algal blooms that negatively affect other fisheries, such as shellfish. Many of the salmon species raised in farms are not native to the waters in which they are raised. Inevitably, some of these fish escape. The introduction of exotic species can have a negative impact on native species. Some people even worry about the genetic stocks of fish that are raised. Many wild salmon populations are in danger of extinction. The introduction of new genetic stocks that escape and interbreed with wild fish can alter the genetic makeup of the wild populations. On an economic front, the increase in the amount of farmed salmon is reducing the prices salmon fishers receive for their catch.

Currently, about 60 percent of all aquaculture production is from freshwater systems, and production is growing rapidly. Aquaculture of freshwater species typically involves the construction of ponds, which allow the close management of the fish, shrimp, or other species. This can be done with a very low level of technology and is easily accomplished in underdeveloped areas of the world. Very little raising of fish occurs in open waters, although immature stages of many species may be raised in special nursery facilities (fish hatcheries) before they are released into the wild.

The environmental impacts of freshwater aquaculture are similar to those of aquaculture in marine systems. Nutrient overloads from concentrations of fish can pollute local bodies of water, and the escape of exotic species may harm native species. In addition, freshwater aquaculture involves the conversion of land to a new use. Often the lands involved are mangrove swamps or other wetlands that many people feel

Farming, Fish Kills, and *Pfiesteria piscicida*

During the summer of 1997, several instances of fish kills by *Pfiesteria piscicida* were suspected along the East Coast of the United States. *Pfiesteria piscicida* is a dinoflagellate that has over 20 stages in its life cycle. Four of the stages release a toxin that causes fish to become lethargic, develop open sores, and often die. *Pfiesteria piscicida* appear to be triggered to transform into the toxic forms when they detect the presence of large numbers of fish. Fish are often caused to concentrate when nutrient-rich waters become depleted of oxygen. The fish move from oxygen-depleted water to the areas that have the best oxygen concentrations. It appears that *Pfiesteria piscicida* detects some component of the excrement of these concentrated populations of fish and develops into its toxic forms. The *Pfiesteria piscicida* organism then feeds on the tissues of the sick and dying fish. Humans working with *Pfiesteria* have been affected by the organism as well. Symptoms include sores, lightheadedness, and asthma-like symptoms, among others.

It appears that the problem occurs in areas of the ocean that are shallow and nutrient rich. Human and swine effluent overflows have been implicated in some cases, as has fertilizer runoff from farming operations.

---

should be protected. Regardless of environmental concerns, the productivity of freshwater aquaculture has the potential to provide for the protein needs of a growing population, so it is likely to continue to increase at a rapid rate.

Managing Ecosystems for Wildlife

Many kinds of terrestrial wild animals are managed as game animals, to protect them from extinction, or for other purposes, and efforts are made to improve conditions for these species. Several techniques, some of which result in ecosystem modifications, are used to enhance certain wildlife populations. These techniques include habitat analysis and management, population assessment and management, predator and competitor control, and establishing refuges.

The economic costs associated with these activities include the salaries of management personnel, the purchase and leasing of land, and the cost of habitat modification activities. The energy costs are primarily involved in the modification of habitat. Often the environmental costs are considered to be minimal because the purpose of management is to enhance the populations of specific desired species. However, it needs to be understood that when actions are taken to benefit one species, usually other species are negatively affected.

Habitat Analysis and Management

Managing a particular species requires an understanding of the habitat needs of that species. An animal's habitat must provide the following: food, water, and cover. **Cover** refers to any set of physical features that conceals or protects animals from the elements or enemies. Several kinds of cover are important and include places where the animal can rest and raise young, where it can escape from enemies, and where it is protected from the elements.

Animals are highly specific in their habitat requirements. For example, bobwhite quail must have a winter supply of food that protrudes above the snow. They require a supply of small rocks called grit, which the birds use in their gizzards to grind food. During most of the year, they need a field of tall grass and weeds as cover from natural predators. However, such protection is not available during the winter, so a thicket of brush is necessary. The thicket serves as cover from predators and shelter from the cold winter weather. Most areas provide suitable cover for resting, but for sleeping, quail prefer an open, elevated location, which allows the birds to quickly take flight if attacked at night. When raising young, they require grassy areas with some patches of bare ground where the young can sun themselves and dry out if they get wet. All of these requirements must be available within a radius of 400 meters (1300 feet), because this is the extent of a quail's normal daily travels. (See figure 12.15.)

Once the critical habitat requirements of a species are understood, steps can be taken to alter the habitat to improve the success of the species. The habitat modifications made to enhance the success of a species are known as **habitat management.** The endangered Kirtland's warbler builds nests near the ground in dense stands of young jack pine. The density is important since their nests are vulnerable to predators and are better hidden in dense cover. Jack pine is a fire-adapted species that releases seeds following forest fires and naturally reestablishes dense stands following fire. Therefore, planned fires to regenerate new dense jack pine stands is a technique that has been used to provide appropriate habitat for this endangered bird.

figure 12.15 **Cover Requirements for Quail** Quail need several kinds of cover to be successful in an area. When raising young, they need areas of tall grass and weeds to provide protection from predators. In the winter, they need thickets to provide cover from weather and predators.

Quail

	Adults	Young	Total
1st year	2	14	16
2nd year	16	112	128
3rd year	128	896	1024

Deer

	Adults	Yearlings	Fawns	Total
1st year	2	0	2	4
2nd year	2	2	2	6
3rd year	4	2	4	10
4th year	6	4	6	16
5th year	10	6	10	26

figure 12.16 **Reproductive Potential** If we assume no mortality, animals have a reproductive capacity far above what is required to just keep the population stable. In the real world, mortality generally keeps populations from growing beyond the ability of their habitat to sustain them.

Source: Aldo Leopold, *Game Management*. Copyright © 1933 by Allyn and Bacon. Reprinted by permission.

est management and deer management may have to be integrated in this case because some other species of animals, such as squirrels, will be excluded if mature trees are cut, since they rely on the seeds of trees as a major food source.

Population Assessment and Management

Population management is another important activity that requires planning. Species must be managed so that they do not exceed the carrying capacity of their habitat. Wildlife managers use several techniques to establish and maintain populations at an appropriate level. Population censuses are used to check the population regularly to see if it is within acceptable limits. These census activities include keeping records of the number of animals killed by hunters, recording the number of singing birds during the breeding season, counting the number of fecal pellets, direct counting of large animals from aircraft, and a variety of other techniques.

Given suitable habitats and protection, most wild animals can maintain a sizable population. In general, organisms produce more offspring than can survive. Figure 12.16 illustrates the reproductive potential of both quail and white-tailed deer. High reproductive potential, protection from hunting, and the management and restoration of suitable habitats have resulted in large populations of once rare animals. In Pennsylvania, where deer were once extinct, the number is now about 1.5 million. The wild turkey population in the United States has increased from about 20,000 birds in 1890 to 5 million birds today. In Zimbabwe, unlike in most African countries, the number of elephants increased from about 200 in 1900 to 35,000 in 2000. This is probably not a sustainable population, and the government would like to reduce the number to about 30,000.

Since wildlife management often involves the harvesting of animals by hunting for sport and for meat, regulation of hunting activity is an important population management technique. Seasons are usually regulated to ensure adequate reproduction and provide the

Habitat management also may take the form of encouraging some species of plants that are the preferred food of the game species. For example, habitat management for deer may involve encourag-ing the growth of many young trees, saplings, and low-growing shrubs by cutting the timber in the area and allow-ing the natural regrowth to supply the food and cover the deer need. Both for-

environmental CLOSE-UP

Native American Fishing Rights

Throughout many parts of the United States, particularly in the Pacific Northwest and the Great Lakes states, a controversy continues over fishing rights of Native Americans. This conflict is unique because it involves treaties that were made 100 to 150 years ago between the U.S. government and Native American nations. It has become a major political, economic, social, and legal issue in some states, such as Washington, Michigan, Wisconsin, and Minnesota, and has involved the entire court system, from local courts to the U.S. Supreme Court. The controversy revolves around the interpretation of treaty language. It has, on several occasions, turned violent and has divided many communities.

According to the wording of many treaties entered into in the 1800s, the rights of Native Americans to fish would not be infringed upon by the states. Native Americans claim the treaties give them the legal authority to engage in commercial fishing enterprises, even when such fishing may be restricted or banned altogether for the general public.

On the other side of the argument are many state officials and sport fishers who believe that Native Americans are seriously endangering populations of such fish as salmon and trout by their uncontrolled harvesting for commercial purposes. They further argue that many species being taken by Native Americans belong to the entire state and not only to a certain group, because the fish are stocked or planted by the state. Another concern is the fishing techniques used by Native Americans. In the 1800s, when the treaties were signed, commercial fishing technology was limited. Today, however, Native American commercial fishers use nylon nets, power boats, depth finders, and other technological aids that enable them to catch much larger quantities of fish than they could with traditional fishing practices.

In the early 1970s, when sport fishers complained that stocks of fish in Lake Michigan were being depleted because of Native Americans' gill-net fishing, Michigan tried to regulate Native American fishing. The issue ended up in court, and, in 1978, a U.S. district court judge in Grand Rapids upheld fishing rights granted under 1836 and 1855 treaties between two Chippewa tribes and the U.S. government.

The federal judge ruled that the state had no authority in the matter because the issue in question involved a federal treaty and that the state did not have the right to make regulations contrary to the treaty; only Congress had such power. Subsequent to the decision, the tribes and the state negotiated changes in the areas in which the tribes could fish in an effort to reduce the potential for conflict between Native American fishers and sport or other commercial fishers.

A similar case was decided by a U.S. district judge in Tacoma, Washington, who interpreted treaties signed in 1854 and 1855 and ruled that Native Americans could catch up to 50 percent of the salmon that passed through their tribal land on the way to other parts of the state. In the face of protests by the non-native commercial fishing industry, the federal judge took over the regulation of salmon fishing in the state. Commercial fishers, who outnumbered Native American fishers by a ratio of approximately eight to one, feared for their livelihood. The case eventually went to the U.S. Supreme Court. In 1979, the Court ruled, in a six-to-three decision, to uphold the federal treaties that entitled Native Americans to half the salmon caught in the area of Puget Sound in Washington. It further held that the 50 percent figure had to include the salmon that Native Americans caught on their lands for home consumption and religious ceremonies as well as the fish they caught for commercial purposes. The high court also voted to uphold the right of the district court to continue supervising the fishing industry because of the state's resistance to the interpretation of the treaties.

Clearly, the issues surrounding Native American fishing rights are complex and broad in scope. There are purely biological questions involving a resource and its wise use, economic issues involving families' livelihoods, cultural and religious concerns pertaining to Native Americans' use of a resource, legal questions involving federal and state conflicts, and moral questions relating to the unfair treatment of Native Americans in the past. In such conflicts, there is seldom one right answer. While the courts have ruled on the cases in Michigan and Washington, and the states and Native Americans are trying to use the fishing resource wisely, the problem still exists and is likely to continue for years to come.

---

largest possible healthy population during the hunting season. Hunting seasons usually occur in the fall so that surplus animals are taken before the challenges of winter. Winter taxes the animal's ability to stay warm and is also a time of low food supplies in most temperate regions. A well-managed wildlife resource allows for a large number of animals to be harvested in the fall and still leaves a healthy population to survive the winter

and reproduce during the following spring. (See figure 12.17.)

In some cases, when a population of a species is below the desired number, organisms may be artificially introduced. Introductions are most often made when a species is being reintroduced to an area where it had become extinct. Wolves have been successfully reintroduced to Yellowstone National Park and large tracts of wooded land in the northern

Midwest (Michigan, Wisconsin, Minnesota). Similarly an exchange between the state of Michigan and the province of Ontario reintroduced moose to Michigan and turkeys to parts of Ontario.

Some game species are non-native species that have been introduced for sport hunting purposes. The ring-necked pheasant was originally from Asia but has been introduced into Europe, Great Britain, and North America. All of the

PART FOUR   Human Influences on Ecosystems

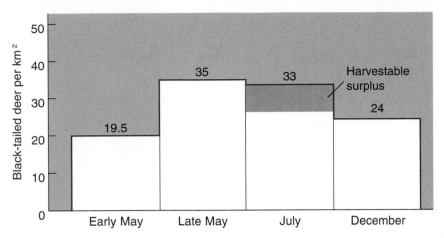

figure 12.17 **Managing a Wildlife Population** The seasonal changes in this population of black-tailed deer are typical of many game species. The hunting season is usually timed to occur in the fall so that surplus animals will be harvested before winter, when the carrying capacity is lower.

Source: Data from R.D. Taber and R.F. Dasmann, "The Dynamics of Three Natural Populations of the Deer *Odocoileus hemionus columbianus," Ecology* 38, no. 2 (1957):233–46.

figure 12.18 **Habitat Protection** The area on the left side of the fence has been protected from cattle grazing. This area provides a haven for many native species of plants and animals that cannot survive in heavily grazed areas.

large game animals of New Zealand are introduced species, since there were none in the original biota of these islands. Several species of deer have been introduced into Europe from Asia. Many of these are raised in deer parks, where the animals are similar to free-ranging domestic cattle. They may be hunted for sport or slaughtered to provide food. Introductions of exotic species were once common. However, today it is recognized that these introductions often result in the decreased viability of native species or in the introduced species becoming pests.

Predator and Competitor Control

At one time, it was thought that populations of game species could be increased substantially if predators were controlled. In Alaska, for example, the salmon-canning industry claimed that bald eagles were reducing the salmon population, and a bounty was placed on the bald eagle. From 1917 to 1952 in Alaska, 128,000 eagles were killed for bounty money. This theory of predator control to increase populations of game species has not proven to be valid in most cases, however, since the predators do not normally take the prime animals anyway. They are more likely to capture sick or injured individuals not suitable for game hunting.

Although predator and competitor control activities have become less common recently, they are still used in some special situations. In intensely managed European systems, gamekeepers are often charged with the responsibility of killing predators or unwanted competitors.

For some species, such as ground nesting birds, it may make some sense to control predators that eat eggs or capture the young, but, in most cases, humans have a greater impact through habitat modification and hunting than do the natural predators. Although at one time they were thought to be helpful, bounties and other forms of predator management have been largely eliminated in North America. In fact, the pendulum has swung the other way, and predator control is not considered to be cost effective in most cases. One exception to this general trend is the hunting and trapping of wolves in Alaska and Canada. One rationale for the taking of wolves is that they kill moose and caribou. Since many of the people who live in these areas rely on wild game as a significant part of their food source, controlling wolf populations is politically popular. However, the number of wolves taken is regulated. Since, in many parts of the world, the major predators of game species are humans, regulation of hunting is a form of predator control.

The control of cowbird populations has been used to enhance the breeding success of the Kirtland's warbler. Cowbirds lay their eggs in the nests of other birds, including Kirtland's warblers. When the cowbird egg hatches, the hatchling pushes the warbler chicks out of the nest. Thus, trapping and killing of cowbirds in the vicinity of nesting Kirtland's warblers has been used to enhance the reproduction of this endangered species.

Many species may require refuges where they are protected from competing introduced species or human interference. Many native rangeland species of wildlife benefit from the exclusion of introduced grazing animals such as cattle and sheep because the absence of grazing livestock allows a more natural grassland community to be reestablished. (See figure 12.18.)

Special Issues with Migratory Waterfowl Management

Waterfowl (ducks, geese, swans, rails, etc.) present some special management problems because they are migratory. **Migratory birds** can fly thousands of kilometers and, therefore, can travel north in the spring to reproduce during the summer months and return to the south when cold weather freezes the ponds, lakes, and streams that serve as

Cinnamon teal

Black duck

Mallard—male and female

Pintail (drake)

Atlantic flyway
Mississippi flyway
Central flyway
Pacific flyway

figure 12.19 **Migration Routes for North American Waterfowl** Migratory waterfowl follow traditional routes when they migrate. These have become known as the Atlantic, Mississippi, Central, and Pacific flyways. Many of these waterfowl are hatched in Canada, migrate through the United States, and winter in the southern United States or Mexico.

their summer homes. (See figure 12.19.) Because many waterfowl nest in Canada and the northern United States and winter in the southern United States and Central America, an international agreement between Canada, the United States, and Mexico is necessary to manage and prevent the destruction of this wildlife resource. Habitat management has taken several forms. In Canada, where much of the breeding occurs, government and private organizations such as Ducks Unlimited have worked to prevent the draining of small ponds and lakes that provide nesting areas for the birds. In addition, new impoundments have been created where it is

practical. Because birds migrate southward during the fall hunting season, a series of wildlife refuges provide resting places, food, and protection from hunting. In addition, these refuges may be used to raise local populations of birds. During the winter, many of these birds congregate in the southern United States. Refuges in these areas are important overwintering areas where the waterfowl can find food and shelter.

A similar management problem exists in Africa and Eurasia. Many birds that rely on wetlands migrate between Eurasia and Africa. In 1995, an international Agreement on the Conservation of African-Eurasian Waterbirds was de-

veloped. Its development was a complicated process, since it involves about 40 percent of the world's land surface and includes 117 countries. Several important countries have signed the agreement, but many have yet to do so.

Extinction and Loss of Biodiversity

Extinction is the death of a species, the elimination of all the individuals of a particular kind. It is a natural and common event in the long history of biological

Table 12.3 Probability of Becoming Extinct

Most Likely to Become Extinct	Least Likely to Become Extinct
Low population density	High population density
Found in small area	Found over large area
Specialized niche	Generalized niche
Low reproductive rates	High reproductive rates

evolution. In addition to complete extinction, we commonly observe local extinctions of populations. While not as final, a local extinction is an indication that the future of the species is not encouraging. Furthermore, as a population is reduced in size, some of the genetic variety in the population is likely to be lost.

Studies of modern local extinctions suggest that certain kinds of species are more likely than others to become extinct. (See table 12.3.) Species that have small populations of dispersed individuals are more prone to extinction because successful breeding is more difficult than in species that have large populations of relatively high density. Some kinds of organisms, such as carnivores at higher trophic levels in food chains, typically have low populations but also have low rates of reproduction compared to their prey species. Organisms in small, restricted areas are also prone to extinction because an environmental change in their locale can eliminate the entire species at once. Organisms scattered over large areas are much less likely to be negatively affected by one event. Specialized organisms are also more likely to become extinct than are generalized ones. Since specialized organisms rely on a few key factors in the environment, anything that negatively affects these factors could result in their extinction, whereas generalists can use alternate resources.

Rabbits and rats are good examples of animals that are not likely to become extinct soon. They have high population densities and a wide geographic distribution. In addition, they have high reproductive rates and are generalists that can live under a variety of conditions and use a variety of items as food. The cheetah is much more likely to become extinct because it has a low population density, is restricted to certain parts of Africa, has low reproductive rates, and has very specialized food habits. It must run down small antelope, in the open, during daylight, by itself. Similarly, the entire wild whooping crane species consists of about 180 individuals that are restricted to small winter and summer ranges that must have isolated marshes. (Captive and experimental populations bring the total number to about 350 individuals.) In addition, their rate of reproduction is low.

Human-Accelerated Extinction

Humans are among the most successful organisms on the face of the Earth. We are adaptable, intelligent animals with few enemies. As our population increases, we displace other kinds of organisms. This has resulted in an accelerated rate of extinction. Wherever humans become the dominant organism,

extinctions occur. Sometimes, we use other animals directly as food. In doing so, we reduce the population of our prey species. Since our population is so large and because we have an advanced technology, catching or killing other animals for food is relatively easy. In some cases, this has led to extinctions. The passenger pigeon in North America, the moas (giant birds) of New Zealand, and the bison and wild cattle of Europe were certainly helped on their way to extinction by people who hunted them for food.

We use organisms for a variety of purposes in addition to food. Many plants and animals are used as ornaments. Flowers are picked, animals skins are worn, and animal parts are used for their purported aphrodisiac qualities. In the United States, many species of cactus are being severely reduced because people like to have them in their front yards. In other parts of the world, rhinoceros horn is used to make dagger handles or is powdered and sold as an aphrodisiac. Because some people are willing to pay huge amounts of money for these products, unscrupulous people are willing to take the chance of poaching these animals for the quick profit they can realize. Table 12.4 shows some of the organisms, or their parts, that are highly prized by buyers. The World Wildlife Fund estimates that illegal trade in wild animals globally produces $2 billion to $30.5 billion per year. These activities have already resulted in local extinctions of some plants and animals and may be a contributing factor to the future extinction of some species.

Some organisms are extinct because they were regarded as pests. Many large predators have been locally exterminated because they preyed on the domestic animals that humans use for food. Mountain lions and grizzly bears in North America have been reduced to small, isolated populations, in part because they were hunted to reduce livestock loss. Tigers in Asia and the lion and wolf in Europe were reduced or eliminated for similar reasons. Even though commercial hunters killed thousands of passenger pigeons, their ultimate extinction was caused primarily by

Table 12.4 Typical Species Prices (1990 Rates)

International Species	Price ($U.S.)	North American Species	Price ($U.S.)
Olive python	1500	Bald eagle	2500
Rhinoceros horn	12,500/pound	Golden eagle	200
Tiger skin (Siberian)	3500	Gila monster	200
Tiger meat	130/pound	Peregrine falcon	10,000
Cockatoo	2000	Grizzly bear	5000
Leopard	8500	Grizzly bear claw necklace	2500
Snow leopard	14,000	Polar bear	6000
Elephant tusk	250/pound	Black bear paw pad	150
Walrus tusk	50/pound	Reindeer antlers	35/pound
Mountain gorilla	150,000	Mountain lion	500
Giant panda	3700	Mountain goat	3500
Ocelot	40,000/coat	Saguaro cactus	15,000
Imperial Amazon macaw	30,000		

Source: Data from U.S. Fish and Wildlife Service.

the increased agricultural use of the forests. Passenger pigeons ate the acorns of oaks and the beechnuts from beech, and also relied on the forests for communal nesting sites. When the forests were cleared, the pigeons became pests to farmers, who shot them to protect their crops from being eaten by the birds. Some extinctions of pest species are considered desirable. Most people would not be disturbed by the extinction of black widow spiders, mosquitoes, rats, or fleas. In fact, people work hard to drive some species to extinction. For example, in the *Morbidity and Mortality Weekly Report* (October 26, 1979), the U.S. Centers for Disease Control triumphantly announced that the virus that causes smallpox was extinct in the human population after many years of continuous effort to eliminate it.

The most important cause of extinctions related to human activity is habitat alteration. (See figure 12.20.) Whenever humans populate an area, they change it by converting the original ecosystem into something that supports themselves. Forests and grasslands have been converted to agricultural and grazing lands. In addition, humans have introduced new species to grow or graze on these lands. These new plants and animals compete with the native organisms for nutrients and living space, and often

the native organisms lose. (See Global Perspective: The History of the Bison.) The African elephant and various species of rhinoceros are in danger because they are hunted for ivory or horn, but it was the original fragmentation of their habitat for agricultural purposes that first placed the animals in conflict with humans.

In much of the tropical world today, rainforests are being cleared to provide grazing land or agricultural land for an expanding human population. This activity destroys existing forests and fragments them into small islands. Scientists studying the effects of this activity have noticed that, as a forest is reduced to small patches, many species of birds disappear from the area. The same kinds of activities certainly happened in Europe, where little of the original forest is left. In North America, the eastern deciduous forests were reduced, which probably resulted in the extinctions of some animals and plants. Almost all of the original prairie in the United States has been replaced with agricultural land, resulting in the loss of some species.

Humans also cause extinction in less direct ways. The building of dams changes the character of rivers, making them less suitable for some species. Air and water pollution may kill all life in an area. For example, acid precipitation

has lowered the pH of some lakes so much that all life has been eliminated. In other cases, indirect human activity may selectively eliminate species that are less tolerant to the pollutants being released, or the introduction of exotic species may eliminate existing species. The accidental introduction of the chestnut blight fungus into the United States resulted in the loss of the American chestnut tree over much of the Appalachian region. Some species that were dependent on these trees probably were also eliminated.

Why Worry About Extinction?

If extinction is a natural event and if the species that become extinct are those that cannot effectively cope with the activities of humans, why should we worry about them? There are several answers to that question. First of all, strictly from a selfish point of view, many species we know little about may be useful to us. Many plants have chemicals in them that can be used as medicines. If we drive them to extinction, we may be eliminating a potentially useful product. Also, as the human population increases, the need for different kinds of food plants and animals will increase. Once they are eliminated, we have lost the opportunity to use them for our own ends. Most of the wild ancestors of our most important food grains, like maize (corn), wheat, and rice, are thought to be extinct. What would our world be like if these plants had not been domesticated before they went extinct as wild populations? Because of this concern, many agricultural institutes and universities maintain "gene banks" of wild and primitive stocks of crop plants.

Another interesting answer to the question is that certain organisms in some ecosystems appear to play pivotal roles. Accidental extinction of one of these species could be devastating to the ecosystem and the humans that use it. For example, sea otters feed on sea urchins, which feed on giant kelp. When harvesting of sea otters reduced their numbers, more sea urchins survived and destroyed kelp forests, which significantly

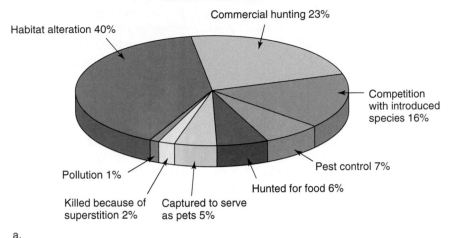

Causes of Extinction

Commercial hunting 23%

Habitat alteration 40%

Competition with introduced species 16%

Pollution 1%

Pest control 7%

Killed because of superstition 2%

Captured to serve as pets 5%

Hunted for food 6%

a.

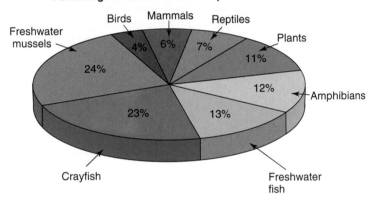

Percentage of North American Species at Risk

Birds 4%

Mammals 6%

Reptiles 7%

Plants 11%

Freshwater mussels 24%

Amphibians 12%

Crayfish 23%

Freshwater fish 13%

b.

figure 12.20 **Causes of Extinction** Many organisms have become extinct as a result of human activities. The most important of these activities is the indirect destruction of habitat by humans modifying the environment for their own purposes. Part (*b*) shows the species at risk in North America (percentage by group).

Source: (*b*) Precious Heritage @ 2000 The Nature Conservancy and NatureServe.

modified the nature of shoreline communities. Similarly, prairie dogs are important components of grasslands in North America. Their burrows are used by many species of animals as temporary or permanent dwellings, they create disturbed sites required by many species of plants, and they are the primary food source of the endangered black-footed ferret. Programs designed to reduce prairie dog numbers negatively affect many other animals as well. Often when a species is eliminated or significantly reduced in numbers, other organisms filled their niche, making it difficult for the population of the endangered species to return to its original size.

A third philosophical answer, according to many people, is that all species have an intrinsic value and a fundamental right to exist without being needlessly eliminated by the unthinking activity of the human species. This is an ethical position that is unrelated to social or economic considerations. According to those who support this philosophical position, extinction by itself is not bad, but human-initiated extinction is. This contrasts with the philosophical position that humans are simply organisms that have achieved a preeminent position on Earth and, therefore, extinctions we cause are no different from extinctions caused by other

forces. Regardless, significant efforts have been made to prevent the human-initiated extinction of species.

What Is Being Done to Prevent Extinction and Protect Biodiversity?

Efforts to prevent human-caused extinctions are difficult to assess. Some countries have enacted legislation to protect species that are in danger of becoming extinct. Species that receive special protection usually are given some sort of designation, such as endangered or threatened. **Endangered species** are those that have such small numbers that they are in immediate jeopardy of becoming extinct. **Threatened species** could become extinct if a critical factor in their environment were changed.

Most of the interest in preventing human-caused extinctions comes from developed countries. There, however, the problem is less acute because vulnerable species have already been eliminated. Extinction is a greater potential problem in tropical, less-developed countries. Many biologists estimate that there may be as many species in the tropical rainforests of the world as in the rest of the world combined. Unfortunately, extinction prevention is not a major issue in many less-developed countries. This difference in level of interest is understandable since the developed world has surplus food, higher disposable income, and higher education levels, while people in many of the less-developed countries, where population growth is high, are most concerned with immediate needs for food and shelter, not with long-term issues like extinction. However, the developed world should not be held blameless because our activities may indirectly be involved in extinction. When we purchase lumber, agricultural products, and fish from the less-developed world at low prices, we often are indirectly encouraging people in the developing world to over-exploit their resources and are thus having an impact on the rates of extinction.

Nevertheless, many of the governments of less-developed countries have responded to pressures and suggestions from outside sources and have established preserves and parks to protect species in danger of extinction. This does not solve the problem, however, since the areas must be protected from poachers and unauthorized agricultural activity of people who trespass on these protected areas. These people are responding to basic biological and economic pressures to provide food for their families. In many countries, hunting for bush meat is an important source of income for people, and the meat provides much needed protein for the people of the area. Furthermore, protected lands generally do not generate any income for the government. Hiring security forces to patrol such areas is expensive for many of these countries, so protection is often inadequate to prevent trespass and poaching. Providing tourism opportunities that generate income for local people and the government is one way that some countries try to offset the costs of protection.

Even in countries where interest in extinction prevention is relatively high, there is a cultural bias in favor of protecting certain kinds of organisms. Most endangered or threatened species are birds, mammals, some insects (particularly butterflies), a few mollusks and fish, and certain categories of plants. Bacteria, fungi, most insects, and many other inconspicuous organisms rarely show up on endangered species lists, even though they play vital roles in the nitrogen cycle, in the carbon cycle, and as decomposers. For example, the U.S. Fish and Wildlife Service lists 233 vertebrates (birds, mammals, amphibians, reptiles, and fish) and 565 flowering plants as endangered in the United States but only 30 insects and no fungi or bacteria.

Several international organizations work to prevent the extinction of organisms. The World Conservation Union (IUCN) estimates that, by the year 2000, at least 500,000 species of plants and animals may have been exterminated. (The International Union for the Conservation of Nature changed its name to the World Conservation Union but kept IUCN as its acronym.) The IUCN classifies species in danger of extinction into four categories: endangered, vulnerable, rare, and indeterminate. Endangered species are those whose survival is unlikely if the conditions threatening their extinction continue. These organisms need action by people to preserve them, or they will become extinct. Vulnerable species are those that have decreasing populations and will become endangered unless causal factors, such as habitat destruction, are stopped. Rare species are primarily those that have small worldwide populations and that could be at risk in the future. Indeterminate species are those that are thought to be extinct, vulnerable, or rare, but so little is known about them that they are impossible to classify.

Although the IUCN is a highly visible international conservation organization, it has very little power to effect change. It generally seeks to protect species in danger by encouraging countries to complete inventories of plants and animals within their borders. It also encourages the training of plant and animal biologists within the countries involved. (There is currently a critical shortage of plant and animal biologists who are familiar with the organisms of the tropics.) The IUCN also encourages the establishment of preserves to protect species in danger of extinction.

The U.S. Endangered Species Act was passed in 1973. This legislation gave the federal government jurisdiction over any species that were designated as endangered. About 300 U.S. species and subspecies have been so designated by the Office of Endangered Species of the Department of the Interior. (See figure 12.21.) The Endangered Species Act directs that no activity by a governmental agency should lead to the extinction of an endangered species and that all governmental agencies must use whatever measures are necessary to preserve these species.

The key to preventing extinctions is preservation of the habitat required by the endangered species. Consequently, many U.S. governmental agencies and private organizations have purchased sensitive habitats or have managed areas to preserve suitable habitats for endangered species. Setting aside certain land areas or bodies of water forces government and private enterprise to confront the issue of endangered species. The question eventually becomes one of assigning a value to the endangered species. This is not an easy task; often politics become involved, and the endangered species does not always win.

Since the protection of a species requires the preservation of its habitat, inevitably landowners are in a position to lose the use of their land. Many lawsuits have been files by landowners claiming that the federal government has taken their land from them by preventing them from using it for development purposes. As a result of these pressures, the rules were changed to allow some of the land to be used if a species conservation plan were in place. Another modification to the Endangered Species Act occurred in 1978, when Congress amended it so that exemptions to the act could be granted for federally declared major disaster areas or for national defense, or by a seven-member Endangered Species Review Committee. Because this group has the power to sanction the extinction of an organism, it has been nicknamed the "God Squad." If the committee found that the economic benefits of a project outweighed the harmful ecological effects, it would exempt a project from the Endangered Species Act.

The amendments to the Endangered Species Act also weakened the ability of the U.S. government to add new species to the endangered and threatened lists. Before a species can be listed, it is now necessary to determine the boundaries of its critical habitat, prepare an economic impact study, and hold public hearings—all within two years of the proposal of the listing. The political aspects of endangered species protection continue to be important in the U.S. Congress. Intense lobbying by environmental and business interests seeks to shape the reauthorization of the act to suit their interests.

Blackfooted ferret

Whooping crane

Mission blue butterfly

Galápagos tortoise

Giant panda

Persistent trillium

figure 12.21 **Endangered Species** Many species have been placed on the endangered species list. These are plants and animals that are present in such low numbers that they are in immediate danger of becoming extinct.

CHAPTER 12   Human Impact on Resources and Ecosystems

The California Condor

The California condor (*Gymnogyps californianus*) is thought to have been adapted to feed on the carcasses of large mammals found in North America during the Ice Age. With the extinction of the large, Ice Age mammals, the condors' major food source disappeared. By the 1940s, their range had shrunk to a small area near Los Angeles, California. Further fragmentation of their habitat caused by human activity, death by shooting, and death by eating animals containing lead shot reduced the wild population to about 17 animals by 1986.

A low reproductive potential makes it difficult for the species to increase in numbers. They do not become sexually mature until six years of age, and females typically lay one egg every two years. Because of concerns about the survival of the species, in 1987, all the remaining wild condors were captured to serve as breeding populations. The total population was 27 individuals. The plan was to raise young condors in captivity, leading to a large population that would ultimately allow for the release of animals back into the wild. Captive breeding is a very involved activity. Extra eggs can be obtained by removing the first egg laid and incubating it artificially. The female will lay a second egg if the first is removed. Raising the young requires careful planning. Although the young are fed by humans, they must not associate food with humans. This would result in inappropriate behaviors in animals eventually released to the wild. Therefore, puppets that resemble the parent birds are used to feed the young birds.

These efforts increased the number of offspring produced per female and resulted in a captive population of 54 individuals by 1991, when two condors were released into the wild north of Los Angeles. Six more were released in 1992. In 1996, a second population of condors was introduced in Arizona north of the Grand Canyon near the Utah border. The goal is to have two populations of up to 150 individuals each. In addition to natural mortality from predation, there has been unanticipated mortality from contact with power lines and lead poisoning from ingesting lead shot with food.

To help reduce these mortality problems, the birds that are currently being released go through a period of training in which they are taught to avoid power lines. In addition, several condors have been recaptured to be treated for lead poisoning. These will eventually be released back to the wild.

In 2002, there were 183 California condors in existence. Twenty-five were in the wild in Arizona. Thirty-two were in the wild in southern and central California, 15 were awaiting release, and the rest were in captive breeding programs at three locations. The behavior of the animals seems natural, and there is hope that they may eventually reproduce in the wild.

Fire As a Forest Management Tool

Fire has been used as a forest management tool for many years. When fires are absent from ecosystems for long periods, the amount of fuel present builds up so that the intensity of any fire that occurs is more destructive than when fires are more frequent. Large destructive fires damage timber and, therefore, reduce the value of the forest for lumber. Frequent small fires that primarily burn the litter on the ground are not as hot, consume the fuel before it reaches dangerous levels, and can be used in such a way that they do little damage to trees. Many private and public organizations that manage forest lands periodically use fire to remove debris and reduce the likelihood of major damaging fires.

In many wilderness areas, natural fires are allowed to burn as long as they are not "too large" and do not threaten human habitation. Yellowstone National Park was burned extensively in 1988. At the time, a "let burn" policy was in effect for wilderness areas. However, the fires eventually threatened buildings within the park, and control measures were initiated. Subsequent studies of the effects of the fires have substantiated that the park recovered and that a natural succession is taking place.

Hundreds of fires are set purposely each year as a normal part of forest management activities. Most of these fires go unnoticed

since they do exactly what they are supposed to do. However, in 2000, a purposely set fire (prescribed burn) in a national forest got out of control and burned beyond the specified area, destroyed hundreds of homes, and threatened the Los Alamos National Laboratory. The U.S. Forest Service maintains that managed use of fire is still an important forest management technique, but some members of Congress believe the practice should be stopped.

- Should fires that are caused by lightning in wilderness areas be treated differently from fires started accidentally by humans?
- Should fire be used as a management tool?
- Should Congress or forestry professionals make decisions about what should be appropriate management practices in national forests?

Summary

Pollution is the result of technological advancements and increased population density. It is defined as any addition of matter or energy that degrades the environment for humans and other organisms.

Natural resources are structures and processes that can be used by people but cannot be created by them. Renewable resources can be regenerated or repaired; nonrenewable resources are consumed.

Mineral resources must be extracted from ores, a process that requires energy and money, and changes the environment. The major steps of mineral exploitation are exploration, mining, processing the ore, transportation, and manufacturing the finished product. Ultimately, all costs of mineral exploitation are reduced to monetary costs. Recycling reduces the demand for new sources of mineral deposits but does not necessarily save money.

As the human population increases, we change natural ecosystems by replacing them with agricultural ecosystems, we alter species mixtures by introducing plants and animals, and we reduce populations by harvesting trees and animals for our use. Forests must be cut in a manner that allows for regrowth so that the soil is not exposed to the erosional effects of wind and water. Cutting small areas and reforestation help to prevent these problems. However, some remote areas have

been changed very little by humans. Tropical rainforests represent some of the last large wilderness areas, but they are being rapidly converted to other uses by the constant pressure of growing populations. Many of these wilderness areas should be protected because of the rich diversity of species they contain.

Grazing of arid and semiarid lands can be a valuable way to provide food for people. However, the land is often overgrazed and then may be degraded to a desert that is not capable of supporting the animals needed to feed growing populations.

Aquatic ecosystems are modified by pollution and activities that occur on the land adjacent to the water. Land that is devoid of vegetation erodes and fills streams and lakes with sediment. Warming of the water is also likely to occur if trees are removed from the stream side. Many exotic species of fish have been introduced into the freshwater ecosystems of the world. This alters normal food chains and reduces the populations of native species.

Most marine fisheries are being overfished or are at capacity. One of the problems associated with marine fisheries management is the enforcement of regulations. Managing for wildlife involves careful planning and habitat manipulation to provide the best possible population for hunting. Some areas are intensely managed, as in many European game parks, while in other parts of the world, the game animals lead a more normal wild life. Waterfowl present a unique problem because they migrate across international boundaries.

Extinction is a normal consequence of not being able to adapt to changes in the environment. However, since humans have such a great influence on nearly every ecosystem in the world, they have been the cause of increased rates of extinction. Many people recognize the value of species, both as possible helpers of humans and for their own intrinsic worth, and are trying to preserve sensitive habitats so that species will not be driven to extinction because of the appropriation of their habitats for other uses.

Key Terms

biodegradable *249*	energy cost *251*	patchwork clear-cutting *256*
biodiversity *253*	environmental cost *251*	pollution *249*
clear-cutting *255*	extinction *270*	reforestation *256*
cover *266*	habitat management *266*	renewable resources *251*
desertification *258*	migratory birds *269*	selective harvesting *256*
economic cost *251*	natural resources *250*	threatened species *273*
endangered species *273*	nonrenewable resources *251*	wilderness *259*

Review Questions

1. Name three ways humans directly alter ecosystems.
2. Why is the impact of humans greater today than at any time in the past?
3. Define pollution. Has the definition changed over time?
4. What are three kinds of costs associated with resource exploitation?
5. Why is recycling usually more energy efficient than mining new raw materials?
6. List three problems associated with forest exploitation.
7. What is desertification? What causes it?
8. What effects do increased temperature and increased organic matter have on aquatic ecosystems?
9. List six techniques utilized by wildlife managers.
10. What special problems are associated with waterfowl management?
11. What is extinction, and why does it occur?
12. Why should humans worry about extinction? Do you?
13. List three actions that can be taken to prevent extinctions.

Critical Thinking Questions

1. Imagine you are a mining company executive trying to determine the costs of your mining operation. Create a list of your costs. How important are environmental costs to your company? Why?
2. Logging, mineral extraction, and grazing on federal lands have been subsidized by the federal government to stimulate economic growth and support rural communities. What place should ecological value have in determining the way land is used? Should subsidies be continued? Justify your stand on this issue.
3. Perhaps 98 to 99 percent of all species that have ever existed are extinct. Nearly all went extinct long before humans arrived on the scene. Why should we be concerned about extinction of organisms today?
4. Would you support clearing of forests and plowing of grasslands that have significant ecological importance in order to support agriculture in countries that have significant hunger? Where do you draw the line between preserving ecosystems and human interest?
5. A Pacific Northwest Native American tribe has been permitted to hunt an endangered whale species to preserve its traditional culture and to uphold its treaty rights. Now there is fear that other countries will hunt the whales, too. Do you think the tribe should be denied its rights? Why, or why not?
6. Pharmaceutical companies are helping some developing countries preserve their rainforests so these companies can look for organisms with possible pharmacological value. How do you feel about these arrangements? What limits would you place on the pharmaceutical companies, if any? Why?

Concept Map

Construct a map to show relationships among the following concepts:

biodiversity

endangered species

extinction

natural resources

nonrenewable resources

reforestation

renewable resources

selective harvesting

threatened species

Interactive Exploration

Check out the website at **http://www.mhhe.com/environmentalscience** and click on the cover of this textbook for quizzing, career information, case studies, and hot links for information on the following topics:

Recycling

Nontropical Forests and Land-Use Issues

Resources from the Sea

Fisheries

Impact of Humans on the Sea; Harvesting

Impact of Humans on the Sea; Pollution

Pfisteria

Biodiversity

Extinction Issues

Extinction and Marine Habitats

Tropical Forests and Extinction

North American Habitats and Extinction Issues

Conservation and Management of Habitats and Species

Global Warming

Air Pollution

Acid Deposition

Water Pollution

CHAPTER 13

Land-Use Planning

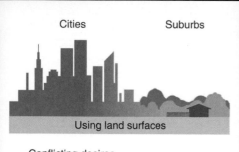

Using land surfaces

Conflicting desires
- Industry
- Housing
- Transportation
- Commercial development
- Recreation

Planning resolves conflicts

Objectives

After reading this chapter, you should be able to:

- Explain why most major cities are located on rivers, lakes, or the ocean.
- Describe the forces that result in farmland adjacent to cities being converted to urban uses.
- Explain why floodplains and wetlands are often mismanaged.
- Describe the economic and social values involved in planning for outdoor recreation opportunities.
- Explain why some land must be designated for particular recreational uses, such as wilderness areas, and why that decision sometimes invites disagreement from those who do not desire to use the land in the designated way.
- List the steps in the development and implementation of a land-use plan.
- Describe methods of enforcing compliance with land-use plans.
- Describe the advantages and disadvantages of both local and regional land-use planning.

Chapter Outline

The Need for Planning

A large proportion of the land surface of the world (about one-third to one-half) has been changed by human activity. Most of this change occurred as people converted the land to agriculture and grazing, but, in our modern world, significant amounts have been covered with buildings, streets, highways, and other products of society. In many cases, cities grew without evaluating and determining the most logical use for the land. Los Angeles and Mexico City have severe air pollution problems because of their geographic location, Venice and New Orleans are threatened by high sea levels, and San Francisco and Tokyo are subject to earthquakes. Currently, most land-use decisions are still based primarily on economic considerations or the short-term needs of a growing population rather than on careful analysis of the capabilities and unique values of the land and landscape. Each piece of land has specific qualities based on its location and physical makeup. Some is valued for the unique species that inhabit it, some is valued for its scenic beauty, and some has outstanding potential for agriculture or urban uses. Since land and the resources it supports (soil, vegetation, elevation, nearness to water, watersheds) are not being created today (except volcanos, river deltas, etc.), land should be considered a nonrenewable resource.

Once land has been converted from natural ecosystems or agriculture to intensive human use, it is generally unavailable for other purposes. As the population of the world grows, competition for the use of the land will increase, and systematic land-use planning will become more important. Furthermore, as the population of the world becomes more urbanized and cities grow, urban planning becomes critical.

Historical Forces That Shaped Land Use in North America

Today, most of the North American continent has been significantly modified by human activity. In the United States, about 47 percent of the land is used for crops and livestock, about 45 percent is forests and natural areas (primarily west of the Mississippi River), and nearly 5 percent is used intensively by people in urban centers and as transportation corridors. Canada is 54 percent forested and wooded and uses only 8 percent of its land for crops and livestock. Less than 1 percent of the land is in urban centers and transportation corridors. A large percentage of its remaining land is wilderness in the north.

This pattern of land use differs greatly from the original conditions experienced by the early European colonists who immigrated to the New World. The first colonists converted only small portions of the original landscape to farming, manufacturing, and housing, but as the population increased, more land was converted to agriculture, and settlements and villages developed into towns and cities. Although most of this early development was not consciously planned, it was not haphazard. Several factors influenced where development took place and the form it would take.

The Importance of Waterways

Waterways were the primary method of transportation, which allowed exploration and the development of commerce in early North America. Thus, early towns were usually built near rivers, lakes, and oceans. Typically, cities developed as far inland as rivers were navigable. Where abrupt changes in elevation caused waterfalls or rapids, goods being transported by boat or barge needed to be offloaded, transported around the obstruction, and loaded onto other boats. Cities often developed at these points. Buffalo, New York, and Sault Sainte Marie, Ontario, are examples of such cities. In addition to transportation, bodies of water provided drinking water, power, and waste disposal for growing villages and towns. Those towns and villages with access to waterways that provided easy transportation could readily receive raw materials and distribute manufactured goods. Some of these grew into major industrial or trade centers. Without access to water, St. Louis, Montreal, Chicago, Detroit, Vancouver, and other cities would not have developed. (See figure 13.1.) The availability of other natural resources, such as minerals, good farm land, or forests was also important in determining where villages and towns were established. Industrial development began on the waterfront since water supplied transportation, waste disposal, and power. As villages grew into towns and cities, large factories replaced small gristmills, sawmills, and blacksmith shops. The waterfront became a center of intense industrial activity. As industrial activity increased in the cities, people began to move from rural to urban centers for the job opportunities these centers presented.

The Rural-to-Urban Shift

North America remained essentially rural until industrial growth began in the last third of the 1800s and the population began a trend toward greater urbanization. (See figure 13.2.) Several forces led to this rural-to-urban transformation. First, the Industrial Revolution led to improvements in agriculture that required less farm labor at the same time industrial jobs became available in the city. Thus, people migrated from the farm to the city. The average person was no longer a farmer but, rather, a factory worker, shopkeeper, or clerk (with a regular paycheck) living in a tenement

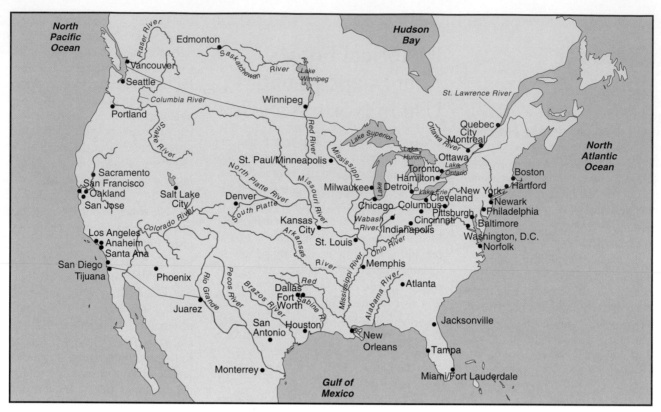

figure 13.1 **Water and Urban Centers** Note that most of the large urban centers are located on water. Water is an important means of transportation and was a major determining factor in the growth of cities. The cities shown have populations of 1 million or more (except Edmonton, Winnipeg, and Quebec City).

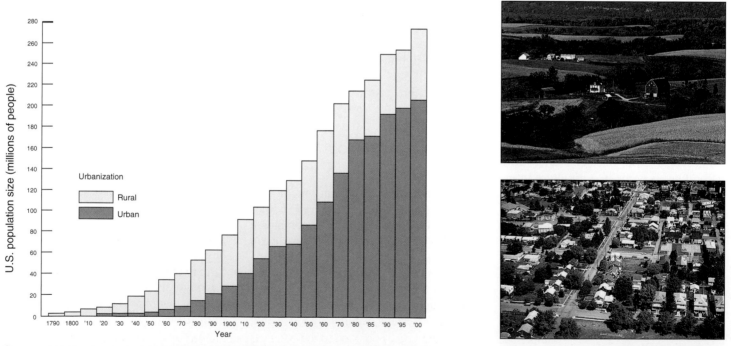

figure 13.2 **Rural-to-Urban Population Shift** In 1800, the United States was essentially a rural country. Industrialization in the late 1800s began the shift to an increasing urban population. In 2000, 75 percent of the U.S. population was urban.

Sources: Data from the *Statistical Abstract of the United States;* 2000 data from Population Reference Bureau, Inc., Washington, D.C.

Global Perspective

Urbanization in the Developing World

Traditionally, most of the population of the developing world has been rural. However, in recent years, the number of people migrating to the cities has grown rapidly. By 2025, about 5 billion people are expected to be living in urban centers. Most of the increase in urban areas will be in the developing world. Many migrate to the cities because they feel they will have greater access to social services and other cultural benefits than are available in rural areas. Many also feel there are more employment opportunities. However, the increase in the urban population is occurring so rapidly that it is very difficult to provide the services needed by the population, and jobs are not being created as fast as the urban population is growing. Thus, many of the people live in poverty on the fringes of the city in shantytowns that lack water, sewer, and other services. Often these shantytowns are constructed without permission only a short distance from affluent urban dwellers. Because the poor lack safe drinking water and sewer services, they pollute the local water sources and disease is common. Because they burn wood and other poor-quality fuels in inefficient stoves, air pollution is common. The additional people also create traffic problems of staggering proportions. The chart shows the seven cities with the largest rate of population increase. All are in the developing world.

Cities Expected to Grow by More Than 50 Percent between 1995 and 2015

City	1995 Population (Millions)	Projected 2015 Population (Millions)
Bombay, India	15.14	26.22
Lagos, Nigeria	10.29	24.61
Delhi, India	9.95	16.86
Karachi, Pakistan	9.73	19.38
Metro Manila, Philippines	9.29	14.66
Jakarta, Indonesia	8.62	13.92
Dhaka, Bangladesh	8.55	19.49

Source: Data from *World Resources 2000–2001*.

or tiny apartment near where they worked. This pattern of rural-to-urban migration occurred throughout North America, and it is still occurring in developing countries today. A second factor that affected the growth of cities was the influx of immigrants from Europe. Although some became farmers, many of these new citizens settled in towns and cities, where jobs were available. A third reason for the growth of cities was that they offered a greater variety of cultural, social, and artistic opportunities than did rural communities. Thus, cities were attractive for cultural as well as economic reasons. San Antonio, Texas, and Las Vegas, Nevada, are examples of cities developing around unique cultural attractions.

Migration from the Central City to the Suburbs

During the early stages of industrial development, there was little control of industry activities, so the waterfront typically became a polluted, unhealthy, undesirable place to live. As roads and rail transport became available, anyone who could afford to do so moved away from the original, industrial city center. The more affluent moved to the outskirts of the city, and the development of suburban metropolitan regions began. Thus, the agricultural land surrounding the towns was converted to housing. Most cities originally had good farmland near them, since the floodplain near rivers typically has a deep, rich soil and agricultural land adjacent to the city was one of the factors that determined whether the city grew or not. This was true because until land transportation systems became well developed, farms needed to be close to the city so that farmers could transport their produce to the markets in the city. This rich farmland adjacent to the city was ideal for the expansion of the city. As the population of the city grew, demand for land increased. As the price of land in the city rose, people and businesses began to look for cheaper land farther away from the city. Developers and real estate

a.

b.

c.

figure 13.3 **Types of Urban Sprawl** Note the three different types of growth depicted in these photos. (*a*) The wealthy suburbs with large lots are adjacent to the city. (*b*) Ribbon sprawl develops as a commercial strip along highways. (*c*) Tract development results in neighborhoods consisting of large numbers of similar houses on small lots.

agents were quick to respond and to help people acquire and convert agricultural land to residential or commercial uses. Land was viewed as a commodity to be bought and sold for a profit, rather than as a nonrenewable resource to be managed. As long as money could be made by converting agricultural land to other purposes, it was impossible to prevent such conversion. There were no counteracting forces strong enough to prevent it.

The conversion of land around cities in North America to urban uses destroyed many natural areas that people had long enjoyed. The Sunday drive from the city to the countryside became more difficult as people had to drive farther to escape the ever-growing suburbs. The unique character of neighborhoods and communities was changed by the erection of shopping malls, apartment complexes, and expressways. Most of these alterations occurred without considering how they would affect the biological community or the lives of the people who lived in the area.

As cities continued to grow, certain sections within each city began to deteriorate. Industrial activity continued to be concentrated near water in the city's center. Industrial pollution and urban crowding turned the core of many cities into undesirable living areas. In the early 1900s, people who could afford to leave began to move to the outskirts. This trend continued after World War II, in the 1940s, 1950s, and 1960s, as a strong economy and government policies that favored new home purchases (tax deductions and low-interest loans) allowed

more people to buy homes. In 1950, about 60 percent of the urban population lived in the central city; by 1990, this number was reduced to about 30 percent. Most of these new single-family homes were in attractive suburbs, away from the pollution and congestion of the central city. These collections of houses were built on large lots that provided outdoor space for family activities. The blocks of homes were added to the periphery of the city in a decentralized pattern in which single-family homes were separated from multifamily structures and both were some distance from places of work, shopping, and other service needs. This pattern also made it very difficult to establish efficient public transportation networks, which decreased energy efficiency and increased the cost of supplying utility services. However, public transport was not considered important because rising automobile ownership and improved highway systems would accommodate the transportation needs of the suburban population. The convenience of a personal automobile escalated decentralized housing patterns, which, in turn, required better highways, which led to further decentralization.

By 1960, unplanned suburban growth had become known as urban sprawl. **Urban sprawl** is a pattern of unplanned, low-density housing and commercial development outside of cities that usually takes place on previously undeveloped land. In addition, blocks of housing are separated from commercial development, and the streets typically form branching patterns and often include cul-de-sacs. These large housing tracts surrounded cities, which made it difficult for people to find open space. A city dweller could no longer take a bus to the city limits and enjoy the open space of the countryside.

Urban sprawl occurs in three ways. (See figure 13.3.) One type of growth involves the development of exclusive, wealthy suburbs adjacent to the city. These homes are usually on large individual lots in the more pleasing geographic areas of the city. Often they are located along water, on elevated sites, or in wooded settings. A second development pattern is tract development. **Tract development** is the construction of similar residential units over large areas. Initially, these tracts are often separated from each other by farmland. New roads are constructed to link new housing to the central city and other suburbs, which stimulates the development of a third form of urban sprawl along transportation routes. This is referred to as **ribbon sprawl** and usually consists of commercial and industrial buildings that line each side of the highway that connects housing areas to the central city and shopping and service areas. Ribbon sprawl results in high costs for the extension of utilities and other public services. It also makes the extent of urbanization seem much larger than it actually is, since the driver cannot see the undeveloped land hidden by the storefronts that face the highway.

As suburbs continued to grow, cities (once separated by farmland) began to merge, and it became difficult to tell

figure 13.4 **Regional Cities in the United States and Canada** The lights from this satellite image show population concentrations. More than 30 major regional cities have developed, with each having more than 1 million people. Many of these cities merge with their neighbors to form huge regional cities. Major urban regions are the northeast coast of the United States (Boston to Washington, D.C.), the region south of the Great Lakes (Chicago to Pittsburgh), south Florida (Jacksonville to Miami), the Toronto and Montreal regions of Canada, and the west coast of California (San Francisco to San Diego).

where one city ended and another began. This type of growth led to the development of regional cities. Although these cities maintain their individual names, they are really just part of one large urban area called a **megalopolis.** (See figure 13.4.) The eastern seaboard of the United States, from Boston, Massachusetts, to Washington, D.C., is an example of a continuous city. In the midwestern United States, the area from Milwaukee, Wisconsin, to Chicago, Illinois, is a further example. Other examples are London to Dover in England, the Toronto-Mississauga region of Canada, and the southern Florida coast from Miami northward.

In some areas throughout North America, the growth of suburbs has been slowed due to the increased cost of housing and transportation. People have migrated back to some cities on a limited scale because of the lower cost of urban houses and the fact that public transportation is generally more efficient in the city than in the suburbs, thus freeing urban residents from the cost of daily commuting. This reverse migration, however, is still greatly offset by the continual growth of the suburban communities.

Factors That Contribute to Sprawl

As we enter the twenty-first century, many, including government policy makers, are beginning to question the prevailing pattern of sprawling urban development. However, powerful social, economic, and policy forces have been behind the development of the current decentralized urban centers typical of North America. If this pattern is to change, these underlying factors must be understood.

Lifestyle Factors

One of the factors that has supported urban sprawl is the relative wealth of the population. This wealth is reflected in material possessions, two of which are automobiles and homes. Based on 2000 census data, in the United States, there are 78 motor vehicles for every 100 people. Since over 20 percent of the population is too young to drive, this means that there is essentially one motor vehicle for every licensed driver in the United States. In addition, about two-thirds of the population live in family-owned homes. With this level of wealth, people are free to choose where and how they want to live. Many are attracted to a lifestyle that includes low-density residential settings, with easy access to open space, isolated from the problems of the city. They are also willing to drive considerable distances to live in these settings. Thus, a decentralized housing pattern is possible because of the high rate of automobile ownership, which allows for ease of movement.

Economic Factors

Several economic forces operate to encourage sprawl development. First of all, it is less expensive to build on agricultural and other nonurban land than it is to build within established cities. The land is less expensive, the regulations and permit requirements are generally less stringent, and there are fewer legal issues to deal with. An analysis of building costs in the San Francisco Bay area of California determined that it costs between 25 percent and 60 percent more to build in the city than it does to build in the suburbs.

Several tax laws also contributed to encouraging home ownership. The interest on home loans is deductible from income taxes, and, in the past, people could avoid paying capital gains taxes on homes they sold if they bought another home of equal or higher value. Since homes usually increased in value, this created a market for increasingly expensive homes.

Planning and Policy Factors

Many planning and policy issues have contributed to sprawl development. First of all, until recently, little coordinated effort has been given to planning how development should occur in metropolitan areas. There are several reasons for this. First of all, most metropolitan areas include hundreds of different political jurisdictions (the New York City metropolitan area includes about 700 separate government units and involves the states of New York, New Jersey, and Connecticut). It is very difficult to integrate the activities of these separate jurisdictions in order to achieve coordinated city planning. In addition, it is very difficult for a small local unit of government to see the "big picture," and many are unwilling to give up their autonomy to a regional governmental body.

Local zoning ordinances have often fostered sprawl by prohibiting the mixing of different kinds of land use. Single-family housing, multiple-family housing, commercial, and light and heavy industry were restricted to specific parts of the community. In addition, many ordinances specify minimum lot sizes and house sizes. This tends to result in a decentralized pattern of development that is supported by the heavy use of automobiles and the roads and parking facilities necessary to support them.

In addition, many government policies actually subsidize the development of decentralized cities. For example, developers and the people who buy the homes and businesses they build are also able to avoid paying for the full cost of extending services to new areas. The local unit of government picks up the cost of these improvements, and the cost is divided among all the taxpayers rather than just those who will benefit. The roads that are needed to support new developments are usually paid for with federal and state monies, so again the cost is not borne by those who benefit most but by the taxpayers of the state or nation. Furthermore, federal and state governments have not supported public transport in an equivalent way.

Problems Associated with Unplanned Urban Growth

As the population increased and metropolitan areas grew, several kinds of problems were recognized. They fall into several broad categories.

Transportation Problems

Most cities experience continual problems with transportation. This is primarily because, as cities grew, little thought was given to how people were going to move around and through the city. Furthermore, when housing patterns and commercial sectors changed, transportation mechanisms had to be changed to meet the shifting needs of the public. This often involved the abandoning of old transportation corridors and the establishment of new ones. Paradoxically, the establishment of new transportation corridors stimulates increased growth in the areas served, and the transportation corridors soon become inadequate. The reliance on the automobile as the primary method of transportation has required the constant building of new highways. A strong automobile-based bias exists in the culture in the United States. In Los Angeles, for example, 70 percent of surface area in the city is dedicated in some way to the automobile (roads, parking garages, etc.), compared with only 5 percent devoted to parks and open space.

According to the U.S. Department of Transportation, the average person in the United States spends nine hours per week traveling in a car. In many metropolitan areas, the amount of time is much greater, and a significant amount of that time involves traffic jams. A study of traffic in the Washington, D.C., area concluded that the average commuter spent 80 hours per year stuck in traffic, in addition to the time normally needed to make the commute. (See figure 13.5.)

Air Pollution

Reliance on the automobile as the primary method of transportation has resulted in significant air pollution problems in many cities. Most of the large industrial sources of air pollution have been contained. However, the individual car with its single occupant going

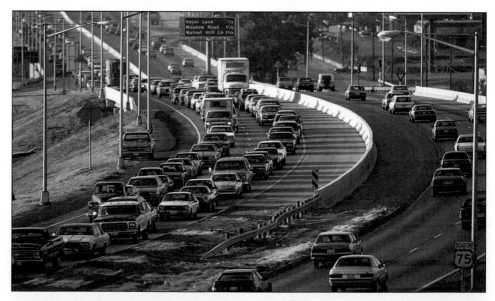

figure 13.5 **Traffic and Suburbia** Traffic congestion is a common experience for people who work in cities but live in the suburbs. The popularity of automobiles and the desire to live in the suburbs are closely tied.

to work, to shop, or to eat a meal is a constant source of air pollution. A simple solution to this problem is a centralized, efficient public transportation system. However, this is difficult to achieve with a highly dispersed population.

Low Energy Efficiency

Energy efficiency is low for several reasons. First of all, automobiles are the least energy-efficient means of transporting people from one place to another. Second, the separation of blocks of homes from places of business and shopping requires that additional distances be driven. Third, congested traffic routes result in hours being spent in stop-and-go traffic, which wastes much fuel. Finally, single-family homes require more energy for heating and cooling than multifamily dwellings.

Loss of Sense of Community

Although it is difficult to measure, general agreement exists that in dispersed suburban developments, there is a loss of a sense of community. In many places, people stay within their homes and yards and do not routinely walk through the neighborhood. When they leave the confines of their home, they get in a car and go somewhere. This pattern of behavior reduces human interaction, isolates people from their neighbors, and greatly reduces the sense of community.

Death of the Central City

Currently, less than 10 percent of people work in the central city. When people leave the city and move to the suburbs, they take their purchasing power and tax payments with them. Therefore, the city has less income to support the services needed by the public. When the quality of services in urban centers drops, the quality of life declines, the flight from the city increases, and a downward spiral of decay begins. An additional problem is the decline of the downtown business district. When shopping malls are built to accommodate the people in the suburbs, the downtown business district de-

clines. Because people no longer need to come to the city to shop, businesses in the city center fail or leave, which deprives the remaining residents of basic services. They must now travel greater distances to satisfy their basic needs.

Higher Infrastructure Costs

Infrastructure includes all the physical, social, and economic elements needed to support the population. Whenever a new housing or commercial development occurs on the outskirts of the city, municipal services must be extended to the area. Sewer and water services, natural gas and electric services, schools and police stations, roads and airports—all are needed to support this new population. Extending services to these new areas is much more costly than supplying services to areas already in the city because most of the basic infrastructure is already present in the city.

Loss of Open Space

One of the important features of a pleasing urban landscape is the presence of open space. Open fields, parks, boulevards, and similar land use allows people to visually escape from the congestion of the city. Unplanned urban growth does not take this important factor into account. Consequently, the buildings must be torn down, and disused spaces must be renovated into parks and other open space at great expense to provide green space in the urban landscape.

Loss of Farmland

Most of the land that has recently been urbanized was previously used for high-value crops. Land that is flat, well drained, accessible to transportation, and close to cities is ideal farmland. However, it is also prime development land. Areas that once supported crops now support housing developments, shopping centers, and parking lots. (See figure 13.6.) For example, a once agricultural San Jose, California, is now known as a high-tech area. The orange groves that were characteristic of Miami

and Los Angeles have long been transformed to suburbs, office complexes, and shopping malls.

Urban development of farmland is proceeding at a rapid pace. Currently, land is being converted to urban uses at a rate of over 400,000 hectares (1 million acres) per year. About one-third of this land is prime agricultural land.

Several states have established programs that provide protection to farmers who do not want to sell their land to developers. The programs may require farmers, in return for lower taxes on the land, to put their land in a conservation easement that prevents the farmer or future owners from using the land for anything other than farming.

Water Pollution Problems

A large impervious surface area results in high runoff and flash flood potential. A typical shopping mall has a paved parking lot that is four times larger than the space taken up by the building. The runoff from paved parking lots carries pollutants (oil, coolant, pieces of rubber) into local streams and ponds.

Floodplain Problems

Because most cities were established along water, many cities are located in areas called **floodplains.** Floodplains are the low areas near rivers and, thus, are subject to periodic flooding. Some floodplains may flood annually, while others flood less regularly. They are generally flat, and so are inviting areas for residential development even though they suffer periodic flooding. A better use of these areas is for open space or for recreation, yet developers continue to build houses and light industry there.

Usually, when a floodplain is developed for residential or commercial use, a retaining wall is built to prevent the periodic flooding natural to the area. This boosts the cost of the development, increases the cost of insurance protection, and creates high-water problems downstream. Frequently, tax monies are used to repair the damage that results from the unwise use of floodplains. Floodplain

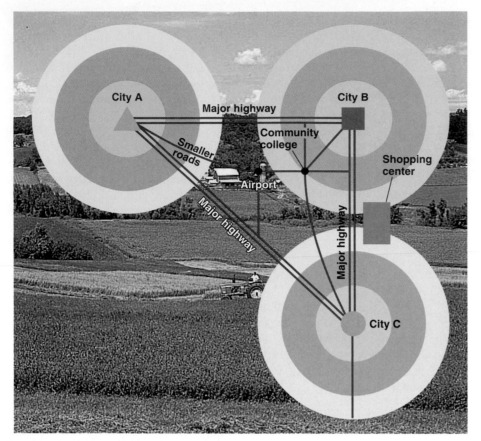

figure 13.6 **Loss of Farmland** As cities grow outward, they eventually grow together to form a regional city. The land between them, once used for farming, becomes developed for residential and commercial purposes. Improved transportation routes and joint facilities (such as airports, shopping centers, and community colleges) hasten this loss of farmland.

figure 13.7 **Flooding in Floodplain** Urban flood control methods (walls, barriers, and levees) often channel flood waters downstream with catastrophic results. This photo shows some of the flooding that occurred along the Mississippi River in 1993.

development is one of the insidious problems associated with a rapidly expanding population. As long as the population continues to increase, these less desirable areas are likely to be used for housing, whether they are subject to annual flooding or less frequent damage.

Floods are natural phenomena. Contrary to popular impressions, no evidence supports the premise that floods are worse today than they were 100 or 200 years ago, except perhaps on small, isolated watersheds. There is little doubt, however, that urban flood control methods (walls, barriers, and levees) have had detrimental downstream effects during high-water periods, because preventing the water from spreading into the floodplain forces increased amounts of water to be channeled to areas downstream. What has increased is the economic loss from the flooding. In 1993, during an extensive flooding event that involved both the Mississippi and Missouri Rivers, the U.S. Army Corps of Engineers estimated over $1.5 billion in damage to residential and commercial property. (See figure 13.7.) This loss reflects the fact that more cities, industries, railroads, and highways have been constructed on floodplains. Sometimes there are no alternatives to floodplain development, but too often risks are simply ignored. Because flooding causes loss of life and property, floodplains should no longer be developed for uses other than agriculture and recreation.

Many communities have enacted **floodplain zoning ordinances** to restrict future building in floodplains. Although such ordinances may prevent further economic losses, what happens to individuals who already live in floodplains? Floodplain building ordinances usually allow current residents to remain. Relocation, usually at a financial loss, is the only alternative. Such situations are unfortunate; perhaps proper planning in the future will prevent these problems.

Wetlands Misuse

Since access to water was and continues to be important to industrial development, many cities are located in areas with extensive wetlands. **Wetlands** are

Wetlands Loss in Louisiana

Louisiana has extensive coastal wetlands along the Gulf of Mexico that constitute about 40 percent of the wetlands in the United States, excluding Alaska and Hawaii. About 3500 square kilometers (1300 square miles) have been lost since 1956. This is an area approximately the size of the state of Rhode Island. Recent rates of loss have ranged from 65 to 90 square kilometers (25–35 square miles) per year. There are several reasons for the loss, but human activity is a primary cause.

Much of the Louisiana coastline consists of poorly compacted muds that are easily eroded if exposed to wind and wave action. Faults in the rock that lies below the mud are causing the land to tilt downward, which allows the ocean to invade. This is the basic cause for about 60 percent of the wetlands loss. In the past, the two actions—loss by tilting and addition of sediments from the Mississippi River—were in balance. The loss due to tilting was replaced by additional sediments from the river. However, two human actions have reduced the amount of sediments being added. Dams and reservoirs upstream on the Missouri and Mississippi Rivers have reduced the amount of sediment delivered to the region, making replacement of lost sediments impossible. In addition, levees on the Mississippi River have prevented the flooding that was a normal part of the behavior of the lower Mississippi River for centuries. The sediments, therefore, are carried out to the Gulf of Mexico and do not contribute to the building of wetlands near the shore. Without the constant addition of sediments, the muds settle and compact, which, along with the subsiding of the land, results in erosion and wetlands loss.

Shipping channels and canals associated with oil and gas production are responsible for about 30 percent of the current wetlands loss. There are nine major shipping channels and 13,000 kilometers (8000 miles) of canals. Typically, production companies cut canals through the marshes to give barges and other equipment easy access to the drilling platforms and production facilities. When oil or gas is found, it must be transported to land through pipelines. Laying pipelines often requires cutting additional canals through the wetlands. These canals break the wall of vegetation that protects the soft muds and allow salt water to penetrate into the marshes, killing the vegetation. Both of these activities open up the wetlands to wave action and accelerate the rate of erosion.

The nutria, a rodent introduced from South America, is also contributing to the problem. These animals reproduce rapidly and form such large populations that they eat all the vegetation in a local area. This exposes the muds to erosion and contributes to the loss of wetlands. Trapping these animals for their fur was at one time profitable for the local people and helped to control the animals. However, a shift in public opinion has reduced the acceptability of fur for clothing and depressed the fur market, making the trapping of nutria uneconomical.

A study led by the Louisiana Department of Natural Resources was released in 1998. It called for billions of dollars of federal investment to protect the wetlands. A logical source of these funds would be royalties paid to the federal government by offshore oil production companies. Major features of the plan include changes to current shipping practices, construction of new shipping facilities downstream from New Orleans, and controlled opening of levees along the lower Mississippi to restore some of the flooding that was at one time a normal part of the ecosystem. Powerful forces are in conflict over this plan. Shipping interests feel changes in shipping will cost them money. People who live in low-lying areas currently protected by levees may need to relocate even if controlled flooding is initiated. Changing the water quality by adding freshwater to a brackish water system will change the kinds of organisms present and hurt some traditional fishing activities. However, there is a plan and Congress will need to decide which elements of the plan to authorize and the amount of money to allocate for their implementation.

areas that periodically are covered with water. They include swamps, tidal marshes, coastal areas, and estuaries. Some wetlands, such as estuaries and marshes, are permanently wet, while others, such as many swamps, have standing water during only part of the year. Many wetlands may have standing water for only a few weeks a year, often in the spring of the year when the snow melts. Because wetlands breed mosquitoes and are sometimes barriers to the free movement of people, they have often been considered useless or harmful.

Most of them have been drained, filled, or used as dumps. Many modern cities have completely covered over extensive wetland areas and may even have small streams running under streets, completely enclosed in concrete. Not including Alaska, the United States has lost about 53 percent of its wetlands from pre-European settlement to the present. That equates to a loss of 89 million to 42 million hectares (220–100 million acres). The current loss rate is about 50,000 hectares (124,000 acres) per year. There are also 68 million hectares

(170 million acres) of wetlands in Alaska. These are mostly peatlands, with almost no loss in the past 200 years.

Each kind of wetland has unique qualities and serves as a home to many kinds of plants and animals. Because most wetlands receive constant inputs of nutrients from the water that drains from the surrounding land, they are highly productive and excellent places for aquatic species to grow rapidly. Wetlands are frequently critical to the reproduction of many kinds of animals. Many fish use estuaries and marshes for spawning. Wetlands also provide nesting sites for many kinds of birds and serve as critical habitats for many other species. Waterfowl hunters and commercial and sport fisheries depend on these habitats to produce and protect the young of the species they harvest. Human impact on wetlands has severely degraded or eliminated these spawning and nursery habitats.

Besides providing a necessary habitat for fish and other organisms, wetlands provide natural filters for sediments and runoff. This filtration process allows time for water to be biologically cleaned before it enters larger bodies of water, such as lakes and oceans, and reduces the sediment load carried by runoff. Wetlands also protect shorelines from erosion. When destroyed, the natural erosion protection provided by wetlands must be replaced by costly artificial measures, such as breakwalls.

Other Land-Use Considerations

The geologic status of an area must also be considered in land-use decisions. Building cities on the sides of volcanos or on major earthquake-prone faults has led to much loss of life and property. Building homes and villages on unstable hillsides or in areas subject to periodic fires is also unwise. Yet, every year more houses slide down California hillsides and wildfires consume homes throughout the dry West.

Another problem in some locations is lack of water. Southern California and metropolitan areas in Arizona must import water to sustain their communities. Wise planning would limit growth to whatever could be sustained by available resources. The risk of serious water shortages in Beijing, China, has forced the government to consider a massive 1200-kilometer (745-mile) diversion of water from the Yangtze River to the city. As the population of cities in such locations continues to grow, the strain on regional water resources will increase.

Water-starved cities often cause land-use dilemmas far from the city boundaries. Supplying water and power to cities often involves the construction of dams that flood valleys that may have significant agricultural, scenic, or cultural value. See chapter 16 on water use for a more extensive development of the topic.

Land-Use Planning Principles

Land-use planning is a process of evaluating the needs and wants of the population, the characteristics and values of the land, and various alternative solutions to the use of a particular land surface before changes are made. Planning land use brings with it the desires of many competing interests. The economic and personal needs of the population are a central driving force that requires land-use decisions to be made. However, the unique qualities of particular portions of the land surface prevent some uses, poorly accommodate others, but are highly suitable for others. For example, the floodplain beside a river is unsuitable for building permanent structures, can easily accommodate recreational uses such as parks, but may be most useful as a nature preserve. Agricultural land near cities can be easily converted to housing but may be more valuable for growing fruits and vegetables that are needed by the people of the city. This is particularly true when agricultural land is in short supply near urban centers. To make good land-use decisions, each piece of land must be evaluated and one of several competing uses assigned to it. When land-use decisions are made, the decision process usually involves the public, private landowners, developers, government, and special interest groups. Each interest has special wants and will argue that its desires are most important. Since this is the case, people must have principles and processes to guide land-use decisions, irrespective of their personal wants and interests.

A basic rule should be to make as few changes as possible, but when changes are suggested or required, several things should be considered.

1. *Evaluate and record any unique geologic, geographic, or biologic features of the land.*

   Some land has unique features that should be preserved because of their special value to society. The Grand Canyon, Yellowstone National Park, and many wilderness areas have been set aside to preserve unique physical structures, scenic characteristics, special ecosystems, or unusual organisms. On a more local level, a stream may provide fishing opportunities near a city, or land may have excellent agricultural potential that should take precedence over other uses. New York City purchased and protects a watershed that provides water for the city at a much lower cost than treating water in the Hudson River.

2. *Preserve unique cultural or historical features.*

   Some portions of the landscape, areas within cities, and structures have important cultural, historic, or religious importance that should not be compromised by land-use decisions. In many cities, historic buildings have been preserved and particular sections set aside as historic districts. Sacred sites, many battlefields, and places of unique historic importance are usually protected from development.

3. *Conserve open space and environmental features.*

   It must be recognized that open space and natural areas are not unused, low-value areas. Many studies of human behavior have shown that when people are given choices, they will choose settings that provide a view of nature or that allow one to see into the distance. Some have argued that this is a deep-seated biological need, while others suggest that it is a culturally derived trait. Regardless of its origin, urban planners know that access to open space and natural areas is an important consideration when determining how to use land. Therefore, it makes sense to protect open space within and near centers of population.

Computer Tools Aid Decision Making— Growing Smarter to Protect Habitats

Picture yourself at a town hall meeting somewhere in America, sitting with dozens of residents debating the local land-use plan. Presentations are given, elected representatives argue, and public comment is heard—a classic exercise in American self-government. What concerns might you bring to this discussion? The need for new jobs? New roads? Protecting forests or farmlands?

Public concern about the loss of our natural heritage is spurring planners and others working at local, regional, and state levels to focus more and more on identifying lands that should be conserved. There are several key sources of environmental information for this effort, including federal agencies, nonprofit groups, and state natural heritage programs, which track the location and status of threatened plants, animals, and ecosystems.

To aid the local land-use planning process, an array of software products—collectively known as *decision-support tools*—are being developed. These tools use sophisticated mapping technology, three-dimensional visualization, modeling, and the capacity to easily incorporate individual and community concerns. Decision-support tools can integrate complex information about biological issues (e.g., where the sensitive habitats are), physical issues (e.g., where the transportation networks are), and socioeconomic issues (e.g., which parcels have the most economic value). This integration provides a comprehensive view of the forces shaping land use. At once powerful and easy to use, these tools will empower citizens to participate more in the development of public policy by giving them access to scientific and technical information about land use previously available only to experts. With this information, perhaps those with different views can work together better to find solutions.

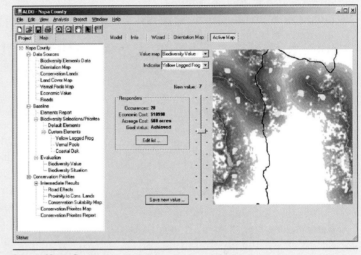

Source: NatureServe.

---

4. *Recognize and calculate the cost of additional changes that will be required to accommodate altered land use.*

Whenever land use is altered, additional modifications will be required to accommodate the change. For example, when a new housing development is constructed, schools and other municipal services will be required, roads will need to be improved, and the former use of the land is lost. Frequently, the cost of these changes is not borne by the developer or the homeowner but becomes the responsibility of the entire community; everyone pays for the additional cost through tax increases. In many areas, basic services, such as provision of water, are severe problems. It makes no sense to build new housing when there is not enough water to support the existing population.

5. *Plan for mixed housing and commercial uses of land in proximity to one another.*

One of the major problems associated with development in North America is segregation of different kinds of housing from one another and from shopping and other service necessities. Mixing various kinds of uses together (single family housing, apartments, shopping and other service areas, and offices) allows easier connection between uses without reliance on the automobile. Walking and biking become possible when these different uses are within a short distance of one another.

6. *Plan for a variety of transportation options.*

Plan for transportation options other than the automobile. Currently, most urban and rural areas do not accommodate bicycles. Although bicycles can legally be ridden on streets, it is generally unsafe because of the high speeds of vehicular traffic and poor road surface conditions near the edge of the pavement. Special bike lanes are available in some areas, but this is rare in much of North America. Walking is a healthy, pleasant way to get from one task to another. However, crossing wide, busy streets is difficult, many areas lack sidewalks, and related service areas are often far apart. This tends to discourage this mode of transport. Clustering housing and service areas allows for easier planning of bus and rail routes to allow people to get from one place to another without relying on automobiles.

7. *Set limits and require managed growth with compact development patterns.*

Much of unplanned growth occurs because there is no plan or the land-use plan is not enforced. One very effective tool that promotes efficient use of the land is to establish an urban growth limit for a municipality. An **urban growth limit** establishes a boundary within which development can occur. Development outside the boundary is severely restricted. One of the most important outcomes of setting urban growth boundaries is that a great deal of planning must precede the establishment of the limit. This lets all in the community know what is going on and can allow development to occur in logical stages that do not stress the community's ability to supply services. This mechanism also stimulates higher density uses of the urban land.

8. *Encourage development within areas that already have a supportive infrastructure so that duplication of resources is not needed.*

Because all development of land for human activity requires that services be provided, it makes sense that housing and commercial development occur where the infrastructure is already present. This includes electric, phone, sewer, water, and transportation systems. It includes service industries, such as shopping, banking, restaurants, hotels, and entertainment. It includes schools, hospitals, and police protection. If development occurs far from these services, it is very costly to extend or duplicate those services in the new location. Furthermore, all large cities and most smaller ones have vacant lots or abandoned

buildings that have outlived their usefulness or are vacant because of changing business or housing needs. This land is already urban, and close to municipal services, and the buildings can be readily renovated or demolished and replaced. Many will argue that this is too expensive, but if there are no options, these spaces will be used, and their redevelopment can be important in revitalizing inner-city spaces.

Mechanisms for Implementing Land-Use Plans

Land-use planning is the construction of an orderly list of priorities for the use of available land. Developing a plan involves gathering data on current use and geological, biological, and sociological information. From these data, projections are made about what human needs will be. All of the data collected are integrated with the projections, and each parcel of land is evaluated and assigned a best use under the circumstances. There are basically three components that contribute to the successful implementation of a land-use development plan: land-use decisions can be assigned to a regional governmental body, the land or its development rights can be purchased, and laws or ordinances can be used to regulate land use.

Establishing State or Regional Planning Agencies

National and regional planning is often more effective than local land-use planning because political boundaries seldom reflect the geological and biological database used in planning. Larger units contain more diverse collections of landscape resources and can afford to hire professional planners. A regional approach is also likely to prevent duplication of facilities and lead to greater efficiency. For example, airport locations should be based on a regional plan that

incorporates all local jurisdictions. Three cities only 30 kilometers (20 miles) apart should not build three separate airports when one regional airport could serve their needs better and at a lower cost to the taxpayers.

In North America, the majority of regional governmental bodies are presently voluntary and lack any power to implement programs. Their only role is to advise the member governments. Members of local governments often are unwilling to give up power. They may view policy from a narrow perspective and put their own interests above the goals of the region.

One way to encourage regional planning is to develop policies at the state, provincial, or national level. The first state to develop a comprehensive statewide land-use program was Hawaii. During the early 1960s, much of Hawaii's natural beauty was being destroyed to build houses and apartments for the increasing population. The same land that attracted tourists was being destroyed to provide hotels and supermarkets for them. Local governments had failed to establish and enforce land-use controls. Consequently, in 1961, the Hawaii State Land-Use Commission was founded. This commission designated all land as urban, agricultural, or conservational. Each parcel of land could be used only for its designated purpose. Other uses were allowed only by special permit. To date, the record for Hawaii's action shows that it has been successful in controlling urban growth and preserving the islands' natural beauty, even though the population continues to grow. (See figure 13.8.)

Several states and cities are attempting to follow Hawaii's lead in land-use regulation. (See tables 13.1 and 13.2.) Some have passed legislation dealing with special types of land use. Examples include wetland preservation, floodplain protection, and scenic and historic site preservation. Although direct state involvement in land-use regulation is relatively new, it is expected to grow. Only large, well-financed levels of government can afford to pay for the growing cost of adequate land-use planning. State, provincial, and regional governments are also more likely to have the power to counter the political and economic influences of land developers, lobbyists, and other special-interest groups when conflicts over specific land-use policies arise.

National governments also have a role to play. Since national governments own and administer the use of large amounts of land, national policy will determine how those lands are used. The designation of lands as wilderness, forests, rangelands, or parks at the federal level often involves a balancing of national priorities with local desires. As with local land-use issues, the conflicts over the use of federal lands are often economic and involve compromise.

Purchasing Land or Use Rights

Probably the simplest way to protect desirable lands is to purchase them. When privately owned land is desired for special purposes, it must be purchased from the owner, and the owner has a right to expect to get a fair price for the property. When it is determined that land has a high public value, then either the land or the rights to use it must be purchased. Many environmental organizations purchase lands of special historic, scenic, or environmental value.

In many cases, the owners may not be willing to sell the land but are willing to limit the uses to which the land can be put in the future. Therefore, landowners may sell the right to develop the land or may agree to place restrictions on the uses any future owners might consider.

Regulating Use

Many communities are not in a financial position to purchase lands; therefore, they attempt to regulate land use by zoning laws.

Zoning is a common type of land-use regulation that restricts the kinds of uses to which land in a specific region can be put. When land is zoned, it is designated for specific potential uses. Common designations are agricultural, commercial, residential, recreational, and industrial. (See figure 13.9.)

Most local zoning boards are elected or appointed and often lack specific training in land-use planning. As a result, zoning regulations are frequently made by people who see only the short-term gain and not the possible long-term loss. Often the land is simply zoned so that its current use is sanctioned and is rezoned when another use appears to have a higher short-term value to the

figure 13.8 **Hawaii Land-Use Plan** Hawaii was the first state to develop a comprehensive land-use plan. This development in an agricultural area was stopped when the plan was implemented in the 1960s.

Table 13.1 Examples of State Land-Use Planning Legislation

State	Legislation
California	Coastal Act (1976)
	Coastal Zone Conservation Act (1972)
	Tahoe Regional Planning Compact (1969)
Florida	Omnibus Growth Management Act (1985)
	State Comprehensive Plan (1985)
	State and Regional Planning Act (1984)
	Environmental Land and Water Management Act (1972)
Georgia	Coordinated Planning Legislation (1989)
Hawaii	Hawaii State Plan (1978)
	Hawaiian Land Use Law (1961)
Maine	Comprehensive Planning and Land-Use Regulation Act (1988)
Maryland	Economic Growth, Resource Protection and Planning Act (1992)
Massachusetts	Cape Cod Commission Act (1989)
	Martha's Vineyard Commission Act (1974)
New Jersey	State Planning Act (1985)
	State Pinelands Protection Act (1974)
New York	Adirondack Park Agency Act (1971)
Oregon	Land Conservation and Development Act (1973)
Rhode Island	Comprehensive Planning and Land Use Regulation Act (1988)
Vermont	Growth Management Act (1988)
Washington	Growth Management Act (1990)
	Amendments to the Growth Management Act (1991)

Table 13.2 Examples of Local Growth Management Actions

Locality	Action
Livermore, California	Established limits on the number of building permits issued
Petaluma, California	Limited the number of building permits to 500 per year and established criteria for awarding permits that emphasized planning
San Diego, California	Adopted a growth management plan
Boulder, Colorado	Established a growth management plan and limited the number of new buildings
Larimer County, Colorado	Established regional urban growth areas
Westminster, Colorado	Limited the number of building permits issued
Boca Raton, Florida	Established a cap on growth at 40,000 dwellings in 1972. The ordinance was eventually overturned by the U.S. Supreme Court.
Broward County, Florida	Public facilities must be developed along with new housing.
Palm Beach County, Florida	Growth control to provide affordable housing
Sanibel, Florida	Established growth limits based on environmental carrying capacity. The city is adjacent to an important wildlife refuge.
Hardin County, Kentucky	Eliminated rural zoning through growth management techniques
Montgomery County, Maryland	Established growth management controls
Minneapolis, Minnesota	Containment of development within city boundaries
Ramapo, New York	Does not permit development until municipal services are established in an area
Austin, Texas	Neighborhood-based conservation overlay districts

figure 13.9 **Zoning** Most communities have a zoning authority that designates areas for particular use. This sign indicates that decisions have been made about the "best" use for the land.

community. Even when well-designed land-use plans exist, they are usually modified to encourage local short-term growth rather than to provide for the long-range needs of the community. The public needs to be alert to variances from established land-use plans, because once the plan is compromised, it becomes easier to accept future deviations that may not be in the best interests of the community. Many times, individuals who make zoning decisions are real-estate agents, developers, or local business people. These individuals wield significant local political power and are not always unbiased in their decisions. Concerned citizens must try to combat special interests by attending zoning commission meetings and by participating in the planning process.

Special Urban Planning Issues

Urban areas present a large number of planning issues. Transportation, open space, and improving the quality of life in the inner city are significant problems.

Urban Transportation Planning

A growing concern of city governments is to develop comprehensive urban transportation plans. While the specifics of such plans might vary from region to region, urban transportation planning usually involves four major goals:

1. Conserve energy and land resources.
2. Provide efficient and inexpensive transportation within the city, with special attention to people who are unable to drive, such as many elderly, young, handicapped, and financially disadvantaged persons.
3. Provide suburban people opportunities to commute efficiently.
4. Reduce urban pollution.

Any successful urban transportation plan should integrate all of these goals, but funding and intergovernmental cooperation are needed to achieve this. The problems associated with current urban transportation will certainly not disappear overnight, but comprehensive planning is the first step to solving them.

Since automobiles are heavily used, transportation corridors and parking facilities must be included in any urban transportation plan.

However, many urban planners recognize that the automobile's disadvantages may outweigh its advantages, so some cities, such as Toronto, London, San Francisco, and New York, have attempted to dissuade automobile use by developing mass transit systems and by allowing automobile parking costs to increase substantially.

The major urban mass transit systems are railroads, subways, trolleys, and buses. In many parts of the world, mass transportation is extremely efficient and effective. However, in the

Land-Use Planning and Aesthetic Pollution

Unpleasant odors, disagreeable tastes, annoying sounds, and offensive sights can be aggravating. Yet it is difficult to get complete agreement on what is acceptable and when some aesthetic boundary has been crossed that is unacceptable. Furthermore, many useful activities generate stimuli that are offensive while the activity itself may be essential or at least very useful. Many of these do not harm us physically but may be harmful from an aesthetic point of view.

Odors are caused by various airborne chemicals. People who live near dairy farms, livestock-raising operations, paper mills, chemical plants, steel mills, and other industries may be offended by the odors originating from these sources, but the products of these activities are needed. Many of these industries discharge wastewater that contains materials that decompose or evaporate and cause odor pollution. People who are constantly exposed to an odor are usually not as offended by it as are people who are newly exposed to it. When an odor is constantly received, the brain ceases to respond to the stimulus. In other words, the person is not aware of the odor.

Chemicals can also affect the taste of things we eat or drink. Some naturally produced chemicals, as well as those discharged into waterways, can affect the taste of food or drinking water. Minute quantities of certain chemicals can affect the taste of our drinking water and our food. Algae in water produce flavors that are offensive to many. Some groundwater sources have sulfur or salts that cause unwanted flavors. In fish, a concentration of 100 ppm of phenol can be tasted. Although this small amount of chemical is not harmful biologically, it does make the fish very unappetizing.

Noise is unwanted sound. It is produced as an incidental by-product of industry, traffic, and other human activity. People vary considerably in their tolerance of unwanted sound. Many in cities adapt to the constant noise of the city, while those who live in more rural areas would find city noises annoying.

Visual pollution is a sight that offends us. This type of pollution is highly subjective and is, therefore, difficult to define or control. To most people, a dilapidated home or building is offensive, especially if located in an area of higher-priced homes. A heavily littered highway or street is aesthetically offensive to most people, and litter along a wilderness trail is even more unacceptable. Some sources of visual pollution are not so clear-cut, however. To many people, roadside billboards are offensive, but they can be helpful to advertisers and to travelers looking for information.

Since aesthetic pollutants—odors, tastes, sounds, and sights—are extremely difficult to define, it is difficult to establish aesthetic pollution standards. One of the simplest ways to eliminate many of these annoyances is to separate the generator of the offensive stimulus from the general public. Proper land-use planning can greatly reduce the amount of annoying aesthetic pollution. If uses that produce annoying stimuli can be clustered together rather than dispersed, the effect on the public is reduced. It does not make sense to allow new home construction in the vicinity of farming operations that will produce odors or industrial operations or airports that will generate noise. Similarly, allowing homes to be built in areas that have poor quality groundwater that will be used for drinking is a problem—unless an additional decision is made to provide a different source of drinking water. The roadside advertising that many find offensive can be regulated by allowing such signs only in specific areas rather than allowing them to be built anywhere along the roadway. None of these land-use remedies eliminates the aesthetic pollutant, but by segregating the public from the source of the annoyance, the impact is reduced.

United States where the automobile is the primary method of transportation, mass transportation systems are often underfunded and difficult to establish because mass transit is:

1. Economically feasible only along heavily populated routes
2. Less convenient than the automobile
3. Extremely expensive to build and operate
4. Often crowded and uncomfortable

Although mass transit meets a substantial portion of the urban transportation needs in some parts of the world, such as in Europe, its use in North America has grown slightly, while automobile and air travel has grown tremendously. One reason for this difference is the size of the region. Unlike North America, Japan and countries in Europe that have extensive public transportation are much smaller geographically. In addition, they have more uniformly dense populations. Thus, it is easier to make public transportation work in those settings. (See figure 13.10.) A variety of forces has contributed to this situation. As people became more affluent, they could afford to own automobiles, which are a convenient, individualized method of transportation. Governments in North America encourage automobile use by financing highways and expressways, by maintaining a cheap energy policy, and by withdrawing support for most forms of mass transportation. Thus, they encourage automobile transportation with hidden subsidies (highway construction and cheap gasoline) but maintain that rail and bus transportation should not be subsidized. North Americans will seek alternatives to private automobile use only when the cost of fuel, the cost of parking, or the inconvenience of driving becomes too high.

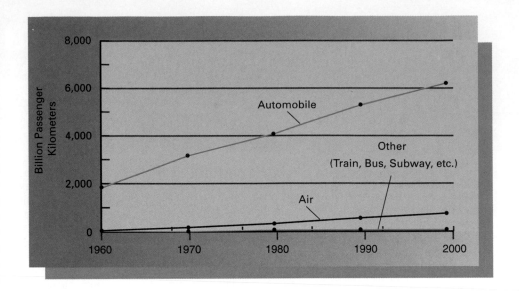

figure 13.10 **Decline of Mass Transportation** Automobile use has increased consistently since 1960, while rail and bus transport use has remained low.

Source: Data from U.S. Department of Transportation, *National Transportation Statistics 2000*.

Urban Recreation Planning

About three-fourths of the population of North America live in urban areas. These urban dwellers value open space because it breaks up the sights and sounds of the city and provides a place for recreation. Inadequate land-use planning in the past has rapidly converted urban open spaces to other uses. Until recently, creating a new park within a city was considered an uneconomical use of the land, but people are now beginning to realize the need for parks and open spaces.

Some cities recognized the need for open space a long time ago and allocated land for parks. London, Toronto, and Perth, Australia, have centrally located and well-used parks. New York City set aside approximately 200 hectares (500 acres) for Central Park in the late 1800s. (See figure 13.11.) Boston has developed a park system that provides a variety of urban open spaces. Other cities have not dealt with this need for open space because they have lacked either the foresight or the funding. Recreation is a basic human need. The most primitive tribes and cultures all engaged in games or recreational activities. New forms of recreation are continually being developed. In the congested urban center, cities often must construct special areas where recreation can take place.

A major problem with urban recreation is locating recreational facilities near residential areas. Facilities that are not conveniently located may be infrequently used. For example, the hundreds of thousands of square kilometers of national parks in Alaska and the Yukon will be visited every year by a tiny proportion of the population of North America. Large urban centers are discovering that they must provide adequate, low-cost recreational opportunities within their jurisdiction. Some of these opportunities take the form of commercial establishments, such as bowling centers, amusement parks, and theaters. Others must be subsidized by the community. (See figure 13.12.) Playgrounds, organized recreational activities, and open space have usually been combined into an arm of the municipal government known as the parks and recreation department. Cities spend millions of dollars to develop and maintain recreation programs. Often, conflict arises over the allocation of financial and land resources. These are closely tied because open land is scarce in urban areas, and it is expensive. Riverfront property is ideal for park and recreational use, but it is also prime land for industry, commerce, or high-rise residential buildings. Although conflict is inevitable, many metropolitan areas are beginning to see that recreational

resources may be as important as economic growth for maintaining a healthy community.

An outgrowth of the trend toward urbanization is the development of **nature centers.** In many urban areas, so little natural area is left that the people who live there need to be given opportunities to learn about nature. Nature centers are basically teaching institutions that provide a variety of methods for people to learn about and appreciate the natural world. Zoos, botanical gardens, and some urban parks, combined with interpretative centers, also provide recreational experiences. Nature centers are usually located near urban centers, in places where some appreciation of the natural processes and phenomena can be developed. They may be operated by municipal governments or by school systems or other nonprofit organizations.

Redevelopment of Inner-City Areas

As people moved to the suburbs during the past 50 years, the inner city was abandoned. Many old industrial sites sit vacant. Businesses have moved to the suburban malls. The quality of housing has declined. Services have been reduced. To improve the quality of life of the residents of the city, special efforts

figure 13.11 **Urban Open Space** These photos show open space in two urban areas, Stanley Park in Vancouver, British Columbia and Echo Park Lake, in Los Angeles, California. If the land had not been set aside, it would have been developed.

must be made to revitalize the city. Although activities that will improve the quality of life in the inner city vary from city to city, several land-use processes can help. One problem that has plagued industrial cities is vacant industrial and commercial sites. Many of these buildings have remained vacant because the cost of cleanup and renovation is expensive. Such sites have been called **brownfields.** Many of these sites involved environmental contamination, and since the EPA required that they be cleaned up to a pristine condition, no one was willing to do so.

A new approach to utilizing these sites is called **brownfields development.** This involves a more realistic approach to dealing with the contamination at these sites. Instead of requiring complete cleanup, the degree of cleanup required is matched to the intended use of the site. Although an old industrial site with specific contamination problems may not be suitable for housing, it may be redeveloped as a new industrial site, since access to the contamination can be controlled. An old industrial site that has soil contamination may be paved to provide parking. (See

figure 13.12 **Urban Recreation** In urban areas, recreation often takes the form of sports programs, playgrounds, and walking. Most cities recognize the need for such activities and facilities, and develop extensive recreation programs for their citizens.

Environmental Close-Up: Rehabilitation of Contaminated Sites in chapter 19.)

Another important focus in urban redevelopment is the remodeling of abandoned commercial buildings for shopping centers, cultural facilities, or high-density housing. Chattanooga, Tennessee, has achieved a reputation for revitalizing its inner-city area. The process of revitalization involved extensive planning activities that included the public, public and private funding of redevelopment activities, establishment of an electric bus system to alleviate air pollution, renovation of existing housing, redevel-

opment of old warehouses into a shopping center, and incorporation of a condemned bridge over the Tennessee River into a portion of a park that is also an important pedestrian connection between a residential area and the downtown business district.

Smart Growth

The current pattern of growth throughout most of North America is commonly referred to as "sprawl." Concern about sprawl is growing because of the sheer pace of land development, which, according to the U.S. Department of Agriculture, is roughly double what it was only a decade ago. Such development has had a number of negative cultural, economic, environmental, and social consequences. In central cities and older suburbs, these include deteriorating infrastructure, poor schools, and a shortage of affordable, quality housing. In newer suburban areas, these problems may include increased traffic congestion, declining air quality, and the loss of open space. Smart growth is an approach that argues that these problems are two sides of the same coin and that the neglect of our central cities is fueling the growth and related problems of the suburbs.

To address these problems, smart growth advocates emphasize the concept of developing "livable" cities and towns. Livability suggests, among other things, that the quality of our built environment and how well we preserve the natural environment both directly affect our quality of life.

Though they support growth, communities are questioning the economic costs of abandoning infrastructure in the city and rebuilding it farther out. They are questioning the necessity of spending increasing time in cars, locked in traffic or traveling miles to the nearest store. They are questioning the practice of abandoning older communities while developing open space and prime agricultural lands at the suburban fringe.

Smart growth recognizes the benefits of growth. It invests time, attention, and resources in restoring a sense of community and vitality to center cities and older suburbs. In new developments, smart growth practices create communities that are more town centered, transit- and pedestrian-oriented, and have a greater mix of housing, commercial, and retail uses. These practices also preserve open space and other environmental amenities. Smart growth recognizes connections between development and quality of life that can be presented as a list of principles.

Smart growth principles

1. Mix land uses.
2. Take advantage of compact building design.
3. Create a range of housing opportunities and choices.
4. Create walkable neighborhoods.
5. Foster distinctive, attractive communities with a strong sense of place.
6. Preserve open space, farmland, natural beauty, and critical environmental areas.
7. Strengthen and direct development toward existing communities.
8. Provide a variety of transportation choices.
9. Make development decisions predictable, fair, and cost-effective.
10. Encourage community and stakeholder collaboration in development decisions.

The features that distinguish smart growth vary by community. No two streets, neighborhoods, or cities are identical. There is no "one-size-fits-all" solution. Smart growth in Portland, Oregon, has different characteristics than smart growth in Austin, Texas. For that reason, smart growth does not prescribe solutions. Rather, it provides choices and seeks to build on proven successes. The smart growth principles listed above reflect the experience of localities that have successfully created smart growth communities. These communities had a vision of where they wanted to go and of what things they valued in their community.

An example of the principles of smart growth being applied is Suisun City, California. In 1989, the *San Francisco Chronicle* rated Suisun City, a town of 25,000 people between San Francisco and Sacramento, the worst place to live in the Bay Area. At that time, Suisun City's historic Main Street was a strip of boarded-up storefronts and vacant lots. Several blocks away, an oil refinery sat at the head of the polluted, silt-laden Suisun Channel. Today, Suisun's harbor is filled with boats and lined with small businesses. A train and bus station that connects the city to the rest of northern California is located nearby. The town is diverse, walkable, and picturesque. Its crime rate is low and its housing affordable. This dramatic change occurred because Suisun City's residents, busi-

nesses, and elected officials agreed on a common vision for their town's future. Cleaning up the polluted Suisun Channel and making the waterfront a focal point of their town was a common goal. The citizens also wanted to reestablish the historic Main Street as a social and retail gathering place. It was recognized that by encouraging tax-generating commercial development such as retail shops and restaurants along Main Street and the waterfront, municipal finances would be strengthened.

In its turnaround, Suisun City avoided large-scale redevelopment projects such as shopping centers and industrial parks that would have altered its historic small-town character. Suisun City is still a work in progress, but it is now invigorated with new businesses and residents, a stronger community spirit, and optimism about its future.

Federal Government Land-Use Issues

Since the federal government manages large amounts of land, the laws and regulations that shape land-use policy are important. For example, the 1960 Multiple Use Sustained Yield Act divided use of national forests into four categories: wildlife habitat preservation, recreation, lumbering, and watershed protection. This act was designed to encourage both economic and recreational use of the forests. However, specific users of this public land are often in conflict, particularly recreational users with timber harvesters.

An 1872 mining law has also been important in federal land management. The law allows anyone to prospect for minerals on public lands and to establish a claim if such minerals are discovered. The miner is then allowed to purchase the rights to extract the mineral for $5 per acre. Many feel that the law is obsolete, but it is still in force and public land is still being sold to mining interests at ridiculously low prices. These laws are examples of past policy decisions that are still in effect. Today, one of the major uses of public lands is outdoor recreation.

Many people want to use the natural world for recreational purposes because nature can provide challenges that may be lacking in their day-to-day lives. Whether the challenge is hiking in the wilderness, exploring underwater, climbing mountains, or driving a vehicle through an area that has no roads, these activities offer a sense of adventure. Look at table 13.3. All of these activities use the out-of-doors but not in the same way. Conflicts develop because some of these activities cannot occur in the same place at the same time. For example, wilderness camping and backpacking often conflict with off-road vehicles.

A basic conflict exists between those who prefer to use motorized vehicles and those who prefer to use muscle power in their recreational pursuits. (See figure 13.13.) This conflict is particularly strong because both groups would like to use the same public land. Both have paid taxes, and both feel that it should be available for them to use as they wish.

Finally, as more rangelands and forests have had vehicular access controlled or eliminated, those who want to use public lands for motorized recreation have become upset.

Land-use conflicts also arise between business interests and recreational users of public lands. Federal and state governments give special use permits to certain users of public lands. Many ski resorts in the West make use of public lands. Grazing is also an important use of federal land. Based on "Animal Use Months" established by the Bureau of Land Management or the Forest Service, ranchers are allowed to graze cattle on certain public lands. Technically, failure to comply can mean a loss of grazing rights. However, since the establishment of regulations is highly political, many maintain that the political influence of ranchers allows them to use a public resource without adequately compensating the government. In addition, the regulatory agencies are understaffed and find it difficult to adequately regulate the actions of individ-

Table 13.3	Number of People Who Participated in Selected Outdoor Recreational Activities in 2000

Activity	Percent of Population over Seven Years of Age Participating
Exercise walking	33
Swimming	27
Bicycling	24
Fishing	15
Camping	18
Golf	14
Hiking	10
Running	8
Hunting	6
Backpacking	4
Skiing	4

Source: Data from *Statistical Abstract of the United States,* 2000.

figure 13.13   **Conflict over Recreational Use of Land**   Land may be used for both motorized and nonmotorized activities. The people who participate in these two kinds of recreation are often antagonists over the allocation of land for recreational use.

ual ranchers. As a result, some lands are overgrazed. Many people who want to use the publicly owned rangelands for outdoor recreation resent the control exercised by grazing interests. On the other hand, ranchers resent the intrusion of hikers and campers on land they have traditionally controlled.

An obvious solution to this problem is to allocate land to specific uses and to

regulate the use once allocations have been made. Several U.S. governmental agencies, such as the National Park Service, the Bureau of Land Management, the Forest Service, and the Fish and Wildlife Service, allocate and regulate the lands they control. However, these agencies have conflicting roles. The Forest Service has a mandate to manage forested public lands for timber

Decision Making in Land-Use Planning—The Malling of America

The following situation has happened thousands of times during the past 20 years:

A developer has just announced plans to build a large shopping mall on the outskirts of your city in what is now prime farmland. Many jobs will be created by the construction and operation of the proposed mall, which will include stores and three new theaters. Presently, your city has some unemployment, only one theater in the downtown area, and little variety in its retail businesses. On the surface, the proposed mall seems to be only good news. Is this the case? Before answering, look at the entire situation.

- What will happen to the downtown area when the mall is opened?
- If you owned a downtown business, would you favor building the mall?
- What will happen to the taxes on the farms near the new mall?
- If you were a farmer, would you favor the project?
- What effect will paving prime farmland have?
- How will storm water runoff be affected?
- How will future housing development be influenced?
- What other problems can be associated with building the complex?

- If you were the mayor or city manager, would you favor construction of the mall? Why?
- Is the proposed project all good news after all?

---

production. This mandate often comes in conflict with recreational uses. Similarly, the Bureau of Land Management has huge tracts of land that can be used for recreation, but it traditionally has been mandated to manage grazing rights.

A particularly sensitive issue is the designation of certain lands as wilderness areas. Obviously, if an area is to be wilderness, human activity must be severely restricted. This means that the vast majority of Americans will never see or make use of it. Many people argue that this is unfair because they are paying taxes to provide recreation for a select few. Others argue that if everyone were to use these areas, their charm and unique character would be destroyed and that, therefore, the cost of preserving wilderness is justifiable.

Areas designated as wilderness make up a very small proportion of the total public land available for recreation. (See figure 13.14.) Such areas, however, are becoming increasingly visible due to their unique character.

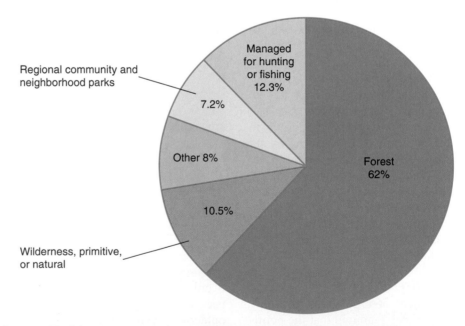

figure 13.14 **U.S. Federal Recreational Lands in 2000** Of the approximately 108 million hectares (267 million acres) of federal recreational lands in the United States, approximately 10 percent is designated as wilderness, primitive, or natural.

Source: Data from the *Statistical Abstract of the United States,* 2000.

Summary

Historically, waterways served as transportation corridors that allowed for the exploration of new land and for the transport of goods. Therefore, most large urban centers began as small towns located near water. Water served the needs of the towns in many ways, especially as transportation. Several factors resulted in the shift of the population from rural to urban. These included the Industrial Revolution, which provided jobs in cities, and the addition of foreign immigrants to the cities. As towns became larger, the farmland surrounding them became suburbs surrounding industrial centers. Unregulated industrial development in cities resulted in the degradation of the waterfront and stimulated the development of suburbs around the city as people sought better places to live and had the money to purchase new homes. The rise in automobile ownership further stimulated the movement of people from the cities to the suburbs.

Many problems have resulted from unplanned growth. Current taxation policies encourage residential development of farmland, which results in a loss of valuable agricultural land. Floodplains and wetlands are often mismanaged. Loss of property and life results when people build on floodplains. Wetlands protect our shorelines and provide a natural habitat for fish and wildlife. Transportation problems and lack of open space are also typical in many large metropolitan areas.

Land-use planning involves gathering data, projecting needs, and developing mechanisms for implementing the plan. Good land-use planning should include assessment of the unique geologic, geographic, biological, and historic and cultural features of the land; the costs of providing additional infrastructure; preservation of open space; provision for a variety of transportation options; a mixture of housing and service establishments; redevelopment of disused urban land; and establishment of urban growth limits. Establishing regional planning agencies, purchasing land or its development rights, and zoning are ways to implement land-use planning. The scale of local planning is often not large enough to be effective because problems may not be confined to political boundaries. Regional planning units can afford professional planners and are better able to withstand political and economic pressures. A growing concern of urban governments is to develop comprehensive urban transportation plans that seek to conserve energy and land resources, provide efficient and inexpensive transportation and commuting, and help to reduce urban pollution. Urban areas must also provide recreational opportunities for their residents and seek ways to rebuild decaying inner cities.

Federal governments own and manage large amounts of land, therefore national policy must be developed. This usually involves designating land for particular purposes, such as timber production, grazing land, parks, or wilderness. The recreational use of public land often requires the establishment of rules that prevent conflict between potential users who have different ideas about what appropriate uses should be. Often federal policy is a compromise between competing uses and land is managed for multiple uses.

Key Terms

brownfields 297
brownfields development 297
floodplains 287
floodplain zoning ordinances 288
infrastructure 287

land-use planning 290
megalopolis 285
nature centers 296
ribbon sprawl 284
tract development 284

urban growth limit 292
urban sprawl 284
wetlands 288
zoning 293

Review Questions

1. Why did urban centers develop near waterways? Are they still located near water?
2. Describe the typical changes that have occurred in cities from the time they were first founded until now.
3. Why do people move to the suburbs?
4. Why do some farmers near urban areas sell their land for residential or commercial development? If you were in this position, would you sell?
5. What is a megalopolis?
6. What land uses are suitable on floodplains?
7. What is multiple land use? Can land be used for multiple purposes?
8. Why is it important to provide recreational space in urban planning?
9. How can recreational activities damage the environment? Do you engage in any of those activities?
10. What is the monetary impact of recreational activities?
11. What are some strictly urban-related recreational activities?
12. List some conflicts that arise when an area is designated strictly as wilderness.
13. Describe the steps necessary to develop a land-use plan.
14. What are the advantages of regional or state planning?
15. List three benefits of land-use planning.

Critical Thinking Questions

1. Choose the city where you live. Interview local residents and look at old city maps. What did the city look like 75 years ago? What were the city's boundaries? Where did people do their shopping? How did they get around? How does this compare with the current situation in the city?

2. What historical factors brought members of your family to the city? How does this compare to the factors that are currently contributing to the growth of cities in the developing world?

3. Consider the outer rim of the city closest to you. Which, if any, of the problems associated with unplanned growth are associated with your city? What factors make them a problem? What do you think can be done about them?

4. There has been tremendous development in the arid West of the United States over the past few decades, creating demands for water. How should these demands be met? Should there be limits to this type of development? What kinds of limits, if any?

5. Imagine you are a U.S. Forest Service supervisor who is creating a 10-year plan that is in the public comment stage. What interests would be contacting you? What power would each interest have? How would you manage the competing interests of timber, mining, grazing, and recreation or between motorized and nonmotorized recreation? What values, beliefs, and perspectives helped you form your recommendations?

6. Imagine that you lived in an area of the country that has the potential to be named a wilderness area. What conflicts do you think would arise from such a declaration? Who might be some of the antagonists? Which perspective do you think is most persuasive? How would you answer the objections of the other perspective?

7. Look at the Issues and Analysis section of this chapter and answer the questions. What values, beliefs, and perspectives lead you, as the imaginary city manager, to reach your decision about the shopping mall?

8. After reading the Environmental Close-Up in this chapter concerning wetland loss in Louisiana, what kinds of recommendations would you make to help preserve wetlands? What do you suppose might happen if nothing is done? What resistance might wetland preservation generate?

Concept Map

Construct a map to show relationships among the following concepts:

land-use planning   limits to growth   suburbia

zoning   open space   transportation

sprawl   urban growth

Interactive Exploration

Check out the website at **http://www.mhhe.com/environmentalscience** and click on the cover of this textbook for quizzing, career information, case studies, and hot links for information on the following topics:

Land-use Planning

Urbanization and Sustainable Cities

Environmental Policy and Decision Making

Air Pollution

Refuges and Sanctuaries

Restoration Ecology

Conservation and Management of Habitats and Species

Landscape Ecology

Soil and Its Uses

CHAPTER 14

Objectives

After reading this chapter, you should be able to:

- Describe the geologic processes that build and erode the Earth's surface.
- List the physical, chemical, and biological factors involved in soil formation.
- Explain the importance of humus to soil fertility.
- Differentiate between soil texture and soil structure.
- Explain how texture and structure influence soil atmosphere and soil water.
- Explain the role of living organisms in soil formation and fertility.
- Describe the various layers in a soil profile.
- Describe the processes of soil erosion by water and wind.
- Explain how contour farming, strip farming, terracing, waterways, windbreaks, and conservation tillage reduce soil erosion.
- Understand that the misuse of soil reduces soil fertility, pollutes streams, and requires expensive remedial measures.
- Explain how land not suited for cultivation may still be productively used for other purposes.

Chapter Outline

Geologic Processes
Soil and Land
Soil Formation
Soil Properties
Soil Profile
Soil Erosion
Soil Conservation Practices
    Contour Farming
    Strip Farming
    Terracing
    Waterways
    Windbreaks
Conventional Versus Conservation Tillage
Environmental Close-Up: *Land Capability Classes*
Global Perspective: *Worldwide Soil Degradation*
Protecting Soil on Nonfarm Land
Issues—Analysis: *Soil Erosion in Virginia*

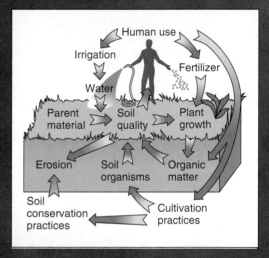

Geologic Processes

We tend to think of the Earth as being stable and unchanging until we recognize such events as earthquakes, volcanic eruptions, floods, and windstorms changing the surface of the places we live. There are forces that build new land and opposing forces that tear it down. Much of the building process involves shifting of large portions of the Earth's surface known as plates. The Earth is composed of an outer crust, a plastic mantle, and a central core. The **crust** is an extremely thin, less dense, solid covering over the underlying mantle. The **mantle** is a layer that makes up the majority of the Earth and surrounds a small core made up primarily of iron. The mantle consists of an inner solid portion and an outer portion that is capable of flow. (See figure 14.1.)

Plate tectonics is the concept that the outer surface of the Earth consists of large plates composed of the crust and the outer portion of the mantle and that these plates are slowly moving over the surface of the liquid outer mantle. This combination of crust and outer mantle is known as the **lithosphere.** The heat from the Earth causes slow movements of the outer layer of the mantle similar to what happens when you heat a liquid on the stove, only much slower. The movements of the plates on this plastic outer layer of the mantle are independent of each other. Therefore, some of the plates are pulling apart from one another, while others are colliding.

Where the plates are pulling apart from one another, the liquid mantle moves upward to fill the gap and solidifies. Thus, new crust is formed from the liquid mantle. Approximately half of the surface of the Earth has been formed in this way in the past 200 million years.

The bottom of the Atlantic and Pacific Oceans and the Rift Valley and Red Sea area of Africa are areas where this is occurring.

If plates are pulling apart on one portion of the Earth, they must be colliding elsewhere. Where plates collide, several things can happen. (See figure 14.2.) Often, one of the plates slides under the other and is melted. Often, when this occurs, some of the liquid mantle makes its way to the surface and volcanoes are formed, which results in the formation of mountains. The west coasts of North and South America have many volcanoes and mountain ranges where the two plates are colliding. The volcanic activity adds new material to the crust. When a collision occurs between two plates under the ocean, the volcanoes may eventually reach the surface and form a chain of volcanic islands, such as can be seen in the Aleutian Islands and many of the Caribbean Islands. When two continental plates

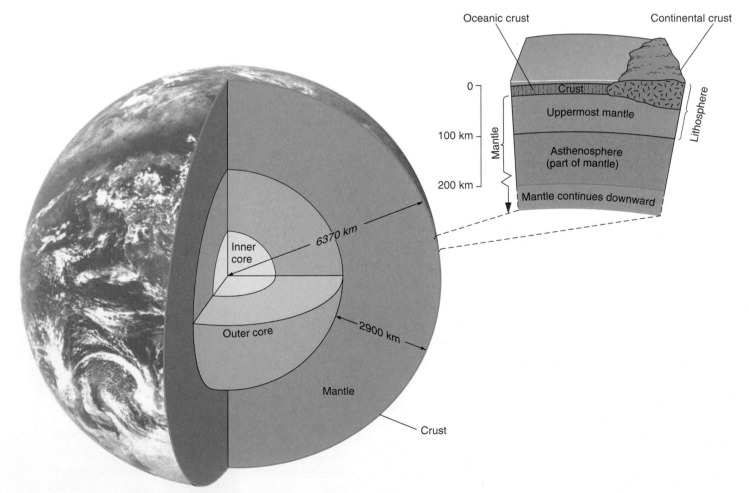

figure 14.1   **Structure of the Earth**   The Earth has a solid outer lithosphere that floats on a plastic mantle.

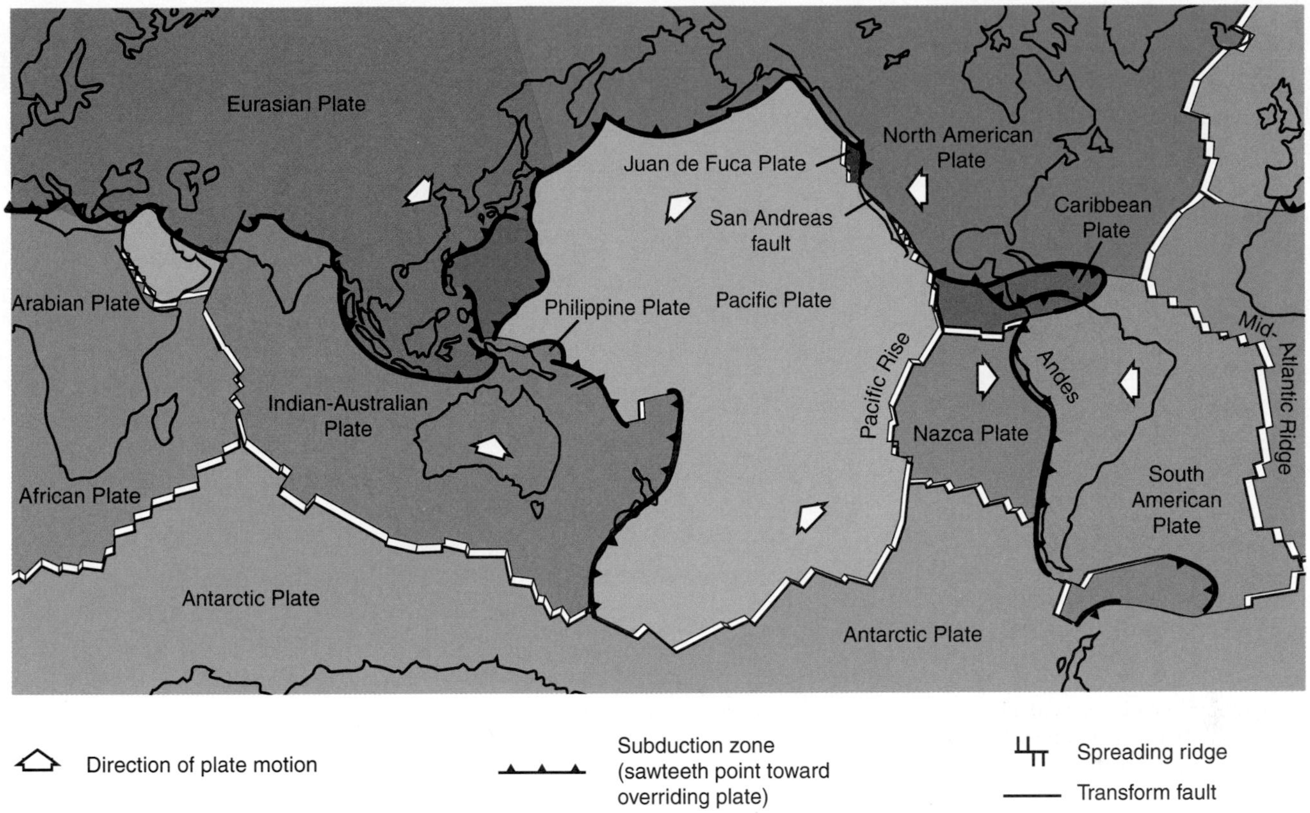
Eurasian Plate
North American Plate
Juan de Fuca Plate
San Andreas fault
Caribbean Plate
Arabian Plate
Philippine Plate
Pacific Plate
Pacific Rise
Andes
Mid-Atlantic Ridge
Indian-Australian Plate
Nazca Plate
African Plate
South American Plate
Antarctic Plate
Antarctic Plate
Direction of plate motion
Subduction zone (sawteeth point toward overriding plate)
Spreading ridge
Transform fault

figure 14.2 **Tectonic Plates** The plates that make up the outer surface of the Earth move with respect to one another. They pull apart from one another in some parts of the world and collide with one another in other parts of the world.

collide, neither plate slides under the other and the crust buckles to form mountains. The Himalayan, Alp, and Appalachian mountain ranges are thought to have formed from the collision of two continental plates. All of these movements of the Earth's surface are associated with earthquakes. The movements of the plates are not slow and steady sliding movements but tend to occur in small jumps. However, what is a small movement between two plates on the Earth is a huge movement for the relatively small structures and buildings produced by humans, so these small movements can cause tremendous amounts of damage.

These building processes are counteracted by processes that tend to make the elevated surfaces lower. Gravity provides a force that tends to wear down the high places. Moving water and ice (glaciers) and wind assist in the process; however, their effectiveness is related to the size of the rock particles. Several kinds of **weathering** processes are important in reducing the size of particles that can then be dislodged by moving water and air. **Mechanical weathering** results from physical forces that reduce the size of rock particles without changing the chemical nature of the rock. Common causes of mechanical weathering are changes in temperature that tend to result in fractures in rock, the freezing of water into ice that expands and tends to split larger pieces of rock into smaller ones, and the actions of plants and animals.

Because rock does not expand evenly, heating a large rock can cause it to fracture, so that pieces of the rock flake off. These pieces can be further reduced in size by other processes, such as the repeated freezing of water and thawing of ice. Water that has seeped into rock cracks and crevices expands as it freezes, causing the cracks to widen. Subsequent thawing allows more water to fill the widened cracks, which are enlarged further by another period of freezing. Alternating freezing and thawing fragments large rock pieces into smaller ones. (See figure 14.3.) The roots of plants growing in cracks can also exert enough force to break rock.

The physical breakdown of rock is also caused by forces that move and rub rock particles against each other (abrasion). For example, a glacier causes rock particles to grind against one another, resulting in smaller fragments and smoother surfaces. These particles are deposited by the glacier when the ice melts. In many parts of the world, the parent material from which soil is formed consists of glacial deposits. Wind and moving water also cause small particles to collide, resulting in further weathering. The smoothness of rocks and pebbles in a stream or on the shore is evidence that moving water has caused them to rub together, removing their sharp edges. Similarly, particles carried by wind collide with objects, fragmenting both the objects and the wind-driven particles.

Wind and moving water also remove small particles and deposit them at new locations, exposing new surfaces to the weathering process. For example, the landscape of the Painted Desert in the southwest United States was created

figure 14.3 **Physical Fragmentation by Freezing and Thawing** The crack in the rock fills with water. As the water freezes and becomes ice, it expands. The pressure of the ice enlarges the crack. The ice melts, and water again fills the crack. The water freezes again and widens the crack. Alternate freezing and thawing splits the rock into smaller fragments.

figure 14.4 **The Painted Desert—An Eroded Landscape** This landscape was created by the action of wind and moving water. The particles removed by these forces were deposited elsewhere and may have become part of the soil in that new location.

by a combination of wind and moving water that removed easily transported particles, while rocks more resistant to weathering remained. (See figure 14.4.)

The activities of organisms can also assist mechanical weathering. The roots of plants can exert considerable force and move particles apart from one another. The burrows of animals expose new surfaces that can be altered by freezing and thawing.

Chemical weathering involves the chemical alteration of the rock in such a manner that it is more likely to fragment or to be dissolved. Some small rock fragments exposed to the atmosphere may be oxidized; that is, they combine with oxygen from the air and chemically change to different compounds. Other kinds of rock may combine with water molecules in a process known as hydrolysis. Often, the oxidized or hydrolyzed molecules are more readily soluble in water and, therefore, may be removed by rain or moving water. Rain is normally slightly acid, and the acid content helps dissolve rocks.

Because of gravity, the prevailing movement of particles is from high elevations to lower ones. This process of loosening and redistributing particles is known as **erosion.** Wind can move sand and dust and can cause the wearing away of rocky surfaces by sandblasting their surfaces. Glaciers can move large rocks and cause their surfaces to be rounded by being rubbed against each other and the surface of the Earth. Moving water transports much material in streams and rivers. In addition, wave action along the shores of lakes and the coasts of oceans constantly wears away and transports particles.

Soil and Land

The geologic processes just discussed are involved in the development of both soil and land; however, soil and land are not the same. **Land** is the part of the world not covered by the oceans. **Soil** is a thin covering over the land consisting of a mixture of minerals, organic material, living organisms, air, and water that together support the growth of plant life. The proportions of the soil components vary with different types of soils, but a typical, "good" agricultural soil is about 45 percent mineral, 25 percent air, 25 percent water, and 5 percent organic matter. (See figure 14.5.) This combination provides good drainage, aeration, and organic matter. Farmers are particularly concerned with soil because the nature of the soil determines the kinds of crops that can be grown and which farming methods must be employed. Urban dwellers should also be concerned about soil because its health determines the quality and quantity of food they will eat. If the soil is so abused that it can no longer grow crops or if it is allowed to erode, degrading air and water quality, both urban and rural residents suffer. To understand how soil can be protected, we must first understand its properties and how it is formed.

Soil Formation

A combination of physical, chemical, and biological events acting over time is responsible for the formation of soil. Soil building begins with the fragmentation

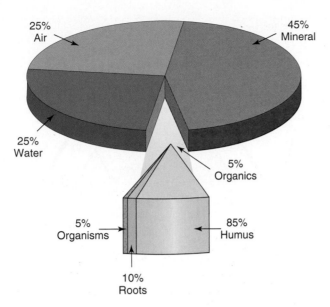

figure 14.5 **The Components of Soil** Although soils vary considerably in composition, they all contain the same basic components: mineral material, air, water, and organic material. The organic material can be further subdivided into humus, roots, and other living organisms. The percentages shown are those that would be present in a good soil.

of the **parent material,** which consists of ancient layers of rock or more recent geologic deposits from lava flows or glacial activity. The kind and amount of soil developed depends on the kind of parent material present, the plants and animals present, the climate, the time involved, and the slope of the land. As was discussed earlier, the breakdown of parent material is known as weathering. The climate and chemical nature of the rock material greatly influence the rate of weathering. Similarly, the size and chemical nature of the particles have a great impact on the nature of the soil that will develop in an area.

The role of organisms in the development of soil is also very important. The first organisms to gain a foothold in this modified parent material also contribute to soil formation. Lichens often form a pioneer community that grows on the surface of rocks and traps small particles. The decomposition of dead lichens and other organic matter releases acids that chemically alter the underlying rock, causing further fragmentation. The release of chemicals from the roots of plants causes further chemical breakdown of rock particles. As other kinds of organisms, such as plants and small animals, become established, they contribute, through their death and decay,

increasing amounts of organic matter, which are incorporated with the small rock fragments.

The organic material resulting from the decay of plant and animal remains is known as **humus.** It is a very important soil component that accumulates on the surface and ultimately becomes mixed with the top layers of mineral particles. This material contains nutrients that are taken up by plants from the soil. Humus also increases the water-holding capacity and the acidity of the soil so that inorganic nutrients, which are more soluble under acidic conditions, become available to plants. Humus also tends to stick other soil particles together and helps to create a loose, crumbly soil that allows water to soak in and permits air to be incorporated into the soil. Compact soils have few pore spaces, so they are poorly aerated, and water has difficulty penetrating, so it runs off.

Burrowing animals, soil bacteria, fungi, and the roots of plants are also part of the biological process of soil formation. One of the most important burrowing animals is the earthworm. One hectare (2.47 acres) of soil may support a population of 500,000 earthworms that can process as much as 9 tonnes (about 10 U.S. tons) of soil a year. These animals literally eat their way through the

soil, resulting in further mixing of organic and inorganic material, which increases the amount of nutrients available for plant use. They often bring nutrients from the deeper layers of the soil up into the area where plant roots are more concentrated, thus improving the soil's fertility. Soil aeration and drainage are also improved by the burrowing of earthworms and other small soil animals, such as nematodes, mites, pill bugs, and tiny insects. They also help to incorporate organic matter into the soil by collecting dead organic material from the surface and transporting it into burrows and tunnels. When the roots of plants die and decay, they release organic matter and nutrients into the soil and provide channels for water and air.

Fungi and bacteria are decomposers and serve as important links in many mineral cycles. (See chapter 5.) They, along with animals, improve the quality of the soil by breaking down organic material to smaller particles and releasing nutrients.

The position on the slope also influences soil development. Soil formation on steep slopes is very slow because materials tend to be moved downslope with wind and water. Conversely, river valleys often have deep soils because they receive materials from elsewhere by these same erosive forces.

Climate and time are also important in the development of soils. In general, extremely dry or cold climates develop soils very slowly, while humid and warm climates develop them more rapidly. Cold and dry climates have slow rates of accumulation of organic matter needed to form soil. Furthermore, chemical weathering proceeds more slowly at lower temperatures and in the absence of water. Under ideal climatic conditions, soft parent material may develop into a centimeter (less than ½ inch) of soil within 15 years. Under poor climatic conditions, a hard parent material may require hundreds of years to develop into that much soil. In any case, soil formation is a slow process.

The amount of rainfall and the amount of organic matter influence the pH of the soil. In regions of high rainfall, basic ions such as calcium, magnesium, and potassium are leached from the soils, and more acid materials are left behind. In addition, the decomposition of organic matter tends to increase the soil's acidity. Soil pH is important since it influences the availability of nutrients, which affects the kinds of plants that will grow, which affects the amount of organic matter added to the soil. Since calcium, magnesium, and potassium are important plant nutrients, their loss by leaching reduces the fertility of the soil. Excessively acidic soils also cause aluminum ions to become soluble, which in high amounts are toxic to many plants. (See the discussion of acid rain in chapter 17.) Most plants grow well in soils with a pH between 6 and 7, although some plants such as blueberries and potatoes grow well in acidic soils. In most agricultural situations, the pH of the soil is usually adjusted by adding chemicals to the soil. Lime can be added to make soils less acid, and acid-forming materials such as sulfates can be added to increase acidity.

Soil Properties

Soil properties include soil texture, structure, atmosphere, moisture, biotic content, and chemical composition. **Soil texture** is determined by the size of the mineral particles within the soil. The largest soil particles are gravel, which consists of fragments larger than 2.0 millimeters in diameter. Particles between 0.05 and 2.0 millimeters are classified as sand. Silt particles range from 0.002 to 0.05 millimeters in diameter, and the smallest particles are clay particles, which are less than 0.002 millimeters in diameter.

Large particles, such as sand and gravel, have many tiny spaces between them, which allow both air and water to flow through the soil. Water drains from this kind of soil very rapidly, often carrying valuable nutrients to lower soil layers, where they are beyond the reach of plant roots. Clay particles tend to be flat and are easily packed together to form layers that greatly reduce the movement

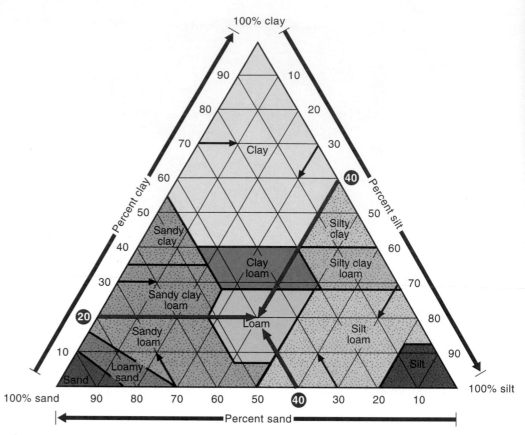

figure 14.6 **Soil Texture** Texture depends on the percentage of clay, silt, and sand particles in the soil. A loam soil has the best texture for most crops. As shown in the illustration, if a soil were 40 percent sand, 40 percent silt, and 20 percent clay, it would be a loam.

Source: Data from Soil Conservation Service.

of water through them. Soils with a lot of clay do not drain well and are poorly aerated. Because water does not flow through clay very well, clay soils tend to stay moist for longer periods of time and do not easily lose minerals to percolating water.

However, rarely does a soil consist of a single size of particle. Various particles are mixed in many different combinations, resulting in many different soil classifications. (See figure 14.6.) An ideal soil for agricultural use is a **loam,** which combines the good aeration and drainage properties of large particles with the nutrient-retention and water-holding ability of clay particles.

Soil structure is different from its texture. **Soil structure** refers to the way various soil particles clump together. The particles in sandy soils do not attach to one another; therefore, sandy soils have a granular structure. The particles in clay soils tend to stick to one another

to form large aggregates. Other soils that have a mixture of particle sizes tend to form smaller aggregates. A good soil is **friable,** which means that it crumbles easily. The soil structure and its moisture content determine how friable a soil is. Sandy soils are very friable, while clay soils are not. If clay soil is worked when it is too wet, it can stick together in massive blocks that will be difficult to break up.

A good soil for agricultural use will crumble and has spaces for air and water. In fact, the air and water content depends on the presence of these spaces. (See figure 14.7.) In good soil, about one-half to two-thirds of the spaces contain air after the excess water has drained. The air provides a source of oxygen for plant root cells and all the other soil organisms. The relationship between the amount of air and water is not fixed. After a heavy rain, most of the spaces may be filled with water and less

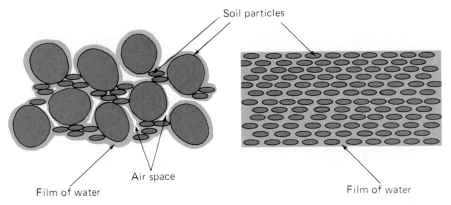

figure 14.7 **Pore Spaces and Particle Size** The soil on the left, which is composed of particles of various sizes, has spaces for both water and air. The particles have water bound to their surfaces (represented by the colored halo around each particle), but some of the spaces are so large that an air space is present. The soil on the right, which is composed of uniformly small particles, has less space for air. Since roots require both air and water, the soil on the left would be better able to support crops than would the soil on the right.

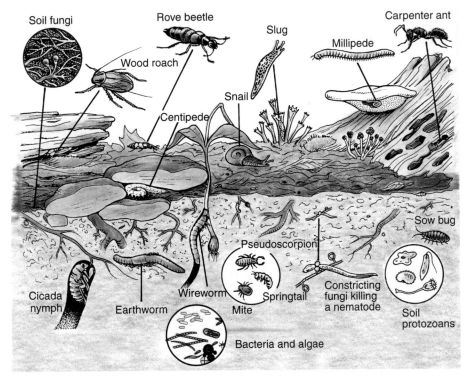

figure 14.8 **Soil Organisms** All of these organisms occupy the soil and contribute to it by rearranging soil particles, participating in chemical transformation, and recycling dead organic matter.

Source: *Ecology and Field Biology,* 5th ed., by Robert Leo Smith; copyright © 1996 by Harper Collins Publishers. Reprinted by permission.

oxygen is available to plant roots and other organisms. If some of the excess water does not drain from the soil, the plant roots may die from lack of oxygen. They are literally drowned. On the other hand, if there is not enough soil moisture, the plants wilt from lack of water. Soil moisture and air are also important

in determining the numbers and kinds of soil organisms.

Protozoa, nematodes, earthworms, insects, algae, bacteria, and fungi are typical inhabitants of soil. (See figure 14.8.) The role of protozoa in the soil is not firmly established, but they seem to act as parasites and predators on other

forms of soil organisms and, therefore, help to regulate the populations of those organisms. Nematodes, which are often called wireworms or roundworms, may aid in the breakdown of dead organic matter. Some nematodes are parasitic on the roots of plants. Insects and other soil arthropods contribute to the soil by forming burrows and consuming and fragmenting organic materials, but they are also major crop pests that feed on plant roots. Several kinds of bacteria are able to fix nitrogen from the atmosphere. Algae carry on photosynthesis and are consumed by other soil organisms. Bacteria and fungi are particularly important in the decay and recycling of materials. Their chemical activities change complex organic materials into simpler forms that can be used as nutrients by plants. For example, some of these microorganisms can convert the nitrogen contained in the protein component of organic matter into ammonia or nitrate, which are nitrogen compounds that can be utilized by plants. The amount of nitrogen produced varies with the type of organic matter, type of microorganisms, drainage, and temperature. Finally, it is important to recognize that the soil contains a complicated food chain in which all organisms are subject to being consumed by others. All of these organisms are active within distinct layers of the soil, known as the soil profile.

Soil Profile

The **soil profile** is a series of horizontal layers in the soil that differ in chemical composition, physical properties, particle size, and amount of organic matter. Each recognizable layer is known as a **horizon.** (See figure 14.9.) There are several systems for describing and classifying the horizons in soils. In general, the uppermost layer of the soil contains more nutrients and organic matter than do the deeper layers. The top layer is known as the *A* horizon, or topsoil. The *A* horizon consists of small mineral particles mixed with organic matter. Because of the relatively high organic

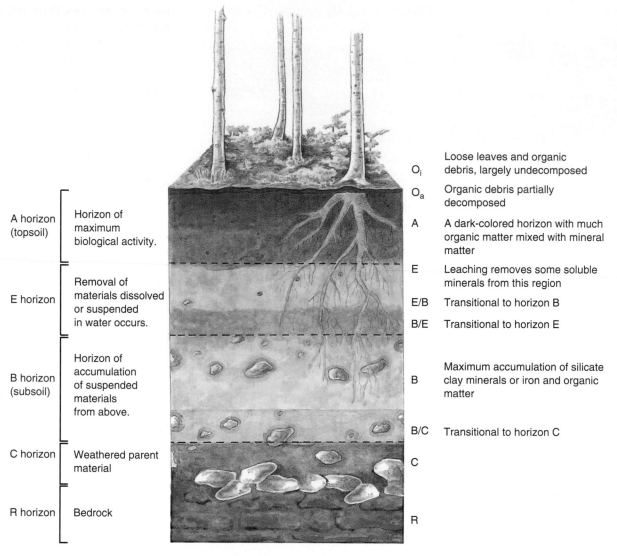

		Loose leaves and organic debris, largely undecomposed
	O_i	
A horizon (topsoil)	Horizon of maximum biological activity.	O_a	Organic debris partially decomposed
		A	A dark-colored horizon with much organic matter mixed with mineral matter
E horizon	Removal of materials dissolved or suspended in water occurs.	E	Leaching removes some soluble minerals from this region
		E/B	Transitional to horizon B
		B/E	Transitional to horizon E
B horizon (subsoil)	Horizon of accumulation of suspended materials from above.	B	Maximum accumulation of silicate clay minerals or iron and organic matter
		B/C	Transitional to horizon C
C horizon	Weathered parent material	C	
R horizon	Bedrock	R	

figure 14.9 **Soil Profile** A soil has layers that differ physically, chemically, and biologically. The top layer is known as the *A* horizon and contains most of the organic matter. Organic matter that collects on the surface is known as the *O* horizon. Many soils have a light-colored *E* horizon below the *A* horizon. It is light in color because dark-colored materials are leached from the layer. The *B* horizon accumulates minerals and particles as water carries dissolved minerals downward from the *A* and *E* to the *B* horizon. The *B* horizon is often called the subsoil. Below the *B* horizon is a *C* horizon of weathered parent material.

content, it is dark in color. If there is a layer of **litter** (undecomposed or partially decomposed organic matter) on the surface, it is known as the *O* horizon. Forest soils typically have an *O* horizon. Many agricultural soils do not, since the soil is worked to incorporate surface crop residue. As the organic matter decomposes, it becomes incorporated into the *A* horizon. The thickness of the *A* horizon may vary from less than a centimeter (less than ½ inch) on steep mountain slopes to over a meter (over 40 inches) in the rich grasslands of central North America. Most of the living or-

ganisms and nutrients are found in the *A* horizon. As water moves down through the *A* horizon, it carries dissolved organic matter and minerals to lower layers. This process is known as **leaching.** Because of the leaching away of darker materials such as iron compounds, a lighter-colored layer develops below the *A* horizon that is known as the *E* horizon. Not all soils develop an *E* horizon. This layer usually contains few nutrients because water flowing down through the soil dissolves and transports nutrients to the underlying *B* horizon. The *B* horizon, often called the subsoil, contains less or-

ganic material and fewer organisms than the *A* horizon. However, it contains accumulations of nutrients that were leached from higher levels. Often, clay minerals that are leached from the topsoil are deposited in this layer. Because nutrients are deposited in this layer, the *B* horizon in many soils is a valuable source of nutrients for plants, and such subsoils support a well-developed root system. Because the amount of leaching depends on the available rainfall, grasslands soils, which develop under low rainfall, often have a poorly developed *B* horizon, while soils in woodlands that receive higher rainfall usually have a well-developed *B* horizon.

The area below the subsoil is known as the *C* horizon, and it consists of weathered parent material. This parent material contains no organic materials, but it does contribute to some of the soil's properties. The chemical composition of the minerals of the *C* horizon helps to determine the pH of the soil. If the parent material is limestone, the soil will tend to neutralize acids; whereas, if the parent material is granite rock, the soil will not be able to do so. The characteristics of the parent material in the *C* horizon may also influence the soil's rate of water absorption and retention. Ultimately, the *C* horizon rests on bedrock, which is known as the *R* horizon.

Soil profiles and the factors that contribute to soil development are extremely varied. Over 15,000 separate soil types have been classified in North America. However, most of the cultivated land in the world can be classified as either grassland soil or forest soil. (See figure 14.10.)

Because the amount of rainfall in grassland areas is relatively low, it does not penetrate into the soil layers very far. Most of the roots of the grasses and other plants remain near the surface, and little leaching of minerals from the topsoil to deeper layers occurs. Since the roots of the plants rot in place when the grasses die, a deep layer of topsoil develops. This lack of leaching also results in a thin layer of subsoil, which is low in mineral and organic content and supports little root growth.

Forest soils develop in areas of more abundant rainfall. Water moves down through the soil so that deeper layers of the soil have a great deal of moisture. The roots of the trees penetrate to this layer and extract the water they need. The leaves and other plant parts that fall to the soil surface form a thin layer of organic matter on the surface. This organic matter decomposes and mixes with the mineral material of the top layers of the soil. The water that moves through the soil tends to carry material from the topsoil to the subsoil, where many of the roots of the plants are located. One of the materials that ac-

cumulates in the *B* horizon is clay. In some soils, particularly forest soils, clay or other minerals may accumulate and form a relatively impermeable "hardpan" layer that limits the growth of roots and may prevent water from reaching the soil's deeper layers.

Desert soils have very poorly developed horizons. Since there is little rainfall, deserts do not support a large amount of plant growth, and much of the soil is exposed. Therefore, little organic matter is added to the soil, and little leaching of materials occurs from upper layers to lower layers. Since much of the soil is exposed to wind and water erosion, much of the organic material and smaller particles are carried away by wind or by flash floods when it rains.

In cold, wet climates, typical of the northern parts of Europe, Russia, and Canada, there may be considerable accumulations of organic matter since the rate of decomposition is reduced. The extreme acidity of these soils also reduces the rate of decomposition. Hot, humid climates also tend to have poorly developed soil horizons since the organic matter decays very rapidly, and soluble materials are carried away by the abundant rainfall.

Because tropical rainforests support such a vigorous growth of plants and an incredible variety of plant and animal species, it is often assumed that tropical soils must be very fertile. Consequently, many people have tried to raise crops on tropical soils. It is possible to grow certain kinds of crops that are specially adapted to tropical soils, but raising of most traditional crop species is not successful. To understand why this is so, it is important to understand the nature of tropical rainforest soils. Two features of the tropical rainforest climate have a great influence on the nature of the soil. High temperature results in rapid decomposition of organic matter, so that the soils have very little litter and humus. High rainfall tends to leach nutrients from the upper layers of the soil, leaving behind a soil that is rich in iron and aluminum. The high iron content results in a reddish color for most of these soils. Because the nutrients are quickly removed, these soils are very infertile.

Furthermore, when the vegetation is removed, the soil is quickly eroded.

In addition to the differences caused by the kind of vegetation and rainfall, topography influences the soil profile. (See figure 14.11.) On a relatively flat area, the topsoil formed by soil-building processes will collect in place and gradually increase in depth. The topsoil formed on rolling hills or steep slopes is often transported down the slope as fast as it is produced. On such slopes, the accumulation of topsoil may not be sufficient to support a cultivated crop. The topsoil removed from these slopes is eventually deposited in the flat floodplains. These regions serve as collection points for topsoil that was produced over extensive areas. As a result, these river-bottom and delta regions have a very deep topsoil layer and are highly productive agricultural land.

Soil Erosion

Erosion is the wearing away and transportation of soil by water, wind, or ice. The Grand Canyon of the Colorado River, the floodplains of the Nile in Egypt, the little gullies on hillsides, and the deltas that develop at the mouths of rivers all attest to the ability of water to move soil. Anyone who has seen muddy water after a rainstorm has observed soil being moved by water. (See figure 14.12.) The force of moving water allows it to carry large amounts of soil. While erosion is a natural process, it is greatly accelerated by agricultural practices that leave the soil exposed. Each year, the Mississippi River transports over 325 million tonnes (360 million U.S. tons) of soil from the central regions of North America to the Gulf of Mexico. This is equal to the removal of a layer of topsoil approximately 1 millimeter (0.04 inches) thick from the entire region. Although the rate of erosion varies from place to place, movement of soil by water occurs in every stream and river in the world. Dry Creek, a small stream in California, has only 500 kilometers (310 miles) of mainstream and

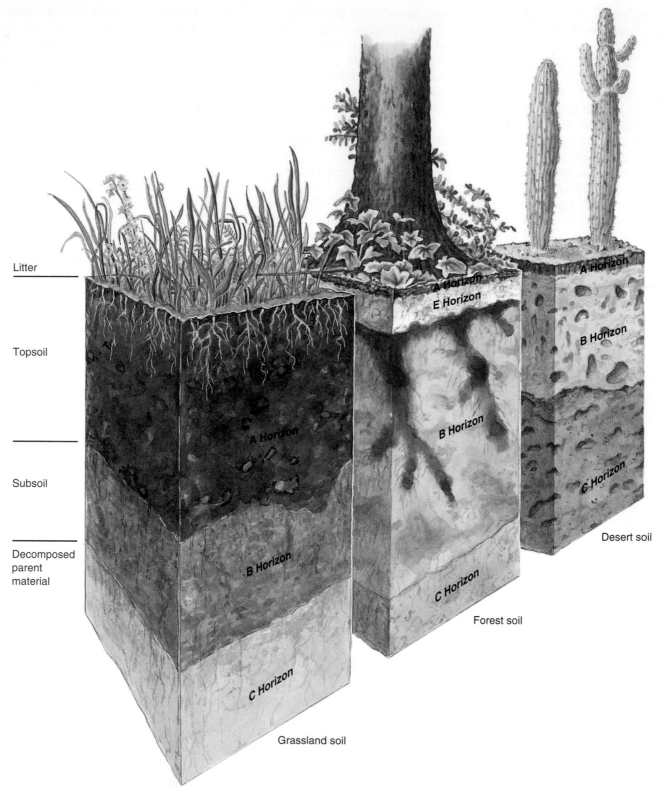

Litter

Topsoil

Subsoil

Decomposed
parent
material

A Horizon

B Horizon

C Horizon

Grassland soil

A Horizon

E Horizon

B Horizon

C Horizon

Forest soil

A Horizon

B Horizon

C Horizon

Desert soil

figure 14.10 **Major Soil Types** There are thousands of different soil types, but many of them can be classified into three broad categories. Soils formed in the grasslands have a deep *A* horizon. The shallow *B* horizon does not have sufficient nutrients to support root growth. In forest soils, the *A* horizon is thinner, and leaching transfers many nutrients to the *B* horizon. Thus, roots are found in both the *A* and *B* horizons. Desert soils have very thin *A* horizons.

A horizon	}	15 cm (6 inches)
B horizon	}	25 cm (10 inches)
A horizon	}	150 cm (60 inches)
B horizon	}	15 cm (6 inches)

figure 14.11 **The Effect of Slope on a Soil Profile** The topsoil formed on a large area of the hillside is continuously transported down the slope by the flow of water. It accumulates at the bottom of the slope and results in a thicker *A* horizon. The resulting "bottomland" is highly productive because it has a deep, fertile layer of topsoil, while the soil on the slope is less productive.

figure 14.12 **Water Erosion** The force of moving water is able to pick up soil particles and remove them. In cases of prolonged erosion, gullies (such as the one shown here) are likely to form.

tributaries; however, each year, it removes 180,000 tonnes (200,000 U.S. tons) of soil from a 340-square-kilometer (130-square-mile) area.

Soil erosion takes place everywhere in the world, but some areas are more exposed than others. Erosion occurs wherever grass, bushes, and trees are disappearing. Deforestation and desertification both leave land open to erosion. In deforested areas, water washes down steep, exposed slopes, taking the soil with it. In desertified regions, exposed soils, cleared for farming, building, or mining, or overgrazed by livestock, simply blow away. Wind erosion is most extensive in Africa and Asia. Blowing soil

not only leaves a degraded area behind but can bury and kill vegetation where it settles. It will also fill drainage and irrigation ditches. When high-tech farm practices are applied to poor lands, soil is washed away and chemical pesticides and fertilizers pollute the runoff. Every year, erosion carries away far more topsoil than is created, primarily because of agricultural practices that leave the soil exposed. (See figure 14.13.)

Worldwide, erosion removes about 25.4 billion tonnes (28 billion U.S. tons) of soil each year. In Africa, soil erosion has reached critical levels, with farmers pushing farther onto deforested hillsides. In Ethiopia, for example, soil loss occurs at a rate of between 1.5 billion and 2 billion cubic meters (53–70 billion cubic feet) a year, with some 4 million hectares (about 10 million acres) of highlands considered irreversibly degraded. In Asia, in the eastern hills of Nepal, 38 percent of the land area is fields that have been abandoned because the topsoil has washed away. In the Western Hemisphere, Ecuador is losing soil at 20 times the acceptable rate.

According to the International Fund for Agricultural Development (IFAD), traditional labor-intensive, small-scale soil conservation efforts that combine maintenance of shrubs and trees with

crop growing and cattle grazing work best at controlling erosion. In parts of Pakistan, a program begun by IFAD in 1980 to control rainfall runoff, erosion, and damage to rivers from siltation has increased crop yields and livestock productivity by 20 to 30 percent.

Badly eroded soil has lost all of the topsoil and some of the subsoil and is no longer productive farmland. Most current agricultural practices lose soil faster than it is replaced. Farming practices that reduce erosion, such as contour farming and terracing, are discussed in the Soil Conservation Practices section of this chapter.

Wind is also an important mover of soil. Under certain conditions, it can move large amounts. (See figure 14.14.) Wind erosion may not be as evident as water erosion, since it does not leave gullies. Nevertheless, it can be a serious problem. Wind erosion is most common in dry, treeless areas where the soil is exposed. In the Sahel region of Africa, much of the land has been denuded of vegetation because of drought, overgrazing, and improper farming practices. This has resulted in extensive wind erosion of the soil. (See figure 14.15.) In the Great Plains region of North America, there have been four serious periods of wind erosion since European settlement in the 1800s. If this area receives less than 30 centimeters (12 inches) of rain per year, there is not enough moisture to support crops. When this occurs for several years in a row, it is called a drought. Farmers plant crops, hoping for rain. When the rain does not come, they plow their fields again to prepare them for another crop. Thus, the loose, dry soil is left exposed, and wind erosion results. Because of the large amounts of dust in the air during those times, the region is known as the Dust Bowl. During the 1930s, wind destroyed 3.5 million hectares (over 8.5 million acres) of farmland and seriously damaged an additional 30 million hectares (75 million acres) in the Dust Bowl.

Fortunately, many soil conservation practices have been instituted that protect soil. However, much soil is being lost to erosion, and more protective measures should be taken.

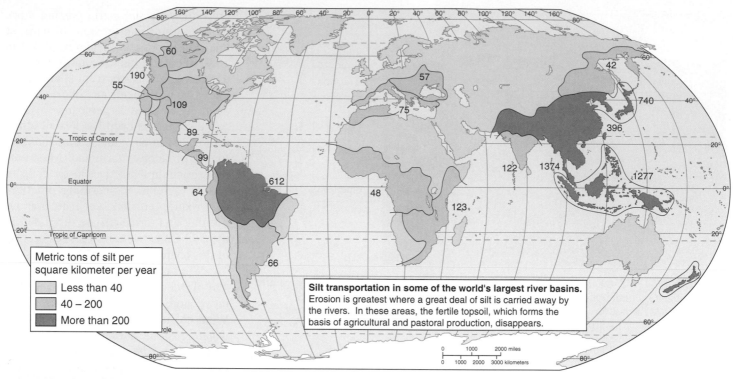

Metric tons of silt per square kilometer per year

- Less than 40
- 40 – 200
- More than 200

Silt transportation in some of the world's largest river basins. Erosion is greatest where a great deal of silt is carried away by the rivers. In these areas, the fertile topsoil, which forms the basis of agricultural and pastoral production, disappears.

Silt-laden river

figure 14.13 **Worldwide Soil Erosion** Soil erosion is widespread throughout the world. Silty rivers are evidence of poor soil conservation practices upstream.

Soil Conservation Practices

The kinds of agricultural activities that land can be used for are determined by soil structure, texture, drainage, fertility, rockiness, slope of the land, amount and nature of rainfall, and other climatic conditions. A relatively large proportion—about 20 percent—of U.S. land is suitable for raising crops. However, only 2 percent of that land does not require some form of soil conservation practice. (See figure 14.16.) This means that nearly all of the soil in the United States must be managed in some way to

reduce the effects of soil erosion by wind or water.

Not all parts of the world are as well supplied as the United States with land that has agricultural potential. (See table 14.1.) For example, worldwide, approximately 11 percent of the land surface is suitable for crops, and an additional 24 percent is in permanent pasture. In the United States, about 20 percent is cropland, and 25 percent is in permanent pasture. Contrast this with the continent of Africa, in which only 6 percent is suitable for crops and 29 percent can be used for pasture. Canada has only 5 percent suitable for crops and 3 percent for pasture. Europe has the highest percentage of cropland with 30 percent, but it has only 17 percent in permanent pasture.

Since very little land is left that can be converted to agriculture, we must use what we have wisely. A study by the Food and Agricultural Organization states that up to 40 percent of the world's agricultural lands are seriously affected by soil degradation. The World

figure 14.14 **Wind Erosion** The dry, unprotected topsoil from this field is being blown away. The force of the wind is capable of removing all the topsoil and transporting it several thousand kilometers.

Resources Institute in 2000 stated that land degradation affects around 70 percent of the world's rangelands, 40 percent of rain-fed agricultural lands, and 30 percent of irrigated lands. According to the United Nations Environment Programme, an estimated 500 million hectares (1.24 billion acres) of land in Africa have been affected by soil degradation since about 1950, including as much as 65 percent of agricultural land.

figure 14.15 **Wind Erosion in the Sahel** The semiarid region just south of the Sahara Desert is in an especially vulnerable position. The rainfall is unpredictable, which often leads to crop failure. In addition, population pressure forces people in this region to try to raise crops in marginal areas. This often results in increased wind erosion.

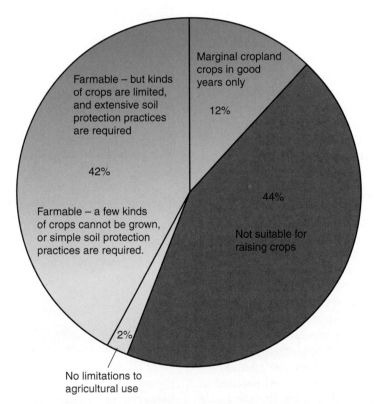

Farmable – but kinds of crops are limited, and extensive soil protection practices are required

42%

Marginal cropland crops in good years only

12%

44%

Not suitable for raising crops

Farmable – a few kinds of crops cannot be grown, or simple soil protection practices are required.

2%

No limitations to agricultural use

figure 14.16 **U.S. Land Used for Agricultural Purposes** Only 2 percent of the land in the United States can be cultivated without some soil conservation practices. This 2 percent is primarily flatland, which is not subject to wind erosion. On 42 percent of the remaining land, some special considerations for protecting the soil are required, or the kinds of crops are limited. Twelve percent of the land is marginal cropland that can provide crops only in years when rainfall and other conditions are ideal. Forty-four percent is not suitable for cultivating crops but may be used for other purposes, such as grazing cattle or as forests.

While the global expansion of agricultural area has been modest in recent decades, intensification has been rapid, as irrigated area increased and fallow time decreased to produce more output per hectare.

Many techniques can protect soil from erosion while allowing agriculture. Some of the more common methods are discussed here. Whenever soil is lost by water or wind erosion, the topsoil, the most productive layer, is the first to be removed. When the topsoil is lost, the soil's fertility decreases, and larger amounts of expensive fertilizers must be used to restore the fertility that was lost. This raises the cost of the food we buy. In addition, the movement of excessive amounts of soil from farmland into streams has several undesirable effects. First, a dirty stream is less aesthetically pleasing than a clear stream. Second, too much sediment in a stream affects the fish population by reducing visibility, covering spawning sites, and clogging the gills of the fish. Fishing may be poor because of unwise farming practices hundreds of kilometers upstream. Third, the soil carried by a river is eventually deposited somewhere. In many cases, this soil must be removed by dredging to clear shipping channels. We pay for dredging with our tax money, and it is a very expensive operation.

For all of these reasons, proper soil conservation measures should be employed to minimize the loss of topsoil. Figure 14.17 contrasts poor soil conservation practices with proper soil protection. When soil is not protected from the effects of running water, the topsoil is removed and gullies result. This can be prevented by slowing the flow of water over sloping land.

Contour Farming

Contour farming, which is tilling at right angles to the slope of the land, is one of the simplest methods for preventing soil erosion. This practice is useful on gentle slopes and produces a series of small ridges at right angles to the slope. (See figure 14.18.) Each ridge acts as a dam to hold water from running down the incline. This allows more of the water to soak into the soil. Contour farming reduces soil erosion by as much as 50 percent and, in drier regions, increases crop yields by conserving water.

Strip Farming

When a slope is too steep or too long, contour farming alone may not prevent soil erosion. However, a combination of contour and strip farming may work. **Strip farming** is alternating strips of closely sown crops like hay, wheat, or other small grains with strips of row crops like corn, soybeans, cotton, or sugar beets. (See figure 14.19.) The closely sown crops retard the flow of water, which reduces soil erosion and allows more water to be absorbed into the ground. The type of soil, steepness, and length of slope dictate the width of the strips and determine whether strip or contour farming is practical.

Terracing

On very steep land, the only practical method of preventing soil erosion is to construct terraces. **Terraces** are level areas constructed at right angles to the slope to retain water and greatly reduce the amount of erosion. (See figure 14.20.) Terracing has been used for centuries in nations with a shortage of level farmland. The type of terracing seen in figure 14.20a requires the use of small machines and considerable hand labor and is not suitable for the mechanized farming typical in much of the world. Terracing is an expensive method of controlling erosion since it requires the moving of soil to construct the level areas, protecting the steep areas between terraces, and constant repair and maintenance. Many factors, such as length and steepness of slope, type of soil, and amount of precipitation, determine whether terracing is feasible.

Waterways

Even with such soil conservation practices as contour farming, strip farming, and terracing, farmers must often provide protected channels for the move-

ment of water. **Waterways** are depressions on sloping land where water collects and flows off the land. When not properly maintained, these areas are highly susceptible to erosion. (See figure 14.21.) If a waterway is maintained with a permanent sod covering, the speed of the water is reduced, the roots tend to hold the soil particles in place, and soil erosion is decreased.

Windbreaks

Contour farming, strip farming, terracing, and maintaining waterways are all important ways of reducing water erosion, but wind is also a problem with certain soils, particularly in dry areas of the world. Wind erosion can be reduced if the soil is protected. The best protection is a layer of vegetation over the surface. However, the process of preparing soil for planting and the method of planting often leave the soil exposed to wind. **Windbreaks** are plantings of trees or other plants that protect bare soil from the full force of the wind. Windbreaks reduce the velocity of the wind, thereby decreasing the amount of soil that it can carry away. (See figure 14.22.)

Table 14.1	Percentage of Land Suitable for Agriculture	
Country	**Percent Cropland**	**Percent Pasture**
World	11.0	26.0
Africa	6.3	28.8
Egypt	2.8	5.0
Ethiopia	12.7	40.7
Kenya	7.9	37.4
South Africa	10.8	66.6
North America	13.0	16.8
Canada	4.9	3.0
United States	19.6	25.0
South America	6.0	28.3
Argentina	9.9	51.9
Venezuela	4.4	20.2
Asia	15.2	25.9
China	10.3	30.6
Japan	12.0	1.7
Europe	29.9	17.1

Source: Data from *World Resources,* 2000.

a.

b.

figure 14.17 **Poor and Proper Soil Conservation Practices** (*a*) This land is no longer productive farmland since erosion has removed the topsoil. (*b*) This rolling farmland shows strip contour farming to minimize soil erosion by running water. It should continue indefinitely to be productive farmland.

figure 14.18 **Contour Farming** Tilling at right angles to the slope creates a series of ridges that slows the flow of the water and prevents soil erosion. This soil conservation practice is useful on gentle slopes.

figure 14.19 **Strip Farming** On rolling land, a combination of contour and strip farming prevents excessive soil erosion. The strips are planted at right angles to the slope, with bands of closely sown crops, such as wheat or hay, alternating with bands of row crops, such as corn or soybeans.

In some cases, rows of trees are planted at right angles to the prevailing winds to reduce their force, while in other cases, a kind of strip farming is practiced in which strips of hay or grains are alternated with row crops that leave large amounts of the soil exposed. In some areas of the world, the only way to protect the soil is to not cultivate it at all but to leave it in a permanent cover of grasses.

Conventional Versus Conservation Tillage

Conventional tillage methods in much of the world require extensive use of farm machinery to prepare the soil for planting and to control weeds. Typically, a field is plowed and then disked or harrowed one to three times before the crop is planted. The plowing, which turns the soil over, has several desirable effects: any weeds or weed seeds are buried, thus reducing the weed problem in the field. Crop residue from previous crops is incorporated into the soil, where it will decay faster and contribute to soil structure. Nutrients that had been leached to deeper layers of the soil are brought near the surface. And the dark soil is exposed to the sun so that it warms up faster. This last effect is most critical in areas with short growing seasons. In many areas, fields are plowed in the fall, after the crop has been harvested, and the soil is left exposed all winter.

After plowing, the soil is worked by disks or harrows to break up any clods of earth, kill remaining weeds, and prepare the soil to receive the seeds. After the seeds are planted, there may still be weed problems. Farmers often must cultivate row crops to kill the weeds that begin to grow between the rows. Each trip over the field costs the farmer money, while at the same time increasing the amount of time the soil is exposed to wind or water erosion.

In recent years, several new systems of tillage have developed, as innovations in chemical herbicides and farm equipment have taken place. These tilling practices protect the soil by leaving the crop residue on the soil surface, thus reducing the amount of time it is exposed to erosion forces. **Reduced tillage** is a method that uses less cultivation to control weeds and to prepare the soil to receive seeds but generally leaves 15 to 30 percent of the soil surface covered with crop residue after planting. **Conservation tillage** methods further reduce the amount of disturbance to the

a.

b.

figure 14.20 **Terraces** Since the construction of terraces requires the movement of soil and the protection of the steep slope between levels, terraces are expensive to build. (*a*) The terraces seen here are extremely important for people who live in countries that have little flatland available. They require much energy and hand labor to maintain but make effective agricultural use of the land without serious erosion. (*b*) This modification of the terracing concept allows the use of the large farm machines typical of farming practices in Canada, Europe, and the United States.

a.

b.

figure 14.21 **Protection of Waterways Prevents Erosion** (*a*) An unprotected waterway has been converted into a gully. (*b*) A well-maintained waterway is not cultivated; a strip of grass retards the flow of water and protects the underlying soil from erosion.

soil and leave 30 percent or more of the soil surface covered with crop residue following planting. Selective herbicides are used to kill unwanted vegetation before planting the new crop and to control weeds afterward. Several variations of conservation tillage are used:

1. Mulch tillage involves tilling the entire surface just before planting or as planting is occurring.
2. Strip tillage is a method that involves tilling only in the narrow strip that is to receive the seeds. The

rest of the soil and the crop residue from the previous crop are left undisturbed.

3. Ridge tillage involves leaving a ridge with the last cultivation of the previous year and planting the crop on the ridge with residue left between the ridges. The crop may be cultivated during the year to reduce weeds.

4. No-till farming involves special planters that place the seeds in slits cut in the soil that still has on its

surface the residue from the previous crop.

Both reduced tillage and conservation tillage methods reduce the amount of time and fuel needed by the farmer to produce the crop and, therefore, represent an economic savings. By 2000, over half of the cropland in the United States was being farmed using reduced or conservation tillage methods. (See table 14.2.) For many kinds of crops, yields are comparable to that produced by conventional tillage methods.

a.

b.

figure 14.22 **Windbreaks** (*a*) In sections of the Great Plains, trees provide protection from wind erosion. The trees along the road protect the land from the prevailing winds. (*b*) In this field, temporary strips of vegetation serve as windbreaks.

Table 14.2 Comparison of Various Tillage Methods		
Tillage Method	**Fuel Use** (Liters per Hectare)	**Time Involved** (Hours per Hectare)
Conventional plowing	8.08	3.00
Reduced tillage	5.11	2.20
Mulch tillage	4.64	2.07
Ridge tillage	4.12	2.25
No tillage	2.19	1.21

Source: Data from University of Nebraska, Institute of Agriculture and Natural Resources, 1997.

Other positive effects of reduced tillage, in addition to reducing erosion, are:

1. The amount of winter food and cover available for wildlife increases, which can lead to increased wildlife populations.

2. Since there is less runoff, siltation in streams and rivers is reduced. This results in clearer water for recreation and less dredging to keep waterways open for shipping.

3. Row crops can be planted on hilly land that cannot be converted to such crops under conventional tilling methods. This allows a farmer to convert low-value pasture land into cropland that gives a greater economic yield.

4. Since fewer trips are made over the field, petroleum is saved, even when the petrochemical feedstocks necessary to produce the herbicides are taken into account.

5. Two crops may be grown on a field in areas that had been restricted to growing one crop per field per year. In some areas, immediately after harvesting wheat, farmers have planted soybeans directly in the wheat stubble.

6. Because conservation tillage reduces the number of trips made over the field by farm machinery, the soil does not become compacted as quickly.

However, there are also some drawbacks to conservation tillage methods:

1. The residue from previous vegetation may delay the warming of the soil, which may, in turn, delay planting some crops for several days.

2. The crop residue reduces evaporation from the soil and the upward movement of water and soil nutrients from deeper layers of the soil, which may retard the growth of plants.

3. The accumulation of plant residue can harbor plant pests and diseases that will require more insecticides and fungicides. This is particularly true if the same crop is planted repeatedly in the same location.

Conservation tillage is not the complete answer to soil erosion problems but may be useful in reducing soil erosion on well-drained soils. It also requires that farmers pay close attention to the condition of the soil and the pests to be dealt with.

In 1972, 12 million hectares (30 million acres) in the United States were under some form of conservation tillage. About 1.3 million hectares (3.2 million acres) were being farmed using no-till methods. By 1992, this had risen to 60 million hectares (150 million acres) of conservation-tilled land, of which 6.2 million hectares (15 million acres) were being farmed using no-till methods. It is

environmental
CLOSE-UP

Land Capability Classes

Not all land is suitable for raising crops or urban building. Such factors as the degree of slope, soil characteristics, rockiness, erodibility, and other characteristics determine the best use for a parcel of land. In an attempt to encourage people to use land wisely, the U.S. Soil Conservation Service has established a system to classify land-use possibilities. The table shows the eight classes of land and lists the characteristics and capabilities of each.

Unfortunately, many of our homes and industries are located on type I and II land, which has the least restrictions on agricultural use.

This does not make the best use of the land. Zoning laws and land-use management plans should consider the land-use capabilities and institute measures to ensure that land will be used to its best potential.

• Can you provide examples from your community where you feel land has not been used properly?

• In your opinion, should there be more or less land-use planning?

• What are some of the consequences of building or farming on land that is not suitable?

	Land Class	Characteristics	Capability	Special Conservation Measures
Land suitable for cultivation	I	Excellent, flat, well-drained land	Cropland	Normal good practices adequate
	II	Good land; has minor limitations, such as slope, sandy soil, or poor drainage	Cropland Pasture	Strip cropping Contour farming
	III	Moderately good land with important limitations of soil, slope, or drainage	Cropland Pasture Watershed	Contour farming Strip cropping Terraces Waterways
	IV	Fair land with severe limitations of soil, slope, or drainage	Pasture Orchards Urban Industry Limited cropland	Crops on a limited basis Contour farming Strip cropping Terraces Waterways
Land not suitable for cultivation	V	Use for grazing and forestry; slightly limited by rockiness, shallow soil, or wetness	Grazing Forestry Watershed Urban Industry	No special precautions if properly grazed or logged; must not be plowed
	VI	Moderate limitations for grazing and forestry because of moderately steep slopes	Grazing Forestry Watershed Urban Industry	Grazing or logging may be limited at times.
	VII	Severe limitations for grazing and forestry because of very steep slopes vulnerable to erosion	Grazing Forestry Watershed Recreation Wildlife Urban Industry	Careful management is required when used for grazing or logging.
	VIII	Unsuitable for grazing and forestry because of steep slope, shallow soil, lack of water, or too much water	Watershed Recreation Wildlife Urban Industry	Not to be used for grazing or logging; steep slope and lack of soil present problems

Global Perspective

Worldwide Soil Degradation

As humans have used the land and its soil for agriculture, grazing, fuelwood production, and forestry, we have had profound effects on the rates at which soils are formed and lost. Removal of the vegetation has accelerated soil erosion so that most of the world is losing soil faster than it is being formed. The degradation is most severe where the need for food is greatest, since starving people are forced to over-exploit their already overused soils. The map and graphs illustrate the magnitude of the soil degradation problem and the relative importance of the causes.

Percent of Soil Degraded by Human Activity

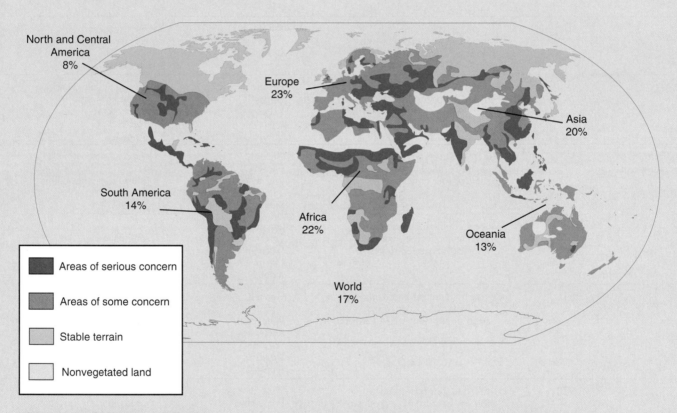

North and Central America
8%

Europe
23%

Asia
20%

South America
14%

Africa
22%

Oceania
13%

World
17%

Areas of serious concern

Areas of some concern

Stable terrain

Nonvegetated land

Causes of Soil Degradation

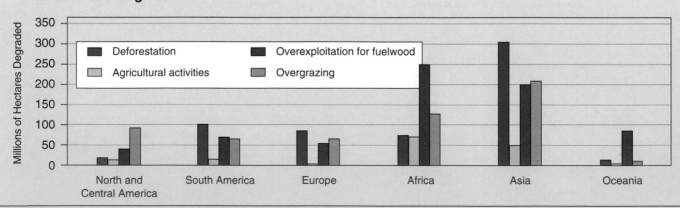

Millions of Hectares Degraded

Deforestation
Agricultural activities
Overexploitation for fuelwood
Overgrazing

North and Central America   South America   Europe   Africa   Asia   Oceania

Source: World Resources Institute in collaboration with the United Nations Environment Programme and the United Nations Development Programme, *World Resources 1992–93* (Oxford University Press, New York, 1992), table 19.4, p. 290.

estimated that by the year 2010, 95 percent of U.S. cropland will be under some form of reduced-tillage practice.

Protecting Soil on Nonfarm Land

Each piece of land has characteristics, such as soil characteristics, climate, and degree of slope, that influence the way it can be used. When all these factors are taken into consideration, a proper use can be determined for each portion of the planet. Wise planning and careful husbandry of the soil is necessary if the land and its soil are to provide food and other necessities of life. Not all land is suitable for crops or continuous cultivation. Some has highly erodible soil and must never be plowed for use as cropland, but it can still serve other useful functions. By using appropriate soil conservation practices, much of the land not usable for crops can be used for grazing, wood production, wildlife production, or scenic and recreational purposes. Figure 14.23 shows land that is not suitable for cultivation because it is too arid but that, if it is used properly, can provide grazing for cattle or sheep.

The land shown in figure 14.24 is not suitable for either crops or grazing because of the steep slope and thin soil. However, it is still a valuable and productive piece of land, since it can be used to furnish lumber, wildlife habitats, and recreational opportunities.

figure 14.23 **Noncrop Use of Land to Raise Food** As long as it is properly protected and managed, this land can produce food through grazing, but it should never be plowed to plant crops because the topsoil is too shallow and the rainfall is too low.

figure 14.24 **Forest and Recreational Use** Although this land is not capable of producing crops or supporting cattle, it furnishes lumber, a habitat for wildlife, and recreational opportunities.

Soil Erosion in Virginia

In Virginia, a city attorney sued a land developer, contending that the developer was responsible for property damages as a result of erosion from his construction site. During the trial, it was shown that the developer had constructed his subdivision without considering the possible effects of soil erosion. As a result, during a heavy rain, mud from his project had covered the yards, sidewalks, and driveways of nearby homes. As much as 10 meters (30 feet) of soil had eroded from the exposed land of his development and been carried downhill, covering portions of the adjacent community.

Speaking to the developer, the city attorney stated: "After gambling and principally winning over a period of fifteen months, it is apparent that you are attempting to go with your winning streak; and flood, storms, and weather hold no terror for you." The city claimed that the developer had acted in a negligent manner and that he was, therefore, responsible for the damages caused by the deposited mud and should be ordered to pay for the cleanup. The developer claimed no responsibility. His position was that grading large tracts of land was a common construction procedure and that the weather was "an act of God" for which he could not be held accountable.

- If you were a member of the jury hearing the case, would you think the developer acted in a negligent manner by leaving the soil exposed for a long period of time?
- Do you believe that property owners can hold a person responsible for actions taken in an area some distance from their property?
- Do you believe that the erosion and subsequent deposit of mud was "an act of God"?
- Would you find the developer guilty or not guilty? Why?

Summary

The surface of the Earth is in constant flux. The movement of tectonic plates results in the formation of new land as old land is worn down by erosive activity. Soil is an organized mixture of minerals, organic material, living organisms, air, and water. Soil formation begins with the breakdown of the parent material by such physical processes as changes in temperature, freezing and thawing, and movement of particles by glaciers, flowing water, or wind. Oxidation and hydrolysis can chemically alter the parent material. Organisms also affect soil building by burrowing into and mixing the soil, releasing nutrients, and decomposing.

Topsoil contains a mixture of humus and inorganic material, both of which supply soil nutrients. The ability of soil to grow crops is determined by the inorganic matter, organic matter, water, and air spaces in the soil. The mineral portion of the soil consists of various mixtures of sand, silt, and clay particles.

A soil profile typically consists of the *O* horizon of litter; the *A* horizon, which is rich in organic matter; an *E* horizon from which materials have been leached; the *B* horizon, which accumulates materials leached from above; and the *C* horizon, which consists of slightly altered parent material. Forest soils typically have a shallow *A* horizon and an *E* horizon and a deep, nutrient-rich *B* horizon with much root development. Grassland soils usually have a thick *A* horizon containing most of the roots of the grasses. They lack an *E* horizon. There are few nutrients in the thin *B* horizon.

Soil erosion is the removal and transportation of soil by water or wind. Proper use of such conservation practices as contour farming, strip farming, terracing, waterways, windbreaks, and conservation tillage can reduce soil erosion. Misuse reduces the soil's fertility and causes air- and water-quality problems. Land unsuitable for crops may be used for grazing, lumber, wildlife habitats, or recreation.

Key Terms

chemical weathering *306*	leaching *310*	soil *306*
conservation tillage *317*	lithosphere *304*	soil profile *309*
contour farming *315*	litter *310*	soil structure *308*
crust *304*	loam *308*	soil texture *308*
erosion *306*	mantle *304*	strip farming *316*
friable *308*	mechanical weathering *305*	terraces *316*
horizon *309*	parent material *307*	waterways *316*
humus *307*	plate tectonics *304*	weathering *305*
land *306*	reduced tillage *317*	windbreaks *316*

Review Questions

1. How are soil and land different?
2. Name the five major components of soil.
3. Describe the process of soil formation.
4. Name five physical and chemical processes that break parent material into smaller pieces.
5. In addition to fertility, what other characteristics determine the usefulness of soil?
6. How does soil particle size affect texture and drainage?
7. Describe a soil profile.
8. Define erosion.
9. Describe three soil conservation practices that help to reduce soil erosion.
10. Besides cropland, what are other possible uses of soil?

Critical Thinking Questions

1. Minimum tillage soil conservation often uses greater amounts of herbicides to control weeds. What do you think about this practice? Why?
2. As populations grow, should we try to bring more land into food production, or should we use technology to aid in producing more food on the land we already have in production of food? What are the tradeoffs?
3. Given what you know about soil formation, how might you explain the presence of a thick *A* horizon in soils in the North American Midwest?
4. Why should nonfarmers be interested in soil conservation?
5. Imagine that you are a scientist hired to consult on a project to evaluate land-use practices at the edge of a small city. The area in question has deep ravines and hills. What kinds of agricultural, commercial, and logging practices would you recommend in this area to help preserve the environment?
6. Look at your own community. Can you see examples of improper land use (urban or rural)? What are the consequences of these land-use practices? What recommendations would you make to improve land use?

Concept Map

Construct a map to show relationships among the following concepts:

soil	soil conservation practices	reduced tillage
parent material	erosion	contour farming
soil profile	soil degradation	

Interactive Exploration

Check out the website at **http://www.mhhe.com/environmentalscience** and click on the cover of this textbook for quizzing, career information, case studies, and hot links for information on the following topics:

Soil and its uses

Agricultural methods

Nutrient cycling

Sustainable agriculture

Deserts

15
CHAPTER

Agricultural Methods and Pest Management

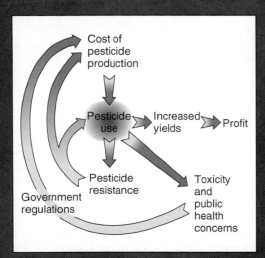

Objectives

After reading this chapter, you should be able to:

- Explain how the invention of new farm machinery encouraged monoculture farming.
- List the advantages and disadvantages of monoculture farming.
- Explain why chemical fertilizers are used.
- Understand how fertilizers alter soil characteristics.
- Explain why modern agriculture makes extensive use of pesticides.
- Differentiate between persistent pesticides and nonpersistent pesticides.
- List four problems associated with pesticide use.
- Define biomagnification.
- Define organic farming.
- Explain why integrated pest management depends on a complete knowledge of the pest's life history.
- Recognize that genetically modified crops are created by using biotechnological techniques to insert genes from one species into another.

Chapter Outline

Different Approaches to Agriculture

Fossil Fuel Versus Muscle Power

The Impact of Fertilizer

Agricultural Chemical Use

Environmental Close-Up: *DDT—A Historical Perspective*
   Insecticides
   Herbicides
   Fungicides and Rodenticides

Environmental Close-Up: *A New Generation of Insecticides*
   Other Agricultural Chemicals

Problems with Pesticide Use
   Persistence
   Bioaccumulation and Biomagnification
   Pesticide Resistance
   Effects on Nontarget Organisms

Global Perspective: *China's Ravenous Appetite*

Global Perspective: *Contaminated Soils in the Former Soviet Union*
   Human Health Concerns

Why Are Pesticides So Widely Used?

Alternatives to Conventional Agriculture

Environmental Close-Up: *Industrial Production of Livestock*

Environmental Close-Up: *Food Additives*
   Techniques for Protecting Soil and Water Resources
   Integrated Pest Management

Issues—Analysis: *Herring Gulls as Indicators of Contamination in the Great Lakes*

Different Approaches to Agriculture

Around the world, people use a variety of agricultural methods to grow food. In some parts of the world with poor soil and low populations, a form of shifting agriculture known as "slash-and-burn" can be used successfully. This method involves cutting down the trees and burning the trees and other vegetation in a small area of the forest. (See figure 15.1.) Burning releases nutrients that were tied up in the biomass and allows a few crops to be raised before the soil is exhausted. Once the soil is no longer suitable for raising crops (within two or three years), the site is abandoned. The surrounding forest recolonizes the area, which will return over time to the original forest through the process of succession. This method is particularly useful on thin, nutrient-poor tropical soils and on steep slopes. The small size of the openings in the forest and their temporary existence prevent widespread damage to the soil, and erosion is minimized. While this system of agriculture is successful when human population densities are low, it is not suitable for large, densely populated areas. When populations become too large, the size and number of the garden plots increase and the time between successive uses of the same plot of land decreases. When a large amount of the forest is disturbed and the time between successive uses is decreased, the forest cannot return and repair the damage done by the previous use of the land, and the nature of the forest is changed.

The traditional practices of the people who engage in small-scale, shifting agriculture have been developed over hundreds of years and often are more effective for their local conditions than other methods of gardening. Typically, these gardens are planted with a mixture of plants; a system known as **polyculture.** Mixing plants together in a garden often is beneficial, since shade-requiring species may be helped by taller plants, or nitrogen-fixing legumes may provide nitrogen for species that require

figure 15.1 **Slash-and-Burn Agriculture** In many areas of the world where the soils are poor and human populations are low, crops can be raised by disturbing small parts of the ecosystem followed by several years of recovery. The burning of vegetation releases nutrients that can be used by crops for one or two years before the soil is exhausted. The return of the natural vegetation prevents erosion and repairs the damage done by temporary agricultural use.

it. In addition, mixing species may reduce insect pest problems because some plants produce molecules that are natural insect repellents. The small, isolated, temporary nature of the gardens also reduces the likelihood of insect plagues. While today we see this form of agriculture practiced most commonly in tropical areas, it is important to note that many Native American cultures used shifting agriculture and polyculture in temperate areas.

In many areas of the world with better soils, more intense forms of agriculture developed. These usually involved a great deal of manual labor. This style of agriculture is still practiced in much of the world today. Three situations favor this kind of farming: (1) when the growing site does not allow for mechanization, (2) when the kind of crop does not allow it, and (3) when the economic condition of the people does not allow them to purchase the tools and machines used for mechanized agriculture. Crops or terrain that requires that fields be small discourages mechanization, since large tractors and other machines cannot be used efficiently on small, oddly shaped fields. Many mountainous areas of the world fit into this category. In addition, some crops require such careful handling in planting, weeding, or harvesting that large amounts of hand labor are required. The planting of paddy rice

and the harvesting of many fruits and vegetables are examples.

However, the primary reason for labor-intensive farming is economic. Many densely populated countries have numerous small farms that can be effectively managed with human labor, supplemented by that of draft animals and a few small gasoline-powered engines. (See figure 15.2.) In addition, in the less-developed regions of the world, the cost of labor is low, which encourages the use of hand labor rather than relatively expensive machines to do planting, weeding, and other activities. Mechanization requires large tracts of land that could be accumulated only by the expenditure of large amounts of money or the development of larger cooperative farms from many small units. Even if social and political obstacles to such large landholdings could be overcome, there is still the problem of obtaining the necessary capital to purchase the machines. Large parts of the developing world fit into this category, including much of Africa, many areas in Central and South America, and many areas in Asia. In countries such as China and India, which have a combined population of over 2 billion people, about 70 percent of the population is rural. Many of these people are engaged in nonmechanized agriculture.

figure 15.2 **Labor-Intensive Agriculture** In many of the less-developed countries of the world, the extensive use of hand labor allows for impressive rates of production with a minimal input of fossil fuels and fertilizers. This kind of agriculture is also necessary in areas that have only small patches of land suitable for farming.

figure 15.3 **Monoculture** This wheat field is an example of monoculture, a kind of agriculture that is highly mechanized and requires large fields for the efficient use of machinery. In this kind of agriculture, machines and fossil-fuel energy have replaced the energy of humans and draft animals.

Mechanized agriculture is typical of North America, much of Europe, the republics of the former Soviet Union, parts of South America, and other parts of the world where money and land are available to support this form of agriculture. In large measure, machines and fossil-fuel energy replace the energy formerly supplied by human and animal muscles. Mechanization requires large expanses of fairly level land for the machines to operate effectively. In addition, large tracts of land must be planted in the same crop, a practice known as **monoculture,** for efficient planting, cultivating, and harvesting. (See figure 15.3.) Small sec-

tions of land with many kinds of crops require many changes of farm machinery, which takes time. Also, many crops interspersed with one another reduces the efficiency of farming operations because farmers must skip parts of the field, which increases travel time and uses expensive fuel.

Even though mechanized monoculture is an efficient method of producing food, it is not without serious drawbacks. When large tracts of land are prepared for planting, they are often left uncovered by vegetation, and soil erosion increases. Because of problems with erosion, many farmers are now using methods that reduce the time the fields are left bare. (See chapter 14.)

Traditionally, mechanized farming has removed much of the organic matter each year when the crop was harvested. This tended to reduce the soil organic matter. As agricultural scientists and farmers have recognized the need to improve the organic-matter content of soils, many farmers have been leaving increased amounts of organic matter after harvest, or they specifically plant a crop that is later plowed under to increase the soil's organic content.

To ensure that a crop can be planted, tended, and harvested efficiently by machines, farmers rely on seeds that are identical genetically and thus provide uniform plants with characteristics suitable for mechanized farming. These special seeds can ensure that all the plants germinate at the same time, resist the same pests, ripen at the same time, and grow to the same height. These are valuable characteristics to the farmer, but these plants have little genetic variety. When all the farmers in an area plant the same genetic varieties, pest control becomes a serious problem. If diseases or pests begin to spread, the magnitude of the problem becomes devastating because all the plants have the same characteristics and, thus, are susceptible to the same diseases. If genetically diverse crops are planted or crops are rotated from year to year, this problem is not as great.

Because farm equipment is expensive, farmers tend to specialize in a few crops. This means that the same crop

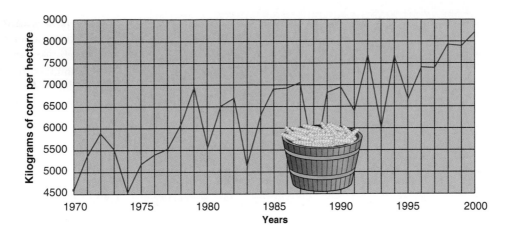

figure 15.4 **Increased Yields Resulting from Modern Technology** The increased crop yields in the United States and many other parts of the world are the result of a combination of factors, including the development of high-yielding varieties, changed agricultural methods, the application of fertilizers and pesticides, and more efficient machinery.

Source: Data from Department of Agriculture, *Agriculture Statistics*.

may be planted in the same field several years in a row. This lack of crop rotation may deplete certain essential soil nutrients, thereby requiring special attention be paid to soil chemistry. In addition, planting the same crop repeatedly encourages the growth of insect and fungus pest populations because they have a huge food supply at their disposal. This requires the frequent use of insecticides and fungicides or some other methods of pest control.

Even though there are problems associated with mechanized, monoculture agriculture, it has greatly increased the amount of food available to the world over the past 100 years. Yields per hectare of land being farmed have increased over much of the world, particularly in the developed world, which includes the United States. (See figure 15.4.) This increase has come about because of improved varieties of crops, irrigation, better farming methods, the use of agricultural chemicals, more efficient machines, and the use of energy-intensive as opposed to labor-intensive technology.

Throughout the 1950s, 1960s, and 1970s, the introduction of new plant varieties and farming methods resulted in increased agricultural production worldwide. This has been called the **Green Revolution.** Both the developed world, which uses highly mechanized farming methods, and the developing world,

where labor-intensive farming is typical, have benefited from these advances, and food production has increased significantly. The Green Revolution has also had some negative affects. For example, many modern varieties of plants require fertilizer and pesticides that the traditional varieties they replaced did not need. In addition, many of the crops require higher amounts of water and increase the demand for irrigation. Farmers also have become more dependent on the industries that provide the specialized seeds. When the positives and negatives are balanced, however, the end result is that food production per hectare has increased. This has not solved the world's food problem, however, because the population of the world continues to increase and more food is needed.

Fossil Fuel Versus Muscle Power

Mechanized agriculture has substituted the energy stored in petroleum products for the labor of humans. For example, in the United States in 1913, it required 135 hours of labor to produce 2500 kilograms (5500 pounds) of corn. In 1980, it required only about 15 hours of labor to produce the same amount of corn. The

energy supplied by petroleum products replaced the equivalent of 120 hours of labor. Energy is needed for tilling, planting, harvesting, and pumping irrigation water. The manufacture of fertilizer and pesticides also requires the input of large amount of fossil fuels, both as a source of energy for the industrial process and as raw material from which these materials are made. For example, about 5 tonnes (5.5 U.S. tons) of fossil fuel are required to produce about 1 tonne (1.1 U.S. tons) of fertilizer. In addition, the pesticides used in mechanized agriculture require the use of oil as a raw material. Since the developed world depends on oil to run machines and manufacture fertilizer and pesticides to support its agriculture, any change in the availability or cost of oil will have a major impact on the world's ability to feed itself.

The Impact of Fertilizer

Various experts estimate that approximately 25 percent of the world's crop yield can be directly attributed to the use of chemical fertilizers. The use of fertilizer has increased significantly over the last few decades and is projected to increase even more. (See figure 15.5.)

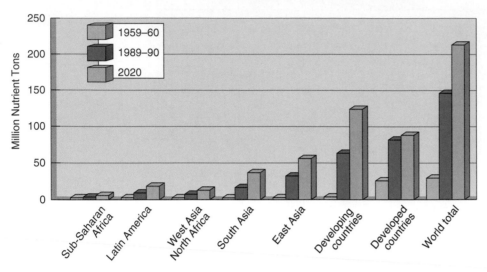

Estimated Growth in Worldwide Fertilizer Use, 1960–2020

Legend: 1959–60, 1989–90, 2020

Y-axis: Million Nutrient Tons

X-axis categories: Sub-Saharan Africa, Latin America, West Asia North Africa, South Asia, East Asia, Developing countries, Developed countries, World total

figure 15.5 **Increasing Fertilizer Use** The use of fertilizer is increasing rapidly. Growth in use will be most rapid in the developing countries over the next 20 years.

Source: Balu Bumb and Carlos Baanaunte, *World Trends in Fertilizer Use and Projections to 2020,* 2020 Brief No. 38 (International Food Policy Research Institute, Washington, D.C., 1996), table 1.

However, since fertilizer production relies on energy from fossil fuels, the price and availability of chemical fertilizers are strongly influenced by world energy prices. If the price of oil increases, the price of fertilizer goes up, as does the cost of food. This is felt most acutely in parts of the world where money is in short supply, since the farmers are unable to buy fertilizer, and crop yields fall accordingly.

Fertilizers are valuable because they replace the soil nutrients removed by plants. Some of the chemical building blocks of plants, such as carbon, hydrogen, and oxygen, are easily replaced by carbon dioxide from the air and water from the soil, but others are less easily replaced. The three primary soil nutrients often in short supply are nitrogen, phosphorus, and potassium compounds. They are often referred to as **macronutrients** and are the common ingredient of chemical fertilizers. Their replacement is important because when the crop is harvested, the chemical elements that are a part of the crop are removed from the field. Since many of those elements originated from the soil, they need to be replaced if another crop is to be grown. Certain other elements are necessary in extremely small amounts and are known as **micronutrients.** Examples are boron,

zinc, and manganese. As an example of the difference between macronutrients and micronutrients, harvesting a tonne (1.1 U.S. tons) of potatoes removes 10 kilograms (22 pounds) of nitrogen (a macronutrient) but only 13 grams (0.03 pounds) of boron (a micronutrient). When the same crop is grown repeatedly in the same field, certain micronutrients may be depleted, resulting in reduced yields. These necessary elements can be returned to the soil in sufficient amounts by incorporating them into the fertilizer the farmer applies to the field.

Although chemical fertilizers replace inorganic nutrients, they do not replace soil organic matter. Organic material is important because it modifies the structure of the soil, preventing compaction and maintaining pore space, which allows water and air to move to the roots. The decomposition of organic matter produces humus, which helps to maintain proper soil chemistry because it tends to loosely bind many soil nutrients and other molecules and modifies the pH so that nutrients are not released too rapidly. Soil bacteria and other organisms use organic matter as a source of energy. Since these organisms serve as important links in the carbon and nitrogen cycles, the presence of organic matter is important to their function.

Thus, total dependency on chemical fertilizers usually reduces the amount of organic matter and can change the physical, chemical, and biological properties of the soil.

As water moves through the soil, it dissolves soil nutrients (particularly nitrogen compounds) and carries them into streams and lakes, where they may encourage the growth of unwanted plants and algae. This is particularly true when fertilizers are applied at the wrong time of the year, just before a heavy rain, or in such large amounts that the plants cannot efficiently remove them from the soil before they are lost. These ideas are covered in greater detail in chapter 16 on water pollution.

Agricultural Chemical Use

In addition to chemical fertilizers, mechanized monoculture requires large amounts of other agricultural chemicals, such as pesticides, growth regulators, and preservatives. These chemicals have specific scientific names but are usually categorized into broad groups based on their effects. A **pesticide** is any chemical used to kill or control populations of unwanted fungi, animals, or plants, often called **pests.** The term *pest* is not scientific but refers to any organism that is unwanted. Insects that feed on crops are pests, while others, like bees, are beneficial for pollinating plants. Unwanted plants are generally referred to as **weeds.**

Pesticides can be subdivided into several categories based on the kinds of organisms they are used to control. **Insecticides** are used to control insect populations by killing them. Unwanted fungal pests that can weaken plants or destroy fruits are controlled by **fungicides.** Mice and rats are killed by **rodenticides,** and plant pests (weeds) are controlled by **herbicides.** Since pesticides do not kill just pests but can kill a large variety of living things, including humans, these chemicals might be more appropriately called **biocides.** They kill many kinds of living things. A perfect pesticide is one

DDT—A Historical Perspective

The first synthetic organic insecticide to be used was DDT [1,1,1-trichloro-2,2-*bis*-(p-chlorophenyl)ethane]. The chemical structure of DDT is shown in the figure. It was originally thought to be the perfect insecticide. It was inexpensive, long-lasting, relatively harmless to humans, and very deadly to insects. After its discovery, it was widely used in agriculture and to control disease-carrying insects. During the first 10 years of its use (1942–52), DDT is estimated to have saved 5 million lives, primarily because of its use in controlling disease-carrying mosquitoes.

However, DDT had several drawbacks. Although most of these drawbacks are not unique to DDT, they were first recognized with DDT because it was the first widely used insecticide. One problem involved the development of resistance in insect populations that were repeatedly subjected to spraying by DDT. Another was the fact that it affected many nontarget organisms, not just the disease-carrying insects that were the original targets. This was a particular problem because DDT is a persistent chemical, which is another major drawback. In temperate regions of the world, DDT has a half-life (the amount of time required for half of the chemical to decompose) of 10 to 15 years. This means that if 1000 kilograms (2200 pounds) of DDT were sprayed over an area, 500 kilograms (1100 pounds) would still be present in the area 10 to 15 years later; 30 years from the date of application, 250 kilograms (550 pounds) would still be present. The half-life of DDT varies depending on soil type, temperature, the kinds of soil organisms present, and other factors. In tropical parts of the world, the half-life may be as short as six months. An additional complication is that persistent pesticides may break down into products that are still harmful. Furthermore, because it is persistent, DDT tends to accumulate and reach higher concentrations in older animals and in animals at higher trophic levels.

When these effects were recognized, the United States and many other developed countries banned the use of DDT. A key force in bringing about this change in thinking was the book *Silent Spring,* by Rachael Carson. (See Environmental Close-Up: Environmental Philosophers on page 23.) The results of this ban are clearly seen in the reduced levels of DDT in the environment and in the recovering populations of such organisms as eagles, cormorants, and pelicans. (See Issues and Analysis: Herring Gulls as Indicators of Contamination in the Great Lakes on page 347.)

Because of these problems, the World Wildlife Fund asked delegates at a United Nations Environment Programme conference on persistent organic pollutants in 1999 to ban DDT from use throughout the world because it was showing up worldwide in the bodies of many kinds of animals that were great distances from sources of DDT and was present in the breast milk of women where DDT is used. This resulted in a great deal of protest from public health people concerned about controlling malaria, since DDT is still widely used in many parts of the world, where it is sprayed on the walls of houses to kill mosquitoes that spread malaria. The use of DDT is still effective and is much less expensive than materials that would be substituted for it. They feared that the total ban of DDT would result in less effective control of mosquitoes and increased deaths from malaria. Where the risk of dying from malaria is high and the risk of possible health affects from exposure to DDT is low, it makes sense to use an inexpensive, effective material such as DDT. In countries where malaria and other serious insect-transmitted diseases are rare (most of the developed world), it is possible to eliminate the use of DDT. Ultimately, continued use of DDT was sanctioned for the control of malaria and other diseases transmitted by insects, and it continues to be used for these purposes today.

---

that kills or inhibits the growth of only the specific pest organism causing a problem. The pest is often referred to as the **target organism.** However, most pesticides are not very specific and kill many **nontarget organisms** as well. For example, most insecticides kill both beneficial and pest species, rodenticides kill other animals as well as rodents, and most herbicides kill a variety of plants, both pests and nonpests.

Many of the older pesticides were very stable and remained active for long periods of time. These are called **persistent pesticides.** Pesticides that break down quickly are called **nonpersistent pesticides.**

Insecticides

If insects are not controlled, they consume a large proportion of the crops produced by farmers. In small garden plots, insects can be controlled by manually removing them and killing them. However, in large fields, this is not practical, so people have sought other ways to control pest insects. In addition, many insects harm humans because they spread diseases, such as sleeping sickness, bubonic plague, and malaria. Mosquitoes are known to carry over 30 diseases harmful to humans. Currently, the World Health Organization estimates that between 300 and 500 million new malaria infections occur each year and that between 1.2 and 2.7 million people die of these infections each year. Most of these deaths are children. Malaria is one of the top five causes of death in children. The discovery of chemicals that could kill insects was celebrated as a major advance in the control of disease and the protection of crops.

Nearly 3000 years ago, the Greek poet Homer mentioned the use of sulfur to control insects. For centuries, it was known that natural plant products could repel or kill insect pests. Plants with insect-repelling abilities were interplanted with crops to help control the pests. Nicotine from tobacco, rotenone from tropical legumes, and pyrethrum from chrysanthemums were extracted and used to control insects. In fact, these compounds are still used today. However, because plant products are difficult to extract and apply and have short-lived effects, other compounds were sought. In 1867, the first synthetic inorganic insecticide, Paris green, was formulated. It was a mixture of acetate and arsenide of copper and was used to control Colorado potato beetles.

DDT was the first synthetic organic insecticide produced. (See Environmental Close-Up: DDT—A Historical Perspective on page 331.) Since then, over 60,000 different compounds that have potential as insecticides have been synthesized. However, most of these have never been put into production because cost, human health effects, or other drawbacks make them unusable. Several categories of these compounds have been developed. Three that are currently used are chlorinated hydrocarbons, organophosphates, and carbamates.

Chlorinated Hydrocarbons

Chlorinated hydrocarbons are a group of pesticides of complex, stable structure that contain carbon, hydrogen, and chlorine. DDT was the first such pesticide manufactured, but several others have been developed. Other chlorinated hydrocarbons are chlordane, aldrin, hep-

tachlor, dieldrin, and endrin. It is not fully understood how these compounds work, but they are believed to affect the nervous systems of insects, resulting in their death.

One of the major characteristics of these pesticides is that they are very stable chemical compounds. This is both an advantage and a disadvantage. They can be applied once and be effective for a long time. However, since they do not break down easily, they tend to accumulate in the soil and in the bodies of animals in the food chain. Thus, they affect many nontarget organisms, not just the original target insects.

Because of their negative effects, most of the chlorinated hydrocarbons are no longer used in many parts of the world. DDT, aldrin, dieldrin, toxaphene, chlordane, and heptachlor have been banned in the United States and many other developed countries. However, many developing countries still use chlorinated hydrocarbons for insect control to protect crops and public health. Because of their persistence and continued use in many parts of the world, chlorinated hydrocarbons are still present in the food chain, although the level of contamination has dropped. These molecules continue to enter parts of the world—where their use has been banned—through the atmosphere and as trace contaminants of imported products.

Organophosphates and Carbamates

Because of the problems associated with persistent insecticides, nonpersistent insecticides that decompose to harmless products in a few hours or days were developed. However, like other insecticides, these are not species specific; they kill beneficial insects as well as harmful ones. Although the short half-life prevents the accumulation of toxic material in the environment, it is a disadvantage for farmers, since more frequent applications are required to control pests. This requires more labor and fuel and, therefore, is more expensive.

Both **organophosphates** and **carbamates** work by interfering with

the ability of the nervous system to conduct impulses normally. Under normal conditions, a nerve impulse is conducted from one nerve cell to another by means of a chemical known as a neurotransmitter. One of the most common neurotransmitters is acetylcholine. When this chemical is produced at the end of one nerve cell, it causes an impulse to be passed to the next cell, thereby transferring the nerve message. As soon as this transfer is completed, an enzyme known as cholinesterase destroys acetylcholine, so the second nerve cell in the chain is stimulated for only a short time. Organophosphates and carbamates interfere with cholinesterase, preventing it from destroying acetylcholine. This results in nerve cells being continuously stimulated, causing uncontrolled spasms of nervous activity and uncoordination that result in death.

Although these pesticides are less persistent in the environment than are chlorinated hydrocarbons, they are generally much more toxic to humans and other vertebrates because these insecticides affect their nerve cells as well. Persons who apply such pesticides must use special equipment and should receive special training because improper use can result in death. Since organophosphates interfere with cholinesterase more strongly than do carbamates, they are considered more dangerous and, for many applications, have been replaced by carbamates.

Common organophosphates are malathion, parathion, and diazinon. Malathion is widely used for such projects as mosquito control, but parathion is a restricted organophosphate because of its high toxicity to humans. Diazinon is widely used in gardens. Sevin, aldicarb, and propoxur are examples of carbamates.

Herbicides

Herbicides are another major class of chemical control agents. In fact, about 60 percent of the approximately 440 million kilograms (about 1 billion pounds) of pesticides used in U.S. agriculture are herbicides. They are widely used to control unwanted vegetation along power-line rights-of-way, railroad rights-of-way, and highways, as well as on lawns and cropland, where they are commonly referred to as weed killers.

Weeds are plants we do not want to have growing in a particular place. Weed control is extremely important for agriculture since weeds take nutrients and water from the soil, making them unavailable to the crop species. In addition, weeds may shade the crop species and prevent it from getting the sunlight it needs for rapid growth. At harvest time, weeds reduce the efficiency of harvesting machines. Also, weeds generally must be sorted from the crop before it can be sold, which adds to the time and expense of harvesting.

Traditionally, farmers have expended much energy trying to control weeds. Initially, weeds were eliminated with manual labor and the hoe. Tilling the soil also helps to control weeds. Once the crop is planted, row crops like corn or sugar beets may be cultivated to remove weeds from between the rows. All of these activities are expensive in time and fuel. Selective use of herbicides can have a tremendous impact on a farmer's profits.

Many of the recently developed herbicides can be very selective if used appropriately. Some are used to kill weed seeds in the soil before the crop is planted, while others are used after the weeds and the crop begin to grow. In some cases, a mixture of herbicides can be used to control several weed species simultaneously. Figure 15.6 shows the effects of using a herbicide that kills grasses but not other kinds of plants.

Several major types of herbicides are in current use. One type is synthetic plant-growth regulators that mimic natural-growth regulators known as **auxins.** Two of the earliest herbicides were of this type: 2,4-dichlorophenoxyacetic acid (2,4-D) and 2,4,5-trichlorophenoxyacetic acid (2,4,5-T). When applied to broadleaf plants, these chemicals disrupt normal growth, causing the death of the plant. 2,4-D has been in use for about 50 years and is still one of the most widely used herbicides. Many

figure 15.6 **The Effect of Herbicides** The grasses in this photograph have been treated with herbicides. The soybeans are unaffected and grow better without competition from grasses.

newer herbicides have other methods of action. Some disrupt photosynthetic activity of plants, causing their death. Others inhibit enzymes, precipitate proteins, stop cell division, or destroy cells directly. Depending on the concentration of the herbicide used, some are toxic to all plants, while others are very selective as to which species of plants they affect. One such herbicide is diuron. In proper concentrations and when applied at the appropriate time, it can be used to control annual grasses and broadleaf weeds in over 20 different crops. However, at higher concentrations, it kills all vegetation in an area. Fenuron is a herbicide that kills woody plants. In low concentration, it is used to control woody weed plants in cropland. In high concentrations, it is used on noncroplands, such as power-line rights-of-way. (See figure 15.7.)

Fungicides and Rodenticides

Fungus pests can be divided into two categories. Some are natural decomposers of organic material, but when the organic material being destroyed happens to be a crop or other product useful to humans,

the fungus is considered a pest. Other fungi are parasites on crop plants; they weaken or kill the plants, thereby reducing the yield. Fungicides are used as fumigants (gases) to protect agricultural products from spoilage, as sprays and dusts to prevent the spread of diseases among plants, and as seed treatments to protect seeds from rotting in the soil before they have a chance to germinate. Methylmercury is often used on seeds to protect them from spoilage before germination. However, since methylmercury is extremely toxic to humans, these seeds should never be used for food. To reduce the chance of a mix-up, treated seeds are usually dyed a bright color.

Like fungi, rodents are harmful because they destroy food supplies. In addition, they can carry disease and damage crops in the field. In many parts of the world, such as India, the government pays a bounty to people who kill rats, because this is an inexpensive way to protect the food supply. Several kinds of rodenticides have been developed to control rodents. One of the most widely used is warfarin, a chemical that causes internal bleeding in animals that

A New Generation of Insecticides

The perfect insecticide would harm only target insect species, not be toxic to humans, not be persistent, and would break down into harmless materials. Most currently used insecticides were developed by screening a wide variety of compounds and modifying compounds that showed insect-killing properties. Most are highly toxic to many kinds of organisms and must be used with great care. However, a new generation of insecticides is being developed that starts with knowledge of insect biology and develops a compound that interferes with some essential aspect of the life cycle. Some mimic normal hormones and prevent insects from maturing into adults. Therefore, the insects do not reproduce, and the populations are greatly reduced.

Other insecticides attack specific chemical processes in cells and cause death. Nicotine is a normal product of some plants such as tobacco, which is toxic to insects. It binds to specific receptors on nerve cells, causing them to fire uncontrollably. However, it is not stable in sunlight. By tinkering with the chemical structure of nicotine, a modified molecule, called imidacloprid, was developed. It is more stable, not toxic to mammals because they do not have as many of the receptors as insects, and very effective against sucking insects.

Another new insecticide works by turning off the energy-producing processes of mitochondria in cells. All cells of higher organisms have mitochondria and would be injured by this insecticide, but a nontoxic form of the molecule was produced that is converted to a toxic form in the bodies of insects but is not converted in mammals. It is highly toxic to birds and some aquatic organisms, however, which would require strict control of its use.

consume it. It is usually incorporated into a food substance so that rodents eat warfarin along with the bait. Because it is effective in all mammals, including humans, it must be used with care to prevent nontarget animals from having access to the chemical. As with many kinds of pesticides, some populations of rodents have become tolerant of warfarin, while others avoid baited areas. In many cases, rodent problems can be minimized by building storage buildings that are rodent proof, rather than relying on rodenticides to control them.

Other Agricultural Chemicals

In addition to herbicides, other agrochemicals are used for special applications. For example, a synthetic auxin sprayed on cotton plants before harvest causes the leaves to drop off, which facilitates the harvesting process by reducing clogging of the mechanical cotton picker.

NAA (naphthaleneacetic acid) is used by fruit growers to prevent apples from dropping from the trees and being damaged. This chemical can keep the apples on the trees for up to 10 extra

figure 15.7 **Herbicide Use to Maintain Rights-of-Way** Power-line rights-of-way are commonly maintained by herbicides that kill the woody vegetation, which might grow so tall that it interferes with the power lines.

days, which allows for a longer harvest period and fewer lost apples.

Under other conditions, it may be valuable to get fruit to fall more easily. Cherry growers use ethephon to promote loosening of the fruit so that the cherries will fall more easily from the tree when shaken by the mechanical harvester. This method lowers the cost of harvesting the fruit. (See figure 15.8.)

figure 15.8 **Chemical Loosening of Cherries to Allow Mechanical Harvesting** By using chemicals and machinery, this farmer can rapidly harvest the cherry crop. This practice reduces the amount of labor required to pick the cherries but requires the application of chemicals to loosen the fruit.

Problems with Pesticide Use

A perfect pesticide would have the following characteristics:

1. It would be inexpensive.
2. It would affect only the target organism.
3. It would have a short half-life.
4. It would break down into harmless materials.

However, the perfect pesticide has not been invented. Many of the more recently developed pesticides have fewer drawbacks than the early pesticides, but none is without problems.

Persistence

Although the trend has been away from using persistent pesticides in North America and much of the developed world, some are still allowed for special purposes, and they are still in common use in other parts of the world. Because of their stability, these chemicals have become a long-term problem. Persistent insecticides become attached to small soil particles, which are easily moved by wind or water to any part of the world. Persistent pesticides and other pollutants have been discovered in the ice of the poles and are present in detectable amounts in the body tissues of animals, including humans, throughout the world. Thus, chemicals originally sprayed to control mosquitoes in Africa or to protect a sugarcane field in Brazil may be distributed throughout the world.

Bioaccumulation and Biomagnification

A problem associated with persistent chemicals is that they may accumulate in the bodies of animals. If an animal receives small quantities of persistent pesticides or other persistent pollutants in its food and is unable to eliminate them, the concentration within the animal increases. This process of accumulating higher and higher amounts of material within the body of an animal is called **bioaccumulation.** Many of the persistent pesticides and their breakdown products are fat soluble and build up in the fat of animals. When affected animals are eaten by a carnivore, these toxins are further concentrated in the body of the carnivore, causing disease or death, even though lower-trophic-level organisms are not injured. This phenomenon of acquiring increasing levels of a substance in the bodies of higher-trophic-level organisms is known as **biomagnification.**

The well-documented case of DDT is an example of how biomagnification occurs. DDT is not very soluble in water but dissolves in oil or fatty compounds. When DDT falls on an insect or is consumed by the insect, the DDT is accumulated in the insect's fatty tissue. Large doses kill insects but small doses do not, and their bodies may contain as much as one part per billion of DDT. This is not very much, but it can have a tremendous effect on the animals that feed on the insects.

If an aquatic habitat is sprayed with a small concentration of DDT or receives DDT from runoff, small aquatic organisms may accumulate a concentration that is up to 250 times greater than the concentration of DDT in the surrounding water. These organisms are eaten by shrimp, clams, and small fish, which are, in turn, eaten by larger fish. DDT concentrations of large fish can be as much as 2000 times the original concentration sprayed on the area. What was a very small initial concentration has now become so high that it could be fatal to animals at higher trophic levels. This has been of particular concern for birds, since DDT interferes with the production of eggshells, making them much more fragile. This problem is more common in carnivorous birds because they are at the top of the food chain. Although all birds of prey have probably been affected to some degree, those that rely on fish for food seem to have been affected most severely. Eagles, osprey, cormorants, and pelicans are particularly susceptible species. (See figure 15.9.)

Other persistent molecules are known to behave in similar fashion. Mercury, aldrin, chlordane, and other chlorinated hydrocarbons, such as poly chlorinated biphenyls (PCBs) used as insulators in electric transformers, are all known to accumulate in ecosystems. PCBs have been strongly implicated in the decline of cormorants in the Great Lakes. As PCB levels have declined, the

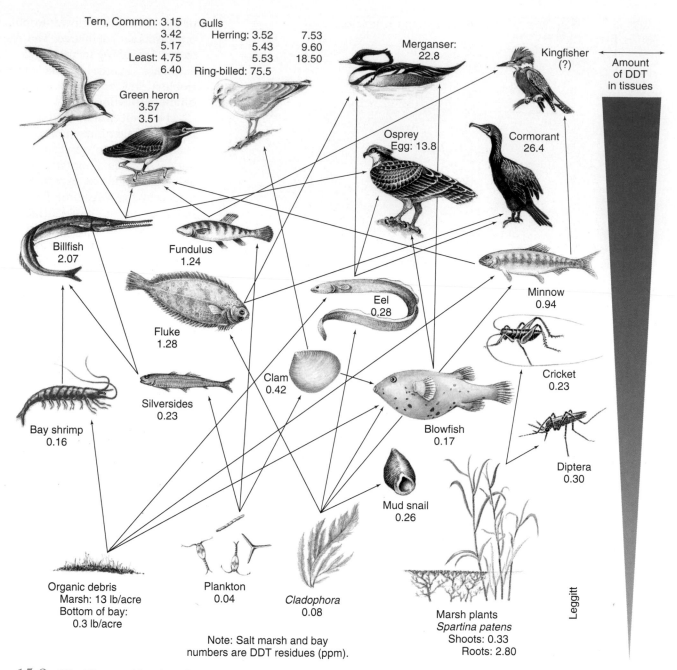

Tern, Common: 3.15
3.42
5.17
Least: 4.75
6.40

Gulls
Herring: 3.52   7.53
5.43   9.60
5.53   18.50
Ring-billed: 75.5

Merganser: 22.8

Kingfisher (?)

Amount of DDT in tissues

Green heron
3.57
3.51

Osprey
Egg: 13.8

Cormorant
26.4

Billfish
2.07

Fundulus
1.24

Minnow
0.94

Fluke
1.28

Eel
0.28

Cricket
0.23

Clam
0.42

Silversides
0.23

Blowfish
0.17

Bay shrimp
0.16

Diptera
0.30

Mud snail
0.26

Organic debris
Marsh: 13 lb/acre
Bottom of bay:
0.3 lb/acre

Plankton
0.04

Cladophora
0.08

Marsh plants
Spartina patens
Shoots: 0.33
Roots: 2.80

Leggitt

Note: Salt marsh and bay
numbers are DDT residues (ppm).

figure 15.9   **The Biomagnification of DDT** All the numbers shown are in parts per million (ppm). A concentration of one part per million means that in a million equal parts of the organism, one of the parts would be DDT. Notice how the amount of DDT in the bodies of the organisms increases as we go from producers to herbivores to carnivores. Because DDT is persistent, it builds up in the top trophic levels of the food chain.

cormorant population has returned to former levels.

Because of their persistence, their effects on organisms at higher trophic levels, and concerns about long-term human health problems, most chlorinated hydrocarbon pesticides have been banned from use in the United States and some other countries. The use of DDT was prohibited in the United States in the early 1970s. Aldrin, diel-

drin, heptachlor and chlordane have also been prohibited from use on crops, although heptachlor and chlordane were still used for termite control until recently. In 1987, Velsicol Chemical Corporation agreed to stop selling chlordane in the United States.

The populations of several species of birds, including the brown pelican, bald eagle, osprey, and cormorant, all of which feed primarily on fish, were se-

verely affected because of biomagnification of persistent organic chemicals. With the control or restriction of most persistent pesticides and several other chemicals, the levels of these chemicals in their body tissues have declined, and their populations have rebounded. The bald eagle has been removed from the endangered species list, and the pelican has been removed from the endangered species list in part of its range.

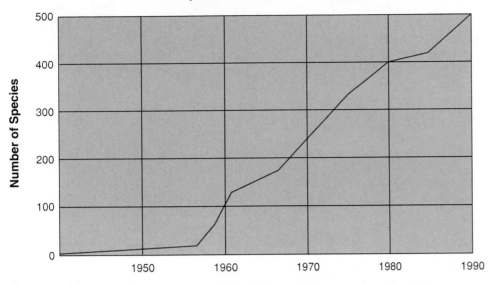

Pest Species Resistant to Insecticides

figure 15.10 **Resistance to Insecticides** The continued use of insecticides has constantly selected for genes that give resistance to a particular insecticide. As a result, many species of insects and other arthropods are now resistant to many kinds of insecticides, and the number continues to increase.

Source: Data from George P. Georghiou, University of California at Riverside.

Table 15.1	Average Dose Necessary to Kill Two Cotton Pests			
Average Dose Necessary to Kill (Milligrams per Gram of Larva)				
	Bollworm		**Tobacco Budworm**	
Compound	**1960**	**1965**	**1961**	**1965**
DDT	0.03	1.000+	0.13	16.51
Endrin	0.01	0.13	0.06	12.94
Carbaryl	0.12	0.54	0.30	54.57
Strobane and DDT	0.05	1.04	0.73	11.12
Toxaphene and DDT	0.04	0.46	0.47	3.52

Source: Reprinted with permission from P.L. Adkisson, "Controlling Cotton's Insect Pests: A New System," *Science* 216 (April 1982):19–22. Copyright © 1982 American Association for the Advancement of Science.

Pesticide Resistance

Another problem associated with pesticides is the ability of pest populations (insects, weeds, rodents, fungi) to become resistant to them. Not all organisms within a given species are identical. Each individual has a slightly different genetic composition and slightly different characteristics. If an insecticide is used for the first time on the population of a particular insect pest, it kills all the individuals that are susceptible. Individuals with characteristics that allow them to tolerate the insecticides may live to reproduce.

If only 5 percent of the individuals possess genes that make them resistant to an insecticide, the first application of the insecticide will kill 95 percent of the population and so will be of great benefit in controlling the size of the insect population. However, the surviving individuals that are tolerant of the insecticide will constitute the majority of the breeding populations. Since these individuals possess genetic characteristics for tolerating the insecticide, so will many of their offspring. Therefore, in the next generation, the number of individuals able to tolerate the insecticide will increase, and the second use of the insecticide will not be as effective as the first. Since some species of insect pests can produce a new generation each month, this process of selecting individuals capable of tolerating the insecticide can result in resistant populations in which 99 percent of the individuals are able to tolerate the insecticide within five years. As a result, that particular insecticide is no longer as effective in controlling insect pests, and increased dosages or more frequent spraying may be necessary. Figure 15.10 indicates that over 500 species of insects have populations resistant to insecticides.

Cotton growers throughout the world are extremely dependent on the use of insecticides. About 40 percent of the insecticides used in the United States are for controlling pests in cotton. Because of this intensive use, many populations of pest insects have become resistant to the commonly used insecticides. Table 15.1 shows the effect of continual use of insecticides on two pests, the bollworm and the tobacco budworm. The size of the dose necessary to kill the pest increased greatly in a five-year period. In some cases, the dose increased tenfold and in others, a hundredfold or more. By 1965, these insecticides were no longer able to control bollworm or tobacco budworm.

Effects on Nontarget Organisms

Most pesticides are not specific and kill beneficial species as well as pest species. With herbicides, this is usually not a problem because a herbicide is chosen that does not harm the desired crop plant, and generally all other plants are competing pests. However, with insecticides, there are several problems associated with the effects on nontarget organisms. The use of insecticides can harm populations of birds, mammals, and harmless insects. Many insecticides that harm vertebrates are restricted in their use. The insecticide may kill

China's Ravenous Appetite

History has shown that when people rise above subsistence level, one of the first things they want is to eat a little better. They want to include more meat in their diet. Therefore, they will feed some of their grains to animals and consume animal products such as meat, fish, eggs, and butter. They eat more vegetables and use more edible oils. In short, they move a bit higher up the food chain.

A sizable percentage of China's nearly 1.3 billion citizens are moving up the food chain. In the early 1980s, the typical urban Chinese diet consisted of rice, porridge, and cabbage. By the middle 1990s, the diet had dramatically changed to include meat, eggs, or fish at least once a day.

Diet is not the only thing changing in China. Even with population control, China's population grows by over 11 million people a year, and young people are leaving the countryside and moving to the cities. China's urban population, currently over 450 million, should double by 2010. China has about 21 percent of the world's population but only 7 percent of the arable land; in other words, it has less arable land than the United States but nearly five times the population. China's farm sector is simply unable to keep up with the surging demand for what people regard as better food.

Total meat consumption in China is growing by 10 percent a year; feed for animal consumption is growing by 15 percent. Demand for poultry, which requires 2 to 3 kilograms of feed per kilogram of bird, has doubled in five years. China's new diet will make it more dependent on the United States, Canada, and Australia for feed grains. By 1997, China had gone from being a net exporter of grain to importing 16 million tonnes (17.6 million U.S. tons). The switch in corn is

KFC and McDonald's in Beijing. Demand for poultry and beef is soaring.

even more dramatic. As recently as 1995, China was the second-largest corn exporter in the world. But with the chickens and pigs eating so much corn, in one year, China moved from exporting 12 million tonnes (13.2 million U.S. tons) of corn to importing 4 million tonnes (4.4 million U.S. tons).

China has also become the world's largest importer of fertilizer. In addition, while the government is converting some marginal lands in the north for agriculture, this is more than offset by the loss of fertile, multiple-cropped farmland in the southern coastal provinces. Overall, China's farmland is shrinking by at least 0.5 percent per year.

China also faces a severe water shortage in the arid north. In the Yellow River Valley's large grain belt, irrigation projects tripled crop yields during the 1950s and 1960s, resulting in severe overpumping of groundwater. Today, the water table is falling, aquifers are vanishing, and farmers now have to compete with industry and households for water. The era of the flush toilet has arrived.

China is only a part—but certainly the biggest part—of a larger story. What's true for China is also happening throughout much of Asia. Large, populous countries such as Indonesia, Thailand, and the Philippines are rapidly urbanizing. They are gaining purchasing power and losing farmland, and the people are adding more animal proteins and processed foods to their diets. Given the magnitude of the changes, the world's ability to feed itself in the future is difficult to predict.

- What changes do you anticipate in the food consumption pattern in China in the next decade?
- Is China capable of maintaining a diet similar to that of North Americans? Is such a diet desirable?

---

predator and parasitic insects that normally control the pest insects. This allows pest species to increase rapidly following the use of a pesticide because there are no natural checks to their population growth. Additional applications of insecticides are necessary to prevent

the pest population from rebounding to levels even higher than the initial one. Once the decision is made to use pesticides, it often becomes an irreversible tactic, because stopping their use would result in rapid increases in pest populations and extensive crop damage.

An associated problem is that the use of insecticides may change the population structures of species present so that a species that was not previously a problem becomes a serious pest. For example, when synthetic organic insecticides came into common use with

Global Perspective

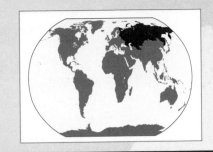

Contaminated Soils in the Former Soviet Union

The independent republics of the former Soviet Union have inherited many kinds of environmental problems generated by the failed policies of the Soviet government. One of these policies was a push for increased agricultural production. New land was brought under cultivation, irrigation projects were developed, and extensive use of fertilizer and pesticides stimulated production.

This single-minded approach has resulted in serious environmental damage. In many areas, the soil has been contaminated by excessive fertilizer and pesticide use, poisoning groundwater and tainting food products. It is estimated that 30 percent of the food produced throughout the former Soviet Union is contaminated and should not be eaten. Forty-two percent of all baby food is contaminated by nitrates and pesticides. Latvian ecologists estimate that excessive pesticide use currently results in 14,000 deaths per year and that 700,000 people become ill from pesticides throughout the former Soviet Union annually.

One problem pesticide is DDT. It was officially banned in the Soviet Union in the 1970s, but the drive to improve agricultural production allowed its continued use. Waivers to the ban were allowed, or DDT was simply renamed by government bureaucrats and used under its new label. It is estimated that 10 million hectares (25 million acres) of agricultural land in the former Soviet Union are contaminated with DDT.

The widespread use of agricultural chemicals has also affected the groundwater. Persistent pesticides, nitrates, and heavy metals have entered the groundwater, which many rely on for drinking water. Some areas have cadmium levels three to 10 times higher than the amount permitted.

It appears that the fall of communism in the former Soviet Union did not solve all the problems of the people. These environmental problems are surfacing after years of secrecy and denial under the shortsighted, production-oriented agricultural policy of the failed political system.

---

cotton in the 1940s, the insect parasites and predators were eliminated, and the bollworm and tobacco budworm became major pests. In the mid-1990s, a similar situation developed when repeated use of malathion removed predator insects and allowed the populations of beet army worms to become a major pest. The repeated use of insecticides caused a different pest problem to develop.

Human Health Concerns

Short-term and long-term health effects to the persons using the pesticide and the public that consumes the food grown by using pesticides are also concerns. If properly applied, most pesticides can be used with little danger to the applicator. However, in many cases, people applying pesticides are unaware of how they work and what precautions should be used in their application. In many parts of the world, farmers may not be able to read the caution labels on the packages or do not have access to the protective gear specified for use with the pesticide. Therefore, many incidences of acute

poisoning occur each year. In most cases, the symptoms disappear after a period free from exposure. Estimates of the number of poisonings are very difficult to obtain since many go unreported, but in the United States, pesticide poisonings requiring medical treatment are in the thousands per year. The World Health Organization estimates that each year, there are between 3.5 million and 5 million acute pesticide poisonings and that about 40,000 deaths occur from pesticide poisonings.

For most people, however, the most critical health problem related to pesticide use is inadvertent exposure to very small quantities of pesticides. Many pesticides have been proven to cause mutations, produce cancers, or cause abnormal births in experimental animals. Studies of farmers who were occupationally exposed to pesticides over many years show that they have higher levels of certain kinds of cancers than the general public. There also are questions about the effects of chronic, minute exposures to pesticide residues in food or through contamination of the environment.

Although the risks of endangering health by consuming tiny amounts of pesticides are very small compared to other risk factors, such as automobile accidents, smoking, or poor eating habits, many people find pesticides unacceptable and seek to prohibit their sale and use. The U.S. Environmental Protection Agency requires careful studies of the effectiveness and possible side effects of each pesticide licensed for use. It has banned several pesticides from further use because new information suggests that they are not as safe as originally thought. For example, it banned the use of dinoseb, a herbicide, because tests by the German chemical company Hoechst AG indicated that dinoseb causes birth defects in rabbits. Other studies indicate that dinoseb causes sterility in rats. Similarly, in 1987, Velsicol Chemical Corporation signed an agreement with the EPA to stop producing and distributing chlordane in the United States. Chlordane had been banned previously for all applications except for termite control. After it was shown that harmful levels of chlordane could exist in treated homes, its use was finally curtailed. Also in 1987, the

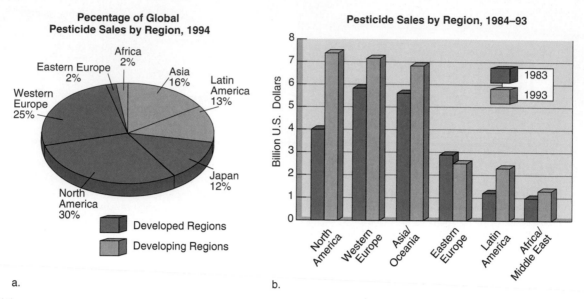

Pecentage of Global Pesticide Sales by Region, 1994

Africa 2%
Eastern Europe 2%
Western Europe 25%
Asia 16%
Latin America 13%
Japan 12%
North America 30%

Developed Regions
Developing Regions

a.

Pesticide Sales by Region, 1984–93

1983
1993

Billion U.S. Dollars

North America
Western Europe
Asia/Oceania
Eastern Europe
Latin America
Africa/Middle East

b.

figure 15.11 **Pesticide Sales** The use of pesticides is greatest in the developed world, with about 67 percent of the total. However, all parts of the world except for Eastern Europe are experiencing increased use of pesticides.

Source: (*a*) "Upturn in World Agrochemical Sales in 1994," *AGRO: World Crop Protection News,* 238 (August 1995): 20; (*b*) The Freedonia Group, *World Pesticides Report No. 636* (The Freedonia Group, Cleveland, Ohio, 1994) as cited in *AGRO,* 225 (1995):16.

Dow Chemical Company announced that it would stop producing the controversial herbicide 2,4,5-T. In both of these cases, new pesticides were developed to replace the older types, so that discontinuing them did not result in an economic hardship for farmers or other consumers.

Finally, during the 1990s, the EPA required that many pesticides be reregistered. During this process, manufacturers needed to justify the continued production of the product. In many cases, companies simply did not want to spend the money to go through the registration process; therefore, the product will be phased out. In other cases, the variety of uses was modified. For example, chlopyrifos, one of the most widely used insecticides, had its use altered. Through negotiations with the producers, its use in homes and schools will be eliminated, while it will still be used in most agricultural settings.

Why Are Pesticides So Widely Used?

Figure 15.11 shows changes in the use of pesticides throughout the world. Use

is projected to continue to increase significantly. If pesticides have so many drawbacks, why are they used so extensively? There are three primary reasons. First, the use of pesticides has increased, at least in the short term, the amount of food that can be grown in many parts of the world. In the United States, pests are estimated to consume 33 percent of the crops grown. On a worldwide basis, pests consume approximately 35 percent of crops. This represents an annual loss of $18.2 billion in the United States alone. Farmers, grain-storage operators, and the food industry continually seek to reduce this loss. A retreat from dependence on pesticides would certainly reduce the amount of food produced. Agricultural planners in most countries are not likely to suggest changes in pesticide use that would result in malnutrition and starvation for many of their inhabitants.

The economic value of pesticides is the second reason they are used so extensively. The cost of pesticides is more than offset by increased yields and profits for the farmer. In addition, the production and distribution of pesticides is big business. Companies that have spent millions of dollars developing a pesticide are going to argue very strongly for its

continued use. Since farmers and agrochemical interests have a powerful voice in government, they have successfully lobbied for continued use of pesticides.

A third reason for extensive pesticide use is that many health problems are currently impossible to control without insecticides. This is particularly true in areas of the world where insect-borne diseases would cause widespread public health consequences if insecticides were not used.

Alternatives to Conventional Agriculture

Before the invention of synthetic fertilizers, herbicides, fungicides, and other agrochemicals, animal manure and crop rotation provided soil nutrients; a mixture of crops prevented regular pest problems; and manual labor killed insects and weeds. With the development of mechanization, larger areas could be farmed, draft animals were no longer needed, and many farmers changed from mixed agriculture, in which animals were an

Industrial Production of Livestock

Over the last couple of decades, the trend has been toward the concentration of livestock production into large factorylike operations. Dairy cattle, beef cattle, pigs, and chickens for both eggs and meat are raised in concentrations that are causing environmental problems. Many of these commercial operations feed thousands of animals, which produce equally large amounts of manure. (See photo.) Although such livestock-raising operations are supposed to contain and dispose of their animal waste properly, problems often result. Holding tanks or lagoons may fail and release animal waste into local streams.

One of the standard ways of dealing with the waste is to use it as fertilizer. This is a relatively simple and effective technology. However, the amount of waste produced is often greater than the land suitable and available for application of manure as fertilizer. The manure may not be applied on all crops, cannot be applied when the ground is frozen, and may not be applied at other times of the year for a variety of reasons. However, the animals produce manure every day.

Groundwater contamination is an additional problem. When concentrated animal wastes are in direct contact with the soil in feedlots or unlined sewage lagoons, or when they are applied to fields at too great a concentration, they can penetrate the soil and enter the groundwater. Other issues involve odors and dusts associated with large industrial livestock operations.

An additional concern related to this form of agriculture is the frequent use of antibiotics. Raising large numbers of animals together creates a ready environment for the spread of disease. Consequently, antibiotics are used regularly to prevent the spread of disease and foster faster growth of the animals. When antibiotics are

used so freely, it is highly likely that the disease organisms being targeted will become resistant to the antibiotics. Since some of these disease organisms (particularly those of pigs and chickens) can spread to humans, resistant disease organisms in livestock can become a human health problem as well.

There are also economic considerations. Since small farmers cannot compete with the large commercial operations, many have gone out of business or had to switch to other agricultural activities to replace the loss of income from raising livestock. While several organizations are concerned about the ethics, health implications, and social implications of such large livestock operations, the average consumer in the supermarket knows little about where the meat products originated or under what conditions the animals were raised.

important ingredient, to monoculture. Chemical fertilizers replaced manure as a source of soil nutrients, and crop rotation was no longer as important since hay and grain were no longer grown for draft animals and cattle. The larger fields of crops like corn, wheat, and cotton presented opportunities for pest problems to develop, and chemical pesticides were used to "solve" this problem.

Today, many people feel current agricultural practices are not sustainable and that we should look for ways to reduce reliance on fertilizer and pesticides, while ensuring good yields and controlling the pests that compete with us for the food that we raise. Several

alternative approaches have somewhat different but overlapping goals. Some of the terms used to describe these approaches are: *alternative agriculture, sustainable agriculture,* and o*rganic agriculture.* **Alternative agriculture** is the broadest term. It includes all nontraditional agricultural methods and encompasses sustainable agriculture, organic agriculture, alternative uses of traditional crops, alternative methods for raising crops, and producing crops for industrial use. The **sustainable agriculture** movement maintains that current practices are degrading natural resources and seeks methods to produce adequate, safe food in an economically

viable manner while enhancing the health of agricultural land and related ecosystems. **Organic agriculture** advocates avoiding the use of chemical fertilizers and pesticides in the production of food, thus preventing damage to related ecosystems and the food-consuming public. A portion of the proposed definition of organic agriculture by the National Organic Standards Board established by the 1990 Organic Foods Production Act points out a central theme of these three movements.

Organic agriculture is an ecological production management system that promotes and enhances biodiversity,

Food Additives

Food additives are chemicals added to food before its sale. They have several purposes:

1. To prolong the storage life of the food.
2. To make the food more attractive by adding color or flavor.
3. To modify nutritive value.

Many kinds of molecules are added to foods to prolong shelf life. Calcium propionate is added to baked goods to retard the growth of airborne spores of molds and bacteria on the food, which can spoil it. BHT, TBHQ, and other commonly seen alphabetic mixtures have a similar function.

Sometimes additives are used just to make the food appear more attractive. For years, red dye II was used to color a variety of foods. Its use was discontinued when it was found to be carcinogenic. Other food colorings are still widely used to increase appeal to the consumer. Many kinds of artificial flavors are added to products as well. Commonly used flavor enhancers are monosodium glutamate, table salt, and citric acid.

Other food additives are used to modify the nutritive value of the food product. Iodine in table salt is a good example. Iodine is a trace element required for proper thyroid functioning. Individuals suffering from a lack of iodine often develop an enlargement of the thyroid gland known as goiter. The addition of iodine to table salt has eliminated goiter in the United States. Most cereals and baked goods and many other products have various vitamins and minerals added to improve their nutritive value. Some additives, such as Nutrasweet™, reduce the calories while giving the consumer the sensation of tasting something sweet.

Some additives are an unavoidable residue of some step of the food production process and could more properly be called contaminants. Pesticide residues are an example. Pesticides are used to grow foods, but they are also used to eliminate pests in the storing, processing, and transportation steps of food production.

Diethylstilbestrol (DES) was at one time used in the poultry industry to produce fatter birds. Because there were indications that DES is carcinogenic, the U.S. Food and Drug Administration banned the use of DES in chickens and declared that it was potentially hazardous to humans. Further studies were conducted to determine if DES was safe to use in raising beef. Cattle gain weight more rapidly with DES in their feed. Because DES was eventually linked to breast cancer in women, it has been banned from all animal feed use in the United States and Europe.

biological cycles, and soil biological activity. It is based on minimal use of off-farm inputs and management practices that restore, maintain, and enhance ecological harmony.

In order to reduce negative impacts of conventional agriculture while making a profit, farmers must deal with the twin problems of protecting the quality and fertility of the soil while controlling pests of their crops. A wide variety of techniques can be used to reduce the negative impact of conventional agricultural activity and maintain economic viability for the farmer.

Techniques for Protecting Soil and Water Resources

Conventional farming practices have several negative effects on soil and water. Soil erosion is a problem throughout the world. (See chapter 14.) However, two other problems are also important: compaction of the soil and reduction in soil organic matter. Several changes in agricultural production methods can help to reduce these problems. Reducing the number of times farm equipment travels over the soil will reduce the degree of compaction. Leaving crop residue on the soil and incorporating it into the soil reduces erosion and increases soil organic matter. In addition, introducing organic matter into the soil makes compaction less likely.

Conventional agriculture has several negative impacts on watersheds: fertilizer runoff stimulates aquatic growth and degrades water resources; pesticides can accumulate in food chains; and groundwater resources can be contaminated by fertilizer, pesticides, or animal waste. Reducing or eliminating these sources of contamination would enhance ecological harmony and reduce a threat to human health. Fertilizer runoff can be lessened by reducing the amount of fertilizer applied and the conditions under which it is applied. Applying fertilizer as plants need it will ensure that more of it is taken up by plants and less runs off. Increased organic matter in the soil also tends to reduce runoff. More careful selection, timing, and use of pesticides would decrease the extent to which these materials become environmental contaminants.

Precision agriculture is a new technique that addresses many of these concerns. With modern computer technology and geographic information systems, it is now possible, based on the soil and topography, to automatically vary the chemicals applied to the crop at different places within a field. Thus, less fertilizer is used, and it is used more effectively.

True organic agriculture that uses neither chemical fertilizers nor pesticides is the most effective in protecting soil and water resources from these forms of pollution, but it requires several adjustments in the way in which farming is done. Crop rotation is an effective way to enhance soil fertility, reduce erosion, and control pests. The use of nitrogen-fixing legumes, such as clover, alfalfa, beans, or soybeans, in crop rotation increases soil nitrogen but places other demands on the farmer. For example, it typically requires that cattle be a part of the farmer's operation in order to make use of forage crops and provide organic fertilizer for subsequent crops. Crop rotation also requires a greater investment in farm machinery, since certain crops require specialized equipment. Also, the raising of cattle requires additional expenditures for feed supplements and veterinary care. Critics say that organic farming cannot produce the amount of food required for today's population and that it can be economically successful only in specific cases. Proponents disagree and stress that when the hidden costs of soil erosion and pollution are included, organic agriculture or some modification of it is a viable alternative approach to conventional means of food production. Furthermore, organic farmers are willing to accept lower yields because they do not have to pay for expensive chemical fertilizers and pesticides. In addition, organic farmers often receive premium prices for products that are organically grown. Thus, even with lower yields, they can still make a profit.

Integrated Pest Management

Integrated pest management uses a variety of methods to control pests rather than relying on pesticides alone. Integrated pest management is a technique that depends on a complete understanding of all ecological aspects of the crop and the particular pests to which it is susceptible to establish pest control strategies. It requires information about the metabolism of the crop plant, the biological interactions between pests and their predators or parasites, the climatic conditions that favor certain pests, and techniques for encouraging beneficial insects. It may involve the selective use of pesticides. Much of the information necessary to make integrated pest management work goes beyond the knowledge of the typical farmer. The metabolic and ecological studies necessary to pinpoint weak points in the life cycles of pests can usually be carried out only at universities or government research institutions. These studies are expensive and must be completed for each kind of pest, since each pest has a unique biology. Once a viable technique has been developed, an educational program is necessary to provide farmers with the information they need in order to use integrated pest management rather than the "spray and save" techniques that they used previously and that pesticide salespeople continually encourage.

Several methods are employed in integrated pest management. These include disrupting reproduction, using beneficial organisms to control pests, developing resistant crops, modifying farming practices, and selectively using pesticides.

Disrupting Reproduction

In some species of insects, a chemical called a **pheromone** is released by females to attract males. Males of some species of moths can detect the presence of a female from a distance of up to 3 kilometers (nearly 2 miles). Since many moths are pests, synthetic odors can be used to control them. Spraying an area with the pheromone confuses the males and prevents them from finding females, which results in a reduced moth population the following year. In a similar way, a synthetic sex attractant molecule known as Gyplure is used to lure gypsy moths into traps, where they become stuck. Since the females cannot fly and the males are trapped, the reproductive rate drops, and the insect population may be controlled.

Another technique that reduces reproduction is male sterilization. In the southern United States and Central America, the screwworm fly weakens or kills large grazing animals, such as cattle, goats, and deer. The female screwworm fly lays eggs in open wounds on these animals, where the larvae feed. However, it was discovered that the female mates only once in her lifetime. Therefore, the fly population can be controlled by raising and releasing large numbers of sterilized male screwworm flies. Any female that mates with a sterile male fails to produce fertilized eggs and cannot reproduce. In Curaçao, an island 65 kilometers (40 miles) north of Venezuela, a program of introducing sterile male screwworm flies eliminated this disease from the 25,000 goats on the island. In parts of the southwestern United States, the sterile male technique has also been very effective. The screwworm fly has been eliminated from the United States and northern Mexico, and much of Central America may become free of them as well. In 1990, sterile males were released in Libya to begin eliminating screwworm flies that had been introduced with a South American cattle shipment.

During an epidemic of Mediterranean fruitflies in southern California and northern Mexico in the early 1980s, a similar technique was employed. Unfortunately, the X-ray technique used to sterilize the males was ineffective, and most of the flies released were not sterile, which made the problem worse rather than better. Pesticides were eventually used to control the fruitflies. Recent concern about the Mediterranean fruitfly in California has resulted in the controversial aerial spraying of malathion.

Using Beneficial Organisms to Control Pests

The manipulation of predator-prey relationships can also be used to control pest populations. For instance, the ladybird beetle, commonly called a ladybug, is a natural predator of aphids and scale insects. (See figure 15.12.) Artificially increasing the population of ladybird beetles reduces aphid and scale populations. In California during the late 1800s, scale insects on orange trees damaged the trees and reduced crop yields. The introduction of a species of ladybird beetle from Australia quickly brought the pests under control. Years later, when chemical pesticides were first used in the area, so many ladybird beetles were accidentally killed that scale insects once again became a serious problem. When pesticide use was discontinued, ladybird beetle populations rebounded, and the scale insects were once again brought under control. (See figure 15.13.)

Herbivorous insects can also be used to control weeds. Purple loosestrife (*Lythrum salicaria*) is a wetlands plant accidentally introduced from Europe in the mid-1800s. It takes over sunny wetlands and eliminates native vegetation such as cattails, therefore, eliminating many species that rely on cattails for food, nesting places, or hiding places. (See figure 15.14.) The plant exists in most states and provinces in the United States and Canada. In Europe, the plant is not a pest because there are several insect species that attack it in various stages of the plant's life cycle. Since purple loosestrife is a European plant, a search was made to identify candidate species of insects that would control it. The criteria were that the insects must live only on purple loosestrife and not infest other plants and must have the capacity to do major damage to purple loosestrife. Five species of beetles have been identified that will attack the plant in various ways. After extensive studies to determine if introductions of these insects from Europe would be likely to cause other problems, several were selected as candidates to help control purple loosestrife in the United States and Canada. Some combination of these beetles has been released in a large proportion of the states and provinces infested with purple loosestrife. Two species (*Galerucella calmariensis,* and *Galerucella pusilla*) feed on the leaves, shoots, and flowers of the newly growing purple loosestrife; one (*Hylobius transversovittatus*) has larvae that feed on the roots. These species have been introduced and appear to be forming colonies. Two other species (*Nanophyes brevis* and *Nanophyes marmoratus*) feed on the flowers. Their release was scheduled for the summer of 1998. This multipronged attack by several species of beetles is projected to reduce the number of purple loosestrife plants by 90 percent.

The use of specific strains of the bacterium *Bacillus thuringiensis* to control mosquitoes and moths is another

figure 15.12 **Beneficial Insect** The ladybird beetle is a predator of many kinds of pest insects, including aphids.

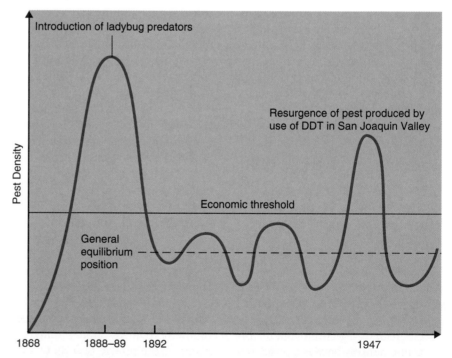

figure 15.13 **Insect Control with Natural Predators** In 1889, the introduction of ladybird beetles (ladybugs) brought the cottony cushion scale under control in the orange groves of the San Joaquin Valley. In the 1940s, DDT reduced the ladybird beetle population and the cottony cushion scale population increased. Stopping the use of DDT allowed the ladybird population to increase, reducing the pest population and allowing the orange growers to make a profit.

Source: *Man and the Environment,* 2d ed. by Arthur S. Boughey, © 1975. Reprinted by permission of Prentice-Hall, Inc., Upper Saddle River, N.J.

PART FOUR Human Influences on Ecosystems

a. Purple loosestrife

b. *Galerucella calmariensis*

c. *Galerucella pusilla*

d. *Hylobius transversovittatus*

figure 15.14 **Biological Control of Purple Loosestrife** Purple loosestrife is a European plant that invades wetlands and prevents the growth and reproduction of native species of plants. Several kinds of European beetles that feed on purple loosestrife have been released as biological control agents. Two species (b. *Galerucella calmariensis;* c. *Galerucella pusilla*) feed on the leaves, shoots, and flowers of the newly growing purple loosestrife, one (d. *Hylobius transversovittatus*) has larvae that feed on the roots. It appears that they are actually slowing the spread of purple loosestrife.

example of the use of one organism to control another. A crystalline toxin produced by the bacterium destroys the lining of the gut of the feeding insect, resulting in its death. One strain of *B. thuringiensis* is used to control mosquitoes, while another is primarily effective against the caterpillars of leaf-eating moths, including the gypsy moth.

Naturally occurring pesticides found in plants also can be used to control pests. For example, marigolds are planted to reduce the number of soil nematodes, and garlic plants are used to check the spread of Japanese beetles.

Developing Resistant Crops

One outcome of the major advances in molecular genetics is the development of specialized organisms that have had specific, valuable gene inserted into their genetic makeup. Although this is a relatively new technology, it is an extension of an ancient art. Farmers have been involved in manipulating the genetic makeup of their plants and animals since these organisms were first domesticated. Initially, farmers either consciously or accidentally chose to plant specific seeds or breed certain animals that had specific desirable characteristics. This resulted in local varieties with particular characteristics. When the laws of genetics began to be understood in the early 1900s, scientists began to make precise crosses between carefully selected individuals to enhance the likelihood that their offspring would have certain highly desirable characteristics. This led to the development of hybrid seeds and specific breeds of domesticated animals. Controlled plant and animal breeding resulted in increased yields and better disease resistance in domesticated plants and animals. These activities are still the major driving force for developing improved varieties of domesticated organisms.

When the structure of DNA was discovered and it was determined that the DNA of organisms could be manipulated, an entirely new field of plant and animal breeding arose. **Genetic engineering** or **biotechnology** involves inserting specific pieces of DNA into the genetic makeup of organisms. The organism with the altered genetic makeup is known as a **genetically modified organism.** The DNA inserted could be from any source, even an entirely different organism. In agricultural practice, two kinds of genetically modified organisms have received particular attention. One involves the insertion of genes from a specific kind of bacterium called *Bacillus thuringiensis israeliensis* (Bti). *Bacillus thuringiensis israeliensis* produces a material that causes the destruction of the lining of the gut of insects that eat it. It is a natural insecticide. To date, the gene has been inserted into the genetic makeup of several crop plants, including corn. In field tests, the genetically engineered corn was protected against some of its insect pests, but there was some concern that pollen grains from the corn might be blown to neighboring areas and affect nontarget insect populations. In particular, a study of monarch butterflies indicated that populations of butterflies adjacent to fields of this genetically engineered corn were negatively affected. One could argue that since the use of Bti corn results in less spraying of insecticides in cornfields, this is just a trade-off.

A second kind of genetically engineered plant involves inserting a gene for herbicide resistance into the genome of certain crop plants. The value of this to farmers is significant. For example, a farmer could plant cotton with very little preparation of the field to rid it of weeds. When both the cotton and the weeds began to grow, the field could be sprayed with a specific herbicide that would kill the weeds but not harm the herbicide-resistant cotton. This has been field tested and it works. Critics have warned that the genes possibly could escape from the crop plants and become part of the genome of the weeds that we are trying to control, thus creating superweeds.

Many groups oppose the use of genetically modified organisms. They argue that this technology is going a step too far, that no long-term studies have been done to ensure their safety, that there are dangers we cannot anticipate, and that if such crops are grown, they should be labeled so that the public knows when they are consuming products from genetically modified organisms. The European Union (EU) will not buy genetically modified grains from the United States. This has resulted in farmers having to segregate their stores of grains so that they can guarantee to an EU buyer that a crop is not genetically modified and at the same time sell a genetically modified crop to other buyers.

Supporters argue that all plant and animal breeding involves genetic manip-ulation and that this is just a new kind of genetic manipulation. A great deal of evidence exists that genes travel between species in nature and that genetic engineering simply makes a common, natural process more frequent. Over the next 20 to 30 years, scientists hope to use biotechnology to produce high-yield plant strains that are more resistant to insects and disease, thrive on less fertilizer, make their own nitrogen fertilizer, do well in slightly salty soils, withstand drought, and use solar energy more efficiently during photosynthesis. These new kinds of genetically modified organisms will continue to be developed and tested, and the political arguments about their appropriateness will continue as well.

Modifying Farming Practices

Modification of farming practices can often reduce the impact of pests. In some cases, all crop residues are destroyed to prevent insect pests from finding overwintering sites. For example, shredding and plowing under the stalks of cotton in the fall reduces overwintering sites for boll weevils and reduces their numbers significantly, thereby reducing the need for expensive insecticide applications. Many farmers are also returning to crop rotation, which tends to prevent the buildup of specific pests that typically occurs when the same crop is raised in a field year after year.

Selective Use of Pesticides

Pesticides can also play a part in integrated pest management. Identifying the precise time when the pesticide will have the greatest effect at the lowest possible dose has these advantages: it reduces the amount of pesticide used and may still allow the parasites and predators of pests to survive. Such precise applications often require the assistance of a trained professional who can correctly identify the pests, measure the size of the population, and time pesticide applications for maximum effect. In several instances, pheromone-baited traps capture insect pests from fields, and an assessment of the number of insects caught can be a guide to when insecticides should be applied.

Integrated pest management will become increasingly popular as the cost of pesticides rises and knowledge about the biology of specific pests becomes available. However, as long as humans raise crops, there will be pests that will outwit the defenses we develop. Integrated pest management is just another approach to a problem that began with the dawn of agriculture.

Herring Gulls as Indicators of Contamination in the Great Lakes

Herring gulls nest on islands and other protected sites throughout the Great Lakes region. Since they feed primarily on fish, they are near the top of aquatic food chains and tend to accumulate toxic materials from the food they eat. Eggs taken from nests can be analyzed for a variety of contaminants.

Since the early 1970s, the Canadian Wildlife Service has operated a monitoring program to assess trends in the levels of contaminants in the eggs of herring gulls. In general, the contaminant levels have declined as both the Canadian and U.S. governments have taken action to stop new contaminants from entering the Great Lakes. The figure shows the trends for PCBs and DDE. PCBs are a group of organic compounds that were used as fire retardants, lu-

bricants, insulation fluids in electrical transformers, and in some printing inks. Some forms of PCB are much more toxic than others. Both Canada and the United States have eliminated most uses of PCBs, reducing the levels found in herring gull eggs. DDE is a breakdown product of DDT. DDT use was discontinued in much of the world in the early 1970s. The illustration shows a general decline in the amount of DDE present following the ban on DDT.

• Should the health of a bird species be used to develop policy? Why or why not?

• If PCBs and DDT are no longer being used, why are herring gulls still being contaminated?

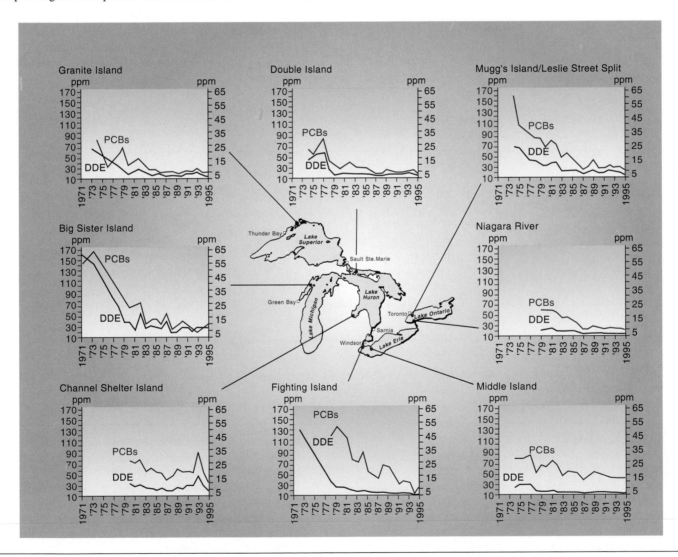

Summary

Although small slash-and-burn garden plots are common in some parts of the world, most of the food in the world is raised on more permanent farms. In countries where population size is high and money is in short supply, much of the farming is labor intensive, using human labor for many of the operations necessary to raise crops. However, much of the world's food is grown on large, mechanized farms that use energy rather than human muscle for tilling, planting, and harvesting crops, and for producing and applying fertilizers and pesticides.

Monoculture involves planting large areas of the same crop year after year. This causes problems with plant diseases, pests, and soil depletion. Although chemical fertilizers can replace soil nutrients that are removed when the crop is harvested, they do not replace the organic matter necessary to maintain soil texture, pH, and biotic richness.

Mechanized monoculture depends heavily on the control of pests by chemical means. Persistent pesticides are stable and persist in the environment, where they may biomagnify in ecosystems. Consequently, many of the older persistent pesticides have been quickly replaced by nonpersistent pesticides that decompose much more quickly and present less of an environmental hazard. However, most nonpersistent pesticides are more toxic to humans and must be handled with greater care than the older persistent pesticides.

Pesticides can be divided into several categories based on the organism they are used to control. Insecticides are used to control insects, herbicides are used for plants, fungicides for fungi, and rodenticides for rodents. Because of the problems of persistence, biomagnification, resistance of pests to pesticides, and effects on human health, many people are seeking pesticide-free alternatives to raising food. Several different philosophies that seek the same ends—less use of chemicals and better stewardship of soil—are alternative agriculture, sustainable agriculture, and organic agriculture. One ingredient in all of these approaches is the use of integrated pest management, which employs a complete understanding of an organism's ecology to develop pest-control strategies that use no or few pesticides.

Key Terms

alternative agriculture *341*
auxins *333*
bioaccumulation *335*
biocides *330*
biomagnification *335*
biotechnology *345*
carbamates *332*
chlorinated hydrocarbons *332*
fungicides *330*
genetically modified organisms *345*
genetic engineering *345*

Green Revolution *329*
herbicides *330*
insecticides *330*
integrated pest management *343*
macronutrient *330*
micronutrient *330*
monoculture *328*
nonpersistent pesticides *331*
nontarget organisms *331*
organic agriculture *341*
organophosphates *332*

persistent pesticides *331*
pests *330*
pesticides *330*
pheromone *343*
polyculture *327*
precision agriculture *342*
rodenticides *330*
sustainable agriculture *341*
target organism *331*
weeds *330*

Review Questions

1. What is monoculture?
2. List three reasons why fossil fuels are essential for mechanized agriculture.
3. Describe why pesticides are commonly used in mechanized agriculture.
4. Why are fertilizers used? What problems are caused by fertilizer use?
5. How do persistent and nonpersistent pesticides differ?
6. What is biomagnification? What problems does it cause?
7. How do organic farms differ from conventional farms?
8. Name three nonchemical methods of controlling pest populations.
9. What are the advantages and disadvantages of integrated pest management?
10. List three uses of food additives.
11. List three actions farmers could take to reduce the effect of pesticides on the environment.

Critical Thinking Questions

1. If you were a public health official in a developing country, would you authorize the spraying of DDT to control mosquitoes that spread malaria? What would be your reasons?

2. Look at table 15.1. What caused the changes in the effectiveness of the insecticides? If you were an agricultural extension agent, what alternatives to pesticides might you recommend?

3. Imagine that you are a scientist examining fish in Lake Superior and you find toxaphene in the fish you are studying. Toxaphene was used primarily in cotton farming and has been banned since 1982. How can you explain its presence in these fish?

4. Are the risks of pesticide use worth the benefits? What values, beliefs, and perspectives lead you to this conclusion?

5. Do you think that current agricultural practices are sustainable? Why or why not? What changes in agriculture do you think will need to happen in the next 50 years?

6. Imagine you are an EPA official who is going to make a recommendation about whether an agricultural pesticide can remain on the market or should be banned. What are some of the facts you would need to make your recommendation? Who are some of the interest groups interested in the outcome of your decision? What arguments might they present regarding their positions? What political pressures might they be able to bring to bear on you?

7. Why are few consumers demanding alternative methods of crop production, and why are farmers not using those methods?

Concept Map

Construct a map to show relationships among the following concepts:

bioaccumulation

biomagnification

genetically modified organisms

herbicides

insecticides

integrated pest management

monoculture

nonpersistent pesticides

organophosphates

persistent pesticides

pests

sustainable agriculture

target organism

weeds

Interactive Exploration

Check out the website at **http://www.mhhe.com/environmentalscience** and click on the cover of this textbook for quizzing, career information, case studies, and hot links for information on the following topics:

Agricultural Methods

Food and Agriculture

Sustainable Agriculture

The Informed Consumer

Genetically Modified Foods

Economic and Ecological Importance of Angiosperms

Hazardous and Toxic Wastes

Food Webs

Pest Control

Integrated Pest Management

Water Management

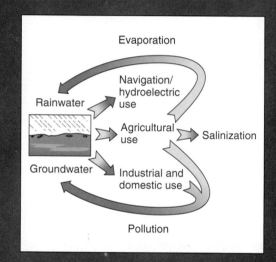

Evaporation

Navigation/hydroelectric use

Rainwater

Agricultural use

Salinization

Groundwater

Industrial and domestic use

Pollution

Objectives

After reading this chapter, you should be able to:

- Explain how water is cycled through the hydrologic cycle.
- Explain the significance of groundwater, aquifers, and runoff.
- Explain how land use affects infiltration and surface runoff.
- List the various kinds of water use and the problems associated with each.
- List the problems associated with water impoundment.
- List the major sources of water pollution.
- Define biochemical oxygen demand (BOD).
- Differentiate between point and nonpoint sources of pollution.
- Explain how heat can be a form of pollution.
- Differentiate between primary, secondary, and tertiary sewage treatments.
- Describe some of the problems associated with storm-water runoff.
- List sources of groundwater pollution.
- Explain how various federal laws control water use and prevent misuse.
- List the problems associated with water-use planning.
- Explain the rationale behind the federal laws that attempt to preserve certain water areas and habitats.
- List the problems associated with groundwater mining.
- Explain the problem of salinization associated with large-scale irrigation in arid areas.
- List the water-related services provided by local governments.

Chapter Outline

The Water Issue

The Hydrologic Cycle

Human Influences on the Hydrologic Cycle

Kinds of Water Use
  Domestic Use of Water
  Agricultural Use of Water
  Industrial Use of Water
  In-Stream Use of Water

Global Perspective: *Comparing Water Use and Pollution in Industrialized and Developing Countries*

Kinds and Sources of Water Pollution

Environmental Close-Up: *Is It Safe to Drink the Water?*
  Municipal Water Pollution

Global Perspective: *The Cleanup of the Holy Ganges*
  Agricultural Water Pollution
  Industrial Water Pollution
  Thermal Pollution
  Marine Oil Pollution
  Groundwater Pollution

Water-Use Planning Issues
  Water Diversion
  Wastewater Treatment

Environmental Close-Up: *Restoring the Everglades*
  Salinization
  Groundwater Mining
  Preserving Scenic Water Areas and Wildlife Habitats

Global Perspective: *Death of a Sea*

Global Perspective: *The Dead Zone of the Gulf of Mexico*

Issues—Analysis: *The California Water Plan*

The Water Issue

Water in its liquid form is the material that makes life possible on Earth. All living organisms are composed of cells that contain at least 60 percent water. Furthermore, their metabolic activities take place in a water solution. Organisms can exist only where they have access to adequate supplies of water. Water is also unique because it has remarkable physical properties. Water molecules are polar; that is, one part of the molecule is slightly positive and the other is slightly negative. Because of this, the water molecules tend to stick together, and they also have a great ability to separate other molecules from each other. Water's ability to act as a solvent and its capacity to store heat are a direct consequence of its polar nature. These abilities make water extremely valuable for societal and industrial activities. Water dissolves and carries substances ranging from nutrients to industrial and domestic wastes. A glance at any urban sewer will quickly point out the importance of water in dissolving and transporting wastes. Because water heats and cools more slowly than most substances, it is used in large quantities for cooling in electric power generation plants and in other industrial processes. Water's ability to retain heat also modifies local climatic conditions in areas near large bodies of water. These areas do not have the wide temperature-change characteristic of other areas.

For most human as well as some commercial and industrial uses, the quality of the water is as important as its quantity. Water must be substantially free of dissolved salts, plant and animal waste, and bacterial contamination to be suitable for human consumption. The oceans, which cover approximately 70 percent of the Earth's surface, contain over 97 percent of its water. However, saltwater cannot be consumed by humans or used for many industrial processes. Freshwater is free of the salt found in ocean waters. Of the freshwater found on Earth, only a tiny fraction is available for use. (See figure 16.1.) Unpolluted freshwater that is suitable for

drinking is known as **potable water.** Early human migration routes and settlement sites were influenced by the availability of drinking water. At one time, clean freshwater supplies were considered inexhaustible. Today, despite advances in drilling, irrigation, and purification, the location, quality, quantity, ownership, and control of potable waters remain important concerns.

Only recently have we begun to understand that we will probably exhaust our usable water supplies in some areas of the world because of increases in human populations and limitations on the supply available. Some areas of the world have abundant freshwater resources, while others have few. In addition, demand is increasing for freshwater for industrial, agricultural, and personal needs.

Shortages of potable freshwater throughout the world can also be directly attributed to human abuse in the form of pollution. Water pollution has negatively affected water supplies throughout the world. In many parts of

the developing world, people do not have access to safe drinking water. The World Health Organization estimates that about 25 percent of the world's people do not have access to safe drinking water. Even in the economically advanced regions of the world, water quality is a major issue. According to the United Nations Environment Programme, 5 million to 10 million deaths occur each year from water-related diseases. The illnesses include cholera, malaria, dengue fever, and dysentery. The United Nations also reports that these illnesses have been increasing over the past decade and that without large economic investments in safe drinking-water supplies, the rate of increase will continue.

Unfortunately, the outlook for the world's freshwater supply is not very promising. According to studies by the United Nations and the International Joint Commission (a joint U.S.-Canadian organization that studies common water bodies between the two nations), many sections of the world are

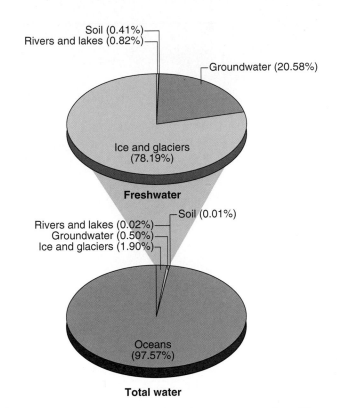

Soil (0.41%)
Rivers and lakes (0.82%)
Groundwater (20.58%)
Ice and glaciers (78.19%)

Freshwater

Soil (0.01%)
Rivers and lakes (0.02%)
Groundwater (0.50%)
Ice and glaciers (1.90%)

Oceans (97.57%)

Total water

figure 16.1 **Freshwater Resources** Although water covers about 70 percent of the Earth's surface, over 97 percent is saltwater. Of the less than 3 percent that is freshwater, only a tiny fraction is available for human use.

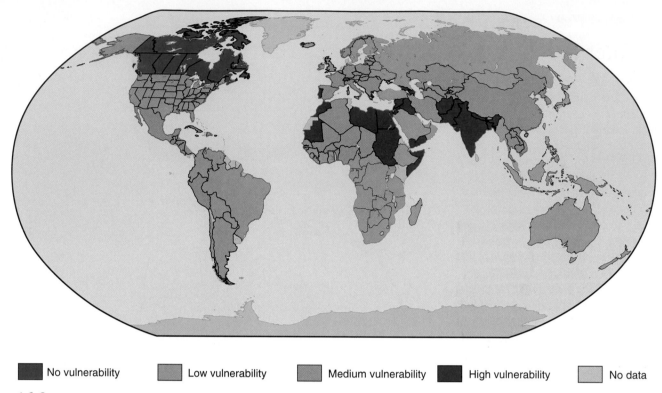

■ No vulnerability	■ Low vulnerability	■ Medium vulnerability	■ High vulnerability	■ No data

figure 16.2 **Areas of the World Experiencing Water Shortage** Many areas of the world are experiencing water shortages. This map indicates that north and central Africa and parts of Asia are particularly vulnerable.

Source: Data from World Resources Institute.

currently experiencing a shortage of freshwater and the problem will only intensify. A study by the United Nations in 2002 said that by 2025, one in three people globally could be threatened with a shortage of freshwater. The United Nations Third World Water Forum stated that currently 450 million people in 29 countries lack water. The report went on to say that the Middle East, India, Pakistan, and China would struggle with serious water shortages. It was stated that water could become as important as oil as a major source of world conflict. Increasing scarcity, competition, and arguments over water in the first quarter of the twenty-first century could dramatically change the way we value and use water and the way we mobilize and manage water resources. Furthermore, changes in the amount of rain from year to year result in periodic droughts for some areas and devastating floods for others. (See figure 16.2.) However, rainfall is needed to regenerate freshwater and, therefore, is an important link in the cycling of water.

The Hydrologic Cycle

All water is locked into a constant recycling process called the **hydrologic cycle.** (See figure 16.3.) Two important processes involved in the cycle are the evaporation and condensation of water. Evaporation involves adding energy to molecules of a liquid so that it becomes a gas in which the molecules are farther apart. Condensation is the reverse process in which molecules of a gas give up energy, get closer together, and become a liquid. Solar energy provides the energy that causes water to evaporate from the ocean surface, the soil, bodies of freshwater, and from the surfaces of plants. The water evaporated from plants comes from two different sources. Some is water that has fallen on plants as rain, dew, or snow. In addition, plants take up water from the soil and transport it to the leaves, where it

evaporates. This process is known as **evapotranspiration.**

The water vapor in the air moves across the surface of the Earth as the atmosphere circulates. As warm, moist air cools, water droplets form and fall to the land as precipitation. Although some precipitation may simply stay on the surface until it evaporates, most will either sink into the soil or flow downhill and enter streams and rivers, which eventually return the water to the ocean. Surface water that moves across the surface of the land and enters streams and rivers is known as **runoff.** Water that enters the soil and is not picked up by plant roots moves slowly downward through the spaces in the soil and subsurface material until it reaches an impervious layer of rock. The water that fills the spaces in the substrate is called **groundwater.** It may be stored for long periods in underground reservoirs.

The porous layer that becomes saturated with water is called an **aquifer.** There are two basic kinds of aquifers: unconfined and confined. An

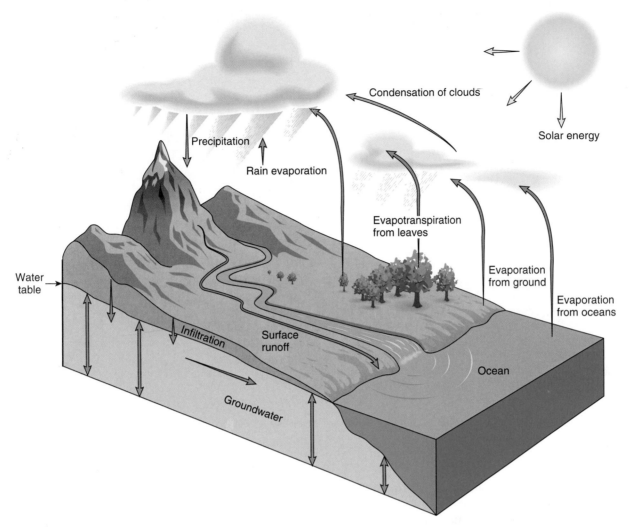

figure 16.3 **The Hydrologic Cycle** The cycling of water through the environment follows a simple pattern. Moisture in the atmosphere condenses into droplets that fall to the Earth as rain or snow, supplying all living things with its life-sustaining properties. Water, flowing over the Earth as surface water or through the soil as groundwater, returns to the oceans, where it evaporates back into the atmosphere to begin the cycle again.

unconfined aquifer usually occurs near the land's surface where water enters the aquifer from the land above it. The top of the layer saturated with water is called the **water table.** The lower boundary of the aquifer is an impervious layer of clay or rock that does not allow water to pass through it. Unconfined aquifers are replenished (recharged) primarily by rain that falls on the ground directly above the aquifer and infiltrates the layers below. The water in such aquifers is at atmospheric pressure and flows in the direction of the water table's slope, which may or may not be similar to the surface of the land above it. Above the water table and below the land surface is a layer known as the **vadose zone** (also known as the unsaturated zone or zone of aeration) that is not saturated with water. (See figure 16.4.)

A **confined aquifer** is bounded on both the top and bottom by layers that are impervious to water and is saturated with water under greater-than-atmospheric pressure. An impervious confining layer is called an **aquiclude.** If water can pass in and out of the confining layer, the layer is called an **aquitard.** A confined aquifer is primarily replenished by rain and surface water from a recharge zone (the area where water is added to the aquifer) that may be many kilometers from where the aquifer is tapped for use. If the recharge area is at a higher elevation than the place where an aquifer is tapped, water will flow up the pipe until it reaches the same elevation as the recharge area. Such wells are called **artesian wells.** If the recharge zone is above the elevation of the top of the well pipe, it is called a flowing artesian well because water will flow from the pipe.

The nature of the substrate in the aquifer influences the amount of water the aquifer can hold and the rate at which water moves through it. **Porosity** is a measure of the size and number of the spaces in the substrate. The greater the porosity, the more water it can contain. The rate at which water moves through an aquifer is determined by the size of the pores, the degree to which they are connected, and any cracks or channels present in the substrate. The rate at which the water moves through the

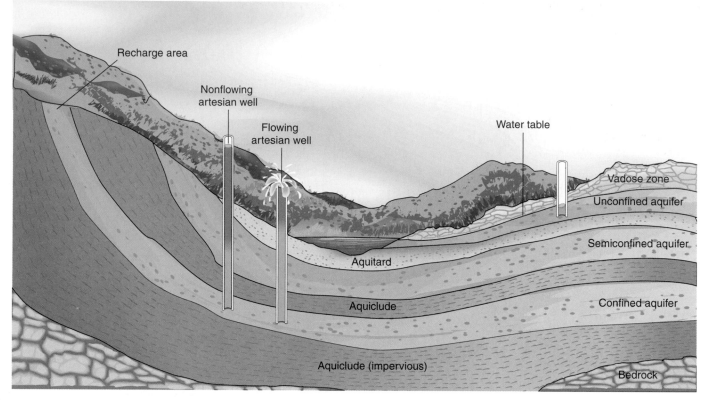

Labels in figure: Recharge area, Nonflowing artesian well, Flowing artesian well, Water table, Vadose zone, Unconfined aquifer, Semiconfined aquifer, Aquitard, Aquiclude, Confined aquifer, Aquiclude (impervious), Bedrock

figure 16.4 **Aquifers and Groundwater** Groundwater is found in the pores in layers of sediment or rock. The various layers of sediment and rock determine the nature of the aquifer and how it can be used.

aquifer determines how rapidly water can be pumped from a well per minute.

Human Influences on the Hydrologic Cycle

Human activities can significantly impact evaporation, runoff, and infiltration. When water is used for cooling in power plants or to irrigate crops, the rate of evaporation is increased. Water impounded in reservoirs also evaporates rapidly. This rapid evaporation can affect local atmospheric conditions. Runoff and the rate of infiltration also are greatly influenced by human activity. Removing the vegetation by logging or agriculture increases runoff and decreases infiltration. Because there is more runoff, there is more erosion of soil. Urban complexes with a high percentage of impervious, paved surfaces have increased runoff and reduced infiltration. A major concern in urban areas

is providing ways to carry storm water away rapidly. This involves designing and constructing surface waterways and storm sewers. Often cities have significant flooding problems when heavy rains overtax the ability of their storm-water management systems to remove excess water. Many cities combine their storm-sewer water with wastewater at their treatment plants, which can cause serious pollution problems after heavy rains. The treatment plants cannot process the increased volume of water and must discharge the combined untreated wastewater and sewer water into the receiving body of water.

Cities also have problems supplying water for industrial and domestic use. In many cases, cities rely on surface water sources for their drinking water. The source may be a lake or river, or an impoundment that stores water. In addition, aquifers are extremely important in supplying water. Many large urban areas in the United States depend on underground water for their water supply. This groundwater supply can be tapped

as long as it is not used faster than it can be replaced. Determining how much groundwater or surface water can be used and what the uses should be is a major concern, especially in water-poor areas of the world.

There are several ways to monitor water use from surface and groundwater sources. Water withdrawals are measurements of the amount of water taken from a source. This water may be used temporarily and then returned to its source and used again. For example, when a factory withdraws water from a river for cooling purposes, it returns most of the water to the river; thus, the water can be used later. Water that is incorporated into a product or lost to the atmosphere through evaporation or evapotranspiration cannot be reused in the same geographic area and is said to be consumed. Much of the water used for irrigation is lost to evaporation and evapotranspiration or is removed with the crop when it is harvested. Therefore, much of the water withdrawn for irrigation is consumed.

Kinds of Water Use

Water use varies considerably around the world, depending on availability of water and degree of industrialization. However, use can be classified into four broad categories: (1) domestic use, (2) agricultural use, (3) industrial use, and (4) in-stream use. It is important to remember that some uses of water are consumptive, while others are nonconsumptive.

Domestic Use of Water

Over 90 percent of the water used for domestic purposes in North America is supplied by municipal water systems, which typically include complex, costly storage, purification, and distribution facilities. Many rural residents, however, can obtain safe water from untreated private wells. Nearly 37 percent of municipal water supplies come from wells.

Regardless of the water source (surface or groundwater), water supplied to cities in the developed world is treated to ensure its safety. Treatment of raw water before distribution usually involves some combination of the following processes. The raw water is filtered through sand or other substrates to remove particles. Chemicals may be added to the water that will cause some dissolved materials to be removed. Then, before the water is released for public use, it is disinfected with chlorine, ozone, or ultraviolet light to remove any organisms that might still be present. Where no freshwater is available, expensive desalinization of saltwater may be the only option available.

Domestic activities in highly developed nations require a great deal of water. This domestic use includes drinking, air conditioning, bathing, washing clothes, washing dishes, flushing toilets, and watering lawns and gardens. On average, each person in a North American home uses about 400 liters (about 100 gallons) of **domestic water** each day. About 70 percent is used as a solvent to carry away wastes (nonconsumptive use), about 30 percent is used for lawns

and gardens, (consumptive use), and only a tiny amount (about 2 percent) is actually consumed by drinking or in cooking. Yet all water that enters the house has been purified and treated to make it safe for drinking. (See figure 16.5.) Natural processes cannot cope with the highly concentrated wastes typical of a large urban area. The unsightly and smelly wastewater also presents a potential health problem, so cities and towns must treat it before returning it to a local water source. Until recently, the cost of water in almost every commu-

nity has been so low that there was very little incentive to conserve, but shortages of water and increasing purification costs have raised the price of domestic water in many parts of the world, and it is becoming evident that increased costs do tend to reduce use. (See figure 16.6.)

Although domestic use of water is a relatively small component of the total water-use picture (see figure 16.7), urban growth has created problems in the development, transportation, and maintenance of quality water supplies. In regions experiencing rapid population

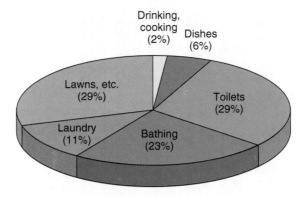

Water Use by a Typical North American Family of Four

figure 16.5 **Urban Domestic Water Uses** Over 150 billion liters (40 billion gallons) of water are used each day for urban domestic purposes in North America.

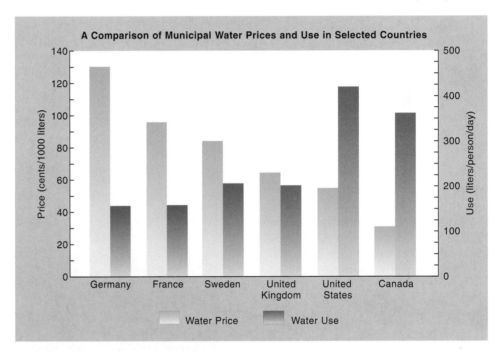

figure 16.6 **Water Use Decreases As Water Price Increases** A general correlation exists between the amount of water that is used and its price.

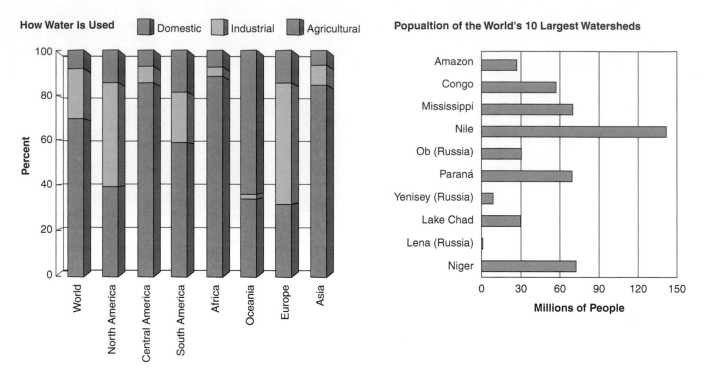

How Water Is Used   ■ Domestic   ■ Industrial   ■ Agricultural

Popualtion of the World's 10 Largest Watersheds

figure 16.7   **World Uses of Water**   Domestic, industrial, and agricultural uses dominate the allocation of water resources. However, there is considerable variety in different parts of the world in how these resources are used, as well as in the population density of the world's watersheds.

Source: World Resources 2000–2001.

growth, such as Asia, domestic use is expected to increase sharply. In North America, fast-growing cities in the West are experiencing water shortages. (See figure 16.8.) Demand for water in urban areas sometimes exceeds the immediate supply, particularly when the supply is local surface water. This is especially true during the summer, when water demand is high and precipitation is often low. Many communities have begun public education campaigns designed to help reduce water usage. (See figure 16.9.) Mexico City, with a population of nearly 20 million and minimal access to surface water, has one of the most serious water management problems in the world. The water supply for the Mexico Basin is severely stressed by water management practices that do not ensure a sustainable supply of clean water. Since extractions from the Valley of Mexico aquifer began

figure 16.8   **Urban Expansion**
Maintaining a suitable supply of water for growing metropolitan areas can pose major problems, especially in arid areas such as Phoenix, Arizona.

Water Savings Guide

Conservative use will save water		Normal use will waste water
Wet down, soap up, rinse off 15 liters (4 gal)	**Shower**	Regular shower 95 liters (25 gal)
May we suggest a shower?	**Tub bath**	Full tub 135 liters (36 gal)
Minimize flushing Each use consumes 20-25 liters (5-7 gal) New toilets use 6 liters (1.6 gal)	**Toilet**	Frequent flushing is very wasteful
Fill basin 4 liters (1 gal)	**Washing hands**	Tap running 8 liters (2 gal)
Fill basin 4 liters (1 gal)	**Shaving**	Tap running 75 liters (20 gal)
Wet brush, rinse briefly 2 liters (1/2 gal)	**Brushing teeth**	Tap running 38 liters (10 gal)
Take only as much as you require	**Ice**	Unused ice goes down drain
Please report immediately	**Leaks**	A small drip wastes 95 liters (25 gal) per week
Turn off light, TV, heaters, and air conditioning when not in room	**Energy**	Wasting energy also wastes water

Thank you for using this column..and not this one

figure 16.9 **Ways to Conserve Water** Minor changes in the way people use water could significantly reduce domestic water use. Note: 1 gallon equals approximately 3.785 liters.

nearly 100 years ago, groundwater levels have dropped significantly. As the city's population continues to grow, managers have been forced to look at alternative sources of water from more than 160 kilometers (100 miles) away.

In addition to encouraging the public to conserve water, municipalities need to pay attention to losses that occur within the distribution system. Leaking water pipes and mains account for significant losses of water. Even in the developed world, such losses may be as high as 20 percent. Poorer countries typically exceed this, and some may lose over 50 percent of the water to leaks. Another major cause of water loss has been public attitudes. As long as water is considered a limitless, inexpensive resource, people will make little effort to conserve it. As the cost of water rises and attitudes toward water change, so will usage and efforts to conserve.

Agricultural Use of Water

In North America, groundwater accounts for about 37 percent of the water used in agriculture and surface water about 63 percent. **Irrigation** is the major consumptive use of water in most parts of the world and accounts for about 80 percent of all the water consumed in North America. About 500,000 million liters per day (134,000 million gallons per day or 150 million acre-feet per year) are used in irrigation in the United States. The amount of water used for irrigation and livestock continues to increase throughout the world. Future agricultural

demand for water will depend on the cost of water for irrigation; the demand for agricultural products, food, and fiber; governmental policies; the development of new technology; and competition for water from a growing human population.

Since irrigation is common in arid and semiarid areas, local water supplies are often lacking, and it is often necessary to transport water great distances to water crops. This is particularly true in the western United States, where about 14 million hectares (35 million acres) of land are irrigated.

Four methods of irrigation are commonly used. Surface or flood irrigation involves supplying the water to crops by having the water flow over the field or in furrows. This requires extensive canals and is not suitable for all kinds of crops. Spray irrigation involves the use of pumps to spray water on the crop. Trickle irrigation uses a series of pipes with strategically placed openings so that water is delivered directly to the roots of the plants. Subirrigation involves supplying water to plants through underground pipes. Often this method is used where soils require draining at certain times of the year. The underground pipes can be used to drain excess water at one time of the year and supply water at others. Each of these methods has its drawbacks and advantages as well as conditions under which it works well.

Construction and maintenance of irrigation structures, such as dams, canals, pipes, and pumps, are expensive. Costs for irrigation water have traditionally been low since many of the dams and canals were constructed with federal assistance, and farmers have often used water wastefully. As there has been greater competition between urban areas and agriculture for scarce water resources, there has been pressure to raise the cost of water used for irrigation. Increasing the cost of water will stimulate farmers to conserve, just as it does homeowners. Another way to reduce the demand for irrigation water is to reduce the quantity of water-demanding crops grown in dry areas or change from high water-demanding to lower water-demanding crops. For example, wheat

a.

b.

c.

figure 16.10 **Types of Irrigation** Many arid areas require irrigation to be farmed economically. (*a*) Surface or flood irrigation uses irrigation canals and ditches to deliver water to the crops. The land is graded so that water flows from the source into the fields. Water is siphoned from a canal into ditches between rows of crops. (*b*) Spray irrigation uses a pump to spray water into the air above the plants. This is an example of central-pivot spray irrigation, in which a long pipe on wheels slowly rotates about a central point. (*c*) Trickle irrigation conserves water by delivering water directly to the roots of the plants but requires an extensive network of pipes.

or soybeans require less water than do potatoes or sugar beets. It is also becoming increasingly important to modify irrigation practices to use less water. (See figure 16.10.) For example, the use of trickle irrigation and some variations of spray irrigation use water more efficiently than the more traditional flood irrigation methods.

Many forms of irrigation require a great deal of energy. This is particularly true when pumps are used to deliver the water to the crop. It is estimated that 40 percent of the energy devoted to agriculture in Nebraska is used for irrigation. Increasing energy costs may force some farmers to reduce or discontinue irrigation. In addition, much of western Nebraska relies on groundwater for irrigation, and the water table is dropping rapidly. If a water shortage develops, land values will decline. Land use and water use are interrelated and cannot be viewed independently. In the

Salinas Valley, along the central coast of California, farmers are irrigating their fields with recycled water from the Monterey Regional Water Pollution Control Agency's (MRWPCA) local wastewater treatment plant. Because seawater had entered the groundwater as far as 10 kilometers (6 miles) inland, well water was becoming too salty for agricultural use. MRWPCA, aware that this problem may be caused by overpumping of groundwater, in the 1980s began studying the possibility of irrigating with treated municipal wastewater. Although California regulations allow the use of treated water for irrigation, health officials expressed concerns that its use on cold-season crops such as broccoli—many of which are consumed raw—may contaminate the crops with pathogens. The MRWPCA conducted a study that showed that crops irrigated with recycled water were not contaminated with pathogens. California then

approved the plan, which required an upgrade of the local wastewater treatment plant. Sixty kilometers (40 miles) of pipeline were constructed to deliver the water to some 75 users throughout the system, which has the capacity to provide 24 billion liters (19,500 acre-feet) of water per year and irrigate 4900 hectares (12,000 acres) of coastal farmland. Such systems may become a common solution to the problem of deteriorating and diminishing groundwater supplies in California and perhaps throughout the world.

Industrial Use of Water

Industrial water use accounts for nearly half of total water withdrawals in the United States, about 70 percent in Canada, and about 23 percent worldwide. Since most industrial processes involve heat exchanges, 90 percent of the water used by industry is for cooling and

a.   b.

figure 16.11 **Dams Interrupt the Flow of Water** The flow of water in most large rivers is controlled by dams. Most of these dams provide electricity. In addition, they prevent flooding and provide recreational areas. (*a*) Large dams, however, destroy the natural river system. Smaller dams (*b*), however, do not always have a significant impact.

is returned to the source, so only a small amount is actually consumed. Industrial use accounts for less than 20 percent of the water consumed in the United States. For example, electric-power generating plants use water to cool steam so that it changes back into water. Many industries, especially power plants, actually can use saltwater for cooling purposes. About 30 percent of the cooling water used by power plants is saltwater. If the water heated in an industrial process is dumped directly into a watercourse, it significantly changes the water temperature. This affects the aquatic ecosystem by increasing the metabolism of the organisms and reducing the water's ability to hold dissolved oxygen.

Industry also uses water to dissipate and transport waste materials. In fact, many streams are now overused for this purpose, especially in urban centers. The use of watercourses for waste dispersal degrades the quality of the water and may reduce its usefulness for other purposes. This is especially true if the industrial wastes are toxic.

Historically, industrial waste and heat were major causes of pollution. However, most industrialized nations have passed laws that severely restrict industrial discharges of wastes or heated water into watercourses. In the United States, the federal government's role in maintaining water quality began in 1972, with the passage of the Federal Water Pollution Control Act (PL 92-500). This act provided federal funds and technical assistance to strengthen local, state, and interstate water-quality programs. The act (and subsequent amendments in 1977, 1981, 1987, and 1993) is commonly referred to as the "Clean Water Act." The Clean Water Act is a comprehensive and technically rigorous piece of legislation. It seeks to protect the waters of the United States from pollution. To do this, the act specifically regulates pollutant discharges into "navigable waters" by implementing two concepts: setting water-quality standards for surface water and limiting effluent discharges into the water. The policy objectives of the Clean Water Act are to restore and maintain the "chemical, physical and biological integrity of the nation's waters." Enforcement of this act has been extremely effective in improving surface-water quality. However, many countries in the developing world have done little to control industrial pollution, and water quality is significantly reduced by careless use. (See Global Perspective: Comparing Water Use and Pollution in Industrialized and Developing Countries on page 361.)

In-Stream Use of Water

In-stream water use does not remove water but makes use of it in its channels and basins. Therefore, all in-stream uses are nonconsumptive. Major in-stream uses of water are for hydroelectric power, recreation, and navigation. Although in-stream uses do not remove water, they may require modification of the direction, time, or volume of flow and can negatively affect the watercourse.

Electricity from hydroelectric power plants is an important energy resource. Presently, hydroelectric power plants produce about 13 percent of the total electricity generated in the United States. (See figure 16.11.) They do not consume water and do not add waste products to it. However, the dams needed for the plants have definite disadvantages, including the high cost of construction and the resulting destruction of the natural habitat in streams and surrounding lands.

Comparing Water Use and Pollution in Industrialized and Developing Countries

Water Use

Characteristic	Industrialized Countries	Developing Countries
Domestic water use per capita	1. Heavy per capita use 2. Highest usage in Australia, New Zealand, United States, and Canada 3. Usage is stabilizing.	1. Small per capita use 2. Water usage increases as living standards go up.
Where water is used	1. Irrigation and industry total about 85 percent. 2. Domestic about 15 percent	1. Irrigation is over 80 percent and particularly high in Asia and Africa. 2. Industry and domestic uses are each less than 10 percent.
Access to safe drinking water and wastewater treatment	1. Safe drinking water generally available 2. Wastewater treatment generally available 3. Only small population increases expected	1. Large numbers of people lack safe drinking water. 2. Effective wastewater treatment generally *not* available 3. Rapidly growing urban population will create greater need for safe drinking water and wastewater treatment.

Pollution Control

	Industrialized Countries	Developing Countries
Domestic wastewater treatment	1. Most countries treat domestic waste. 2. Central and Eastern European countries lack effective sewage treatment.	1. Almost all sewage is discharged without effective treatment.
Industrial wastes	1. Industrial discharges strictly regulated in most countries 2. Some accidental discharges 3. Discharges from disused industrial sites are a problem. 4. Industrial air pollution has caused acidification of lakes in parts of North America and Europe. 5. Eastern and Central Europe have serious industrial waste problems from historically unregulated industry. Black and Baltic Seas are heavily polluted.	1. Largely untreated 2. Little attention to regulating industrial waste 3. Acidification of lakes becoming important in China and tropical Africa
Land-use runoff	1. Fertilizers and pesticides are a continuing problem. 2. Soil conservation practices are used, but agricultural runoff is still a significant problem. 3. Runoff from urban areas causes some particles, oil, and other chemicals to enter water.	1. Heavy fertilizer and pesticide use causes serious water-quality problems. 2. Deforestation and poor farming practices cause soil erosion and degrade water quality. 3. Runoff from urban areas causes serious pollution problems; trash, human waste, animal waste, chemical waste are all present.

figure 16.12 **Recreational Use of Water** Marinas provide recreation; however, wetlands destruction and large dredging operations may be necessary to build them.

The sudden discharge of impounded water from a dam can seriously alter the downstream environment. If the discharge is from the top of the reservoir, the stream temperature rapidly increases. Discharging the colder water at the bottom of the reservoir causes a sudden decrease in the stream's water temperature. Either of these changes is harmful to aquatic life. The impoundment of water also reduces the natural scouring action of a flowing stream. If water is allowed to flow freely, the silt accumulated in the river is carried downstream during times of high water. This maintains the river channel and carries nutrient materials to the river's mouth. But if a dam is constructed, the silt is deposited behind the dam, eventually filling the reservoir. Many other dams were constructed to control floodwaters. While dams reduce flooding, they do not eliminate it. In fact, the building of a dam often encourages people to develop the floodplain. As a result, when flooding occurs, the loss of property and lives may be greater.

Because dams create lakes that have a large surface area, evaporation is increased. In arid regions, the amount of water lost can be serious. This is particularly evident in hot climates. Furthermore, flow is often intermittent below the dam, which alters the water's oxygen content and interrupts fish migration. The populations of algae and other small organisms are also altered. Because of all these impacts, dam construction requires careful planning. It should also be noted that the ecological impact of dams varies considerably, depending on the size of the dam. Large-scale dams have significantly greater impact than small-scale dams.

Dam construction creates new recreational opportunities because reservoirs provide sites for boating, camping, and related recreation. (See figure 16.12.) However, these opportunities come at the expense of a previously free-flowing river. Some recreational pursuits, such as river fishing, are lost. Sailing, waterskiing, swimming, fishing, and camping all

require water of reasonably good quality. Water is used for recreation in its natural setting and often is not physically affected. Even so, it is necessary to plan for recreational use, because overuse or inconsiderate use can degrade water quality. For example, waves generated by powerboats can accelerate shoreline erosion and cause siltation. When large numbers of powerboats are used on water, they contribute significantly to water pollution since their exhaust contains unburned hydrocarbons. Newer boat motors are designed to produce less pollution.

Most major rivers and large lakes are used for navigation. North America currently has more than 40,000 kilometers (25,000 miles) of commercially navigable waterways. These waterways must have sufficient water depth to ensure passage of ships and barges. Canals, locks, and dams are used to ensure that adequate depths are provided. Often, dredging is necessary to maintain the proper channel depth. Dredging can resuspend contaminated sediments. Another problem is determining where to deposit the contaminated sediments when they are removed from the bottom. In addition, the flow within the hydrologic system is changed, which, in turn, affects the water's value for other uses.

Most large urban areas rely on water to transport resources. During recent years, the inland waterway system in the United States has carried about 10 percent of goods such as grain, coal, ore, and oil. In North America, expenditures for the improvement of the inland waterway system have totaled billions of dollars.

In the past, almost any navigation project was quickly approved and funded, regardless of the impact on other uses. Today, however, such decisions are not made until ecological impacts are analyzed.

Kinds and Sources of Water Pollution

Water pollution occurs when something enters water that changes the natural ecosystem or interferes with water use

environmental CLOSE-UP

Is It Safe to Drink the Water?

Roughly 1000 contaminants have been detected in the public water supply in the United States, and virtually every major water source is vulnerable to pollution. About 60 percent of the U.S. population relies on surface water from rivers, lakes, and reservoirs that may contain industrial and agricultural wastes and pesticides washed off fields by rain. The other 40 percent uses groundwater that may be tainted by chemicals slowly seeping in from toxic-waste dumps, agricultural activities, and leaking sewage and septic systems. In some areas where groundwater supplies are being gradually depleted, the chemical pollutants are becoming more concentrated.

Most pollutants are probably not concentrated enough to pose significant health hazards; however, there are exceptions. The most widespread danger in water is lead, which can cause high blood pressure and an array of other health problems. Lead is especially hazardous to children, since it impairs the development of brain cells.

The U.S. Environmental Protection Agency estimates that at least 42 million Americans are exposed to unacceptably high levels of lead, and the U.S. Public Health Service estimates that perhaps 9 million children are at least slightly affected by it.

The contamination comes from old lead pipes and solder that have been used in plumbing for years. These materials are gradually being replaced in homes and water systems. Individuals may want to have their water tested for lead by an official lab. If the level is too high, they can investigate ways to deal with the problem or switch to bottled water for drinking and cooking. Even then, caution is called for—some bottled waters contain many of the same contaminants that tap water does.

The other four types of contamination in the U.S. water supply, along with their source and risk, are shown in the table. Regardless of the problems, however, the water supply in the United States is among the cleanest in the world.

Common Contaminants of Drinking Water

Substance	Source	Health Effects
Persistent chlorinated organic compounds	Used as solvents in industry; past use as pesticides	Various, including reproductive problems and cancer
Trihalomethanes	Produced by chemical reactions when water is disinfected by chlorination	Liver and kidney damage and possible cancer
Nitrates	Primarily from fertilizer and effluent from concentrated livestock raising	Can react to reduce oxygen uptake by blood; particularly a problem for children
Lead	Old piping and solder in public water distribution systems, homes, and other buildings	Nerve damage, learning difficulties in children, birth defects, possible cancer
Pathogenic bacteria, protozoa, and viruses	Leaking septic tanks and sewers; contamination of water supply from birds and mammals; inadequate disinfection	Acute gastrointestinal illness and other serious health problems

Source: Data, in part, from the U.S. Environmental Protection Agency.

by segments of society. In an industrialized society, maintaining completely unpolluted water in all drains, streams, rivers, and lakes is probably impossible. (See table 16.1.) But we can evaluate the water quality of a body of water and take steps to preserve or improve its quality by eliminating sources of pollution. Some pollutants seriously affect the quality and possible uses of water. In general, water pollutants can be divided into several broad categories.

Toxic chemicals or acids may kill organisms and make the water unfit for human use. If these chemicals are persistent, they may bioaccumulate in individual organisms and biomagnify in food chains.

Dissolved organic matter is a significant water pollution problem because it decays in the water. As the microorganisms naturally present in water break down the organic matter, they use up available dissolved oxygen from the wa-

ter. If too much dissolved oxygen is removed, aquatic organisms die. The amount of oxygen required to decay a certain amount of organic matter is called the **biochemical oxygen demand (BOD).** (See figure 16.13.) Measuring the BOD of a body of water is one way to determine how polluted it is. If too much organic matter is added to the water, all of the available oxygen will be used up. Then, anaerobic (not requiring oxygen) bacteria begin to break down

Table 16.1 Sources and Impacts of Selected Pollutants

Pollutant	Source	Effects on Humans	Effects on Aquatic Ecosystem
Acids	Atmospheric deposition; mine drainage; decomposing organic matter	Reduced availability of fish and shellfish Increased heavy metals in fish	Death of sensitive aquatic organisms; increased release of trace metals from soils, rock, and metal surfaces, such as water pipes
Chlorides	Runoff from roads treated for removal of ice or snow; irrigation runoff; brine produced in oil extraction; mining	Reduced availability of drinking water supplies; reduced availability of shellfish	At high levels, toxic to freshwater organisms
Disease-causing organisms	Dumping of raw and partially treated sewage; runoff of animal wastes from feedlots	Increased costs of water treatment; death and disease; reduced availability and contamination of fish, shellfish, and associated species	Reduced survival and reproduction of aquatic organisms due to disease
Elevated temperatures	Heat trapped by cities that is transferred to water; unshaded streams; solar heating of reservoirs; warm-water discharges from power plants and industrial facilities	Reduced availability of fish	Elimination of cold-water species of fish and shellfish; less oxygen; heat-stressed animals susceptible to disease; inappropriate spawning behavior
Heavy metals	Atmospheric deposition; road runoff; discharges from sewage treatment plants and industrial sources; creation of reservoirs; acidic mine effluents	Increased costs of water treatment; disease and death; reduced availability and healthfulness of fish and shellfish; biomagnification	Lower fish population due to failed reproduction; death of invertebrates leading to reduced prey for fish; biomagnification
Nutrient enrichment	Runoff from agricultural fields, pastures, and livestock feedlots; landscaped urban areas; dumping of raw and treated sewage and industrial discharges; phosphate detergents	Increased water treatment costs; reduced availability of fish, shellfish, and associated species; color and odor associated with algal growth; impairment of recreational uses	Algal blooms occur. Death of algae results in low oxygen levels and reduced diversity and growth of large plants. Reduced diversity of animals; fish kills
Organic molecules	Runoff from agricultural fields and pastures; landscaped urban areas; logged areas; discharges from chemical manufacturing and other industrial processes; combined sewers	Increased costs of water treatment; reduced availability of fish, shellfish, and associated species; odors	Reduced oxygen; fish kills; reduced numbers and diversity of aquatic life
Sediment	Runoff from agricultural land and livestock feedlots; logged hillsides; degraded stream banks; road construction; and other improper land use	Increased water treatment costs; reduced availability of fish, shellfish, and associated species; filling in of lakes, streams, and artificial reservoirs and harbors, requiring dredging	Covering of spawning sites for fish; reduced numbers of insect species; reduced plant growth and diversity; reduced prey for predators; clogging of gills and filters
Toxic chemicals	Urban and agricultural runoff; municipal and industrial discharges; leachate from landfills and mines; atmospheric deposits	Increased costs of water treatment; increased risk of certain cancers; reduced availability and healthfulness of fish and shellfish	Reduced growth and survivability of fish eggs and young; fish diseases; death of carnivores due to biomagnification in the food chain

Source: Data, in part, from *World Resources* 1994–95.

wastes. Anaerobic respiration produces chemicals that have a foul odor and an unpleasant taste and generally interfere with the well-being of humans.

Disease-causing organisms are a very important pollution problem in most of the world. Untreated or inadequately treated human or domesticated animal waste is most often the source of these organisms. In the developed world, sewage treatment and drinking-water treatment plants greatly reduce this public health problem.

Nutrients are also a pollution problem. Additional nutrients in the form of nitrogen and phosphorus compounds increase the rate of growth of aquatic plants and algae. However, phosphates and nitrates are generally present in very limited amounts in unpolluted freshwater and, therefore, are a limiting factor on the growth of aquatic plants and algae. (A **limiting factor** is a necessary material that is in short supply, and because of the lack of it, an organism cannot reach its full potential growth. See chapter 5.) Thus, when phosphates or nitrates are added to the surface water, they act as a fertilizer and promote the growth of undesirable algae populations. The excessive growth of algae and aquatic plants due to added nutrients is called **eutrophication.** Algae and larger aquatic plants may interfere with the use of the water by fouling boat propellers, clogging water-intake pipes, changing the taste and odor of water, and causing the buildup of organic matter on the bottom. As this organic matter decays, oxygen levels decrease, and fish and other aquatic species die.

Physical particles also can negatively affect water quality. Particles alter the clarity of the water, can cover spawning sites, act as abrasives that injure organisms, and carry toxic materials.

Ideally, we should think in terms of eliminating all pollution, but any human use is going to have at least some minor negative impact. Even activities such as swimming and boating add particles and

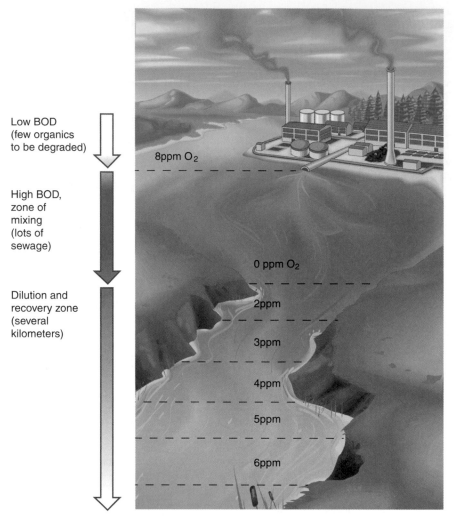

Low BOD
(few organics
to be degraded)

8ppm O₂

High BOD,
zone of
mixing
(lots of
sewage)

0 ppm O₂

2ppm

Dilution and
recovery zone
(several
kilometers)

3ppm

4ppm

5ppm

6ppm

figure 16.13 **Effect of Organic Wastes on Dissolved Oxygen** Sewage contains a high concentration of organic materials. When these are degraded by organisms, oxygen is removed from the water. This is called the biochemical oxygen demand (BOD). An inverse relationship exists between the amount of organic matter and oxygen in the water. The greater the BOD, the more difficult it is for aquatic animals to survive and the less desirable the water is for human use. The more the organic pollution, the greater the BOD.

chemicals to the water. So, determining acceptable water quality involves economic considerations. Removing the last few parts per million of some materials from the water may not significantly improve water quality and may not be economically justifiable. This is certainly true of organic matter, which is biodegradable. However, radioactive wastes and toxins that may accumulate in living tissue are a different matter. Vigorous attempts to remove these materials are often justified because of the materials' potential harm to humans and other organisms.

Sources of pollution are classified as either point sources or nonpoint sources. When a source of pollution can be readily identified because it has a definite source and place where it enters the water, it is said to come from a **point source.** Municipal and industrial discharge pipes are good examples of point sources. Diffuse pollutants, such as from agricultural land and urban paved surfaces, acid rain, and runoff, are said to come from **nonpoint sources** and are much more difficult to identify and control. Initial attempts to control water pollution were focused on point sources of pollution, since these were readily identifiable and economic pressure and adverse publicity could be brought to bear on companies that continued to

pollute from point sources. In North America, most point sources of water pollution have been identified and are regulated. In the United States, the EPA is responsible for identifying point sources of pollution, negotiating the permissible levels of pollution allowed from each source, and enforcing the terms of the permits.

Nonpoint sources of water pollution are being addressed, but this is much more difficult to do since regulating many small, individual human acts is necessary. In 1998, President Clinton announced the formation of the Clean Water Action Plan. This plan requires the cooperation of several levels of government and focuses on watersheds and nonpoint sources of pollution, particularly runoff from urban areas and agricultural land. In addition, plans have been developed to address several long-term water-quality issues, such as acid runoff from mines, contaminated sediments, the presence of toxic chemicals in fish, and restoring degraded wetlands.

Municipal Water Pollution

Municipalities are faced with the double-edged problem of providing suitable drinking water and disposing of wastes. These wastes consist of stormwater runoff, wastes from industry, and wastes from homes and commercial establishments. Wastes from homes consist primarily of organic matter from garbage, food preparation, cleaning of clothes and dishes, and human wastes. Human wastes are mostly undigested food material and a concentrated population of bacteria, such as *Escherichia coli* and *Streptococcus faecalis.* These particular bacteria normally grow in the large intestine (colon) of humans and are present in high numbers in the feces of humans; therefore, they are commonly called **fecal coliform bacteria.** Fecal coliform bacteria are also present in the wastes of other warm-blooded animals, such as birds and mammals. Low numbers of these bacteria in water are not harmful to healthy people. However, because they can be easily identified, their presence in the water is used to indicate the amount of pollution from the

Global Perspective

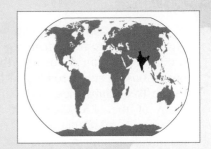

The Cleanup of the Holy Ganges

Every day, thousands of Hindus flock to the banks of the Ganges River in India. There, they drink and bathe in what they believe to be holy water, as partially cremated corpses float past them and nearby drains emit millions of liters of raw sewage. Clean water is one of India's most scarce resources, but, as in many developing nations, the money and technology needed to properly treat sewage are not available in most cities and villages. With the country's population of slightly more than 1 billion people expected to double in 39 years, officials are concerned that Indians will have no choice but to continue dumping raw waste into local waterways, contributing to epidemics of diarrhea and other diseases that kill thousands of people annually.

Keeping the Ganges clean is made especially difficult because faith in the river's incorruptible purity has generated complacency and ambivalence about its pollution among many of the 300 million people who live in the Ganges Basin. More than 1600 million liters (425 million gallons) of untreated municipal sewage, industrial waste, agricultural runoff, and other pollutants are discharged into the river every day. At the same time, officials estimate that more than 1 million people a day

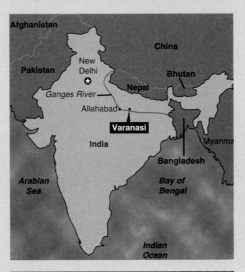

bathe or take a "holy dip" in the Ganges, and thousands drink straight from its banks.

The Hindu belief in cremation has led to several environmental problems. A wood cremation takes more than 50 pounds of wood, costing two week's wages for the typical Indian. There are complaints that wood cremations are helping to devour India's forests. Given the high cost of wood, many bodies are not completely cremated, and the partially burned corpses are disposed of in the Ganges.

One solution to this problem has been the introduction of 25,000 specially raised snapping turtles to devour the corpses. While this solution may seem extreme, there is another that is more acceptable in the long run. In 1992, as part of the Ganges Action Plan, an electric crematorium was built at the city of Varanasi that charges less than $2 per body and does a thorough job of turning the bodies to ash. There is some concern, however, that many Hindus will not want to abandon the traditional ritual of a wood cremation for the more efficient and less costly electric cremation.

Although this entire scenario may seem somewhat unusual to North Americans, it is important to keep in mind how culture and religion affect our environment. To many Indians, the cultural and religious practices of North Americans are equally puzzling.

---

fecal wastes of humans and other warm-blooded animals. The numbers of these types of bacteria present in water are directly related to the amount of fecal waste entering the water.

When human wastes are disposed of in water systems, potentially harmful bacteria from humans may be present in amounts too small to detect by sampling. Even in small numbers, these harmful bacteria may cause disease epidemics. It is estimated that some 1.5 million people in the United States become ill each year from infections caused by fecal contamination. In 1993, for example, a protozoan pathogen called *Cryptosporidium* was identified in the Milwaukee, Wis-

consin, public water system. This resulted in over 400,000 people becoming ill and at least 100 deaths. The total costs of such diseases amount to billions of dollars per year in the United States alone. The greater the amount of wastes deposited in the water, the more likely it is that there will be populations of disease-causing bacteria. Therefore, the presence of fecal coliform bacteria is used as an indication that other more harmful organisms may be present as well.

Wastewater from cleaning dishes and clothing contains some organic material along with the soap or detergent, which helps to separate the contaminant

from the dishes or clothes. Soaps and detergents are useful because one end of the molecule dissolves in dirt or grease and the other end dissolves in water. When the soap or detergent molecules are rinsed away by the water, the dirt or grease goes with them.

At one time, many detergents contained phosphates as a part of their chemical makeup, which contributed to eutrophication. However, because of the environmental effects of phosphate on aquatic environments, since 1994 most major detergent manufacturers in North America and other developed countries have eliminated phosphates from most of their formulations. Today, the majority of phosphate entering water in North America is from human waste and runoff from farm fields and livestock operations. A study conducted in 2000 by the Toxic Substances Hydrology Program of the U.S. Geological Survey (USGS) revealed that a broad range of chemicals found in residential, industrial, and agricultural wastewaters commonly occurs at low concentrations downstream from areas of intense urbanization and animal production. The chemicals include human and veterinary drugs (including antibiotics), natural and synthetic hormones, detergent metabolites, plasticizers, insecticides, and fire retardants. The report found that in addition to caffeine, the most frequently detected compounds were cholesterol and coprostanol, which is a by-product of DEET, a common insect repellent. The compounds found in the water are sold on supermarket shelves and are in virtually every medicine cabinet and broom closet, as well as farms and factories. Although they are flushed or rinsed down the drain every day, they do not disappear.

In a study of 139 streams throughout the United States, the USGS found that one or more of these chemicals were in 80 percent of the streams sampled. Half of the streams contained seven or more of these chemicals, and about one-third contained 10 or more. This was the first national-scale examination of streams for these organic wastewater contaminants. The chemicals identified largely escape regulation and are not removed by municipal wastewater treatment. At this point, the long-term effects of exposure to such chemicals are not clear; however, further study is anticipated.

Agricultural Water Pollution

Agricultural activities are the primary cause of water pollution problems. Excessive use of fertilizer results in eutrophication in many aquatic habitats, because precipitation carries dissolved nutrients (nitrogen and phosphorus compounds) into streams and lakes. In addition, groundwater may become contaminated with fertilizer and pesticides. The exposure of land to erosion results

in increased amounts of sediment being added to water courses. Runoff from animal feedlots carries nutrients, organic matter, and bacteria. Water used to flush irrigated land to get rid of excess salt in the soil carries a heavy load of salt that degrades the water body. And the use of agricultural chemicals results in contamination of sediments and aquatic organisms. One of the largest water pollution problems is agricultural runoff from large expanses of open fields. See chapter 14 for a general discussion of methods for reducing runoff and soil erosion.

Farmers can reduce runoff in several ways. One is to leave a zone of undisturbed, permanently vegetated land, called a conservation buffer, near drains or stream banks. This retards surface runoff because soil covered with vegetation tends to slow the movement of water and allows the silt to be deposited on the surface of the land rather than in the streams. This can be costly because farmers may need to remove valuable cropland from cultivation. One goal of the Clean Water Action Plan is to establish 3.2 million kilometers (2 million miles) of conservation buffer strips. By 2001, the plan was about halfway to its goal. Another way to retard runoff is keep the soil covered with a crop as long as possible. Careful control of the amount and the timing of fertilizer application can also reduce the amount of nutrients lost to streams. This makes good economic sense because any fertilizer that runs off or leaches out of the soil is unavailable to crop plants and results in less productivity.

Industrial Water Pollution

Factories and industrial complexes frequently dispose of some or all of their wastes into municipal sewage systems. Depending on the type of industry involved, these wastes contain organic materials, petroleum products, metals, acids, toxic materials, organisms, nutrients, or particulates. Organic materials and oil add to the BOD of the water. The metals, acids, and specific toxic materials need special treatment, depending on their nature and concentration. In these cases, a municipal wastewater treatment

plant will require that the industry pretreat the waste before sending it to the wastewater treatment plant. If this is not done, the municipal sewage treatment plants must be designed with their industrial customers in mind. In most cases, cities prefer that industries take care of their own wastes. This allows industries to segregate and control toxic wastes and design wastewater facilities that meet their specific needs.

Since industries are point sources of pollution, they have been relatively easy to identify as pollution sources, have been vigorously regulated, and have responded to mandates that they clean up their effluent. Most companies, when they remodel their facilities, include wastewater treatment as a necessary part of an industrial complex. However, some older facilities continue to pollute. These companies discharge acids, particulates, heated water, and noxious gases into the water. While industrial water pollution in the industrialized world has been significantly regulated, in much of the developing world this is not the case, and many lakes, streams, and harbors are severely polluted with heavy metals and other toxic materials, organic matter, and human and animal waste.

A special source of industrial water pollution is mining. By its very nature, mining disturbs the surface of the Earth and increases the chances that sediment and other materials will pollute surface waters. Hydraulic mining is practiced in some countries and involves spraying hillsides with high-pressure water jets to dislodge valuable ores. Often, chemicals are used to separate the valuable metals from the ores, and the waste from these processes is released into streams as well. Water that drains from current or abandoned coal mines is often very acid. Pyrite is a mineral associated with many coal deposits. It contains sulfur, and when exposed to weathering, the sulfur reacts with oxygen, resulting in the formation of sulfuric acid. In addition, fine coal-dust particles are suspended in the water, which makes the water chemically and physically less valuable as a habitat. Dissolved ions of iron, sulfur, zinc, and copper also are present in mine drainage. Control involves containing mine drainage and treating it before it is released to surface water. While federal legislation was passed in the 1990s requiring backfilling and land restoration after a mine is closed down, issues of responsibility and compliance with the law persist.

Thermal Pollution

Amendments to the Federal Water Pollution Control Act of 1972 mandated changes in how industry treats water. Industries were no longer allowed to use water and return it to its source in poor condition. One of the standards regulates the temperature of the water that is returned to its source. Because many industries use water for cooling, thermal pollution can be a problem. **Thermal pollution** occurs when an industry removes water from a source, uses the water for cooling purposes, and then returns the heated water to its source.

Power plants heat water to convert it into steam, which drives the turbines that generate electricity. For steam turbines to function efficiently, the steam must be condensed into water after it leaves the turbine. This condensation is usually accomplished by taking water from a lake, stream, or ocean to absorb the heat. This heated water is then discharged. The least expensive and easiest method of discharging heated water is to return the water to the aquatic environment, but this can create problems for aquatic organisms. Although an increase in temperature of only a few degrees may not seem significant, some aquatic organisms are very sensitive to minor temperature changes. Many fish are triggered to spawn by increases in temperature, while others may be inhibited from spawning if the temperature rises. For example, lake trout will not spawn in water above 10°C (50°F). If a lake has a temperature of 8°C (46°F), the lake trout will reproduce, but an increase of 3°C (5°F) would prevent spawning and result in this species' eventual elimination from that lake. Another problem associated with elevated water temperature is that it results in a decrease in the amount of oxygen dissolved in the water.

Ocean estuaries are very fragile. The discharge of heated water into an estuary may alter the type of plants present. As a result, animals with specific food habits may be eliminated because the warm water supports different food organisms. The entire food web in the estuary may be altered by only slight temperature increases.

Cooling water used by industry does not have to be released into aquatic ecosystems. Today in the industrialized world, most cooling water is not released in such a way that aquatic ecosystems are endangered. Three other methods of discharging the heat are commonly used. One method is to construct a large shallow pond. Hot water is pumped into one end of the pond, and cooler water is removed from the other end. The heat is dissipated from the pond into the atmosphere and substrate.

A second method is to use a cooling tower. In a cooling tower, the heated water is sprayed into the air and cooled by evaporation. The disadvantage of cooling towers and shallow ponds is that large amounts of water are lost by evaporation. The release of this water into the air can also produce localized fogs.

The third method of cooling, the dry tower, does not release water into the atmosphere. In this method, the heated water is pumped through tubes, and the heat is released into the air. This is the same principle used in an automobile radiator. The dry tower is the most expensive to construct and operate.

Marine Oil Pollution

Marine oil pollution has many sources. One source is accidents, such as oil-drilling blowouts or oil tanker accidents. The *Exxon Valdez,* which ran aground in Prince William Sound, Alaska, in 1989, released over 42 million liters (11 million gallons) of oil and affected nearly 1500 kilometers (930 miles) of Alaskan coastline. The event had a great effect on the algae and animal populations of the sound, and the economic impact on the local economy was severe. However, by 1994, little visible sign remained of the oil that had covered the coastline in 1989. A U.S. National Oceanic and Atmospheric Administra-

tion study estimates that 50 percent of the oil biodegraded on beaches or in the water; 20 percent evaporated; 14 percent was recovered; 12 percent is at the bottom of the sea, mostly in the Gulf of Alaska; 3 percent lies on shorelines; and less than 1 percent still drifts in the water column. Furthermore, it appears that some of the animal populations are recovering. This points out the tremendous resilience of natural ecosystems to recover from disastrous events.

Although accidents such as the *Exxon Valdez* are spectacular events, much more oil is released as a result of small, regular discharges from other, less-visible sources. Nearly two-thirds of all human-caused marine oil pollution comes from three sources: (1) runoff from streets, (2) improper disposal of lubricating oil from machines or automobile crankcases, and (3) intentional oil discharges that occur during the loading and unloading of tankers. With regard to the latter, pollution occurs when the tanks are cleaned or oil-contaminated ballast water is released. Oil tankers use seawater as ballast to stabilize the craft after they have discharged their oil. This oil-contaminated water is then discharged back into the ocean when the tanker is refilled. In addition to human-caused oil pollution, oil naturally seeps into the water from underlying oil deposits in many places.

As the number of offshore oil wells and the number and size of oil tankers have grown, the potential for increased oil pollution has also grown. Many methods for controlling marine oil pollution have been tried. Some of the more promising methods are recycling and reprocessing used oil and grease from automobile service stations and from industries, and enforcing stricter regulations on the offshore drilling, refining, and shipping of oil. As a result of oil spills from shipping tankers, an international agreement was reached in 1992 that required that all new oil tankers be constructed with two hulls—one inside the other. Such double-hulled vessels would be much less likely to rupture and spill their contents. Today, approximately 15 percent of oil tankers are double hulled.

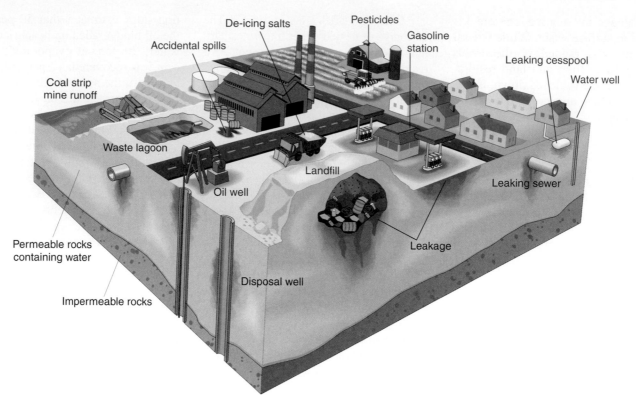

figure 16.14 **Sources of Groundwater Contamination** A wide variety of activities have been identified as sources of groundwater contamination.

Groundwater Pollution

A wide variety of activities, some once thought harmless, have been identified as potential sources of groundwater contamination. In fact, possible sources of human-induced groundwater contamination span every facet of social, agricultural, and industrial activities. (See figure 16.14.) Once groundwater pollution has occurred, it is extremely difficult to remedy. Pumping groundwater and treating it is very slow and costly, and it is difficult to know when all of the contaminated water has been removed. A much better way to deal with the issue of groundwater pollution is to work very hard to prevent the pollution from occurring in the first place.

Major sources of groundwater contamination include:

1. **Agricultural products.** Pesticides contribute to unsafe levels of organic contaminants in groundwater. Seventy-three different pesticides have been detected in the groundwater in Canada and the United States. Accidental spills or leaks of pesticides pollute groundwater sources with 10 to 20 additional pesticides. Other agricultural practices contributing to groundwater pollution include animal-feeding operations, fertilizer applications, and irrigation practices.

2. **Underground storage tanks.** For many years in North America, a large number of underground storage tanks containing gasoline and other hazardous substances have leaked. Four liters (1 gallon) of gasoline can contaminate the water supply of a community of 50,000 people. A major program of replacing leaking underground storage tanks recently was completed in the United States. However, the effects of past leaks and abandoned tanks will continue to be a problem for many years.

3. **Landfills.** Even though recently constructed landfills have special liners and water collection systems, approximately 90 percent of the landfills in North America have no liners to stop leaks to underlying groundwater, and 96 percent have no system to collect the leachate that seeps from the landfill. Sixty percent of landfills place no restrictions on the waste accepted, and many landfills are not inspected even once a year.

4. **Septic tanks.** Poorly designed and inadequately maintained septic systems have contaminated groundwater with nitrates, bacteria, and toxic cleaning agents. Over 20 million septic tanks are in use in the United States, and up to a third have been found to be operating improperly.

5. **Surface impoundments.** Over 225,000 pits, ponds, and lagoons are used in North America to store or treat wastes. Seventy-one percent are unlined, and only 1 percent use a plastic or other synthetic, nonsoil liner. Ninety-nine percent of these impoundments have no leak-detection systems. Seventy-three percent have no restriction on the waste placed in the impoundment.

Sixty percent are not even inspected annually. Many of these ponds are located near groundwater supplies.

Other sources of groundwater contamination include mining wastes, salting for controlling road ice, land application of treated wastewater, open dumps, cemeteries, radioactive disposal sites, urban runoff, construction excavation, fallout from the atmosphere, and animal feedlots.

Water-Use Planning Issues

In the past, wastes were discharged into waterways with little regard for the costs imposed on other users by the resulting decrease in water quality. Furthermore, as the population has grown and the need for irrigation and domestic water has intensified, in many parts of the world, there has not been enough water to satisfy everyone's needs. With today's increasing demands for high-quality water, unrestrained waste disposal and unlimited withdrawal of water could lead to serious conflicts about water uses, causing social, economic, and environmental losses at both local and international levels. (See table 16.2.)

Metropolitan areas must deal with a variety of issues and maintain an extensive infrastructure to provide three basic water services:

1. Water supply for human and industrial needs
2. Wastewater collection and treatment
3. Storm-water collection and management

Water sources must be identified and preserved for use. Some cities obtain all their municipal water from groundwater and must have a thorough understanding of the size and characteristics of the aquifer they use. Some cities, such as New York City, obtain potable water by preserving a watershed that supplies the water needed by the population. Other cities have abundant water in the large rivers that flow by them but must deal

Table 16.2 International Water Disputes		
River/Lakes	**Countries Involved**	**Issues**
Asia		
Brahmaputra, Ganges, Farakka	Bangladesh, India, Nepal	Alluvial deposits, dams, floods, irrigation, international quotas
Mekong	Cambodia, Laos, Thailand, Vietnam	Floods, international quotas
Salween	Tibet, China (Yunan), Burma	Alluvial deposits, floods
Middle East		
Euphrates, Tigris	Iraq, Syria, Turkey	International quotas, salinity levels
West Bank Aquifer, Jordan, Litani, Yarmuk	Israel, Jordan, Lebanon, Syria	Water diversion, international quotas
Africa		
Nile	Mainly Egypt, Ethiopia, Sudan	Alluvial deposits, water diversion, floods, irrigation, international quotas
Lake Chad	Nigeria, Chad	Dam
Okavango	Namibia, Angola, Botswana	Water diversion
Europe		
Danube	Hungary, Slovak Republic	Industrial pollution
Elbe	Germany, Czech Republic	Industrial pollution, salinity levels
Meuse, Escaut	Belgium, Netherlands	Industrial pollution
Szamos (Somes)	Hungary, Romania	Water allocation
Tagus	Spain, Portugal	Water allocation
Americas		
Colorado, Rio Grande	United States, Mexico	Chemical pollution, international quotas, salinity levels
Great Lakes	Canada, United States	Pollution
Lauca	Bolivia, Chile	Dams, salinity
Paraná	Argentina, Brazil	Dams, flooding of land
Cenepa	Ecuador, Peru	Water allocation

Sources: Walter H. Corson (ed.), *The Global Ecology Handbook* (Boston: Beacon Press, 1990), pp. 160–1; and Peter H. Fleick (ed.), *Water in Crisis: A Guide to the World's Freshwater Resources* (New York: Oxford University Press, 1993).

with pollution problems caused by upstream users. In many places where water is in short supply, municipal and industrial-agricultural needs for water conflict.

Water for human and industrial use must be properly treated and purified. It is then pumped through a series of pipes to consumers. After the water is used, it flows through a network of sewers to a wastewater treatment plant, where it is treated before it is released. Maintaining the infrastructure of pipes, pumps, and treatment plants is expensive.

Metropolitan areas must also deal with great volumes of excess water during storms. This water is known as **storm-water runoff.** Because urban areas are paved and little rainwater can be absorbed into the ground, managing storm water is a significant problem. Cities often have severe local flooding because the water is channeled along streets to storm sewers. If these sewers are overloaded or blocked with debris, the water cannot escape and flooding occurs.

The Water Quality Act of 1987 requires that municipalities obtain permits for discharges of storm-water runoff so that nonpoint sources of pollution are controlled. In the past, many cities had a single system to handle both sewage and storm-water runoff. During heavy precipitation or spring thaws, the runoff from streets could be so large that the wastewater treatment plant could not handle the volume. The wastewater was then diverted directly into the receiving body of water without being treated. Because of these new requirements, some cities have created areas in which to store this excess water until it can be treated. This is expensive and, therefore, is done only if federal or state funding is available. Many cities have also gone through the expensive process of separating their storm sewers from their sanitary sewers. A good example of the long and costly process of separation of sewers is Portland, Oregon. By 2001, Portland had completed only half of a 20-year project to separate its storm and sanitary sewer systems. The final costs are estimated to exceed the project's $1 billion budget.

Providing water services is expensive. We must understand that water supplies are limited. We must also understand that water's ability to dilute and degrade pollutants is limited and that proper land-use planning is essential if metropolitan areas are to provide services and limit pollution.

In pursuing these objectives, city planners encounter many obstacles. Large metropolitan areas often have hundreds of local jurisdictions (governmental and bureaucratic areas) that divide responsibility for management of basic water services. The Chicago metropolitan area is a good example. This area is composed of six counties and approximately 2000 local units of government. It has 349 separate water-supply systems and 135 separate wastewater disposal systems. Efforts to implement a water-management plan, when so many layers of government are involved, can be complicated and frustrating.

To meet future needs, urban, agricultural, and national interests will need to deal with a number of issues, such as the following:

- Increased demand for water will generate pressure to divert water to highly populated areas or areas capable of irrigated agriculture.

- Increased demand for water will force increased treatment of wastewater and reuse of existing water supplies.

- In many areas where water is used for irrigation, evaporation of water from the soil over many years results in a buildup of salt in the soil. When the water used to flush the salt from the soil is returned to a stream, the quality of the water is lowered.

- In some areas, wells provide water for all categories of use. If the groundwater is pumped out faster than it is replaced, the water table is lowered.

- In coastal areas, seawater may intrude into the aquifers and ruin the water supply.

- The demand for water-based recreation is increasing dramatically and requires high-quality water, especially for activities involving total body contact, such as swimming.

In 2002, the National Research Council issued a report entitled *Envisioning the Agenda for Water Resources Research in the 21st Century*. The report addresses the future of U.S. water resources and the research needed to support efforts to manage those resources in a sustainable fashion. The purpose of the report is to draw attention to the urgency and complexity of the water-resources issues facing the United States in the twenty-first century and to inform decision makers, researchers, and the public about these issues and the challenges they bring.

The report is organized around the general categories of water availability, water use, water institutions, and water research organization. It makes some 43 recommendations, which can be summarized into four key themes:

1. The challenge of solving the nation's water problems requires a renewed national commitment, which must entail changes in the way research agendas and priorities are established and a significant infusion of new federal funding.

2. Water quality and water quantity need to be viewed as interrelated.

Research priorities in these areas, therefore, should be developed in an integrated fashion.

3. More attention must be given to water-related research in the social sciences and research focused on the development of innovative institutions. Such institutions must translate public preferences into water-management goals and objectives, and facilitate the application of science and technology to achieve them.

4. Research on water-related environmental issues needs to become a major and continuing part of the water research agenda.

Water Diversion

Water diversion is the physical process of transferring water from one area to another. The aqueducts of ancient Rome are early examples of water diversion. Thousands of diversion projects have been constructed since then. New York City, for example, receives 90 percent of its water supply from the Catskill Mountains (Schoharie and Ashokan Reservoirs) and from four reservoirs collecting water from tributaries to the Delaware River west of the Catskills. About 10 percent of its supply comes from the Croton watershed east of the Hudson River. The most distant watershed serving New York City is some 200 kilometers (125 miles) away. Los Angeles is another example. It began importing water from the Owens Valley, 400 kilometers (250 miles) to the north, as early as 1913. (See Issues—Analysis: The California Water Plan on page 380.)

While diversion is seen as a necessity in many parts of the world, it often generates controversy. An example of this is the Garrison Diversion Unit in North Dakota, which was originally envisioned as a way to divert water from the Missouri River system for irrigation. (See figure 16.15.) The initial plan to irrigate portions of the Great Plains was developed during the Dust Bowl era of the 1930s. The Federal Flood Control Act of 1944 authorized the construction of the Garrison Diversion Unit. However, one of the original intents of the plan—to

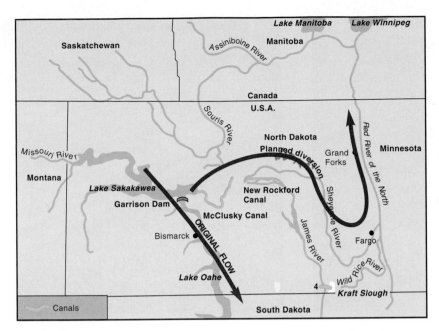

figure 16.15 **The Garrison Diversion Unit** The original intent of this plan was to divert water from the Missouri River to the McClusky Canal and the Sheyenne River and eventually into Lake Winnepeg by way of the Red River. In the process, additional land could be irrigated, and growing populations in the Red River Valley could be served with adequate water. Opposition has stalled the project, and only portions of the canal system and a pumping station have been completed.

Table 16.3 Percent of Sewage Treated in Selected Areas	
Area	**Percent**
North America	90
Europe	72
Mediterranean Sea	30
Caribbean Basin	Less than 10
Southeast Pacific	Almost zero
South Asia	Almost zero
South Pacific	Almost zero
West and Central Africa	Almost zero

Source: Data from World Resources Institute, 2000–01.

divert water from the Missouri River to the Red River—has not been met. Since the Red River flows north into Canada, the project requires international cooperation. The Canadian government has concerns about the effects on water quality of releasing additional water into the Red River. There are also concerns about environmental consequences. Portions of wildlife refuges, native grasslands, and waterfowl breeding marshes would be damaged or destroyed. Some states also have expressed concern about diverting water from the Missouri River, since any water diverted is not available for those downstream on the Missouri River.

While the plan has been modified several times, two sections of canal and a pumping station at Lake Sakakawea have been completed but not used. Proponents continue to push for legislation to complete the connection between the two canals that would allow water to be diverted to the Red River for municipal use. The original intent of diverting water for irrigation has been eliminated from the most recent proposal.

One major consequence of diverting water for irrigation and other purposes is that the water bodies downstream of the diversion are deprived of their source of water. This often has major ecological consequences. Water levels in Lake Chad in Africa are falling due to drought and increased demand for irrigation water. This has affected the populations of fish and other wildlife. In Mexico, 2500 kilometers (1550 miles) of rivers have dried up because the water was diverted for other purposes, resulting in the extinction of 15 species of fish and threatening half of the remaining species.

As people recognize the significance of wildlife habitat, plans are being worked out to balance societal needs with the need for water to maintain wildlife habitat. For example, Mono Lake in the Sierra Mountains in central California began to shrink in 1941 when much of the water that fed the lake was diverted to supply Los Angeles. Since Mono Lake has no outlet, its size is determined by the balance between the water flowing into the lake and evaporation from the lake surface. Consequently, the water level in the lake dropped 13 meters (43 feet), the lake's volume decreased by half, and its salinity doubled. These changes resulted in a loss of habitat important for many ducks and other water-

fowl. In 1994, an agreement was reached to increase the amount of water flowing to the lake so that the level of the lake would rise. The plan is to increase the level of the lake by 5 meters (17 feet) over a 20-to-30-year period.

In New South Wales, Australia, an area known as the Macquarie Marshes was affected by water diversion. The original extent of the marshes was reduced by about 50 percent because the Macquarie River, which feeds the marsh, was dammed to provide irrigation water. In the mid-1990s, because of concerns about the loss of wildlife, an agreement was reached to provide additional water to the area to maintain the marsh and protect the breeding areas of waterfowl.

Wastewater Treatment

Because water must be cleaned before it is released, most companies and municipalities in the developed world maintain wastewater treatment facilities. The percentage of sewage that is treated, however, varies greatly throughout the world. (See table 16.3.) Treatment of sewage is usually classified as primary, secondary, or tertiary. **Primary sewage treatment** is primarily a physical process that removes larger particles by filtering water through large screens and then allowing smaller particles to settle in ponds or lagoons. Water is removed

a.

c.

b.

figure 16.16 **Primary and Secondary Wastewater Treatment** Primary treatment is physical; it includes filtrating and settling of wastes. Photograph (*a*) is of a settling tank in which particles settle to the bottom. Secondary treatment is mostly biological and includes the concentration of dissolved organics by microorganisms. Two major types of secondary treatment are trickling filter and activated sludge methods. Photograph (*b*) shows a trickling filter system, and photograph (*c*) shows an activated sludge system.

from the top of the settling stage and either released to the environment or to a subsequent stage of treatment. If the water is released to the environment, it does not have any sand or grit; but it still carries a heavy load of organic matter, dissolved salts, bacteria, and other microorganisms. The microorganisms use the organic material for food, and as long as there is sufficient oxygen, they will continue to grow and reproduce. If the receiving body of water is large

enough and the organisms have enough time, the organic matter will be degraded. In crowded areas, where several municipalities take water and return it to a lake or stream within a few kilometers of each other, primary water treatment is not adequate and major portions of the receiving body of water are affected.

Secondary sewage treatment is a biological process that usually follows primary treatment. It involves holding the wastewater until the organic material has been degraded by the bacteria and other microorganisms. Secondary treatment facilities are designed to promote the growth of microorganisms. To encourage this action, the wastewater is mixed with large quantities of highly oxygenated water, or the water is aerated directly, as in a trickling filter system or an activated sludge system. In a **trickling filter system,** the wastewater is sprayed over the surface of rock or other substrate to increase the amount of dissolved oxygen. The rock also provides a place for a film of bacteria and other microbes to attach so they are exposed simultaneously to the organic material in the water and to oxygen. These

microorganisms feed on the dissolved organic matter and small suspended particles, which then become incorporated into their bodies as part of their cell structure. The bodies of the microorganisms are larger than the dissolved and suspended organic matter, so this process concentrates the organic wastes into particles that are large enough to settle out. This mixture of organisms and other particular matter is called **sewage sludge.** The sludge that settles consists of living and dead microorganisms and their waste products.

In **activated-sludge sewage treatment** plants, the wastewater is held in tanks and has air continuously bubbled through it. The sludge eventually is moved to settling tanks where the water and sludge can be separated. To make sure that the incoming wastewater has appropriate kinds and amounts of decay organisms, some of the sludge is returned to aeration tanks, where it is mixed with incoming wastewater. This kind of process uses less land than a trickling filter. (See figure 16.16.) Both processes produce a sludge that settles out of the water.

CLOSE-UP

Restoring the Everglades

Everglades National Park is a unique, subtropical, freshwater wetland visited by about a million people per year. This unique ecosystem exists because of an unusual set of conditions. South Florida is a nearly flat landscape with a slight decrease in elevation from the north to the south. In addition, the substrate is a porous limestone that allows water to flow through rather easily. Originally, water drained in a broad sheet from Lake Okeechobee southward to Florida Bay. This constant flow of water sustained a vast, grassy wetland interspersed with patches of trees.

After Everglades Park was established in 1947, about 800,000 hectares (2 million acres) of wetlands to the north were converted to farms and urban development. South Florida boomed. Some 4.5 million people currently live in the horseshoe crescent around the Everglades region, and new residents arrive each day. Conversion of land to agriculture and urban development required changes in the natural flow of water. Dams, drainage canals, and water diversion supported and protected the human uses of the area but cut off the essential, natural flow of freshwater to the Everglades. Changes in the normal pattern of water flow to the Everglades resulted in periods of drought and a general reduction in the size of this unique wetland region. Populations of wading birds in the southern Everglades fell drastically as their former breeding and nesting areas dried up. Other wildlife, such as alligators, Florida panthers, snail kites, and wood storks, also were negatively affected because drought reduces suitable habitat during parts of the year and the animals are forced to congregate around the remaining sources of water. The quality of the water is also important. The original wetland ecosystem was a nutrient-poor system. The introduction of nutrients into the water from agricultural activities encouraged the growth of exotic plants that replaced natural vegetation.

As people recognized that the key element necessary to preserve the Everglades was a constant, reliable source of clean freshwater, several steps were taken to modify water use to preserve the Everglades. The Kissimmee River, which flows into Lake Okeechobee, was channelized into an arrow-straight river in the late 1960s. This project destroyed the marshes and allowed nutrients from dairy farms and other agricultural activities to pollute the lake and Everglades Park, to which the water from Lake Okeechobee eventually flowed. To help alleviate this problem, in 1990, the U.S. Army Corps of Engineers began to return the Kissimmee to its natural state, with twisting oxbow curves and extensive wetlands. This allows the plants in the natural wetlands to remove much of the nutrient load before it enters the lake. To reduce the likelihood of further development near the park, in 1989, Congress approved the purchase of 43,000 hectares (100,000 acres) for an addition to the east section of the park. Florida obtained an additional 60,000 hectares (150,000 acres) as an additional buffer zone for the park.

For several years, the U.S. Army Corps of Engineers and the South Florida Water Management District cooperated in the development of a comprehensive restoration plan that was finished in 2000. The development of the plan involved the input of scientists, politicians, various business interests, and environmentalists. Key components of the plan are:

1. Developing facilities to store surface water and pump water into aquifers so that it can be released when needed

2. Developing wetlands to treat municipal and agricultural runoff so that nutrient loads are reduced

3. Using clean wastewater to recharge aquifers and supply water to wetlands in the Miami area

4. Reducing the amount of water lost through levees and redirecting water to the Everglades

5. Removing barriers to the natural flow of water through the Everglades

The sludge that remains is concentrated and often dewatered (dried) before disposal. Sludge disposal is a major problem in large population centers. In the San Francisco Bay area, 2500 tonnes (2750 U.S. tons) of sludge are produced each day. Most of this is carried to landfills and lagoons, and some is composted and returned to the land as fertilizer. Some municipalities incinerate their sludge. In other areas, if the sewage sludge is free of heavy metals and other contaminants that might affect plant growth or the quality and safety of food products, it is applied directly to agricultural land as a fertilizer and soil conditioner.

In North America and much of the developed world, wastewater receives both primary and secondary sewage treatment. While the water has been cleansed of its particles and dissolved organic matter, it still has microorganisms that might be harmful. Therefore, the water discharged from these sewage treatment plants must be disinfected. The least costly method of disinfection is chlorination. However, many people feel that the use of chlorine should be discontinued, since chlorination may be responsible for the creation of chlorinated organic compounds that are harmful. Therefore, other methods of killing microorganisms are being explored. The chemical ozone also kills microorganisms and has been substituted for chlorine by some facilities. Ultraviolet light and ultrasonic energy can also be used. Vancouver, Washington, for example, has been using ultraviolet light treatment since 1998 as a final treatment before returning water to the Columbia River. However, chlorine is inexpensive and very effective, so it continues to be the primary method used.

CHAPTER 16 Water Management

environmental CLOSE-UP

The plan received strong support from Congress in the fall of 2001, when it allocated $1.4 billion to begin implementing the plan. It will require many more billions of dollars and up to 30 years to accomplish all aspects of the plan, but if the plan continues to be implemented over the next few decades, the Everglades ecosystem will be restored to a more stable condition and will have a more hopeful future.

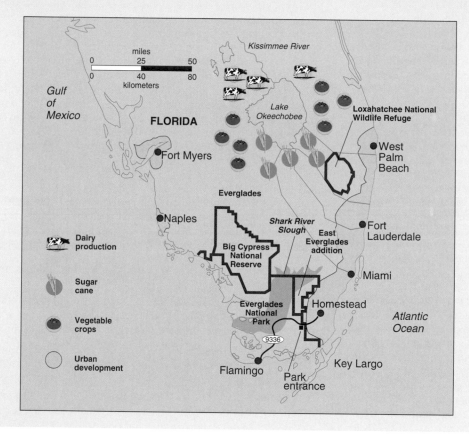

Everglades wading birds once numbered more than a million, but pollution and drought brought on by water diversion have decimated their number.

Construction threatens to destroy the fragile ecosystem of the Everglades.

A growing number of larger sewage treatment plants use additional processes called tertiary sewage treatment. **Tertiary sewage treatment** involves a variety of techniques to remove dissolved pollutants left after primary and secondary treatments. (See table 16.4.) The tertiary treatment of municipal wastewater is often used to remove phosphorus and nitrogen that could increase aquatic plant growth. Some municipalities are using natural or constructed wetlands to serve as tertiary sewage treatment systems. In other cases, the effluent from the treatment facility is used to irrigate golf courses, roadside vegetation, or cropland. The vegetation removes excess nutrients and prevents them from entering streams and lakes where they would present a pollution problem. Tertiary treatment of industrial and other specialized wastewater streams is very costly because it requires specific chemical treatment of the water to eliminate specific problem materials. Many industries maintain their own wastewater facilities and design specific tertiary treatment processes to match the specific nature of their waste products.

As water has become scarce in many parts of the world, people have looked at wastewater as a source of water for other purposes. Water from wastewater plants can be used for irrigation, industrial purposes, cooling, and many other activities. Ultimately, it is possible to have a closed loop system for domestic water in which the output of the wastewater plant becomes the input for the drinking water supply.

Salinization

Another water-use problem results from **salinization,** an increase in salinity caused by growing salt concentrations

Table 16.4 Tertiary Treatment Methods

Kind of Tertiary Treatment	Problem Chemicals	Methods
Biological	Phosphorus and nitrogen compounds	1. Large ponds are used to allow aquatic plants to assimilate the nitrogen and phosphorus compounds from the water before the water is released. 2. Columns containing denitrifying bacteria are used to convert nitrogen compounds into atmospheric nitrogen.
Chemical	Phosphates and industrial pollutants	1. Water can be filtered through calcium carbonate. The phosphate substitutes for the carbonate ion, and the calcium phosphate can be removed. 2. Specific industrial pollutants, which are nonbiodegradable, may be removed by a variety of specific chemical processes.
Physical	Primarily industrial pollutants	1. Distillation 2. Water can be passed between electrically charged plates to remove ions. 3. High-pressure filtration through small-pored filters 4. Ion-exchange columns

figure 16.17 **Salinization** As water evaporates from the surface of the soil, the salts it was carrying are left behind. In some areas of the world, this has permanently damaged cropland.

in soil. This is primarily a problem in areas where irrigation has been practiced for several decades. When water evaporates from soil or plants extract the water they need, the salts present in all natural waters become concentrated. Since irrigation is most common in hot, dry areas that have high rates of evaporation, there is generally an increase in the concentration of salts in the soil and in the water that runs off the land. (See figure 16.17.) Every river increases in salinity as it flows to the ocean. The salinity of the Colorado River water increases 20 times as it passes through irrigated cropland between Grand Lake in north-central Colorado and the Imperial Dam in southwest Arizona. The problem of salinity will continue to increase as irrigation increases.

Groundwater Mining

Groundwater mining means that water is removed from an aquifer faster than it is replaced. When this practice continues for a long time, the water table eventually declines. Groundwater mining is common in areas of the western United States and throughout the world. In North America, it is a particular problem due to growing cities and increasing irrigation. In aquifers with little or no recharge, virtually any withdrawal constitutes mining, and sustained withdrawals will eventually exhaust the supply. This problem is particularly serious in communities that depend heavily on groundwater for their domestic needs.

Groundwater mining can also lead to problems of settling or subsidence of the ground surface. Removal of the water allows the ground to compact, and large depressions may result. For example, in the San Joaquin Valley of California, groundwater has been withdrawn for irrigation and cultivation since the 1850s, and groundwater levels have fallen over 100 meters (300 feet). More than 1000 hectares (approximately 2500 acres) of ground have subsided, some as much as 6 meters (20 feet). Currently, the ground surface in that area is sinking 30 centimeters (12 inches) per year. London, Mexico City, Venice, Houston, and Las Vegas are some other cities that have experienced subsidence as a result of groundwater withdrawal. (See table 16.5.) As people recognize the severity of the problem, public officials are beginning to develop water conservation plans for their cities. Albuquerque, New Mexico, which relies on groundwater for its water supply, has an extensive public education program to encourage people to reduce their water consumption. Since grass demands water, people are encouraged to use desert plants for landscaping or collect rainwater to water their lawns. Finding and correcting leaks, reducing the amount of water used in bathing, and recycling water from swimming pools are other conservation strategies.

Groundwater mining poses a special problem in coastal areas. As the fresh groundwater is pumped from wells

Table 16.5 Groundwater Depletion in Major Regions of the World

Region/Aquifer	Estimates of Depletion
California	Groundwater overdraft exceeds 1.7 billion cubic meters (60 billion cubic feet) per year. The majority of the depletion occurs in the Central Valley, which is referred to as the vegetable basket of the United States.
Southwestern United States	In parts of Arizona, water tables have dropped more than 120 meters (400 feet). Projections for parts of New Mexico indicate that water tables will drop an additional 22 meters (70 feet) by 2020.
High Plains aquifer system, United States	The Ogallala aquifer underlies nearly 20 percent of all the irrigated land in the United States. To date, the net depletion of the aquifer is in excess of 350 billion cubic meters (12 trillion cubic feet), or roughly 15 times the average annual flow of the Colorado River. Most of the depletion has been in the Texas high plains, which have witnessed a 26 percent decline in irrigated land from 1979 to 1989. Current depletion is estimated to be in excess of 13 billion cubic meters (450 billion cubic feet) a year.
Mexico City and Valley of Mexico	Use exceeds natural recharge by 60 to 85 percent, causing land subsidence and falling water tables.
African Sahara	North Africa has vast nonrecharging aquifers where current depletion exceeds 12 billion cubic meters (425 billion cubic feet) a year.
India	Water tables are declining throughout much of the most productive agricultural land in India. In parts of the country, groundwater levels have declined 90 percent during the past two decades.
North China	The water table underneath portions of Beijing has dropped 40 meters (130 feet) during the past 40 years. A large portion of northern China has significant groundwater overdraft.
Arabian peninsula	Groundwater use is nearly three times greater than recharge. At projected depletion rates, exploitable groundwater reserves could be exhausted within the next 50 years. Saudi Arabia depends on nonrenewable groundwater for roughly 75 percent of its waters. This includes irrigation of 2 million to 4 million tonnes (2.2–4.4 million U.S. tons) of wheat per year.

along the coast, the saline groundwater moves inland, replacing fresh groundwater with unusable saltwater. (See figure 16.18.) Saltwater intrusion is a serious problem in heavily populated coastal areas throughout the world.

Preserving Scenic Water Areas and Wildlife Habitats

Some bodies of water have unique scenic value. To protect these resources, the way in which the land adjacent to the water is used must be consistent with preserving these scenic areas.

The U.S. Federal Wild and Scenic Rivers Act of 1968 established a system to protect wild and scenic rivers from development. All federal agencies must consider the wild, scenic, or recreational values of certain rivers in planning for the use and development of the rivers and adjacent land. The process of designating a river or part of a river as wild or scenic is complicated. It often encounters local opposition from businesses dependent on growth. Following reviews by state and federal agencies, rivers may be designated as wild and scenic by action of either Congress or the secretary of the Interior. Sections of over 150 streams comprising about 12,000 kilometers (7700 miles) in the

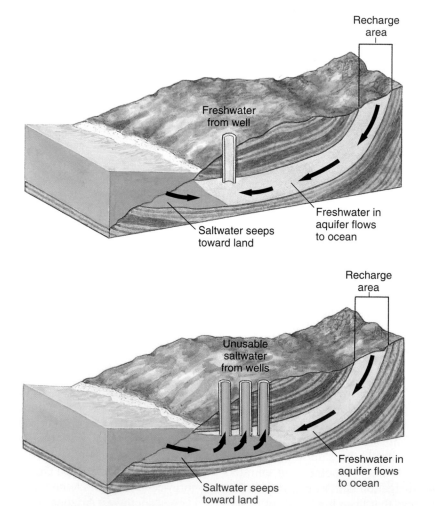

figure 16.18 **Saltwater Intrusion** When saltwater intrudes on fresh groundwater, the groundwater becomes unusable for human consumption and for many industrial processes.

Global Perspective

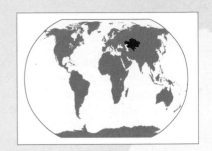

Death of a Sea

The Aral Sea lies on the border between Uzbekistan and Kazakhstan in the former Soviet Union. It was once larger than any of the Great Lakes except Superior, and now it is disappearing. Since the 1920s, Soviet agricultural planners have used up the Aral Sea, diverting its waters for irrigation. The two rivers feeding the Aral were drawn off to irrigate millions of hectares of cotton. An irrigation canal, the world's longest, stretches over 1300 kilometers (800 miles) into Turkmenistan, paralleling the boundaries of Afghanistan and Iran. The cotton production plan worked and by 1937 the Soviet Union was a net exporter of cotton. The success of the cotton program, however, spelled the end for the Aral Sea.

For a long time, the ecological impact on the sea and surrounding area was largely hidden from public view. Since the 1960s, however, the Aral has lost 75 percent of its volume and over 50 percent of its surface area, or over 22,000 square kilometers (8500 square miles) of what are now largely dry, salt-encrusted wastelands. The once-thriving fishing industry that depended on the water is all but gone. Some 20 of the 24 fish species there have disappeared. The fish catch, which totaled 44,000 tonnes (48,000 U.S. tons) a year in the 1950s and supported some 60,000 jobs, has dropped to zero. Abandoned fishing villages dot the sea's former coastline.

Another apparent consequence of the dried-up sea is a host of human illnesses. A high rate of throat cancer is attributed to dust from the drying sea. Each year, winds pick up a million tonnes of a toxic dust-salt mixture from the dry sea bed and deposit it on the surrounding farmland, harming or killing crops. The low river flows have concentrated salts and toxic chemicals, making water supplies hazardous to drink and contributing to disease. In the northwest part of the Republic of Uzbekistan, the infant mortality rate is the highest in the former Soviet Union.

The former fishing center of the sea was a town named Muynak. The town is now landlocked more than 30 kilometers (20 miles) from the water. Less than 25 years ago, Muynak was a seaport. The population of Muynak is down from 40,000 in 1970 to 12,000 today. In 1990, the mayor and last harbormaster of Muynak commented:

> The water continued to go away while the salinity increased. The weather changed for the worse, with the summers getting hotter and the winters colder. The people feel salt on their lips and in their eyes all the time. It's getting harder to open your eyes here.

In 1992 the central Asian republics of Uzbekistan, Kazakhstan, Turkmenistan, Tajikistan, and Kaygyzstan signed an international agreement to save the Aral Sea. Additional water is now flowing to the sea, but it will take years of cooperative efforts to return the sea to its original size.

Death of a sea

United States have been designated as wild or scenic.

Many unique and scenic shorelands have also been protected from future development. Until recently, estuaries and shorelands have been subjected to significant physical modifications, such as dredging and filling, which may improve conditions for navigation and construction but destroy fish and wildlife habitats. Recent actions throughout North America have attempted to restrict the development of shorelands. Development has been restricted in some particularly scenic areas, such as Cape Cod National Seashore in Massachusetts and the Bay of Fundy in the Atlantic provinces of Canada.

Historically, poorly drained areas were considered worthless. Subsequently, many of these wetlands were filled or drained and used for building sites. At the time of European settlement, the area that is now the conterminous United States contained an estimated 89 million hectares (221 million acres) of wetlands. Over time, wetlands have been drained, dredged, filled, leveled, and flooded to the extent that less than half of the original wetland acreage remains.

Today, 95 percent of the remaining 20 million hectares (50 million acres) of wetlands in the United States are inland freshwater wetlands. The remaining 5 percent are in saltwater estuarine environments. Freshwater forested wetlands make up the single largest category of all wetlands in the conterminous United States (see figure 16.19). The natural and economic importance of wetlands has been recognized only recently. In addition to providing spawning and breeding habitats for many species of wildlife, wetlands act as natural filtration systems by trapping nutrients and pollutants and preventing them from entering adjoining lakes, streams, or estuaries. Wetlands also slow down floodwaters and permit nutrient-rich particles to settle out. In addition, wetlands can act as reservoirs and release water slowly into lakes, streams, or aquifers, thereby preventing floods. (See figure 16.20.) Coastal estuarine zones

Causes of Wetlands Loss (1986–2000)

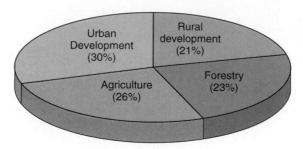

figure 16.19   **Wetlands Conversion** Urban development accounted for the greatest portion of wetlands loss from 1986 to 2000.

Source: U.S. Geological Survey

The value of coastal wetlands

figure 16.20   **The Value of Wetlands** Wetlands are areas covered with water for most of the year that support aquatic plant and animal life. Wetlands can be either fresh or saltwater and may be isolated potholes or extensive areas along rivers, lakes, and oceans. We once thought of wetlands as only a breeding site for mosquitoes. Today, we are beginning to appreciate their true value.

and adjoining sand dunes also provide significant natural flood control. Sand dunes act as barriers and absorb damaging waves caused by severe storms. In recent years, public appreciation of the ecological, social, and economic values of wetlands has increased substantially.

The increased awareness of how much wetland acreage has been lost or damaged since the time of European settlement and the consequences of those losses has led to the development of many federal, state, and local wetland protection programs and laws.

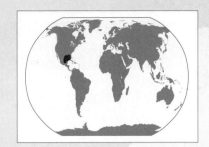

Global Perspective

The Dead Zone of the Gulf of Mexico

Each summer, a major "dead zone" of about 18,000 square kilometers (7000 square miles) develops in the Gulf of Mexico off the mouth of the Mississippi River. This dead zone contains few fish and bottom-dwelling organisms. It is caused by low oxygen levels (hypoxia) brought about by the rapid growth of algae and bacteria in the nutrient-rich waters. The nutrients can be traced to the extensive use of fertilizer in the major farming areas of the central United States, farming areas drained by the Mississippi River and its tributaries. About 1.6 million tonnes (1.75 million U.S. tons) of nitrogen-mostly fertilizer runoff from Midwestern farms flows out of the Mississippi River every year.

The hypoxia problem begins when nitrogen and other nutrients wash down the Mississippi River to the Gulf of Mexico, where they trigger a bloom of microscopic plants and animals. The dead cells and fecal matter of the organisms then fall to the seafloor. As growing colonies of bacteria digest this waste, they consume dissolved oxygen faster than it can be replenished. The flow of oxygen-rich water from the Mississippi cannot rectify the problem, because differences in temperature and density cause the warm freshwater to float above the colder, salty ocean water. Crustaceans, worms, and any other animals that cannot swim out of the hypoxic zone will die.

According to the EPA, nutrient pollution has degraded more than half of U.S. estuaries. In 2002, the National Research Council named nutrient pollution and the sustainability of fisheries as the most important problems facing the U.S. coastal waters in the next decade.

Gulf of Mexico fisheries, which generate some $2.8 billion a year in revenue, are one potential casualty of hypoxia. Hypoxia can block crucial migration of shrimp, which must move from inland nurseries to feed and spawn offshore. In other places in the world, such as the Black and Baltic Seas, hypoxia has been responsible for the collapse of some commercial fisheries.

Several approaches could reduce the amount of hypoxia-causing nitrogen released into the Mississippi River Basin. These include:

- Reduce use of nitrogen-based fertilizers and improve storage of manure. Reduce runoff from feedlots.
- Plant perennial crops instead of fertilizer-intensive corn and soybeans on 10 percent of the acreage.

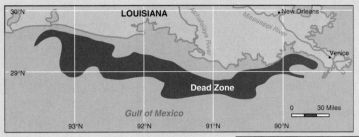

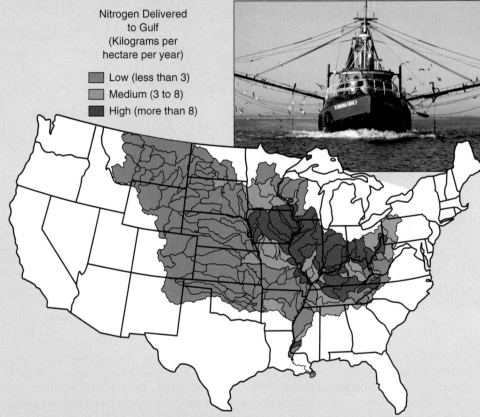

Nitrogen Delivered to Gulf (Kilograms per hectare per year)

- Low (less than 3)
- Medium (3 to 8)
- High (more than 8)

- Remove nitrogen and phosphorus from domestic wastewater.
- Restore 2 million to 4 million hectares (5–10 million acres) of wetlands, which absorb nitrogen runoff.

National Science and Technology Committee on Environment and Natural Resources.

Note: A considerable amount of total nitrogen originates from watersheds in the Mississippi River Basin very distant from the Gulf of Mexico.

Source: Modified from R. B. Alexander, R. A. Smith, and G. E. Schwarz, "Effect of Stream Channel Size on the Delivery of Nitrogen to the Gulf of Mexico," *Nature,* 17 (February 2000): 761.

The California Water Plan

The management of freshwater is often a controversial subject, involving social, ecological, and economic aspects. A good example is the California Water Plan.

In the early 1900s, it became clear that the growth of Los Angeles, which was then a small coastal town, would be encouraged by irrigating the surrounding land. Los Angeles looked to the Owens Valley, near Bishop, 400 kilometers (250 miles) north, for a source of water. The Los Angeles Aqueduct connecting these two areas was completed in 1913.

After the Owens Valley project, California developed a statewide water program known as the California Water Plan. This plan is necessary because most of the state's population and irrigated land are found in the central and southern regions, but most of the water is in the north. The plan details the construction of aqueducts, canals, dams, reservoirs, and power stations to transport water from the north to the south. In addition to supplying water for southern regions, the aqueducts provide irrigation for the San Joaquin Valley. Eventually, the new land made available for agriculture amounted to about 400,000 hectares (1 million acres).

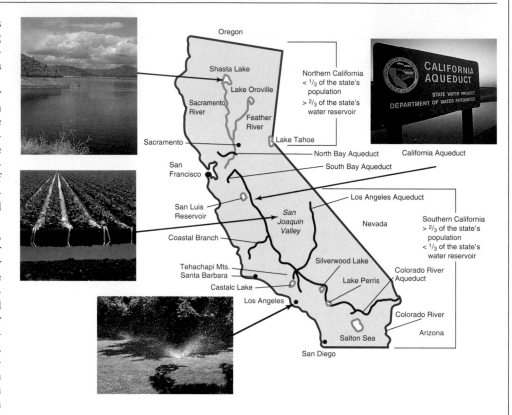

The California Water Plan has been one of the most controversial programs ever undertaken in California. Its adoption instigated a sectional feud between the moist "north" and the dry "south." Southern California was accused of trying to steal northern water, but because the large population in the southern part of the state carried the vote, the plan was adopted. Environmentalists still claim that the project has irreparably scarred the countryside and upset natural balances of streams, estuaries, vegetation, and wildlife. They argue that providing water to southern California promotes population growth, which leads to further urbanization and land development.

Many questions and controversies center on whether the water is really needed. Ninety percent of the water used in southern California is for irrigation; there is evidently abundant water for domestic and industrial use. A few of the crops raised in the San Joaquin Valley account for a large percentage of the irrigation water. Most notable is rice. Rice is not a native crop and demands intensive irrigation. The cost is borne by all the rate payers, most of whom are urban. California is now one of the nation's most productive agricultural areas.

Allocation of water resources is a matter of economics as well as technology. The California water project has been criticized for using public funds to increase the value of privately held farmland. Furthermore, technological advances in desalination plants may give a new dimension (unforeseen when the water plan was devised) to the problem of water resources. Although more aqueducts, canals, and pumping plants are planned, whether they will be completed is uncertain.

Southern California has been in a drought emergency for several years. Water rates have increased, certain municipal uses have been curtailed, and irrigation has been reduced. Increasing population creates a demand for water that is now causing people to recognize that they will need to choose between using water for municipal purposes or irrigation. There is not enough water to increase both.

- What are the major advantages of having a water plan?
- What problems develop when water is transported to arid regions?
- Should water from northern California be sent to southern California?
- Should crops like rice be raised in California?
- Aside from lack of rainfall, why do you think southern California is in a water crisis?

Summary

Water is a renewable resource that circulates continually between the atmosphere and the Earth's surface. The energy for the hydrologic cycle is provided by the sun. Water loss from plants is called evapotranspiration. Water that infiltrates the soil and is stored underground in the tiny spaces between rock particles is called groundwater, as opposed to surface water that enters a river system as runoff. There are two basic kinds of aquifers. Unconfined aquifers have an impervious layer at the bottom and receive water that infiltrates from above. The top of the layer of water is called the water table. A confined aquifer is sandwiched between two impervious layers and is often under pressure. The recharge area may be a great distance from where the aquifer is tapped for use. The way in which land is used has a significant impact on rates of evaporation, runoff, and infiltration.

The four human uses of water are domestic, agricultural, in-stream, and industrial. Water use is measured by either the amount withdrawn or the amount consumed. Domestic water is in short supply in many metropolitan areas. Most domestic water is used for waste disposal and washing, with only a small amount used for drinking. The largest consumptive use of water is for agricultural irrigation. Major in-stream uses of water are for hydroelectric power, recreation, and navigation. Most industrial uses of water are for cooling and for dissipating and transporting waste materials.

Major sources of water pollution are municipal sewage, industrial wastes, and agricultural runoff. Nutrients, such as nitrates and phosphates from wastewater treatment plants and agricultural runoff, enrich water and stimulate algae and aquatic plant growth. Organic matter in water requires oxygen for its decomposition and, therefore, has a large biochemical oxygen demand (BOD). Oxygen depletion can result in fish death and changes in the normal algal community, which leads to visual and odor problems.

Point sources of pollution are easy to identify and resolve. Nonpoint sources of pollution, such as agricultural runoff and mine drainage, are more difficult to detect and control than those from municipalities or industries.

Thermal pollution occurs when an industry returns heated water to its source. Temperature changes in water can alter the kinds and numbers of plants and animals that live in it. The methods of controlling thermal pollution include cooling ponds, cooling towers, and dry cooling towers.

Wastewater treatment consists of primary treatment, a physical settling process; secondary treatment, biological degradation of the wastes; and tertiary treatment, chemical treatment to remove specific components. Two major types of secondary wastewater treatments are the trickling filter and the activated sludge sewage methods.

Groundwater pollution comes from a variety of sources, including agriculture, landfills, and septic tanks. Marine oil pollution results from oil drilling and oil-tanker accidents, runoff from streets, improper disposal of lubricating oil from machines and car crankcases, and intentional discharges from oil tankers during loading or unloading.

Reduced water quality can seriously threaten land use and in-place water use. In the United States and other nations, legislation helps to preserve certain scenic water areas and wildlife habitats. Shorelands and wetlands provide valuable services as buffers, filters, reservoirs, and wildlife areas. Water management concerns of growing importance are groundwater mining, increasing salinity, water diversion, and managing urban water use. Urban areas face several problems, such as providing water suitable for human use, collecting and treating wastewater, and handling storm-water runoff in an environmentally sound manner. Water planning involves many governmental layers, which makes effective planning difficult.

Key Terms

activated-sludge sewage treatment *372*
aquiclude *353*
aquifer *352*
aquitard *353*
artesian well *353*
biochemical oxygen demand (BOD) *362*
confined aquifer *353*
domestic water *355*
eutrophication *363*
evapotranspiration *352*
fecal coliform bacteria *364*
groundwater *352*

groundwater mining *375*
hydrologic cycle *352*
industrial water use *358*
in-stream water use *359*
irrigation *357*
limiting factor *363*
nonpoint source *364*
point source *364*
porosity *353*
potable water *351*
primary sewage treatment *371*
runoff *352*

salinization *374*
secondary sewage treatment *372*
sewage sludge *372*
storm-water runoff *369*
tertiary sewage treatment *374*
thermal pollution *367*
trickling filter system *372*
unconfined aquifer *353*
vadose zone *353*
water diversion *370*
water table *353*

Review Questions

1. Describe the hydrologic cycle.
2. Distinguish between withdrawal and consumption of water.
3. What are the similarities between domestic and industrial water use? How are they different from in-stream use?
4. How is land use related to water quality and quantity? Can you provide local examples?
5. What is biochemical oxygen demand? How is it related to water quality?
6. How can the addition of nutrients such as nitrates and phosphates result in a reduction of the amount of dissolved oxygen in the water?
7. Differentiate between point and nonpoint sources of water pollution.
8. How are most industrial wastes disposed of? How has this changed over the past 25 years?
9. What is thermal pollution? How can it be controlled?
10. Describe primary, secondary, and tertiary sewage treatment.
11. What are the types of wastes associated with agriculture?
12. Why is storm-water management more of a problem in an urban area than in a rural area?
13. Define groundwater mining.
14. How does irrigation increase salinity?
15. What are the three major water services provided by metropolitan areas?

Critical Thinking Questions

1. Leakage from freshwater distribution systems accounts for significant losses. Is water so valuable that governments should require systems that minimize leakage to preserve the resource? Under what conditions would you change your evaluation?
2. Do nonfarmers have an interest in how water is used for irrigation? Under what conditions should the general public be involved in making these decisions along with the farmers who are directly involved?
3. Should the United States allow Mexico to have water from the Rio Grande and the Colorado Rivers, both of which originate in the United States and flow to Mexico?
4. Do you believe that large-scale hydroelectric power plants should be promoted as a renewable alternative to power plants that burn fossil fuels? What criteria do you use for this decision?
5. What are the costs and what are the benefits of the proposed Garrison Diversion Unit? What do you think should happen with this project?
6. How might you be able to help save freshwater in your daily life? Would the savings be worth the costs?
7. Look at the hydrologic cycle in figure 16.3. If global warming increases the worldwide temperature, how should increased temperature directly affect the hydrologic cycle?
8. After reading the Issues—Analysis concerning the California Water Plan, do you believe water should be diverted from northern California to southern California?

Concept Map

Construct a map to show relationships among the following concepts:

hydrologic cycle
water distribution
water use

water pollution
groundwater
water diversion

waste water treatment
salinization

Interactive Exploration

Check out the website at **http://www.mhhe.com/environmentalscience** and click on the cover of this textbook for quizzing, career information, case studies, and hot links for the following topics:

Water Use and Management
Water Pollution

Groundwater
Impact of Humans on the Sea: Pollution

PART

V

Pollution and Policy

Environmental Policy: Pragmatic or Polluted?

by Frank X. Phillips

McNeese State University/Louisiana

Discussing pollution and policy is complex due to the dynamic interactions of the social and technical aspects that typically compete for the limited resources needed to resolve a problem or challenge. A viewer's cultural, economic, political, or religious viewpoints all impact on the viewer's interpretation of whether a condition may or may not be a problem. When we view a particular situation as not bothering or impacting us, it isn't a problem. However, if we are affected, our point of view changes, and we're more likely to believe that "something should be done about it."

Environmental policies are often seemingly obvious, printed statements that reasonable people believe to be desirable and beneficial for all affected parties. Policies are generally established by three stakeholder groups: the government regulators, the regulated groups, and the general public. These three entities work together to produce a mutually agreed-upon resolution. Stakeholders negotiate or barter trade-offs, or sometimes compromise some portion of their stake, to achieve a desirable state for the greater good. The system is not always equitable; each party may not get everything they want, but, when done properly, the results are acceptable. This is the "TANSTAAFL Principle" in action (There Ain't No Such Thing as a Free Lunch, [Heinlein, 1966]).

Lake Charles, Louisiana, and its surrounding areas are noted for agriculture, commercial and sports fishing, and petroleum-related industries. These activities and the people who live and work in the area impact the region's water quality. The Calcasieu, a major river that serves the area, has been influenced by human activities for over 150 years. Industrial activities designed to support the war effort increased in the early 1940s—primarily petroleum refining, which produces motor fuels and secondary derivatives. The region's population has also grown over the decades, partially in response to the job opportunities provided by these industries and their support services.

Before the 1970s, the release of wastes to the river's drainage system was not regulated. Since then, wastewater discharges, both industrial and municipal, have required permits and have been monitored, resulting in a significant decrease in waste from these facilities. However, we now know, based on modern data, that the bulk of materials that negatively impact the area's water quality originate from nonindustrial sources such as agriculture, private septic systems, stormwater runoff, public health pest control, and so on. These waste inputs are nonpermitted sources of pollutants that ultimately end up in the river. Nationwide, 40 percent of the assessed waterways are still too polluted to meet the *basic* goal of the Clean Water Act ("fishable and swimmable water"). This figure represents about 21,000 polluted river segments, lakes, and estuaries.

Louisiana (as well as many other states) continues to assess the condition of specific river segments to estimate and plan the most practical approach to return impaired water bodies to being fishable and swimmable. The major causes of pollution are excess sediments, nutrients, and harmful microorganisms. A reasonable amount of scientific evidence indicates that a large percentage of these materials are associated with agricultural operations.

Understandably, the agricultural community takes issue with these recent findings because of the probable changes that will be required to achieve local and regional water quality improvements. As with most things in life, all it will take is time and money! The guiding laws and regulations are already in place. All that remains is to have the stakeholders negotiate how to do what needs to be done. This is where you and I come in—by engaging in the political process to *optimize* the desirable results for all parties affected by the changes, while *minimizing* problems.

Remember TANSTAAFL? We can have "fishable and swimmable waters" in exchange for higher prices in the marketplace: perhaps increased license fees for hunting and fishing, higher household water bills, or additional costs for wastewater treatment systems. Farmers will have to comply with the environmental regulations, and the changes in their operations will be very costly. The costs will need to be shared by all of us. And, in turn, we will all share in the benefits.

What do you think?

1. Your town notifies you and your neighbors that you'll have to hook up to the new municipal wastewater treatment system. You're on individual septic systems now. You will be charged for the service provided by the town. What are the pros and cons of this situation?

2. You don't hunt, fish, boat, or swim. In fact, you really hate water. Why should you have to pay more for your groceries because of the changes required of agricultural producers? Isn't this really just an underhanded, political trick to help subsidize the "water lovers" and "eco-freaks"? Discuss.

Air Quality Issues

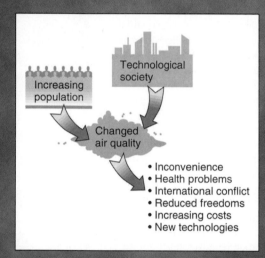

- Increasing population
- Technological society
- Changed air quality
  - Inconvenience
  - Health problems
  - International conflict
  - Reduced freedoms
  - Increasing costs
  - New technologies

Objectives

After reading this chapter, you should be able to:

- Recognize that air can accept and disperse significant amounts of pollutants.
- List the major sources and effects of the six criteria air pollutants.
- Describe how photochemical smog is formed and how it affects humans.
- Explain how acid rain is formed.
- Understand that human activities can alter the atmosphere in such a way that they can change climate.
- Describe the kinds of changes that could occur as a result of global warming.
- Describe the link between chlorofluorocarbon use and ozone depletion.
- Recognize that there are many positive actions that have improved air quality.
- Recognize that enclosed areas can trap air pollutants that are normally diluted in the atmosphere.

Chapter Outline

The Atmosphere

The atmosphere (air) is composed of 78.1 percent nitrogen, 20.9 percent oxygen, and a number of other gases such as argon, carbon dioxide, methane, and water vapor that total about 1 percent. Most of the atmosphere is held close to the Earth by the pull of gravitational force, so it gets less dense with increasing distance from the Earth. Throughout the various layers of the atmosphere, nitrogen and oxygen are the most common gases present, although the molecules are farther apart at higher altitudes. Figure 17.1 shows the various layers of the atmosphere and some of the characteristics of each layer. The atmosphere is divided into four sections. The troposphere extends from the Earth's surface to about 10 kilometers (about 6.2 miles) above the Earth. It actually varies from about 8 to 18 kilometers (5–11 miles), depending on the position of the Earth and the season of the year. The temperature of the troposphere declines by about 6°C (11°F) for every kilometer above the surface. The troposphere contains most of the water vapor of the atmosphere and is the layer in which weather takes place. The stratosphere extends from the top of the troposphere to about 50 kilometers (about 31 miles) and contains most of the ozone. The ozone is in a band between 15 and 30 kilometers (9–19 miles) above the Earth's surface. Because the ozone layer absorbs sunlight, the upper layers of the stratosphere are warmer than the lower layers. The mesosphere is a layer with decreasing temperature from 50 to 80 kilometers (31–50 miles) above the Earth. The thermosphere is a layer with increasing temperature that extends up to about 300 kilometers (186 miles).

Even though gravitational force keeps the majority of the air near the Earth, the air is not static. As it absorbs heat from the Earth, it expands and rises. When its heat content is radiated into space, the air cools, becomes more dense, and flows toward the Earth. As the air circulates due to heating and cooling, it also moves horizontally over the surface of the Earth because the Earth rotates on its axis. The combination of all air movements creates the wind and weather patterns characteristic of different regions of the world. (See figure 17.2.)

As we discussed in chapter 12, pollution is any addition of matter or energy that degrades the environment for humans and other organisms. Because we cause pollution, we may be able to do something to prevent it. There are several natural sources of gases and particles that degrade the quality of the air, including material emitted from volcanoes, dust from wind erosion, and gases from the decomposition of dead plants and animals. Since these events are not controlled by humans, they do not fit our definition of pollution. However, automobile emissions, chemical odors, factory smoke, and similar materials are considered air pollution and will be the focus of this chapter.

The problem of air pollution is directly related to the number of people living in an area and the kinds of activities in which they are involved. When a population is small and its energy use is low, the impact of people is minimal. The pollutants released into the air are diluted, carried away by the wind, washed from the air by rain, or react with oxygen in the air to form harmless materials. Thus, the overall negative effect is slight. However, our urbanized, industrialized civilization has dense concentrations of people that use large quantities of fossil fuels for manufacturing, transportation, and domestic purposes. These activities release large quantities of polluting by-products into our environment.

Gases or small particles released into the atmosphere are likely to stay near the Earth due to gravity. We do not get rid of them; they are just diluted and moved out of the immediate area. When people lived in small groups, the smoke from their fires was diluted. It was produced in such low concentrations that it did not interfere with neighboring groups downwind. In industrialized urban areas, pollutants cannot always be diluted sufficiently before the air reaches another city. The polluted air from

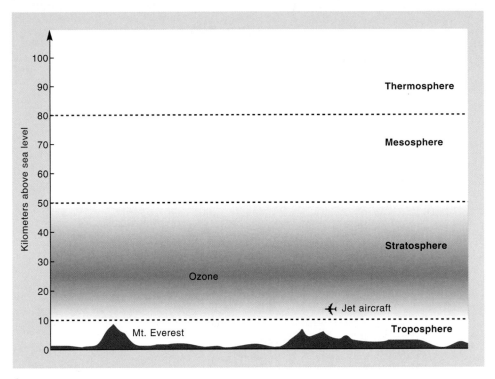

figure 17.1   **The Atmosphere** The atmosphere is divided into the troposphere, the relatively dense layer of gases close to the surface of the Earth; the stratosphere, more distant with similar gases but less dense; the mesosphere; and the thermosphere. Weather takes place in the troposphere, and the important ozone layer is present in the stratosphere.

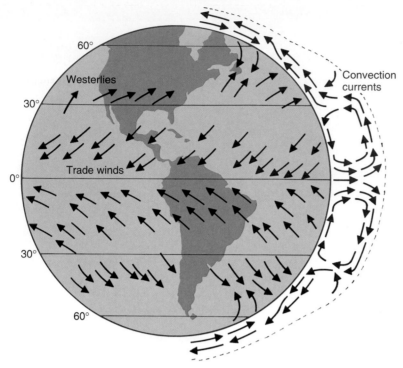

figure 17.2 **Global Wind Patterns** Wind is the movement of air caused by the rotation of the Earth and atmospheric pressure changes brought about by temperature differences. Both of these contribute to the patterns of world air movement. In North America, most of the winds are westerlies (from the west to the east).

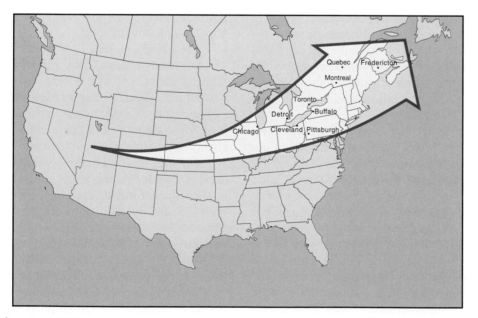

figure 17.3 **Accumulation of Pollutants** As an air mass moves across the continent from west to east, each population center adds its pollutants to the total load in the atmosphere.

Chicago is further polluted when it reaches Gary, Indiana, supplemented by the wastes of Detroit and Cleveland, and finally moves over southeastern Canada and New England to the ocean. (See figure 17.3.) While not every population center adds the same kind or amount of waste, each adds to the total load carried.

Air pollution is not just an aesthetic problem. It also causes health problems. Thousands of deaths have been directly related to poor air quality in cities. A well-documented case of pollution that was harmful to human health occurred in Donora, Pennsylvania. (See figure 17.4.) The city of Donora is located in a valley. In October 1948, the pollutants from a zinc plant and steel mills became trapped in the valley, and a dense smog formed. Within five days, 17 people died and 5910 became ill. The polluted atmosphere affected nearly 50 percent of the city's 12,300 inhabitants. The technology that provided jobs for the people was also killing them.

Many of the megacities of the developing world have extremely poor air quality. Beijing, Seoul, Mexico City, and Cairo exceed World Health Organization guidelines for air quality for at least two pollutants. The causes of this air pollution are open fires, large numbers of poorly maintained motor vehicles, and poorly regulated industrial plants. The World Health Organization estimates that particles in Mexico City contribute to 6400 deaths each year. Not only does poor air quality in such cities increase the death rate, but the general health of the populace is lowered. Chronic coughing and susceptibility to infections are common in these cities. Deaths from air pollution occur primarily among the elderly, the infirm, and the very young. Bronchial inflammations, allergic reactions, and irritation of the mucous membranes of the eyes and nose all indicate that air pollution must be reduced.

Categories of Air Pollutants

Around the world, five major types of substances are released directly into the atmosphere in their unmodified forms in sufficient quantities to pose a health risk and are called **primary air pollutants.** They are carbon monoxide, volatile organic compounds (hydrocarbons), particulate matter, sulfur dioxide, and oxides of nitrogen. Primary air pollutants may interact with one another in the presence of sunlight to form new compounds such as ozone that are known as **secondary air pollutants.** Secondary air pollutants also form from reactions with substances that occur naturally in the atmosphere. In addition, the U.S. Environmental Protection Agency has a category of air pollutants known as criteria air pollutants. **Criteria air pollutants** are those for which specific air quality standards have been set. (See table 17.1.) The criteria air pollutants are carbon monoxide, nitrogen dioxide, ozone, sulfur dioxide, particulate matter, and lead. In addition, certain compounds with high toxicity are known as **hazardous air pollutants** or **air toxics.**

Carbon Monoxide

Carbon monoxide (CO) is produced when organic materials such as gasoline, coal, wood, and trash are burned with insufficient oxygen. When carbon-containing compounds are burned with abundant oxygen present, carbon dioxide is formed ($C + O_2 \rightarrow CO_2$). When the amount of oxygen is restricted, carbon monoxide is formed instead of carbon dioxide ($2C + O_2 \rightarrow 2CO$). Any process that involves the burning of fossil fuels has the potential to produce carbon monoxide. The single largest source of carbon monoxide is the automobile. (See figure 17.5.) About 60 percent of CO comes from vehicles driven on roads and 20 percent comes from vehicles not used on roads. The remainder comes from other processes that involve burning (power plants, industry, burning leaves, etc.). Although increased fuel efficiency and the use of catalytic converters have reduced carbon monoxide emissions per kilometer driven, carbon monoxide remains a problem because the number of automobiles on the road and the number of kilometers driven have risen. In urban areas, as much as 90 percent of carbon monoxide is from motor vehicles. In many parts of the world, automobiles are poorly maintained and may have inoperable pollution control equipment, resulting in even greater amounts of carbon monoxide.

figure 17.4 **Air Pollution** Donora, Pennsylvania, was the scene of a serious air-pollution incident. The pollutants from industry were trapped by an inversion, and almost half of the population was affected.

Table 17.1 National Ambient Air Quality Standards	
Pollutant	**Standard**
Carbon monoxide (CO)	8-hour average (9 ppm)
	1-hour average (35 ppm)
Nitrogen dioxide (NO₂)	Annual mean (0.053 ppm)
Ozone (O₃)	8-hour average (0.08 ppm)
	1-hour average (0.12 ppm)
Lead (Pb)	3-month average (1.5 μg/m³)
Particulate matter (PM₁₀)	Annual mean (50 μg/m³)
	24-hour average (150 μg/m³)
Particulate matter (PM₂.₅)	Annual mean (15 μg/m³)
	24-hour average (65 μg/m³)
Sulfur dioxide (SO₂)	Annual mean (0.03 ppm)
	24-hour average (0.14 ppm)
	3-hour average (0.50 ppm)

Carbon monoxide is dangerous because it binds to the hemoglobin in blood and makes the hemoglobin less able to carry oxygen. Carbon monoxide is most dangerous in enclosed spaces, where it is not diluted by fresh air entering the space. Several hours of exposure to air containing only 0.001 percent of carbon monoxide can cause death. Because carbon monoxide remains attached to hemoglobin for a long time, even small amounts tend to accumulate and reduce the blood's oxygen-carrying capacity. The amount of carbon monox-ide produced in heavy traffic can cause headaches, drowsiness, and blurred vision. Cigarette smoking is also an important source of carbon monoxide because the smoker is inhaling it directly. A heavy smoker in congested traffic is doubly exposed and may experience severely impaired reaction time compared to a nonsmoking driver.

Fortunately, carbon monoxide is not a persistent pollutant. It readily combines with oxygen in the air to form carbon dioxide ($2CO + O_2 \rightarrow 2CO_2$). Therefore, the air can be cleared of its

figure 17.5 **Carbon Monoxide** The major source of carbon monoxide, hydrocarbons, and nitrogen oxides is the internal combustion engine, which is used to provide most of our transportation. The more concentrated the number of automobiles, the more concentrated the pollutants. Carbon monoxide concentrations of a hundred parts per million are not unusual in rush-hour traffic in large metropolitan areas. These concentrations are high enough to cause fatigue, dizziness, and headaches.

carbon monoxide if no new carbon monoxide is introduced into it. Control of carbon monoxide in the United States has been very good. Carbon monoxide levels decreased by about 25 percent between 1970 and 2000, and nearly all communities now meet the standards set by the EPA. This was accomplished with a variety of controls on industry and, in particular, on motor vehicles. Catalytic converters reduce the amount of carbon monoxide released by vehicle engines, and specially formulated fuels that produce less carbon monoxide are used in many cities that have a carbon monoxide problem. Often these special fuels are only used in winter, when car engines run less efficiently and produce more carbon monoxide.

Volatile Organic Compounds

In addition to carbon monoxide, automobiles emit a variety of **volatile organic compounds (VOCs).** Volatile organic compounds are composed primarily of carbon and hydrogen, so they are often referred to as **hydrocarbons.** If they evaporate into the air, they are volatile; hence, airborne organic compounds are volatile organic compounds. They are either evaporated from fuel supplies or are remnants of fuel that did not burn completely. The use of internal combustion engines accounts for the majority of volatile organic compounds released into the air, although refineries and other industries add to the total atmospheric burden. Consumer products such as oil-based paint, charcoal lighter, and many other chemicals are also important. Some of these compounds are toxic and are known as hazardous air pollutants. They are also important because they contribute to the production of the secondary air pollutants found in smog. Smog will be discussed in the Ground-Level Ozone and Photochemical Smog section of this chapter.

Many modifications to automobiles have significantly reduced the amount of volatile organic compounds entering the atmosphere. Recycling some gases through the engine so they burn rather than escape, increasing the proportion of oxygen in the fuel-air mixture to obtain more complete burning of the fuel, and using devices to prevent the escape of gases from the fuel tank and crankcase are three of these modifications. In addition, catalytic converters allow unburned organic compounds in exhaust gases to be oxidized more completely so that fewer volatile organic compounds leave the tail pipe. Because of the many modifications to automobiles and other restrictions, the quantity of volatile organic compounds decreased by 43 percent between 1970 and 2000 in the United States.

Particulate Matter

Particulate matter consists of minute (10 microns and smaller) solid particles and liquid droplets dispersed into the atmosphere. A micron is one millionth of a meter. Many bacteria are about 1 micron in diameter. The U.S. Environmental Protection Agency has set standards for particles smaller than 10 microns (**PM$_{10}$**) and 2.5 microns (**PM$_{2.5}$**). Most of the coarse particles (greater than 2.5 microns) are primary pollutants that are released directly into the air. Travel on roads, agricultural activities, construction sites, industrial processes, and smoke particles from fires are the primary sources of coarse particles. Fine particles (2.5 microns or less) are mostly secondary pollutants that form in the atmosphere from interactions of primary pollutants. Sulfates and nitrates formed from sulfur dioxide and nitrogen oxides are examples. Particulates cause problems ranging from the annoyance of reduced visibility at national parks or soot settling on a backyard picnic table to the **carcinogenic** (cancer-causing) effects of asbestos. Particulates frequently get attention because the coarse particles are so readily detected by the public. Heavy black smoke from a factory or a view obscured by dust can be observed without expensive monitoring equipment and generally causes an outcry.

Particles can accumulate in the lungs and interfere with their ability to exchange gases. Such lung damage usually occurs in people who are repeatedly exposed to large amounts of particulate

Global Perspective

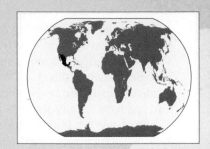

Air Pollution in Mexico City

Mexico City has been labeled the city with the worst air pollution ever recorded. The air over Mexico City exceeded ozone limits set by the World Health Organization on more than 300 days in one year. With about 20 million people, Mexico City is one of the largest cities in the world and grows larger each day. Open sewers and garbage dumps contribute dust and bacteria to the atmosphere. The city has about 35,000 factories and 3.6 million vehicles. Most of these vehicles are older models that are poorly tuned and, therefore, pollute the air with a mixture of hydrocarbons, carbon monoxide, and nitrogen oxides. A large proportion of the air pollution is the result of automobiles. A major source of hydrocarbons is from leakage of LP gas. Fixing the leaks would significantly reduce the hydrocarbon level in the atmosphere. The high altitude of over 2000 meters (6500 feet) results in even greater air pollution from automobiles because automobile engines do not burn fuel efficiently at such high altitudes. Mexico City's location in a valley also allows for conditions suitable for thermal inversions during the winter.

Many foreign companies and governments give employees special "hazard pay" for working in Mexico City because of the air pollution. Pediatricians estimate that 85 percent of childhood illnesses are related to air pollution and say that the only way to improve the health of many of the children is to get them out of the city. The government has responded with a comprehensive program to clean up the air in the city. Twenty-five million trees have been planted to help clean the air, and 4400 hectares (11,000 acres) of land have been purchased to provide green space for the city.

Public information campaigns encourage people to keep their automobiles tuned to reduce air pollution, lead-free gasoline has been made available and is being used, and catalytic converters are now required on all automobiles manufactured after 1991. The use of taxis manufactured before 1985 has been banned, and by 2002, all pre-1991 taxis must be replaced.

Personal automobiles and taxis are prohibited from being driven on the streets one day a week. This encourages people to carpool or use public transportation. The government is planning to improve public transportation to make it more attractive for people to switch from private automobiles to public transport. Both actions should reduce the number of automobiles releasing pollutants into the air. A polluting, government-owned oil refinery was shut down, and power plants and many industries have switched from oil to natural gas, which pollutes less.

These actions have had significant effects on the quality of the air. Although ozone in the air continues to be a major problem, as it is in metropolitan areas around the world, lead levels, carbon monoxide, and sulfur dioxide have been reduced. To further improve air quality, increasingly strong restrictions on polluting industries and the use of private automobiles will be necessary. The Mexican government is considering such actions.

---

matter on the job. Miners and others who work in dusty conditions are most likely to be affected. Droplets and solid particles can also serve as centers for the deposition of other substances from the atmosphere. As we breathe air containing particulates, we come in contact with concentrations of other potentially more harmful materials that have accumulated on the particulates. Sulfuric, nitric, and carbonic acids, which irritate the lining of our respiratory system, frequently are associated with particulates.

The amount of PM_{10} pollution has decreased by about 88 percent between 1970 and 2000. The U.S. Environmental Protection Agency has been setting standards for $PM_{2.5}$ particles for a shorter period, and pollution by these particles decreased by only about 6 percent between 1991 and 2000.

Sulfur Dioxide

Sulfur dioxide (SO_2) is a compound of sulfur and oxygen that is produced when sulfur-containing fossil fuels are burned ($S + O_2 \rightarrow SO_2$). There is sulfur in coal and oil because they were produced from the bodies of organisms that had sulfur as a component of some of their molecules. The sulfur combines with oxygen to form sulfur dioxide when fossil fuels are burned. Today, the major source of sulfur dioxide is from the burning of coal in power plants.

Sulfur dioxide has a sharp odor, irritates respiratory tissue, and aggravates asthmatic and other respiratory conditions. It also reacts with water, oxygen, and other materials in the air to form sulfur-containing acids. The acids can become attached to particles that, when inhaled, are very corrosive to lung tissue. These acid-containing particles are also involved in acid deposition, which is discussed in the Acid Deposition section of this chapter.

In 1306, Edward I of England banned the burning of "sea coles," coal found on the seashore, in London. This coal contained sulfur and was, in part, responsible for the city's poor air quality. Edward's ban might very well be the earliest environmental legislation concerning air quality. London was also the site of one of the earliest killer fogs. In 1952, the city was covered with a dense fog for several days. During this time, air over the city failed to mix with layers of air in the upper atmosphere due to temperature conditions. The factories continued to release smoke and dust into this stagnant layer of air, and it became so full of the fog, smoke, and dust that people got lost in familiar surroundings. This combination of smoke and fog has become known as smog. Many London residents developed respiratory discomfort, headaches, and nausea as a result of breathing this air. Four thousand people died in a few weeks. Their deaths have been associated with the high levels of sulfur compounds in the smog. Thousands of others suffered from severe bronchial irritation, sore throats, and chest pains. The 1948 Donora, Pennsylvania, incident already mentioned also involved symptoms related to the high concentration of particles and sulfur dioxide in the air. Because the major sources of sulfur dioxide are power plants and they are easily monitored, much has been done to reduce the amount of sulfur dioxide released. In the United States between 1970 and 2000, SO_2 levels decreased by 44 percent.

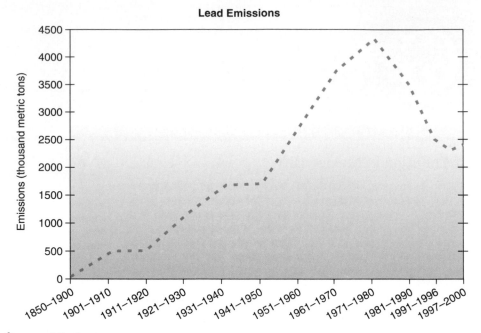

figure 17.6 **Lead Emissions** The amount of lead released into the atmosphere has declined significantly since the early 1980s, when the lead additive was removed from gasolines in North America and much of Europe.

Source: U.S. Environmental Protective Agency, Office of Air and Radiation.

Nitrogen Dioxide

The burning of fossil fuels produces a mixture of nitrogen-containing compounds commonly known as **oxides of nitrogen (NO_X).** These compounds are formed because the nitrogen and oxygen molecules in the air combine with one another when subjected to the high temperatures experienced during combustion. The two most common molecules are **nitrogen monoxide (NO)** and **nitrogen dioxide (NO_2).** The primary molecule produced is nitrogen monoxide ($N_2 + O_2 \rightarrow 2NO$), but nitrogen monoxide can be converted to nitrogen dioxide in the air ($2NO + O_2 \rightarrow 2NO_2$) to produce a mixture of NO and NO_2. Thus, NO_2 is, for the most part, a secondary pollutant. Nitrogen dioxide is a reddish brown, highly reactive gas that is responsible for much of the haze seen over cities, causes respiratory problems, and is a component of acid precipitation.

Nitrogen dioxide also is important in the production of the mixture of secondary air pollutants known as photochemical smog, which is discussed in the section on Ground-Level Ozone and Photochemical Smog in this chapter.

The primary source of nitrogen oxides is the automobile engine. Catalytic converters significantly reduce the amount of nitrogen monoxide released from the internal combustion engine. (About 75 percent of the NO produced by an automobile engine is converted back into N_2 and O_2 by the car's catalytic converter.) However, the increase in the numbers of cars and kilometers driven has resulted in an increase of NO_X levels of about 20 percent in the United States between 1971 and 2000, and NO_2 is a significant pollutant.

Lead

Lead (Pb) can enter the body through breathing airborne particles or consuming lead that was deposited on surfaces. Lead accumulates in the body and causes a variety of health effects, including mental retardation and kidney damage. At one time, the primary source of airborne lead was from additives in gasoline. Lead was added to gasoline to help engines run more effectively. Recognition that lead emissions were hazardous resulted in the lead additives being removed from gasoline in North America and Europe. Many other countries in the world, however, still use leaded gasoline but are moving toward eliminating lead. Since leaded gasoline has been eliminated in much of the developed world, lead levels have fallen. In North America, lead levels fell by 98 percent between 1970 and 2000, and they have fallen by about 50 percent throughout the world. (See figure 17.6.)

Another major source of lead is paints. Many older homes have paints that contain lead, since various lead compounds are colorful pigments. Dust from flaking paint, remodeling, or demolition is released into the atmosphere. Although the amount of lead may be small, its presence in the home can result in significant exposure to inhabitants, particularly young children who chew on painted surfaces and often eat paint chips. Today, however, the major sources of lead are from industrial sources where metals are smelted or batteries are manufactured.

Ground-Level Ozone and Photochemical Smog

Ozone (O_3) is a molecule that consists of three oxygen atoms bonded to one another. Ozone is an extremely reactive

figure 17.7 **Photochemical Smog** The interaction among hydrocarbons, oxides of nitrogen, and sunlight produces new compounds that are irritants to humans. The visual impact of smog is shown in these photographs taken in Los Angeles, California.

molecule that irritates respiratory tissues and can cause permanent lung damage. It also damages plants and reduces agricultural yields. Ozone is a secondary pollutant that is formed as a component of photochemical smog.

Photochemical smog is a mixture of pollutants including ozone, aldehydes, and peroxyacetyl nitrates resulting from the interaction of nitrogen dioxide and volatile organic compounds with sunlight in a warm environment. (See figure 17.7.) The two most destructive components of photochemical smog are ozone and peroxyacetyl nitrates.

Both of these secondary pollutants are excellent oxidizing agents that will react readily with many other compounds, including those found in living things, causing destructive changes. Ozone is particularly harmful because it destroys chlorophyll in plants and injures lung tissue in humans and other animals. Peroxyacetyl nitrates, in addition to being oxidizing agents, are eye irritants.

For photochemical smog to develop, several ingredients are required. Nitrogen monoxide, nitrogen dioxide, and volatile organic compounds must be present, and sunlight and warm tempera-

tures are important to support the chemical reactions involved. The chemical reactions that cause the development of photochemical smog are quite complicated, but several key reactions will help you understand the process. In most urban areas, nitrogen monoxide, nitrogen dioxide, and volatile organic compounds are present as a result of cars and industrial processes. In the presence of sunlight, nitrogen dioxide breaks down to nitrogen monoxide and atomic oxygen ($NO_2 \xrightarrow{sunlight} NO + O^*$). Atomic oxygen is extremely reactive and will react with molecular oxygen in the air to form ozone ($O^* + O_2 \rightarrow O_3$).

Both ozone and atomic oxygen will react with volatile organic compounds to produce very reactive organic free radicals (O^* or O_3 + organic molecule $\rightarrow$ organic free radical). Free radicals are very reactive and cause the formation of additional nitrogen dioxide from nitrogen monoxide (NO + free radical $\rightarrow NO_2$ + other organic molecules). This step is important because the presence of additional NO_2 results in the further production of ozone. The organic free radicals also react with nitrogen dioxide to form peroxyacetyl nitrates and aldehydes. (Figure 17.8 depicts these events.) The development of photochemical smog in an area involves the interaction of climate, time of day, and motor vehicle emissions in the following manner.

During morning rush-hour traffic, the amounts of nitrogen monoxide and volatile organic compounds increase. This leads to an increase in the amount of NO_2, and NO levels fall as NO is converted to NO_2. During this same time, ozone levels rise as NO_2 is broken down by sunlight, and levels of peroxyacetyl nitrates rise as well. As the sun sets, production of ozone lessens. In addition, ozone and other smog components react with their surroundings and are destroyed, so smog conditions improve in the evening. Figure 17.9 shows how the concentration of these various molecules changes during the day.

While smog can develop in any area, some cities have greater problems because of their climate, traffic, and geographic features. Cities that are located

Source	Necessary conditions	Reactions take place in atmosphere		Products
Primarily automobiles	volatile organic compounds (VOC) present	VOC + O* or O₃ → highly reactive organic radicals + NO₂		→ Peroxyacetyl nitrates Aldehydes
Primarily automobiles	Nitrogen monoxide (NO) present	NO + organic radicals → NO₂		
From automobiles and formed from NO	Nitrogen dioxide (NO₂) present	NO₂ → NO + O*(atomic oxygen) O* + O₂ → O₃ (ozone) sunlight		→ Ozone
Sun	Sunlight			
Sun (summer temperatures)	Heat	Reactions take place more rapidly at higher temperatures.		

figure 17.8 **Major Steps in the Development of Photochemical Smog**
Photochemical smog develops when specific reactants and conditions are present. The necessary reactants are volatile organic compounds, nitrogen monoxide, and nitrogen dioxide. The conditions that cause these compounds to react are warm temperatures and the presence of sunlight. When these conditions exist, ozone and other components of smog are formed as secondary pollutants.

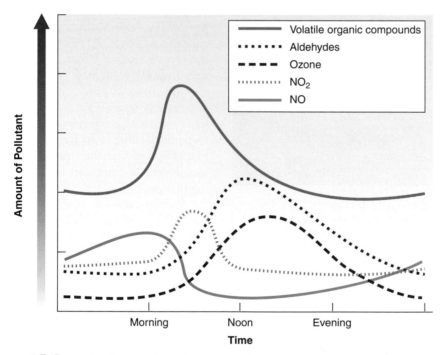

figure 17.9 **Daily Changes in Pollutants During a Photochemical Smog Incident**
The development of photochemical smog begins with a release of nitrogen oxides and volatile organic compounds associated with automobile use during morning traffic. As the sun rises and the day warms up, these reactants interact to form ozone and other secondary pollutants. These peak during the early afternoon and decline as the sun sets.

adjacent to mountain ranges or in valleys have greater problems because the pollutants can be trapped by thermal inversions. Normally, air is warmer at the surface of the Earth and gets cooler at higher altitudes. In some instances, a layer of warmer air may be above a layer of cooler air at the Earth's surface. This condition is known as a **thermal inversion.** In the case of cities like Los Angeles that have mountains to the east and the ocean to the west, cool air from the ocean may push in under a layer of warm air and create a thermal inversion. In cities located in valleys, as the surface of the Earth cools at night, the cooler air on the sides of the valley can flow down into the valley and create a similar situation. As cool air flows into these valleys, it pushes the warm air upward. In either case, the warm air becomes sandwiched between two layers of cold air and acts like a lid on the valley. (See figure 17.10.) The lid of warm air cannot rise because it is covered by a layer of cooler air pushing down on it. It cannot move out of the area because of the mountains. Without normal air circulation, smog accumulates. Harmful chemicals continue to increase in concentration until a major weather change causes the lid of warm air to rise and move over the mountains. Then the underlying cool air can begin to circulate, and the polluted air can be diluted.

Smog problems could be substantially decreased by reducing the NO_X and VOCs associated with the use of internal combustion engines (perhaps eliminating them completely) or by moving population centers away from the valleys where thermal inversions occur. Currently, major advances have been made in reducing the smog problem by reformulating gasoline and installing devices on automobiles that reduce the amount of NO_X and VOCs released.

Hazardous Air Pollutants

While all criteria air pollutants already discussed cause health concerns, hundreds of other chemical compounds are purposely or accidentally released into the air that can cause harm to human health or damage the environment. These compounds are collectively known as

Normal Situation

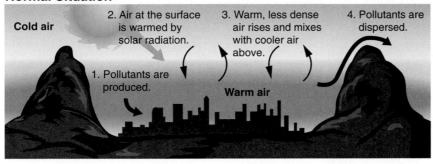

Cold air

2. Air at the surface is warmed by solar radiation.

3. Warm, less dense air rises and mixes with cooler air above.

4. Pollutants are dispersed.

1. Pollutants are produced.

Warm air

a.

Thermal Inversion

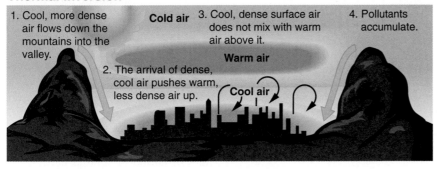

1. Cool, more dense air flows down the mountains into the valley.

Cold air

3. Cool, dense surface air does not mix with warm air above it.

4. Pollutants accumulate.

Warm air

2. The arrival of dense, cool air pushes warm, less dense air up.

Cool air

b.

Thermal Inversion

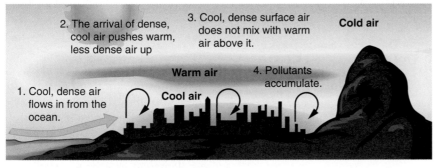

2. The arrival of dense, cool air pushes warm, less dense air up

3. Cool, dense surface air does not mix with warm air above it.

Cold air

Warm air

4. Pollutants accumulate.

1. Cool, dense air flows in from the ocean.

Cool air

c.

figure 17.10   **Thermal Inversion** Under normal conditions, (*a*) the air at the Earth's surface is heated by the sun and rises to mix with the cooler air above it. When a thermal inversion occurs, (*b* and *c*) a layer of heavier cool air flows into a valley and pushes the warmer air up. The heavy cooler air is then unable to mix with the less dense warm air above and cannot escape because of surrounding mountains. The cool air is trapped, sometimes for several days, and accumulates pollutants. If the thermal inversion continues, the levels of pollution can become dangerously high.

hazardous air pollutants (HAP) or **air toxics.** Materials such as pesticides are purposely released to kill insects or other pests. Some are released as a result of consumer activities. Benzene in gasoline escapes when gasoline is put into the tank, and the use of some consumer products such as glues and cleaners releases toxic materials into the air. The majority of air toxics, however, are released as a result of manufacturing processes. Per-

chloroethylene is released from dry cleaning establishments, and toxic metals are released from smelters. The chemical and petroleum industries are the primary sources of hazardous air pollutants. Although air toxics are harmful to the entire public, their presence is most serious for people who are exposed on the job, since they are likely to be exposed to higher concentrations of hazardous substances over longer time periods. Chapter 19

deals with hazardous materials and the issues related to regulating their production, use, and disposal.

Control of Air Pollution

All of the air pollutants we have examined thus far are produced by humans. That means their release into the atmosphere can be controlled. Methods of controlling air pollution depend on the type of pollutant and the willingness or ability of industries, governments, and individuals to make changes. It is important to note that positive change is possible. (See figure 17.11.)

Control of Motor Vehicle Emissions

Motor vehicles are the primary source of several important air pollutants: carbon monoxide, volatile organic compounds, and nitrogen oxides. In addition, ozone is a secondary pollutant of automobile use. Placing controls on the emissions from motor vehicles has resulted in a significant improvement in the air quality in North America. The positive crankcase ventilation valve and gas caps with air-pollution control valves reduce volatile organic compound loss. Better fuel efficiency and specially blended fuels that produce less carbon monoxide and unburned organic compounds have improved air quality. Catalytic converters reduce carbon monoxide, oxides of nitrogen, and volatile organic compounds in emissions and necessitated the use of lead-free fuel. This lead-free fuel requirement, in turn, significantly reduces the amount of lead (and other metal additives) in the atmosphere. Figure 17.11 shows that levels of volatile organic compounds, carbon monoxide, and lead are down significantly. The levels of oxides of nitrogen have risen, but that is the result of more vehicles and more kilometers driven. The actual quantity of nitrogen oxides produced per vehicle is down. In addition, ozone levels decreased by about 12 percent between 1981 and 2000. Table 17.2 shows

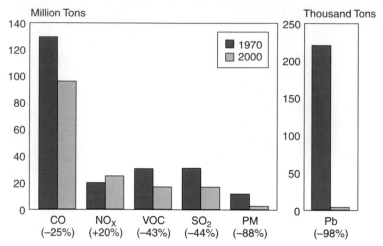

Comparison of 1970 and 2000 Emissions

figure 17.11 **Improvement in Air Quality** The many initiatives to improve air quality have been effective. In the past 30 years, nearly all of the major air pollutants have decreased significantly. A major exception is nitrogen oxides (NO$_x$), which have increased. However, even this is a partial success since the amount of NO$_x$ released per vehicle has gone down. But more vehicles and greater distances driven have resulted in an increase in the total NO$_x$ produced.

Source: Environmental Protection Agency.

Table 17.2 Top 10 Cities for Smog Improvement in United States from 1987 to 2000

	Percent Reduction	Smog Days 1987–1989	Smog Days 1994–2000
1. Seattle	100.0	14.0	0.0
2. Buffalo	100.0	7.7	0.0
3. Tucson	100.0	4.0	0.0
4. Tacoma	95.5	7.3	0.3
5. Albuquerque	93.2	14.7	1.0
6. Denver	92.5	22.3	2.1
7. Miami	92.3	4.3	0.3
8. Boston	91.7	8.0	0.7
9. Raleigh-Durham	90.9	3.7	0.5
10. Charlotte	88.2	11.3	1.6

Source: Data from the U.S. Environmental Protection Agency.

that the smog problem in cities has significantly improved.

Control of Particulate Matter Emissions

Particulate matter comes from a variety of sources. Many industrial activities involve processes that produce dusts. Mining and other earth-moving activities, farming operations, and the transfer of grain or coal from one container to another all produce dust.

Improper land use can also be a major source of airborne particles. Over the past several years, satellites have tracked large amounts of dust and industrial pollution rapidly moving across the Pacific toward North America. In 1997, researchers at the Cheeka Peak Observatory in the Olympic Peninsula in Washington measured carbon monoxide levels that were 10 percent higher than average and fine particulate levels that were 50 percent higher than average. In 2000, large volumes of dust from the

Gobi Desert in Mongolia and other Asian deserts and adjacent overgrazed regions traveled in a cloud across the Pacific, reaching as far east as Texas. In Seattle, Vancouver, and other West Coast cities, the sky turned a visible milky white from the presence of dust particles.

Burning of fossil fuels is another major source of particulate matter. Because of air quality regulations, industries have done much to reduce the amount of particulate matter released from the burning of fossil fuels. Various kinds of devices can be used to trap particles so they do not escape from the smokestack. They are very effective, and particulate matter from burning of fossil fuels is greatly reduced. However, the smaller particles that form from gaseous emissions (SO$_2$, NO$_x$) are still a problem. Diesel engines are a significant source of particulate matter, and the gaseous emissions from motor vehicles contribute to the formation of smaller particles in the same way that industrial sources do.

While industrial activities, motor vehicle use, and land-use practices are major sources of particulate matter, many individual personal activities are also important. Many people in the world use wood as their primary source of fuel for cooking and heating. In developed countries like the United States and Canada, some people use fireplaces and wood-burning stoves as a primary source of heat, but most use them for supplemental heat or for aesthetic purposes. However, the use of large numbers of wood-burning stoves and fireplaces can generate a significant air-pollution problem, called a brown cloud. Many municipalities, such as Boise, Idaho; Salt Lake City, Utah; and Denver, Colorado (and some entire states), impose fines to enforce a ban on wood-burning during severe air-pollution episodes. Many other communities issue pollution alerts and request that people do not use wood-burners. Many of these communities have regulations governing the number and efficiency of wood-burning stoves and fireplaces. Some communities, such as Castle Rock, Colorado, prohibit the construction of houses with fireplaces or wood-burning stoves. Many people readily switch to gas fireplaces or high-efficiency wood-burners when they understand that their enjoyment of a rustic stove may be leading to a degraded environment, but most need to be forced to comply by official regulations and the threat of fines. Obviously, accumulations of many small sources of air pollution may cause as big a problem as one large emission source and are frequently more difficult to control.

Control of Power Plant Emissions

To control sulfur dioxide, which is produced primarily by electric-power generating plants, several possibilities are available. One alternative is to change from high-sulfur fuel to low-sulfur fuel. Switching from a high-sulfur to a low-sulfur coal reduces the amount of sulfur released into the atmosphere by 66 percent. Switching to oil, natural gas, or nuclear fuels would reduce sulfur dioxide emissions even more. However, these are not long-term solutions because low-sulfur fuels are in short supply, and nuclear power plants pose a different set of pollution problems. (See chapter 11.)

A second alternative is to remove the sulfur from the fuel before the fuel is used. Chemical or physical treatment of coal before it is burned can remove nearly 40 percent of the sulfur. This is technically possible, but it increases the cost of electricity to the rate payer.

Scrubbing the gases emitted from a smokestack is a third alternative. The technology is available, but, of course, these control devices are costly to install, maintain, and operate. As with auto emissions, governments have required the installation of these devices, but when industries install them, the cost of construction and operation is passed on to the consumer. The cost of installing scrubbers on a typical power plant is about $200 million. In the United States, an estimated 20 million or more tonnes (25 million U.S. tons) of sulfur oxides are released into the atmosphere each year. Installing scrubbers on just 50 of the largest coal-burning plants would reduce this amount by over one-third.

In the past, a common solution was to build taller smokestacks. Tall stacks release their gases above the inversion layers and, therefore, add the sulfur dioxide to the upper atmosphere, where it is diluted before it comes in contact with the population downwind. This works as long as the stack is tall enough and as long as not so much pollution is added to the air that it cannot be diluted to an acceptable level. Taller smokestacks do not reduce pollutants; they simply make pollutants disperse downwind. However, sulfur dioxide reacts with oxygen and dissolves in water in the atmosphere to form sulfuric acid. This acid is washed from the air when it rains or snows. It damages plants and animals and increases corrosion of building materials and metal surfaces.

Clean Air Act

In the United States, implementation of the requirements of the Clean Air Act has been the primary means of controlling air pollution. Initially enacted in 1967 as the Air Quality Act and extensively amended in 1970, 1977, and 1990, the current version of the Clean Air Act established a series of detailed control requirements that the federal government implements and the states administer.

1. All new and existing sources of air pollution are subject to ambient air quality regulation.
2. New sources are subject to more stringent control technology and permitting requirements.
3. Hazardous air emissions and pollution that affects visibility are regulated.
4. A comprehensive operating permit program was added with the 1990 amendments.

National ambient air quality standards (NAAQS) were established for six pollutants (criteria pollutants) that are widespread pollution problems.

1. Sulfur dioxide (SO_2)
2. Nitrogen dioxide (NO_2)
3. Particulate matter
4. Carbon monoxide (CO)
5. Ozone
6. Lead

The states have primary responsibility for ensuring that air quality is maintained at a level consistent with the NAAQS. State Implementation Plans (SIPs) must include enforceable emission limits, an enforcement program, protection against interstate air pollution, monitoring and emission data requirements, and preconstruction review and notification requirements.

Under the 1990 amendments, 189 substances are regulated, including hazardous organic chemicals and metals. Any source emitting 9.1 tonnes (10 U.S. tons) per year of any listed substance or 22.7 tonnes (25 U.S. tons) combined is considered a major source. With regard to acid rain, the 1990 amendments added new regulatory programs to control sulfur dioxide and nitrogen oxides emissions. The 1990 amendments established a program for the phaseout of ozone-depleting substances (CFCs, halons, carbon tetrachloride, and methyl chloroform).

In 2000, the EPA released a study of how effective the Clean Air Act has been. While stating that there is still considerable progress to be made, the report highlighted the successes of the act. Since passage of the Clean Air Act, the EPA report stated, air pollution has been cut by a third and acid rain by 25 percent. Perhaps most impressive of all: emissions of the six worst air pollutants dropped 33 percent from 1970 to 2000, despite a 31 percent increase in U.S. population, a 114 percent rise in productivity, and a 127 percent jump in the number of kilometers driven by Americans. The EPA estimates that the Clean Air Act's human health, welfare, and environmental benefits have outweighed its costs by 40 to 1. But problems remain. The EPA has targeted two major sources of continuing air pollution: the aging coal-fired power plants in the Midwest and motor vehicles such as sport utility vehicles, diesel trucks, and buses. A report issued by the U.S. Public Interest Research Group in 2000 said that even with the Clean Air Act, smog levels exceeded federal standards in 43 states in 1999. Smog helps cause more than 6 million asthma attacks and sends some 160,000 people to hospital emergency rooms each year.

Acid Deposition

Acid deposition is the accumulation of potential acid-forming particles on a surface. The acid-forming particles can be dissolved in rain, snow, or fog or can be deposited as dry particles. When dry particles are deposited, an acid does not actually form until these materials mix with water. Even though the acids are formed and deposited in different ways, all of these sources of acid-forming particles are commonly referred to as **acid rain.** Acids result from natural causes, such as vegetation, volcanoes, and lightning, and from human activities, such as the burning of coal and use of the internal combustion engine. (See figure 17.12.) These combustion processes produce sulfur dioxide (SO_2) and oxides of nitrogen (NO_X). Oxidizing agents, such as ozone, hydroxide ions, or hydrogen peroxide, along with water, are necessary to convert the sulfur dioxide or nitrogen oxides to sulfuric or nitric acid.

Acid rain is a worldwide problem. Reports of high acid-rain damage have come from Canada, England, Germany, France, Scandinavia, and the United States. Rain is normally slightly acidic, (pH between 5.6 and 5.7) since atmospheric carbon dioxide dissolves in water to produce carbonic acid. But acid rains sometimes have a concentration of acid a thousand times higher than normal. In 1969, New Hampshire had a rain with a pH of 2.1. In 1974, Scotland had a rain with a pH of 2.4. The normal rain in much of the northeastern part of the United States and parts of Ontario has a pH between 4.0 and 4.5.

Acid rain can cause damage in several ways. Buildings and monuments are often made from materials that contain limestone (calcium carbonate, $CaCO_3$), because limestone is relatively soft and easy to work. Sulfuric acid (H_2SO_4), a major component of acid rain, converts limestone to gypsum ($CaSO_4$), which is more soluble than calcium carbonate and is eroded over many years of contact with acid rain. (See figure 17.13.) Metal surfaces can also be attacked by acid rain.

The effects of acid rain on ecosystems are often subtle and difficult to quantify. However, in many parts of the world, acid rain is suspected of causing

the death of many forests and reducing the vigor and rate of growth of others. (See figure 17.14.) In central Europe, many forests have declined significantly, resulting in the death of about 6 million hectares (14.8 million acres) of trees. Northeastern North America has been affected with significant tree death and reduction in vigor, particularly at higher elevations. Some areas have had 50 percent mortality of red spruce trees.

A strong link can be established between the decline of the forests and acid rain. Sulfur dioxide and oxides of nitrogen are the primary molecules that contribute to acid rain. The deposition of acids causes major changes to the soils in areas where the soils are not able to buffer the additional acid. As soil becomes acidic, aluminum is released from binding sites and becomes part of the soil water, where it interferes with the ability of plant roots to absorb nutrients. A recent long-term study in New Hampshire strongly suggests that the many years of acid precipitation have reduced the amount of calcium and magnesium in the soil, which are essential

figure 17.13 **Damage Due to Acid Deposition** Sulfuric acid (H_2SO_4), which is a major component of acid deposition, reacts with limestone ($CaCO_3$) to form gypsum ($CaSO_4$). Since gypsum is water soluble, it washes away with rain. The damage to this monument is the result of such acid reacting with the stone.

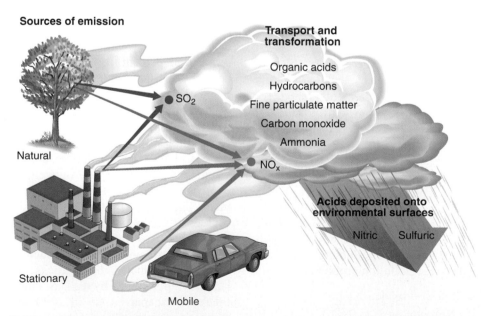

figure 17.12 **Acid Deposition** Molecules from natural sources, power plants, and internal combustion engines react to produce the chemicals that are the source of acid deposition.

for plant growth. Because there are no easy ways to replace the calcium even if acid rain were to stop, it would still take many years for the forests to return to health. Reduction in the pH of the soil may also change the kinds of bacteria in the soil and reduce the availability of nutrients for plants. While none of these factors alone would necessarily result in tree death, each could add to the stresses on the plant and may allow other factors, such as insect infestations, extreme weather conditions (particularly at high elevations), or drought, to further weaken trees and ultimately cause their death.

The effects of acid rain on aquatic ecosystems are much more clear-cut. In several experiments, lakes were purposely made acidic and the changes in the ecosystems recorded. The experiments showed that as lakes became more acidic, there is a progressive loss of many kinds of organisms. The food web becomes less complicated, many organisms fail to reproduce, and many others die. Most healthy lakes have a pH above 6. At a pH of 5.5, many desirable species of fish are eliminated; at a pH of 5, only a few starving fish may be found, and none are able to reproduce. Lakes with a pH of 4.5 are nearly sterile.

Several reasons account for these changes. Many of the early developmental stages of insects and fish are more sensitive to acid conditions than are the adults. In addition, the young often live in shallow water, which is most affected by a flood of acid into lakes and rivers during the spring snowmelt. The snow and its acids have accumulated over the winter, and the snowmelt releases large amounts of acid all at once. Crayfish and other crustaceans need calcium to form their external skeleton. As the pH of the water decreases, the crayfish are unable to form new exoskeletons and so they die. Reduced calcium availability also results in the development of some fish with malformed skeletons. (See figure 17.15.) As mentioned earlier, increased acidity also results in the release of aluminum, which impairs the function of a fish's gills.

About 14,000 lakes in Canada and 11,000 in the United States have been seriously altered by becoming acidic. Many lakes in Scandinavia are similarly affected. The extent to which acid deposition affects an ecosystem depends on the nature of the bedrock in the area and the ecosystem's proximity to acid-forming pollution sources. (See figure 17.16.) Soils derived from igneous rock are not capable of buffering the effects of acid deposition, while soils derived from sedimentary rocks such as limestone release bases that neutralize the effects of acids. Because of this, eastern

figure 17.14 **Forest Decline** Many forests, particularly at high elevations in northeastern North America, have shown significant decline, and dead trees are common.

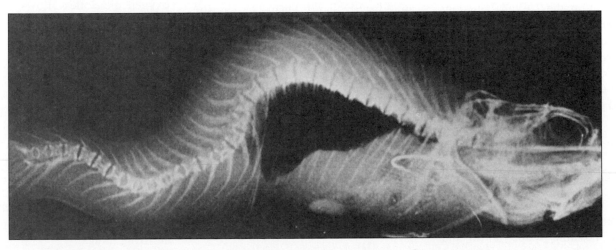

figure 17.15 **Effects of Acid Deposition on Organisms** The low pH of the water in which this fish lived caused the abnormal bone development that ultimately resulted in the death of the fish.

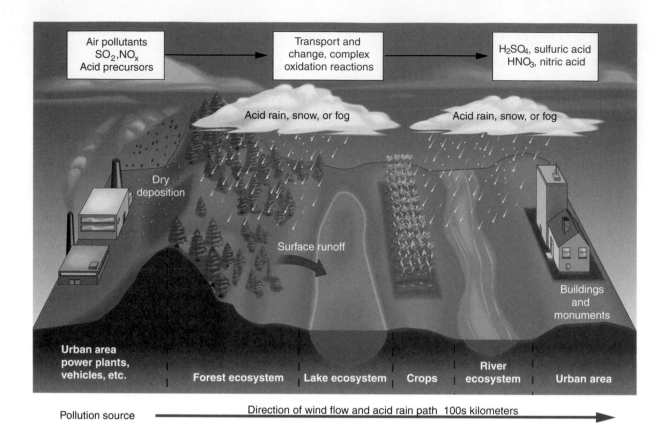

Air pollutants
SO₂, NOₓ
Acid precursors

→

Transport and
change, complex
oxidation reactions

→

H₂SO₄, sulfuric acid
HNO₃, nitric acid

Acid rain, snow, or fog

Acid rain, snow, or fog

Dry
deposition

Surface runoff

Buildings
and
monuments

Urban area
power plants,
vehicles, etc.

Forest ecosystem | Lake ecosystem | Crops | River
ecosystem | Urban area

Pollution source

Direction of wind flow and acid rain path 100s kilometers

figure 17.16 **Factors That Contribute to Acid Rain Damage** In any aquatic ecosystem, the following factors increase the risk of damage from acid deposition: (1) location downwind from a major source of pollution; (2) hard, insoluble bedrock with a thin layer of infertile soil in the watershed; (3) low buffering capacity in the soil of the watershed; (4) a low lake surface area to watershed ratio.

Source: Data from U.S. Environmental Protection Agency, *Acid Rain.*

Canada and the U.S. Northeast are particularly susceptible to acid rain. These areas have a high proportion of granite rock and are downwind from the major air-pollution sources of North America. Scandinavian countries have a similar geology and receive pollution from industrial areas in the United Kingdom and Europe. Thousands of kilometers of streams and up to 200,000 lakes in eastern Canada and the northeastern United States are estimated to be in danger of becoming acidified because of their location and geology.

Ozone Depletion

In the 1970s, various sectors of the scientific community became concerned about the possible reduction in the ozone layer in the Earth's upper atmosphere. **Ozone** is a molecule of three

atoms of oxygen (O_3). In 1985, it was discovered that a significant thinning of the ozone layer over the Antarctic occurred during the Southern Hemisphere spring. Some regions of the ozone layer showed 95 percent depletion. Ozone depletion also has been found to be occurring farther north than previously. Measurements in Arctic regions suggest a thinning of the ozone layer there also. These findings caused several countries to become involved in efforts to protect the ozone layer.

The ozone in the outer layers of the atmosphere, approximately 15 to 35 kilometers (9–21 miles) from the Earth's surface, shields the Earth from the harmful effects of ultraviolet light radiation. Ozone absorbs ultraviolet light and is split into an oxygen molecule and an oxygen atom:

$$O_3 \xrightarrow{\text{Ultraviolet light}} O_2 + O$$

Oxygen molecules are also split by ultraviolet light to form oxygen atoms:

$$O_2 \xrightarrow{\text{Ultraviolet light}} 2O$$

Recombination of oxygen atoms and oxygen molecules allows ozone to be formed again and to be available to absorb more ultraviolet light.

$$O_2 + O \rightarrow O_3$$

This series of reactions results in the absorption of 99 percent of the ultraviolet light energy coming from the sun and prevents it from reaching the Earth's surface. Less ozone in the upper atmosphere would result in more ultraviolet light reaching the Earth's surface, causing increased skin cancers and cataracts in humans and increased mutations in all living things.

Chlorofluorocarbons are strongly implicated in the ozone reduction in the

Secondhand Smoke

Smoking of tobacco has long been associated with a variety of respiratory diseases, including lung cancer, emphysema, and heart disease. Because of these strong links, tobacco products carry warning labels. Many nonsmoking people, however, are exposed to environmental tobacco smoke (secondhand smoke) because they live and work in spaces where people smoke. The U.S. Environmental Protection Agency estimates that approximately 3000 nonsmokers die each year of lung cancer as a result of breathing air that contains secondhand smoke. Young children who are exposed to secondhand smoke are much more likely to have respiratory infections.

In July 1993, the U.S. Environmental Protection Agency recommended several actions to prevent people from being exposed to secondhand indoor smoke, including:

- People not smoke in their homes or permit others to do so
- All organizations that deal with children have policies that protect children from secondhand smoke
- Every company have a policy that protects employees from secondhand smoke
- Smoking areas in restaurants and bars be placed so that the smoke will have little chance of coming into contact with nonsmokers

California has some of the most restrictive smoking laws in the United States. In 1998, California passed a no-smoking law that effectively banned all smoking in indoor public places, including restaurants and bars. The law was designed for employees who did not want to inhale secondhand smoke. While there are no no-smoking police going from building to building if an employee complains, the owner of the establishment could pay a fine of up to $7,000.

The chart lists the various compounds and their toxicity found in secondhand smoke. Note that secondhand smoke is also referred to as environmental tobacco smoke.

Compound	Type of Toxicity
Vapor phase	
Carbon monoxide	T
Carbonyl sulfide	T
Benzene	c
Formaldehyde	c
3-Vinylpyridine	SC
Hydrogen cyanide	T
Hydrazine	c
Nitrogen oxides	T
N-nitrosodimethylamine	C
N-nitrosopyrrolidine	C
Particulate phase	
Tar	C
Nicotine	T
Phenol	TP
Catechol	CoC
o-Toluidine	C
2-Naphthylamine	C
4-Aminobiphenyl	C
Benz(a)anthracene	C
Benzo(a)pyrene	C
Quinoline	C
N'-nitrosonornicotine	C
NNK	C
N-nitrosodiethanolamine	C
Cadmium	C
Nickel	C
Polonium-210	C

Abbreviations: C, carcinogenic; CoC, cocarcinogenic; SC, suspected carcinogen; T, toxic; TP, tumor promoter. Environmental tobacco smoke (ETS) is diluted in the air before it is inhaled and thus is less concentrated than mainstream smoke (MS). However, active inhalation of MS is limited to the time it takes to smoke each cigarette, whereas exposure to ETS is constant over the period spent in the ETS-polluted environment. This fact is reflected in measurements of nicotine uptake by smokers and ETS-exposed nonsmokers.

Sources: Data from Department of Health and Human Services, 1989; State of California, *Report on Secondhand Smoke*, 1999.

upper atmosphere. Chlorine reacts with ozone in the following way to reduce the quantity of ozone present:

$$Cl + O_3 \rightarrow ClO + O_2$$
$$ClO + O \rightarrow Cl + O_2$$

These reactions both destroy ozone and reduce the likelihood that it will be formed because atomic oxygen (O) is removed as well. It is also important to note that it can take 10 to 20 years for chlorofluorocarbon molecules to get into the stratosphere, and then they can react with the ozone for up to 120 years. As a result, the World Meteorological Association predicts ozone depletion will worsen before improvements are seen.

Since the 1970s, when chlorofluorocarbons were linked to the depletion of the ozone layer in the upper atmosphere, their use as propellants in aerosol cans has been banned in the United States, Canada, Norway, and Sweden, and the European Union agreed to reduce use of chlorofluorocarbons in aerosol cans. In other parts of the world, however, chlorofluorocarbons are still widely used as aerosol propellants.

In 1987, several industrialized countries, including Canada, the United States, the United Kingdom, Sweden, Norway, Netherlands, the Soviet Union, and West Germany, agreed to freeze chlorofluorocarbon production at current levels and reduce production by 50 percent by the year 2000. This document, known as the Montreal Protocol, was ratified by the U.S. Senate in 1988. Although the initial concerns related to the problem of ozone depletion, efforts to reduce chlorofluorocarbons have been effective at removing a greenhouse gas as well. As a result of the 1987 Montreal Protocol, chlorofluorocarbon emissions dropped 87 percent from their peak in 1988. In 1990, in London, international agreements were reached to further reduce the use of chlorofluorocarbons. A major barrier to these negotiations was the reluctance of the developed countries of the world to establish a fund to help less-developed countries implement technologies that would allow them to obtain refrigeration and air conditioning

without the use of chlorofluorocarbons. In 1991, DuPont announced the development of new refrigerants that would not harm the ozone layer. These and other alternative refrigerants are now used in refrigerators and air conditioners in many nations, including the United States. In 1996, the United States stopped producing chlorofluorocarbons. As a result of these international efforts and rapid changes in technology, the use of chlorofluorocarbons has dropped rapidly, and concentrations of chlorofluorocarbons in the atmosphere will slowly fall over the next few decades.

Global Warming and Climate Change

In recent years, scientists suspected that the average temperature of the Earth was increasing and looked for causes for the change. It is clear that the Earth has had changes in its average temperature many times in the geologic past before humans were present. So, scientists initially tried to determine if the warming was a natural phenomenon or the result of human activity. Several gases such as carbon dioxide, chlorofluorocarbons, methane, and nitrous oxide are known as **greenhouse gases** because they let sunlight enter the atmosphere but slow the loss of heat from the Earth's surface. Evidence of past climate change going back as far as 160,000 years indicates a close correlation between the concentration of greenhouse gases in the atmosphere and global temperatures. Computer simulations of climate indicate that global temperatures will rise as atmospheric concentrations of greenhouse gases increase, and there are many other effects predicted by an increase in temperature. Since these predictions are based on computer models of climate, some scientists have criticized them as being inaccurate and constructed from sketchy data. As additional, more accurate information is gathered and inserted into the models, however, the general conclusions remain the same.

It is difficult for the general public to relate to these changes or see evidence

of them since each of us experiences only our own local weather and climate, and we observe many short-term changes in weather patterns that seem greater than the predictions generated by the computer models. The models, however, are attempting to predict the long-term trends for large regions of the Earth. Since the potential consequences of a global warming are so great, many are suggesting we alter our lifestyles regardless of whether the phenomenon is true, just to be on the safe side.

Because major disagreements arose over the significance of global warming, the United Nations Environment Programme established an Intergovernmental Panel on Climate Change (IPCC) to study the issue and make recommendations. Its *First Assessment* was published in 1990. In 1996, the IPCC published its *Second Assessment* and concluded that climate change is occurring and that it is highly probable that human activity is an important cause of the change. The IPCC has reached several important conclusions:

1. The average temperature of the Earth has increased 0.3 to 0.6°C (0.5–1.0°F), (1998 was the warmest year on record), and sea level has risen 10 to 25 centimeters (4–10 inches) in the last 100 years. (See figure 17.17.)

2. A strong correlation exists between the increase in temperature and the concentration of greenhouse gases in the atmosphere.

3. Human activity greatly increases the amounts of these greenhouse gases.

Causes of Global Warming and Climate Change

What actually causes global warming? An explanation is relatively straightforward. Several gases in the atmosphere are transparent to ultraviolet and visible light but absorb infrared radiation. These gases allow sunlight to penetrate the atmosphere and be absorbed by the Earth's surface. This sunlight energy is reradiated as infrared radiation (heat), which is absorbed by the greenhouse gases in the atmosphere. Because the effect is similar

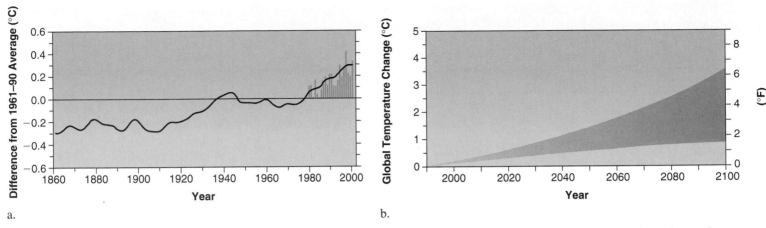

a.

b.

figure 17.17 **Changes in Average Global Temperature** (*a*) Despite considerable variation, there has been a general trend toward increasing temperatures: 1998 was the warmest year on record. (*b*) The possible range of globally averaged surface temperature increase is shown for the period 1990 to 2100.

Source: (*a*) Data from "U.K. Meteorological Office/University of East Anglia," *Science,* 12 January 1999 and annual updates; (*b*) Data from "Global Temperature Change and Range of Globally Averaged Sea Level," in *Climate Changes—State of Knowledge,* Office of Science and Technology Policy, 1997, Washington, D.C.; and *State of the World* 2000.

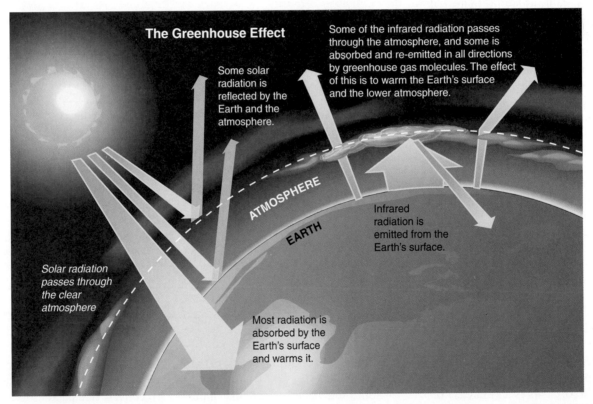

figure 17.18 **Greenhouse Effect** The greenhouse effect naturally warms the Earth's surface. Without it, Earth would be 33°C (60°F) cooler than it is today—uninhabitable for life as we know it.

Source: Data from *Climate Change—State of Knowledge.* October 1997, Office of Science and Technology Policy, Washington D.C.

to what happens in a greenhouse (the glass allows light to enter but retards the loss of heat), these gases are called greenhouse gases, and the warming thought to occur from their increase is called the **greenhouse effect.** (See figure 17.18.) The most important greenhouse gases are carbon dioxide (CO_2), chloro-fluorocarbons (primarily CCl_3F and CCl_2F_2), methane (CH_4), and nitrous ox-ide (N_2O). Table 17.3 lists the relative contribution of each of these gases to the potential for global warming.

Carbon dioxide (CO_2) is the most abundant of the greenhouse gases. It

Table 17.3 Major Greenhouse Gases and Their Characteristics

Gas	Atmospheric Concentration (ppm)	Annual Increase (Percent)	Life Span (Years)	Contribution to Global Warming (Percent)	Principal Sources
Carbon dioxide	369.4	0.3	50–200	67	Coal, oil, natural gas, deforestation
Chlorofluorocarbons	0.00088	stable	50–102	9	Foams, aerosols, refrigerants, solvents
Methane	1.75	stable	12–17	18	Wetlands, rice fields, fossil fuels, livestock
Nitrous Oxide	0.316	0.25	120	6	Fossil fuels, fertilizers, deforestation

Source: Climate Monitoring and Diagnostics Lab, U.S. Department of Commerce.

occurs as a natural consequence of respiration. However, much larger quantities are put into the atmosphere as a waste product of energy production. Coal, oil, natural gas, and biomass are all burned to provide heat and electricity for industrial processes, home heating, and cooking.

Another factor contributing to the increase in the concentration of carbon dioxide in the atmosphere is deforestation. Trees and other vegetation remove carbon dioxide from the air and use it for photosynthesis. Since trees live for a long time, they effectively tie up carbon in their structure. Cutting down trees to convert forested land to other uses releases this carbon, and a reduction in the amount of forest lessens its ability to remove carbon dioxide from the atmosphere. The combination of these factors (fossil-fuel burning and deforestation) has resulted in an increase in the concentration of carbon dioxide in the atmosphere. Measurement of carbon dioxide levels at the Mauna Loa Observatory in Hawaii show that the carbon dioxide level has increased from about 315 parts per million (ppm) in 1958 to about 370 ppm in 2001. (See figures 17.19 and 17.20.) This is an increase of over 17 percent.

Chlorofluorocarbons (CFCs) are entirely the result of human activity. They were widely used as refrigerant gases in refrigerators and air conditioners, as cleaning solvents, as propellants in aerosol containers, and as expanders in foam products. Although they are present in the atmosphere in minute quantities, they are extremely efficient as greenhouse gases (about 15,000 times

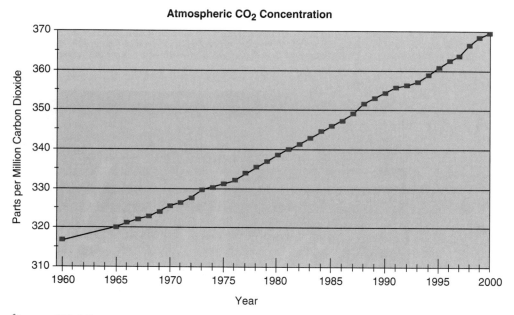

Atmospheric CO₂ Concentration

figure 17.19   **Change in Atmospheric Carbon Dioxide** Since the establishment of a carbon dioxide monitoring station at Mauna Loa Observatory in Hawaii, a steady increase in carbon dioxide levels has been observed. Since 1958, the concentration of carbon dioxide in the atmosphere has increased by about 17 percent.

Source: Data from Scripps Institution of Oceanography.

more efficient at retarding heat loss than is carbon dioxide).

Some methane enters the atmosphere from fossil-fuel sources, but the majority comes primarily from biological sources. Several kinds of bacteria that are particularly abundant in wetlands and rice fields release methane into the atmosphere. Methane-releasing bacteria are also found in large numbers in the guts of termites and various kinds of ruminant animals such as cattle. Control of methane sources is unlikely since the primary sources involve agricultural practices that would be very difficult to change. For example, nations would have to convert rice paddies to other forms of agriculture and drastically reduce the number of animals used for meat production. Neither is likely to occur, since food production in most parts of the world needs to be increased, not decreased.

Nitrous oxide, a minor component of the greenhouse gas picture, enters the atmosphere primarily from fossil fuels and fertilizers. It could be reduced by more careful use of nitrogen-containing fertilizers.

Surface Air Warming

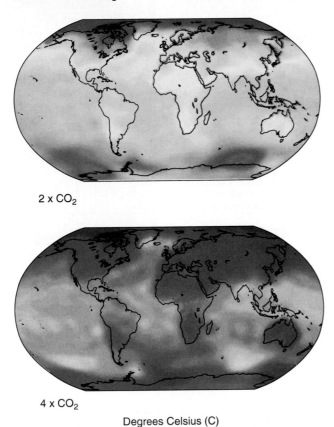

2 x CO$_2$

4 x CO$_2$

Degrees Celsius (C)

-3   0   3   6   9   12   15

figure 17.20   **Predicted Surface Air Temperatures**   The maps show the predicted increases in surface air temperature with a doubling or quadrupling of carbon dioxide levels from preindustrial levels.

Source: Geophysical Fluid Dynamics Laboratory (NOAA).

Potential Consequences of Global Warming and Climate Change

It is important to recognize that although a small increase in the average temperature of the Earth may seem trivial, this increase could set in motion changes that could significantly alter the climate of major regions of the world. Computer models suggest that rising temperature will lead to increased incidences of severe weather and changes in rainfall patterns that would result in more rain in some areas and drought in others. These models suggest that the magnitude and rate of change will differ from region to region. Furthermore, some natural ecosystems or human settlements will be able to withstand or adapt to the changes, while others will not.

Poorer nations are generally more vulnerable to the consequences of global warming. These nations tend to be more dependent on climate-sensitive sectors, such as subsistence agriculture, and lack the resources to buffer themselves against the changes that global warming may bring. The Intergovernmental Panel on Climate Change has identified Africa as "the continent most vulnerable to the impacts of projected changes because widespread poverty limits adaptation capabilities."

In addition to changes in weather, there are many other potential consequences of warmer temperatures and changes in climate. (See figure 17.21.) These include rising sea levels, disruption of the water cycle, potential health concerns, changing forests and natural areas, and challenges to agriculture and the food supply.

Rising Sea Level

A warmer Earth would result in rising sea levels for two different reasons. When water increases in temperature, it expands and takes up more space. In addition, a warming of the Earth would result in the melting of glaciers, which would add more water to the oceans. Rising sea level erodes beaches and coastal wetland, inundates low-lying areas, and increases the vulnerability of coastal areas to flooding from storm surges and intense rainfall. By 2100, sea level is expected to rise by 15 to 90 centimeters (6–35 inches). (See figure 17.22.) A 50-centimeter (20-inch) sea-level rise will result in substantial loss of coastal land in North America, especially along the southern Atlantic and Gulf coasts, which are subsiding and are particularly vulnerable. Many coastal cities would be significantly affected by an increase in sea level. The land area of some island nations and countries such as Bangladesh would change dramatically as flooding occurred. The oceans will continue to expand for several centuries after temperatures stabilize. Because of this, the sea-level rise associated with carbon dioxide levels of 550 ppm (double preindustrial levels) could eventually exceed 100 centimeters (39.4 inches). A CO$_2$ level of 1100 ppm could produce a sea-level rise of 200 centimeters (78.7 inches) or even more, depending on the extent to which the Greenland and Antarctic ice sheets melt.

A 50-centimeter (20-inch) sea-level rise would double the global population at risk from storm surges, from roughly 45 million at present to over 90 million people, and this figure does not account for any increases in coastal populations. A 100-centimeter (39.4 inch) rise would triple the size of the population at risk.

Disruption of the Water Cycle

Among the most fundamental effects of climate change is disruption of the water cycle. Rising temperatures are expected

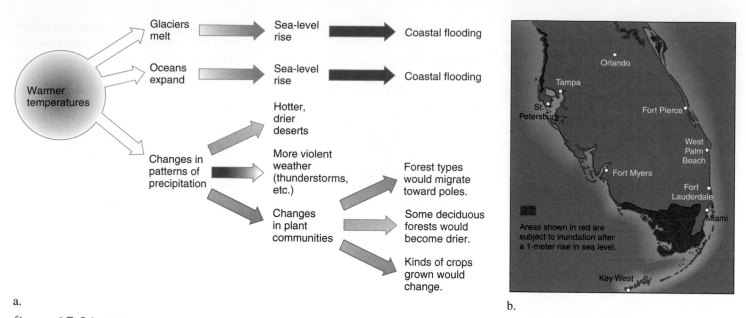

a.

b.

figure 17.21 **Effects of Global Warming** (*a*) Global warming would have several effects on the climate of the world. The climate changes would have important impacts on human and other living things. (*b*) Sea-level rise could inundate many low-lying coastal areas in Florida, and will increase the vulnerability of all such areas to storm surges.

Source: (*b*) Elevations from U.S. Geological Survey digital data. Prepared by the USGS, 2000.

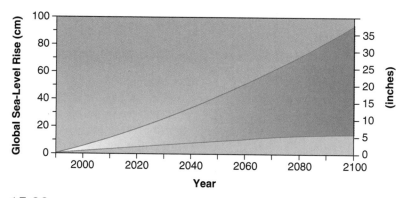

figure 17.22 **Possible Sea-Level Rise** The possible range of globally averaged sea-level rise is shown for the period 1990 to 2100.

Source: Data from *Climate Change—State of Knowledge,* October 1997, Office of Science and Technology Policy, Washington, D.C., and *State of the World* 2000.

to result in increased evaporation, which will cause some areas to become drier, while the increased moisture in the air will result in greater rainfall in other areas. This is expected to cause droughts in some areas and flooding in others. In those areas where evaporation increases more than precipitation, soil will become drier, lake levels will drop, and rivers will carry less water. Lower river flows and lake levels could impair navigation, hydroelectric power generation, and water quality, and reduce the sup-

plies of water available for agricultural, residential, and industrial uses.

Some areas may experience increased flooding during winter and spring, as well as lower supplies during summer. In California's Central Valley, for example, melting snow provides much of the summer water supply; warmer temperatures would cause the snow to melt earlier and thus reduce summer supplies, even if rainfall increased during the spring. More generally, the tendency for rainfall to be more con-

centrated in large storms as temperatures rise would tend to increase river flooding, without increasing the amount of water available. It is very difficult to predict the effects of changes in the water cycle, but several concerns have been raised.

Navigation: Climate change could impair navigation by affecting average water levels in rivers and lakes, increasing the frequency of both floods (during which navigation is hazardous) and droughts (during which passage is difficult), and necessitating changes in navigational infrastructure. On the other hand, warmer temperatures could extend the ice-free season in many parts of the world.

Hydropower: Hydropower depends on the flow of water in rivers. Increases in the amount of water flowing down a river would be beneficial, and decreases would be harmful.

Water supply and demand: In some parts of the world, the most widely discussed potential impact of climate change is the effect on water supply and demand. The potential changes in water supplies would result directly from the changes in runoff and the levels of rivers, lakes, and aquifers.

Flood control: While the impacts of sea-level rise and associated coastal flooding have been widely discussed, global climate change could also change the frequency and severity of inland flooding, particularly along rivers.

Environmental quality and recreation: Decreased river flows and higher temperatures could harm the water quality of rivers, bays, and lakes. In areas where river flows decrease, pollution concentrations will rise because there will be less water to dilute the pollutants. Increased frequency of severe rainstorms could boost the amount of chemicals that run off from farms, lawns, and streets into rivers, lakes, and bays.

Political issues: The areas of the world that currently are experiencing problems with water quantity and quality are likely to see these problems become more severe. This is particularly true in the arid and semiarid regions of the world. Water scarcity in the Middle East and Africa is likely to be aggravated by climate change, which could increase tensions among countries that depend on water supplies originating outside their borders.

Health Effects

Climate change will impact human health in a variety of ways. The most direct effect of climate change would be the impacts of hotter temperatures. Extremely hot temperatures increase the number of people who die (of various causes) on a given day. For example, people with heart problems are vulnerable because the cardiovascular system must work harder to keep the body cool during hot weather. Heat exhaustion and some respiratory problems increase. For example, in July 1995, more than 700 deaths in Chicago were attributed to a heat wave during which temperatures exceeded 32°C (90°F) day and night. Carbon dioxide concentrations of 550 ppm (double preindustrial levels) could cause such heat wave events to occur six times more frequently. The potential increases in the heat index, a calculation combining temperature and humidity, illustrate the magnitude of this threat. Washington, D.C., currently has an average July heat index of 30°C (86°F), but if CO_2 levels reach 550 ppm, this could increase to 35°C (95°F), and if concentrations quadrupled to 1100 ppm, it could increase to 43°C (110°F).

Climate change will also aggravate air-quality problems. Higher air temperature increases the concentration of ozone at ground level, which leads to injury of lung tissue and intensifies the effects of airborne pollen and spores that cause respiratory disease, asthma, and allergic disorders. Because children and the elderly are the most vulnerable, they are likely to suffer disproportionately with both warmer temperatures and poorer air quality.

Throughout the world, the prevalence of particular diseases depends largely on local climate. Several serious diseases appear only in warm areas. As the Earth becomes warmer, some of these tropical diseases may be able to spread to parts of the world where they do not currently occur. Diseases that are spread by mosquitoes and other insects could become more prevalent if warmer temperatures enabled those insects to become established farther north. Such "vector-borne" diseases include malaria, dengue fever, yellow fever, and encephalitis. Some scientists believe that algal blooms could occur more frequently as temperatures rise, particularly in areas with polluted waters, in which case outbreaks of diseases such as cholera that tend to accompany algal blooms could become more frequent.

Changing Forests and Natural Areas

Some of the most dramatic projections regarding global warming involve natural systems. Climate change could dramatically alter the geographic distributions of vegetation types. The composition of one-third of the Earth's forests could undergo major changes as a result of climate changes associated with a carbon dioxide level of 700 ppm. Over the next 100 years, the ideal range for some North American forest species could shift by as much as 500 kilometers (300 miles) to the north, far faster than the forests can migrate naturally. (See figure 17.23.) For example, sugar maples—long a fixture in the northeastern United States—and the Everglades system in Florida both disappear in modeling forecasts. Wetlands and coral reefs may also undergo radical decline due to climate change. The wetlands of the prairie pothole region of Minnesota, North Dakota, South Dakota, and southern Manitoba, Saskatchewan, and Alberta, which support half the waterfowl population of North America, could diminish in area and change dramatically in character in response to climate change. These kinds of changes would result in an accelerated loss of species and an additional challenge to efforts to protect biological diversity. The more gradually climate changes, the easier it could be for species to adapt. As a result, ecologists have proposed that emission constraints be calculated so as to limit the rate of warming to no more than a tenth of a degree Celsius (0.18°F) per decade.

Challenges to Agriculture and the Food Supply

Climate strongly affects crop yields. A carbon dioxide concentration of 550 ppm is likely to increase crop yields in some areas by as much as 30 percent to 40 percent, but it will decrease yields in other places by similar amounts, even for the same crop. A warmer climate would reduce flexibility in crop distribution and increase irrigation demands. Expansion of the ranges of pests could also increase vulnerability and result in greater use of pesticides. Despite these effects, total global food production is not expected to be altered substantially by climate change, but negative regional impacts are likely. Agricultural systems in the developed countries are highly adaptable and can probably cope with the expected range of climate changes without dramatic reductions in yields. It is the poorest countries, where many already are subject to hunger, that are the most likely to suffer significant decreases in agricultural productivity.

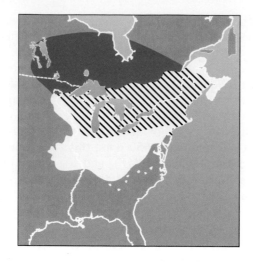

Present range

Overlap

Predicted range

Prediction based on increased temperature

Prediction based on increased temperature and moisture reduction

figure 17.23 **Climatic Shifts** Climatic shifts will force some species to migrate northwards or to higher elevations in order to stay in the appropriate climatic zone. The climatic zone for sugar maple, for example, could shift northward into Canada. This would compromise the maple syrup industry and the fall foliage colors, both of which make New England famous.

Source: Data from the U.S. Environmental Protection Agency and United Nations.

Unanticipated Changes

The greatest risks may be those yet to be discovered. Just as the Antarctic ozone hole was a surprise, scientists have hypothesized many troubling possibilities, including more frequent or severe hurricanes and a shift in ocean currents responsible for moderating the climate of northern Europe. As we learned in the context of ozone depletion, what we don't know can hurt us—the rate of ozone depletion proved to be much greater than what had been predicted. The most serious effects of climate change also may lie outside current calculation. The evolution of a common international scientific understanding of such complex issues has been a critical step toward addressing the problem, just as with the problem of ozone depletion.

Addressing Climate Change

Approaches to dealing with climate change involve technological change coupled with political will and economic realities. A major step toward

slowing global warming would be to increase the efficiency of energy utilization. More efficient use of fuels conserves the shrinking supplies of energy resources. It makes sense to increase energy efficiency, thus reducing carbon dioxide production, even if global warming is not a concern. One way to stimulate a move toward greater efficiency would be to place a tax on the amount of carbon individuals and corporations release into the atmosphere. This would increase the cost of fuels and the demand for fuel efficiency because the cost of fuel would rise. It would also stimulate the development of alternative fuels with a lower carbon content and generate funds for research in many aspects of fuel efficiency and alternative fuel technologies.

Another approach to the problem is to increase the amount of carbon dioxide removed from the atmosphere. If enough biomass is present, the excess carbon dioxide can be used by vegetation during photosynthesis, thereby reducing the impact of carbon dioxide released by fossil-fuel burning. Australia, the United States, and several other countries have announced plans to plant billions of trees to help remove carbon dioxide from the atmosphere.

Many critics argue that this approach will provide only a short-term benefit since, eventually, the trees will mature and die, and their decay will release carbon dioxide into the atmosphere at some later time.

An associated concern is the destruction of vast areas of rainforest in tropical regions of the world. These ecosystems are extremely efficient at removing carbon dioxide and storing the carbon atoms in plant structures. The burning of tropical rainforests to provide farm or grazing land not only adds carbon dioxide to the atmosphere, but it also reduces their ability to remove carbon dioxide from the atmosphere, since the grasslands or farms created do not remove carbon dioxide as efficiently as do the rainforests. Furthermore, the grazing lands and farms in such regions of the world are often abandoned after a few years and do not return to their original forest condition.

With the recognition that chlorofluorocarbons were destroying the stratospheric ozone layer that protects us from ultraviolet radiation, international agreements were reached that have led to a sharp reduction in the amount of chlorofluorocarbons being released. Changes made to protect the ozone layer, therefore, have had the side benefit of reducing the release of a potent greenhouse gas. See the section on ozone depletion in this chapter.

As experience protecting the ozone layer demonstrates, technological change can be very difficult to initiate but extremely rapid once broad consensus is reached. Once a new direction has been firmly agreed upon, the economic incentives are to move as quickly as possible. Few firms seek to be the last left producing a banned substance or failing to comply with a new regulation.

The incentive for change is reinforced by the globalization of the economy, which encourages manufacture for the widest possible market acceptance. Thus, many developing nations that were allowed to continue using chlorofluorocarbons beyond the time allowed more developed nations quickly discovered that this grace period was a mixed blessing. If they continued to produce products that contained chlorofluorocarbons, they were excluded from the largest export markets. Multinational firms reinforced the trend by requiring that products they purchased be "chlorofluorocarbon-free." Financial assistance to countries through the Ozone Multilateral Fund further encouraged a more rapid transition.

It will be more difficult to achieve a similar result with carbon dioxide because it is released as a by-product of manufacturing and is not part of the product itself. However, the fundamental benefit of cleaner and more energy-efficient production potentially could encourage changes to industry, energy production, and transportation that would result in lower carbon dioxide releases. The policy challenge is to find different paths to the same end, channeling the enormous creativity and resources of the private sector to the search for carbon alternatives.

The initial challenge of reducing greenhouse gas emissions is as much political as economic, as much about how we organize our economies as about how much we consume. The Intergovernmental Panel on Climate Change and numerous national reports have shown that there is no shortage of cost-effective

Table 17.4	Technologies to Address Climate Change

1. Fuel cell-powered cars and buses that substantially reduce local pollution, greenhouse gas emissions, and oil consumption
2. Microturbines (building-sized power plants) with lower costs and equivalent efficiency to current-scale coal generation
3. Oxy-fuel firing (combustion using oxygen rather than air) for manufacture of glass, steel, aluminum, and metal casting, reducing energy use up to 45 percent and dramatically reducing local emissions
4. "Zero energy" buildings that minimize energy requirements and produce more on-site energy (e.g., through photovoltaic rooftiles) than is purchased
5. High-strength, lightweight materials that allow reduced costs and improved efficiency in all transportation modes
6. Coal gasification combined with CO_2 capture and sequestration to achieve high efficiency, clean burning of coal with near-zero carbon emissions
7. Wind power systems combined with compressed air energy storage at a total cost competitive with coal generation
8. Generation of electricity directly from sunlight through photovoltaics integrated with roofing materials at a cost competitive with central station power

Source: Energy Research and Development Panel of the President's Committee of Advisors on Science and Technology, *Federal Energy Research and Development for the Challenges of the Twenty-First Century.*

technologies for achieving sizable reductions in greenhouse gas emissions. For example, the Intergovernmental Panel on Climate Change estimates that the potential is great to reduce greenhouse gases by making changes to buildings that conserve energy. These opportunities exist today in virtually all nations, including developing economies and nations that are already relatively energy efficient. For example, China's energy intensity (the ratio of commercial energy consumption per unit of gross national product) has fallen 50 percent since 1980, an unprecedented rate over such a period. (China remains four times as energy intensive as the United States, however, making it among the world's least energy-efficient economies.) An important reason for China's progress is the recognition that its continued economic success requires the gradual elimination of fossil-fuel subsidies, a philosophy increasingly accepted by many developing nations.

Reductions in greenhouse gas emissions are likely to bring considerable ancillary benefits, potentially offsetting much of the costs. For example, the World Bank estimated that in 1999, air pollution in China, primarily attributable to poorly controlled burning of coal, caused 200,000 premature deaths, 1.8 million cases of chronic bronchitis, 1.7 billion restricted-activity days, and

more than 5 billion cases of respiratory illness. Each of these statistics can be converted into monetary terms so that costs of improving fuel efficiency and reducing pollution can be offset by lower health care costs and higher worker productivity.

Improved energy efficiency also reduces the need for new power plants and related energy infrastructure, now estimated at about $100 billion annually in developing nations.

Many candidates for this longer-term technological transition have been identified, and some have already begun to penetrate the market. For example, wind energy systems are improving rapidly and, in ideal conditions, already compare favorably with conventional coal-burning power plants. Direct conversion of sunlight to electricity is now possible with photovoltaic and solar thermal technologies. While relatively expensive today, they are already competitive in areas remote from electric utility grids. The costs of these technologies are likely to decline significantly over time because of their small scale, which allows economies of scale and learning by doing. (See table 17.4.)

The U.S. Department of Energy has concluded that, relying primarily on already proven technology, the United States could reduce its carbon emissions by almost 400 million tonnes (440

The Kyoto Protocol on Greenhouse Gases

On December 10, 1997, 160 nations reached agreement in Kyoto, Japan, on limiting emissions of carbon dioxide and other greenhouse gases. The Kyoto Protocol was a significant agreement in focusing the attention of world leaders on the issues of climate change. The Kyoto Protocol called for the industrialized nations—the so-called Annex 1 countries—to reduce their average national emissions over the period 2008–2012 to about 5 percent below 1990 levels. The United States pledged to achieve a level 7 percent below the 1990 level, slightly less than the European Union's pledge and slightly more than Japan's. None of the developing countries was required to set any limits.

The protocol initially covered only three greenhouse gases: carbon dioxide, methane, and nitrous oxide. Three more compounds—hydrofluorocarbons, perfluorocarbons, and sulfur hexafluoride—were to be added in subsequent years. The protocol also contains the elements of a program for international trading of greenhouse gas emissions. Such trading would employ market incentives to help ensure that the lowest-cost opportunities for reduction are pursued.

While the Kyoto climate agreement was important in bringing a new level of international attention to greenhouse gas emissions, there remain many important issues to be resolved, issues pertaining to the agreement. These include:

1. The rules and institutions that are to govern international trading of greenhouse gas emissions among Annex 1 countries must be better established.
2. The criteria used to judge compliance, and any penalties for noncompliance, must be clearly articulated.
3. To make longer-term objectives more credible, moderate but specific near-term goals should be set for Annex 1 countries, and these countries should be able to use early emissions reductions to meet longer-term requirements.

One thing is for sure, if the goals of the protocol are to be met by 2008, the developed nations of the world will be forced to focus attention on their energy policies, and this in turn will affect all of us.

Key dates in the global warming story

1898—Swedish scientist Svante Ahrrenius warns that Industrial Revolution's carbon dioxide emissions, from coal and oil, could accumulate in atmosphere and lead to global warming.

1961—New observatory atop Hawaii's Mauna Loa volcano detects rise in atmospheric carbon dioxide.

1980s—Computer models of world climate project temperature rises.

1988—UN establishes an authoritative network of climate scientists, the Intergovernmental Panel on Climate Change, to study the problem and make recommendations.

1990—Intergovernmental Panel on Climate Change certifies scientific basis for "greenhouse effect" and global warming predictions.

1992—Climate change treaty is signed, setting voluntary goals for industrial nations to lower greenhouse gas emissions to 1990 levels by 2000. Almost 170 nations eventually ratify. The United States signs the treaty, but the Senate does not ratify it.

1996—The Intergovernmental Panel on Climate Change concludes that climate change is occurring and that it is highly probable that human activity is an important cause of the change.

October 1997—Negotiators end two years of preliminary tasks with major issues, including level of binding targets, unresolved and prepare for final conference in Kyoto.

July 2001—Representatives from 180 countries meet in Bonn, Germany, to continue work on the details of implementing the global warming treaty. The European Union supports the treaty, but few countries within the EU are moving toward ratification. The United States withdraws from the treaty but continues to participate in the discussions.

2002—Meetings will continue, but it is difficult to see full ratification in the near future.

million U.S. tons) in 2010, or enough to stabilize U.S. emissions in that year at 1990 levels, with savings from reduced energy costs roughly equal to the added cost of investment.

Resources and policies to increase investment in renewables and other longer-term technologies will be needed. Good examples from the industrialized nations include a policy in the United Kingdom that reserves a small part of electricity demand for competitive acquisition of designated renewable energy technologies, wind power purchase programs in Denmark and Germany, the "10,000 rooftops" photovoltaic program in Japan, and the evolution of "green marketing" campaigns in the United States and Europe to capture consumer willingness to pay modest premiums for electricity from clean energy technologies.

Indoor Air Pollution

A growing body of scientific evidence indicates that the air within homes and other buildings can be more seriously polluted than outdoor air in even the largest and most industrialized cities. Many indoor air pollutants and pollutant sources are thought to have an adverse

Table 17.5 Indoor Air Pollution Summary

Pollutant	Description	Sources	Effects*
Asbestos	Light, fibrous mineral; fireproof and insulative	Ceilings and floor tiles, insulation, and spackling compounds	Easily inhaled to cause lung damage Cancer
Carbon monoxide (CO)	Odorless, colorless gas	Combustion sources such as charcoal, kerosene heaters, attached garages with poor ventilation, tobacco smoke	Reduces ability of blood to carry oxygen; impairs vision and alertness Dizziness, headaches, fatigue, death by suffocation
Formaldehyde	Pungent gas; preservative and disinfectant	Foam insulation, resin in particleboards, plywood paneling, fiberboard, some carpets, drapery, and upholstery fabrics	Headaches, dizziness, nausea, lethargy, rashes, upper respiratory irritations
Lead (Pb)	Metallic element	House paints manufactured prior to 1976, old plumbing lines and solder, leaded crystal and pottery, old toys	Learning and behavior problems in children, high blood pressure, joint pain, concentration problems, reproductive problems
Biological pollutants and microorganisms	Pollen, dust mites, and pet dander Bacteria, molds, fungi, and viruses	Improperly maintained heating and cooling systems Washrooms, humidifiers, and dehumidifiers Pets and plants Wall-to-wall carpeting	Allergic diseases and skin irritations, influenza and Legionnaires' disease, acute cases of asthma
Nitrogen dioxide (NO$_2$)	Brownish gas	Gas appliances, fireplaces, wood and coal stoves	Eye and respiratory tract irritations Lowers resistance to respiratory infections Chronic bronchitis
Radon (Rn)	Naturally occurring, radioactive gas	Rocks and soils that contain the decaying radioactive elements of uranium and radium Enters through cracks in foundation and basements	Lung cancer
Tobacco smoke	Mixture of several substances including known human carcinogens	Cigarettes, pipes and cigars, secondhand smoke	Eye, nose and throat irritations, headaches and nausea, coughing and chest discomfort, respiratory infections, lung cancer
Volatile organic compounds (VOCs)	Substances produced by synthetic chemical industry and naturally; vaporize at ordinary temperatures	Some furniture, paint, adhesives, solvents, upholstery and drapery fabrics, construction materials, cleaning compounds, deodorizers, felt-tipped markers, correction fluid, dry-cleaned clothes, pesticides, wool preservatives, tobacco smoke, etc.	Eye and respiratory irritations, kidney and liver damage in animals Long-term effects are still being studied.

*Effects from prolonged exposure or high concentrations.

Source: Chart compiled by Earth Force, Inc., (www.earthforce.org) with data from the U.S. Environmental Protection Agency and Wisconsin Department of Natural Resources. All rights reserved. Reprinted by permission.

effect on human health. These pollutants include asbestos; formaldehyde, which is associated with many consumer products, including certain wood products and aerosols; airborne pesticide residues; chloroform; perchloroethylene (associated particularly with dry cleaning); paradichlorobenzene (from mothballs and air fresheners); and many disease-causing or allergy-producing organisms. (See table 17.5.) Smoking is the most important air pollutant source in the United States in terms of human health. The surgeon general estimates that 350,000 people in this country die each year from emphysema, heart attacks, strokes, lung cancer, or other diseases caused by tobacco smoking. Banning smoking probably would save more lives than would any other pollution-control measure.

A recent contributing factor to the concern about indoor air pollution is the weatherizing of buildings to reduce heat loss and save on fuel costs. In most older homes, a complete exchange of air occurs every hour. This means that fresh air leaks in around doors and windows and through cracks and holes in the building. In a weatherized home, a complete air exchange may occur only once every five hours. Such a home is more energy efficient, but it also tends to trap air pollutants.

Even though we spend on average almost 90 percent of our time indoors, the movements to reduce indoor air pollution lag behind regulations governing outdoor air pollution. In the United States, the Environmental Protection Agency is conducting research to identify and rank the human health risks that result from exposure to individual indoor pollutants or mixtures of multiple indoor pollutants.

environmental CLOSE-UP

Radon

In 1985, the clothing worn by an engineer at the Limerick Nuclear Generating Station in Pottstown, Pennsylvania, registered a high radiation level. Initially, the generating station was believed to be the source of the radiation. However, subsequent studies indicated that radon 222 from the engineer's home was the source. Following this incident, interest in radon and its effects has increased.

The source of radon is uranium-238, a naturally occurring element that makes up about 3 parts per million of the Earth's crust. Uranium-238 goes through 14 steps of decay before it becomes stable, nonradioactive lead 206. **Radon,** an inert radioactive gas having a half-life of 3.8 days, is one of the products formed during this process.

Since radon is an inert gas, it does not enter into any chemical reactions within the body, but it can be inhaled. Once in the lungs, it may undergo radioactive decay, producing other kinds of atoms called "daughters" of radon. These decay products (daughters) of radon—plutonium 218, which has a 3-minute half-life; lead 214, which has a 27-minute half-life; bismuth 214, which has a 20-minute half-life; and polonium 214, which has a millisecond half-life—are solid materials that remain in the lungs and are chemically active.

Increased incidence of lung cancer is the only known health effect associated with radon decay products. It is estimated that the decay products of radon are responsible for about 15,000 lung cancer deaths annually in the United States. This is about 10 percent of lung cancer deaths.

As the radon gas is formed in the rocks, it usually diffuses up through the rocks and soil and escapes harmlessly into the atmosphere. It can also diffuse into groundwater. Radon usually enters a home through an open space in the foundation. A crack in the basement floor or the foundation, the gap around a water or sewer pipe, or a crawl space allows the radon to enter the home. It may also enter in the water supply from wells.

About 10 percent of the homes in the United States have a potential radon problem. However, the increased publicity about radon has many people worried. In addition, the Environmental Protection Agency and the U.S. surgeon general recommend that all Americans (other than those living in apartment buildings above the second floor) test their homes for radon. If the tests indicate radon levels at or above 4 picocuries/liter, the EPA recommends that the homeowner take action to lower the level. This is usually not expensive and consists of blocking the places where radon is entering or venting radon sources to the outside. People who are concerned about radon should contact their state's public health department or environmental protection agency.

Radon Risk Evaluation Chart

pCi/L	WL	Estimated Lung-Cancer Deaths Due to Radon Exposure (out of 1000)	Comparable Exposure Levels	Comparable Risk
200	1.0	440-770	One thousand times average outdoor level	More than 60 times nonsmoker risk
100	0.5	270-630	One hundred times average indoor level	Four-pack-a-day smoker
				Two thousand chest X rays per year
40	0.2	120-380		
				Two-pack-a-day smoker
20	0.1	60-120	One hundred times average outdoor level	One-pack-a-day smoker
10	0.05	30-120	Ten times average indoor level	Five times nonsmoker risk
4	0.02	13-50	Level at which EPA suggests remedial action	Two hundred chest X rays per year
2	0.01	7-30	Ten times average outdoor level	Nonsmoker risk of dying from lung cancer
1	0.005	3-13	Average indoor level	
0.2	0.001	1-3	Average outdoor level	Twenty chest X rays per year

Note: Measurement results are reported in one of two ways: (1) pCi/L (picocuries per liter) — measurement of *radon gas*, or (2) WL (working levels) – measurement of *radon decay products*.

Source: Data from Office of Air and Radiation Programs, U.S. Environmental Protection Agency.

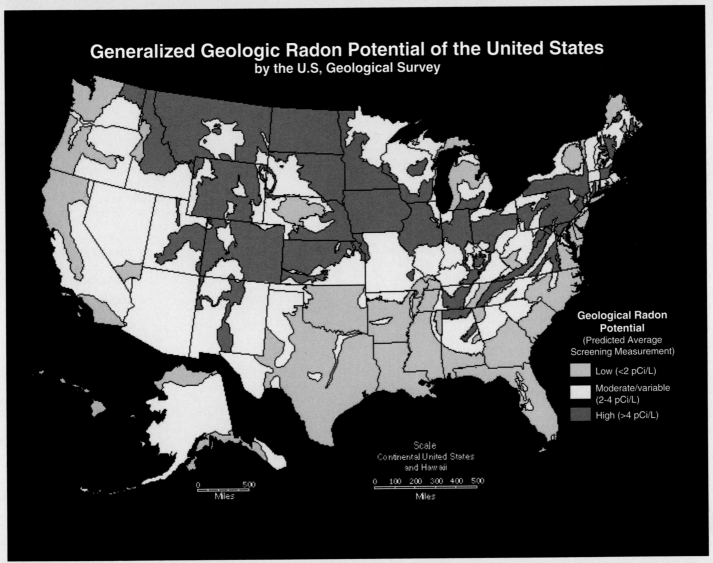

Generalized Geologic Radon Potential of the United States
by the U.S. Geological Survey

Geological Radon Potential
(Predicted Average Screening Measurement)

Low (<2 pCi/L)

Moderate/variable (2-4 pCi/L)

High (>4 pCi/L)

Scale
Continental United States and Hawaii

0 100 200 300 400 500
Miles

0 500
Miles

Source: U.S. Geological Survey.

environmental CLOSE-UP

Noise Pollution

Noise is referred to as unwanted sound. However, noise can be more than just an unpleasant sensation. Research has shown that exposure to noise can cause physical, as well as mental, harm. The loudness of the noise is measured by **decibels** (db). Decibel scales are logarithmic rather than linear. Thus, the change from 40 db (a library) to 80 db (a dishwasher or garbage disposal) represents a ten-thousandfold increase in sound loudness.

The frequency or pitch of a sound is also a factor in determining its degree of harm. High-pitched sounds are the most annoying. The most common sound pressure scale for high-pitched sounds is the A scale, whose units are written "dbA." Hearing loss begins with prolonged exposure (eight hours or more per day) to 80 or 90 dbA levels of sound pressure. Sound pressure becomes painful at around 140 dbA and can kill at 180 dbA. (See the table.)

In addition to hearing loss, noise pollution is linked to a variety of other ailments, ranging from nervous tension headaches to neuroses. Research has also shown that noise may cause blood vessels to constrict (which reduces the blood flow to key body parts), disturbs unborn children, and sometimes causes seizures in epileptics. The U.S. Environmental Protection Agency has estimated that noise causes about 40 million U.S. citizens to suffer hearing damage or other men-

tal or physical effects. Up to 64 million people are estimated to live in homes affected by aircraft, traffic, or construction noise.

The Noise Control Act of 1972 was the first major attempt in the United States to protect the public health and welfare from detrimental noise. This act also attempted to coordinate federal research and activities in noise control, to set federal noise emission standards for commercial products, and to provide information to the public. After the passage of the Noise Control Act, many local communities in the United States enacted their own noise ordinances. While such efforts are a step in the right direction, the United States is still controlling noise less than are many European countries. Several European countries have developed quiet construction equipment in conjunction with strongly enforced noise ordinances. The Germans and Swiss have established maximum day and night noise levels for certain areas. Regarding noise-pollution abatement, North America has much to learn from European countries.

Intensity of Noise

Source of Sound	Intensity in Decibels
Jet aircraft at takeoff	145
Pain occurs	140
Hydraulic press	130
Jet airplane (160 meters overhead)	120
Unmuffled motorcycle	110
Subway train	100
Farm tractor	98
Gasoline lawn mower	96
Food blender	93
Heavy truck (15 meters away)	90
Heavy city traffic	90
Vacuum cleaner	85
Hearing loss after long exposure	85
Garbage disposal unit	80
Dishwasher	65
Window air conditioner	60
Normal speech	60

International Air Pollution

Acid rain originates as a form of air pollution, but it may damage the environment in the form of water pollution. Also, air pollution that originates in one country may result in water pollution in another country. A particular nation may have stringent environmental controls within its own boundaries but have environmental problems because of the actions of a neighboring country.

Recently, the phenomenon of acid rain has underscored the need for international cooperation in dealing with various environmental concerns. For example, an estimated 56 percent of the acid rain falling in Sweden originates outside of that country. The main sources of the pollutants are Germany and the United Kingdom—because the west coast of Sweden receives winds from these countries. When these industrialized countries release sulfur dioxide into the atmosphere, the result is acid rain in Sweden. Since the 1930s, lakes in western Sweden have become more acidic by a value of two pH units. Ten thousand lakes have pHs below 6.0, and 5000 lakes have pHs below 5.0. A pH below 5.5 is too acidic for many species of fish.

In the 1930s, the United Kingdom initiated a program to reduce air pollution. This program has been successful in that the air in the United Kingdom has become cleaner. However, since one of the country's solutions was to build taller smokestacks, more of the pollutants from the United Kingdom are transported to Sweden.

- Should the United Kingdom be permitted to disperse air pollutants in a manner that damages the Swedish environment?
- Should there be a series of international agreements to control and regulate the movement of airborne pollutants across international boundaries?
- In addition to air pollutants, are there other forms of pollutants that can naturally be transported across international boundaries?

Summary

The atmosphere has a tremendous ability to disperse pollutants. Carbon monoxide, hydrocarbons, particulate matter, sulfur dioxide, and oxides of nitrogen compounds are the primary air pollutants. They can cause a variety of health problems. The U.S. Environmental Protection Agency establishes standards for six pollutants known as criteria air pollutants. They are carbon monoxide, nitrogen dioxide, sulfur dioxide, volatile organic compounds (hydrocarbons), ozone, and lead. The EPA also regulates hazardous air pollutants.

Photochemical smog is a secondary pollutant, formed when hydrocarbons and oxides of nitrogen are trapped by thermal inversions and react with each other in the presence of sunlight to form peroxyacetyl nitrates and ozone. Elimination of photochemical smog requires changes in technology, such as more fuel-efficient automobiles, special devices to prevent the loss of hydrocarbons, and catalytic converters to more completely burn hydrocarbons in exhaust gases.

Acid rain is caused by emissions of sulfur dioxide and oxides of nitrogen in the upper atmosphere, which form acids that are washed from the air when it rains or snows or settle as particles on surfaces. Direct effects of acid rain on terrestrial ecosystems are difficult to prove, but changes in many forested areas are suspected of being partly the result of additional stresses caused by acid rain. Recent evidence suggests that loss of calcium from the soil may be a major problem associated with acid rain. The effect of acid rain on aquatic ecosystems is easy to quantify. As waters become more acidic, the complexity of the ecosystem decreases, and many species fail to reproduce. The control of acid rain requires the use of scrubbers, precipitators, and filters—or the removal of sulfur from fuels. However, oxides of nitrogen are still a problem.

Currently, many are concerned about the damaging effects of greenhouse gases: carbon dioxide, methane, and chlorofluorocarbons. These gases are likely to be causing an increase in the average temperature of the Earth and, consequently, are leading to major changes in the climate. Human and ecological systems are already vulnerable to a range of environmental pressures, including climate extremes and variability. Global warming is likely to amplify the effects of other pressures and disrupt our lives in numerous ways. Significant impacts on our health, the vitality of forests and other natural areas, the distribution of freshwater supplies, and the productivity of agriculture are among the probable consequences of climate change. Chlorofluorocarbons also lead to the destruction of ozone in the upper atmosphere, which results in increased amounts of ultraviolet light reaching the Earth. Concern about the effects of chlorofluorocarbons has led to international efforts that have resulted in significant reductions in the amount of these substances reaching the atmosphere. Many commonly used materials release gases into closed spaces (indoor air pollution), where they cause health problems. The most important of these health problems are associated with smoking.

Key Terms

acid deposition *398*
acid rain *398*
air toxics *389*
carbon dioxide (CO$_2$) *403*
carbon monoxide (CO) *389*
carcinogenic *390*
chlorofluorocarbons (CFCs) *404*
criteria air pollutants *389*
decibel *414*
greenhouse effect *403*

greenhouse gas *402*
hazardous air pollutants *389*
hydrocarbons (HC) *390*
nitrogen dioxide (NO$_2$) *392*
nitrogen monoxide (NO) *392*
nitrous oxide(N$_2$O) *404*
oxides of nitrogen (NO$_X$) *392*
ozone(O$_3$) *392*
particulate matter *390*

photochemical smog *393*
PM$_{2.5}$ *390*
PM$_{10}$ *390*
primary air pollutants *388*
radon *412*
secondary air pollutants *389*
sulfur dioxide (SO$_2$) *391*
thermal inversion *394*
volatile organic compounds (VOCs) *390*

Review Questions

1. List the five primary air pollutants commonly released into the atmosphere and their sources.
2. List the six criteria air pollutants and their sources.
3. Define *secondary air pollutants* and give an example.
4. List three health effects of air pollution.
5. Why is air pollution such a large problem in urban areas?
6. What is photochemical smog? What causes it?
7. Describe three actions that can be taken to control air pollution.
8. What causes acid rain? List three probable detrimental consequences of acid rain.
9. Why is carbon dioxide (a nontoxic normal component of the atmosphere) called a "greenhouse gas"?
10. What would the consequences be if the ozone layer surrounding the Earth were destroyed?
11. How does energy conservation influence air quality?

Critical Thinking Questions

1. What could you do to limit the air pollution you create?
2. Do you agree with a ban on smoking, as in California, that includes all indoor public places, even privately owned restaurants and bars? Why or why not?
3. Some developing countries argue that they should be exempt from limits on the production of greenhouse gases and that developed countries should bear the brunt of the changes that appear to be necessary to curb global climate change. What values, beliefs, and perspectives underlie this argument? What do you think about this argument?
4. As a nation, the United States provides many subsidies to make energy cheap because policy makers feel that economic development depends on cheap energy. If these subsidies were withdrawn or taxes on energy were added, what effect would this have on your own energy consumption? Would you be willing to support high gasoline prices, in the $3- to $4-a-gallon range as in many European countries, if it would cut greenhouse gas emissions?
5. Why do you think air pollution is so much worse in developing countries than in developed countries? What should developed countries do about this, if anything?
6. What common indoor air pollutants are you exposed to? What can you do to limit this exposure?
7. What kinds of noise pollution do you encounter? How important is noise pollution to you? What can you do to reduce noise pollution? How can we make cities quieter?
8. Is it possible to have zero emissions of pollutants? What level of risk are you willing to live with?

Concept Map

Construct a map to show relationships among the following concepts:

acid deposition

carbon dioxide (CO$_2$)

chlorofluorocarbons

greenhouse effect

greenhouse gas

nitrogen monoxide

nitrogen dioxide

photochemical smog

sulfur dioxide (SO$_2$)

thermal inversion

volatile organic compounds

Interactive Exploration

Check out the website at **http://www.mhhe.com/environmentalscience** and click on the cover of this textbook for quizzing, career information, case studies, and hot links for the following topics:

Air Pollution

Acid Deposition

Atmosphere, Climate, and Weather

Global Warming

Human Respiratory Topics

18 CHAPTER

Solid Waste Management and Disposal

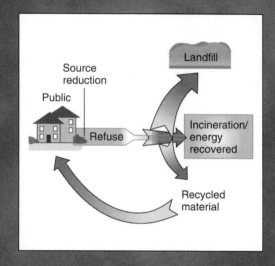

Objectives

After reading this chapter, you should be able to:

- Explain why solid waste has become a problem throughout the world.
- Understand that the management of municipal solid waste is directly affected by economics, changes in technology, and citizen awareness and involvement.
- Describe the various methods of waste disposal and the problems associated with each method.
- Understand the difficulties in developing new municipal landfills.
- Define the problems associated with incineration as a method of waste disposal.
- Describe some methods of source reduction.
- Describe composting and how it fits into solid waste disposal.
- List some benefits and drawbacks of recycling.

Chapter Outline

Introduction to Waste Management

The Nature of the Problem
    The Disposable Decades
    Current Trends

Methods of Waste Disposal
    Landfills
    Incineration

Environmental Close-Up: *Resins Used in Consumer Packaging*
    Composting
    Source Reduction
    Recycling

Environmental Close-Up: *What You Can Do to Reduce Waste and Save Money*

Environmental Close-Up: *Recycling Is Big Business*

Environmental Close-Up: *Recyclables Market Basket*

Issues—Analysis: *Corporate Response to Environmental Concerns*

Introduction to Waste Management

The several kinds of waste produced by a technological society can be categorized in many ways. Some kinds of wastes are released into the air and water. These were discussed in chapters 16 and 17. **Solid waste** is generally made up of objects or particles that accumulate on the site where they are produced, as opposed to water- and air-borne wastes that are carried away from the site of production. Solid waste can be divided into several categories. There are wastes with particularly dangerous characteristics, such as nuclear wastes, medical wastes, industrial hazardous wastes, and household hazardous wastes. Chapter 11 dealt with the nuclear waste issue, and chapter 19 will consider hazardous waste issues. Solid wastes can also be placed in categories based on the sector of the economy responsible for producing them, such as mining, agriculture, manufacturing, and municipalities.

We have very good information about those wastes streams that are tightly regulated (hazardous wastes, municipal solid waste, medical wastes, nuclear wastes) but only general estimates for many of the other kinds of wastes (mining and agricultural). The U.S. Geological Survey estimates that about 1700 million tonnes (1900 million U.S. tons) of mining wastes are produced each year. Most of this is in the form of mine tailings produced when the small amount of valuable ore is extracted from large volumes of rock. Agricultural waste is the second most common form of waste. An estimate of the amount of animal manure produced annually is about 1240 million tonnes (1370 million U.S. tons). Other wastes associated with agriculture might bring the total to about 1500 million tonnes (1650 million U.S. tons) per year. Most agricultural waste can be used as fertilizer or for other soil-enhancement activities. However, when too much waste is produced in one place, there may not be enough farmland available to accept the agricultural waste without causing water pollution problems associated with runoff. Other industrial activity is responsible for about 635 million tonnes (700 million U.S. tons) of waste per year.

Municipal solid waste (MSW) consists of all the materials that people in a region no longer want because they are broken, spoiled, or have no further use. It includes waste from households, commercial establishments, institutions, and some industrial sources, and amounts to about 210 million tonnes (230 million U.S. tons) per year. This chapter will focus on the generation and disposal of municipal solid waste.

The Nature of the Problem

Wherever people exist, waste disposal is a problem. Archaeologists are eager to find the "middens," the waste heaps, of ancient civilizations. The kinds of articles discarded can tell us a great deal about the nature of the society that produced them. In modern society, the production of materials that are designed to be replaced has increased the amount of waste produced.

The Disposable Decades

In 1955, *Life* magazine pictured a happy family in an article entitled "Throwaway Living." A disposable lifestyle was marketed as the wave of the future and as a way to cut down on household chores. "Use it once and throw it away" became a very popular advertising slogan in the 1950s.

The solid waste problems facing us today have their roots in the economic boom that followed World War II. Marketing experts set to work trying new tactics to get consumers to buy and toss or, as economists would say, to "stimulate consumption." In the mid-1950s, marketing consultant Victor Lebow wrote an emotional plea for "forced consumption" in the New York *Journal of Retailing:*

Our enormously productive economy demands that we make consumption a way of life, that we convert the buying and use of goods into rituals, that we seek our spiritual satisfaction in consumption. . . . We need things consumed, burned up, worn out, replaced, and discarded at an ever-growing rate.

Consumers were quick to adapt to the lifestyle that Lebow envisioned. In fact, it was not long until a disposable, throwaway lifestyle was seen as a consumer's right. The idea was to sell "convenience" to the prosperous postwar consumers. What was initially a convenience was soon to become a "necessity"; at least, that was what the advertisements were telling people.

A good example of this way of thinking is the TV dinner, which was first marketed in 1953. The food-packaging industry was evolving into a major service, closely allied with marketing and advertising. From the first TV dinner, we progressed rapidly to microwave meals. One such meal consisted of 340 grams (3/4 pound) of edible material and six separate layers of packaging, five of them plastic.

After several decades of throwaway living, the disposable lifestyle is changing. Many cities and counties face a shortage of space in old landfills. Solid waste has become a major problem, and in a growing list of communities around the world, it has reached crisis proportions. Many states are running out of places to put their garbage. (See figure 18.1.)

Current Trends

The United States produces about 210 million tonnes (230 million U.S. tons) of municipal solid waste each year. This equates to about 2 kilograms (4.4 pounds) of trash per person per day, or 0.73 tonnes (0.8 U.S. tons) per person per year. The amount of municipal solid waste has more than doubled since 1960, and the per capita rate has increased by nearly 70 percent in that same time, although per capita rates began to stabilize about 1990. When

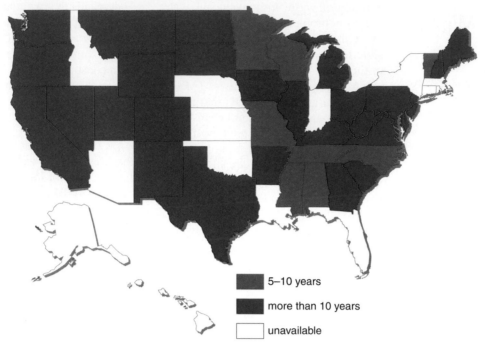

Years of Remaining Landfill Capacity

5–10 years

more than 10 years

unavailable

figure 18.1 **Available Landfill Space** Landfills are the primary method of dealing with municipal solid waste in the United States. Many states in highly urbanized parts of the country have a shortage of suitable landfill space. The development of new landfills is expensive, and in many areas, no suitable land is available for development.

recycling is included, the net waste produced has actually fallen since 1990. (See figure 18.2.)

Nations with high standards of living and productivity tend to have more municipal solid waste per person than less-developed countries. (See figure 18.3.) The United States and Canada, therefore, are world leaders in waste production. Large metropolitan areas have the greatest difficulty dealing with their solid waste because of the large volume and the challenge of finding suitable landfill sites near the city. For example, Toronto, Canada, is running out of places to put its municipal solid waste. Even with a very ambitious plan to reduce waste production by 50 percent, the metropolitan area will run out of space by 2006. The outlying districts are not willing to be the site of a new landfill for Toronto, and the metropolitan area is looking for disposal sites north of the city as well as in the United States. Many states that could serve as disposal sites are beginning to pass laws

that regulate the importation of waste from other countries, and metropolitan Toronto continues to struggle with its waste problem.

In March 2001, New York City closed its Fresh Kills Landfill on Staten Island. (It was reopened for a time following the September 11, 2001, terrorist attack to serve as a place to take the debris from the World Trade Center so that it could be processed.) Before its closure, it was the largest landfill in the world and received about 12,600 tonnes (14,000 U.S. tons) of trash each day. The remainder of the trash was shipped to Pennsylvania and other states. Today, New York City is exporting all of its solid waste to areas outside the city. Less densely populated areas typically have adequate landfill space.

Archaeologists rely on the waste of past societies to tell them about the nature of the culture and lifestyle of ancient civilizations. In the same way today, our municipal solid waste is a reflection of our society. Figure 18.4

shows how the composition of our trash has changed since 1960. Notice particularly the increase in the amount of paper and plastic and the effect that recycling has had on the amount of glass in the trash. In the United States, the two most common items in the waste stream are paper products and yard waste, and other significant segments are wood, metal, glass, plastics, and food waste. (See figure 18.5.) An analysis of the composition of our waste will present us with possible approaches to reducing the amount of waste we generate.

Methods of Waste Disposal

Until recently, the disposal of municipal solid waste did not attract much public attention. From prehistory through the present day, the favored means of disposal was simply to dump solid wastes

PART FIVE Pollution and Policy

Total and Net Municipal Solid Waste Generation—1960 to 2000

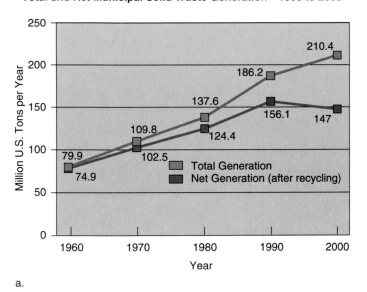

a.

Per Capita Municipal Solid Waste Generation

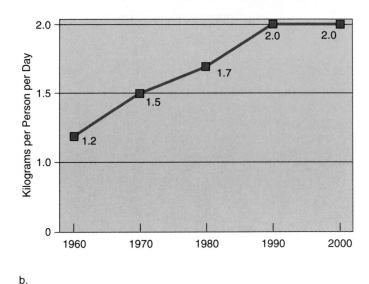

b.

figure 18.2 **Municipal Solid Waste Generation Rates** The generation of municipal solid waste in the United States has increased steadily. However, because of increased recycling rates, the net production rates (after recyclables have been removed) has actually fallen since 1990 (*a*) and the per capita rate has stabilized (*b*).

Source: U.S. Environmental Protection Agency, Washington, D.C.

outside of the city or village limits or in the "back 40." Frequently, these dumps were in wetlands adjacent to a river or lake. To minimize the volume of the waste, the dump was often burned. Unfortunately, this method is still being used in remote or sparsely populated areas in the world. (See figure 18.6.)

As better waste-disposal technologies were developed and as values changed, more emphasis was placed on the environment and quality of life. Simply dumping and burning our wastes is no longer an acceptable practice from an environmental or health perspective. While the technology of waste disposal has evolved during the past several decades, our options are still limited. Realistically, there are no ways of dealing with waste that have not been known for many thousands of years. Essentially, five techniques are used: (1) landfills, (2) incineration, (3) source reduction, (4) composting, and (5) recycling.

Landfills have historically been the primary method of waste disposal because this method is the cheapest and most convenient, and because the threat of groundwater contamination was not initially recognized. As we have recognized some of the problems associated

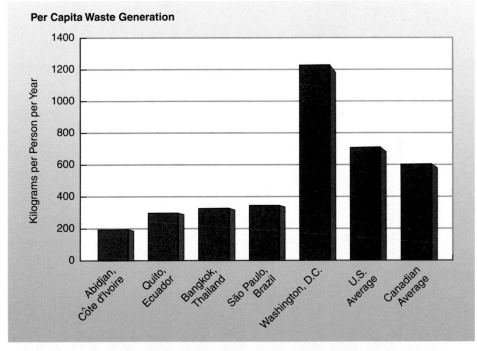

figure 18.3 **Waste Generation and Lifestyle** The waste generation rates of people are directly related to their economic condition. People in richer countries produce more garbage than those in poorer countries.

Source: Data from *World Resources* 1996–1997, and U.S. Environmental Protection Agency.

with poorly designed landfills, efforts to reduce the amount of material placed in landfills have been substantial. Although the amount of waste has in-creased, composting and recycling have removed significant amounts of materials from the waste stream, and the amount of material entering landfills has

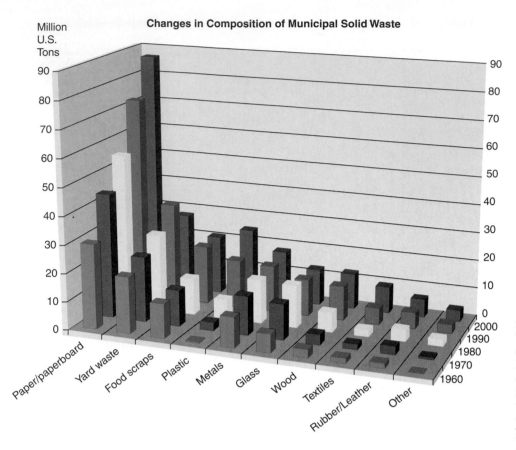

Changes in Composition of Municipal Solid Waste

Million U.S. Tons

figure 18.4 **The Changing Nature of Trash** Paper products are the largest component of the waste stream. Changes in lifestyle and packaging have led to a change in the nature of trash. Note the increase in the amount of plastics in the waste stream. Most of what is currently disposed of could be recycled.

Source: Data from the U.S. Environmental Protection Agency.

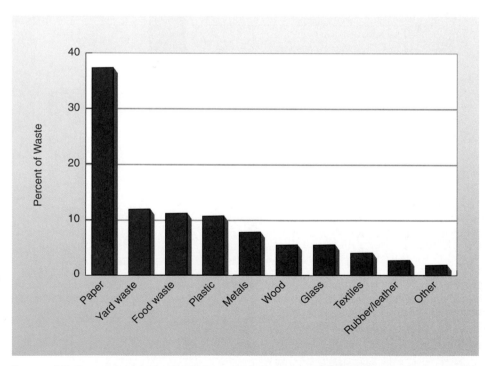

figure 18.5 **Composition of Trash in the United States (2000)** Paper and yard waste are the most common materials disposed of, accounting for about 50 percent of the waste stream.

Source: Data from the U.S. Environmental Protection Agency.

declined. (See figure 18.7.) However, the landfill of today is far different from a simple hole in the ground into which garbage is dumped.

Landfills

A modern **municipal solid waste landfill** is typically constructed above an impermeable clay layer that is lined with an impermeable membrane and includes mechanisms for dealing with liquid and gas materials generated by the contents of the landfill. Each day's deposit of fresh garbage is covered with a layer of soil to prevent it from blowing around and to discourage animals from scavenging for food. Selection of landfill sites is based on an understanding of local geologic conditions such as the presence of a suitable clay base, groundwater geology, and soil type. In addition, it is important to address local citizens' concerns. Once the site is selected, extensive construction activities are necessary to prepare it for use. New landfills have complex bottom layers to trap contaminant-laden water, called **leachate,** leaking through the buried trash. In addition, monitoring systems are necessary to detect methane gas production and groundwater contamination. In some cases, methane produced by decomposing waste is collected and used to produce heat or generate electricity. The water that leaches through the site must be collected and treated. As a result, new landfills are becoming increasingly more complex and expensive. They currently cost up to $1 million per hectare ($400,000 per acre) to prepare. (See figure 18.8.)

Today, about 55 percent of the municipal solid waste from the United States and about 80 percent of Canadian municipal solid waste goes into landfills. The number of landfills is declining. In 1988, there were about 8000 landfills, and in 2000, there were about 2000 active sanitary landfills. (See figure 18.9.) The number of landfills has decreased for two reasons. Many small, poorly run landfills have been closed because they were not meeting regulations. Others have closed because they reached their capacity. The overall capacity, however, has remained relatively constant because new landfills are much larger than the old ones.

A prolonged public debate over how to replace lost landfill capacity is developing where population density is high and available land is scarce. Selecting sites for new landfills in locations like Toronto, New York, and Los Angeles is extremely difficult because of (1) the difficulty in finding a geologically suitable site and (2) local opposition, which is commonly referred to as the NIMBY, or "not-in-my-backyard," syndrome. Resistance by the public comes from concern over groundwater contamination, rodents and other vectors of disease, odors, and truck traffic. Public officials look for alternatives to landfills to avoid problems with the public over landfill site selection. Although sites must be chosen for new landfills, politicians are often unwilling to take strong positions that might alienate their constituents.

figure 18.6 **Burning Landfills** In the past, it was common practice to burn the waste in landfills to reduce the volume. Waste is still being burned in sparsely populated areas in North America and other parts of the world.

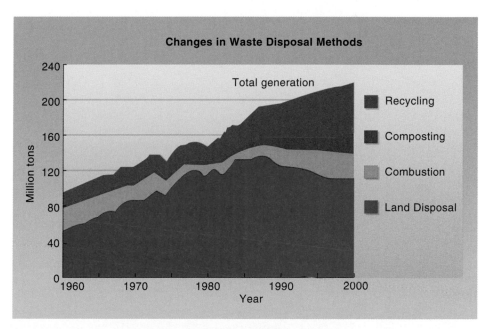

figure 18.7 **Changes in Waste Disposal Methods** The landfill is still the primary method of waste disposal. Historically, landfills have been the cheapest means of disposal. This may not be the case in the future. Notice that recycling and composting have grown over the past decade, while the amount going to landfills has declined somewhat.

Source: Data from the U.S. Environmental Protection Agency.

Japan and many Western European countries have already moved away from landfills as the primary method of waste disposal because of land scarcities and related environmental concerns. Switzerland and Japan dispose of less than 15 percent of their waste in landfills, compared to 55 percent in the United States. Instead, recycling and incineration are the primary methods. (See

How a Modern Landfill Works

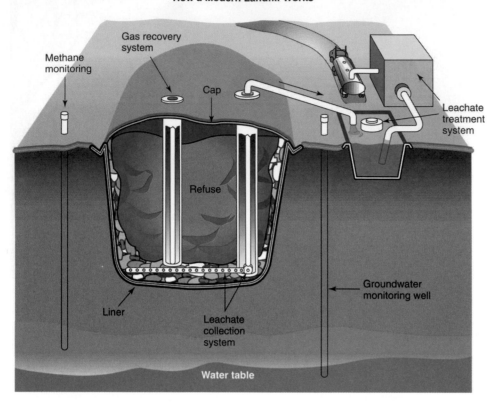

figure 18.8 **A Well-Designed Modern Landfill** A modern sanitary landfill is far different from a simple hole in the ground filled with trash. A modern landfill is a self-contained unit that is separated from the soil by impermeable membranes and sealed when filled. Methane gas and groundwater are continuously monitored to ensure that wastes are not escaping to the air or groundwater.

Source: National Solid Waste Management Association.

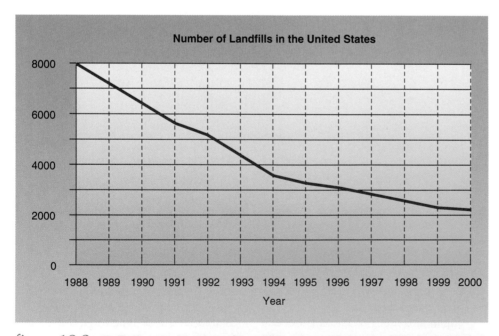

figure 18.9 **Reducing the Number of Landfills** The number of landfills in the United States is declining because they are filling up or because their design and operation do not meet environmental standards.

Source: Data from the U.S. Environmental Protection Agency.

figure 18.10.) In addition, the energy produced by incineration can be used for electric generation or heating.

Incineration

Incineration is the process of burning refuse in a controlled manner. The incineration of refuse was quite common in North America and Western Europe before 1940. However, many incinerators were eliminated because of aesthetic concerns, such as foul odors, noxious gases, and gritty smoke, rather than for reasons of public health. Today, about 15 percent of the municipal solid waste in the United States is incinerated; Canada incinerates about 8 percent. Most incinerators are not used just to burn trash. The heat derived from the burning is converted into steam and electricity. (See figure 18.11.) In 2000, about 100 combustors with energy recovery existed in the United States, with the

environmental CLOSE-UP

Resins Used in Consumer Packaging

Thermoplastics, which account for 87 percent of plastics sold, are the most recyclable form of plastics because they can be remelted and reprocessed, usually with only minor changes in their properties. Thermoplastic resins are commonly used in consumer packaging applications:

1. Polyethylene terephthalate (PET) is used extensively in rigid containers, particularly beverage bottles for carbonated beverages and medicine containers.
2. Polyethylene is the most widely used resin. High-density polyethylene (HDPE) is used for rigid containers, such as milk and water jugs, household-product containers, and motor oil bottles.
3. Polyvinyl chloride (PVC) is a tough plastic often used in construction and plumbing. It is also used in some food, shampoo, oil, and household-product containers.
4. Low-density polyethylene (LDPE) is often used in films and bags.
5. Polypropylene (PP) is used in a variety of areas, from yogurt containers to battery cases to disposable diaper linings. It is frequently interchanged for polyethylene or polystyrene.
6. Polystyrene (PS) is best known as a foam in the form of cups, trays, and food containers. In its rigid form, it is used in cutlery.
7. Other. These usually contain layers of different kinds of resins and are most commonly used for squeezable bottles (for example, for ketchup).

Currently, HDPE and PET are the two most commonly recycled resins. Efforts to recycle PS are being explored, especially within the fast-food industry. The percentages listed below are based on each type in relation to all plastics.

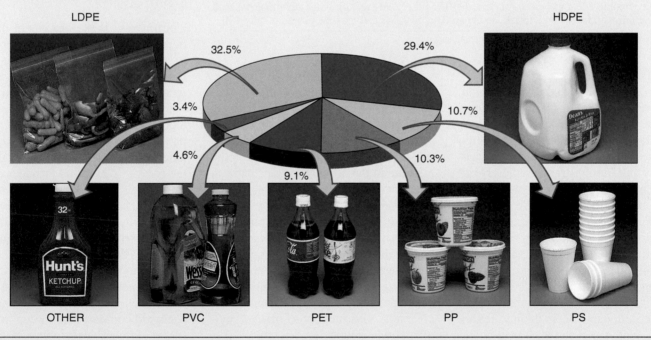

LDPE   32.5%   29.4%   HDPE

3.4%   10.7%

4.6%   10.3%

9.1%

OTHER   PVC   PET   PP   PS

Source: Graph data from Center for Plastics Recycling Research.

capacity to burn up to 100,000 tonnes (110,000 U.S. tons) of MSW per day.

Most incineration facilities burn unprocessed municipal solid waste. This is not as efficient as some other technologies. About one-fourth of the incinerators use refuse-derived fuel—collected refuse that has been processed into pellets prior to combustion.

The newest means of incineration, a European concept, is called **mass burn.** In the mass-burn technique, municipal solid waste is fed into a furnace, where it falls onto moving grates and is burned at temperatures up to 1300°C (2400°F). The burning waste heats water, and the steam drives a turbine to generate electricity, which is sold to a utility.

Incinerators drastically reduce the amount of municipal solid waste—up to 90 percent by volume and 75 percent by

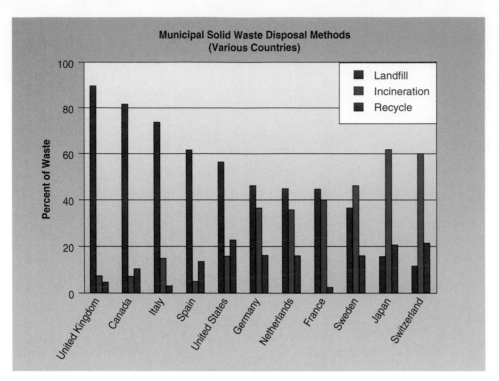

Municipal Solid Waste Disposal Methods (Various Countries)

Legend:
- Landfill
- Incineration
- Recycle

Percent of Waste (y-axis, 0 to 100)

Countries (x-axis): United Kingdom, Canada, Italy, Spain, United States, Germany, Netherlands, France, Sweden, Japan, Switzerland

figure 18.10 **Disposal Methods Used in Various Countries** Many countries have difficulty finding adequate space for landfills. Therefore, they rely on other technologies, such as incineration and recycling, to reduce the amount of waste that must be placed in a landfill.

Source: Data from the U.S. Environmental Protection Agency.

weight. Primary risks of incineration, however, involve air-quality problems and the toxicity and disposal of the ash.

Though mass-burn technology works efficiently in Europe, the technology is not easily transferable. North American municipal solid waste contains more plastic and toxic materials than European waste, thus creating air-pollution and ash-toxicity concerns. Modern incinerators have many pollution control devices that trap nearly all of the pollutants produced. However, tiny amounts of pollutants are released into the atmosphere, including certain metals, acid gases, and classes of chemicals known as dioxins and furans, which have been implicated in birth defects and several kinds of cancer. The long-term risks from the emissions are still a subject of debate.

Ash from incineration is also an important issue. Small concentrations of heavy metals are present both in the air emissions (fly ash) and residue (bottom ash) from these facilities. Because the ash contains lead, cadmium, mercury, and ar-

senic in varying concentrations from such items as batteries, lighting fixtures, and pigments, this ash may need to be treated as hazardous waste. The toxic substances are more concentrated in the ash than in the original garbage and can seep into groundwater from poorly sealed landfills. Many cities have had difficulty disposing of incinerator ash, and there is still considerable debate about what is the best method of disposal.

The cost of the land and construction for new incinerators are also major concerns facing many communities. Incinerator construction is often a municipality's single largest bond issue. Incinerator construction costs in North America in 2000 ranged from $45 million to $350 million, and the costs are not likely to decline.

Incineration is also more costly than landfills in most situations. Figure 18.12 shows the cost of landfills in comparison with solid waste incinerators. As long as landfills are available, they will have a cost advantage. When cities are unable to dispose of their trash locally in a land-

fill and must begin to transport the trash to distant sites, incinerators become more cost effective. The U.S. Environmental Protection Agency has not looked favorably on the construction of new waste-to-energy facilities and has encouraged recycling and source reduction as more effective ways to reduce the solid waste problem. Critics have argued that cities and towns have impeded waste reduction and recycling efforts by putting a priority on incinerators and committing resources to them. Proponents of incineration have been known to oppose source reduction. They argue that incinerators need large amounts of municipal solid waste to operate and that reducing the amount of waste generated makes incineration impractical. Many communities that have opposed incineration say that they support a vigorous waste-reduction and recycling effort.

Composting

Composting is the process of harnessing the natural process of decomposition to transform organic materials—anything from manure and corncobs to grass and soiled paper—into compost, a humuslike material with many environmental benefits. In nature, leaves and branches that fall to the forest floor form a rich, moist layer of mulch that protects the roots of plants and provides a home for nature's most fundamental recyclers: worms, insects, and a host of bacteria, fungi, and other microorganisms.

Properly managing air and moisture provides ideal conditions for these organisms to transform large quantities of organic material into compost in a few weeks. A good small-scale example is a backyard compost pile. Green materials (grass, kitchen vegetable scraps, and flower clippings) mixed with brown materials (twigs, dry leaves, and soiled paper towels) at a ratio of 1:3 provide a balance of nitrogen and carbon that helps microbes efficiently decompose these materials.

Large-scale municipal composting uses the same principles of organic decomposition to process large volumes of organic materials. Composting facilities of various sizes and technological

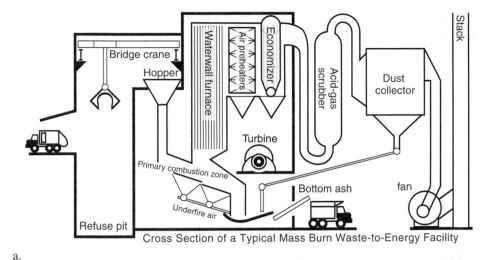

a.

b.

figure 18.11 **A Mass Burn Incineration System** (*a*) The diagram shows how a typical mass burn, waste-to-energy incinerator operates. (*b*) Incineration of municipal solid waste reduces its weight and volume significantly. However, there are concerns about air-quality problems and the toxicity and disposal of the ash.

Source: (*a*) Data from the U.S. Environmental Protection Agency.

sophistication accept materials such as yard trimmings, food scraps, biosolids from sewage treatment plants, wood shavings, unrecyclable paper, and other organic materials. These materials undergo processing—shredding, turning, and mixing—and, depending on the materials, can be turned into compost in a period ranging from 8 to 24 weeks. About 3800 composting facilities are in use in the United States. In 2000, 57 percent of yard trimmings were composted in the United States through municipal programs. (See figure 18.13.) Most municipal programs involve one of three composting methods: windrows, aerated piles, or enclosed vessels.

- *Windrow* systems involve placing the compostable materials into long piles or rows called windrows. Tractors with front end loaders or other kinds of specialized machinery are used to turn the piles periodically. Turning mixes the different kinds of materials and aerates the mixture.

- *Aerated piles* are large piles of material that can be composted if they have air pumped through them (aeration), and a layer of mature compost or other material is used to insulate the pile to keep it at an optimal temperature. Air is forced through the pile, but no mechanical turning or agitation is done.

- *Enclosed vessels* can also be used to compost materials very rapidly (within days). However, these systems are much more technologically complex. In such systems, compostable material is fed into a drum, silo, or other structure where the environmental conditions are closely controlled, and the material is mechanically aerated and mixed.

In addition to keeping wastes from entering a landfill, composting has other significant benefits. The addition of compost to soil will improve it by making clay soils more porous or increasing the water-holding capacity of sandy soils. Nitrogen, potassium, iron, phosphorus, sulfur, and calcium are all common in compost and are beneficial to plant growth. Microorganisms are an important component of compost and play a valuable role in organic matter decomposition, which, in turn, leads to humus formation and nutrient availability.

Source Reduction

The simplest way to reduce waste is to prevent it from ever becoming waste in the first place. Waste prevention, also known as **source reduction,** is the practice of designing, manufacturing, purchasing, or using materials (such as products and packaging) in ways that reduce the amount or toxicity of trash created. Reusing items is another way to stop waste at the source because it delays or prevents the entry of those items into the waste collection and disposal system.

Source reduction, including reuse, can help reduce waste disposal and handling costs, because it avoids the costs associated with recycling, municipal composting, landfilling, and combustion. Soft drink bottles are an example of source reduction. Since 1977, the weight

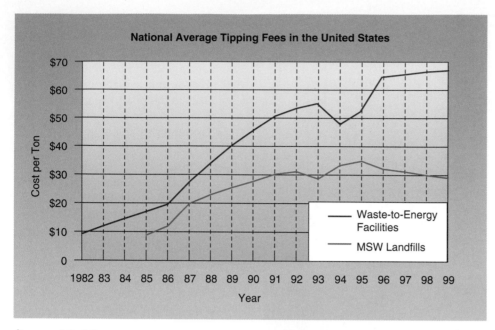

figure 18.12 **Relative Costs of Landfills and Incinerators** In general, the cost of disposing waste by incineration is greater than that of putting it in a landfill. However, in areas where landfills are not available and transportation costs for land disposal are high, incineration is a logical alternative.

Source: Data from the U.S. Environmental Protection Agency.

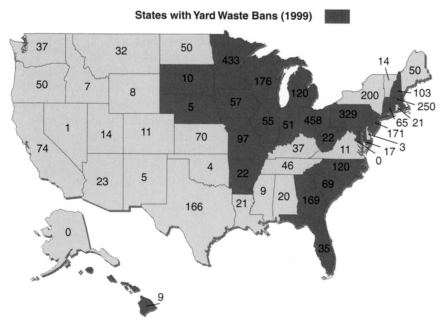

figure 18.13 **States with Yard Waste Bans** Since yard waste is such an important segment of the solid waste stream, many states have passed laws that prohibit the deposition of yard waste in landfills. The states shown in green have laws that ban yard waste from going into a landfill. This will extend the useful life of the landfill. To accommodate their citizens, many communities still collect yard waste but have instituted composting programs to deal with that waste. Even states that do not ban yard waste from landfills can extend the life of landfills by diverting yard waste to composting programs. The numbers on the map indicate the number of composting programs in each state.

Source: Data from the U.S. Environmental Protection Agency.

of a 2-liter plastic soft drink bottle has been reduced from 68 grams (2.4 ounces) to 51 grams (1.8 ounces). That translates to 114 million kilograms (250 million pounds) of plastic per year that has been kept out of the waste stream. Source reduction also conserves resources and reduces pollution, including greenhouse gases that contribute to global warming.

Source reduction and reuse have many benefits, including saving natural resources, reducing the toxicity of wastes, and reducing costs. Waste is not just created when consumers throw items away. Throughout the life cycle of a product—from extraction of raw materials, to transportation, to processing and manufacturing facilities, to manufacture and use—waste is generated. Reusing items or making them with less material decreases waste significantly. Ultimately, fewer materials will need to be recycled or sent to landfills or waste combustion facilities.

Selecting nonhazardous or less hazardous items is another important component of source reduction. Using less hazardous alternatives for certain items (e.g., cleaning products and pesticides), sharing products that contain hazardous chemicals instead of throwing out leftovers, following label directions carefully, and using the smallest amount necessary are ways to reduce waste toxicity.

Preventing waste also can mean economic saving for communities, businesses, schools, and individual consumers.

- **Communities.** In the United States, over 4000 communities have instituted "pay-as-you-throw" programs in which citizens pay for each can or bag of trash they set out for disposal rather than through the tax base or a flat fee. When these households reduce waste at the source, they dispose of less trash and pay lower trash bills.

For example, in 1994, Gainesville, Florida, entered into a contract with two companies for the collection of residential solid waste and recyclable materials (glass, plastic, paper, metal cans). The new contract for solid waste service included a variable rate for residential collections: residents paid $13.50, $15.90, or $19.75 per month according to whether they placed 35, 64, or 96 gallons of solid waste at the curb for collection.

Recycling service is unlimited. While residents have had curbside collection of recyclables since 1989, the implementation of this program added brown paper bags, corrugated cardboard, and phone books to the list of items recycled.

The results of the first five years of the program were very positive. The amount of solid waste collected decreased 20 percent, and the recyclables recovered increased 25 percent. This resulted in annual savings of $200,000 to the residents of Gainesville.

- **Businesses.** Industry also has an economic incentive to practice source reduction. When businesses market products with less packaging, they are buying fewer raw materials. A decrease in manufacturing costs can mean a larger profit margin, with savings that could be passed on to the consumer.
- **Consumers.** The buying habits of consumers can result in source reduction. When consumers buy in bulk or select products that are reusable or have less packaging, they reduce the amount of waste they will generate and frequently save money as well.

figure 18.14 **Recycling Rates for Various Materials** Recycling rates for materials that have high value such as automobile batteries are extremely high. Other materials are more difficult to market. But recycling rates today are much higher than in the past as technology and markets have found uses for materials that once were considered valueless.

Source: (*a*) Data from the U.S. Environmental Protection Agency, *Characterization of Municipal Solid Waste in the United States,* 2000. (*b*) *Characterization of Municipal Solid Waste in the United States; Update.* U.S. Environmental Protection Agency, Washington, D.C.

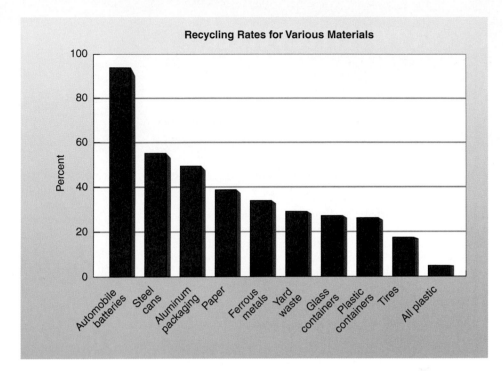

a.

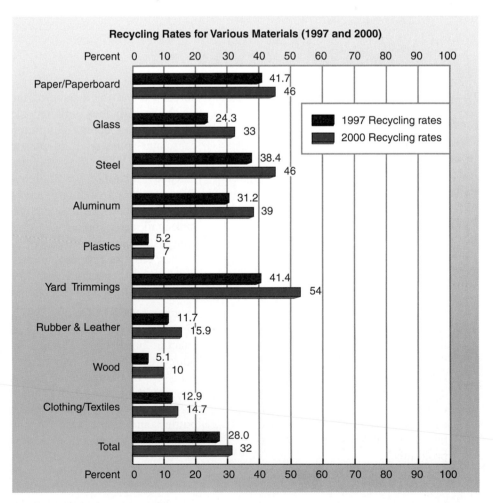

b.

environmental CLOSE-UP

What You Can Do to Reduce Waste and Save Money

You can make a difference. While this statement is sometimes overused, it does speak the truth when it comes to your ability to lessen the stream of solid waste being generated every day. Here are a few ideas that are easy to follow, will save you money, and will help reduce waste.

- Buy things that last, keep them as long as possible, and have them repaired, if possible.
- Buy things that are reusable or recyclable, and be sure to reuse and recycle them.
- Buy beverages in refillable glass containers instead of cans or throwaway bottles.
- Use plastic or metal lunch boxes and metal or plastic garbage containers without throwaway plastic liners.
- Use rechargeable batteries.

- Skip the bag when you buy only a quart of milk, a loaf of bread, or anything you can carry with your hands.
- Buy recycled goods, especially those made by primary recycling, and then recycle them.
- Recycle all newspaper, glass, and aluminum, and any other items accepted for recycling in your community.
- Reduce the amount of junk mail you get. This can be accomplished by writing to Mail Preference Service, Direct Marketing Association Inc., 1120 Avenue of the Americas, New York, NY 10036-6700, or by calling (212) 768-7277. Ask that your name not be sold to large mailing-list companies. Of the junk mail you do receive, recycle as much of the paper as possible.
- Push for mandatory trash separation and recycling programs in your community and schools.
- Choose items that have the least packaging or, better yet, no packaging ("nude products").
- Compost your yard and food wastes, and pressure local officials to set up a community composting program.

Recycling

Recycling is one of the best environmental success stories of the late twentieth century. (See figure 18.14.) In the United States, recycling, including composting, diverted about 30 percent of the solid waste stream from landfills and incinerators in 2000, up from about 16 percent in 1990. Several kinds of programs have contributed to the increase in recycling rate.

Curbside recycling—In 1990, a thousand U.S. cities had curbside recycling programs. By 2000, the number had grown to about 9000 cities. By 1999, mandatory recycling laws for all materials had been passed in 15 states. In Canada, Toronto, Mississauga, and the province of Ontario have comprehensive recycling programs.

Some local communities—from Portland, Oregon, to Chatham, New Jersey—also have achieved recycling rates of more than 50 percent, in part by expanding the types of trash they will recycle beyond newspapers, aluminum cans, and some types of plastic bottles. In 1999, Los Angeles reached a record 46 percent recycling rate by expanding its curbside pickup program to include junk mail, cereal boxes, and yard compost. The city also made recycling easy for residents by providing each household with one large recycling bin on wheels.

Container laws—In October 1972, Oregon became the first state to enact a "bottle bill." The law required a deposit of two to five cents on all beverage containers that could be reused. It banned the sale of one-time-use beverage bottles and cans. One of the primary goals of the law was to reduce the amount of litter, and it worked. Within two years of when it went into effect, beverage-container litter decreased by about 49 percent.

Nine other states—Vermont, Maine, Connecticut, New York, Iowa, Rhode Island, Michigan, Delaware, and California—have enacted legislation requiring deposits on bottles, specifically beverage containers. In 1991, Maine expanded its bottle law to all nondairy beverage containers, including all noncarbonated juice containers holding a gallon or less. Deposits are five cents except on liquor and wine bottles, which carry a 15-cent deposit. Excluded are containers for milk and other dairy products, cough syrup, baby formula, soap, and vinegar.

Many argue that a national bottle bill is long overdue. A national bottle bill would reduce litter, save energy and money, and create jobs. It would also help to conserve natural resources. But the lobbying efforts of the soft drink and brewing industries are very strong, and the U.S.

environmental CLOSE-UP

Recycling Is Big Business

In 1994, Weyerhaeuser Company, one of the world's largest forest products producers, began building a new paper mill in Cedar Rapids, Iowa, an area of North America not known for its abundant forests. The Cedar River Paper Company is a joint venture between Weyerhaeuser and Midwest Recycling Company and is the largest paper recycling plant in the United States. Even with a lack of local trees, the new mill will not be hurting for raw material to make paper and corrugated cardboard.

The mill is supplied with old paper, including scrap boxes from K-Mart. K-Mart signed an agreement with Weyerhaeuser in 1992 to have Weyerhaeuser purchase its cardboard. The agreement provides Weyerhaeuser with a supply of paper for its recycled paper plants and solves a waste disposal problem for K-Mart. The fact that a supply of recycled paper can determine where new paper-making factories will be built indicates just how important the recycling business is becoming. It is projected that by the turn of the century, over a third of the raw material that Weyerhaeuser uses for paper making will come from recycled materials. Weyerhaeuser is the third largest paper recycler and collector in the United States. It has 37 wastepaper processing plants in both Canada and the United States. According to a company spokesperson, the recycling business is becoming as significant as the millions of hectares of forests the company owns.

Congress currently has failed to pass a national container law.

Mandatory recycling laws—Many states and cities have passed mandatory recycling laws. Some of these laws simply require that residents separate their recyclables from other trash. Other laws are aimed at particular products such as beverage containers and require that they be recycled. Some are aimed at businesses and require them to recycle certain kinds of materials such as cardboard or batteries. Finally, some laws forbid the disposal of certain kinds of materials in landfills. Therefore, the materials must be recycled or dealt with in some other way. For example, banning yard waste from landfills has resulted in extensive composting programs.

Benefits of Recycling

Some benefits of recycling are readily recognizable, such as conservation of resources and pollution reduction. These perceived benefits provide the primary motivation for participation in recycling programs. The following examples illustrate the resource-conservation and pollution-reducing benefits of recycling:

One Sunday edition of the *New York Times* consumes 62,000 trees. Currently, about 40 percent of all paper that enters the waste stream in North America is recycled.

The United States imports nearly all of its aluminum and recycles over 60 percent of its aluminum beverage cans.

There are no sources of tin within the United States; however, about 2 kilograms (4.4 pounds) of tin

can be reclaimed from each 1000 kilograms (2200 pounds) of metal food cans.

Crushed glass (cullet) reduces the energy required to manufacture new glass by 50 percent. Cullet lowers the temperature requirements of the glassmaking process, thus conserving energy and reducing air pollution. In 2000, 23 percent of the glass in the solid waste stream was recycled.

While recycling is a viable alternative to landfilling or the incineration of municipal solid waste, recycling does present several problems.

Recycling Concerns

Problems associated with recycling tend to be either technical or economic. Technical questions are of particular concern when recycling plastics. (See

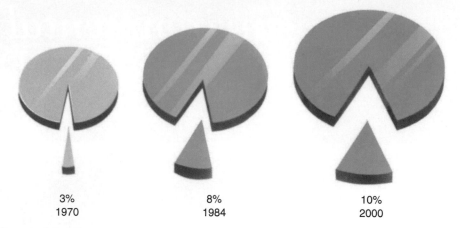

3%	8%	10%
1970	1984	2000

figure 18.15   **Increasing Amounts of Plastics in Trash** Plastics are a growing component of municipal solid waste in North America. Increased recycling of plastics could reverse this trend.

Source: Data from Franklin Associates, Ltd.

Recycling Composite Prices		1999	2000
Metal			
Ferrous ($/ton)	Used steel cans	71	69
Nonferrous (¢/lb.)	Aluminum cans	50	99
	Auto batteries	6	6
Plastic (¢/lb.)			
	Green PET	7	11
	Clear PET	6	13
	Mixed HDPE	9	14
Paper ($/ton)			
	Corregated	92	62
	Newspaper	31	53
	High-grade office	93	158
	Computer laser	163	193
Glass ($/ton)			
	Clear	39	39
	Green	14	14
	Brown	24	24

figure 18.16   **Recycling Composite Prices** Prices for materials can vary widely from year to year, depending on demand.

figure 18.15.) While the plastics used in packaging are recyclable, the technology to do so differs from plastic to plastic. There are many different types of plastic polymers. Since each type has its own chemical makeup, different plastics cannot be recycled together. In other words, a milk container is likely to be high-density polyethylene (HDPE), while an egg container is polystyrene (PS), and a soft-drink bottle is polyethylene terephthalate (PET).

Plastic recycling is still a relatively new field. Industry is researching new technologies that promise to increase the quality of plastics produced from recycled materials and that will allow mixing of different plastics. Until such technology is developed, separation of different plastics before recycling will be necessary.

The economics of recycling are also a primary area of concern. The stepped-up commitment to recycling in many developed nations has produced a glut of certain materials on the market. Markets for collected materials fill up just like landfills. Unless the demand for recycled products keeps pace with the growing supply, recycling programs will face an uncertain future. The prices for selected recycled materials are listed in figure 18.16. Prices for materials can vary widely from year to year, depending on demand.

Markets for materials collected in recycling programs grew dramatically during the 1990s. The establishment of a recyclables exchange on the Chicago Board of Trade allows for consumers and producers of recyclable materials to participate in an efficient market so that it is less likely that recyclable materials will be left unclaimed. In 1995, the American Forest and Paper Association stated that it was planning to spend $10 billion by 2002 to retool and build new mills to produce recycled paper.

The long-term success of recycling programs is also tied to other economic incentives, such as taxing issues and the development of and demand for products manufactured from recycled material. Government tax policy needs to be readjusted to encourage recycling efforts. Currently in the United States, it is still cheaper to transport virgin material, such as fresh-cut pulp wood, than to transport collected paper for recycling. Such taxing policy severely inhibits the cost-effectiveness of paper recycling. In addition, on an individual level, we can have an impact by purchasing products made from recycled materials. The demand for recycled products must grow if recycling is to succeed on a large scale. (See Environmental Close-Up: Recyclables Market Basket on page 433.)

environmental
CLOSE-UP

Recyclables Market Basket

Annually in the United States, about 75 percent of scrap used tires are recovered. Scrap tires are difficult to dispose of in landfills and waste incinerators. An estimated 800 million are currently stockpiled. These stockpiles can provide convenient habitats for rodents, serve as breeding grounds for mosquitoes, and pose fire hazards. Of the scrap tires that are used, about 40% are burned for energy. Scrap tires also are used for rubberized asphalt paving, molded rubber products, and athletic surfaces.

About 96 percent of automotive batteries are recovered each year in North America. Although these lead-acid batteries constitute a small portion of the waste stream, they contain metals that may be a concern when disposed of in landfills and incinerators. All three components of automotive batteries are recyclable: the lead, the acid, and the plastic casing.

About seventy percent of all used oil is recovered in North America. Only 10 percent of the amount generated by people who change their own motor oil is returned to collection programs. If disposed of improperly, such as being poured down sewage drains, used oil can contaminate soil, groundwater, and surface water. In some communities, used motor oil is collected at service stations, corporate or municipal collection sites, or at curbside.

General Motors REP Car
Recycling Examples in Production

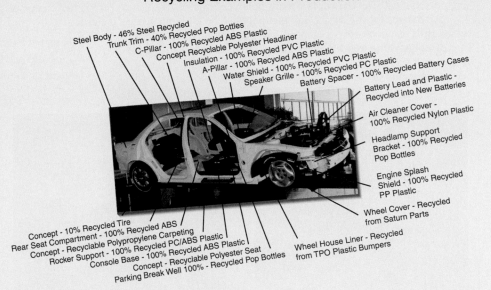

Steel Body - 46% Steel Recycled
Trunk Trim - 40% Recycled Pop Bottles
C-Pillar - 100% Recycled ABS Plastic
Concept Recyclable Polyester Headliner
Insulation - 100% Recycled PVC Plastic
A-Pillar - 100% Recycled ABS Plastic
Water Shield - 100% Recycled PVC Plastic
Speaker Grille - 100% Recycled PC Plastic
Battery Spacer - 100% Recycled Battery Cases
Battery Lead and Plastic - Recycled into New Batteries
Air Cleaner Cover - 100% Recycled Nylon Plastic
Headlamp Support Bracket - 100% Recycled Pop Bottles
Engine Splash Shield - 100% Recycled PP Plastic
Wheel Cover - Recycled from Saturn Parts

Concept - 10% Recycled Tire
Rear Seat Compartment - 100% Recycled ABS
Concept - Recyclable Polypropylene Carpeting
Rocker Support - 100% Recycled PC/ABS Plastic
Console Base - 100% Recycled ABS Plastic
Concept - Recyclable Polyester Seat
Parking Break Well 100% - Recycled Pop Bottles
Wheel House Liner - Recycled from TPO Plastic Bumpers

Corporate Response to Environmental Concerns

In the 1990s, McDonald's Corporation announced it would switch from polystyrene to paper for packaging its food products. In making this announcement, McDonald's stated that it was responding to consumer pressure to become more environmentally conscientious. Is using paper to wrap fast-food better for the environment than using polystyrene? The answer is not a simple yes or no.

Advocates for the switch say that polystyrene takes up space in landfills and does not decompose. They also argue that the burning of polystyrene foam in incinerators might release harmful air pollutants.

Opponents of the switch argue that polystyrene can be recycled into useful products, such as insulation board or playground equipment. They further argue that using paper means cutting forests and that since the paper used to wrap the food is coated with wax, it cannot be recycled.

Many grocery chains now offer several choices of carryout bags. Some encourage reuse of bags by deducting a small amount from the customer's bill. Others provide a choice of paper or plastic.

- What are the pros and cons of paper and plastic?
- Which alternative do you prefer? Why?
- Is there an alternative to both?

Which is better?

Which do you use?

Is there an alternative?

Summary

Beginning with the post–World War II era, increased consumption of consumer goods became a way of life. Products were designed to be used once and then thrown away. By the 1980s, a disposal lifestyle began to cause problems. There simply were no places to dispose of waste.

Municipal solid waste is managed by landfills, incineration, composting, waste reduction, and recycling. Landfills are the primary means of disposal; however, a contemporary landfill is significantly more complex and expensive than the simple holes in the ground of the past. The availability of suitable landfill land is also a problem in large metropolitan areas.

About 15 percent of the municipal solid waste in the United States is incinerated. While incineration does reduce the volume of municipal solid waste, the problems of ash disposal and air quality continue to be major concerns. There are several forms of composting that can keep organic wastes from entering a landfill.

The most fundamental way to reduce waste is to prevent it from ever becoming waste in the first place. Using less material in packaging, producing consumer products in concentrated form, and composting yard waste are all examples of source reduction. On an individual level, we can all attempt to reduce the amount of waste we generate.

About 30 percent of the waste generated in North America is handled through recycling. Recycling initiatives have grown rapidly in North America during the past several years. As a result, the markets for some recycled materials have become very volatile. Recycling of municipal solid waste will be successful only if markets exist for the recycled materials. Another problem in recycling is the current inability to mix various plastics.

Future management of municipal solid waste will be an integrated approach involving landfills, incineration, composting, source reduction, and recycling. The degree to which any option will be used will depend on economics, changes in technology, and citizen awareness and involvement.

Key Terms

composting *426*
incineration *424*
leachate *423*

mass burn *425*
municipal solid waste landfill *422*
municipal solid waste (MSW) *419*

recycling *430*
solid waste *419*
source reduction *427*

Review Questions

1. How is lifestyle related to our growing municipal solid waste problem?
2. What four methods are incorporated under integrated waste management?
3. Describe some of the problems associated with modern landfills.
4. What are four concerns associated with incineration?

5. Describe examples of source reduction.
6. Describe the importance of recycling household solid wastes.
7. Name several strategies that would help to encourage the growth of recycling.
8. Describe the various types of composting and the role of composting in solid waste management.

Critical Thinking Questions

1. Why do you suppose consumers were so quick to adapt the "use it once, then throw it away" lifestyle after World War II? What values, beliefs, and perceptions does this reveal?
2. How can you help solve the solid waste problem?
3. Given that you have only so much time, should you spend your time acting locally, as a recycling coordinator, for example, or advocating for larger political and economic changes at the national level, changes that would solve the waste problems? Why? Or should you do nothing? Why?
4. How does your school or city deal with solid waste? Can solid waste production be limited at your institution or city? How?

What barriers exist that might make it difficult to limit solid waste production?
5. It is possible to have a high standard of living, as in North America and Western Europe, and not produce large amounts of solid waste. How?
6. Often environmental costs are hidden from view, and the "correct" response to an environmental problem is not readily apparent. Read the Issues—Analysis section of this chapter. Which alternative do you prefer? Are there other alternatives?
7. Incineration of solid waste is controversial. Do you support solid waste incineration in general? Would you support an incineration facility in your neighborhood?

Concept Map

Construct a map to show relationships among the following concepts:

composting
leachate

municipal solid waste landfill
recycling

solid waste
source reduction

Interactive Exploration

Check out the website at **http://www.mhhe.com/environmentalscience** and click on the cover of this textbook for quizzing, career information, case studies, and hot links for information on the following topics:

Solid, Toxic, and Hazardous Waste

Recycling

Solid Waste

19 CHAPTER

Regulating Hazardous Materials

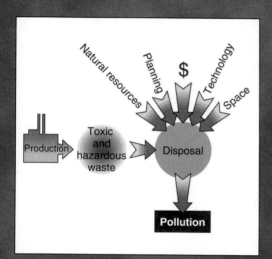

Objectives

After reading this chapter, you should be able to:

- Distinguish between hazardous substances and hazardous wastes.
- Distinguish between hazardous and toxic substances.
- Explain the complexity in regulating hazardous materials.
- Describe the four characteristics by which hazardous materials are identified.
- Describe the environmental problems of hazardous and toxic materials.
- Understand the difference between persistent and nonpersistent pollutants.
- Describe the health risks associated with hazardous wastes.
- Explain the problems associated with hazardous-waste dump sites and how such sites developed.
- Describe how hazardous wastes are managed, and list five technologies used in their disposal.
- Describe the importance of source reduction with regard to hazardous wastes.

Chapter Outline

Hazardous and Toxic Materials in Our Environment

Hazardous and Toxic Substances—Some Definitions

Defining Hazardous Waste

Issues Involved in Setting Regulations

Environmental Close-Up: *Determining Toxicity*

    Identification of Hazardous and Toxic Materials

    Setting Exposure Limits

    Acute and Chronic Toxicity

    Synergism

    Persistent and Nonpersistent Pollutants

Global Perspective: *Lead and Mercury Poisoning*

Environmental Problems Caused by Hazardous Wastes

Health Risks Associated with Hazardous Wastes

Hazardous-Waste Dumps—A Legacy of Abuse

Environmental Close-Up: *Computers— A Hazardous Waste*

    Toxic Chemical Releases

Hazardous-Waste Management Choices

    Reducing the Amount of Waste at the Source

    Recycling of Wastes

    Treatment of Wastes

    Disposal Methods

International Trade in Hazardous Wastes

Global Perspective: *Hazardous Wastes and Toxic Materials in China*

Hazardous-Waste Management Program Evolution

Issues—Analysis: *Love Canal*

Hazardous and Toxic Materials in Our Environment

Our modern technological society makes use of a large number of substances that are hazardous or toxic. The benefits gained from using these materials must be weighed against the risks associated with their use. When these materials are released into the environment in an inappropriate way through accident or neglect, significant environmental and human health consequences result.

- Pesticides thought to degrade in soils turn up in rural drinking-water wells.

- Underground plumes of toxic chemicals emanating from abandoned waste sites contaminate city water supplies.

- A gas leak at a chemical production plant in Bhopal, India, killed more than 2000 people.

- Pesticides spilled into the Rhine River from a warehouse near Basel, Switzerland, destroyed a half million fish, disrupted water supplies, and caused considerable ecological damage.

- The collapse of an oil storage tank in Pennsylvania spilled over 2 million liters (500,000 gallons) of oil into the Monongahela River and threatened the water supply of millions of residents.

- Release of PCBs from electrical parts manufacturing plants before the 1970s contaminated hundreds of hectares of marine sediments in New Bedford Harbor, Massachusetts, resulting in marine life being reduced and the area being closed to fishing.

Toxic and hazardous products and by-products are becoming a major issue of our time. At sites around the world, accidental or purposeful releases of hazardous and toxic chemicals are contami-

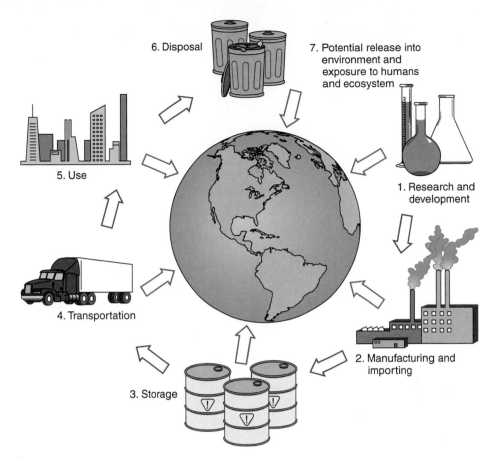

figure 19.1 **The Life Cycle of Toxic Substances** Controlling the problems of hazardous substances is complicated because of the many steps involved in a substance's life cycle.

nating the land, air, and water. The potential health effects of these chemicals range from minor, short-term discomforts, such as headaches and nausea, to serious health problems, such as cancers and birth defects (that may not manifest themselves for years), to major accidents that cause immediate injury or death. Today, names like Love Canal, New York, and Times Beach, Missouri, in the United States; Lekkerkerk in the Netherlands; Vác, Hungary; and Minamata Bay, Japan, are synonymous with the problems associated with the release of hazardous and toxic wastes into the environment.

Increasingly, governments and international agencies are attempting to control the growing problem of hazardous substances in our environment. Controlling the release of these substances is difficult since there are so many places in their cycles of use at which they may be released. (See figure 19.1.)

Hazardous and Toxic Substances—Some Definitions

To begin, it is important to clarify various uses of the words *hazardous* and *toxic,* as well as to distinguish between things that are wastes and those that are not. **Hazardous substances** or **hazardous materials** are those that can cause harm to humans or the environment. Different government agencies have slightly different definitions for what constitutes a hazardous substance. The U.S. Environmental Protection Agency defines hazardous substances as having one or more of the following characteristics:

1. **Ignitability**—Describes materials that pose a fire hazard during routine management. Fires not only

present immediate dangers of heat and smoke but also can spread harmful particles over wide areas. Common examples are gasoline, paint thinner, and alcohol.

2. **Corrosiveness**—Describes materials requiring special containers because of their ability to corrode standard materials or requiring segregation from other materials because of their ability to dissolve toxic contaminants. Common examples are strong acids and bases.

3. **Reactivity** (or explosiveness)—Describes materials that, during routine management, tend to react spontaneously, to react vigorously with air or water, to be unstable to shock or heat, to generate toxic gases, or to explode. Common examples are gunpowder, which will burn or explode; the metal sodium, which reacts violently with water; and nitroglycerine, which explodes under a variety of conditions.

4. **Toxicity**—Describes materials that, when improperly managed, may release toxicants (poisons) in sufficient quantities to pose a substantial hazard to human health or the environment. Almost everything that is hazardous is toxic in high enough quantities. For example, tiny amounts of carbon dioxide in the air are not toxic, but high levels are.

Some hazardous materials fall into several of these categories. Gasoline, for example, is ignitable, can explode, and is toxic. It is even corrosive to certain kinds of materials. Other hazardous materials meet only one of the criteria. Polychlorinated biphenyls (PCBs) are toxic but will not burn, explode, or corrode other materials. While the terms *toxic* and *hazardous* are often used interchangeably, there is a difference. **Toxic** commonly refers to a narrow group of substances that are poisonous and cause death or serious injury to humans and other organisms by interfering with normal body physiology. **Hazardous,** the broader term, refers to all dangerous materials, including toxic ones, that present an immediate or long-term human health risk or environmental risk.

Another important distinction is the difference between hazardous substances and hazardous wastes. Although the health and safety considerations regarding hazardous substances and hazardous wastes are similar, the legal and regulatory implications are quite different. Hazardous substances are materials that are used in business and industry for the production of goods and services. Typically, hazardous substances are consumed or modified in industrial processes. **Hazardous wastes** are by-products of industrial, business, or household activities for which there is no immediate use. These wastes must be disposed of in an appropriate manner, and there are stringent regulations pertaining to their production, storage, and disposal.

There are numerous types of hazardous waste, ranging from materials contaminated with dioxins and heavy metals (such as mercury, cadmium, and lead) to organic wastes. These wastes can also take many forms, such as barrels of liquid waste or sludge, old computer parts, used batteries, and incinerator ash. In industrialized countries, industry and mining are the main sources of hazardous wastes, though small-scale industry, hospitals, military establishments, transport services, and small workshops contribute to the generation of large quantities of such wastes in both the industrialized and developing worlds.

Improper handling and disposal of hazardous wastes can affect human health and the environment through the leakage of toxins into groundwater, soil, waterways, and the atmosphere. The environmental and health effects can be immediate (acute), as when exposure to toxins at a particular site causes sudden illness, or long-term (chronic), as when contaminated waste leaches into groundwater or soil and then works its way into the food chain. The damage caused by hazardous wastes also takes an economic toll, and cleaning up contaminated sites can be costly for local authorities, particularly in poor communities. The handling of hazardous waste requires special training to prevent exposure of workers.

While exact figures regarding the amounts of hazardous waste generated internationally are quite difficult to obtain, some information does exist. The United Nations Environment Programme estimates total annual international generation of hazardous wastes to be between 300 million tonnes (330 million U.S. tons) and 500 million tonnes (550 million U.S. tons), with Organization for Economic Cooperation and Development (OECD) countries accounting for 80 to 90 percent of this quantity. Data from the EPA indicate that the United States generated about 243 tonnes (270 U.S. tons) of hazardous waste in 1999, out of a total of 9 billion tonnes (10 billion U.S. tons) of solid waste. However, in some of the rapidly developing countries of Southeast Asia, the technical and regulatory structures required for proper hazardous-waste management have not kept pace with industrialization. Thailand, for example, generated approximately 2 million tonnes (2.2 million U.S. tons) of hazardous waste in 1990, a figure that is expected to quadruple by 2002.

Defining Hazardous Waste

The definition of hazardous waste varies from one country to another. One of the most widely used definitions, however, is contained in the U.S. **Resource Conservation and Recovery Act** of 1976 (RCRA). The RCRA considers wastes toxic and/or hazardous if they:

cause or significantly contribute to an increase in mortality or an increase in serious irreversible, or incapacitating reversible, illness; or pose a substantial present or potential hazard to human health or the environment when improperly treated, stored, transported, disposed of, or otherwise managed.

This definition gives one an appreciation for the complexity of hazardous-waste regulation.

Working from the RCRA definition, the EPA compiled a list of hazardous

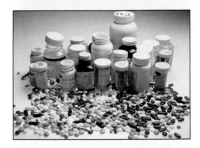

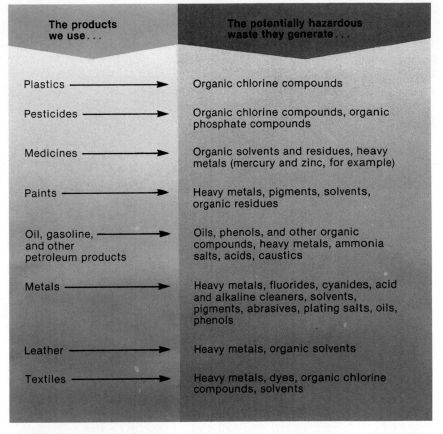

The products we use...	The potentially hazardous waste they generate...
Plastics	Organic chlorine compounds
Pesticides	Organic chlorine compounds, organic phosphate compounds
Medicines	Organic solvents and residues, heavy metals (mercury and zinc, for example)
Paints	Heavy metals, pigments, solvents, organic residues
Oil, gasoline, and other petroleum products	Oils, phenols, and other organic compounds, heavy metals, ammonia salts, acids, caustics
Metals	Heavy metals, fluorides, cyanides, acid and alkaline cleaners, solvents, pigments, abrasives, plating salts, oils, phenols
Leather	Heavy metals, organic solvents
Textiles	Heavy metals, dyes, organic chlorine compounds, solvents

figure 19.2 **Common Materials Can Produce Hazardous Wastes** Many commonly used materials can release toxic or hazardous wastes if not properly disposed of.

wastes. Listing is the most common method for defining hazardous waste in European countries and in some state laws. The EPA has also required that a hazardous waste be identified by testing it to determine if it possesses any one of the four characteristics discussed earlier: ignitability, corrosiveness, reactivity, and toxicity. If it does, it is subject to regulation under RCRA.

Issues Involved in Setting Regulations

Whether a hazardous substance is a raw material, an ingredient in a product, or a waste, problems are associated with determining regulations that pertain to it. In the industrialized countries of Europe and North America, chemical and petro-chemical industries produce nearly 70 percent of all hazardous wastes; in developing countries, the figure is 50 to 66 percent. These industries produce many useful materials that are converted into the everyday products we use. Most toxic and hazardous wastes come from chemical and related industries that produce plastics, soaps, synthetic rubber, fertilizers, medicines, paints, pesticides, herbicides, and cosmetics. (See figure 19.2.)

environmental CLOSE-UP

Determining Toxicity

We are all exposed to potentially harmful toxins. The question is, at what levels is such exposure harmful? One measure of toxicity of a substance is its LD_{50}, the dosage of a substance that will kill (lethal dose) 50 percent of a test population (LD_{50} = lethal dose 50 percent). Toxicity is measured in units of poisonous substance per kilogram of body weight. For example, the deadly chemical that causes botulism, a form of food poisoning, has an LD_{50} in adult human males of 0.0014 milligrams per kilogram. This means that if each of 100 human adult males weighing 100 kilograms consumed a dose of only 0.14 milligrams—about the equivalent of a few grains of table salt—approximately 50 of them will die.

Lethal doses are not the only danger from toxic substances. During the past decade, concern has been growing over minimum harmful dosages, or threshold dosages, of poisons, as well as their sublethal effects.

The length of exposure further complicates the determination of toxicity values. Acute exposure refers to a single exposure lasting from a few seconds to a few days. Chronic exposure refers to continuous or repeated exposure for several days, months, or even years. Acute exposure usually is the result of a sudden accident, such as the tragedy at Bhopal, India, mentioned at the beginning of the chapter.

Acute exposures often make disaster headlines in the press, but chronic exposure to sublethal quantities of toxic materials presents a much greater hazard to public health. For example, millions of urban residents are continually exposed to low levels of a wide variety of pollutants. Many deaths attributed to heart failure or such diseases as emphysema may actually be brought on by a lifetime of exposure to sublethal amounts of pollutants in the air.

Identification of Hazardous and Toxic Materials

In attempting to regulate the use of toxic and hazardous substances and the generation of toxic and hazardous wastes, most countries simply draw up a list of specific substances that have been scientifically linked to adverse human health or environmental effects. However, since many potentially harmful chemical compounds have yet to be tested adequately, most lists include only the known offenders. Historically, we have often identified toxic materials only after their effects have shown up in humans or other animals. Asbestos was identified as a cause of lung cancer in humans who were exposed on the job, and DDT was identified as toxic to birds when robins began to die and eagles and other fish-eating birds failed to reproduce. Once these substances are identified as toxic, their use is regulated.

Countries contemplating regulation of hazardous and toxic materials and wastes must consider not only how toxic each one is but also how flammable, corrosive, and explosive it is, and whether it will produce mutations or cause cancer.

Setting Exposure Limits

Even after a material is identified as hazardous or toxic, there are problems in determining appropriate exposure limits. Nearly all substances are toxic in sufficiently high doses. The question is, When does a chemical cross over from safe to toxic? There is no easy way to establish acceptable levels. Several government agencies set limits for different purposes. The Occupational Safety and Health Administration, the Food and Drug Administration, the U.S. Public Health Service, the Environmental Protection Agency, and others publish

guidelines or set exposure limits for hazardous substances in the air, water, and soil. Furthermore, it is important to recognize that people can be exposed in three primary ways: by breathing the material, by consuming the material through the mouth, or by absorbing the material through the skin. Each of these routes may require different exposure classifications. Regardless, for any new compounds that are to be brought on the market, extensive toxicology studies must be done to establish their ability to do harm. Usually these are tests on animals. (See Environmental Close-Up: Determining Toxicity.) Typically, the regulatory agency will determine the level of exposure at which none of the test animals is affected (**threshold level**) and then set the human exposure level lower to allow for a safety margin. This safety margin is important because it is known that threshold levels vary significantly among species, as well as among members of the same species. Even when concentrations are set, they may vary considerably from country to country. For example, in the Netherlands, 50 milligrams of cyanide per kilogram of waste is considered hazardous; in neighboring Belgium, the toxicity standard is fixed at 250 milligrams per kilogram.

Acute and Chronic Toxicity

Regulatory agencies must look at both the effects of one massive dose of a substance (**acute toxicity**) and the effects of exposure to small doses over long periods (**chronic toxicity**). Acute toxicity is readily apparent because organisms respond to the toxin shortly after being exposed. Chronic toxicity is much more difficult to determine because the effects may not be seen for years. Furthermore, an acute exposure may make an organism ill but not kill it, while chronic exposure to a toxic material may cause death. A good example of this effect is alcohol toxicity. Consuming extremely high amounts of alcohol can result in death (acute toxicity and death). Consuming moderate amounts may result in illness (acute toxicity and full recovery). Consuming moderate amounts over a number of years may result in liver damage and death (chronic toxicity and death).

Another example of chronic toxicity involves lead. Lead has been used in paints, gasoline, and pottery glazes for many years, but researchers discovered that it has harmful effects. The chronic effects on the nervous system are most noticeable in children, particularly when children eat paint chips.

Synergism

Another problem in regulating hazardous materials is assessing the effects of mixtures of chemicals. Most toxicological studies focus on a single compound, even though industry workers may be exposed to a variety of chemicals; and in waste dumps, the compounds are usually found in mixtures. Although the materials may be relatively harmless as separate compounds, once mixed, they may become highly toxic and cause more serious problems than do individual pollutants. This is referred to as **synergism.** For example, all uranium miners are exposed to radioactive gases, but those who smoke tobacco and thus are exposed to the toxins in tobacco smoke have unusually high incidences of lung cancer. Apparently, the radioactive gases found in uranium mines interact synergistically with the carcinogens found in tobacco smoke.

Persistent and Nonpersistent Pollutants

The regulation of hazardous and toxic materials is also influenced by the degree of persistence of the pollutant. **Persistent pollutants** are those that remain in the environment for many years in an unchanged condition. Most of the persistent pollutants are human-made materials. An estimated 30,000 synthetic chemicals are used in the United States. They are mixed in an endless variety of combinations to produce all types of products used in every aspect of daily life. They are part of our food, transportation, clothing, building materials, home appliances, medicine, recreational equipment, and many other items. Our way of life depends heavily on synthetic materials.

An example of a persistent pollutant is DDT. It was used as an effective pesticide worldwide and is still used in some countries because it is so inexpensive and very effective in killing pests. However, once released into the environment, it accumulates in the food chain and causes death when its concentration is high enough. (See chapter 15 for a discussion of DDT as a pesticide.)

Another widely used group of synthetic compounds of environmental concern are polychlorinated biphenyls. PCBs are highly stable compounds that resist changes from heat, acids, bases, and oxidation. These characteristics make PCBs desirable for industrial use but also make them persistent pollutants when released into the environment. At one time, these materials were commonly used in transformers and electrical capacitors. Other uses included inks, plastics, tapes, paints, glues, waxes, and polishes. Although the manufacture of PCBs in the United States stopped in 1977, these persistent chemicals are still present in the soil and sediments and continue to do harm. PCBs are harmful to fish and other aquatic forms of life because they interfere with reproduction. In humans, PCBs produce liver ailments and skin lesions. In high concentration, they can damage the nervous system, and they are suspected carcinogens.

In addition to synthetic compounds, our society uses heavy metals for many purposes. Mercury, beryllium, arsenic, lead, and cadmium are examples of heavy metals that are toxic. These metals are used as alloys with other metals, in batteries, and have many other special applications. In addition, these metals may be released as a by-product of the extraction and use of other metals. When released into the environment, they enter the food chain and become concentrated. In humans, these metals can produce kidney and liver disorders, weaken the bone structure, damage the central nervous system, cause blindness, and lead to death. Because these materials are persistent, they can accumulate in the environment even though only small amounts might be released each year. When industries use these materials in a concentrated form, it presents a hazard not found naturally.

A **nonpersistent pollutant** does not remain in the environment for very long. Most nonpersistent pollutants are biodegradable. Others decompose as a result of inorganic chemical reactions. Still others quickly disperse to concentrations that are too low to cause harm. A biodegradable material is chemically changed by living organisms and often serves as a source of food and energy for decomposer organisms, such as bacteria and fungi. Phenol and many other kinds of toxic organic materials can be destroyed by decomposer organisms.

Other toxic materials, such as many insecticides, are destroyed by sunlight or reaction with oxygen or water in the atmosphere. These include the "soft biocides." For example, organophosphates usually decompose within several weeks. As a result, organophosphates do not accumulate in food chains because they are pollutants for only a short period of time.

Lead and Mercury Poisoning

Lead and mercury are naturally present in rock, and as the rock weathers, they are released into the soil and water. They have been present in the environment and probably have been a source of pollution for centuries. For example, the lead drinking and eating vessels used by the wealthy Romans may have caused the death of many of them. Another example, believe it or not, comes from *Alice in Wonderland,* published in 1865. One of the characters was the Mad Hatter. At that period in history, mercury was widely used in the treatment of beaver skins for making hats. As a result of exposure to mercury, hat makers often suffered from a variety of mental problems; hence, the phrase "mad as a hatter."

In 1953, a number of physical and mental disorders in the Minamata Bay region of Japan were diagnosed as being caused by mercury: 52 people developed symptoms of mercury poisoning, 17 died, and 23 became permanently disabled. In 1970, an outbreak of mercury poisoning in North America was traced to mercury in the meat of swordfish and tuna. In both incidents, the toxic material was not metallic mercury but a mercurous compound, methylmercury. Metallic mercury is converted to methylmercury by bacteria in the water. Methylmercury enters the food chain and may become concentrated as the result of biological amplification. Sufficient amounts in humans can cause brain damage, kidney damage, or birth defects. Today, regulations reduce the release of mercury into the environment and set allowable levels in foods. The problem persists, however, because it is impossible to eliminate the large amounts of mercury already present in the environment, and it is difficult to prevent the release of mercury in all cases. For example, burning coal releases 3200 tonnes (3500 U.S. tons) of mercury into the Earth's atmosphere each year, and mercury is still "lost" when it is used for various industrial purposes.

Like mercury, lead is a heavy metal and has been a pollutant for centuries. Studies of the Greenland Ice Cap indicate a 1500 percent increase in the lead content today as compared to 800 B.C. These studies reveal that the first large increase occurred during the Industrial Revolution, and the second great increase occurred after the invention of the automobile. Oil companies added lead to gasoline to improve performance, and burning gasoline is a major source of lead pollution. There has been a reduction in airborne lead as a result of North America and Europe and many other countries eliminating lead from gasoline. Another source of lead pollution is older paints. Before 1940, indoor and outdoor paints often contained lead.

Disagreement still exists over what levels of lead and mercury can cause human health problems. Although ingested lead from paint can cause death or disability, such a strong correlation cannot be made for atmospheric lead.

There are concerns about fish contamination and public health as a result of methylmercury.

Other toxic and hazardous materials such as carbon monoxide, ammonia, or hydrocarbons can be dispersed harmlessly into the atmosphere (as long as their concentration is not too great), where they eventually react with oxygen.

Because persistent materials can continue to do harm for a long time (chronic toxicity), their regulation is particularly important. Nonpersistent materials need to be kept below threshold levels to protect the public from acute toxicity. They are not likely to present a danger of chronic toxicity since they either disperse or decompose.

Environmental Problems Caused by Hazardous Wastes

Hazardous wastes contaminate the environment in several ways. Many hazardous materials are released directly to the environment. Many molecules that evaporate readily are vented directly to the atmosphere or escape from faulty piping and valves. These materials are often not even thought of as being hazardous waste. Once hazardous wastes are produced, they must be stored. Improper storage or even poor bookkeeping may inadvertently result in a release. Uncontrolled or improper incineration of hazardous wastes, whether on land or at sea, can contaminate the atmosphere and the surrounding environment. The discharge of hazardous substances into the sea, lakes, or rivers often kills fish and other aquatic life. Further, disposal on land in dumps that are later abandoned or in improperly controlled landfills can pollute both the soil and the groundwater as materials leach below the site.

Table 19.1 Top 15 Hazardous Substances, 2001

Substance	Source	Toxic Effects
Arsenic	From elevated levels in soil or water	Multiple organ systems affected. Heart and blood vessel abnormalities, liver and kidney damage, impaired nervous system function
Lead	Lead-based paint Lead additives in gasoline	Neurological damage. Affects brain development in children. Large doses affect brain and kidneys in adults and children.
Mercury	Air or water at contaminated sites Methylmercury in contaminated fish and shellfish	Permanent damage to brain, kidneys, developing fetus
Vinyl chloride	Plastics manufacturing Air or water at contaminated sites	Acute effects: dizziness, headache, unconsciousness, death Chronic effects: liver, lung, and circulatory damage
Polychlorinated biphenyls (PCBs)	Eating contaminated fish Industrial exposure	Probable carcinogens; acne and skin lesions
Benzene	Industrial exposure Glues, cleaning products, gasoline	Acute effects: drowsiness, headache, death at high levels Chronic effects: damage to blood-forming tissues and immune system; also carcinogenic
Cadmium	Released during combustion Living near a smelter or power plant Picked up in food	Probable carcinogen; kidney damage, lung damage, high blood pressure
Benzo[a]pyrene	Product of combustion of gasoline or other fuels In smoke and soot	Probable carcinogen; possible birth defects
Polycyclic aromatic hydrocarbons	Exposure to smoke from a variety of sources	Probable carcinogen; possible birth defects
Benzo[b]fluoranthene	Product of combustion of gasoline and other fuels Inhaled in smoke	Probable carcinogen
Chloroform	Contaminated air and water Many kinds of industrial settings	Affects central nervous system, liver, and kidneys; probable carcinogen
DDT	From food with low levels of contamination Still used as pesticide in parts of world	Probable carcinogen; possible long-term effect on liver; possible reproductive problems
Aroclor 1254 (a mixture of PCBs)	From food and air	Probable carcinogens; acne and skin lesions
Aroclor 1260 (a mixture of PCBs)	From food and air	Probable carcinogens; acne and skin lesions
Trichloroethylene	Used as a degreaser, evaporates into air	Dizziness, numbness, unconsciousness, death
Dibenz[a,h]anthracene	Product of combustion in smoke	Probable carcinogen

Source: Data from Agency for Toxic Substances and Disease Registry.

Because most hazardous wastes are disposed of on or in land, the most serious environmental effect is contaminated groundwater. In the United States alone, an estimated 100,000 active industrial landfill sites may be possible sources of groundwater contamination, along with 200 special facilities for disposal of both liquid and solid hazardous wastes, and some 180,000 surface impoundments (ponds) for all types of waste. (See chapter 15.) Since many kinds of hazardous materials have been stored in underground storage tanks, leaking tanks have been a major source of pollution of soil and groundwater. Nearly 2 percent of North America's underground aquifers could be contaminated with such chemicals as chlorinated solvents, pesticides, trace metals, and PCBs. Once groundwater is polluted with hazardous wastes, the cost of reversing the damage is prohibitive. In fact, if an aquifer is contaminated with organic chemicals, restoring the water to its original state is seldom physically or economically feasible.

Health Risks Associated with Hazardous Wastes

Because most hazardous wastes are chemical wastes, controlling chemicals and their waste products is a major issue in most developed countries. Every year, roughly 1000 new chemicals join the nearly 70,000 in daily use. Many of these chemicals are toxic, but they pose little threat to human health unless they are used or disposed of improperly. For example, many insecticides are extremely toxic to humans. However, if they are stored, used, and disposed of properly, they do not constitute a human health hazard. Unfortunately, at the center of the hazardous-waste problem is the fact that the products and by-products of industry are often handled and disposed of improperly. Table 19.1 is a list of 15 top toxic materials as identified by the Federal Agency for Toxic Substances and Disease Registry.

Establishing the medical consequences of exposure to toxic chemicals is extremely complicated. The problem of linking a particular chemical to specific injuries or diseases is further compounded by the lack of toxicity data on most hazardous substances.

Although assessing environmental contamination from toxic substances and determining health effects is extremely difficult, what little is known is cause for concern. Most older hazardous-waste dump sites, for example, contain dangerous and toxic chemicals along with heavy-metal residues and other hazardous substances.

Hazardous-Waste Dumps—A Legacy of Abuse

In the United States prior to the passage of the Resource Conservation and Recovery Act (RCRA) in 1976, hazardous waste was essentially unregulated. Similar conditions existed throughout most of the industrial nations of the world. The solution to hazardous-waste disposal was simply to bury or dump the wastes without any concern for potential environmental or health risks. Such uncontrolled sites included open dumps, landfills, bulk storage containers, and surface impoundments. These sites were typically located convenient to the industry and were often in environmentally sensitive areas, such as floodplains or wetlands. Rain and melting snow soaked through the sites, carrying chemicals that contaminated underground waters. When these groundwaters reached streams and lakes, they were contaminated as well. When the sites became full or were abandoned, they were frequently left uncovered, thus increasing the likelihood of water pollution from leaching or flooding and of people having direct contact with the wastes. At some sites, specifically the uncovered ones, the air was also contaminated as toxic vapors rose from evaporating liquid wastes or from un-

figure 19.3 **Toxic Chemical Storage** This is a site where toxic wastes were improperly stored. Local governments often assume the problems and costs of correcting bankrupt companies' errors.

controlled chemical reactions. (See figure 19.3.) In North America alone, the number of abandoned or uncontrolled sites is over 25,000, and the list grows yearly. The costs involved in cleaning up the sites are high. Nearly all industrialized countries are faced with costly, even massive, cleanup bills.

The United States has the highest number of hazardous-waste dumps needing immediate attention. Europeans are also paying a heavy price for their negligence. Every country in Europe is plagued by an abundance of toxic-waste sites—both old and new—needing urgent attention. Holland is a good example. Authorities estimate that up to 8 million tonnes (8.8 million U.S. tons) of hazardous chemical wastes may be buried in Holland. Estimates for cleaning up those wastes run as high as $7 billion. In the republics of the former Soviet Union and Eastern Europe, many

hazardous waste sites have been identified recently, and there is no money to pay for cleanup.

In the United States, the federal government has become the principal participant in the cleanup of hazardous-waste sites. The program that deals with the cleanup has popularly become known as **Superfund.** Superfund was established when Congress responded to public pressure to clean up hazardous-waste dumps and protect the public against the dangers of such wastes. The **Comprehensive Environmental Response, Compensation, and Liability Act (CERCLA)** (Superfund) was enacted in 1980. CERCLA had several key objectives:

1. To develop a comprehensive program to set priorities for cleaning up the worst existing hazardous-waste sites.

environmental CLOSE-UP

Computers—A Hazardous Waste

They once cost thousands of dollars and were considered a major investment meant to last. Now computes sell for hundreds of dollars and are obsolete 18 months after they are out of the bubble wrap. Rapid innovation in computer hardware is dramatically cutting the cost and the useful life of modern computers, creating a solid waste problem in the process.

Computers are more than just household waste. They contain large amounts of substances such as lead, cadmium, mercury, and chromium that can leach into soil and contaminate groundwater or, if incinerated, can be released into the air. The average PC contains 5 to 8 pounds of lead (to protect the user from radiation) in the cathode ray tube screen alone. Circuit boards typically contain several pounds of cadmium, mercury, and chromium. In 1998, more than 20 million computers became obsolete in the United States, but only 11 percent were recycled. It will no doubt get worse as computers get increasingly cheaper, faster, and more disposable. According to the National Safety Council, by 2005, 350 million machines in the United States will have reached obsolescence, with at least 55 million of them expected to end up in landfills. Currently, about 75 percent of all computers ever bought in the United States are idling in

attics, basements, and office closets, thus creating a major backlog. The good news is that most computers are highly reusable, and up to 97 percent of the parts can be recycled, either as upgraded components for use in other computers or melted down as scrap.

In 2000, Massachusetts became the first state in the country to initiate a ban on the disposal of household computer screens, TV sets, and other glass picture tubes in landfills and incinerators. The state set up six collection centers to handle the items, and cities and towns must now transport the items to those centers. From there, they will either be refurbished or sent on for recycling. Perhaps, in the not-too-distant future, your used computer will be treated like your old car battery or used car oil—when your computer dies or you upgrade, you will simply take it to your local facility for proper disposal.

2. To make responsible parties pay for those cleanups whenever possible.

3. To set up a $1.6 billion Hazardous Waste Trust Fund—popularly known as Superfund—to support the identification and cleanup of abandoned hazardous-waste sites.

4. To advance scientific and technological capabilities in all aspects of hazardous-waste management, treatment, and disposal.

A **National Priority List** of hazardous-waste dump sites requiring urgent attention was drawn up for Superfund action. Different government entities have different estimates of the extent of the hazardous-waste problem. The U.S. Office of Technology Assessment estimates that 10,000 sites may eventually be placed on the National Priority List and that cleanup could take 50 years and cost up to $100 billion because of the complex mixture of contaminants in most sites. (See table 19.2.) The Gov-

Table 19.2	Common Contaminants Found at Superfund Sites (in Order of Occurrence)
Chemical	
Lead	Ethylbenzene
Cadmium	Benzo[a]anthracene
Toluene	Bromodichloromethane
Mercury	Polychlorinated biphenyls (PCBs)
Benzene	Toxaphene
Trichloroethylene	

Source: Data from U.S. Environmental Protection Agency, Office of Emergency and Remedial Resource, 1999.

ernment Accounting Office believes that the National Priority List could reach more than 4000 sites with cleanup costs of around $40 billion. The EPA has the shortest list of sites (2500) that it says should be placed on the National Priority List at a cost of nearly $30 billion.

By the late 1990s, the Superfund program was still controversial. Millions of dollars have been spent by both the

federal government and industry, but most of the money was spent on litigation and technical studies to support or disprove the claims of the parties. The money has paid for lawyers but has not paid for cleanup. One of the primary reasons for this is the way CERCLA was written. It provided that anyone who contributed to a specific hazardous-waste site could be required to pay for the cleanup of the entire site, regardless of the degree to which they contributed to the problem. Since many industries that contributed to the problem had gone out of business or could not be identified, those that could be identified were asked to pay for the cleanup. Most businesses found it cost-effective to hire lawyers to fight their inclusion in a cleanup effort rather than to pay for the cleanup. Consequently, cleanup has been slow.

In 1997, the U.S. Congress attempted to reauthorize CERCLA. Unable to reach an agreement, CERCLA has been authorized on a year-to-year basis, referred to as a "continuing resolution." However, after a slow start, the Superfund program is showing significant results. By the end of 2000, 1450 sites were on the National Priorities List, and another 59 were being considered for addition to the list. Of the sites listed, 757 (over 50 percent) have had all cleanup activities completed. A further 417 (nearly 30 percent) were in the process of being cleaned up. Most of the remaining sites were either about ready to have cleanup activities begin or were under study to determine the best way to proceed. This has been an expensive undertaking. The EPA estimates that the settlements it has reached with responsible parties amount to over $16 billion. Furthermore, the parties responsible for creating the problem are doing about 70 percent of the cleanup activity.

Toxic Chemical Releases

In 1987, as the result of EPA requirements, certain industries in the United States had to begin reporting toxic chemicals released into the environment. Any industrial plant that released 23,000 kilograms (50,000 pounds) or more of toxic pollutants was required to file a report. These were primarily manufacturing industries. These industries reduced emissions by about 45 percent from 1988 to 1999. In the intervening years, changes have been made in who must report. The quantity that required reporting was reduced to 11,400 kilograms (25,000 pounds). In addition, industries that were originally exempt

from reporting now must report toxic releases. About 3.3 billion kilograms (7.3 billion pounds) of toxic chemicals were reported released into the environment by industry in 1999. (See figure 19.4.)

Today, the primary industries involved in producing these wastes are the mining, metal manufacturing, power generation, chemical, and paper industries. The top 10 toxic wastes released are listed in figure 19.4.

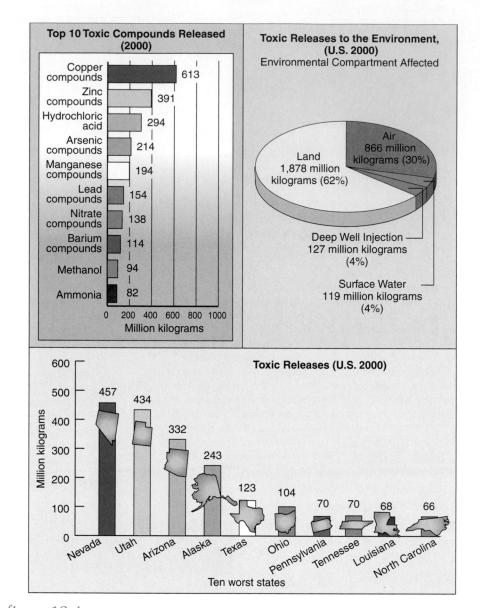

figure 19.4 **Toxic Releases** Toxic substances commonly released to the environment in the United States and the amount for selected states.

Source: Data from the U.S. Environmental Protection Agency.

Hazardous-Waste Management Choices

Today, the two most common methods for disposing of hazardous wastes are land disposal and incineration. The choice between these two methods involves both economic decisions and acceptance by the public. In North

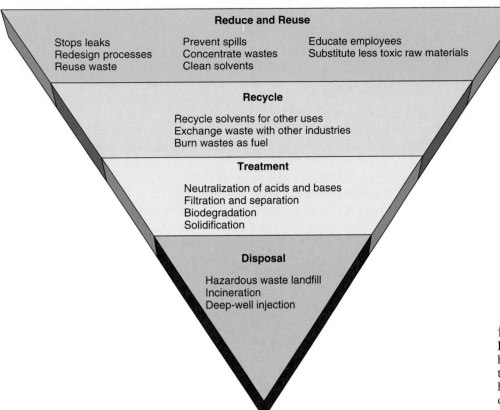

Reduce and Reuse

Stops leaks	Prevent spills	Educate employees
Redesign processes	Concentrate wastes	Substitute less toxic raw materials
Reuse waste	Clean solvents	

Recycle

Recycle solvents for other uses
Exchange waste with other industries
Burn wastes as fuel

Treatment

Neutralization of acids and bases
Filtration and separation
Biodegradation
Solidification

Disposal

Hazardous waste landfill
Incineration
Deep-well injection

figure 19.5 **Pollution-Prevention Hierarchy** The simplest way to deal with hazardous wastes is to not produce them in the first place. The pollution-prevention hierarchy stresses reductions in the amount of hazardous waste produced by employing several different strategies.

America, abundant land is available for land disposal, making it the most economical and most widely used method. In Europe and Japan, where land is in short supply and expensive, incineration is more economical and a major method for dealing with hazardous waste. Because of concerns about the emissions from incinerators, significant amounts of hazardous waste are incinerated at sea on specially designed ships.

Laws and regulations dealing with hazardous-waste disposal are driving industrial behavior toward pollution prevention and waste minimization. In addition, because the costs of safe disposal are mounting, waste-handling firms—both private and public—are looking for better and cheaper ways to treat and dispose of hazardous wastes. Strong, enforceable laws have eliminated the economic incentives to pollute. Unfortunately, as strong laws have been enacted, some small companies that were unable or, more likely, unwilling to properly dispose of their wastes have turned to illegal nighttime dumping.

The environmental costs of not managing hazardous wastes, as witnessed in virtually every industrialized country, are astronomical. And because major generators of hazardous wastes remain liable for past mistakes, economic and regulatory incentives for complying with hazardous-waste regulations should continue to encourage responsible management.

In the past, the management of hazardous waste was always added on to the end of the industrial process. The effluents from pipes or smokestacks were treated to reduce their toxicity or concentration.

In recent years, it has become obvious that a better way to deal with the problem of hazardous waste is to not produce it in the first place. To this end, the EPA and regulatory agencies in other countries have emphasized pollution prevention and waste minimization. Strong regulatory control requires that industries report the hazardous wastes they produce and that the wastes be stored, transported, and disposed of properly.

The EPA now fosters a **pollution-prevention hierarchy** that emphasizes reducing the amount of hazardous waste produced. This involves the following strategy:

First—reduce the amount of pollution at the source.

Second—recycle wastes wherever possible.

Third—treat wastes to reduce their hazard or volume.

Fourth—dispose of wastes on land or incinerate them as a last resort. (See figure 19.5.)

Reducing the Amount of Waste at the Source

Pollution prevention encourages changes in the operations of business and industry that prevent hazardous wastes from being produced in the first place. Many of these actions are simple to perform and cost little. Primary among them are activities that result in fewer accidental spills, leaks from pipes and valves, loss from broken containers, and similar mishaps. These reductions often can be achieved at little cost through better housekeeping and awareness training for employees. Many industries actually save money because they need to buy less raw material because less is being lost.

Pollution prevention can be applied in unusual ways. In 2000, the U.S. Army announced that it would begin issuing an environmentally friendly "green bullet" that contains no lead. The bullet contains a nonpolluting tungsten core instead of lead, which contaminates the soil and air around firing ranges. The U.S. military uses between 300 million and 400 million rounds of small-caliber ammunition each year. The military plans to phase out all use of lead by 2003. Lead contamination has closed hundreds of outdoor firing ranges on military bases across the United States. In 1998, when lead concentrated in firing berms was found to be leaching into Cape Cod's water supply, the Environmental Protection Agency ordered the Massachusetts Military Reservation to stop live-fire training.

Waste minimization involves changes that industries could make in the way they manufacture products, changes that would reduce the amount of waste produced. For example, it may be possible to change a process so that a solvent that is a hazardous material is replaced with water, which is not a hazardous material. This is an example of source reduction: any change or strategy that reduces the amount of waste produced.

Another strategy is to use the waste produced in a process in another aspect of the process, thus reducing the amount of waste produced. For example, water used to clean equipment might be included as a part of the product rather than being discarded as a contaminated waste.

Another technique that can be used to reduce the amount of waste produced is to clean solvents used in processes. Using a still to purify solvents results in a lower total volume of hazardous waste being produced because the same solvent can be used over and over again.

The simple process of allowing water to evaporate from waste can reduce the total amount of waste produced. Obviously, the hazardous components of the waste are concentrated by this process.

Recycling of Wastes

Often it is possible to use a waste for another purpose and thus eliminate it as a waste. Many kinds of solvents can be burned as a fuel in other kinds of operations. For example, waste oils can be used as fuels for power plants, and other kinds of solvents can be burned as fuel in cement kilns. Care needs to be taken that the contaminants in the oils or solvents are not released into the environment during the burning process, but the burning of these wastes destroys them and serves a useful purpose at the same time.

Similarly, many kinds of acids and bases are produced as a result of industrial activity. Often these can be used by other industries. Ash or other solid wastes can often be incorporated into concrete or other building materials and therefore do not require disposal. Thus, the total amount of waste is reduced.

Treatment of Wastes

Wastes can often be treated in such a way that their amount is reduced or their hazardous nature is modified. Dangerous acids and bases can be reacted with one another to produce materials that are not hazardous.

Hazardous wastes that are biodegradable can be subjected to the actions of microorganisms that destroy the hazardous chemicals. Many kinds of organic molecules can be handled in this way.

Air stripping is sometimes used to remove volatile chemicals from water. Volatile chemicals, which have a tendency to vaporize easily, can be forced out of liquid when air passes through it. Steam stripping works on the same principle, except that it uses heated air to raise the temperature of the liquid and force out volatile chemicals that ordinary air would not. The volatile compounds can be captured and reused or disposed of.

Carbon absorption tanks contain specifically activated particles of carbon to treat hazardous chemicals in gaseous and liquid waste. The carbon chemically combines with the waste or catches hazardous particles just as a fine wire mesh catches grains of sand. Contaminated carbon must then be disposed of or cleaned and reused.

Precipitation involves adding special materials to a liquid waste. These bind to hazardous chemicals and cause them to precipitate out of the liquid and form large particles called floc. Floc that settles can be separated as sludge; floc that remains suspended can be filtered, and the concentrated waste can be sent to a hazardous-waste landfill.

Disposal Methods

Incineration (thermal treatment) can be used to destroy a variety of kinds of wastes, although many people are skeptical of this technique because they feel that toxic materials may be escaping from the smokestacks. A hazardous waste incinerator can be used to burn organic wastes but is unable to destroy inorganic waste. A well-designed and well-run incinerator can destroy 99.9999 percent of the hazardous materials that go through it. The relatively high costs of incineration (compared with landfills) and concerns for the safety of surrounding areas in case of accidents have kept incineration from becoming a major method of treatment or disposal.

Incineration accounts for the disposal of only about 2 percent of the hazardous wastes in North America. In Europe, the amount of hazardous wastes destroyed by incineration is higher, but it still amounts to less than 50 percent.

When all other options have been exhausted, any remaining hazardous wastes are typically disposed of on land. (See table 19.3.) For over 80 percent of their hazardous wastes, North America, Europe, and Japan still rely principally on six methods of disposal:

1. Deep-well injection into porous geological formations or salt caverns
2. Discharge of treated and untreated liquids into municipal sewers, rivers, and streams

Table 19.3 Hazardous-Waste Management Methods, United States

Management Method	Share of Total Waste Managed (Percent)	
Land disposal	67	
Injection wells		25
Surface impoundments		19
Hazardous-waste landfills		13
Sanitary landfills		10
Discharge to sewers, rivers, streams	22	
Distillation for recovery of solvents	4	
Burning in industrial boilers	4	
Chemical treatment by oxidation	1	
Land treatment of biodegradable waste	1	
Incineration	1	
Recovery of metals through ion exchange	less than 1	
Total	**100**	

Source: Data from the U.S. Environmental Protection Agency, Office of Solid Waste.

3. Placement of liquid wastes or sludges in surface pits, ponds, or lagoons

4. Storage of solid wastes in specially lined dumps covered by soil

5. Storage of liquid and solid wastes in underground caverns and abandoned salt mines

6. Sending wastes to sanitary landfills not designated for toxic or hazardous wastes

There are techniques that reduce the chance that hazardous materials will escape from these locations and become a problem for the public. Immobilizing a waste puts it into a solid form that is easier to handle and less likely to enter the surrounding environment. Waste immobilization is useful for dealing with wastes, such as certain metals, that cannot be destroyed. Two popular methods of immobilizing waste are fixation and solidification. Engineers and scientists mix materials such as fly ash or cement with hazardous wastes. This either "fixes" hazardous particles, in the sense of immobilizing them or making them chemically inert, or "solidifies" them into a solid mass. Solidified waste is sometimes made into solid blocks that can be stored more easily than can a liquid.

International Trade in Hazardous Wastes

The growth in uncontrolled movement of hazardous wastes between countries has been one of the most contentious environmental issues on the international political agenda. There is particular concern about rich, industrialized countries exporting such wastes to poorer, developing countries lacking the administrative and technological resources to safely dispose of or recycle the waste. For example, in 1999, between 3000 and 4000 tonnes (3300–4400 U.S. tons) of mercury-contaminated concrete waste packed in plastic bags was found in an open dump in a small town in Cambodia. The waste, labeled as "construction waste" on import documents, came from a Taiwanese petrochemical company. In this case, the waste was tracked down and returned to its point of origin. Unfortunately, most such cases are not reported or detected. International awareness of the problems associated with the trade in hazardous wastes has increased noticeably owing to several factors: the growing amounts of such wastes being generated; closure of old waste disposal facilities and political opposition to the development of new ones; and the dramatically higher costs associated with the disposal of hazardous wastes in industrialized countries (and thus the potential to earn profits by exporting such wastes to developing countries with low disposal costs). The debate over controlling hazardous-waste movements between countries culminated in 1989 with the creation of the Basel Convention.

The Basel Convention was negotiated under the auspices of the United Nations Environmental Programme between 1987 and 1989. The objectives of the convention are to minimize the generation of hazardous wastes and to control and reduce their transboundary movements to protect human health and the environment. To achieve these objectives, the convention prohibits exports of hazardous waste to Antarctica, to countries that have banned such imports as a national policy, and to nonparties to the convention (unless those transactions are subject to an agreement that is as stringent as the Basel Convention). Though not part of the original agreement, there is now a broad ban on the export of hazardous wastes from the Northern to the Southern Hemisphere. The waste transfers that are permitted under the Basel regime are subject to the mechanism of prior notification and consent, which requires parties to not export hazardous wastes unless a "competent authority" in the importing country has been properly informed and has consented to the trade.

While the Basel regime may not be perceived as being as successful or significant as some other multilateral environmental agreements, it remains an important part of the international community's attempt to protect the global environment and human health from hazardous materials. Now that the convention has been in place for more than a decade, parties to the convention are beginning to focus on assisting parties with the environmentally sound management of hazardous wastes and with reducing the amount of wastes generated.

The exportation of hazardous wastes for recycling from industrialized nations to developing nations raises other

Global Perspective

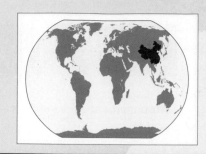

Hazardous Wastes and Toxic Materials in China

Large areas of land are currently used in China for the uncontrolled disposal of industrial wastes. Approximately 600 million tonnes (660 U.S. million tons) of industrialized waste, of which 50 to 70 percent is hazardous, are generated in China annually. Estimates indicate that in recent years, a total of 5.9 billion tonnes (6.5 billion U.S. tons) of industrial waste, occupying 540 million cubic meters (19,000 million cubic feet), have been improperly stored or discarded. The majority of the waste is simply piled on unprotected areas, which causes leaching to surface and groundwater bodies. As a result, environmental accidents are prevalent. For instance, a chromium-residue disposal site in Jinzhou caused groundwater pollution in a 12.5-square-kilometer (4.8-square-mile) area: as a result, water from 1800 wells in nine villages is no longer safe to drink.

The production and use of chemicals is developing rapidly in China. More than 30,000 classes of chemicals are now produced, of which many are toxic. During production, transportation, storage, and use, many releases and spills occur. For example, in 1995, the toxic chemical storage in Shenzhen exploded, causing significant damage to life, property, and the environment.

Management of hazardous and toxic materials in China is still in its early stages; therefore, treatment and disposal technologies are primitive and equipment is poor. None of China's current hazardous-waste disposal sites meets environmental standards. In part, this is due to a lack of funds and management capabilities. A demonstration project, however, is under way in China. It is hoped that this will assist China in developing a management system for hazardous and toxic materials.

The "Law of Pollution Prevention and Control of Solid Wastes" has been made a priority item of legislation by the National People's Congress. Research on hazardous-waste management and disposal has been classified as key in national scientific and technological development plans. Experiments with solid waste declaration and registration, and waste exchanges are being performed, and standard protocols for chemical testing, toxicity evaluation of synthetic chemicals, laboratory analysis, and risk assessment have begun to be formulated. The long- and short-term objectives of the new project include the following:

- Formulate and strengthen China's hazardous- and toxic-materials control laws and regulations, and establish criteria for sound environmental management of wastes.
- Establish technical support system for hazardous-waste management.
- Design demonstration projects for hazardous-waste treatment and disposal.
- Formulate laws and regulations, criteria, and policies for toxic-chemicals management.
- Construct in Beijing a hazardous-waste incineration plant with an annual capacity of 3000 tonnes (3300 U.S. tons).

This project will introduce hazardous- and toxic-materials control laws and regulations, antipollution criteria, and mitigation measures. The capabilities for the management and control of hazardous materials will be developed, and regulatory enforcement will be initiated. It is hoped that the establishment of declaration, registration, licensing, and wastes exchange for hazardous-waste procedures will result in a reduction in the volume of waste generated in China while providing an incentive for recovery and reuse.

The hazardous-waste disposal demonstration project will provide a model, and increased capabilities and experience can be used by China in establishing regional central-disposal facilities for hazardous wastes. Dissemination of the experience gained in the demonstration project should produce a significant improvement in waste-disposal management strategies, resulting in greater awareness, regulation, and environmental protection.

---

questions. Those who support the export of hazardous substances for recycling argue that this practice offers two major benefits: reducing the quantity of such substances that get into the environment through final disposal and slowing down the depletion of natural resources. This argument is undoubtedly correct, provided the receiving country has the proper recycling facilities and adequate environmental standards. Environmentally beneficial trade in hazardous wastes ordinarily requires that there be an established market for these wastes and that the trade be economically viable.

Hazardous-Waste Management Program Evolution

Fundamentally, the goal of a hazardous-waste management program is to change the behavior of those who generate hazardous wastes so that they routinely store, transport, treat, and dispose of them in an environmentally safe manner. The focus on hazardous-waste management typically comes in the second phase of countries' environmental programs, after efforts to address more immediate threats to public health, such as safe drinking water. In the early years of many countries' hazardous-waste management programs, uncontrolled disposal of hazardous waste is the norm. Few, if any, proper treatment and disposal facilities exist. Information on who is generating waste, what types are being generated, and where it is being disposed of is meager or nonexistent.

The transition from an unregulated environment to a regulated one is complex, but hazardous-waste management

Love Canal

Love Canal, near Niagara Falls, New York, was constructed as a waterway in the nineteenth century. It was subsequently abandoned and remained unused for many years. In the 1930s, it became an industrial dump. The Hooker Chemical Company purchased the area in 1947 and also used it as a burial site for 20,000 tonnes (22,000 U.S. tons) of chemicals.

Hooker, later to become part of Occidental Petroleum, then sold the property to the local government for one dollar. A housing development and an elementary school were constructed on the site. Soon after the houses were constructed, people began to complain about chemicals seeping into their basements. In 1978, 80 different chemicals were found in this seepage. Approximately one dozen probable carcinogens were identified among these chemicals.

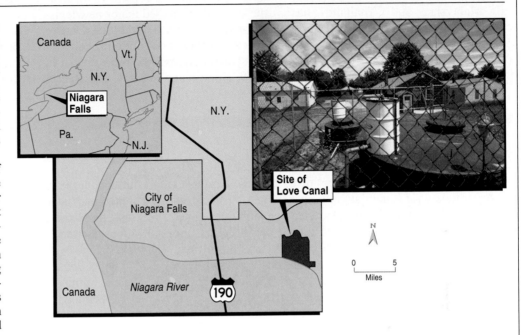

That same year, as a result of these findings, $27 million in government funds were appropriated to purchase homes and permanently relocate 237 families. The funds also provided for the construction of a series of ditches to contain the chemicals and a clay cap to prevent the fumes from entering the atmosphere. But the problems continued.

The remaining 710 families in the Love Canal area were not satisfied with the government's approach to the problem. They cited the fact that women in the area had a 50-percent-higher rate of miscarriages. Of 17 reported pregnancies in the area during 1979, two children were born normal, nine had defects, two were stillborn, and four pregnancies ended in miscarriage. In addition to the abnormalities in birth, there are other biological problems in the Love Canal area.

Neurologists determined that the speed of the nerve impulses in 37 residents who were examined was slower than normal. They stated that chemical exposure could have caused this damage. In 1980, the EPA released the findings of a study that found that 11 out of 36 residents tested in the Love Canal area had broken chromosomes, which are linked to cancer and birth defects. As a result, the federal government released $5 million to temporarily relocate Love Canal residents to motels or other quarters.

By 1990, a $150 million cleanup effort had sealed off the leaky dump, demolished 238 homes nearest the chemical graveyard, and scoured toxins from neighborhood storm sewers and streams (the two major sources of danger to area homes).

In early 1991, some families began to move back into the area. The Love Canal Area Revitalization Agency is planning to sell 236 homes in Love Canal (renamed Black Creek Village). One of the incentives is that the houses can be purchased at a low cost. The families are confident that the houses are safe, but many environmental groups oppose them. In March 1998, the last of 2300 families received compensation for medical claims. Individual compensation ranged from $400,000 to as little as $83.

- Who is responsible for providing treatment for the physical and mental problems experienced by the residents in this community?
- There is a question no one ever seems to ask: Why were permits ever awarded to construct a thousand-unit housing development on top of a site known to contain 20,000 tonnes (22,000 U.S. tons) of toxic wastes?

programs typically evolve through the following major stages: identifying the problem and enacting legislation, designating a lead agency, establishing rules and regulations, developing treatment and disposal capacity, and creating a compliance and enforcement program.

Each of these stages takes a number of years, and at each stage, there are many difficult issues to be resolved. Denmark, Germany, and the United States began this process by passing their first major hazardous waste laws between 1972 and 1976; Canada did so in 1980. In the

decade that followed, all four of these countries developed hazardous-waste regulations and requirements, so that by the end of the 1980s, their regulatory systems were largely operational. Laws and policies developed during the 1990s have focused mainly on waste minimization and recycling, as well as on harmonization with international standards and the cleanup of contaminated sites.

The more recent evolution of regulatory programs in Hong Kong, Indonesia, Malaysia, and Thailand has followed a similar pattern. By the early 1980s, all of these countries had enacted some form of environmental legislation providing at least limited authority to regulate hazardous waste. However, hazardous-waste management received little attention until the late 1980s, after periods of rapid economic growth and the expansion of the countries' manufacturing sectors. From 1989 to 1998, all of these countries passed major new legislation that addressed hazardous waste or developed regulations outlining comprehensive programs: Malaysia did so in 1989, Hong Kong in 1991, Thailand in 1992, and Indonesia in 1995–98. All four countries now have at least one modern hazardous waste treatment, storage, and disposal facility.

While all countries pass through the same stages of program development, no two countries follow precisely the same path. Differences in geography, demographics, industrial profile, politics, and culture lead countries to make different choices at each stage.

Summary

Public awareness of the problems of hazardous substances and hazardous wastes is relatively recent. The industrialized countries of Europe and North America began major regulation of hazardous materials only during the past 30 years, and most developing countries exercise little or no control over such substances. As a result, many countries are living with serious problems from prior uncontrolled dumping practices, while current systems for management of hazardous and toxic waste remain incomplete and incapable of even identifying all hazardous waste.

A number of fundamental problems are involved in hazardous-waste management. First, there is no agreement as to what constitutes a hazardous waste. Moreover, little is known about the amounts of hazardous wastes generated throughout the world. The issue is further complicated by our limited understanding of the health effects of most hazardous wastes and the fact that large numbers of potentially hazardous chemicals are being developed faster than their health risks can be evaluated.

Hazardous-waste management must move beyond burying and burning. Industries need to be encouraged to generate less hazardous waste in their manufacturing processes. Although toxic wastes cannot be entirely eliminated, technologies are available for minimizing, recycling, and treating wastes. It is possible to enjoy the benefits of modern technology while avoiding the consequences of a poisoned environment. The final outcome rests with governmental and agency policy makers, as well as with an educated public.

Key Terms

acute toxicity *441*

chronic toxicity *441*

Comprehensive Environmental Response, Compensation, and Liability Act (CERCLA) *444*

corrosiveness *438*

hazardous *438*

hazardous materials *437*

hazardous substances *437*

hazardous wastes *438*

ignitability *437*

incineration *448*

LD$_{50}$ *440*

National Priority List *445*

nonpersistent pollutant *441*

persistent pollutant *441*

pollution prevention *447*

pollution-prevention hierarchy *447*

reactivity *438*

Resource Conservation and Recovery Act (RCRA) *438*

Superfund *444*

synergism *441*

threshold level *440*

toxic *438*

toxicity *438*

waste minimization *448*

Review Questions

1. Explain the problems associated with hazardous-waste dump sites and how such sites developed.
2. Distinguish between acute and chronic toxicity.
3. Give two reasons why regulating hazardous wastes is difficult.
4. In what ways do hazardous wastes contaminate the environment?
5. Describe how hazardous wastes contaminate groundwater.
6. Why is it often a problem to link a particular chemical or hazardous waste to a particular human health problem?
7. Describe what is meant by the U.S. National Priority List.
8. Describe five technologies for managing hazardous wastes.
9. What is meant by pollution prevention and waste minimization?
10. Describe the pollution-prevention hierarchy.
11. What are RCRA and CERCLA? Why is each important for managing hazardous wastes?

Critical Thinking Questions

1. Scientists at the EPA have to make decisions about thresholds in order to identify which materials are toxic materials. What thresholds would you establish for various toxic materials? What is your reasoning for establishing the limits you do?

2. Go to the EPA's web site (www.epa.gov/enviro/html/ef_overview.html) and identify the major releasers of toxic materials in your area. Were there any surprises? Are there other releasers of toxic materials that might not be required to list their releases?

3. In North America alone, there are over 25,000 abandoned and uncontrolled hazardous-waste dumps. Many were abandoned before the RCRA of 1976 was passed. Who should be responsible for cleaning up these dumps?

4. Look at this chapter's section called "Hazardous-Waste Dumps—A Legacy of Abuse." Do the authors present the information from a particular point of view? What other points of view might there be on this issue? What information do you think these other viewpoints would provide?

5. Many economically deprived areas, Native American reservations, and developing countries that need an influx of cash have agreed, over significant local opposition, to site hazardous-waste facilities in their areas. What do you think about this practice? Should "outsiders" have a say in what happens within these sovereign territories?

6. After reading about the problem with hazardous wastes and toxic materials in China, do you think that the United States or any other country should have the right to intervene if another country is creating significant environmental damage? Why?

Concept Map

Construct a map to show relationships among the following concepts:

CERCLA
corrosiveness
hazardous substances
hazardous wastes

ignitability
national priority list
pollution-prevention hierarchy
reactivity

superfund
toxic
waste minimization

Interactive Exploration

Check out the website at **http://www.mhhe.com/environmentalscience** and click on the cover of this textbook for quizzing, career information, case studies, and hot links for information on the following topics:

Solid, Toxic, and Hazardous Waste

Hazardous and Toxic Wastes

Environmental Policy and Decision Making

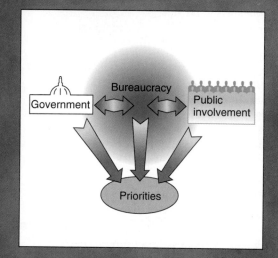

Objectives

After reading this chapter, you should be able to:

- Explain how the executive, judicial, and legislative branches of the U.S. government interact in forming policy.
- Understand how environmental laws are enforced in the United States.
- Describe the forces that led to changes in environmental policy in the United States during the past three decades.
- Understand the history of the major U.S. environmental legislation.
- Understand why some individuals in the United States are concerned about environmental regulations.
- Understand what is meant by "green" politics.
- Describe the reasons environmentalism is a growing factor in international relations.
- Understand the factors that could result in "eco-conflicts."
- Understand why it is not possible to separate politics and the environment.
- Explain how citizen pressure can influence governmental environmental policies.

Chapter Outline

New Challenges for a New Century

We live in remarkable times. This is an era of rapid and often bewildering alterations in the forces and conditions that shape human life. This is evident both in the altered nature of geopolitics in the post–Cold War era and in the growing understanding of the relationship between human beings and the natural world. This relationship varies, however, between the developing and developed countries of the world. One of the major challenges of the foreseeable future will be centered on worldwide environmental impact as developing nations evolve economically.

The end of the Cold War has been accompanied by the swift advance of democracy in places where it was previously unknown and an even more rapid spread of market-based economies. The authority of central governments is eroding, and power has begun to shift to local governments and private institutions. In some countries, freedom and opportunity are flourishing, while in others, these changes have unleashed the violence of old conflicts and new ambitions.

Internationally, trade, investment, information, and even people flow across borders largely outside of governmental control. Domestically, deregulation and the shift of responsibilities from federal to state and local governments are changing the relationships among levels of government and between government and the private sector.

Communications technology has enhanced people's ability to receive information and influence events that affect them. This has sparked explosive growth in the number of organizations, associations, and networks formed by citizens, businesses, and communities seeking a greater voice for their interests. As a result, society outside of government—civil society—is demanding a greater role in governmental decisions, while at the same time impatiently seeking solutions outside government's power to decide.

But technological innovation is changing much more than communication. It is changing the ways in which we live, work, produce, and consume. Knowledge has become the economy's most important and dynamic resource. It has rapidly improved efficiency as those who create and sell goods and services substitute information and innovation for raw materials. During the past 25 years, the amount of energy and natural resources the U.S. economy uses to produce each constant dollar of output has steadily declined, as have many forms of pollution. When U.S. laws first required industry to control pollution, the response was to install cleanup equipment. The shift to a knowledge-driven economy has emphasized the positive connection among efficiency, profits, and environmental protection and helped launch a trend in profitable pollution prevention. More and more people today now understand that pollution is waste, waste is inefficient, and inefficiency is expensive.

Even as their access to information and to means of communication has increased, citizens of the more developed nations are becoming cynical about, and frustrated with, traditional political arrangements that no longer seem responsive to their needs. The confidence of many citizens in the large institutions that affect their lives—such as business, government, the media, and environmental, labor, and civic organizations—is eroding. Individual citizens have lost faith in their ability to influence events and have surrendered to apathy, or worse, to anger.

Since the end of World War II, the world's economic output has increased substantially, allowing widespread improvements in health, education, and opportunity but also creating growing disparities between rich and poor. Even in the highly developed nations, the gap between rich and poor is widening.

Tomorrow's world will be shaped by the aspirations of a much larger global population. The number of people living on Earth has doubled in the last 50 years. Growing populations demand more food, goods, services, and space. Where there is scarcity, population increase aggravates it. Where there is conflict, rising demand for land and natural resources exacerbates it. Struggling to survive in places that can no longer sustain them, growing populations overfish, overharvest, and overgraze. (See figure 20.1.)

As we begin the new century, it is important that we recognize that economic, environmental, and social goals are integrally linked and that we develop policies that reflect that interrelationship. Thinking narrowly about jobs, energy, transportation, housing, or ecosystems—as if they were not connected—creates new problems even as it attempts to solve old ones. Asking the wrong questions is a sure way to get misleading answers that result in short-term remedies for symptoms, instead of cures for long-term problems.

All of this will require new modes of decision making, ranging from the local to the international level. While trend

figure 20.1 **Population Increase Aggravates Scarcity** Struggling to survive in places that can no longer sustain them, growing populations overfish, overharvest, and overgraze.

figure 20.2 **Trend Is Not Destiny** Scenes such as these were very common in North America only a short time ago. Fortunately, for the most part, such photos are today only historic in nature. Positive change is possible.

is not always destiny, the trend that has been evolving over the past several years has been toward more collaborative forms of decision making. Perhaps such collaborative structures will involve more people and a broader range of interests in shaping and making public policy. It is hoped that this will improve decisions, mitigate conflict, and begin to counteract the corrosive trends of cynicism and civic disengagement that seem to be growing.

More collaborative approaches to making decisions can be arduous and time-consuming, and all of the players must change their customary roles. For government, this means using its power to convene and facilitate, shifting gradually from prescribing behavior to supporting responsibility by setting goals, creating incentives, monitoring performance, and providing information.

For their part, businesses need to build the practice and skills of dialogue with communities and citizens, participating in community decision making and opening their own values, strategies, and performance to their community and the society.

Advocates, too, must accept the burdens and constraints of rational dialogue built on trust, and communities must create open and inclusive debate about their future.

Does all of this sound too idealistic? Perhaps it is; however, without a vision for the future, where would we be? As was stated previously, trend is not destiny. In other words, we are capable of change regardless of the status quo. This is perhaps nowhere more important than in the world of environmental decision making. (See figure 20.2.)

Learning from the Past

For the past quarter century, the basic pattern of environmental protection in economically developed nations has been to react to specific crises. Institutions have been established, laws passed, and regulations written in response to problems that already were posing substantial ecological and public health

risks and costs, or that already were causing deep-seated public concern.

The United States is no exception. The U.S. Environmental Protection Agency (EPA) has focused its attention almost exclusively on present and past problems. The political will to establish the agency grew out of a series of highly publicized, serious environmental problems, such as the fire on the Cuyahoga River in Ohio, smog in Los Angeles, and the near extinction of the bald eagle. During the 1970s and 1980s, Congress enacted a series of laws intended to solve these problems, and the EPA, which was created in 1970, was given the responsibility for enforcing most environmental laws.

Despite success in correcting a number of existing environmental problems, there has been a continuing pattern of not responding to environmental problems until they pose immediate and unambiguous risks. Such policies, however, will not adequately protect the environment in the future. People are recognizing that the agencies and organizations whose activities affect the environment must begin to anticipate future environmental problems and then take steps to avoid them. One of the most important lessons learned during the past quarter century of environmental history is that the failure to think about the future environmental consequences of prospective social, economic, and technological changes may impose substantial and avoidable economic and environmental costs on future generations.

Thinking about the Future

Thinking about the future is more important today than ever before, because the accelerating rate of change is shrinking the distance between the present and the future. Technological capabilities that seemed beyond the horizon just a few years ago are now outdated. Scientific developments and the flow of information are accelerating. For example, who would have envisioned cellular phones, voice-activated computers, or a widely used Internet only 20 years ago? Similarly, the environmental effects of changes in global economic activity are being felt more rapidly by both nations and individuals. Examples include the collapse of the cod-fishing industry in the Maritime provinces of Canada and the northeastern United States; the threatened salmon-fishing industry in the Pacific northwest of the United States and British Columbia, Canada; the severe air pollution problems in Mexico; and the shortage of safe drinking water supplies in Russia.

Initiating thought and analysis well in advance of anticipated change can shorten the time needed to respond to such change or allow us to avoid the problem entirely. Because some damage is irreversible, response time is critical.

Thinking about the future is valuable also because the cost of avoiding a problem is often far less than the cost of solving it later. The U.S. experience with hazardous-waste disposal provides a compelling example. Some private companies and federal facilities un-

figure 20.3 **Environmental Debt** The plight of these fishermen, resulting from the collapse of the cod fishery in the northeast United States and Canada and the salmon fishery in the northwest United States and Canada, is an environmental debt inherited from past generations of abuse and misuse.

doubtedly saved money in the short term by disposing of hazardous wastes inadequately, but those savings were dwarfed by the cost of cleaning up hazardous-waste sites years later. In that case, foresight could have saved private industry, insurance companies, and the federal government (i.e., taxpayers) billions of dollars, while reducing exposure to pollutants and public anxieties in the affected communities.

Thinking about the future has another value, one that goes beyond the immediate costs and benefits of environmental protection. Environmental foresight can preserve the environment for future generations. When one generation's behavior necessitates environmental remediation in the future, an environmental debt is bequeathed to future generations just as surely as unbalanced government budgets bequeath a burden of financial debt. (See figure 20.3.)

By anticipating environmental problems and taking steps now to prevent them, the present generation can minimize the environmental and financial debts that its children will incur.

Today, we face new classes of environmental problems that are more diffuse

than those of the past and thus demand different approaches. Since the first Earth Day in 1970, the vast majority of the significant "point" sources of air and water pollution—large industrial facilities and municipal sewage systems—that once spewed untreated wastes into the air, rivers, and lakes have been controlled. The most important remaining sources of pollution are diffuse and widespread: sediment, pesticides, and fertilizers that run off farmland; oil and toxic heavy metals that wash off city streets and highways; and air pollutants from automobiles, outdoor grills, and woodstoves. Pollution from these sources cannot always be controlled with sewage treatment plants or the same regulatory techniques used to check emissions from large industries. Furthermore, we now have the global environmental problems of biodiversity loss, ozone depletion, and climate change. These problems will require cooperative international responses. We are also recognizing that controlling pollutants alone, no matter how successful, will not achieve an environmentally sustainable economy, since many global concerns are related to the size of the human population and the unequal distribution of resources.

Many of the challenges facing human development at the beginning of the twenty-first century are well known. For the most fortunate, the past 50 years have produced a quality of life unprecedented in human history. For another 3 billion people, it has brought about marked improvements in living standards, including significant increases in life expectancy, infant survival, literacy, and access to safe drinking water. This progress notwithstanding, however, more than 20 percent of today's human population still lives in poverty, 15 percent experience persistent hunger, and at least 10 percent is homeless. Moreover, the gap between the very rich and very poor is widening, with whole regions of the world clearly losing ground.

Through both its successes and failures, modern human development has transformed the planet. Human activities have doubled the planet's rate of nitrogen fixation, tripled the rate of invasion by exotic organisms, increased sediment loads in rivers fivefold, and vastly increased natural rates of species extinctions. Clearly, in the future, we will need a different vision from that which shaped our past.

Defining the Future

We are progressing from an environmental paradigm based on cleanup and control to one including assessment, anticipation, and avoidance. Expenditures to develop technologies that prevent environmental harm are beginning to pay off. Agricultural practices are becoming less wasteful and more sustainable, manufacturing processes are becoming more efficient in the use of resources, and consumer products are being designed with the environment in mind. The infrastructures that supply energy, transportation services, and water supplies are becoming more resource efficient and environmentally benign. Remediation efforts are cleaning up a large portion of existing hazardous-waste sites. Our ability to respond to emerging problems is being aided by more advanced monitoring systems and data analysis tools that continually assess the state of the local, regional, and global environment. Finally, we are developing effective ways of restoring or recreating severely damaged ecosystems to preserve the long-term health and productivity of our natural resource base.

This trend will continue if we are willing to develop strong environmental policies, develop new strategies, and closely coordinate the actions of the public and private sectors toward a set of shared, long-term goals. Such a strategy should include three aims. The first is to articulate a vision of the future and the role environmental technology will play in shaping that future. This vision must be built on an understanding of the strengths and weaknesses of past policies and actions. The second is to define the roles of the many individuals and organizations that are needed to implement the goals. The third is to chart a course by offering suggestions for strategic goals for all the partners in this endeavor.

In the long run, environmental quality is not determined solely by the actions of government, regulated industries, or nongovernment organizations. It is largely a function of the decisions and behavior of individuals, families, businesses, and communities everywhere. Consequently, the extent of environmental awareness and the strength of environmental institutions will be two critical factors driving changes in environmental quality in the future.

A concerned, educated public, acting through responsive local, national, and international institutions, will serve as effective agents for avoiding future environmental problems, no matter what they are. Environmental institutions, strengthened by informed public support, will play a critical role in devising and implementing effective national and international responses to emerging issues.

Looking ahead, consensus is growing that the next 50 years will see a world in which people are more crowded, more connected, and more consuming than at any time in human history. The natural environment in which those people live will be stressed as never before. It will be almost certainly warmer, more polluted, and less species-rich. Many of these trends have generated a good deal of discussion recently, under headings ranging from "globalization" to "climate change." Less remarked upon is the fundamental transition under way in the growth of human populations: Rates of increase are now falling almost everywhere in the world, with the result that the number of people on the planet is expected to level off at 10 or 11 billion by the end of the twenty-first century, reaching around 9 billion—still half again as many as today's count—by 2050.

This transition toward a stable human population, if brought to fruition, could fundamentally transform the challenges of environmentally sustainable human development and how people think about them. It could allow people to focus, for the first time in modern history, on questions of sustaining increases in the *quality*, not just the *quantity*, of human life. Can the transition to a stabilizing human population also be a transition to sustainability in which the people living on Earth over the next half century meet their needs while nurturing and restoring the planet's life support systems? These are the questions yet to be answered.

The Development of Environmental Policy in the United States

Public **policy** is the general principal by which government **branches**—the **legislative, executive,** and **judicial**—are guided in their management of public affairs. The legislature (Congress) is directed to declare and shape national policy by passing legislation, which is the same as enacting law. The executive (president) is directed to enforce the law while the judiciary (the court system) interprets the law when a dispute arises. (See figure 20.4.)

When Congress considers certain conduct to be against public policy and against the public good, it passes

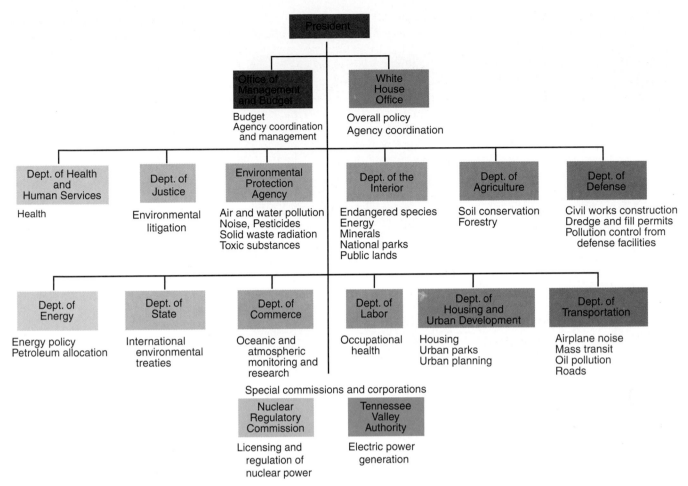

figure 20.4   **Major Agencies of the Executive Branch** Major agencies of the executive branch are shown with their environmental responsibility.

legislation in the form of acts or statutes. Congress specifically regulates, controls, or prohibits activity in conflict with public policy and attempts to encourage desirable behavior.

Through legislation, Congress regulates behavior, selects agencies to implement new programs, and sets general procedural guidelines. When Congress passes environmental legislation, it also declares and shapes the national environmental policy, thus fulfilling its policy-making function. (See figure 20.5.)

Over 90 years ago, President Teddy Roosevelt declared that nothing short of defending your country in wartime "compares in leaving the land even better land for our descendants than it is for us." The environmental issues that Roosevelt strongly believed in, however, did not become major political issues until the early 1970s.

While the publication of Rachel Carson's *Silent Spring* in 1962 is considered to be the beginning of the modern environmental movement, the first Earth Day on April 22, 1970, was perhaps the single event that put the movement into high gear. In 1970, as a result of mounting public concern over environmental deterioration—cities clouded by smog, rivers on fire, waterways choked by raw sewage—many nations, including the United States, began to address the most obvious, most acute environmental problems.

Public opinion polls indicate that a permanent change in national priorities followed Earth Day 1970. When polled in May 1971, 25 percent of the U.S. public declared protecting the environment to be an important goal—a 2500 percent increase over the proportion in 1969.

During the 1970s, many important pieces of environmental legislation were enacted in the United States. (See figure 20.5.) Many of the identified environmental problems were so immediate, so obvious, that it was relatively easy to see what had to be done and to summon the political will to do it. (See table 20.1.)

Just as it was beginning to gain momentum, however, the environmental movement began to decline. When the energy crisis threatened to stall the North American economy in the early 1970s, environmental concerns quickly faded. By 1974, President Gerald Ford had proposed accelerating his administration's leasing program for offshore gas and oil drilling. A turnaround in environmental policy was even more pronounced in the 1980s during the Reagan administration. Former Vice President Walter Mondale was fond of noting that President Ronald

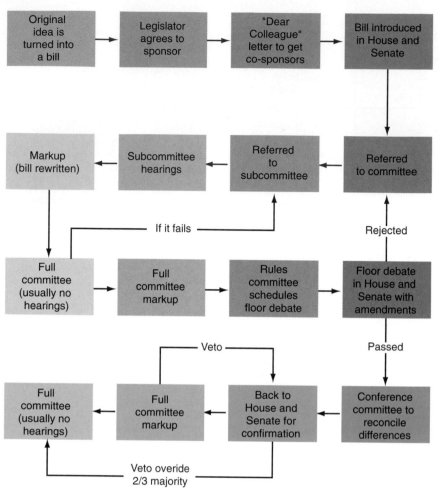

The Legislative Process

figure 20.5 **Passage of a Law** This figure illustrates the path of a bill in the U.S. Congress from organization to becoming a law. As we can see, the process is not a quick one.

presidential election, the environment was established as a major campaign issue, a trend that continues today.

In many respects, the environmental movement of the 1970s and 1980s came of age in the 1990s. The linking of politics and science, and emotionalism and logic in a new environmental movement represented a significant integration of human thinking. It has been said that politics have always forged science. Prioritization of issues and political will determine where money will be spent. By the mid-1990s, it appeared that the political will in the United States to address environmental concerns was on the rise. This was, however, a backlash to environmental policies developing in the United States.

Environmental Backlash— The Wise Use Movement

During the late 1980s, there began to develop a backlash or anti-environmental attitude among sectors of the U.S. populace. A loose-knit organization of several hundred of these groups became known as the Wise Use Movement. The majority of funding for this coalition comes from various interests, including the timber, oil, and coal industries, real estate developers, and ranchers.

In his 1988 book, *The Wise Use Agenda,* Ron Arnold, an early leader of the movement, laid out some of its goals, including:

Elimination of the National Park Service—to be replaced with privately operated parks

Removal of government restrictions on development in wetlands

The opening of all national parks, wilderness areas, and wildlife refuges to off-road vehicles, commercial development, mining, and drilling for oil

Cutting of all the remaining old-growth forests in national forests and replacing them with tree plantations

Recognition of private property rights to water, grazing permits, and mining claims on public lands

Reagan "would rather take a polluter to lunch than to court." During the mid-1980s, the environment was not a priority in the Reagan administration.

The period from 1970 to 1990, however, did bring forth some very tangible accomplishments in environmental policy. Among the most visible and quantifiable is the expansion of protected areas. During this period, federal parklands in the United States—excluding Alaska—increased 800,000 hectares (2 million acres), to 10.5 million (25.9 million acres). In Alaska, 18.3 million additional hectares (45 million acres) were protected, bringing the state's total to over 232 million hectares (573 million acres). Also, the extent of the waterways included in the National Wild and Scenic Rivers System increased by more than 12 times, to some 15,000 kilometers (9300 miles).

By the late 1980s, however, a new environmental awareness and concern began to surface as a major political issue. This was in part due to a number of highly visible environmental problems that appeared nightly on the evening news. Images of toxic waste (including hospital waste, such as used syringes) washing up on the nation's beaches and of the pristine waters of Alaska covered in oil from the *Exxon Valdez* spill made an impact on the public. Once again, the public reacted by organizing and putting pressure on the political system, and, as in 1970, the politicians began to respond. For the first time in the history of the United States, the environment became a key issue in a presidential campaign. In 1988, the environmental records of the two major candidates were hotly debated. Environmentalism was evolving as a major public issue. By the 1992 U.S.

Table 20.1 Major U.S. Environmental and Resource Conservation Legislation

Wildlife conservation
Anadromous Fish Conservation Act of 1965
Fur Seal Act of 1966
National Wildlife Refuge System Act of 1966, 1976, 1978
Species Conservation Act of 1966, 1969
Marine Mammal Protection Act of 1972
Marine Protection, Research, and Sanctuaries Act of 1972
Endangered Species Act of 1973, 1982, 1985, 1988, 1995
Fishery Conservation and Management Act of 1976, 1978, 1982, 1996
Whale Conservation and Protection Study Act of 1976
Fish and Wildlife Improvement Act of 1978
Fish and Wildlife Conservation Act of 1980 (Nongame Act)
Fur Seal Act Amendments of 1983

Land use and conservation
Taylor Grazing Act of 1934
Wilderness Act of 1964
Multiple Use Sustained Yield Act of 1968
Wild and Scenic Rivers Act of 1968
National Trails System Act of 1968
National Coastal Zone Management Act of 1972, 1980
Forest Reserves Management Act of 1974, 1976
Forest and Rangeland Renewable Resources Act of 1974, 1978
Federal Land Policy and Management Act of 1976
National Forest Management Act of 1976
Soil and Water Conservation Act of 1977
Surface Mining Control and Reclamation Act of 1977
Antarctic Conservation Act of 1978
Endangered American Wilderness Act of 1978
Alaskan National Interests Lands Conservation Act of 1980
Coastal Barrier Resources Act of 1982
Food Security Act of 1985
Emergency Wetlands Resources Act of 1986
North American Wetlands Conservation Act of 1989
Coastal Development Act of 1990
California Desert Protection Act of 1994
Federal Agriculture Improvement and Reform Act of 1996

General
National Environmental Policy Act of 1969 (NEPA)
International Environmental Protection Act of 1983

Energy
Energy Policy and Conservation Act of 1975
National Energy Act of 1978, 1980
Northwest Power Act of 1980
National Appliance Energy Conservation Act of 1987
Energy Policy Act of 1992

Water quality
Refuse Act of 1899
Water Quality Act of 1965
Water Resources Planning Act of 1965
Federal Water Pollution Control Acts of 1965, 1972
Ocean Dumping Act of 1972
Safe Drinking Water Act of 1974, 1984, 1996
Clean Water Act of 1977, 1987
Great Lakes Toxic Substance Control Agreement of 1986
Great Lakes Critical Programs Act of 1990
Oil Spill Prevention and Liability Act of 1990

Air quality
Clean Air Act of 1963, 1965, 1970, 1977, 1990

Noise control
Noise Control Act of 1965
Quiet Communities Act of 1978

Resources and solid waste management
Solid Waste Disposal Act of 1965
Resources Recovery Act of 1970
Resource Conservation and Recovery Act of 1976
Waste Reduction Act of 1990

Toxic substances
Toxic Substances Control Act of 1976
Resource Conservation and Recovery Act of 1976
Comprehensive Environmental Response, Compensation, and Liability
    (Superfund) Act of 1980, 1986, 1990
Nuclear Waste Policy Act of 1982

Pesticides
Food, Drug, and Cosmetics Act of 1938
Federal Insecticide, Fungicide, and Rodenticide Control Act of 1972,
    1988
Food Quality Protection Act of 1996

In addition, the Wise Use Movement has been in the forefront of advocating private property rights. They argue that regulations protecting environmentally sensitive areas on private property are unconstitutional "takings." They cite the Fifth Amendment to the U.S. Constitution, which states in part: "nor shall private property be taken for public use, without just compensation." That clause is the basis for the concept of eminent domain, which allows government entities to take land for public projects by paying property owners the land's fair market value.

Wise Use pamphlets argue that extinction is a natural process and that some species were not meant to survive. It has been argued that the movement's signature public relations tactic is to frame complex environmental and economic issues in simple, scapegoating terms that benefit its corporate backers. In the Pacific Northwest, for example, the movement has continually dwelled on a supposed "battle" for survival between spotted owls and the families of the workers involved in harvesting the old-growth timber. "Jobs versus owls" is a good sound bite, but it is not the issue. The real issue is much more complicated.

As would be expected, the environmental community has come out strongly against the Wise Use Movement. Environmentalists have renamed it the "earth- and people-abuse movement" and stated that the basic principles of Wise Users are: "If it grows, cut it; if it's swampy, fill it; if it moves, kill it."

The Changing Nature of Environmental Policy

A 2002 survey revealed that most Americans would prefer a federal government more actively involved in environmental protection. Ninety percent desired stronger active participation by leaders in government and business in environmental concerns. Forty percent believed the general public should bear the primary responsibility for a clean environment, followed by industry at 34 percent and government at 23 percent. The poll also showed that 55 percent of Americans consider the environment to be a serious problem facing the country. Only 24 percent said that a great deal of progress had been made in the past few decades in dealing with environmental problems. It should be noted, however, that these polls, like most polls, often reflect attitudes and not necessarily actions. For example, saying that you support stronger environmental laws does not always translate into individual actions such as purchasing environmentally "friendly" products, active recycling, or support of higher taxes for

environmental CLOSE-UP

Shaping U.S. Environmental Policy as the New Century Begins

During the presidential campaign of 2000, the economy, health care, and education were primary issues, and environmental issues were secondary. However, as the new administration began to formulate policy, national and global environmental issues became increasingly visible. Public debates about drilling in the Arctic National Wildlife Refuge, global warming, the Kyoto Protocol, and the environmental concerns surrounding world trade made the environment front-page news.

Rather than attempting to pass judgment on the Bush administration's policies, it is perhaps more beneficial to look at the polarization that exists over the policies. The statements that follow present widely different perspectives on the significance of policy decisions of the Bush administration. One presents the views of the administration, the other the views of environmental organizations.

The Bush Administration's Environmental Record
(Published by the Bush Administration)

Preamble

President Bush has articulated a vision for environmental protection that focuses on results: Cleaner air, water and land, and healthier people and ecosystems. To achieve these goals, we need a strong and growing economy. Our environmental policies thus recognize the importance of a robust economy, which provides the public and private resources needed to make new investments in environmental conservation. The President sees these goals as complementary, rather than competing—strong economic growth and strong environmental protection can and must go hand-in-hand. In his first year, President Bush made significant progress toward achieving each of these goals.

Brownfields Cleanup—Bringing New Life to Abandoned Sites in Our Cities and Towns: Fulfilling an important campaign commitment, President Bush signed historic legislation that will result in more cleanup and redevelopment of contaminated industrial sites, improving the environment, protecting public health, creating jobs, and revitalizing communities.

Clear Skies—A Clean Air Act for the 21st Century: President Bush's initiative would dramatically improve air quality by cutting power plant's emissions of three critical pollutants by 70 percent—more than any other presidential clean air initiative. This historic legislative proposal would bring clean air to American communities faster, more reliably, and more effectively than the current Clean Air Act.

Energy Bill—Promoting Clean, Affordable, Reliable Energy for America: President Bush has prepared the first national energy policy in years, and is working with Congress to pass legislation that will promote affordable, reliable, and clean energy that is essential to America's security, environmental quality, and economic growth.

Land Conservation—Working in Partnership with States: President Bush has pushed to fully fund the Land and Water Conservation Fund, and worked successfully with Congress to significantly increase its funding. President Bush has also used the Land and Water Conservation Fund to increase support for partnerships for cooperative conservation, and has requested $100 million in FY '03 funding for a new Cooperative Conservation Initiative.

Global Environment—A Realistic, Growth-Oriented Approach to Climate Change: The President has committed America to a new strategy to meet the challenge of long-term global climate change by reducing the greenhouse gas intensity of our economy by 18 percent over the next 10 years. This goal is supported by a broad range of domestic and international climate change initiatives, including $4.5 billion in FY '03 funding for climate change, as well as $178 million for the Global Environment Facility and $50 million to help conserve tropical forests through programs like debt-for-nature swaps.

Budget: President Bush's $44.4 billion FY '03 environment and natural resources budget request is the highest ever—$1.4 billion, or 3 percent, higher than FY '02 enacted. The President's budget proposal provides $4.1 billion, the highest level ever for EPA's operating program, and provides the highest level ever for EPA state program grants, $1.2 billion.

The Bush Budget: Bad News for Our Environment and Our Health
(Published by the Sierra Club, with comments from the Natural Resources Defense Council)

Preamble

As this is written on the eve of Earth Day 2002, our nation's environmental landscape is changing for the worse. Agencies throughout the Bush administration are taking explicit directions from big corporate polluters, allowing these corporations to rewrite the agency rules that give life to America's environmental laws.

It is not news that the Bush administration has an anti-environmental tilt. In fact, the early months of this presidency were defined in part by overwhelming public disapproval of the administration's positions on arsenic in drinking water, drilling in the Arctic National Wildlife Refuge, and carbon dioxide pollution from power plants. Since September 11, however, the environmental assault has a quietly intensified, bolstered by a growing critical mass of presidential appointees at key federal agencies actively pursuing an anti-environment agenda, emboldened by the president's surge in popularity, and unchecked by news media distracted by the war on terrorism.

Energy Research Cuts: The Bush Administration has proposed cutting energy efficiency research and development by 27% overall, with over 50% cut in some specific programs in FY '03. These cuts would hamstring efforts to improve efficiency in homes, vehicles, businesses, and industry.

President Bush has proposed cutting renewable energy research and development programs by 36% in FY '03. This cut would slow the development of key renewable energy technologies.

Environmental Protection Agency Cuts: Overall President Bush is cutting $500 million from the EPA's budget including a cut of $158 million from the EPA's efforts to enforce laws that keep polluters from fouling the air we breathe and the water we drink. In addition to these cuts, his budgetary sleight of hand shifts money to states, crippling the federal government's ability to enforce fair and consistent environmental standards.

Interior Department Cuts: The President's budget includes numerous examples where he shifts money away from conserving landscapes and wildlife and instead uses the money for mining and oil drilling on our public lands. In addition, it tilts the balance from experienced federal oversight and lets individual states decide whether to protect wildlife and open space.

The President cuts the U.S. Fish and Wildlife Service budget by $168 million, slashing money dedicated to protecting wildlife habitat, wetlands restoration, and endangered species.

more governmental regulation of the environment.

While the public was supportive of environmental protection, the 1990s also witnessed a decline in membership of the major environmental organizations. By 1998, membership declines in many of the large environmental organizations was forcing the layoff of staff and the closure of offices. A major part of the problem with the larger environmental groups may be that they have grown into large bureaucracies and, in the process, have lost the trust of many grassroots environmentally concerned individuals. For example, the National Wildlife Federation has an annual budget in excess of $100 million. This includes profits on merchandise of nearly $25 million and nearly $12 million spent on magazines. By any definition, the National Wildlife Federation is a large bureaucracy.

While many of the larger, established environmental organizations are still declining somewhat, many newer, smaller, local, and grassroots organizations are being created or are expanding, often in response to environmental threats in their own communities. Estimates are that some 7000 local environmental organizations are active in the United States. On college campuses, environmental studies programs are expanding at both the undergraduate and graduate levels. Students are not just concerned with local recycling programs or nuclear power plants but are focusing on the broader issues to be faced in the new century. The primary issue is the achievement of long-term sustainability and the fundamental changes in society that this will entail.

Environmental Policy and Regulation

Environmental laws are not a recent phenomenon. As early as 1306, London adopted an ordinance limiting the burning of coal because of the degradation of local air quality. Such laws became more common as industrialization cre-

ated many sources of air and water pollution throughout the world. In the United States, environmental laws often evolved from ordinances passed by local governments. Interested in protecting public health, officials of towns and cities enacted local laws to limit activities of private citizens for the common good. For example, to have "healthy air," many communities enacted ordinances in the 1880s to regulate rubbish burning within city limits. Public health issues were the foundations on which the environmental laws of today were built.

Environmental law in the United States is governed by administrative law. Administrative law is a relatively new concept, having been developed only during the twentieth century. In 1946, Congress passed the Federal Administrative Procedure Act. This act designated general procedures to be used by federal agencies when they exercised their rule-making, adjudicatory, and enforcement powers. This is a rapidly expanding area of law and defines how governmental organizations such as agencies, boards, and commissions develop and implement the regulatory programs they are legislatively authorized to create. Some of the many U.S. federal agencies that impact environmental issues include the Environmental Protection Agency (EPA), the Council on Environmental Quality, the National Forest Service, and the Bureau of Land Management.

Administrative law applies to government agencies and to those that are affected by agency actions. In the United States, many federal environmental programs are administered by the states under the authority of federal and related state laws. States often differ from both the federal government and each other in the way they interpret, implement, and enforce federal laws. In addition, each state has its own administrative guidelines that govern and define how state agencies act. All actions of federal agencies must comply with the 1946 Administrative Procedure Act.

The National Environmental Policy Act of 1969 was enacted in 1969 and signed into law by President Richard

Nixon on New Year's Day, 1970. It is a short, general statute designed to institutionalize within the federal government a concern for the "quality of the environment." NEPA helps encourage environmental awareness among all federal agencies, not just those that prior to NEPA had to consider environmental factors in their planning and decision making. Until 1970, most federal agencies acted within their delegated authority without considering the environmental impacts of their actions. However, in the 1960s, Congress seriously began to study pollution problems. Because Congress has found that the federal government is both a major cause of environmental degradation and a major source of regulatory activity, all actions of the federal government now fall under NEPA.

NEPA forces federal agencies to consider the environmental consequences of their actions before implementing a proposal or recommendation. NEPA has two purposes: first, to advise the president on the state of the nation's environment; and second, to create an advisory council called the Council on Environmental Quality (CEQ). The CEQ outlines NEPA compliance guidelines. The CEQ also provides the president with consistent expert advice on national environmental policies and problems.

Between 1970 and 1977, the CEQ served only as an advisory council to the president. However, in 1977, through an executive order, President Jimmy Carter granted the CEQ authority to issue binding regulations. These regulations set out details for matters that are broadly addressed by NEPA. The CEQ applies to all federal agencies except Congress, the judiciary, and the president.

NEPA has been interpreted narrowly by the federal courts. As a result, many states have passed much stronger state environmental protection acts (SEPAs) as well. Today, NEPA analysis is undertaken as part of almost every recommendation or proposal for federal action. This includes not only actions by agencies of the federal government but also actions of states, local municipalities, and private corporations. NEPA is

Congress's mission statement that mandates the means by which the federal government, through the guidance of the CEQ, will achieve its national environmental policy.

In addition to the passage of NEPA, the 1970s saw a series of new environmental laws passed, including the Resource Conservation and Recovery Act (RCRA), the Comprehensive Environmental Response, Compensation and Liability Act (CERCLA), and the Clean Air Act (CAA). All of these acts are broadly worded to identify existing problems that Congress believes can be corrected to protect human health, welfare, and the environment.

Protecting human health, welfare, and the environment is the national policy that Congress has chosen to encourage. For example, under NEPA, the national policy is to "promote efforts which will prevent or eliminate damage to the environment and biosphere and stimulate the health and welfare of man [humans]." Similarly, under the Clean Air Act, the policy is to "protect and enhance the quality of the Nation's air resources so as to promote the public

health and welfare and the productive capacity of its population."

Each of the above statutes declares national policy on environmental issues and addresses distinct problems. These statutes also authorize the use of some or all of the administrative functions discussed above, such as rule making, adjudication, administrative, civil and criminal enforcement, citizen suits, and judicial review. (See figure 20.6.)

Congress established the Environmental Protection Agency in 1970 as the primary agency to implement the statutes. Administrative functions empower EPA, the states, and private citizens to take responsibility for enforcing the various authorized programs. These administrative functions not only shape environmental law but also control the daily operations of both the regulated industry and agencies authorized to protect the environment.

To date, much environmental law has reflected the perception that environmental problems are localized in time, space, and media (i.e., air, water, soil). For example, many hazardous-waste sites in the United States have been

"cleaned" by simply shipping the contaminated dirt someplace else, which not only does not solve the problem but creates the danger of incidents during removal and transportation. Environmental regulation has focused on specific phenomena and adopted the so-called "command-and-control" approach, in which restrictive and highly specific legislation and regulation are implemented by centralized authorities and used to achieve narrowly defined ends.

Such regulations generally have very rigid standards, often mandate the use of specific emission-control technologies, and generally define compliance in terms of "end-of-pipe" requirements. (See figure 20.7.) Examples in the United States include the Clean Water Act (which applied only to surface waters), the Clean Air Act (urban air quality), and the Comprehensive Environmental Response, Compensation, and Liability Act (Superfund), which applied to specific landfill sites.

If properly implemented, command-and-control methods can be effective in addressing specific environmental problems. For example, rivers such as the Potomac and Hudson in the United States are much cleaner as a result of the Clean Water Act. (See figure 20.8.) Moreover, where applied against particular substances, such as the ban on tetraethyl lead in gasoline in the United States, the command-and-control approach has clearly worked well.

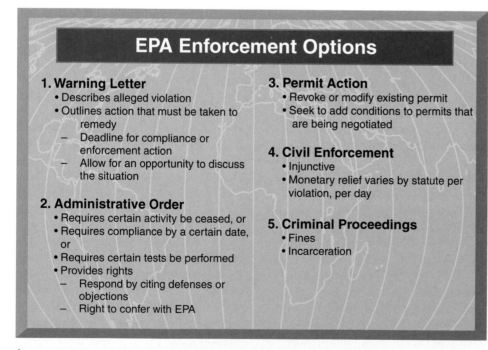

figure 20.6 **Enforcement Options of the U.S. Environmental Protection Agency**
The enforcement options of the U.S. Environmental Protection Agency range from a warning letter to a jail sentence.

Source: U.S. Environmental Protection Agency, Office of Enforcement, Washington, D.C.

The Greening of Geopolitics

Environmental or "green" politics have emerged from minority status and become a political movement in many nations. Issues such as transboundary water supply and pollution, acid precipitation, and global climate change have served to bolster the emergence of green politics. Even in Russia and Eastern Europe, the public has demanded more environmental protection, and leaders seem to be listening.

Concern about the environment is not limited to developed nations. A 1989 treaty signed in Switzerland limits what poorer nations call toxic terrorism—use of their lands by richer countries as dumping grounds for industrial waste. In 1990, more than 100 developing nations called for a "productive dialogue with the developed world" on "protection of the environment." As was covered in chapter 2, the Earth Summit in Rio brought together nearly 180 governments to address world environmental concerns, and the 1997 conference on global warming in Kyoto brought together some 120 nations.

Environmental concern is also a growing factor in international relations. Many world leaders see the concern for environment, health, and natural resources as entering the policy mainstream. A sense of urgency and common cause about the environment is leading to cooperation in some areas. Ecological degradation in any nation is now understood almost inevitably to impinge on the quality of life in others. Drought in Africa and deforestation in Haiti have resulted in large numbers of refugees, whose migrations generate tensions both within and between nations. From the Nile to the Rio Grande, conflicts flare over water rights. The growing megacities of the developing world are areas of potential civil unrest. Sheer numbers of people overwhelm social services and natural resources. The government of the Maldives has pleaded with the industrialized nations to reduce their production of greenhouse gases, fearing that the polar ice caps may melt and inundate the island nation.

Economic progress in developing nations could also bring the possibility for environmental peril and international tension. China, which accounts for 21 percent of the world's population, has the world's third-largest recoverable

Criticism of Environmental Regulation

- **Unrealistic nature of regulations**

- **Failure to use market incentives**

- **Inadequate assignment of responsibility for pollution**

- **Disproportionate harm to small businesses**

- **Technology-based rather than performance-based standards**

- **"Business versus Environment" mentality**

figure 20.7 **Criticisms of Environmental Regulation** While some of the above criticisms are being addressed, others still need to be studied and discussed.

Source: U.S. Environmental Protection Agency, Office of Enforcement, Washington, D.C.

figure 20.8 **Effect of the Command-and-Control Approach** The environmental quality of rivers has dramatically improved over the past 30 years as a direct result of the Clean Water Act. The photo on the left shows a visibly impaired river in New England in the 1970s. The photo on the right shows an environmentally healthy Hudson River in the late 1990s.

environmental CLOSE-UP

Changing the Nature of Environmental Regulation—The Safe Drinking Water Act

According to the American Water Works Association, the 1996 amendments to the Safe Drinking Water Act (SDWA) amount to a radical rewrite of the 1986 amendments. The amendments were written with little substantive input from the regulated community. The 1996 amendments, however, were developed with significant contributions from water suppliers and state and local officials. The new amendments embody a partnership approach that includes major new infusions of federal funds to help water utilities, especially the thousands of smaller systems, comply with the law.

What all this means is that the amended SDWA is more focused on what it was initially intended to achieve. For example, rather than require the EPA to set standards for 25 new contaminants every three years, regardless of need or actual threat to public health (in some cases, it has been argued, this requirement impeded public health protection by diverting resources from the most critical needs), the new law focuses the EPA's efforts on regulating contaminants known to pose health risks and requires cost-benefit analyses and risk assessments to take place before new standards are set. The act also allows the EPA to adopt interim regulations for contaminants on an emergency basis if they pose urgent health threats. The act provides millions of dollars in new funding for crucial research; links source water to drinking water (and sets out a state-administered system designed to protect drinking water from becoming contaminated in the first place); authorizes a new system of federal grants to states, that will enable water utilities to borrow funds to upgrade their systems; and requires water systems to provide customers with annual updates delineating the sources and quality of the water they provide. It has been stated that the 1996 amendments are good for the public and drinking water industry alike. The law has more flexibility, more state responsibility, more cooperative approaches, and is achieved through partnerships.

All too often we take clean, safe drinking water for granted.

coal reserves. If China's current "modernization" campaign succeeds, the boom will be fueled by coal, to the possible detriment of the planet as a whole. (See figure 20.9.)

Some experts estimate that the developing world, which today produces one-fourth of all greenhouse gas emissions, could be responsible for nearly two-thirds by the middle of this next century. Developing nations have repeatedly indicated that they are not prepared to slow down their own already weak economic growth to help compensate for decades of environmental problems caused largely by the industrialized world.

Some developing countries may resist environmental action because they see a chance to improve their bargaining leverage with foreign aid donors and international bankers. Where before the poor nations never had a strategic advantage, they now may have an ecological edge. Ecologically, there could be more parity than there ever was economically or militarily.

A former U.S. ambassador to the United Nations stated that just as the Cold War between the East and West seems to be winding down, "eco-conflicts" between the industrialized North and the developing South may pose a comparable challenge to world peace. National security may no longer be about fighting forces and weaponry alone. It also relates increasingly to wa-

tersheds, croplands, forests, climate, and other factors rarely considered by military experts and political leaders but that, when taken together, deserve to be viewed as equally crucial to a nation's security as are military factors. It is interesting to note that the North Atlantic Treaty Organization (NATO) has developed an office for Scientific and Environmental Affairs, and the U.S. Department of Defense has created an Office of Environmental Security. Environmental offices within traditional military organizations would have been viewed as very unusual until only recently.

In 2000, a dike holding millions of liters of cyanide-laced wastewater gave

figure 20.9 **Developing Concerns** If China's current "modernization" continues, the boom will be fueled by coal to the possible detriment of the planet as a whole.

way at a gold-extraction operation in northwestern Romania, sending a deadly waterborne plume across the Hungarian border and down the nation's second largest river. Cyanide separates gold from ore, and mining operations often store cyanide-laced sludge in diked-off lagoons. After devastating the upper Tisza River, the 50-kilometer (30-mile) long pulse of cyanide and heavy metals spilled into the Danube River in northern Yugoslavia. The diluted plume finally filtered into the Danube delta at the Black Sea, more than 1000 kilometers (620 miles) from and three weeks after the spill. Scientists across Europe warned that the accident could leave the upper Tisza and the nearby Somes Rivers a poisonous legacy for several years if heavy metals are left to linger in the river sediments.

The Hungarian Academy of Sciences has pointed out that 90 percent of the water entering the tributaries of Hungary's rivers flows through Romania, Ukraine, Slovakia, and Austria. Numerous mines, chemical plants, oil refineries, and other sources of pollutants line those tributaries. The Hungarian government stated that the management of en-

vironmental security cannot be stopped at the borders. The government has filed lawsuits seeking monetary damage against the operators of the Baia-Mare gold extraction lagoon, and it has threatened to sue the Romanian government to help recover the cleanup costs. The accident stoked bilateral tensions: Romanian officials accused the Hungarian side of exaggerating the extent of the damage, while Hungarians asserted that the Romanians are downplaying the spill. The issue of environmental security continues to expand in the dialogue between the two nations.

The increased attention to the environment as a foreign policy and national security issue is only the beginning of what will be necessary to avert problems in the future. The most formidable obstacle may be the entrenched economic and political interests of the world's most advanced nations. If the United States, for example, asks others not to cut their forests, then it will have to be more judicious about cutting its own. If North Americans wish to stem the supply of hardwood from a fragile jungle or furs from endangered species, then they will have to stem demand for fancy furniture and fur coats. If they wish to preserve wilderness from the intrusions of the oil industry, then they will have to find alternative sources of energy and use all fuels more efficiently.

What may be needed is self-discipline on the part of the world's haves and increased assistance to the have-nots. In the world today, a billion people live in a degree of poverty that forces them to deplete the environment without regard to its future. Their governments often are too crippled by international debt to afford the short-term costs of environmental safeguards.

William Ruckelshaus, a former administrator of the U.S. Environmental Protection Agency, believes a historical watershed may be at hand. If the industrialized and developing countries did everything they should, he says, the resulting change would represent a "modification of society comparable in scale to the agricultural revolution of the Neolithic age and to the Industrial Revolution of the past two centuries."

Let's put all of this into another perspective that may ring closer to home. By even conservative estimates, the population of the United States is expected to double in the next 50 years. It could be argued then that with twice as many people projected for 2050, it will be necessary to double the total U.S. infrastructure in the next 50 years. Potentially then, this translates to the following: twice as many cars, trucks, planes, airports, parking lots, streets, and freeways; twice as many houses and apartment buildings; twice as many landfills, wastewater treatment plants, hazardous-waste treatment facilities, and chemicals (pesticides and herbicides) for agriculture; in short, twice as much of everything. Under this scenario, it will be necessary to take over and develop in the next 50 years an amount of farmland and scenic countryside equal to the total already developed in the past 200 years. (See figure 20.10.) Megacities such as New York and Los Angeles will double in size, and many new megacities will develop. In short, this vision is not very appealing. What happens to wilderness areas, remote and quiet places, habitat for songbirds, waterfowl, and other wild creatures? What happens to our quality of life?

Another way of looking at the future would be along the lines of what was previously mentioned about the analogy to the agricultural and industrial revolutions. This translates into a future of profound change. A future in which virtually everything we do will change. This future of profound change will also be one of phenomenal opportunity and excitement. It is a future in which we will farm, build, and transport in entirely new ways. This vision of the future could well become known as the environmental revolution.

Terrorism and the Environment

A study of environmental terrorism requires understanding motivations, identifying vulnerabilities and risks, and

figure 20.10 **Options and Tradeoffs for the New Century** Unless we become more creative in areas such as transportation and land use, it will be necessary to develop in the United States alone, in the next 50 years, an amount of farmland and scenic countryside as was developed over the past 200 years.

working on effective solutions. At a time when world populations are increasing, the existing resource base (water, energy, soils) is being stretched to provide for more people and consumed at a faster rate. As the value and vulnerability of these resources increase, so does their attractiveness as terrorism targets. Terrorists often choose targets because of what they represent, for example, skyscrapers and government buildings. Rivaling both of those, however, for the amount of long-term damage that can be inflicted on a country, environmental resources should be considered as being at risk. **Environmental terrorism** is defined as the unlawful use of force against environmental resources so as to deprive populations of their benefits or destroy other property. Over the past several decades, several types of environmental terrorism have been waged, from acts of war to acts by individuals or small groups.

The effects of war on the natural environment are catastrophic. By its nature, war is destructive, and the environment is an innocent victim. While the effects of war on the environment have been with us over the centuries, the 1990–91 con-

flict in the Persian Gulf witnessed a new role for the environment in war: its degradation as a weapon.

The term *eco-terrorism* was applied to Iraqi leader Saddam Hussein's use of oil as a weapon in the war. Iraq dumped an estimated 1.1 billion liters (290 million gallons) of crude oil into the Persian Gulf from Kuwait's Sea Island terminal. The oil spill that resulted from the dumping was the world's largest—over 30 times the size of the 1989 *Exxon Valdez* disaster in Alaska. (See figure 20.11a.)

Saddam Hussein may have engineered the spill to block allied plans for an amphibious invasion of Kuwait, but he was also probably trying to shut down seaside desalination plants that provide much of the freshwater for Saudi Arabia's Eastern Province. Whatever the military or political effect, the environmental effect was much greater.

The Persian Gulf waters, shores, and islands are dotted with coral reefs, mangrove swamps, and beds of sea grass, alive with many unique species of birds, fish, and marine mammals. The gulf is also rather isolated, with only one narrow outlet—the Strait of Hormuz,

just 55 kilometers (34 miles) across. The gulf takes up to five years to flush out. Many of the region's species, such as the bottlenose dolphin, dugong, green turtle, and Caspian tern, were already classified as threatened. The 1991 Gulf War oil spill provided opportunities to monitor the effects of the oil slick on marine organisms and biotic communities. Despite the devastating effects on coastal habitats such as salt marshes in Saudi Arabia, postwar coastal surveys showed that the recovery of these habitats had been relatively rapid. By 1997, several coastal habitats had attained normal to 60 percent diversity. Despite considerable population fluctuations, postwar coastal bird surveys also suggest no overall decline of coastal bird populations when compared to prewar population levels. The long-term fate and possible ecological effects of sunken oil and oil buried under sediments still are uncertain.

Another act of eco-terrorism that took place during the Iraq-Kuwait war was the intentional burning of hundreds of oil wells in Kuwait. (See figure 20.11b.) The air pollution resulting from

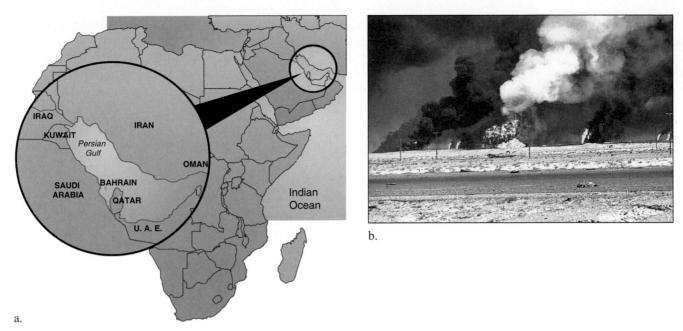

figure 20.11 **Eco-terrorism** (*a*) The Persian Gulf was the victim of environmental terrorism during the 1991 Gulf War. (*b*) Oil well fires burning in Kuwait.

Source: United Nations Development Program.

the burning oil covered a large area. It took several months to extinguish the fires following the war. Using environmental damage as a weapon of war is a frightening concept. The environmental consequences of such acts of eco-terrorism will not be fully understood for years. Though the war in the Persian Gulf lasted less than one year, the environmental effects on some species could be irreversible.

Not all acts of environmental terrorism are as dramatic as those of the 1991 Gulf War. Individual acts affecting a more localized region are also documented, such as in July 2000, when workers at the Cellatex chemical plant in northern France dumped 3000 liters (793 gallons) of sulfuric acid into the Meuse River when they were denied health benefits.

Concerns over the use of biological and chemical weapons as a means of terrorism became, unfortunately, all too real following the events of September 11, 2001. The use of such weapons, however, is not new. During the Vietnam War, a chemical defoliant called Agent Orange was used extensively. During the First World War, several chemicals were used as poison gas. His-

tory is full of examples of both chemical and biological weapons being used. There is a difference, however, between biological and chemical weapons.

Biological weapons are naturally occurring organisms that cause disease. The two most common examples are the bacterium *Bacillus anthracis* (anthrax), which produces a toxin, and the virus that causes smallpox, a highly infectious disease. Anthrax bacteria produce spores that allow them to live in a dormant state in soil. When used as a weapon, the spores enter the lungs, where they are carried into the blood and the immune system. The spores become active, reproduce in large numbers, and release a devastating toxin that is lethal to cells. If enough spores are inhaled, anthrax can be fatal.

The history of biological weapons is a long one. Almost as soon as humans figured out how to make arrows, they were dipping them in animal feces to poison them. The Roman Empire used animal carcasses to contaminate their enemies' water wells. This practice was used again in Europe's many wars, in the American Civil War, and even into the twentieth century. In the fifteenth century, Spanish conquistador Francisco Pizarro gave clothing contaminated with

the smallpox virus to natives in South America. Britain's Lord Amherst continued the practice into the late eighteenth century, spreading smallpox among Native Americans during the French-Indian War by giving them blankets that had been used at a hospital treating smallpox victims.

Chemical weapons are poisons such as mustard gas and nerve gases such as sarin. In the First World War, poison gas was used on both the Eastern and Western fronts from 1915 to 1918. Chlorine gas was the most widely used of the poison gases. Chlorine gas burns and destroys lung tissue. Chlorine is not an exotic chemical. Most municipal water systems use it to kill bacteria.

Modern chemical weapons tend to be made with agents having much greater power, meaning that it takes a lot less of the chemical to kill the same number of people. Many of them use the types of chemicals found in insecticides. When you spray your lawn or garden with a chemical to control aphids, you are, in essence, waging a chemical war on aphids.

In 1972, 103 countries signed the Biological Weapons Convention (BWC), which prohibited the development and use of biological and chemical weapons. Even so, several countries are known to have developed biological weapons since the convention. The BWC still allows research for defense, such as vaccines, against biological weapons. In 2001, the United States announced it was developing a new form of anthrax for defensive research.

Controlling a biological agent is the most troublesome part of using biological weapons and one of the most important reasons they have not been widely used. Bacteria and viruses do not discriminate between ally and foe, and the so-called boomerang effect, the biological agent affecting those who released it, is a potential problem. Still, the use of biological and chemical agents for terrorist activities in the recent past has been alarming.

Fortunately, turning chemical, biological, or nuclear materials into usable weapons is not that easy. First, you have to acquire or manufacture sufficient quantities of the lethal agent. Second, you have to deliver it to the target. And third, you have to either detonate it or spread it around in a way that will actually harm a lot of people.

In 1995, a Japanese cult called Aum Shinrikyo released the nerve agent sarin on the Tokyo subway. The intention was to kill thousands; however, only 12 people died. The cult researchers had spent more than $30 million attempting to develop sarin-based weapons, yet they failed to overcome three major obstacles. They could not produce the chemical in the purity required. Their delivery mechanism was simply to carry plastic bags of sarin onto the trains. And their idea of a distribution system was to pierce those bags with umbrella tips to release the liquid, which would then evaporate. Thus, while 12 people did die, the attack was viewed as unsuccessful.

Of the three classes of weapons of mass destruction, those based on chemicals should be the easiest to make. Their ingredients are often commercially available, and their manufacturing techniques are well known. As mentioned, these weapons have been used from time to time in real warfare, so their deployment is also understood. Biological weapons are trickier; and nuclear weapons trickier still. Germs need to be coddled and are hard to spread. The manufacture and detonation of a nuclear device are extremely complicated.

Nevertheless, there may be ways around these obstacles. One way would be to hire unemployed weapons specialists from the former Soviet Union. Some of these people are known to have left Russia for Iran, Iraq, and North Korea, but none has yet been directly associated with any terrorist group. Numerous attempts have been made to smuggle nuclear materials out of the former Soviet Union, and there are unconfirmed suspicions that Iran may have obtained a Russian nuclear warhead. To date, police and customs officers have seized mostly low-grade nuclear waste. This could not be turned into a proper atomic bomb, but with enough of it, a terrorist group might hope to build a device to spread radioactive contamination in an area.

International Environmental Policy

If there must be a war, let it be against environment contamination, nuclear contamination, chemical contamination; against the bankruptcy of soil and water systems; against the driving of people away from the lands as environmental refugees. If there must be war, let it be against those who assault people and other forms of life by profiteering at the expense of nature's capacity to support life. If there must be war, let the weapons be your healing hands, the hands of the world's youth in defense of the environment.

Mustafa Tolba
Former Secretary General
United Nations Environment
Programme

There are many institutions that address the global environment. In the United Nations system, there are 21 separate agencies that deal with environmental issues. In addition to the United Nations, the World Bank and other institutions charged with economic development play an important role in the implementation of policies and projects that affect the global environment. Institutions that deal with trade such as the World Trade Organization (WTO) and the North American Free Trade Agreement (NAFTA) also affect the global environment.

While there have been examples of successful initiatives, global organizations have not been able to achieve significant progress in reversing global environmental degradation. There are several reasons for this. Some fail because they are controlled by a disparate membership with competing interests who are unable to reach consensus on complex and difficult issues. An example of this is the UN Food and Agriculture Organization (FAO), under whose policies only limited progress has been made in dealing with global fishing and agriculture issues. Although the FAO has pursued the collection of statistics on fish stocks and it is clear that most fisheries are being fished at or above capacity, it has not acted to develop systems that would conserve fish stocks, and individual governments continue to set catch quotas above sustainable levels.

Other institutions do not succeed because they are responsible only for specific functions or activities and are unable to address whole issues on their own. The World Bank, for example, can provide guidelines for dealing with air pollution or biodiversity preservation only for development projects that use World Bank funds. Still other institutions have failed to take even initial steps to weave the environment into their policies and programs as a permanent concern. For example, the WTO focuses on trade issues but has yet to confront the nexus of trade—sustainable economic growth and the environment.

International coordination and political resolve are necessary if the goals of preserving and protecting the global environment are to be realized. A step in that direction was the 1972 United Nations Conference in Stockholm, Sweden. This was the first international conference

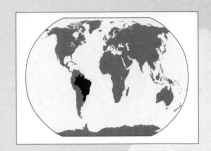
Earth Summit on Environment and Development

In June 1992, representatives from 178 countries, including 115 heads of state, met in Rio de Janeiro, Brazil, at the Earth Summit. Officially, the meeting was titled the United Nations Conference on Environment and Development (UNCED), and it was the largest gathering of world leaders ever held. The first Earth Summit had been held 20 years earlier in Stockholm, Sweden. At that time, the planet was divided into rival East and West blocs and was preoccupied with the perils of the nuclear arms race. With the collapse of the East bloc and the thawing of the Cold War, a fundamental shift in the global base of power had occurred.

Today, the more important division, especially on environmental issues, is not between East and West but between North (Europe, North America, and Japan) and South (most of Asia, Africa, and Latin America). And, though the immediate threat of nuclear destruction has lifted, the planet is still at risk.

The idea behind the Earth Summit was that the relaxation of Cold War tensions, combined with the growing awareness of ecological crises, offered a rare opportunity to persuade countries to look beyond their national interests and agree to some basic changes in the way they treat the environment. The major issues are clear: The developed countries of the North have grown accustomed to lifestyles that are consuming a disproportionate share of natural resources and generating the bulk of global pollution. Many of the developing countries of the South are consuming irreplaceable global resources to provide for their growing populations.

The Earth Summit was intended to promote better integration of nations' environmental goals with their economic aspirations. Although the hopes of some developing nations for large commitments of new foreign assistance did not fully materialize, much was accomplished during the summit.

- The *Rio Declaration on Environment and Development* sets out 27 principles to guide the behavior of nations toward more environmentally sustainable patterns of development. The declaration, a compromise between developing and industrialized countries that was crafted at preparatory meetings, was adopted in Rio without negotiation due to fears that further debate would jeopardize any agreement.

- States at UNCED also adopted a voluntary action plan called *Agenda 21,* named because it is intended to provide an agenda for local, national, regional, and global action into the twenty-first century. UNCED Secretary General Maurice Strong called Agenda 21 "the most comprehensive, the most far-reaching and, if implemented, the most effective program of international action ever sanctioned by the international community." Agenda 21 includes hundreds of pages of recommended actions to address environmental problems and promote sustainable development. It also represents a process of building consensus on a "global work-plan" for the economic, social, and environmental tasks of the United Nations as they evolve over time.

- The third official product of UNCED was a "*non-legally binding authoritative statement of principles for a global consensus on the management, conservation, and sustainable development of all types of forests.*" Negotiations on the forest statement, begun as negotiations for a legally binding convention on forests, were among the most difficult of the UNCED process. Many states and experts, dissatisfied with the end result, came away from UNCED seeking further negotiations toward agreement on a framework convention on forests.

specifically dealing with global environmental concerns. The UN Environment Programme, a separate department of the United Nations that deals with environmental issues, developed out of that conference. The United Nations Conference on the Law of the Sea produced a comprehensive convention that addressed many of the issues concerning jurisdiction over ocean waters and use of ocean resources. This treaty is viewed by many as a model for international environmental protection. Positive results are already evident from application of the agreements on pollution control, marine mammal protection, navigation safety, and other aspects of the marine environment. The issue of deep ocean mining, primarily of manganese nodules, has to date kept many industrial nations, such as Germany and the United States, from ratifying the treaty.

Over 150 global environmental treaties have been negotiated since the start of the twentieth century. In addition, at least 500 bilateral agreements are now in effect dealing with cross-border environmental issues. These agreements generally address single issues, such as the Framework Convention on Climate Change or the Convention on International Trade in Endangered Species of Wild Fauna and Flora. History has shown that most of the existing agreements are lacking in ambition or are framed in such a way that they result in little concrete action.

However, there have been several successful international conventions and treaties. The Antarctic Treaty of 1961 reserves the Antarctic continent for peaceful scientific research and bans all military activities in the region. The 1979 Convention on Long-Range Transboundary Air Pollution was the first multilateral agreement on air pollution and the first environmental accord involving all the nations of Eastern and Western Europe and North America. The 1987 Montreal Protocol on Substances That Deplete the Stratospheric

Overview of an International Organization—The International Whaling Commission

The first records of whaling can be traced back hundreds of years to the twelfth century in the Bay of Biscay of Spain. During whaling's heydays of the 1930s, 85 percent of the world's catch of whales was coming from the Antarctic, with 95 percent of these catches dominated by Norway and the United Kingdom.

With the possible exhaustion of whale stock, the problem of whales as a common property resource came to the forefront. When a resource is common property, no single user has a right to the resource, nor can they prevent others from sharing in its exploitation. These doctrines of freedom of the seas and common property resources date back to Roman times and were constantly used by states as an argument to justify the overfishing of whale stocks in previous centuries. Seeing a need to keep statistical data on the populations of different whale species, Norway, at the request of the International Council for the Exploration of the Seas, set up the Bureau of International Whaling Statistics in 1930 to deal with whaling problems. This eventually led to the drafting and signing of the Convention for the Regulation of Whaling, in 1931. The convention "applied to all waters . . . but was only applicable to baleen whale," and provided exemptions for aboriginal subsistence whaling. This convention had little effect on the amount of whaling, because important nations such as Germany and Japan failed to adhere to the rules. A new draft of the convention was brought into effect in 1937 and included such countries as Norway, the United Kingdom, and Germany. In 1946, the International Whaling Conference (IWC) was convened by the United States to address the post–World War II whaling issues.

The measures to be taken by this commission were to completely protect certain whale species, specify certain whale sanctuaries, set limits on the number of whales taken, decide when open and closed hunting seasons would begin, and specify the size and age of those whales that could be taken. It is important to point out that the majority of the nations who were original members of the IWC were whaling states and had no real interest in protecting the whale in order to preserve biologic diversity. The formation and actions of the IWC in its early years were purely economically motivated.

The ineffectiveness of the IWC to control the slaughter of whale populations continued into the 1970s until environmentally oriented nongovernmental organizations began to champion the struggle of the whale. Soon millions of people in the world were chanting "save the

whale" and began demanding that the IWC pass a moratorium to end whaling. But a powerful veto coalition made up of Japan, Norway, the Soviet Union, Iceland, Chile, and Peru was able to control the vote in the IWC and defeat any measures that were harmful to the whaling industry as a whole.

Seeing that they could not influence the voting of the IWC alone, the United States began to form a new strategy. To ensure a moratorium on whaling in the IWC, nonwhaling states from the developing nations were brought in to outnumber the whaling states. The state leading this group of new members was the small island nation of Seychelles. Despite economic threats from Japan, Seychelles held strong and was able to push for a moratorium that was eventually passed by the IWC in 1982. Although many believed that this was the final victory in the struggle to end whaling, the war waged on. Nations such as Norway and Iceland threatened to leave the IWC and eventually did so. Other nations found loopholes in the moratorium and continued to kill large numbers of whales. Today, the future of both the moratorium and the IWC are unclear. Japan rallies developing nations onto its side with promises of economic aid. Other nations such as Norway are refusing to abide by the ruling handed down by the IWC and state that they will resume whaling. With a decline in the United States' ability to influence the IWC, perhaps the only hope for the whale in the future is the changing attitude of the Japanese public toward the killing of whales.

Ozone Layer addresses the ozone protective shield problem.

Agreed to in 1987, the Montreal Protocol's objective was to phase out the manufacture and use of chemicals depleting the Earth's protective ozone layer (see chapter 17). Since its adoption over a decade ago, 160 countries are now parties, representing over 95 percent of the Earth's population. Production and consumption of chlorofluorocarbons (CFCs), carbon tetrachloride, halons, and methyl chloroform have been phased out in developed countries, with reduction schedules set for their use in developing countries. In 1999, developing countries ended the production and consumption of CFCs, with phaseout scheduled for 2010.

By the late 1990s, the concentration of some CFCs in the atmosphere had started to decline, and predictions are

$\mathcal{G}$lobal Perspective

Eco-Labels

Green consumerism is the concept of rational consumption of our scarce resources for the benefit of the environment and future generations. The old saying "the world is enough for everyone's need but not for everyone's greed" calls for a change in our behavior and lifestyle in favor of a sustainable future. Eco-labels have been introduced in a number of countries to help consumers choose products with a proven environmental edge, determined by the product's choice of raw materials, production process, product life cycle, and associated disposal problems. Eco-labels provide evidence that prod-

ucts have met the safety, quality, and environmental protection requirements of the authority that issues the label.

The first eco-label was introduced in Germany in 1978. This distinctive Blue Angel label was followed by other environmental certification, including the White Swan of Northern European countries, Environmental Choice of Canada, Green Label of Singapore, and Environmental Label of China. By 1995, over a dozen countries had adopted this system of providing consumers with information enabling them to become responsible green consumers. The eco-label of the European Union is significant because it is the world's first regional scheme to apply the same minimum standards across national markets.

1992
European Union - 'Eco-label'

1992
Singapore - 'Green Label'

1993
China - 'Environmental Label'

1978
Germany - 'Blue Angel'

1988
Canada - 'Environmental Choice'

1989
Nordic Council - 'White Swan'

Source: ECCO, *Bulletin of the Environmental Campaign Committee.*

that the ozone layer could recover by the middle of this century. Among the many reasons for these achievements, three merit attention:

- *Global agreement on the nature and seriousness of the threat.* Even the strongest skeptics could not deny the Antarctic ozone hole, which was first brought to international attention by British scientists in 1985. It

was understood for the first time that emissions of ozone-depleting substances were in reality putting our lives and the lives of future generations at risk. Decisive action was required.

- *A cooperative approach, especially between developed and developing countries.* In the developed countries, it was recognized that their in-

dustries had contributed significantly to this global problem and that they had to take the lead in stopping emissions and finding alternatives. It was also recognized that solving the ozone problem required a global solution, with all countries committed to eliminating ozone-depleting substances. Innovative partnerships, including early controls for developed countries and a grace period, funding, and technology transfer for developing countries, were set in place.

- *Policy based on expert and impartial advice.* The parties to the Montreal Protocol were able to receive impartial advice from their science, technical, and economic committees. These drew together experts from around the world to evaluate the need for further action and propose options that are technically and economically feasible.

Put these three factors together—acknowledgment of the threat, agreement to cooperate, and commitment to take effective action based on expert advice—and you have a potentially strong recipe to solve global issues, such as climate change and biodiversity loss.

Even in cases where countries have ratified treaties that have entered into force, parties to the treaties do not always comply with their provisions. An example is Russia's noncompliance with the Montreal Protocol. There are few penalties for noncompliance other than public anger, in part because countries are unwilling to give up their sovereignty.

Why is the implementation of global environmental agreements so difficult? One answer is that for most of these accords to function effectively, participation must be truly global. Yet because of the unique nature of individual nations and their differing economies, it is difficult to reach meaningful agreement among the necessary participants. Another obstacle to implementing global environmental agreements is poorer nations' lack of capacity to comply with international treaty requirements. Some treaties have addressed this problem by offering financial assistance to countries that are unable to comply, for example, the Multilateral Fund of the Montreal Protocol. Other agreements have established more relaxed timetables for compliance for these countries. In many cases, however, the financing has not been forthcoming, and differentiating among countries based on their ability to act has become politically controversial. The Kyoto Protocol, for example, deals with emissions reductions from developed countries only, which has opened the treaty to criticism in the United States. The challenge of developing accords that are both effective and fair has proved to be a major obstacle to progress.

Despite these problems, treaties are the best traditional tools of global environmental governance. Frameworks for action that are negotiated among different nations are key to leveling the playing field and establishing the rules under which governments, businesses, nongovernmental organizations, and citizens can work together toward a common goal.

In 1997, the General Assembly of the United Nations held a special session and adopted a comprehensive document entitled Programme for the Further Implementation of Agenda 21 prepared by the Commission on Sustainable Development. It also adopted the program of work of the Commission for 1999–2003. The Commission on Sustainable Development (CSD) was created in 1992 to ensure effective follow-up of the United Nations Conference on Environment and Development held in Rio de Janeiro, Brazil.

The Commission on Sustainable Development consistently generates a high level of public interest. Over 50 national leaders attend the CSD each year, and more than 1000 nongovernmental organizations are accredited to participate in the commission's work. The commission ensures the high visibility of sustainable development issues within the UN system and helps to improve the UN's coordination of environment and development activities. The CSD also encourages governments and international organizations to host workshops and conferences on different environmental and cross-sectoral issues. The results of these expert-level meetings enhance the work of CSD and help it work better with national governments and various nongovernmental partners in promoting sustainable development worldwide.

There is no international legislature with authority to pass laws; nor are there international agencies with power to regulate resources on a global scale. An international court at The Hague in the Netherlands has no power to enforce its decisions. Nations can simply ignore the court if they wish. However, a growing network of multilateral environmental organizations have developed a greater sense of their roles and a greater incentive to work together. These include not only the United Nations Environment Programme but also the Environment Committee of the Organization for Economic Cooperation and Development (OECD) and the Senior Advisors on Environmental Problems of the Economic Commission for Europe. Such institutions perform unique functions that cannot be carried out by governments acting alone or bilaterally.

Is the goal of an environmentally healthy world realistic? There is growing optimism that the community of nations is slowly maturing with regard to our common environment. We have all suffered a loss of innocence about "earth management." Laissez-faire may be good economics, but it can be a prescription for disaster in ecology.

This environmental "coming of age" is reflected in the broadening of intellectual perspective. Governments used to be preoccupied with domestic environmental affairs. Now, they are beginning to broaden their scope to confront problems that cross international borders, such as transboundary air and water pollution, and threats of a planetary nature, such as stratospheric ozone depletion and climatic warming. It is becoming increasingly evident that only decisive mutual action can secure the kind of world we seek.

Environmental Policy and the European Union

The environment knows no frontiers" was the slogan of the 1970s, when the European Community—known today as the European Union—began to develop its first environmental legislation.

The early laws focused on the testing and labeling of dangerous chemicals, testing drinking water, and controlling air pollutants such as sulfur dioxide, oxides of nitrogen, and particulates from power plants and automobiles. Many of the directives from the 1970s and 1980s were linked to Europe's desire to improve the living and working conditions of its citizens.

In 1987, the Single European Act gave this growing body of environmental legislation a formal legal basis and set three objectives: protection of the environment, protection of human health, and prudent and rational use of natural resources.

The treaty reflected what many governments had already understood: that countries are part of an interconnected and interdependent world of people who are bound together by the air they breathe, the food they eat, the products they use, the wastes they throw away, and the energy they consume.

Similarly, a factory in one European nation may import supplies and raw material from several other neighboring nations; it may consume energy produced from imported gas, produce wastes that affect the air and water quality across the border or downstream, and export products whose wastes become the risk and responsibility of governments and peoples several hundred or many thousand kilometers away.

The 1992 Maastricht Treaty formally established the concept of sustainable development in European Union law. Then, in 1997, the Amsterdam Treaty made sustainable development one of the overriding objectives of the European Union. The treaty considerably strengthened the commitment to the principle that the European Union's future development must be based on the principle of sustainable development and a high level of protection of the environment. The environment must be integrated into the definition and implementation of all of the Union's other economic and social policies, including trade, industry, energy, agriculture, transport, and tourism.

Under certain conditions, the member states may maintain or introduce environmental standards and requirements that are stricter than those of the Union. The European Commission verifies the compatibility of these stricter national programs and the potential effects on other member states.

New International Instruments

Over the past few decades, the global community has responded to emerging environmental problems with unprecedented international agreements— notably the Montreal Protocol, the Framework Convention on Climate Change, the Convention on Biological Diversity, and the Convention to Combat Desertification. There are many others.

In the course of crafting these global agreements, many important lessons have been learned. As the experience with the Montreal Protocol shows, the scientific community can play a crucial role in two ways: first, confirming the links between human activities and global environmental problems; and second, showing what could happen to human health and the global environment if nothing is done.

When the evidence is in hand, an international consensus to act can emerge quickly. The same process is underway in the current international debate on climate change, but the process has been more difficult because the linkages between human activities and global environmental impacts are more complex and still not completely understood. Nevertheless, a global consensus for action is emerging. Another important lesson learned has to do with the structure of international agreements and the elements that can contribute to an effective structure. In the case of the Montreal Protocol, the agreement was not punitive and favored incentives and results-oriented approaches. All nations participated in the agreement but at different levels of responsibility in recognition of their differing conditions. The Kyoto Protocol to the Climate Change Convention has benefited from the experience with the Montreal Protocol and included many of the same elements.

A third vital lesson is that, to the extent possible, every interested party must have an opportunity to participate as full partners in the process and to voice their concerns. It is particularly helpful for environmental advocacy groups and the business community to be part of this process. International agreements need to provide incentives to foster public-private partnerships and to provide a role for business leaders to seek innovative technical solutions.

A fourth lesson concerns the role of governments in implementing these conventions. Government actions need to be consistent and predictable; provide sufficient lead times; favor government-led incentives over direct industry subsidies; and use flexible, market-based solutions where they are appropriate.

Finally, a fifth lesson is that agreements mark the beginning of a process, not the end. Scientists and nongovernmental organizations must continue to further global understanding of environmental problems and communicate what they have learned to the public and policy makers. Policy makers, in turn, must be flexible and respond to changing circumstances with new or modified policy solutions.

It All Comes Back to You

No set of policies, no system of incentives, no amount of information can substitute for individual responsibility when it comes to ensuring that our grandchildren will enjoy a quality of life that comes from a quality environment. Information can provide a basis for action. Vision and ideas can influence perceptions and inspire change. New ways to make decisions can empower those who seek a role in shaping the future. However, all of this will be meaningless unless individuals acting as citizens, consumers, investors, managers, workers, and professionals decide that it is important to them to make choices on the basis of a broader, longer view of their self-interest; to get involved in turning those choices into action; and, most important, to be held accountable for their actions.

The combination of political will, technological innovation, and a very large investment of resources and human ingenuity in pursuit of environmental goals has produced enormous benefits over the past two decades. This is an achievement to celebrate, but in a world that steadily uses more materials to make more goods for more people, we must recognize that we will have to achieve more in the future for the sake of the future. We must move toward a world in which zero waste will become the ideal for society that zero defects has become for manufacturing. Finally, we must recognize that the pursuit of one set of goals affects others and that we must pursue policies that integrate economic, environmental, and social goals.

Summary

Politics and the environment cannot be separated. In the United States, the government is structured into three separate branches, each of which impacts environmental policy.

The increase in environmental regulations in the United States over the past 30 years has caused concerns in some sectors of society.

The late 1980s and early 1990s witnessed a new international concern about the environment, both in the developed and developing nations of the world. Environmentalism is also seen as a growing factor in international relations. This concern is leading to international cooperation where only tension had existed before.

While there exists no world political body that can enforce international environmental protection, the list of multilateral environmental organizations is growing.

It remains too early to tell what the ultimate outcome will be, but progress is being made in protecting our common resources for future generations. Several international conventions and treaties have been successful. In the final analysis, however, each of us has to adjust our lifestyle to clean up our own small part of the world.

Key Terms

executive branch *458*
judicial branch *458*

legislative branch *458*

policy *458*

Review Questions

1. What are the major responsibilities of each of the three branches of the U.S. government?
2. What are some of the enforcement options in U.S. environmental policy?
3. What role does administrative law play in U.S. environmental policy?
4. What are some of the criticisms of U.S. environmental policy?
5. In the past 10 years, how has public opinion in the United States changed concerning the protection of the environment?
6. Why is environmentalism a growing factor in international relations?
7. Give some examples of international environmental conventions and treaties.

Critical Thinking Questions

1. Does Chapter 20 have an overall point of view? If you were going to present the problems of environmental policy making and enforcement to others, what framework would you use?

2. The authors of this text say that "we are progressing from an environmental paradigm based on cleanup and control to one including assessment, anticipation, and avoidance." Do you agree with this assessment? Are there environmental problems that are harder to be proactive about than others?

3. Does a command-and-control approach to environmental problems, an approach that emphasizes regulation and remediation, make sense with global environmental problems such as global climate change, habitat destruction, and ozone depletion?

4. What values, perspectives, and beliefs does the Wise Use Movement exhibit in their response to environmental legislation? How are they similar to or different from your own?

5. How is it best, as a global society with many political demarcations, to preserve the resources that are held in common? What special problems does this kind of preservation entail?

6. Do you agree with William Ruckelshaus that current environmental problems require a change on the part of industrialized and developing countries that would be "a modification in society comparable in scale to the agricultural revolution . . . and Industrial Revolution"? What kinds of changes might that mean in your life? Would these be positive or negative changes?

7. New treaties regarding free trade might enable some nations to argue that other nations' environmental legislation is too restrictive, thereby imposing a barrier to trade that is subject to sanction. What special problems and possibilities might the new global economy present for environmental preservation? What do you think about that?

Concept Map

Construct a map to show relationships among the following concepts:

environmental policy
environmental regulation
green politics

activism
environmental politics

globalization
individual action

Interactive Exploration

Check out the website at **http://www.mhhe.com/environmentalscience** and click on the cover of this textbook for quizzing, career information, case studies, and hot links for information on the following topics:

Environmental Policy and Decision Making

The Ultimate Ecological Answers

Individual Contributions to Environmental Issues

Environmental Organizations

Miscellaneous Environmental Resources

Environmental and Ecological Organization Sites

Environmental Philosophy

Miscellaneous Environmental Resources

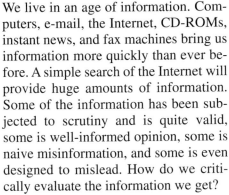

We live in an age of information. Computers, e-mail, the Internet, CD-ROMs, instant news, and fax machines bring us information more quickly than ever before. A simple search of the Internet will provide huge amounts of information. Some of the information has been subjected to scrutiny and is quite valid, some is well-informed opinion, some is naive misinformation, and some is even designed to mislead. How do we critically evaluate the information we get?

Critical thinking involves a set of skills that helps us to evaluate information, arguments, and opinions in a systematic and thoughtful way. Critical thinking also can help us better understand our own opinions as well as the points of view of others. It can help us evaluate the quality of evidence, recognize bias, characterize the assumptions behind arguments, identify the implications of decisions, and avoid jumping to conclusions.

Characteristics of Critical Thinking

Critical thinking involves skills that allow us to sort information in a meaningful way and discard invalid or useless information while recognizing that which is valuable. Some key components of critical thinking are:

Recognize the importance of context.
All information is based on certain assumptions. It is important to recognize what those assumptions are.

Critical thinking involves looking closely at an argument or opinion by identifying the historical, social, political, economic, and scientific context in which the argument is being made. It is also important to understand the kinds of bias contained in the argument and the level of knowledge the presenter has.

Consider alternative views.
A critical thinker must be able to understand and evaluate different points of view. Often these points of view may be quite varied. It is important to keep an open mind and to look at all the information objectively and try to see the value in alternative points of view. Often people miss obvious solutions to problems because they focus on a certain avenue of thinking and unconsciously dismiss valid alternative solutions.

Expect and accept mistakes.
Good critical thinking is exploratory and speculative, tempered by honesty and a recognition that we may be wrong. It takes courage to develop an argument, engage in debate with others, and admit that your thinking contains errors or illogical components. By the same token, be willing to point out what you perceive to be shortcomings in the arguments of others. It is always best to do this with good grace and good humor.

Have clear goals.
When analyzing an argument or information, keep your goals clearly in mind. It is often easy to get sidetracked.

A clear goal will allow you to quickly sort information into that which is pertinent and that which may be interesting but not germane to the particular issue you are exploring.

Evaluate the validity of evidence.
Information comes in many forms and has differing degrees of validity. When evaluating information, it is important to understand that not all the information from a source may be of equal quality. Often content about a topic is a mix of solid information interspersed with less certain speculations or assumptions. Apply a strong critical attitude to each separate piece of information. Often what appears to be a minor, insignificant error or misunderstanding can cause an entire argument to unravel.

Critical thinking requires practice.
As with most skills, you become better if you practice. At the end of each chapter in the text, there are a series of questions that allow you to practice critical thinking skills. Some of these questions are straightforward and simply ask you to recall information from the chapter. Others ask you to apply the information from the chapter to other similar contexts. Still others ask you to develop arguments that require you to superimpose the knowledge you have gained from the chapter on quite different social, economic, or political contexts from your own.

Practice, practice, practice.

appendix 2

The Periodic Table of the Elements

Traditionally, elements are represented in a shorthand form by letters. For example, the formula for water, H_2O, shows that a molecule of water consists of two atoms of hydrogen and one atom of oxygen. These chemical symbols for each of the atoms can be found on any periodic table of the elements. Using the periodic table, we can determine the number and position of the various parts of atoms.

Notice that atoms number 3, 11, 19, and so on are in column one. The atoms in this column act in a similar way since they all have one electron in their outermost layer. In the next column, Be, Mg, Ca, and so on act alike because these metals all have two electrons in their outermost electron layer. Similarly, atoms number 9, 17, 35, and so on all have seven electrons in their outer layer.

Knowing how fluorine, chlorine, and bromine act, you can probably predict how iodine will act under similar conditions. At the far right in the last column, argon, neon, and so on all act alike. They all have eight electrons in their outer electron layer. Atoms with eight electrons in their outer electron layer seldom form bonds with other atoms.

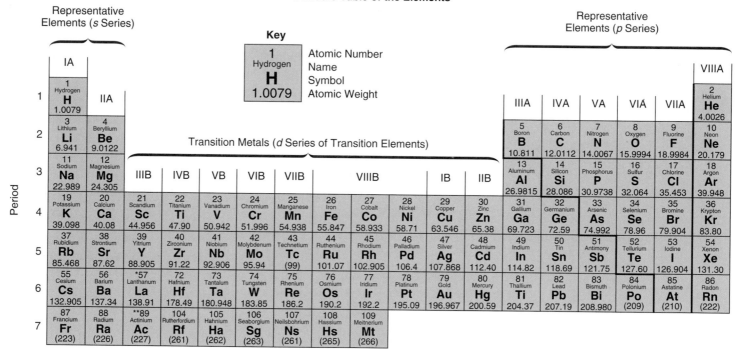

Periodic Table of the Elements

glossary

A

abiotic factors   Nonliving factors that influence the life and activities of an organism.

abyssal ecosystem   The collection of organisms and the conditions that exist in the deep portions of the ocean.

acid   Any substance that, when dissolved in water, releases hydrogen ions.

acid deposition   The accumulation of potential acid-forming particles on a surface.

acid mine drainage   A kind of pollution, associated with coal mines, in which bacteria convert the sulfur in coal into compounds that form sulfuric acid.

acid rain (acid precipitation)   The deposition of wet acidic solutions or dry acidic particles from air.

activated sludge sewage treatment   Method of treating sewage in which some of the sludge is returned to aeration tanks, where it is mixed with incoming wastewater to encourage degradation of the wastes in the sewage.

activation energy   The initial energy input required to start a reaction.

active solar system   A system that traps sunlight energy as heat energy and uses mechanical means to move it to another location.

acute toxicity   A serious effect, such as a burn, illness, or death, that occurs shortly after exposure to a hazardous substance.

age distribution   The comparative percentages of different age groups within a population.

agricultural products   Any output from farming: milk, grain, meat, etc.

agricultural runoff   Surface water that carries soil particles, nutrients, such as phosphate, nitrates, and other agricultural chemicals, as it runs off agricultural land to lakes and streams.

air stripping   The process of pumping air through water to remove volatile materials dissolved in the water.

alpha radiation   A type of radiation consisting of a particle with two neutrons and two protons.

alpine tundra   The biome that exists above the tree line in mountainous regions.

alternative agriculture   All nontraditional agricultural practices.

anthropocentric   Human-centered; a theory of moral responsibility that views the environment as a resource for humankind.

aquiclude   An impervious confining layer of an aquifer.

aquifer   A porous layer of earth material that becomes saturated with water.

aquitard   A partially permeable layer in an aquifer.

artesian well   The result of a pressurized aquifer being penetrated by a pipe or conduit, within which water rises without being pumped.

atom   The basic subunit of elements, composed of protons, neutrons, and electrons.

auxin   A plant hormone that stimulates growth.

B

base   Any substance that, when dissolved in water, removes hydrogen ions from solution; forms a salt when combined with an acid.

benthic   Describes organisms that live on the bottom of marine and freshwater ecosystems.

benthic ecosystem   A type of marine or freshwater ecosystem consisting of organisms that live on the bottom.

beta radiation   A type of radiation consisting of electrons released from the nuclei of many fissionable atoms.

bioaccumulation   The buildup of a material in the body of an organism.

biocentric   Life-centered; a theory of moral responsibility that states that all forms of life have an inherent right to exist.

biochemical oxygen demand (BOD)   The amount of oxygen required by microbes to degrade organic molecules in aquatic ecosystems.

biocide   A kind of chemical that kills many different types of living things.

biodegradable   Able to be broken down by natural biological processes.

biodiversity   A measure of the variety of kinds of organisms present in an ecosystem.

biomagnification   The increases in the amount of a material in the bodies of organisms at successively higher trophic levels.

biomass   Any accumulation of organic material produced by living things.

biome   A kind of plant and animal community that covers large geographic areas. Climate is a major determiner of the biome found in a particular area.

biotechnology   Inserting specific pieces of DNA into the genetic makeup of organisms.

biotic factors   Living portions of the environment.

biotic potential   The inherent reproductive capacity.

birthrate   The number of individuals born per thousand individuals in the population per year.

black lung disease   A respiratory condition resulting from the accumulation of large amounts of fine coal dust particles in miners' lungs.

boiling-water reactor (BWR)   A type of light water reactor in which steam is formed directly in the reactor, and is used to generate electricity.

boreal forest   A broad band of mixed coniferous and deciduous trees that stretches across northern North America (and also Europe and Asia); its northernmost edge is integrated with the Arctic tundra.

brownfields   Buildings and land that have been abandoned because they are contaminated and the cost of cleaning up the site is high.

brownfields cleanup   Cleaning a contaminated industrial site to the point that it is safe to use for specific purposes.

brownfields development   The concept that abandoned contaminated sites can be cleaned up sufficiently to allow some specified uses without totally removing all of the contaminants

C

carbamate   A class of soft pesticides that work by interfering with normal nerve impulses.

carbon absorption   The use of carbon particles to treat chemicals by having the chemicals attach to the carbon particles.

carbon cycle   The cyclic flow of carbon from the atmosphere to living organisms and back to the atmospheric reservoir.

carbon dioxide (CO_2)   A normal component of the Earth's atmosphere that in elevated concentrations may interfere with the Earth's heat budget.

carbon monoxide (CO)   A primary air pollutant produced when organic materials, such as gasoline, coal, wood, and trash, are incompletely burned.

carcinogen   A substance that causes cancer.

carcinogenic   The ability of a substance to cause cancer.

carnivores   Animals that eat other animals.

carrying capacity   The optimum number of individuals of a species that can be supported in an area over an extended period of time.

catalyst   A substance that alters the rate of a reaction but is not itself changed.

chemical bond   The physical attraction between atoms that results from the interaction of their electrons.

chemical weathering   Processes that involve the chemical alteration of rock in such a manner that it is more likely to fragment or to be dissolved.

chlorinated hydrocarbon   A class of pesticide consisting of carbon, hydrogen, and chlorine; these pesticides are very stable.

chlorofluorocarbons (CFC)   Stable compounds containing carbon, hydrogen, chlorine, and fluorine. They were formerly used as refrigerants, propellants in aerosol containers, and expanders in foam products. They are linked to the depletion of the ozone layer.

chronic toxicity   A serious effect, such as an illness or death, that occurs after prolonged exposure to small doses of a toxic substance.

clear-cutting   A forest harvesting method in which all the trees in a large area are cut and removed.

climax community   Last stage of succession; a relatively stable, long-lasting, complex, and interrelated community of plants, animals, fungi, and bacteria.

coevolution   Two or more species of organisms reciprocally influencing the evolutionary direction of the other.

combustion The process of releasing chemical bond energy from fuel.

commensalism The relationship between organisms in which one organism benefits while the other is not affected.

community Interacting groups of different species.

competition An interaction between two organisms in which both require the same limited resource, which results in harm to both.

composting The process of harnessing the natural process of decomposition to transform organic materials into compost, a humuslike material with many environmental benefits.

compound A kind of matter composed of two or more different kinds of atoms bonded together.

Comprehensive Environmental Response, Compensation, and Liability Act (CERCLA) The 1980 U.S. law that addressed the cleanup of hazardous-waste sites.

confined aquifer An aquifer that is bounded on the top and bottom by impermeable confining layers.

conservation To use in the best possible way so that the greatest long-term benefit is realized by society.

conservation ethic An environmental ethic that stresses a balance between total development and absolute preservation.

consumers Organisms that use other organisms as food.

contour farming A method of tilling and planting at right angles to the slope, which reduces soil erosion by runoff.

controlled experiment An experiment in which two groups are compared. One, the control, is used as a basis of comparison and the other, the experimental, has one factor different from the control.

coral reef ecosystem A tropical, shallow-water, marine ecosystem dominated by coral organisms that produce external skeletons.

corporation A business structure that has a particular legal status.

corrosiveness Ability of a chemical to degrade standard materials.

cost-benefit analysis A method used to determine the feasibility of pursuing a particular project by balancing estimated costs against expected benefits.

cover A term used to refer to any set of physical features that conceals or protects animals from the elements or their enemies.

crust The thin, outer, solid surface of the Earth.

D

death phase The portion of the population growth curve of some organisms that shows the population declining.

death rate The number of deaths per thousand individuals in the population per year.

decibel A unit used to measure the loudness of sound.

decommissioning Decontaminating and disassembling a nuclear power plant and safely disposing of the radioactive materials.

decomposers Small organisms, such as bacteria and fungi, that cause the decay of dead organic matter and recycle nutrients.

demand Amount of a product that consumers are willing and able to buy at various prices.

demographic transition The hypothesis that economies proceed through a series of stages, beginning with growing populations with high birth and death rates and low economic development and ending with stable populations with low birth and death rates and high economic development.

demography The study of human populations, their characteristics, and their changes.

denitrifying bacteria Bacteria that convert nitrogen compounds into nitrogen gas.

density-dependent limiting factors Those limiting factors that become more severe as the size of the population increases.

density-independent limiting factors Those limiting factors that are not affected by population size.

desert A biome that receives less than 25 centimeters (10 inches) of precipitation per year.

desertification The conversion of arid and semiarid lands into deserts by inappropriate farming practices or overgrazing.

detritus Tiny particles of organic material that result from fecal waste material or the decomposition of plants and animals.

development ethic Philosophy that states that the human race should be the master of nature and that the Earth and its resources exist for human benefit and pleasure.

dispersal Migration of organisms from a concentrated population into areas with lower population densities.

domestic water Water used for domestic activities, such as drinking, air conditioning, bathing, washing clothes, washing dishes, flushing toilets, and watering lawns and gardens.

E

ecocentrism An approach to environmental responsibility that maintains that the environment deserves direct moral consideration rather than consideration derived merely from human interests.

ecology A branch of science that deals with the interrelationship between organisms and their environment.

economic cost Those monetary costs that are necessary to exploit a natural resource.

economic growth The perceived increase in monetary growth within a society.

ecosystem A group of interacting species along with their physical environment.

ectoparasite A parasite that is adapted to live on the outside of its host.

electron The lightweight, negatively charged particle that moves around at some distance from the nucleus of an atom.

element A form of matter consisting of a specific kind of atom.

emergent plants Aquatic vegetation that is rooted on the bottom but has leaves that float on the surface or protrude above the water.

emigration Movement out of an area that was once one's place of residence.

endangered species Those species that are present in such small numbers that they are in immediate jeopardy of becoming extinct.

endoparasite A parasite that is adapted to live within a host.

endothermic reaction Chemical reaction in which the newly formed chemical bonds contain more energy than was present in the compounds from which they were formed.

energy The ability to do work.

energy cost The amount of energy required to exploit a resource.

environment Everything that affects an organism during its lifetime.

environmental cost Damage done to the environment as a resource is exploited.

environmental justice Fair application of laws designed to protect the health of human beings and ecosystems; that no groups suffer unequal environmental harm.

Environmental Protection Agency (EPA) U.S. government organization responsible for the establishment and enforcement of regulations concerning the environment.

environmental resistance The combination of all environmental influences that tend to keep populations stable.

environmental science An interdisciplinary area of study that includes both applied and theoretical aspects of human impact on the world.

environmental terrorism The unlawful use of force against environmental resources so as to deprive populations of their benefits or destroy other property.

enzyme Protein molecules that speed up the rate of specific chemical reactions.

erosion The processes that loosen and move particles from one place to another.

estuaries Marine ecosystems that consist of shallow, partially enclosed areas where freshwater enters the ocean.

ethics A discipline that seeks to define what is fundamentally right and wrong.

euphotic zone The upper layer in the ocean where the sun's rays penetrate.

eutrophication The enrichment of water (either natural or cultural) with nutrients.

eutrophic lake A usually shallow, warm-water lake that is nutrient rich.

evapotranspiration The process of plants transporting water from the roots to the leaves where it evaporates.

evolution A change in the structure, behavior, or physiology of a population of organisms as a result of some organisms with favorable characteristics having greater reproductive success than those organisms with less favorable characteristics.

executive branch The office of the president of the United States.

exothermic reaction Chemical reaction in which the newly formed compounds have less chemical energy than the compounds from which they were formed.

experiment An artificial situation designed to test the validity of a hypothesis.

exponential growth phase The period during population growth when the population increases at an ever-increasing rate.

external costs Expenses, monetary or otherwise, borne by someone other than the individuals or groups who use a resource.

extinction The death of a species; the elimination of all the individuals of a particular kind.

F

fecal coliform bacteria   Bacteria found in the intestines of humans and other animals, often used as an indicator of water pollution.

first law of thermodynamics   A statement about energy that says that under normal physical conditions, energy is neither created nor destroyed.

fissionable   The property of the nucleus of some atoms that allows them to split into smaller particles.

fixation   A form of waste immobilization in which materials, such as fly ash or cement, are mixed with hazardous waste to prevent the waste from dispersing.

floodplain   Lowland area on either side of a river that is periodically covered by water.

floodplain zoning ordinances   Municipal laws that restrict future building in floodplains.

food chain   The series of organisms involved in the passage of energy from one trophic level to the next.

food web   Intersecting and overlapping food chains.

fossil fuels   The organic remains of plants, animals, and microorganisms that lived millions of years ago that are preserved as natural gas, oil, and coal.

free-living nitrogen-fixing bacteria   Bacteria that live in the soil and can convert nitrogen gas (N_2) in the atmosphere into forms that plants can use.

freshwater ecosystem   Aquatic ecosystems that have low amounts of dissolved salts.

friable   A soil characteristic that describes how well a soil crumbles.

fungicide   A pesticide designed to kill or control fungi.

G

gamma radiation   A type of electromagnetic radiation that comes from disintegrating atomic nuclei.

gas-cooled reactor (GCR)   A type of nuclear reactor that uses graphite as a moderator and carbon dioxide or helium as a coolant.

genetically modified organisms   Organisms that have had their genetic makeup modified by biotechnology.

genetic engineering   Inserting specific pieces of DNA into the genetic makeup of organisms.

geothermal energy   The heat energy from the Earth's molten core.

Global Reporting Initiative   Guidelines for reporting on the economic, environmental, and social performance of corporations.

grasslands   Areas receiving between 25 and 75 centimeters (10-30 inches) of precipitation per year. Grasses are the dominant vegetation, and trees are rare.

greenhouse effect   The property of carbon dioxide (CO_2) that allows light energy to pass through the atmosphere but prevents heat from leaving; similar to the action of glass in a greenhouse.

greenhouse gas   Gas in the atmosphere that allows sunlight to enter but retards the outward flow of heat from the Earth.

Green Revolution   The introduction of new plant varieties and farming practices that increased agricultural production worldwide during the 1950s, 1960s, and 1970s.

gross national income (GNI)   An index that measures the total goods and services generated within a country as well as income earned by citizens of the country who are living in other countries.

gross national product (GNP)   An index that measures the total goods and services generated annually within a country.

groundwater   Water that infiltrates the soil and is stored in the spaces between particles in the earth.

groundwater mining   Removal of water from an aquifer faster than it is replaced.

H

habitat   The specific kind of place where a particular kind of organism lives.

habitat management   The process of changing the natural community to encourage the increase in populations of certain desirable species.

hard pesticide   A pesticide that persists for long periods of time; a persistent pesticide.

hazardous   All dangerous materials, including toxic ones, that present an immediate or long-term human health risk or environmental risk.

hazardous substances   Substances that can cause harm to humans or the environment.

hazardous-waste dump   A site in which hazardous-waste is disposed in a dump, landfill, or surface impoundment without any concern for potential environmental or health risks.

hazardous wastes   Substances that could endanger life if released into the environment.

heavy-water reactor (HWR)   A type of nuclear reactor that uses the hydrogen isotope deuterium in the molecular structure of the coolant water.

herbicide   A pesticide designed to kill or control plants.

herbivores   Primary consumers; animals that eat plants.

horizon   A horizontal layer in the soil. The top layer (A horizon) has organic matter. The lower layer (B horizon) receives nutrients by leaching. The C horizon is partially weathered parent material.

host   The organism a parasite uses for its source of food.

humus   Partially decomposed organic matter typically found in the top layer of the soil.

hydrocarbons (HC)   Group of organic compounds consisting of carbon and hydrogen atoms that are evaporated from fuel supplies or are remnants of the fuel that did not burn completely and that act as a primary air pollutant.

hydrologic cycle   Constant movement of water from surface water to air and back to surface water as a result of evaporation and condensation.

hydroxyl ion   A negatively charged particle consisting of a hydrogen and an oxygen atom, commonly released from materials that are bases.

hypothesis   A logical statement that explains an event or answers a question that can be tested.

I

ignitability   Characteristic of materials that results in their ability to combust.

immigration   Movement into an area where one has not previously resided.

incineration   Method of disposing of solid waste by burning.

industrial ecology   A concept that stresses cycling resources rather than extracting and eventually discarding them.

Industrial Revolution   A period of history during which machinery replaced human labor.

industrial uses   Uses of water for cooling and for dissipating and transporting waste materials.

insecticide   A pesticide designed to kill or control insects.

in-stream uses   Use of a stream's water flow for such purposes as hydroelectric power, recreation, and navigation.

integrated pest management   A method of pest management in which many aspects of the pest's biology are exploited to control its numbers.

interspecific competition   Competition between members of different species for a limited resource.

intraspecific competition   Competition among members of the same species for a limited resource.

ion   An atom or group of atoms that has an electric charge because it has either gained or lost electrons.

irrigation   Adding water to an agricultural field to allow certain crops to grow where the lack of water would normally prevent their cultivation.

isotope   Atoms of the same element that have different numbers of neutrons.

J

judicial branch   That portion of the U.S. government that includes the court system.

K

keystone species   One that has a critical role to play in the maintenance of specific ecosystems.

kinetic energy   Energy of moving objects.

kinetic molecular theory   The widely accepted theory that all matter is made of small particles that are in constant movement.

K-strategists   Large organisms that have relatively long lives, produce few offspring, provide care for their offspring, and typically have populations that stabilize at the carrying capacity.

L

lag phase   The initial stage of population growth during which growth occurs very slowly.

land   The surface of the Earth not covered by water.

landfill   A method of disposing of solid wastes that involves burying the wastes in specially constructed sites.

land-use planning   The process of evaluating the needs and wants of the population, the characteristics and values of the land, and various alternative solutions before changes in land use are made.

law of conservation of mass   States that matter is not gained or lost during a chemical reaction.

LD$_{50}$ A measure of toxicity; the dosage of a substance that will kill (lethal dose) 50 percent of a test population.

leachate   Contaminant-laden water that flows from landfills or other contaminated sites.

leaching   The movement of minerals from the top layers of the soil to the B horizon by the downward movement of soil water.

legislative branch   That portion of the U.S. government that is responsible for developing laws.

light-water reactor   A nuclear reactor that uses ordinary water as a coolant.

limiting factor   The one primary condition of the environment that determines population size of an organism.

limnetic zone   Region that does not have rooted vegetation in a freshwater ecosystem.

liquefied natural gas Natural gas that has been converted to a liquid by cooling to −162°C (−260°F).

liquid metal fast-breeder reactor (LMFBR) Nuclear fission reactor using liquid sodium as the moderator and heat transfer medium; produces radioactive plutonium-235, which can be used as a nuclear fuel.

lithosphere A combination of the crust and outer layer of the mantle that forms the plates that move over the Earth's surface.

litter A layer of undecomposed or partially decomposed organic matter on the soil surface.

littoral zone Region with rooted vegetation in a freshwater ecosystem.

loam A soil type with good drainage and good texture that is ideal for growing crops.

M

macronutrient A nutrient, such as nitrogen, phosphorus, and potassium, that is required in relatively large amounts by plants.

management ethic An environmental ethic that stresses a balance between total development and absolute preservation.

mangrove swamp ecosystems Marine shoreline ecosystems dominated by trees that can tolerate high salt concentrations.

mantle The layer of the Earth between the crust and the core.

marine ecosystems Aquatic ecosystems that have high salt content.

marsh Area of grasses and reeds that is flooded either permanently or for a major part of the year.

mass burn A method of incineration of solid waste in which material is fed into a furnace on movable metal grates.

matter Substance with measurable mass and volume.

mechanical weathering Physical forces that reduce the size of rock particles without changing the chemical nature of the rock.

megalopolis A large, regional urban center.

micronutrient A nutrient needed in extremely small amounts for proper plant growth; examples are boron, zinc, and magnesium.

migratory birds Birds that fly considerable distances between their summer breeding areas and their wintering areas.

mixture A kind of matter consisting of two or more kinds of matter intermingled with no specific ratio of the kinds of matter.

moderator Material that absorbs the energy from neutrons released by fission.

molecule Two or more atoms chemically bonded to form a stable unit.

monoculture A system of agriculture in which large tracts of land are planted with the same crop.

morals Predominant feeling of a culture about ethical issues.

mortality The number of deaths per year.

multiple land use Land uses that do not have to be exclusionary, so that two or more uses of land may occur at the same time.

municipal solid waste landfill A waste storage site constructed above an impermeable clay layer that is lined with an impermeable membrane and includes mechanisms for dealing with liquid and gas materials generated by the contents of the landfill.

municipal solid waste All the waste produced by the residents of a community.

mutualism The association between organisms in which both benefit.

N

natality The number of individuals added to the population through reproduction.

National Priority List A listing of hazardous-waste dump sites requiring urgent attention as identified by Superfund legislation.

natural resources Those structures and processes that can be used by humans for their own purposes but cannot be created by them.

natural selection A process that determines which individuals within a species will reproduce more effectively and therefore results in changes in the characteristics within a species.

nature centers Teaching institutions that provide a variety of methods for people to learn about and appreciate the natural world.

negligible risk A point at which there is no significant health or environmental risk.

neutralization Reacting acids with bases to produce relatively safe end products.

neutron Neutrally charged particle located in the nucleus of an atom.

niche The total role an organism plays in its ecosystem.

nitrifying bacteria Bacteria that are able to convert ammonia to nitrite, which can be converted to nitrate.

nitrogen cycle The series of stages in the flow of nitrogen in ecosystems.

nitrogen dioxide A compound composed of one atom of nitrogen and two atoms of oxygen; a secondary air pollutant.

nitrogen-fixing bacteria Bacteria that are able to convert the nitrogen gas (N_2) in the atmosphere into forms that plants can use.

nitrogen monoxide A compound composed of one atom of nitrogen and one atom of oxygen; a primary air pollutant.

nitrous oxide N_2O, one of the oxides of nitrogen.

nonpersistent pesticide A pesticide that degrades in a short period of time.

nonpersistent pollutants Those pollutants that do not remain in the environment for long periods.

nonpoint source Diffuse pollutants, such as agricultural runoff, road salt, and acid rain, that are not from a single, confined source.

nonrenewable energy sources Those energy sources that are not replaced by natural processes within a reasonable length of time.

nonrenewable resources Those resources that are not replaced by natural processes, or those whose rate of replacement is so slow as to be noneffective.

nontarget organism An organism whose elimination is not the purpose of pesticide application.

northern coniferous forest See boreal forest.

nuclear breeder reactor Nuclear fission reactor designed to produce radioactive fuel from nonradioactive uranium and at the same time release energy to use in the generation of electricity.

nuclear chain reaction A continuous process in which a splitting nucleus releases neutrons that strike and split the nuclei of other atoms, releasing nuclear energy.

nuclear fission The decomposition of an atom's nucleus with the release of particles and energy.

nuclear fusion The union of smaller nuclei to form a heavier nucleus accompanied by the release of energy.

nuclear reactor A device that permits a controlled nuclear fission chain reaction.

nucleus The central region of an atom that contains protons and neutrons.

O

observation Ability to detect events by the senses or machines that extend the senses.

oligotrophic lakes Deep, cold, nutrient-poor lakes that are low in productivity.

omnivores Animals that eat both plants and other animals.

organic agriculture Agricultural practices that avoid the use of chemical fertilizers and pesticides in the production of food, thus preventing damage to related ecosystems and consumers.

organophosphate A class of soft pesticides that work by interfering with normal nerve impulses.

outdoor recreation Leisure activities carried out in the natural out-of-doors.

overburden The layer of soil and rock that covers deposits of desirable minerals.

oxides of nitrogen (NO, N_2O, and NO_2) Primary air pollutants consisting of a variety of different compounds containing nitrogen and oxygen.

ozone (O_3) A molecule consisting of three atoms of oxygen, which absorb much of the sun's ultraviolet energy before it reaches the Earth's surface.

P

parasite An organism adapted to survival by using another living organism (host) for nourishment.

parasitism A relationship between organisms in which one, known as the parasite, lives in or on the host and derives benefit from the relationship while the host is harmed.

parent material Material that is weathered to become the mineral part of the soil.

particulate matter Minute solid particles and liquid droplets dispersed into the atmosphere.

particulates Small pieces of solid materials, such as smoke particles from fires, bits of asbestos from brake linings and insulation, dust particles, or ash from industrial plants, that are dispersed into the atmosphere.

passive solar system A design that allows for the entrapment and transfer of heat from the sun to a building without the use of moving parts or machinery.

patchwork clear-cutting A forest harvest method in which patches of trees are clear-cut among patches of timber that are left untouched.

peat The first stage in the conversion of organic material into coal.

pelagic Those organisms that swim in open water.

pelagic ecosystem A portion of a marine or freshwater ecosystem that occurs in open water away from the shore.

periphyton Attached organisms in freshwater streams and rivers, including algae, animals, and fungi.

permafrost Permanently frozen ground.

persistent pesticide A pesticide that remains unchanged for a long period of time; a hard pesticide.

persistent pollutant A pollutant that remains in the environment for many years in an unchanged condition.

pest An unwanted plant or animal that interferes with domesticated plants and animals or human activity.

pesticide A chemical used to eliminate pests; a general term used to describe a variety of different kinds of pest killers, such as insecticides, fungicides, rodenticides, and herbicides.

pH The negative logarithm of the hydrogen ion concentration; a measure of the number of hydrogen ions present.

pheromone A chemical produced by one animal that changes the behavior of another.

photochemical smog A yellowish-brown haze that is the result of the interaction of hydrocarbons, oxides of nitrogen, and sunlight.

photosynthesis The process by which plants manufacture food. Light energy is used to convert carbon dioxide and water to sugar and oxygen.

photovoltaic cell A means of directly converting light energy into electricity.

phytoplankton Free-floating, microscopic, chlorophyll-containing organisms.

pioneer community The early stages of succession that begin the soil-building process.

plankton Tiny aquatic organisms that are moved by tides and currents.

plate tectonics The concept that the outer surface of the Earth consists of large plates that are slowly moving over the surface of a plastic layer.

plutonium-239 (Pu-239) A radioactive isotope produced in a breeder reactor and used as a nuclear fuel.

pm$_{10}$ Particulate matter that is 10 microns or less than in diameter.

pm$_{2.5}$ Particulate matter that is 2.5 microns or less than in diameter.

point source Pollution that can be traced to a single source.

policy Planned course of action on a question or a topic.

pollution Any addition of matter or energy that degrades the environment for humans and other organism.

pollution costs The private or public expenditures undertaken to avoid pollution damage once pollution has occurred and the increased health costs and loss of the use of public resources because of pollution.

pollution prevention Action to prevent either entirely or partially the pollution that would otherwise result from some production or consumption activity.

pollution-prevention hierarchy Regulatory controls that emphasize reducing the amount of hazardous waste produced.

polyculture A system of agriculture that mixes different plant species in the same plots of land.

population density A measure of how close organisms are to one another, generally expressed as the number of organisms per unit area.

porosity A measure of the size and number of spaces in an aquifer.

postwar baby boom A large increase in the birthrate immediately following World War II.

potable waters Unpolluted freshwater supplies suitable for drinking.

potential energy The energy of position.

prairies Grasslands.

precipitation Removal of materials by mixing with chemicals that cause the materials to settle out of the mixture.

precision agriculture The use of computer technology and geographic information systems to automatically vary the chemicals applied to a crop at different places within a field.

predator An animal that kills and eats another organism.

preservation Action to keep from harm or damage; to maintain in its original condition.

preservation ethic Philosophy that considers nature to be so special that it should remain intact.

pressurized-water reactor (PWR) A type of light-water reactor in which the water in the reactor is kept at high pressure and steam is formed in a secondary loop.

prey An organism that is killed and eaten by a predator.

primary air pollutants Types of unmodified materials that, when released into the environment in sufficient quantities, are considered hazardous.

primary consumer An animal that eats plants (producers) directly.

primary sewage treatment Process that removes larger particles by settling or filtering raw sewage through large screens.

primary succession Succession that begins with bare mineral surfaces or water.

probability A mathematical statement about how likely it is that something will happen.

producer An organism that can manufacture food from inorganic compounds and light energy.

profitability The extent to which economic benefits exceed the economic costs of doing business.

proton The positively charged particle located in the nucleus of an atom.

public resources Those parts of the environment that are owned by everyone.

R

radiation Energy that travels through space in the form of waves or particles.

radioactive Describes unstable nuclei that release particles and energy as they disintegrate.

radioactive half-life The time it takes for half of the radioactive material to spontaneously decompose.

radon Radioactive gas emitted from certain kinds of rock; can accumulate in very tightly sealed buildings.

range of tolerance The ability organisms have to succeed under a variety of environmental conditions. The breadth of this tolerance is an important ecological characteristic of a species.

reactivity The property of materials that indicates the degree to which a material is likely to react vigorously to water or air, or to become unstable or explode.

recycling The process of reclaiming a resource and reusing it for another or the same structure or purpose.

reforestation The process of replanting areas after the original trees are removed.

rem A measure of the biological damage to tissue caused by certain amounts of radiation.

renewable energy sources Those energy sources that can be regenerated by natural processes.

renewable resources Those resources that can be formed or regenerated by natural processes.

replacement fertility The number of children per woman needed just to replace the parents.

reproducibility A characteristic of the scientific method in which independent investigators must be able to reproduce the experiment to see if they get the same results.

reserves The known deposits from which materials can be extracted profitably with existing technology under present economic conditions.

Resource Conservation and Recovery Act (RCRA) The 1976 U.S. law that specifically addressed the issue of hazardous waste.

resource exploitation The use of natural resources by society.

resources Naturally occurring substances that can be utilized by people but may not be economic.

respiration The process that organisms use to release chemical bond energy from food.

ribbon sprawl Development along transportation routes that usually consists of commercial and industrial building.

risk assessment The use of facts and assumptions to estimate the probability of harm to human health or the environment that may result from exposures to specific pollutants, toxic agents, or management decisions.

risk management Decision-making process that uses input such as risk assessment, technological feasibility, economic impacts, public concerns, and legal requirements.

rodenticide A pesticide designed to kill rodents.

r-strategist Typically, a small organism that has a short life span, produces a large number of offspring, and does not reach a carrying capacity.

runoff The water that moves across the surface of the land and enters a river system.

S

salinization An increase in the amount of salt in soil due to the evaporation of irrigation water.

savanna Tropical biome having seasonal rainfall of 50 to 150 centimeters (20-60 inches) per year. The dominant plants are grasses, with some scattered fire- and drought-resistant trees.

science A method for gathering and organizing information that involves observation, asking questions about observations hypothesis formation, testing hypothesis, critically evaluating the results, and publishing information so that others can evaluate the process and the conclusions.

scientific law A uniform or constant fact of nature that describes *what* happens in nature.

scientific method A way of gathering and evaluating information. It involves observation, hypothesis formation, hypothesis testing, critical evaluation of results, and the publishing of findings.

secondary air pollutants Pollutants produced by the interaction of primary air pollutants in the presence of an appropriate energy source.

secondary consumers Animals that eat animals that have eaten plants.

secondary recovery Techniques used to obtain the maximum amount of oil or natural gas from a well.

secondary sewage treatment Process that involves holding the wastewater until the organic material has been degraded by bacteria and other microorganisms.

secondary succession Succession that begins with the destruction or disturbance of an existing ecosystem.

second law of thermodynamics A statement about energy conversion that says that whenever energy is converted from one form to another, some of the useful energy is lost.

selective harvesting A forest harvesting method in which individual high-value trees are removed from the forest, leaving the majority of the forest undisturbed.

septic tank Underground holding tank into which sewage is pumped and where biological degradation of organic material takes place; used in places where sewers are not available.

seral stage A stage in the successional process.

sere A stage in succession.

sewage sludge A mixture of organic material, organisms, and water in which the organisms consume the organic matter.

sex ratio Comparison between the number of males and females in a population.

smart growth Land development that emphasizes the concept of livable cities and towns.

soft pesticide A nonpersistent pesticide that breaks down into harmless products in a few hours or days.

soil A mixture of mineral material, organic matter, air, water, and living organisms; capable of supporting plant growth.

soil profile The series of layers (horizons) seen as one digs down into the soil.

soil structure Refers to the way that soil particles clump together. Sand has little structure because the particles do not stick to one another.

soil texture Refers to the size of the particles that make up the soil. Sandy soil has large particles, and clay soil has small particles.

solidification The conversion of liquid wastes to a solid form to allow for more safe storage or transport.

solid waste Unwanted objects or particles that accumulate on the site where they are produced.

source reduction Reducing the amount of solid waste generated by using less, or converting from heavy packaging materials to lightweight ones.

speciation The process of developing a new species.

species A group of organisms that can interbreed and produce offspring capable of reproduction.

stable equilibrium phase The phase in a population growth curve in which the death rate and birthrate become equal.

standard of living The necessities and luxuries essential to a level of existence that is customary within a society.

steam stripping The use of heated air to drive volatile compounds from liquids.

steppe A grassland.

storm-water runoff Storm water that runs off of streets and buildings and is often added directly to the sewer system and sent to the municipal wastewater treatment facility.

strip farming The planting of crops in strips that alternate with other crops. The primary purpose is to reduce erosion.

submerged plants Aquatic vegetation that is rooted on the bottom and has leaves that stay submerged below the surface of the water.

subsidy A gift given to private enterprise by government when the enterprise is in temporary economic difficulty but is viewed as being important to the public.

succession Regular and predictable changes in the structure of a community, ultimately leading to a climax community.

successional stage A stage in succession.

sulfur dioxide (SO₂) A compound containing sulfur and oxygen produced when sulfur-containing fossil fuels are burned. When released into the atmosphere, it is a primary air pollutant.

Superfund The common name given to the U.S. 1980 Comprehensive Environmental Response, Compensation, and Liability Act, which was designed to address hazardous-waste sites.

supply Amount of a good or service available to be purchased.

supply/demand curve The relationship between available supply of a commodity or service and its demand. The supply and demand change as the price changes.

surface impoundment Pond created to hold liquid materials. Some may hold only water, while others may be used to contain polluted water or liquid contaminants.

surface mining (strip mining) A type of mining in which the overburden is removed to procure the underlying deposit.

sustainable development Using renewable resources in harmony with ecological systems to produce a rise in real income per person and an improved standard of living for everyone.

sustainable agriculture Agricultural methods used to produce adequate, safe food in an economically viable manner while enhancing the health of agricultural land and related ecosystems.

swamp Area of trees that is flooded either permanently or for a major part of the year.

symbiosis A close, long-lasting physical relationship between members of two different species.

symbiotic nitrogen-fixing bacteria Bacteria that grow within a plant's root system and that can convert nitrogen gas (N₂) from the atmosphere to nitrogen compounds that the plant can use.

synergism The interaction of materials or energy that increases the potential for harm.

T

taiga Biome having short, cool summers and long winters with abundant snowfall. The trees are adapted to winter conditions.

target organism The organism a pesticide is designed to eliminate.

technological advances Increasing use of machines to replace human labor.

temperate deciduous forest Biome that has a winter-summer change of seasons and that typically receives 75 to 150 centimeters (30–60 inches) or more of relatively evenly distributed precipitation throughout the year.

terrace A level area constructed on steep slopes to allow agriculture without extensive erosion.

tertiary sewage treatment Process that involves a variety of different techniques designed to remove dissolved pollutants left after primary and secondary treatments.

theory A unifying principle that binds together large areas of scientific knowledge.

thermal inversion The condition in which warm air in a valley is sandwiched between two layers of cold air and acts like a lid on the valley.

thermal pollution Waste heat that industries release into the environment.

thermal treatment A form of hazardous-waste destruction involving heating waste.

threatened species Those species that could become extinct if a critical factor in their environment were changed.

threshold level The minimum amount of something required to cause measurable effects.

total fertility rate The number of children born per woman per lifetime.

toxic A narrow group of substances that are poisonous and cause death or serious injury to humans and other organisms by interfering with normal body physiology.

toxicity A measure of how toxic a material is.

toxic waste Substances that are poisonous and cause death or serious injury to humans and animals when released into the environment.

tract development The construction of similar residential units over large areas.

transuranic waste Nuclear wastes of the U.S. weapons program that consist primarily of isotopes of plutonium.

trickling filter system A secondary sewage treatment technique that allows polluted water to flow over surfaces harboring microorganisms.

trophic level A stage in the energy flow through ecosystems.

tropical rainforest A biome with warm, relatively constant temperatures where there is no frost. These areas receive more than 200 centimeters (80 inches) of rain per year in rains that fall nearly every day.

tundra A biome that lacks trees and has permanently frozen soil.

U

unconfined aquifer An aquifer that usually occurs near the land's surface, receives water by percolation from above, and may be called a water table aquifer.

underground mining A type of mining in which the deposited material is removed without disturbing the overburden.

underground storage tank Tank located below ground level for the storage of materials, such as oil, gasoline, or other chemicals.

uranium-235 (U-235) A naturally occurring radioactive isotope of uranium used as fuel in nuclear reactors.

urban growth limit A boundary established by municipal government that encourages development within the boundary and prohibits it outside the boundary.

urban sprawl A pattern of unplanned, low-density housing and commercial development outside of cities that usually takes place on previously undeveloped land.

V

vadose zone A zone above the water table and below the land surface that is not saturated with water.

variable Things that change from time to time.

vector An organism that carries a disease from one host to another.

volatile organic compounds Airborne organic compounds; primary air pollutants.

W

waste destruction Destruction of a portion of hazardous waste with harmful residues still left behind.

waste immobilization Putting hazardous wastes into a solid form that is easier to handle and less likely to enter the surrounding environment.

waste minimization   A process that involves changes that industries could make in the way they manufacture products that would reduce the waste produced.

waste separation   Separating one hazardous waste from another or from nonhazardous material that it has contaminated.

water diversion   The physical process of transferring water from one area to another.

water table   The top of the layer of water in an aquifer.

waterways   Low areas that water normally flows through.

weathering   The physical and chemical breakdown of materials; involved in the breakdown of parent material in soil formation.

weed   An unwanted plant.

wetlands   Areas that include swamps, tidal marshes, coastal wetlands, and estuaries.

wilderness   Designation of land use for the exclusive protection of the area's natural wildlife; thus, no human development is allowed.

windbreak   The planting of trees or strips of grasses at right angles to the prevailing wind to reduce erosion of soil by wind.

Z

zero population growth   The stabilized growth stage of human population during which births equal deaths and equilibrium is reached.

zoning   Type of land-use regulation in which land is designated for specific potential uses, such as agricultural, commercial, residential, recreational, and industrial.

zooplankton   Weakly swimming microscopic animals.

credits

Text Credits

Chapter 2
p. 29: *Environmental Justice Highlights*, Reprinted with permission of The Detroit News.

Chapter 3
Fig. 3.02: Reprinted with permission from "Great Lakes Fish Consumption Study," 1989, Great Lakes National Resource Center, Ann Arbor, MI.

Chapter 7
Fig. 7.02: From *Concepts and Applications*, 2nd ed. by Manuel C. Molles Jr. Copyright © 2002 The McGraw-Hill Companies. Reprinted by permission.

Chapter 10
Fig. 10.12: From *Man, Energy, Society* by Earl Cook, W. H. Freeman and Company. Copyright © 1976. Reprinted with permission.

Chapter 12
Fig. 12.16: From *Game Management* by Aldo Leopold. Copyright © 1933. Reprinted by permission of Pearson Education, Inc. **Fig. 12.20:** Precious Heritage @ 2000 The Nature Conservancy and NatureServe.

Chapter 14
Fig. 14.01: From *Physical Geology*, 8th ed. by Charles C. Plummer, David McGeary, and Diane Carlson. Copyright © 1999 The McGraw-Hill Companies. All Rights Reserved. Reprinted by permission. **Fig. 14.02:** From *Fundamentals of Geology*, 5th ed. by Carla W. Montgomery. Copyright © 1999 The McGraw-Hill Companies. All Rights Reserved. Reprinted by permission. **Fig. 14.08:** From *Ecology and Field Biology*, 5th ed. by Robert Leo Smith. Copyright © 1996 by Addison-Wesley Educational Publishers, Inc. Reprinted by permission of Pearson Education, Inc.

Chapter 15
Table 15.01: Reprinted with permission from *Science*, 216: 19-22, April 1982 by Adkisson. Copyright © 1982 American Association for the Advancement of Science. **Fig. 15.13:** From Man and the *Environment*, 2/e by Arthur S. Boughey, © 1975. Reprinted by permission of Prentice-Hall, Inc., Upper Saddle River, NJ. **I—A Box, PCB's p. 347:** From Government of Canada, 1991, *The State of Canada's Environment*, 1991. Reproduced with permission of the Minister of Public Works and Government Services Canada, 1998.

Chapter 16
Fig. 16.09: Reprinted by permission of the San Francisco Convention & Visitors Bureau. **ECU BA 1,** Everglades map, **p. 373:** EVERGLADES MAP Copyright, April 2, 1990, U.S. News & World

Report **GP BA 1**, Nitrogen delivered to Gulf, **p. 379:** Reprinted by permission from *Nature*, 403: 761, copyright © 2000 Macmillan Publishers Ltd.

Chapter 17
Fig. 17.12: From *Environment*, Vol. 25, Issue 4, pp. 6–9. Published by Heldref Publications, 1319 18th St., NW, Washington DC 20036-1802. **Table 17.12:** Source: Chart compiled by Earth Force, Inc. (www.earthforce.org) with data from the Environmental Protection Agency and Wisconsin Department of Natural Resources. All Rights Reserved. Reprinted by permission of Earth Force, Inc.

Photo Credits

Part Openers
One: © Gary Griffin/Animals Animals/ Earth Scenes; **Two:** © CORBIS R-F Website; **Three:** © CORBIS R-F Website; **Four:** © Phil Schermeister/ CORBIS; **Five:** © Charles Rotkin/CORBIS.

Chapter 1
Figure 1.4 (left): © Steve McCutcheon/Visuals Unlimited; **1.4** (middle): © Vol. 1/ PhotoDisc; **1.4** (right): © Vol. 86/CORBIS; **1.5** (left): © Vol. 102/ CORBIS; **1.5** (middle): © Red Diamond Stock Photos/Bob Coyle, photographer; **1.5** (right): © Vol. 39/PhotoDisc; **1.6** (left): © Vol. 15/CORBIS; **1.6** (top middle): © J. Eastcott and Y. Momatiuk/ The Image Works; **1.6** (bottom middle): © Bayard H. Brattstrom/Visuals Unlimited; **1.6** (right): © Vol. 98/ CORBIS; **1.7** (top left): © Vol. 14/ CORBIS; **1.7** (bottom left): © William J. Weber/ Visuals Unlimited; **1.7** (top right): © Vol. 72/ CORBIS; **1.8** (left): © Vol. 25/PhotoDisc; **1.8** (middle): © Matt Bradley/Tom Stack & Associates; **1.8** (right): © Vol. 16/PhotoDisc; **1.9** (left): © Vol. 98/ CORBIS; **1.9** (middle): © Vol. 16/PhotoDisc; **1.9** (right): © Martin G. Miller/ Visuals Unlimited.

Chapter 2
Figure 2.1: © Vol. 34/PhotoDisc; **2.2** (top left): © Vol. 80/CORBIS; **2.2** (top right): © Vol. 74/ CORBIS; **2.2** (bottom left): © Vol. 80/CORBIS; **2.2** (bottom right): © Vol. 10/PhotoDisc; **p. 23** (Emerson): © Granger Collection; **p. 23** (Thoreau): © Bettmann/CORBIS; **p. 23** (Muir): © Bettmann/CORBIS; **p. 23** (Leopold): © AP/Wide World Photos; **p. 23** (Carson): © AP/Wide World Photos; **2.3** (top left): © Toni Michaels; **2.3** (top right): © Vol. 154/ CORBIS; **2.3** (bottom): © CORBIS R-FWebsite; **2.4a:** © Greg Vaughn/Tom Stack & Associates; **2.4b:** © Natalie Fobes/ Stone/Getty Images; **p. 32:** © AP/Wide World Photos; **2.5** (top left): © Nigel J.H. Smith/Animals Animals/Earth Scenes; **2.5** (top right): © Vol. 74/CORBIS; **2.5** (middle left): © Vol. 5/PhotoDisc; **2.5** (middle right): © Vol. 102/CORBIS; **2.5** (bottom left): © George Gainsburgh Bernard Photo Productions/

Animals Animals/Earth Scenes; **2.5** (bottom right): © Vol. 102/CORBIS.

Chapter 3
Figure 3.4 (top): © William E. Ferguson, photographer; **3.4** (middle right): © Vol. 38/ CORBIS; **3.4** (bottom right): © Scott Blackman/Tom Stack & Associates; **3.4** (bottom middle): © Vol. 38/CORBIS; **3.4** (bottom left): © McGraw-Hill Companies, Inc./ Bob Coyle, photographer; **3.4** (top left): © Vol. 14/ CORBIS; **3.7** (top left): © Cedric Max Dunham/ Photo Researchers, Inc.; **3.7** (top middle): © Vol. 10/ PhotoDisc; **3.7** (top right): © Jan Halaska/The Image Works; **3.7** (bottom left): © Didier Givois/ Vandystadt/Photo Researchers, Inc.; **3.7** (bottom middle): © Vol. 10/PhotoDisc; **3.7** (bottom right): © George E. Jones III/ Photo Researchers, Inc.; **3.8** (right): © AP/Wide World Photos; **3.9** (right): © Vol. 31/PhotoDisc; **p. 60** (left): © C. Allan Morgan/ Peter Arnold Inc.; **p. 60** (right): © John R. MacGregor/Peter Arnold, Inc.

Chapter 4
Page 70 (left & right): © McGraw-Hill Companies, Inc./Bob Coyle, photographer; **4.9:** © Carl Purcell/ Photo Researchers, Inc.

Chapter 5
Figure 5.3: © Joanne Lotter/Tom Stack & Associates; **5.5:** © Vol. 26/PhotoDisc; **5.7:** © Stephen Krasemann/Photo Researchers, Inc.; **5.8:** © Fritz Polking/Peter Arnold, Inc.; **5.9** (top): © SPL/Photo Researchers, Inc.; **5.9** (bottom): © Manfred Kage/ Peter Arnold, Inc.; **5.10:** © Noble Proctor/Photo Researchers, Inc.; **5.11:** © J. Burgess/SPL/Photo Researchers, Inc.; **5.12:** © K. Maslowski/Visuals Unlimited; **p. 99:** © Marc A. Blovin, National Biological Survey/U.S. Dept. of Interior Fish and Wildlife Service; **5.17** (beans): © Alexander Lowry/ Photo Researchers, Inc.; **5.17** (fawn): © Vol. 6/ PhotoDisc; **5.17** (fox): © Paul Souders/Stone/Getty Images; **5.17** (duck): © Knolan Benfield/Visuals Unlimited; **5.17** (decomposers): © R. Kessel and C. Shih/Visuals Unlimited; **5.17** (nitrate No2): © David Phillips/Visuals Unlimited; **5.17** (nitrate No3): Fred Hossler/Visuals Unlimited.

Chapter 6
Figure 6.1: © William E. Ferguson, photographer; **6.4:** © Larry Mellichamp/Visuals Unlimited; **6.10b:** © Leonard Lee Rue, Jr./ Photo Researchers, Inc.; **6.11b:** © William E. Ferguson, photo by Stephanie E. Ferguson; **p. 118** (left & middle): © Harold Hungerford/University of Southern Illinois-Carbondale; **p. 118** (right): © Carl Bollwinkel/University of Northern Iowa; **6.12b:** © Vol. 6/CORBIS; **6.13b:** © Vol. 37/CORBIS; **6.14b:** © Doug Wechsler/Animals Animals/Earth Scenes; **p. 121:** © Michael J. Balick/Peter Arnold, Inc.; **6.15b:** © Eldon Enger; **p. 123:** © Jerry Franklin; **6.16b:** © Vol. 44/PhotoDisc; **6.17b:** © Vol. 36/ PhotoDisc; **6.18b:** © John Shaw/Tom Stack &

20.3: © John Neubauer/PictureQuest; **20.8** (left): Courtesy of New York State Department of Environmental Conservation; **20.8** (right): © Paul David Mozell/Stock Boston; **p. 466** (left): © C. Osbourne/ Photo Researchers; **p. 466** (right): © Mary Kate Denny/Photo Edit; **20.9:** © Forrest Anderson/ Liaison Agency/Getty News Service; **20.10** (left): © Vol. 19/PhotoDisc; **20.10** (right): © A. Ramey/ Stock Boston; **20.11b:** © Peter Turnley/CORBIS; **p. 472:** © Greenpeace/Morgan.

Image Credits

Title Page / Half Title Page Icon

Bird, Digital Vision; Pine Cone, Corbis; Quartz, Corbis; Stream, Corbis; Earth, Digital Vision.

Dedication Page

Earth, Digital Vision.

Guided Tour

Earth, Digital Vision.

Preface

Ferret, Corbis.

Brief TOC

Bird in Nest, Digital Vision.

Contents

Water Lilies, Corbis.

Chapter Opening Icons

Brown Icon

Bird, Digital Vision; Pine Cone, Corbis; Quartz, Corbis; Stream, Corbis; Earth, Digital Vision.

Red Icon

Butterfly, Corbis; Bear Grass, Digital Vision; Desert, Digital Vision; Lake, Digital Vision.

Green Icon

Fox, Corbis; Palm, Corbis; Marsh, Corbis; Glacier, Corbis.

Purple Icon

Fish, Corbis; Mushroom, Digital Vision; Ocean Coast, Digital Vision; Volcano, Digital Stock.

Blue Icon

Turtle, Digital Vision; Corn, Corbis; Water Lilies, Corbis; Canyon, Corbis.

Environmental Close-Up Banner

River, Corbis.

Global Perspective Banner

Earth, Digital Vision.

Issues & Analysis Banner

Earth, Digital Vision.

Appendix 1

Clouds, Digital Vision.

Appendix 2

Whales, Digital Stock.

Glossary

Snake, Digital Vision.

Credits

Fern Frond, Corbis.

Index

Water Lilies, Corbis.

Part References

Part II References

Cohn, J. P. 1994. Restoring the Everglades. *BioScience* 44:579–583.; Craighead, F. C. 1971. *The Trees of South Florida. Vol. 1 The Natural Environments and Their Succession.* Coral Gables, Fla; *University of Miami Press.*; Dalrymple, G. H., N. K. Dalrymple, and K. A. Fanning. 1993. Vegetation of restored rock-plowed wetlands of the East Everglades. *Restoration Ecology* 1:220–25.; Davis, S. M., and J. C. Ogden. 1994. Everglades: *The Ecosystem and Its Restoration.* Boca Raton, Fla.: St. Lucie Press; Douglas, M. S. 1997. *The Everglades: River of Grass; 50th Anniversary Edition.* Sarasota, Fla.: Pineapple Press; Enserink, M. 1999. Plan to quench the Everglades thirst. *Science* 285:180a.; Lodge, T.E. 1994. *The Everglades Handbook: Understanding the Ecosystem.* Delray Beach, Fla.: St. Lucie Press; Myers, R. L., and J. J. Ewel. 1990. *Ecosystems of Florida.*
Orlando: University of Central Florida Press.

Part III References:

KUED documentary on Skull Valley: http://www.kued.org/skullvalley/index_flash.html.; Skull Valley goshutes web site: http://www.skullvalleygoshutes.org.

Related articles: http://www.skullvalleygoshutes.org/newspaper.; Utah Department of Environmental Quality, comments and concerns regarding the Skull Valley site: http://www.deq.state.ut.us/EQOAS/ no_high_level_waste/DEIScomments and http://www.deq.state.ut.us/EQOAS/ no_high_level_waste/concerns.htm.; Environmental Justice Database: http://www.msue.msu.edu/msue/imp/modej/ masterej.html.

Part IV References

Cohen, M., J. Morrison, and E. Glenn. 1999. *Haven or Hazard: The Ecology and Future of the Salton Sea.* Oakland, Calif.: Pacific Institute for Studies in Development, Environment, and Security.; Kaiser, J. 1999. Battle over a dying sea. *Science* 284: 28–30.; Matsui, M., P. Garrahan, and J. Hose. 1992. Development defects in fish embryos from Salton Sea, California. *Bulletin of Environmental Contamination and Toxicology* 48: 914–20.; Ohlendorf, H., and K. Marois. 1990. Organochlorines and selenium in California USA night-heron and egret eggs. *Environmental Monitoring and Assessment* 15: 91–104.; Robbins, Jim. Farms and growth threaten a sea and its creatures. *New York Times*, 2 April 2002.; Sanchez, Rene. New California water law seeks to curb runaway sprawl. *Washington Post*, 23 December 2001.; Vessey, K. B. 2000. Salton: A sea of controversy. *Journal of College Science Teaching* 30: 67–69.; Map of the Salton Sea, U.S. Geological Survey Geographic Names Information System Map Server: http://mapping.usgs.gov:8888/ gnis/owa/MapServer?f_name=Salton+Sea& f_state=CA&f_latlong=332000N1155000W& f_ht=2&server=TIGER;
Salton Sea home page (numerous links): http://www.sci.sdsu.edu/salton/ SaltonSeaHomePage.html;
History of the Salton Sea and information on bird die-offs: http://www.desertusa.com/salton/ salton.html;
"Dr. Milton Friend to Chair Salton Sea Science Subcommittee," news release: http://www.nbs.gov/pr/newsrelease/1998/3-19.html; The Salton Sea: http://visearth.ucsd.edu/VisE_Int/ aralseahtml/SaltSea.intro.html;
Vessey, K. B. 1999. "Salton, A Sea of Controversy." http://ublib.buffalo.edu/libraries/projects/cases/ salton.html

index